Self-Organised Criticality
Theory, Models and Characterisation

Giving a detailed overview of the subject, this book takes in the results and methods that have arisen since the term 'self-organised criticality' was coined twenty years ago.

Providing an overview of numerical and analytical methods, from their theoretical foundation to actual application and implementation, the book is an easy access point to important results and sophisticated methods. Starting with the famous Bak–Tang–Wiesenfeld sandpile, ten key models are carefully defined, together with their results and applications. Comprehensive tables of numerical results are collected in one volume for the first time, making the information readily accessible to readers.

Written for graduate students and practising researchers in a range of disciplines, from physics and mathematics to biology, sociology, finance, medicine and engineering, the book gives a practical, hands-on approach throughout. Methods and results are applied in ways that will relate to the reader's own research.

Gunnar Pruessner is a Lecturer in Mathematical Physics in the Department of Mathematics at Imperial College London. His research ranges from complexity, through field theoretic methods and applications, to synchronisation and the application of statistical mechanics in the medical sciences.

Self-Organised Criticality

Theory, Models and Characterisation

GUNNAR PRUESSNER

Imperial College London

CAMBRIDGE
UNIVERSITY PRESS

Shaftesbury Road, Cambridge CB2 8EA, United Kingdom

One Liberty Plaza, 20th Floor, New York, NY 10006, USA

477 Williamstown Road, Port Melbourne, VIC 3207, Australia

314–321, 3rd Floor, Plot 3, Splendor Forum, Jasola District Centre, New Delhi – 110025, India

103 Penang Road, #05–06/07, Visioncrest Commercial, Singapore 238467

Cambridge University Press is part of Cambridge University Press & Assessment,
a department of the University of Cambridge.

We share the University's mission to contribute to society through the pursuit of
education, learning and research at the highest international levels of excellence.

www.cambridge.org
Information on this title: www.cambridge.org/9780521853354

ISBN   978-0-521-85335-4   Hardback

In memory of our friend Holger Bruhn.

To see the world in a grain of sand,
and to see heaven in a wild flower,
hold infinity in the palm of your hands,
and eternity in an hour.

William Blake, *Auguries of Innocence*, ca 1803

Contents

Tables

Foreword by Henrik J. Jensen

When Self-Organised Criticality (SOC) was first introduced in 1987 by Bak, Tang, and Wiesenfeld, it was suggested to be *the* explanation of the fractal structures surrounding us everywhere in space and time. The very poetic intuitive appeal of the combination of terms self-organisation and criticality, meant that the field gained immediate attention. The excitement was not lowered much by the fact that the claimed $1/f$ and fractal behaviour were soon realised in reality not to be present in the sandpile model used by the authors to introduce their research agenda. Nor did the lack of power laws in experiments on real piles of sand deter investigators from interpreting pieces of power laws observed in various theoretical models and physical systems as evidence of SOC being essentially everywhere. This led rapidly to a strong polarisation between two camps. On the one side there was the group of researchers who did not worry about the lack of a reasonably precise exclusive definition of the SOC concept and therefore tended to use SOC as synonymous with snippets of power laws, rendering the term fairly meaningless. The other camp maintained that SOC was not to be taken seriously. They arrived at this conclusion through a mixture of factors including the observation that SOC was ill defined, not demonstrated convincingly in models, and absent from experiments on sandpiles. The debate sometimes reflected a reaction in response to bruises received during fierce exchanges at meetings as much as a reaction to scientific evidence.

All in all this was a difficult, somewhat overwrought and unfortunate situation. Science proceeds through gradual uncovering of hierarchical insights and it is of course most unlikely that one simple mechanism is able to explain *how nature works*. Nevertheless, we are certainly surrounded by abundant power laws and fractals of more or less pure forms. They are typical features of non-equilibrium systems. But we obviously need to go beyond the classification of the world into only two categories: equilibrium and non-equlibrum. There is structure among the non-equilibrium systems – exactly as the classification of zoology into elephants and non-elephants does not represent the rich substructures within the class of non-elephants, such as tigers, parrots, monkeys and tortoises, to mention a few. When SOC is scrutinised at a seriously scientific level it will most definitely help us in our quest to understand the substructures within the non-equilibrium class.

The present book does exactly this. It seeks to lay out carefully what is certain and the uncertain aspects about SOC. Gunnar Pruessner has produced a comprehensive, detailed and authoritative overview of a quarter of a century's intensive experimental, observational and theoretical investigations. He carefully discusses the main achievements, the still unsolved problems and some of the compelling reasons why SOC remains an important concept helping us to place some demarcation lines within the vast field of non-equilibrium systems. The book is a masterpiece in clarity surveying a huge literature and helping to extract the essential durable lessons uncovered by the many scientists working in the field. The book will play an essential rôle in turning SOC into a serious and specific research field. It is worth mentioning how the book compares with Bak's (1996) *How Nature Works* and with my own (1998) book, *Self-Organized Criticality*. Bak's book is not really scientific. It is an entertaining and readable, enthusiastic and very optimistic document about how SOC came into being, what Bak hoped it would achieve and how Bak experienced fellow researchers in the field. My own book was a very brief and simple attempt to make a status report of what SOC might mean scientifically after the first 10 years of research. Not only does Pruessner's book present 15 more years of research, it also represents a significantly more detailed and mathematically careful discussion.

Like many other fields of statistical mechanics SOC now has its own reference book that helps us as we continue to develop our understanding of what SOC really is – and is not. By the clear exposition of the established understanding of models and the relationships between them, the book will help research aimed at developing further our theoretical comprehension. Furthermore, by the circumspect discussion of the relation between theoretical insights and expectations on the one side and experiments and observations on the other, the book will act as a guide when we continue our attempts to place experimental and observational findings from physics, biology, neuroscience, geophysics, astrophysics, sociology etc. in relation to SOC.

Henrik Jeldtoft Jensen

Preface

Self-organised criticality (SOC) is a very lively field that in recent years has branched out into many different areas and contributed immensely to the understanding of critical phenomena in nature. Since its discovery in 1987, it has been one of the most active and influential fields in statistical mechanics. It has found innumerable applications in a large variety of fields, such as physics, chemistry, medicine, sociology, linguistics, to name but a few. A lot of progress has been made over the last 20 years in understanding the phenomenology of SOC and its causes. During this time, many of the original concepts have been revised a number of times, and some, such as complexity and emergence, are still very actively discussed. Nevertheless, some if not most of the original questions remain unanswered. Is SOC ubiquituous? How does it work?

As the field matured and reached a widening audience, the demand for a summary or a commented review grew. When Professor Henrik J. Jensen asked me to write an updated version of his book on self-organised criticality six years ago, it struck me as a great honour, but an equally great challenge. His book is widely regarded as a wonderfully concise, well-written introduction to the field. After more than 24 years since its conception, self-organised criticality is in a process of consolidation, which an up-to-date review has to appreciate just as much as the many new results discovered and the new directions explored. Very soon the idea was born to deviate from Henrik's plan and try to be more comprehensive. It proved a formidable task to include all the material I thought deserved mentioning, and a daunting and precarious one to decide what not to include. There is a degree of uncertainty that I did not expect and for which I sincerely apologise.

Leaving all difficulties aside, SOC is regarded in the following as *scale invariance without external tuning of a control parameter, but with all the features of the critical point of an ordinary phase transition, in particular long range (algebraic) spatiotemporal correlations.* In contrast to (other) instances of generic scale invariance it displays a separation of time scales, avalanches, is interaction dominated and contains non-linearities. The ideal SOC model differs from a corresponding model displaying an ordinary phase transition solely by a self-tuning mechanism, which is absent in the latter. Defining SOC primarily as a phenomenon rather than a class of systems allows the question which natural phenomena, experiments, computer models and theories display the desired behaviour. This is the central theme of the present book. It was written with the aim of giving an overview of the vast literature, the many theories and disparate methods used to analyse SOC phenomena. It is my hope that it will be of as much use for the theoretician working in statistical mechanics, as for the practitioner trying to apply ideas of SOC to his or her particular field.

Style

With this broad scope, I have tried to make the material as accessible as possible, by placing it in a wider context or a narrative, whenever suitable, by retracing the historic development of the field. I have aimed for a balanced view on contentious questions and have tried to flag up where my personal opinion and convictions have taken over. I hope the occasional casual remark is never mistaken as an offence. I have tried to use notation, symbols and names consistently throughout, but only as long as this did not clash with established usage. Where I could not establish a particular custom, I used the notation of the original works. A table of commonly used symbols can be found on pp. xvii–xxii.

Structure

The book is divided in three parts. The first part gives an introductory overview of the subject: the basic concepts and terminology, the principles of scaling and the huge range of experimental evidence.

In the second part, the key SOC models [1] are discussed, divided in three categories: sandpile-like, deterministic models, dissipative models and stochastic sandpiles. I have tried to pick models that have received a large amount of attention and had a great impact on the field as a whole. These difficult choices are, admittedly, arbitrary and thus bound to be imperfect. Each model is discussed in some detail, starting with a short history followed by

> its definition, as well as its key features and observables set apart in a Box, which provides the reader with an overview of the dynamics of the model.

Because every such box follows the same structure, they can serve as a reference and are easily turned into a concrete implementation.

After a few introductory remarks, the remainder is dedicated to the particular characteristics of the model, focusing on those that have evoked the most wide-ranging discourse in the literature. The material has been brought together with the intention of covering as much of the relevant findings as possible, even when they are sometimes contradictory. The last chapter of the second part is dedicated to various numerical techniques used for the characterisation and analysis of SOC models. I mention implementation details for particular models whenever I see fit, unfortunately I was not able to present a complete computer program for each model. In the appendix, an actual C implementation of an SOC model is discussed in great detail using the example of the OFC Model.

---

[1] To call these systems 'models', as is traditionally done, is misleading if one accepts that they are not intended to model much more than themselves. A similar concern applies to the term 'simulation', see footnote 1 on p. 210.

The final, third part reviews many theoretical concepts, such as mean-field theories and random walker approaches, which have proved invaluable for the analysis of SOC models. Such theories have led to various proposed mechanisms, which are presented in Ch. 9. Again, I had to make an arbitrary choice and decided to focus on two theories of SOC, summarising others only very briefly. The very last chapter might be the most interesting for many readers and might even prove a bit contentious. Here I have collected facts and fallacies, as ignored and embraced in the literature (although I have refrained from stating bibliographical references). From my point of view, such erroneous beliefs are remarkably instructive and do more good than harm. As the chapter draws much of its arguments from the preceding ones, one might regard it as a form of summary – the book ends with a very brief outlook.

Inevitably, I myself suffer from such fallacies, ignore the truth, misinterpret findings, overrate my own work and overlook important contributions by others. I have tried very hard to avoid such mistakes and I sincerely apologise for the failure to do so. I would like to invite readers to share their insights with me and contact me whenever corrections are due.

Keywords, notes and indices

Throughout the text, important **concepts and keywords** are shown in bold, as are their page numbers in the index. *Key statements* are printed in italics, and slanted text in a figure caption refers to a particular feature in the figure. Footnotes are indicated by arabic superscript numbers. [2] Where more extensive notes could not be presented in a footnote, endnotes are used, marked in the text by a number in square brackets. [0.0] Endnotes can be found on pp. 391–398, just before the references. Each is preceded by the page number referring to it.

The accessibility of the material is facilitated by the indices in the back of the book. Each bibliographical entry contains, after an arrow ($\rightarrow$), the page numbers where the work is cited, with italics signifying *quotations*. Publications that deserve special attention are marked as follows: ● marks selected reviews and overview articles, ▶ designates the definition of some characteristic or important models and ⇒indicates a proposed mechanism of SOC. To simplify sequencing, in all indexes name prefixes, such as 'van' and 'de', are regarded part of the name, for example 'van Kampen' can be found under 'V'. The references are followed by an author index and a subject index. In the former, bold page numbers refer to **bibliographical entries** by that author and italicised ones, again, refer to *quotes*. The subject index is rather detailed and has a large number of entries; where entire sections are dedicated to a subject, the page numbers are underlined. Page numbers of **keywords** are highlighted in bold. **Key SOC models**, discussed in Part II of the book, are printed in bold. EPONYMS and NAMESAKES of SOC models are highlighted in small capitals throughout. I have used established abbreviations and names as much as possible, but inevitably some authors use different ones.

---

[2] See footnote 1 on p. xiv for an example.

Numerics

A pitfall Henrik Jensen wisely avoided in his introduction to SOC is the inclusion of numerical results. What should be included? The most up-to-date results, which soon will be out-dated? The historic values, which have been built upon? I decided to include as much as possible and necessary to allow the reader to judge which results can be relied upon. Unfortunately, it is not always immediately clear which numbers have been derived from numerical data under which assumptions. I have tried to avoid making additional ones. Where it occurred to me as being particularly important, I point out that exponents have been derived from each other, not least because that puts an apparent consistency in perspective.

Acknowledgments

I would like to thank Henrik J. Jensen for his kindness and sincerity and the opportunity he has given me by suggesting that I write this book. I am deeply grateful for all he taught me about science and SOC in particular and look forward to the many projects we enjoy together. This is what Science should be like. I am indebted to Kim Christensen, Nicholas Moloney, Andy Parry and Ole Peters, who allowed me to learn so much in the many discussions and collaborations we have had. I would like to extend my gratitude to everyone who helped me to write this book and contributed to it. Attempting to name everybody is a task bound to fail and I apologise for that. People I would like to mention, because they so kindly shared their insights with me include: Alvin Chua, Deepak Dhar, Ronald Dickman, Vidar Frette, Peter Grassberger, Christopher Henley, Satya Majumdar, Robert McKay, Zoltan Racz, Richard Rittel, Beate Schmittmann, Paolo Sibani, Alessandro Vespignani, Alistair Windus and Royce Zia. I would also like to thank my friends at Imperial College London, in particular Seng Cheang, Nguyen Huynh and Ed Sherman, who make working here such a wonderful experience. We have the good fortune of a great library staff, who tirelessly tracked down and provided me with the most elusive material. I am truly grateful for their efforts. The same goes for our fantastic computing support, in particular Dan Moore and Andy Thomas, who provided me with a seemingly endless amount of CPU time, even when the demands of modern university life make that all the more difficult.

The cover of this book shows 'Work No. 370: Balls' by M. Creed (2004), which is a stunning example for the need of a reference scale to judge correctly the size of the objects shown – without the parquet in the background it would be difficult to tell whether the baseball or football is out of proportion. I am very grateful to the artist and the gallery Hauser and Wirth for giving me access to this wonderful work.

Finally, I would like to thank Anton, Ida and Sonja without whom nothing would be much fun. It is impossible to make up for all the support and patience I received from them, as I did from our friends and family, in particular from both our parents. One day I wish to have the opportunity to pass on to my children all this kind, patient help to build a home, a family and a life. This book is dedicated to our friend Holger Bruhn, who is so greatly missed.

Symbols

An attempt was made to use symbols consistently throughout the book, at the same time keeping them in line with common usage and tradition as much as possible. Most of them are carefully introduced when they are used for the first time in a chapter. A few have more than one or a more specific meaning, depending on the particular context, or are used briefly for a very different purpose than the one listed below. Many have derived forms with a more specialised meaning not listed below. The page numbers refer to the earliest occurrence of the variable or its most detailed discussion. Further references can be found in the index.

Symbol	Description	Page
α	Avalanche duration exponent in SOC models	14
α	Scaling exponent of the power spectrum ($1/f^{\alpha}$ noise)	15
α	Level of conservation in the OFC Model	126
α_E	Scaling exponent of the scaling function of the energy dissipated in the ricepile experiment	191
α_R	Asymptotic power law of the residence time distribution in the OSLO Model	195
β	Order parameter exponent in ordinary critical phenomena	337
β_E	Finite size scaling exponent of the energy dissipated in the ricepile experiment	191
β_R	Finite size scaling exponent of the residence time distribution in the OSLO Model	194
$\beta^{(int)}$	Growth (roughening) exponent of an interface *Superscripts in brackets generally denote the observable the variable characterises.*	310
$\beta^{(AS)}$	Order parameter exponent in an absorbing state phase transition	337
Γ	Noise amplitude (of various types of noise)	205
$\Gamma(\cdot)$	Gamma function	57
γ	Critical exponent of the average avalanche size in the BS Model	148
γ	Susceptibility exponent in ordinary critical phenomena	337
γ_{xy}	Exponent characterising scaling of observable x conditional to y	42
$\gamma^{(DP)}$	Susceptibility exponent in directed percolation	297
Δp	Distance from the critical point in percolation (and similar for other variables)	30

Symbol	Description	Page
$\Delta_{\mathbf{nm}}$	Toppling matrix	91
$\Delta(g)$	Correlator of the noise in the quenched Edwards–Wilkinson equation	205
$\delta_{ij}, \delta(\cdot)$	Kronecker and Dirac δ function	
$\delta^{(AS)}$	Survival exponent in absorbing state phase transitions	305
ϵ	Small quantity, correction etc.	
ϵ	Dissipation rate (often in the bulk)	169
ϵ	Mass term	323
ζ	Particle density	330
η	Correlation exponent	47
θ	Tuning variable in the DS-FFM	116
θ_m, θ_r	Angle at which avalanching must occur, angle of repose	54
$\theta(\cdot)$	Heaviside step function	32
Λ	Total rate of multiple Poisson processes	389
λ	Scaling factor	32
λ	Scaling exponent of the characteristic burnt cluster size in the DS-FFM	119
λ	Coupling of the non-linearity	324
λ	Control parameter in the contact process	341
$\lambda, \lambda_i, \ldots$	Rate of a Poisson process	388
λ_i	ith eigenvalue of eigenvector $|e_i\rangle$	201
μ	Generic scaling exponent	28
ν	Surface tension	47
ν	Correlation length exponent in the FFM	112
ν	Correlation length exponent in ordinary critical phenomena	337
ν'	Scaling of the characteristic burning duration in the DS-FFM	118
ν_E	Finite size scaling exponent of the energy dissipated in the ricepile experiment	191
ν_R	Finite size scaling exponent of the residence time distribution in the Oslo Model	194
$\nu_\perp^{(AS)}$	Correlation length exponent in absorbing state phase transitions	305
ξ	Correlation length	112
$\xi, \tilde{\xi}$	Noise (usually Gaussian, white, vanishing mean)	204
ρ	Tree density in the DS-FFM	118
ρ_n	Moment ratio	40
ρ_a	Activity	299
Σ	Finite size scaling exponent	43
σ	Branching ratio (effective)	130
σ	Critical exponent of the characteristic avalanche size in the BS Model	160
$\sigma^2(\cdot)$	Variance	

Symbol	Description	Page
$\overline{\sigma}^2(\cdot)$	Numerical estimate of the variance	
	Overbars generally denote numerical estimates.	
$\sigma_n^{(s)}, \sigma_b^{(T)}$	Finite size scaling exponent of the nth moment of the avalanche size (or duration) distribution	38
τ	Avalanche size exponent in SOC models	13
τ	Correlation time on the microscopic time scale	212
$\tilde{\tau}$	Apparent avalanche size exponent in SOC models	31
τ_r	Avalanche radius exponent	43
Φ	Number of potential topplings in the OSLO Model	202
$\chi^{(pile)}$	Roughness exponent of the surface of a sandpile	197
$\chi^{(int)}$	Roughness exponent of an interface	308
$\Omega(f(x))$	Asymptotically bounded from below by $f(x)$	292
ω	Frequency (angular velocity)	15
$\omega, \omega_1, \ldots$	Correction to scaling exponent	30
A	Avalanche area	12
$A, A_n, B, \ldots$	Generic amplitudes	31
A_h, A_t	Areas two distributions differ by	34
$a, a(t)$	Number of active sites, position of a random walker	252
$a^{(s)}, b^{(s)}, \ldots$	Metric factors	12
$a(t)$	Time series	15
a_0	Initial distance of a random walker from an absorbing wall	36
$a_{\mathbf{n}}$	Evolution operator in deterministic sandpiles	95
$\hat{a}_{\mathbf{n}}$	Markov matrix	272
b	Exponent in the Gutenberg–Richter law	66
b_j	Bin boundary	231
$\mathcal{C}$	Stable configuration (state) on the lattice	95
$C, C^\dagger, C_{\mathbf{n}} \ldots$	Creation and annihilation operators in the MANNA Model (matrix representation of its dynamics), acting on site $\mathbf{n}$	173
$\mathcal{C}(\cdot)$	Cumulant generating function	46
C_t	Catalan number	253
$c(t)$	Time correlation function	15
$c(\mathbf{x}, \mathbf{x}'\, t, t')$	Connected correlation function	47
$c_{\mathbf{n}}$	Probability for site $\mathbf{n}$ to be charged	97
c_0	Density of empty sites in the sticky sandpile	295
D	Avalanche dimension (finite size scaling exponent) in SOC models	12
D	Diffusion constant	36
$D_t(t)$	Probability for the activity to cease at time t	253
d	Spatial dimension	27
E	Energy release in an earthquake	66
E	Energy release in the ricepile experiment	191
$E_{\mathbf{n}}$	Energy at site $\mathbf{n}$ in the ZHANG Model	105

Symbol	Description	Page
$E(t), E(\mathbf{n}, t)$	External drive (boundary condition or source term) in the OSLO Model	205
$\langle e_i |, | e_i \rangle, \ldots$	Left and right eigenvectors, respectively, with eigenvalue λ_i	201
$\mathcal{F}, \mathcal{G}, \tilde{\mathcal{G}}, \mathcal{K}, \ldots$	Scaling and cutoff functions	12
$\mathsf{F}(\cdot)$	Cumulative distribution function	26
$F_{\mathbf{n}}, F_{\mathbf{th}}$	Force at $\mathbf{n}$ and threshold in the OFC Model	127
f	Frequency	15
f	Lightning probability in the DS-FFM	116
$f(x), g(x)$	Generic function or functional	
$f(s; s_{\mathbf{c}})$	Non-universal part of the probability density function of s	28
$f_n^{(s)}$	Moment of the non-universal part of the probability density function	29
$f_{\mathbf{n}}$	Fitness at site $\mathbf{n}$ in the BS Model	142
f_0	Fitness threshold to define avalanches in the BS Model	142
$G(t)$	Gap function in the BS Model	144
$G(\mathbf{n}, t)$	Number of topplings performed by site $\mathbf{n}$ up to and including time t	176
$g_n^{(s)}$	Moment of the scaling function	29
$g(\mathbf{n}, t)$	Number of charges received by site $\mathbf{n}$ up to and including time t	176
$g, g_{\mathbf{n}}$	Generational time scale, local generational time scale	257
$H(\mathbf{x}; l)$	Coarse grained local dynamical variable	49
h	External driving field in the AS mechanism	331
$h(\mathbf{x}, t)$	Height field in continuum, generally local dynamical variable	47
$h_{\mathbf{n}}$	Height of particle pile or particle number at lattice site $\mathbf{n}$ in BTW and MANNA Models	86
h^c	Critical height or particle number in the MANNA Model	164
h_i	Normalised histogram count for slot i	230
$I_n(\mathbf{n})$	Indicator function of site $\mathbf{n}$ being in state n	99
$i, j, \ldots$	Generic indices and counters	
$\mathbf{1}$	Imaginary unit, $\sqrt{-1}$	
$\mathbf{j}$	Current	324
K	Number of new fitnesses drawn per site updated in the BS Model	392
$\mathcal{K}(\cdot)$	Scaling function of the two point correlation function	47
K_L	Spring constant in the OFC Model	125
$\mathbf{k}$	Generic vector in Fourier space	
L	Linear system size	86
l	Block size	49
$M, N, \ldots$	Generic number of items (N is often the number of sites)	
$\mathcal{M}(\cdot)$	Moment generating function	45
$\mathbf{m}, \mathbf{n}, \ldots$	Generic position on the lattice	
m_i	Arithmetic mean of a subsample	216
$\langle m \rangle (\mathbf{n})$	Average number of moves (escape time) from $\mathbf{n}$	245
$m(g)$	Size of the gth generation in a branching process	258

Symbol	Description	Page
$N_i, \mathcal{N}$	Size of a subsample, total sample size	216
$n(s)$	Site normalised cluster size distribution	103
n_{cov}	Number of distinct sites updated in the BS Model (avalanche area)	392
$\mathcal{O}(f(x))$	Asymptotically bounded from above by $f(x)$	
$P(\cdot)$	Generic probability of an event to occur, a quantity (usually discrete) to be observed etc.	
$\mathcal{P}^{(\mathrm{s})}(s')$	Probability density function of observable s evaluated at s'	12
p	Number of boundary sites toppling in an avalanche	42
p	Period of an Abelian operator	95
p	Tree growth probability in the FFM	113
p	Parameter in the ricepile model	185
p	Toppling probability in the sticky sandpile	295
p_r, p_l, p_0	Toppling probabilities in the generalised AMM	174
$p_c^{(\mathrm{DP})}$	Directed percolation critical point	295
Q	Linear observable	275
q	Coordination number on a lattice	
$\tilde{q}$	Removals per toppling	245
$q'(\mathbf{x})$	Local ratio of number of charges and topplings	307
R	Residence time in the Oslo Model	186
R	Binning range when coarse graining	233
$R_{1/2}$	Characteristic length for the decay of correlations	48
r	Radius of gyration	12
r, r_0	Distance between two points in space	48
r_i	Binning range	230
S	Total area encapsulated in two interface representations of an Oslo Model configuration	395
$S(\omega)$	Power spectrum	15
S_1, S_2	Parameter in the ricepile model	185
$S, S_{\mathbf{n}}, T, \ldots$	Evolution matrices acting on site $\mathbf{n}$ in the matrix representation of the Abelian BTW Model and the (totally asymmetric) Oslo Model	200
S_N	Sum of the sizes of N consecutive avalanches	212
s	Avalanche size	11
s_0	Lower cutoff of the avalanche size distribution	12
$s_{\mathbf{c}}$	Upper cutoff of the avalanche size distribution (characteristic size)	12
T	Avalanche duration	14
T	Number of subsamples	216
T	Longitudinal direction in the DR Model	286
T_0	Lower cutoff of the avalanche duration distribution	14
$T_{\mathbf{c}}$	Upper cutoff of the avalanche duration distribution (characteristic time scale)	14
T_t, T_g	Termination probability (random walk and branching process)	253
$\mathcal{T}(\mathcal{C} \rightarrow \mathcal{C}')$	Probability for a transition from $\mathcal{C}$ to $\mathcal{C}'$	273

Symbol	Description	Page
t	Generic time	
$\mathcal{U}(\cdot)$	Diagonal matrix generating the potential number of topplings in the OSLO Model	202
$u(s_0; s_c)$	Fraction of events governed by the universal part of the probability density function	29
V	Volume of the system	118
v	Drift velocity	206
W	Double charging Markov matrix of the AMM	174
w, w^2	Width and roughness of an interface, surface etc.	313
x, y	Generic real variable, components of $\mathbf{x}$	
$\mathbf{x}$	Generic position in continuum space	
z	Dynamical exponent in SOC models	14
$z_{\mathbf{n}}$	Slope at $\mathbf{n}$	86
$z^c, z_{\mathbf{n}}^c$	Critical slope, at site $\mathbf{n}$	87
$z^{(AS)}$	Dynamical exponent in absorbing state phase transitions	305
$z^{(int)}$	Dynamical exponent in interface growth phenomena	309

PART I

INTRODUCTION

1   Introduction

When Bak, Tang, and Wiesenfeld (1987) coined the term **Self-Organised Criticality** (SOC), it was an explanation for an unexpected observation of scale invariance and, at the same time, a programme of further research. Over the years it developed into a subject area which is concerned mostly with the analysis of computer models that display a form of **generic scale invariance**. The primacy of the **computer model** is manifest in the first publication and throughout the history of SOC, which evolved with and revolved around such computer models. That has led to a plethora of computer 'models', many of which are not intended to model much except themselves (also Gisiger, 2001), in the hope that they display a certain aspect of SOC in a particularly clear way.

The question whether SOC exists is empty if SOC is merely the title for a certain class of computer models. In the following, the term SOC will therefore be used in its original meaning (Bak *et al.*, 1987), to be assigned to systems

> with spatial degrees of freedom [which] naturally evolve into a self-organized critical point.

Such behaviour is to be juxtaposed to the traditional notion of a **phase transition**, which is the singular, **critical point** in a phase diagram, where a system experiences a breakdown of symmetry and **long-range spatial** and, in non-equilibrium, also **temporal correlations**, generally summarised as (power law) **scaling** (Widom, 1965a,b; Stanley, 1971). The paradigmatic example for such a **critical phenomenon** is the Ising Model, which displays scaling only at a specific critical temperature, the value of which depends on the type and dimension of the lattice considered, as well as the details of the interaction, and is generally not known analytically.

Bak *et al.* (1987) found a simple computer model which seemed to develop into a non-trivial scale invariant state, displaying long-range spatiotemporal correlations and self-similarity, without the need of any tuning of a temperature-like control parameter to a critical value. **Universality** of the asymptote of the correlation function is immediately suspected by analogy with ordinary scale invariance, i.e. critical phenomena. More than twenty years later, one can review the situation. Does SOC exist? A host of models has been studied in great detail with mixed results. While some eventually turned out not to display scaling, a range of models displays the expected behaviour. The situation is less encouraging for experimental evidence, which suggests that there are very few systems with solid scaling behaviour. The most pessimistic perspective on the numerical and experimental evidence is that none of the systems displays asymptotic scaling behaviour and SOC does not exist. With the theoretical, numerical and experimental evidence presented in the following, this point of view is difficult to maintain. SOC exists, very convincingly, in flux avalanches in superconductors and certainly in the MANNA Model and the OSLO Model. Yet, it is probably not as common as originally envisaged so that the enormous body of work reporting long-range behaviour, which is close to but not exactly a power law, still awaits an explanation beyond SOC.

There is a number of reasons why SOC is important independent of how broadly it applies. Initially, it was a solution of the riddle of $1/f$ noise (Sec. 1.3.2), the frequently found scaling of the **power spectrum** of a time series. Press (1978) popularised the notion of $1/f$ noise and its ubiquity (also Dutta and Horn, 1981; Hooge, Kleinpenning, and Vandamme, 1981; Weissman, 1988). SOC provided a promising explanation, which was long overdue. Perhaps more importantly, it answered Kadanoff's (1986) call for a 'physics of fractals' with a theory that epitomises Anderson's (1972) credo 'more is different': 'The aim of the science of self-organized criticality is to yield insight into the fundamental question of why nature is complex, not simple, as the laws of physics imply' (Bak, 1996, p. xi). As Gisiger (2001) summarised: with the advent of SOC '[. . .] the important question which arose from the work of Mandelbrot [. . .] has shifted from "Why is there scale invariance in nature?" to "Is nature critical?" [. . .]'. Thirdly and more concretely, if SOC were generally to be found across a large class of systems, then **universality** would unify these systems, and their internal interactions could be identified and studied. They dominate the long time and large scale behaviour, just like in ordinary critical phenomena with a broken symmetry and could be studied in a simplified experiment on the laboratory scale or a simple computer model, rather than 'attempt[ing] to model the detailed, and perhaps insuperably complex[,] microphysics' (Dendy, Helander, and Tagger, 1999). Even if universality does not apply to SOC, scaling still means that some large scale phenomena can be studied on the laboratory scale, provided only that both are governed by the asymptotic behaviour of the system. Finally, even if SOC is less common than initially expected, it might still help to elucidate the nature of the critical state in traditional critical phenomena. It might even serve as a recipe to make these systems self-tune to a transition.

To appreciate its impact, it is interesting to retrace the historical context of SOC. In the 1960s statistical mechanics gained the ability to incorporate large fluctuations in a meaningful way and moved from the physics of gases and liquids with small, local perturbations, to the physics of long-range correlations as observed at phase transitions. Kadanoff (1966) explained the scaling ideas brought forward by Ben Widom using the concept of renormalisation. In the 1970s Wilson's renormalisation group led to a deep understanding of phase transitions and symmetry breaking. The 1980s made fractals popular; their physical manifestations are collected in Mandelbrot's (1983) famous 'manifesto and casebook'. By the end of that decade, SOC was born and helped to promote the notion of complexity. Since then, the focus has shifted to the more general idea of **emergence**, which summarises in a single word the phenomenon that many interacting degrees of freedom can bring about **cooperative phenomena**, i.e. the whole is more than the sum of its parts and effective long-range interaction looks fundamentally different from the microscopic interaction it is caused by.

Like the 1980s theme of fractals, SOC reaches far beyond its original realm. Other subject areas that tap into the results in SOC often regard it as an *explanation* for the observed emergent phenomenon, as originally envisaged by Bak *et al.* (1987). Given, however, that SOC itself is not fully understood, its explanatory power stretches only so far as to assume that a given *scale-free phenomenon is caused by the system **self-organising** to an underlying critical point.* Whether that leads to any further insight depends on the nature of the phenomenon in question. For example, if evolution is self-organised critical (Sec. 3.5), the hunt for external causes of mass extinction is put into question. Claiming that the Barkhausen effect (Sec. 3.3) is self-organised critical adds much less to the understanding of the phenomenon and might ultimately be seen merely as a change of perspective.

Apart from identifying systems that indisputably display SOC, a lot of research is dedicated to the question how SOC works, i.e. its necessary and sufficient conditions. Among the many proposed mechanisms, the Absorbing State Mechanism (AS mechanism, Sec. 9.3) has gained so much popularity that SOC is considered by some as explained. The fact that many of its key ideas were laid out already by Bak *et al.* (1987, also Tang and Bak, 1988b) is a striking testimony of their genius. According to the AS mechanism, SOC is due to a very simple feedback loop, leading to the **self-tuning** of a parameter that controls a non-equilibrium phase transition to its critical point. As discussed in detail elsewhere (Sec. 9.3.4), some important ingredients of the AS mechanism might still be missing.

The big challenges in SOC on a more technical level thus have remained the same for almost twenty years: on the most basic level, the identification of universality classes containing models that display solid scaling behaviour. This is mostly numerical work and significant progress has been made in particular with respect to the Manna universality class (Sec. 6.2.1.2). Other universality classes displaying robust scaling are very scarce, virtually non-existent. A much bigger challenge is to develop a full understanding of the underlying mechanism including the link to ordinary critical phenomena, which includes absorbing state phase transitions. Such a mechanism might be applicable to ordinary critical phenomena. Finally, the big question: why are so many natural phenomena so broadly distributed, (almost) resembling a power law? The answer to this question might, of course, lie far outside the realm of SOC.

The questions where SOC can be found, which systems are governed by it, what that implies, what the general features of SOC are, its necessary and sufficient conditions, all of this is still a very active research field. After a short overview of other reviews on SOC, in the remainder of this chapter the key concepts of SOC are introduced as well as its basic ingredients and observables. Two more chapters in Part I introduce the concepts and technicalities of scaling and a review of the many SOC experiments and observations. Part II is dedicated to a number of widely studied (computer) models displaying SOC, while Part III discusses various analytical approaches to the understanding of SOC.

1.1 Reviews

Over the years, a large number of reviews of SOC has been published. Of those with very broad scope, Bak's (1996) famous book with the equally ambitious as teasing title 'how nature works', popularised the subject, but also antagonised a few. The very readable, succinct review by Jensen (1998) soon became the most cited academic reference, providing a comprehensive and comprehensible overview of the main results and central themes in SOC. Sornette's (2nd edition 2006) book places SOC in a broader context with a wider background in probability theory and physics. Christensen and Moloney's (2005) textbook takes a similar approach and is aimed mainly at the audience of researchers in other subjects as well as undergraduates. A light and much shorter introduction can be found in the beautifully illustrated article by Bak and Chen (1991) and an even shorter one in Bak and Tang (1989b), which is probably relevant only for historic reasons. A broad but more technical overview is given by Bak and Paczuski (1993) and Bak and Paczuski (1995), which place SOC in the context of complexity and vice versa (Vicsek, 2002). As for the relation between complexity and statistical mechanics, Kadanoff (2000) is an invaluable collection of original articles and commentary.

Shorter, often more specialised and much more technical, but also more up-to-date reviews have frequently been published over the last fifteen years or so. Dhar (1999a,c, 2006) concentrated mostly on exact results, whereas Alava (2004) discussed the relation of SOC to (ordinary) non-equilibrium phase transitions, growth phenomena and interfaces in random media. This approach can be traced back to the highly influential review by Dickman, Muñoz, Vespignani, and Zapperi (2000), which presented some of the key models and experiments on SOC in the light of absorbing state phase transitions. Volume 340, issue 4 of *Physica A* (Alstrøm, Bohr, Christensen, *et al.*, 2004) contains short and very diverse reviews from the participants of a symposium in memory of Per Bak, a valuable resource to retrace the history of SOC and its impact on current research. At least two other proceedings volumes are similarly useful. The first half of the volume edited by Riste and Sherrington (1991) gives an overview of SOC and its context only a few years after its conception. The volume edited by McKane, Droz, Vannimenus, and Wolf (1995), which contains Grinstein's (1995) highly influential review, shows a more mature field, with links to growth, generic scale invariance and cellular automata. An early, detailed and quite technical review of SOC with special attention to extremal dynamics and interface depinning was published by Paczuski, Maslov, and Bak (1996). The recent development regarding the relation between SOC and absorbing states is discussed in Muñoz, Dickman, Pastor-Satorras *et al.* (2001). On the more applied side, Hergarten (2002) published a monograph on SOC in earth systems.

A highly critical assessment of SOC on a philosophical level can be found in Frigg (2003), who also presented an introduction to some of the frequently used SOC models. The epistomology of emergence and thus complexity is discussed in Batterman (2002). Earlier, Horgan (1995) wrote a more entertaining piece on complexity, including SOC, questioning to what extent it is delivering on the promises it apparently made (but Vicsek, 2002). Complexity might have replaced the oversimplified interpretation of reductionism as a way of reconstructing the world, as criticised by Anderson (1972), by an equally naïve notion of universality (Batterman, 2002), which takes away too many details of the phenomena science and technology are interested in. As much as psychology is not applied biology and chemistry is not applied particle physics, the universal phenomena shared by them might be pretty meaningless to them all.

1.2 Basic ingredients

A precise definition of the term 'self-organised criticality' is hampered by its indiscriminant use, which has blurred its meaning. Sometimes it is used to label models and natural systems with particular design features, sometimes for models with a particular (statistical) phenomenology, sometimes for the phenomenon itself. The literal meaning of the term refers to its supposed origin: a critical, i.e. scale invariant, phenomenon in a system with many degrees of freedom reminiscent of a phase transition, yet not triggered by gentle tuning of a **temperature-like control parameter** [1] to the critical point, as for example in the Ising Model, but by **self-organisation** to that critical point. This self-organisation might involve a control parameter which is subject to an equation of motion, or it might affect directly the statistical ensemble sampled by the system.

As SOC became more popular, the meaning of SOC switched from labelling the cause to labelling the phenomenon, i.e. scale invariance, characterised by power laws in spatiotemporal observables and their distributions. [2] As SOC was intended to explain $1/f$ noise, the latter became indicative or even synonymous with SOC. This is the beginning of a time when the question 'Is it SOC?' became meaningless (Sornette, 1994). It broke down completely when it became common to label anything SOC that bore some resemblance to its typical features, such as avalanches, thresholds or simply a broad distribution of some observable, regardless of whether or not there was any signature of scale invariant behaviour – some disputes in the literature are down to this simple confusion, as not all SOC models would display SOC.

It can be difficult to keep these different perspectives on SOC apart. Ideally, SOC should be reserved for the first meaning, a supposed underlying mechanism which keeps a system at or drives it to a (more or less) ordinary critical point. There is no need to apply the term SOC to the statistical phenomena, since plenty of terminology from the theory of phase transitions is available. Characterising certain models as 'SOC models', however, is so undeniably widespread that it would be unrealistic to propose to bar the usage of the term here, on the understanding that some 'SOC models' might ultimately not be governed by SOC.

There are a few basic ingredients that can be found in every SOC model, summarised by Jensen (1998, p. 126) as 'slowly driven, interaction-dominated threshold [(SDIDT)] systems'. First of all, there are **many (discrete) interacting degrees of freedom**, usually structured in space by nearest neighbour or at least local interaction, for example as **sites** on a lattice. Apart from some exotic exceptions, all models have a **finite number of degrees of freedom**, most suitably captured in the **finite size** of the lattice. Secondly, the interaction involves a **threshold**, which represents a very strong non-linearity. In many models the degrees of freedom can be thought of as a **local (dynamical) variable** indicating the amount of local **energy, force or particles**, which is redistributed among neighbours once it exceeds a threshold. In other models, sites interact in a certain way only if they are in

---

[1] Sometimes called the 'tuning parameter'.

[2] 'Distribution' is used synonymously with 'probability density function' in the following, p. 26.

particular states. Sites for which the local degree of freedom is below the threshold are said to be **stable**, those for which the local degree of freedom is above the threshold are called **active** or **unstable** and those which become unstable once charged are sometimes called **susceptible**.

The models are subject to an **external drive** or **driving**, which changes or **charges** the local variables, continuously or discretely, either everywhere or at one particular point. The former is often called **uniform driving** or **homogeneous driving** (more generally sometimes **global driving**) in particular when the dynamical variable is continuous, whereas no general terminology is established for the latter. Some authors refer to it as **local driving**, some call it **point driving** or **boundary driving** when it takes place at the boundary of the lattice. When the driving takes place with uniform probability throughout the lattice it is often called **(stochastic or random) bulk driving**, in particular when the driving is discrete. The amount of charge transferred from the external source is usually chosen to be small and can be fixed or random. If it involves only one site (or, for that matter, a few sites), the driving position can be fixed or chosen at random. **Stochastic driving** and **deterministic driving** thus both require further specification as to whether that applies to position and/or amount. Finally, if the driving does *not* respect the separation of time scales (see below), it is normally referred to as **continuous driving**.

The external driving plays two different rôles, which are distinguished most clearly when the charges transferred are not conserved under the relaxation described below. On the one hand, the external drive acts as a supply or a **loading** mechanism, which allows the system to respond strongly and very sensitively to an external perturbation. On the other hand, the external driving acts as such a perturbation, **triggering** a potentially very large response. In conserved models like the BTW, Manna and Oslo Models, but also in the non-conserved OFC Model, the external driving does both loading and triggering. Otherwise, in most non-conserved models like the Drossel–Schwabl Forest Fire Model the two are distinct, whereas in the BS Model loading does not exist as such, because there is nothing to be loaded and there is no quantity, certainly not a conserved one, being transported in response to the triggering event.

Once the external charges trigger the threshold, interaction occurs in the form of **toppling**, which reduces or, more generally, changes the local variable but in turn can lead to the threshold being triggered at neighbouring sites provided they are susceptible. This **relaxation process** defined by the **microscopic dynamics** follows a set of **update rules**, which specifies how degrees of freedom are updated as they interact. Analytical approaches to many models start with an attempt to capture these rules in a mathematical formalism, which can be analysed using established tools of statistical mechanics. These rules can be either **deterministic** or **stochastic**. The most common form of a stochastic relaxation is that interacting sites are picked at random among nearest neighbours. Locally and temporarily the relaxation is therefore **anisotropic**, even when it is on average (in time, space and/or across an ensemble) isotropic. Other forms of stochastic relaxation involve the amount of charge transferred or random values of the degree of freedom at updated sites. The totality of such interaction or **relaxation events** is called an **avalanche**. Avalanching is the archetypal relaxation mechanism in SOC models and is often considered as the signature of **metastability**, as small perturbations due to the external driving can lead to catastrophic responses involving the entire system.

An SOC model is often described as being driven to the brink of stability (also Sec. 1.3.3). Some authors distinguish **stable states**, which respond to external perturbations with small, local changes, **metastable states**,[3] which respond with avalanches of possibly system spanning size, and **unstable states**, which are still changing under the dynamics. Sometimes the distinction between stable and metastable is dropped and both are subsumed under stable. To differentiate the two, the stable states might be labelled as **transient** and the metastable ones as **recurrent** (also Sec. 4.2.1 and Sec. 8.3).

Being so enormously **susceptible**, the system might be considered as being in a **critical state**. Avalanches can be tallied and analysed for their size, duration, the number of sites involved etc. The resulting histograms can be probed for **power laws**, the hallmark of (full) scale invariance. Observables are discussed further in Sec. 1.3.

Any scaling observed in SOC models is usually **finite size scaling**, since the finiteness of the lattice is supposed to be the *only* scale that controls the statistics of the observables. If there is another tunable scale dominating and cutting off the statistical features of avalanches, then the system apparently requires explicit tuning and therefore cannot be called 'self-organised critical'. However, some widely accepted SOC models, e.g. the DS-FFM (Sec. 5.2) possess such a second scale, even when it diverges in some 'trivial' limiting procedure. There is widespread confusion as to what amounts to a second scale and what its consequences are. Generally, *competing* scales are a necessary condition for non-trivial scaling, which can occur only in the presence of dimensionless quantities (Sec. 2.3, p. 48 and Sec. 8.1.3, in particular p. 257). In ordinary critical phenomena, such a second scale needs to be tuned to its particular critical value. If, however, a second scale *dominates* the behaviour of a system over its finite size, it generally is not self-organised critical. The competing scales ultimately succumb to the dominant scale and play a (reduced) rôle in only a finite range of the observable.

While an avalanche is running, the external driving is stopped, known as the **separation of time scales** of driving and relaxation and generally regarded as the key *cause* of SOC. Some authors refer to it as **slow drive**, alluding also to the smallness of the perturbations caused by the external driving. Separation of time scales is achieved as long as the system is driven *slowly enough*, but there is no lower limit, i.e. an SOC system can be driven more slowly without changing its statistical properties. Separation of time scales is akin to the thermodynamic limit, in that it does not require tuning of a control parameter to a particular value that could provide a characteristic scale. Nevertheless, some regard the separation of time scales as a form of tuning.

The time scale on which an avalanche is resolved into individual events of interacting sites governed by the microscopic dynamics is the **microscopic time scale**. As no external charges arrive during an avalanche, these triggering events can be regarded as infinitely far apart on the microscopic time scale. On the **macroscopic time scale**, on the other hand, the external drive has finite (Poissonian) frequency and avalanches are instantaneous, i.e.

---

[3] Bak *et al.* (1987) called a metastable state (locally) **minimally stable**.

collapse into a point. Such a **complete separation of time scales** where one explodes to infinity from the point of view of the other, which in turn implodes into a point, can often be realised exactly in a computer implementation of the model, simply by not attempting to trigger a new avalanche, while one is running. This way, the driving frequency on the microscopic time scale can be regarded as anything between 0 and the average frequency, i.e. the inverse of the average duration.

To measure these frequencies and other temporal observables, a **microscopic time** needs to be defined explicitly, which is very often not fixed in the rules of the model. In the case of **parallel updating**, sites due for an update are ordered in generations and each generation is dealt with separately, the current generation giving rise to the next. In this case, the microscopic time simply counts the generations and naturally advances by one unit for each parallel update. In the case of **(random) sequential update**, a site is picked at random from all those due to be updated, of which there are, say, N, and time advances by $1/N$, corresponding to the average time spent on each site in a parallel update. If sites are updated as if each were subject to a local **Poisson process**, waiting times between updates can be drawn at random and added to the microscopic time (Sec. A.6.3). Most models remain well defined if the updating scheme is changed, and many models that were originally defined with parallel update are more elegantly defined with random sequential or Poissonian update (in particular the MANNA Model, Sec. 6.1, and OSLO Model, Sec. 6.3), which now is very widely used.

Some authors regard the separation of time scales as a form of global supervision or interaction (Dickman, Vespignani, and Zapperi, 1998) and 'fire the babysitter' (Dickman *et al.*, 2000) by considering a driving frequency that is finite but asymptotically vanishing on the *microscopic time scale*. In a finite system, avalanches (normally) have finite duration, so that a maximum frequency can be found below which the separation of time scales is realised without global supervision, possibly accepting its occasional but very rare violation. This maximum frequency is essentially the inverse of the characteristic duration and diverges with system size much faster than the average duration.

Many models obey an **Abelian symmetry** which is often understood as an invariance of the microscopic dynamics under change of the updating order (Sec. 8.3). Consequently, such models are not uniquely defined on the microscopic time scale and temporal observables might differ in different implementations. **Conservation** during updates of interacting sites *in the bulk* (known as **bulk conservation** or local conservation) is a second important symmetry, which was considered as a necessary ingredient of SOC models very early, but whose rôle was questioned with the arrival of non-conservative models. The quantity conserved is normally the totality of the local dynamical variable, such as the total of the energy or force. Given the external drive, **dissipation** is necessary for a **stationary state** to exist in the presence of bulk conservation. This is often realised by **boundary dissipation** at **open boundaries**, i.e. loss of the otherwise conserved quantity when a boundary site topples. Bulk conservation in conjunction with boundary dissipation leads to **transport** forcing a current through the system, even when it is **isotropic**. If **periodic boundary conditions** apply, for example in order to restore translational variance in a finite lattice, **bulk dissipation** can be implemented explicitly and conservation therefore destroyed.

Many **anisotropic** systems, which have a preferred direction at relaxation, can be dealt with analytically. In **totally asymmetric** models charges occur in only one direction, so that a site that charges another one does not become charged in return when the charged site topples. This effectively suppresses correlations and usually prevents a site from performing **multiple topplings**. Such **directed models** are often exactly solvable.

As mentioned above, SOC was intended to be more than generic scale invariance. To establish the link to ordinary critical phenomena at a transition between phases with different symmetries, in some systems a control parameter can be identified, often the energy, force or particle density (Sornette, 1994, p. 210):

> SOC requires that, as a function of a tunable control parameter, one has a phase transition at some critical point, and that the dynamics of the system brings this parameter automatically to its critical point without external fine-tuning.

For example, many directed models can be mapped to random walks, eliminating initial bias (drift) in a process of self-organisation. If it were not for that process or if the system were placed at criticality *by definition* there would be no point talking about *self-organised* criticality.

In summary, *the most basic design elements of an SOC model are: many interacting degrees of freedom, a local energy or force, a slow external drive and thresholds triggering a fast internal relaxation mechanism (separation of time scales), giving rise to avalanching.* As Jensen (1998) succinctly sums up: SDIDT – 'slowly driven, interaction-dominated threshold systems'. Ideally, an underlying phase transition can be identified which has a temperature-like control parameter tuned to the critical value by the dynamics of the system. Whether or not they exhibit (self-organised) criticality is a matter of observables, to be discussed in the next section.

1.3 Basic observables and observations

The characterisation of avalanches, by size (mass), duration, area covered and radius of gyration, was at the centre of SOC from the very beginning. If avalanches were compiled from (almost) independent patches of toppling sites, their distribution would tend to a Gaussian. Their (non-trivial) power law distribution therefore signals underlying spatiotemporal correlations (Sec. 2.1.3, Sec. 2.3), the direct measurement of which, unfortunately, is technically difficult (and rather unpopular). The distribution also determines the power spectrum, the scaling of which was used to characterise models especially in the early years of SOC. On a more immediate level, power laws do not allow for the definition of a characteristic scale *from within*; to half the probability of a certain event size, it has to be *multiplied* by a *dimensionless* constant. If the distribution is, say, exponential, a *dimensionful* constant, which allows the definition of a characteristic scale, has to be *added* (also p. 355).

The **size s of an avalanche**, sometimes also called the 'mass', is usually defined as the **number of topplings** that occur in the system between an external charge and complete

quiescence, i.e. when all activity has ceased. If the external drive does not lead to an avalanche, its size is defined as $s = 0$. Especially in analytic approaches, it can be advantageous to define the avalanche size as the **number of charges** received throughout the system, including the external drive, so that the smallest possible avalanche size is greater than 0. Depending on the details of the model, the number of topplings and the number of charges have (asymptotically in large avalanches) a fixed ratio, which is complicated by the convention of whether or not to discount the initial charge (external drive) and by dissipation at boundary sites or in the bulk.

Kadanoff, Nagel, Wu, and Zhou (1989) introduced a different measure for the avalanche size, called the **drop number** as opposed to the **flip number** introduced above. The drop number counts the number of particles, energy or force units, that are dissipated at the boundary. Experimentally easier to capture, it was identified early to be somewhat problematic (Jensen, Christensen, and Fogedby, 1989) and has since fallen out of common usage.

The **duration** T of an avalanche is the microscopic time span from the external charge by the driving to complete quiescence. Characteristics of the avalanche duration are riddled with ambiguity when the microscopic time is not defined explicitly. The number of distinct sites toppling (sometimes sites charged) in an avalanche is its **area**, A, and the average distance between every distinct pair of sites toppling during the course of the avalanche is the **radius of gyration**, r. The latter has various alternative definitions, which are borrowed from percolation theory (Stauffer and Aharony, 1994) and polymer science. Of the four features, the avalanche size is by far the most studied, followed by the duration. Much rarer is the analysis of the avalanche area and rarer still analysis of the radius of gyration.

1.3.1 Simple scaling

Avalanche features are typically either tallied into histograms or averaged directly in moments. Measured in an experiment or computer implementation of an SOC model, they can only **estimate** characteristics of the full, 'true' population average. For example, the histogram of the avalanche sizes estimates their **probability density function** (PDF), denoted[4] by $\mathcal{P}^{(s)}(s L)$. Such PDFs and derived quantities are subject to a **finite size scaling analysis**, as commonly used for ordinary critical phenomena (Barber, 1983; Privman and Fisher, 1984; Cardy, 1988).

Under the **finite size scaling** (FSS) **hypothesis** the histogram of the avalanche size s is expected to follow **simple scaling** asymptotically for $s \gg s_0$

$$\mathcal{P}^{(s)}(s; L) = a^{(s)} s^{-\tau} \mathcal{G}^{(s)}(s/s_c(L)) \quad \text{with} \quad s_c(L) = b^{(s)} L^D. \tag{1.1}$$

The two **amplitudes** $a^{(s)}$ and $b^{(s)}$ are non-universal **metric factors**, s_c is the **upper cutoff** or **characteristic avalanche size** and s_0 is the fixed **lower cutoff**. Simple scaling is an asymptote, which approximates the observed histogram increasingly well with increasing ratio s/s_0. Below the constant lower cutoff s_0 the histogram follows a non-universal function, which often depends on the details of the implementation. As opposed to the lower cutoff,

---

[4] Henceforth, a superscript indicates which observable a particular quantity describes.

the upper cutoff s_c increases with system size. It is in fact the *only L* dependent quantity in Eq. (1.1) and indicates the scale on which the features of the *dimensionless* **scaling function** $\mathcal{G}^{(s)}$ are visible. In most finite systems observables are expected to be finite as well, so that all moments have an upper bound and $\mathcal{G}^{(s)}(x)$ vanishes in large arguments x faster than any power of x.

The PDF in Eq. (1.1) is continuous, even when the observable is discrete. Some authors prefer the cumulative distribution function, which is the integral of the probability density, because it is a proper probability, namely that for the observable to be larger (or smaller) than a particular value. It is also more common in the mathematical literature.

The two **critical exponents** τ and D, *uniquely defined* in Eq. (1.1), are known as the **avalanche size exponent** and the **avalanche dimension** respectively. By dimensional consistency, the exponents are fixed to the **canonical dimension** (i.e. the engineering dimension determined by dimensional analysis) in the absence of another scale. They can obtain non-trivial values only if another scale is available to make expressions dimensionally consistent. For example, if $a^{(s)}$ in Eq. (1.1) is dimensionless, then $\tau = 1$ since $\mathcal{P}^{(s)}(s)$ is a density. The presence of another scale, say s^*, allows the creation of a dimensionless quantity s/s^*, which can be raised to any power, for example $\mathcal{P}^{(s)}(s) = s^{-3/2}\sqrt{s^*}\mathcal{G}^{(s)}(s/s_c)$. As illustrated in Sec. 2.3, such a (power law) dependence indicates **scale invariance** – the statistical appearance of a system is self-similar under rescaling of observables such as s and parameters such as L. It means that no part of the system, no matter how distant, can ever be fully self-contained, as *all* constituents, regardless how far, can contribute to a global event: 'More is different' (Anderson, 1972).

Scale invariance means the presence of a scaling symmetry, which might be due to the absence of *any* **dominant scale** (**full scale invariance** or **pure power law scaling**, see below), or, as in case of FSS, controlled by only one dominant scale, namely the system size L, which determines essentially all long-range features. As illustrated above, scale invariance does not imply the complete absence of any scale. On the contrary, additional **competing scales** are necessary for non-trivial scaling.[5]

Power laws appear on an intermediate scale, away from any cutoffs, which ultimately dominate the behaviour on the very large and the very small scales. Power laws also govern these cutoffs themselves. Not least because of that, the (very widely used) term **scale invariant** is normally applied even in the presence of a cutoff, provided that a continuous **scaling symmetry** like Eq. (1.1) is present, which allows for a **data collapse** (Sec. 2.1) to occur. Strictly, a data collapse probes only for the scaling symmetry but not (directly) for the presence of a power law. **Full scale invariance** is achieved in the absence of all cutoffs in the relevant (normally the large) length (or time, size, etc.) scale, at which point all scales can be regarded as equally important and pure power laws emerge.[6] The

---

[5] The term 'competing' might be a bit misleading, as competing scales are not meant to struggle for control of the large scale behaviour of the system. Yet, they (often) enter with non-trivial powers and are instrumental in the creation of non-trivial long-range scaling, by *cancelling* some of the dimensionality of the dominant scale, possibly by controlling the short-range scaling.

[6] Sometimes the distinction made here between 'scale invariance' and 'full scale invariance', namely the presence or absence of a scaling function like $\mathcal{G}^{(s)}(x)$ in Eq. (1.1), is made between 'scaling' and 'scale invariance'. In the following, the context should resolve any such ambiguity.

presence of an intermediate power law in a data collapse can therefore be interpreted as the onset of full scale invariance in a system with a scaling symmetry.

The exponents τ and D, characterising the self-similar power law asymptote of $\mathcal{P}^{(s)}(s)$, are **universal** and thus expected to be largely independent of implementation details, characterising an entire class of models. The scaling function $\mathcal{G}^{(s)}(x)$ is also universal across different models (Privman and Fisher, 1984, but Fig. 6.5), but like τ can depend on boundary and driving conditions.

The avalanche duration T is also expected to follow simple scaling for $T \gg T_0$

$$\mathcal{P}^{(T)}(T; L) = a^{(T)} T^{-\alpha} \mathcal{G}^{(T)}(T/T_c(L)) \quad \text{with} \quad T_c(L) = b^{(T)} L^z, \tag{1.2}$$

with two more exponents α and z, the **avalanche duration exponent** and the **dynamical exponent** (also 'dynamic exponent') respectively. The rôles of the quantities in Eq. (1.2) correspond to those in Eq. (1.1), in particular $T_c(L)$ is the **characteristic time scale** and is therefore expected to be proportional to the **correlation time**. Expressions similar to Eq. (1.1) and Eq. (1.2) are expected to apply to other quantities, such as the avalanche area or its radius of gyration, defining more exponents. A priori exponents are independent, however, constraints and symmetries, such as bulk conservation and boundary conditions, can lead to **scaling relations** or, more accurately, **scaling laws**, which normally are identities between exponents. The two most common scaling laws are exact values for the scaling of the average avalanche size, say $D(2 - \tau) = 1$ or $D(2 - \tau) = 2$, and a relation between avalanche size and duration $D(1 - \tau) = z(1 - \alpha)$. This and other consequences of the FSS hypothesis are discussed further in Ch. 2.

Critical exponents depend on the spatial dimension of the system studied. A significant simplification takes place above the **upper critical dimension**, where many systems behave as if all their constituents interact directly even though they might be many lattice spacings apart. The effect of one degree of freedom on all others can then be captured in a **molecular** or **mean-field theory** (MFT), where the molecular field 'felt locally' is the average over the entire system. In this case, exponents and scaling functions can be calculated in exact form (Sec. 8.1).

The *key observations in SOC models are thus (simple) finite size scaling of the distributions of various geometric and temporal features of avalanches, such as their size, duration, radius of gyration and area, as well as long-range spatiotemporal correlations like those observed in ordinary phase transitions, yet without the need of tuning a control parameter to its critical value.* If such features are present, a system is deemed to exhibit SOC, irrespective of the details of its setup – in the following, SOC is primarily regarded as the phenomenon, not as a class of systems or models.

1.3.2 $1/f$ noise

Particularly in the early days of SOC, power spectra of temporal signals were analysed for their power law behaviour in low frequencies f, either by analysing a temporal signal directly, or by transforming, under suitable assumptions, a distribution into a (supposed) power spectrum. The latter procedure has led to some inconsistencies in the early SOC literature – proposed as an explanation of $1/f$ noise, SOC's most prominent model seemed to produce rather $1/f^2$ noise (Jensen *et al.*, 1989, also Davidsen and Lüthje, 2001).

$1/f$ noise is widely regarded as indicating **algebraic** or **infinite correlations**, even when that might hold only in some limited sense to be discussed below. Generally, any time series with an algebraic power spectrum $S(f) \propto f^{-\alpha}$ is called $1/f$ noise, although some authors restrict the usage of the term to values of α close to 1. As SOC was intended to explain $1/f$ noise, this was seen as the hallmark of SOC. It has grown out of fashion for various reasons (but Zhang, 2000): it is somewhat cumbersome to handle numerically, few models display it solidly (but Jensen, 1990), and it is shown to exist in a variety of settings, even without accompanying spatial correlations (Montroll and Shlesinger, 1982; Marinari, Parisi, Ruelle, and Windey, 1983; Lowen and Teich, 1993; Grinstein, 1995; De Los Rios and Zhang, 1999, also Christensen, Olami, and Bak, 1992).

In the present context, the **auto-correlation function** of a (real) stationary time series $a(t)$ is normally defined as (e.g. Gardiner, 1997)

$$c(t) = \lim_{T \to \infty} \frac{1}{T} \int_{-T/2}^{T/2} \mathrm{d}t'\, a(t')a(t'+t), \tag{1.3}$$

whose convergence is guaranteed by the normalisation and by $a(t)$ being bounded, even when $a(t)$ has non-vanishing average. As $a(t)$ is recorded in the stationary state, the time average above can often be supplemented or even replaced by an ensemble average.[7] The power spectrum $S(\omega)$ of the time series is defined through its Fourier transform

$$a(\omega) = \lim_{T \to \infty} \frac{1}{\sqrt{T}} \int_{-T/2}^{T/2} \mathrm{d}t\, a(t)\mathrm{e}^{-\imath\omega t} \tag{1.4}$$

with $S(\omega) = a(\omega)a^*(\omega)$ and $\omega = 2\pi f$. Provided the Fourier transform

$$c(\omega) = \int_{-\infty}^{\infty} \mathrm{d}t\, c(t)\mathrm{e}^{-\imath\omega t} \tag{1.5}$$

exists, it equals the power spectrum by the Wiener–Khinchin theorem (Gardiner, 1997).

The significance of an algebraic power spectrum lies in the implications for its inverse Fourier transform; if $S(\omega) \propto \omega^{-\alpha}$, then[8]

$$c(t) = \int_{-\infty}^{\infty} đ\omega\, c(\omega)\mathrm{e}^{\imath\omega t} \propto t^{\alpha-1}, \tag{1.6}$$

which, one might argue, holds even when $S(\omega)$ is not a power law everywhere, provided that it decays algebraically for small frequencies, i.e. there is no characteristic long-time cutoff. This canonical result, however, has some unexpected limitations. For example, the Ornstein–Zernike-type correlation function $c(t) = \exp(-|t|/\tau)$ with correlation time τ has a Lorentzian as Fourier transform, $c(\omega) = 2\tau/(1 + \omega^2\tau^2)$. In the small frequency region $c(\omega)$ is essentially constant, so $\alpha = 0$ and one might thus conclude $c(t) \propto t^{-1}$,

---

[7] The normalisation $1/T$ is somewhat unfortunate, as it invades all subsequent calculations, but can be omitted only when $a(t)$ dies off sufficiently quickly, which is not the case at stationarity (also Bonabeau and Lederer, 1994).

[8] Here and in the following $đ\omega = \mathrm{d}\omega/(2\pi)$.

which is clearly incorrect and in more than one way. Not only does that power law deny the exponential decay of the correlations in the limit of large t as $c(t) = \exp(-|t|/\tau)$, but the correct result of the Fourier transform of $c(\omega) = $ constant is a Dirac δ function, $\delta(t)$, which has the same dimension as t^{-1}, but a completely different character.

The result $c(t) \propto t^{\alpha-1}$ is based on dimensional analysis and ignores the potential presence of other time scales, such as τ. Nevertheless, $\alpha = 1$ is widely regarded as a sign of 'infinite correlations' or 'long-range temporal correlations', as it seems to indicate an essentially constant correlation function $c(t)$. Even though this argument seems flawed, a lack of a characteristic frequency in the slow modes of the power spectrum undoubtedly is significant in its own right.

The link to SOC was originally made in the following way. The auto-correlation of a Markov process can be shown to be a sum of exponentials, and by considering the leading term only, $\exp(-t/\tau)$, a power spectrum of of the form $2\tau/(1 + \omega^2\tau^2)$ is a natural choice. To explain the frequent occurrence of $1/f$ power spectra, van der Ziel (1950) suggested their superposition with varying correlation time (also Hooge and Bobbert, 1997). Bak, Tang, and Wiesenfeld (1987, 1988) assumed that in avalanching systems these correlation times were essentially given by the avalanche durations, so that the overall spectrum is given by

$$S(\omega) = \int_0^\infty d\tau\, D(\tau) \frac{2\tau}{1 + \omega^2\tau^2}, \qquad (1.7)$$

where $D(\tau)$ is the weight, with which the relaxation time τ enters; if $D(\tau) \propto \tau^{-a}$ then $S(\omega) \propto \omega^{a-2}$, i.e. $\alpha = 2 - a$ and accepting the dimensional analysis above $c(t) \propto t^{1-a}$. Bak *et al.* (1988) argued that $D(\tau)$ was in fact the re-weighted distribution of avalanche durations. The re-weighting incorporated the different amplitudes with which avalanches of different size and duration enter into the correlation function. This was confirmed numerically by measuring $D(\tau)$ and comparing the derived scaling of the power spectrum with that calculated directly from the activity time series.

These results were soon revised, numerically as well as analytically. Jensen *et al.* (1989) proposed not only a different weighting, but also a different correlation function of individual avalanches, arriving at $S(\omega) \propto \omega^{-2}$. The relation of the exponent α to exponents characterising avalanche size and duration distribution depends on that choice, and different assumptions therefore lead to different results (e.g. Kertész and Kiss, 1990; Christensen, Fogedby, and Jensen, 1991, briefly reviewed in Jensen, 1998). The variety of the underlying assumptions ties in well with the wider discussion about the origin of $1/f$ noise. The review by Weissman (1988) contains a comprehensive overview of many such approaches.

1.3.3 Edge of chaos

Historically, SOC was introduced at a time when chaos, fractals and dynamical systems were very widely studied (Kadanoff, 1986), so that the link between SOC and chaos seemed a natural one to make. Bak (1990, also Bak *et al.*, 1987) established it explicitly by considering the SOC 'state' as a dynamical system with many degrees of freedom, operating at the 'border of chaos'. Around the same time, Grassberger (1986) suggested 'a quantitative theory of self-generated complexity' and Langton (1990) introduced his concept of 'computation at the edge of chaos', investigating cellular automata similar to the models considered in SOC. Later Bak (1996, also Bak and Sneppen, 1993) made a clear distinction between SOC and chaotic dynamical systems, as the latter produce white noise for all values of a control parameter, except for one critical value, at the transition to chaos, where the behaviour is complex. Like ordinary critical phenomena, it requires tuning and thus 'cannot explain complexity'. Stapleton, Dingler, and Christensen (2004) illustrated the meaning of 'the edge of chaos' by showing that in the OSLO Model perturbations never heal, but never get out of control either. Rather, the system remembers them indefinitely.

The language and some concepts of dynamical systems are still commonly used to describe SOC and complex systems, such as the 'SOC attractor' (e.g. Mehta and Barker, 1994; Corral, 2004a) or sensitivity to initial conditions and perturbations (e.g. Pinho and Andrade, 2004; Stapleton *et al.*, 2004). Yet, there are some important differences from chaotic dynamical systems. First of all, SOC models have *many* degrees of freedom and are normally even considered in the limit of infinitely many degrees of freedom, the thermodynamic limit. As a consequence, **phase space** is so high dimensional, it often makes little sense to study the phase portrait. In fact, most of the statistical analysis is based on an incredibly small subsection of phase space that it explores, and relies heavily on the assumption that it is in a sense representative of the whole accessible space. That is normally *very* different in low dimensional dynamical systems. Secondly, SOC models display scaling with fractal dimensions in spatiotemporal correlations, as well as global observables, whereas it is the phase portrait of dynamical systems that is probed for such fractal features. Even when they are often mentioned in the context of fractals, dynamical systems are generally not thought to be at the origin of well-known natural fractals, such as the iconic picture of the coastline of Southern Norway (Feder, 1988, p. 7). This is where SOC comes in as the 'physics of fractals'.

Thirdly, as pointed out by Bak (1996), dynamical systems might have a critical point, but are chaotic in a broad region of their parameter space. SOC is concerned with systems that are thought to organise themselves towards this critical point. Even if chaos were a relevant perspective on SOC, it would not explain this (supposed) behaviour. If chaos and dynamical systems are applicable, then at most this is to the same extent as critical phenomena. Finally, SOC is frequently stochastic, whereas chaotic systems are normally deterministic.

This last point, however, hints at another link between SOC and chaos: the origin of noise. A deterministic world, whose equation of motion preserves many symmetries, can still give rise to a noisy environment by virtue of deterministic chaos. Some SOC models, such as the OLAMI–FEDER–CHRISTENSEN earthquake model (Sec. 5.3) are, except for their initial condition, deterministic and, nevertheless, produce time series seemingly 'as random as' those from stochastic processes.

1.3.4 The signature of SOC

The difficulty of producing a clear definition of SOC and its distinctive features is frequently bemoaned in the literature. If SOC means self-organisation to some underlying ordinary critical point, which is the perspective to be taken here, its most important sign is **avalanching**, which is an expression of the **separation of time scales** of driving and relaxation. Secondly, the event size distribution should display **simple finite size scaling**. This restriction is very dramatic and excludes many models that are commonly believed to display **multiscaling** (or **multifractal scaling**) and were proposed early enough to be regarded by now as classic or even paradigmatic SOC models (Bak *et al.*, 1987; Kadanoff *et al.*, 1989). In fact, there are even exactly solvable models that display multiscaling (e.g. Welinder, Pruessner, and Christensen, 2007) and whose scale invariance is well understood. However, the absence of simple finite size scaling is difficult to reconcile with the traditional understanding of scale invariant behaviour as found at ordinary phase transitions. If **multiscaling** is admitted as indicative of underlying scale *invariance*, other types of broad distributions with **fat tails** (also called 'heavy tails') are equally legitimate indicators, which would leave the concept of SOC too loosely defined.

It is difficult to fit a system in the framework of SOC without being able to identify avalanches, yet the absence of extensive statistics can be tolerated as long as moments can be shown to scale systematically with system size. For example, an average avalanche size which diverges in the thermodynamic limit might be regarded as a basic hint of SOC, without referring to the scaling of the full avalanche size distribution.

Ideally, in an SOC model, a control parameter can be identified, which allows a phase transition to be triggered if the control parameter is freely manipulated (in a modified version of the SOC model) rather than being left to self-organisation. Clearly, no tuning should be necessary to obtain the scaling in any supposed SOC system. Strictly, the DROSSEL–SCHWABL Forest Fire Model (Sec. 5.2) and the BAK–SNEPPEN Model (Sec. 5.4) violate this requirement, but in these cases the tuning might be seen as trivial or irrelevant to the dynamics, affecting only the definition of the observables.

1.4 Universality

Universality is the observation first made in the context of ordinary critical phenomena, that certain **universal quantities** are independent of many details of a system, such as the specifics of the interaction or the type of lattice. They are *necessarily dimensionless*. These universal quantities are first and foremost critical exponents, but include also the scaling function as well as many dimensionless moment and amplitude ratios, discussed detail in Ch. 2. As SOC is concerned with finite size scaling, boundary conditions can alter the universal behaviour (Privman, Hohenberg, and Aharony, 1991), as do driving conditions, an effect that is well understood in the presence of conservation laws (Sec. 8.1.1.1) and is in line with the observation in ordinary critical phenomena, that universal quantities lose their dependence on boundary conditions, lattice shape and aspect ratio only if the thermodynamic limit is taken before the critical point is reached.

At the heart of universality is the insight that the features of asymptotes are determined by a few characteristics of the Hamiltonian or more generally the equation of motion of a system, such as symmetries and types of interaction. Systems that share universal features ought to share these characteristics and vice versa. They are said to be in the same

universality class. Often very few features of the universality class suffice to determine it uniquely, implying that certain other features and therefore the characteristics of the equation of motion *must* be present as well. In this sense, universality is a form of reductionism (Anderson, 1972). Even when it is not possible to 'see' the universal features merely by inspecting the interaction, the latter is nevertheless intricately linked to the former and determines it. If the interaction is known, the universal features that follow from it can at least be 'looked up', even when they cannot be (easily) derived.

Universality therefore allows the study of universal phenomena on different scales and in simplified models, provided they incorporate the important interactions. Maybe more crucially, the inverse approach allows determination of the relevant interactions solely on the basis of the universal features of the asymptotes. For example, inspecting the cluster size distribution of a forest could therefore reveal by what basic type of interaction this population evolves in time and space.

Universality justifies the simplified models in SOC, which ignore all but a few details of what they are modelling, such as ricepiles or evolution, and allows them to display the 'right' behaviour provided they incorporate the few relevant interactions. However, it is found only very rarely. As reviewed below (Sec. 1.4.2), there is only one widely accepted non-trivial universality class, the MANNA universality class (Sec. 6.2.1.2, p. 177). Apart from that, directed models generally belong to one of a well-understood set of trivial universality classes (Sec. 8.4 and Ben-Hur and Biham, 1996) and variants of the BTW Model have been reported to form universality classes (Kadanoff *et al.*, 1989), based, however, on a multifractal analysis. It is a common critique of SOC, that each new model is a representative of a new universality class.

Unfortunately, experiments have not yielded much evidence for universality in SOC (Ch. 3) either. In summary, universality is very important but largely missing (Kardar, 1996) in SOC: 'The continuing challenge to both theorists and experimentalists is to go beyond observation of power law distributed events to the identification of the exponent and demonstration of its universality in different systems – that is, to go from qualitative to quantitative agreement.'

The taxonomy of universality classes in **non-equilibrium**, i.e. in systems lacking **detailed balance**, was famously introduced by Hohenberg and Halperin (1977). Their work focused on systems governed by a Hamiltonian, which relax back to the equilibrium state, where detailed balance is restored. These systems might thus be called **out-of-equilibrium**. Their features derive largely from their equilibrium counterparts, but are enriched by the additional presence of a dynamics, giving rise to a notion of time. Owing to the transport of particles, energy or force taking place in SOC systems, they clearly lack detailed balance, and do not even relax back to an equilibrium state. Rather, they are permanently out of equilibrium. Phase transitions in this regime, also known as **far-from-equilibrium** critical phenomena, have received increasing attention over recent years. Far-from-equilibrium systems have been studied for a long time in the form of growth phenomena (Sec. 9.1), where, however, the anomalous scaling is often generic, i.e. it does not require tuning, as, for example, in the case of the KPZ equation. In that respect they resemble SOC, however, they do not normally incorporate a separation of time scales and (thus) do not display avalanching. Where they do, they normally require tuning, as, for example, in the qEW equation. Universality in far-from-equilibrium phenomena different from growth, in particular in **absorbing state phase transitions**, was recently reviewed by Ódor (2004) as well as Lübeck (2004) and Henkel, Hinrichsen, and Lübeck (2008). In the following, some elements of universality in SOC are reviewed, most importantly appropriate observables and the various universality classes of SOC.

1.4.1 Universal quantities

The central limit theorem states that the suitably rescaled sum of independent random variables is distributed like a Gaussian. This is a form of universality, because the limiting distribution does not depend on the distribution the random variables are drawn from, provided only it has finite variance. On the other hand, it is clear that different initial distributions generally result in distributions of the sum, which differ widely in appearance, in particular have different mean and variance. The apparent clash is resolved by 'suitable rescaling' or by deriving universal quantities. In the case of the central limit theorem, for example, the $2n$th central moment normalised by the variance, i.e. divided by the variance raised to the nth power, is $(2n - 1)!!$. In this section, universal quantities relevant in SOC are briefly introduced.

The most important universal observables in the study of SOC systems are the **exponents** D, τ and slightly less so z and α, as *defined* through simple finite size scaling, Eq. (1.1) and Eq. (1.2) respectively. The consequences of simple scaling, the corrections to it and the pitfalls are introduced in Ch. 2 and the measurement of suitable observables is discussed in Ch. 7. Very often, only these exponents are analysed and reported. There are more exponents, however, for example characterising the scaling of the avalanche area and its radius of gyration.

In addition to exponents, **scaling functions** are universal up to their normalisations. Being functions, they are not as easy to compare across models as exponents. Agreement or disagreement is difficult to quantify, even when facilitated by a **data collapse**. This is further exacerbated by the presence of non-universal contributions in the event size distribution. Derived characteristics of the scaling function, such as **moment ratios**, can be used to circumvent such problems. A moment ratio is a form of **amplitude ratio**, as established in ordinary critical phenomena (e.g. Privman et al., 1991; Salas and Sokal, 2000) and traditionally involves powers of two moments of the relevant observable, for example, the second and the fourth moment of the avalanche size.[9] In SOC, where the distribution more often than not follows power laws different from unity, i.e. $\tau \neq 1$ and $\alpha \neq 1$, at least three moments are needed, such as $\left(\langle s \rangle \langle s^3 \rangle \right) / \langle s^2 \rangle^2$, where the moments are defined via

$$\langle s^n \rangle = \int_0^\infty \mathrm{d}s\, s^n \mathcal{P}^{(s)}(s). \tag{1.8}$$

The major advantage of using moments to characterise distributions is that contributions from the non-universal part of the distribution, which enter primarily through small values

---

[9] Although 'amplitude' often refers to a metric factor, amplitude ratios are often set up in a way that metric factors do not enter at all, or at least combine to a dimensionless, universal ratio.

of the observable, are effectively masked if the moments are of sufficiently high order, e.g. $n > \tau - 1$.

Correlation functions are of great conceptual importance to SOC but are rarely measured directly, because they are computationally very expensive and often turn out to be too noisy. In translationally invariant systems they are expected to follow simple scaling and thus can be analysed along the same lines as a distribution function, determining exponents as well as the scaling function or amplitude ratios. In many SOC models, however, at least some boundaries have to be open, which introduces additional scales, so that correlations are not only a function of the distance between two points, but also of the distance to the boundary. These technical difficulties might be responsible for correlation functions being virtually absent in the literature. Block scaling is based on the integral of the correlation function and is often better behaved (Pruessner, 2008), yet is equally absent in the literature.

All universal quantities together characterise a **universality class**, but very few such quantities normally suffice to determine it uniquely. Universal quantities are generally a function of the spatial dimension d. Below the lower critical dimension, normally $d = 1$ in SOC, the systems do not display scale invariant behaviour. Above the upper critical dimension d_c they display their mean-field values (Sec. 8.1.3), which are often identical across universality classes. At d_c the mean-field values are amended by logarithmic corrections. Often, some scaling laws break down above d_c. A universality class is thus normally specific to a certain spatial dimension.

The question what basic ingredients a system needs in order to belong to a particular universality class remains one of the unanswered, fundamental questions in SOC. The answer would allow much more reliable modelling of complex systems, at least with respect to their universal features, without getting caught up in the details. In turn, it would allow the identification of such fundamental interactions, which determine the universality class, without actually analysing a given system in detail.

The number of universal quantities generally required to determine the SOC universality class is still unknown and might not be unique. As in out-of-equilibrium critical phenomena, the details of the dynamics and therefore the microscopic time scale might affect dynamical universal features, such as the dynamical exponent z or the avalanche duration exponent α. While this is not widely studied, because the dynamics of many models is actually fixed, the choice of boundary conditions on the other hand is known to affect universal exponents (normally τ but not D, Nakanishi and Sneppen, 1997). This observation is in line with the finding that SOC models can be associated with an underlying absorbing state phase transition, an interfacial or a growth phenomenon, which immediately sets D, but not τ. In that sense, some universal quantities are more universal than others. Accepting such a hierarchy of exponents, D seems to determine the wider universality class, z the dynamics and τ reflects the boundary conditions; where appropriate scaling laws (Sec. 8.1.1.1) are known, D determines τ and vice versa. From the scaling laws discussed in Sec. 2.2.2 it seems that every independent observable is characterised by only one new exponent, but what it depends on remains an open question. If universality prevails, any such additional exponent is determined by the universality class.

Unfortunately, the quantity most frequently reported in the SOC literature is τ and, even more unfortunately, it is very often based on a 'straight line fit' of $\mathcal{P}^{(s)}(s)$, Sec. 2.1.2.1.

The next most common exponents are the avalanche dimension D, the avalanche duration exponent α and finally the dynamical exponent z. Together, D, z as well as τ and α make up what should be regarded the **standard set of exponents** of an SOC model. Few authors report exponents for avalanche areas and their radius of gyration, which can be explained by the computational effort these measurements often require. Although rarely reported, it takes hardly any additional computational effort during the simulation and in the data analysis to derive a few, independent universal moment ratios (Sec. 2.2.1).

The significance of the particular value of a critical exponent or any other universal quantity lies mainly in its power to determine the universality class. Applications of critical exponents, i.e. 'real life' situations where an exponent is used to calculate another quantity, are very rare. [10] Even then, it seems its significance does not extend far beyond its ability to summarise a particular dependence of an observable from a control parameter.

1.4.2 Universality classes of SOC

The first comprehensive comparative study of SOC models was undertaken by Kadanoff *et al.* (1989), considering models with different rules in one and two dimensions. All being derived from the original model by Bak *et al.* (1987, the BTW Model), they differ in the range of interaction and details of that interaction, which, in two dimensions, involves many different choices for the neighbourhood sites interact with. None of the models displayed very robust scaling and only models that had similar rules seemed to belong to the same universality class, with universal as well as non-universal quantities coinciding. This is an interesting outcome in its own right, but not one that is testimony to the power of universality. The only models that showed clear signs of universality in the traditional sense, i.e. the same scaling behaviour without being identical in a trivial sense, were two directed ones.

Directed SOC models and their universality are very well understood (Sec. 8.4). Dhar and Ramaswamy (1989) introduced a directed version of the BTW Model and solved it in all dimensions. It has deterministic updating rules (but stochastic driving) and is solved by mapping it to an annihilating random walk. Pastor-Satorras and Vespignani (2000e) analysed numerically several directed models with stochastic rules. Paczuski and Bassler (2000) as well as Kloster, Maslov, and Tang (2001) placed them in a wider theoretical framework, using again a mapping to a random walk. These are undoubtedly very encouraging results, yet the mapping to a random walk is often based on the insight that there are no (relevant) correlations in the bulk (but Hughes and Paczuski, 2002), rendering these models in some sense trivial.

Many of the early SOC models were designed to be in the same universality class as the BTW Model, to support the case of universality in SOC. The ZHANG Model (Zhang, 1989) is one of these early developments, but in face of the difficulties with deriving exponents for the BTW Model and the ZHANG Model, this case is still not settled. The MANNA Model (Manna, 1991b) was another such attempt to reproduce the scaling of the BTW Model in

---

[10] Apparently, the pressure dependence of the flow rate of fuel nozzles follows a non-trivial power law over several decades, and this exponent *is* actually in use in the industry (Fenger, 1976; Schwoll, 2004), even when its meaning is reduced to hardly more than a good fit.

a simplified model. Ironically, the scaling of the MANNA Model turns out to be far more robust than that of the BTW Model. When the OSLO Model came along half a decade later (Christensen, Corral, Frette, *et al.*, 1996), nobody noticed at first that it was in the same universality class as the MANNA Model, because the two models were 'at home' in different spatial dimensions and driven in a way that made their exponent τ appear very different. Nakanishi and Sneppen (1997) were the first to note that these models are in fact in the same universality class. Their update rules, on the other hand, are different enough to justify the notion of real universality in SOC. Many more models, many of which are listed in Sec. 6.2.1.2, followed. The **MANNA universality class**, also known as the **OSLO universality class** among others, is by far the largest universality class in SOC and contains most of the SOC models that display robust scaling. In terms of its level of theoretical understanding, it is second only to the universality classes of directed models.

The three classes, BTW Model, MANNA universality class and directed models, were first suggested by Ben-Hur and Biham (1996). Later attempts to categorise SOC models focused on more general features rather than on specific models along some natural dividing lines such as stochastic/deterministic (discussed in Sec. 6.2.1.2), Abelian/non-Abelian (Sec. 8.3), directed/undirected (Sec. 8.4) (Biham, Milshtein, and Solomon, 1998; Milshtein, Biham, and Solomon, 1998; Biham, Milshtein, and Malcai, 2001; Hughes and Paczuski, 2002). By that time, it had become fashionable to invent ever more models that were apparently *not* in the same universality class as any of the known models, which has left a plethora of models which are not easily classified. Models that are paradigmatic in their features are detailed in Part II.

Only a few models have been studied systematically with respect to fundamental changes in the microscopic dynamics, such as changes of the lattice type, the interaction range, or model-specific parameters. In the work mentioned above, Kadanoff *et al.* (1989) investigated the effect of changing the interaction range and other parameters of the dynamics ('limited' versus 'un-limited' models) for variants of the BTW Model, which was repeated for variants of the OSLO Model by Markošová (2000b). Universality in the former class was restricted mainly to trivial identity, whereas the latter class displayed visible changes in the macroscopic behaviour, preserving universal quantities, however. Hughes and Paczuski (2002) changed the **Abelian symmetry** in directed models and found a significant change in spatial correlations, without affecting avalanche exponents.

Very few studies have considered lattices different from the usual square and hyper-cubic ones. This might come as a surprise, given that universality across different lattices is generally perceived as a cornerstone of equilibrium critical phenomena (Fisher and Sykes, 1959; Gaunt, Fisher, Sykes, and Essam, 1964; Stanley, 1971; Syozi, 1972). Duarte (1990), cited by Manna (1990), was probably the first to consider this issue by studying the BTW Model on a triangular lattice (later by Manna, 1991a). Much later, Hu and Lin (2003) extended the study to many more lattice types and found the known universal features of the BTW Model, based on the wave decomposition, reproduced. Azimi-Tafreshi, Dashti-Naserabadi, Moghimi-Araghi, and Ruelle (2010) performed analytical calculations, based on conformal field theory and Schramm–Loewner evolution, for the BTW Model on a hexagonal lattice. The effect of disorder on the BTW Model was studied by Karmakar, Manna, and Stella (2005), who found, depending on certain symmetries, either the scaling of the MANNA universality class, or multiscaling as known for the BTW Model. Some numerical simulations exist for SOC models on fractal lattices, mostly for the BTW Model (Kutnjak-Urbanc, Zapperi, Milošević, and Stanley, 1996; Daerden and Vanderzande, 1998, for the MANNA Model, Huynh, Chew, and Pruessner, 2010). Analytical results are available for directed models on triangular (Dhar and Ramaswamy, 1989) and partially directed square lattices (Kloster *et al.*, 2001), reproducing those of the usual directed square lattice.

The following chapter discusses scaling and universality in greater technical detail. It forms the basis required to appreciate fully the scaling behaviour found in experiments and observations (Ch. 3), as well as numerical models which are discussed in Part II.

2 Scaling

In broad terms, the aim of the analysis of a supposed self-organised critical system is to determine whether the phenomenon is merely the sum of independent local events, or is caused by interactions on a global scale, i.e. cooperation, which is signalled by algebraic correlations and non-Gaussian event distributions.[1] Self-organised criticality therefore revolves around scaling and scale invariance, as it describes the asymptotic behaviour of large, complex systems and hints at their universality (Kadanoff, 1990). Numerical and analytical work generally concentrates on the scaling features of a model. Understanding their origin and consequences is fundamental to the analysis as well as to the interpretation of SOC models, beginning at the very motivation of a particular model and permeating down to the level of the presentation of data.

During the last fifteen years or so, the understanding of scaling in complex systems has greatly improved and some standard numerical techniques have been established, which allow the comparison of different models, assumptions and approaches. Yet, there is still noticeable confusion regarding the implications of scaling as well as its quantification.

Most concepts, such as universality and generalised homogeneous functions, are taken from or are motivated by the equilibrium statistical mechanics of phase transitions (Stanley, 1971; Privman *et al.*, 1991), and were first applied to SOC in a systematic manner by Kadanoff *et al.* (1989). Yet, what appears to be rather natural in the context of *equilibrium* statistical mechanics, might not be so for complex systems. Scaling is a continuous symmetry (Barenblatt, 1996) which in equilibrium has a long-standing tradition and a beautiful 'exegesis' in terms of the free energy and the structure of the Hamiltonian entering the Boltzmann–Gibbs factor. For example, the lower cutoff in distributions can be associated with the (crystal) lattice constant or with an ultraviolet cutoff in an action, and the scaling of observables can be traced back to the scaling of the free energy density under change of length scales (Kadanoff, 1966). After all, equilibrium statistical mechanics is an approach to thermodynamics and therefore an expression of the real world. Complex systems and SOC models, on the other hand, are more often than not **spatially extended stochastic processes**, conceived on the lattice and originally involving only dimensionless parameters to describe them. A priori, one cannot appeal to the physical world to derive symmetries under rescaling of time and space.

This chapter introduces scaling in SOC, often referring to corresponding concepts in equilibrium statistical mechanics. The key observables, most importantly the avalanche

---

[1] In the following, it is normally assumed that variances are finite, which is guaranteed whenever observables are bounded, so that the central limit theorem of suitably defined observables applies. Otherwise, non-Gaussian distributions can also be the result of *independent* events.

size, have already been introduced in Sec. 1.3. Even though the latter is used in almost all examples below, all results apply equally to any other observable, except where indicated. In the first section, **simple scaling** will be discussed in a broader context, followed by a section on moment scaling, which is the key concept used in the modern analysis of SOC systems. The chapter concludes with two short sections on correlation functions and on multiscaling.

2.1 Distribution functions

The probability density for an event of size s to occur is given by the **probability density function** (PDF) $\mathcal{P}^{(s)}(s)$. It is a function of a continuous argument $s \in \mathbb{R}$ and strictly speaking therefore not applicable to discrete observables, such as the avalanche size in discrete systems. Yet, none of the following changes significantly if the condition of continuous s is lifted. Similarly, event sizes s will be assumed to be non-negative.

A remark on the terminology is in order. The **cumulative distribution function** (CDF), $F^{(s)}(s) = \int_0^s ds' \, \mathcal{P}^{(s)}(s')$, is in many respects better behaved than the density $\mathcal{P}^{(s)}(s')$: $F^{(s)}(s)$ is the probability that the avalanche size is less than s, which can be continuous, even if avalanche sizes are discrete. Any quantity derived from the PDF can equally well be derived from the CDF. The CDF is more widely used among mathematicians, while the PDF is more common in physics, as it can be 'measured' in the form of a histogram. Unfortunately, the PDF is normally referred to as the 'distribution' in the literature, even when mathematicians reserve this term for the CDF. To avoid this confusion, the usage of the CDF is confined in the following to very few special cases and 'distribution' is normally used synonymously with PDF.

The moments of the PDF are defined as

$$\langle s^n \rangle = \int_0^\infty ds \, s^n \mathcal{P}^{(s)}(s), \tag{2.1}$$

which do not necessarily all exist. Throughout the following, the normalisation condition $\langle s^0 \rangle = \langle 1 \rangle = 1$ is assumed to apply to the PDF. It is generally advisable to include zero-size events in the moments, as otherwise exact relations, in particular for the first moment (Sec. 8.1.1.1), can break down.[2] Excluding events of size 0 amounts to rescaling all moments by the same factor, namely essentially $1/(1 - \mathcal{P}^{(s)}(o))$, and therefore does not affect either the appropriate moment ratios or the scaling of the moments, unless $1 - \mathcal{P}^{(s)}(o)$ is a power law. Nevertheless, results on the basis of conservation lose validity, if the effect of some of the particles added is discounted.

If s is a global event, made up from smaller, individual contributions throughout a spatially extended system and subject to some local variability, its PDF is expected to be a Gaussian, unless the local contributions are correlated. If the PDF does not even possess

---

[2] When vanishing event sizes are collected into the PDF, $\mathcal{P}^{(s)}(0) \neq 0$, the condition $\langle s^0 \rangle = 1$ can lead to inconsistencies, since $0 = \lim_{n \to 0^+} 0^n \neq \lim_{s \to 0} s^0 = 1$.

a **characteristic scale**, then these correlations are equally expected to be scale invariant, which is discussed further in Sec. 2.3 (also Sec. 2.1.3).

In most of what follows, it will be assumed that the PDF is subject to **simple scaling**, sometimes referred to more specifically as **(simple) finite size scaling** (Barber, 1983; Privman and Fisher, 1984; Cardy, 1988). Yet, the origin of the scaling, the finite linear size L of the (d-dimensional) system considered, is not strictly relevant for the following arguments. It is, however, fundamental to SOC that the *only* dominant scale is the finite size of the system. As for the terminology, **finite size scaling** (FSS) is a well behaved **finite size effect**, whereas **finite size corrections** refer to undesired effects due to finite size, such as corrections to scaling (Sec. 2.1.1, Eq. (2.7) and Sec. 2.2, Eq. (2.26)), which vanish asymptotically.

Simple scaling of a PDF means that it obeys, asymptotically in large $s \gg s_0$,

$$\mathcal{P}^{(\mathrm{s})}(s; L) = a^{(\mathrm{s})} s^{-\tau} \mathcal{G}^{(\mathrm{s})}(s/s_{\mathrm{c}}(L)) \quad \text{with} \quad s_{\mathrm{c}}(L) = b^{(\mathrm{s})} L^{\mathrm{D}}, \tag{2.2}$$

see Eq. (1.1) on p. 12. Without changing $s_{\mathrm{c}}(L)$ and thus L it is impossible to probe for simple scaling, even if $\mathcal{P}^{(\mathrm{s})}(s; L)$ can be approximated by a power law $\propto s^{-\tau}$; FSS can be revealed and the relevant quantities be determined only if the finite size is changed; simple (finite size) scaling is the invariance under a continuous symmetry, which can be probed *only* by changing the scale.

The nature of simple scaling as an asymptote for $s \gg s_0$ is more than a technicality.[3] If Eq. (2.2) applied for all s, regardless how small, i.e. $s_0 = 0$, then τ would be bound to unity (Christensen, Farid, Pruessner, and Stapleton, 2008). It is difficult to pin down a precise definition of the asymptote, though. One fruitful interpretation is that for every $\delta > 0$ an s_0 exists, so that the difference of the right and left hand sides of Eq. (2.2) is smaller than δ for all $s > s_0$ and all L. Most authors do not dwell on these details, but those who do sometimes include $s_{\mathrm{c}}(L) \gg s_0$ as a condition, which must be included once s is integrated over, Eq. (2.4). A more pragmatic perspective is to impose similar conditions on the asymptotes of moments of $\mathcal{P}^{(\mathrm{s})}(s)$ as they follow from the simple scaling form Eq. (2.2).

The ingredients of Eq. (2.2) have already been introduced in Sec. 1.3.1. The **metric factors** $a^{(\mathrm{s})}$ and $b^{(\mathrm{s})}$ do not depend on L and allow the (universal finite size) **scaling function** (Privman and Fisher, 1984) $\mathcal{G}^{(\mathrm{s})}$ to be a dimensionless function of a dimensionless quantity. The metric factor $a^{(\mathrm{s})}$ is not needed if $\tau = 1$, which implies that if a system is parameterised by so few quantities that no dimensionful metric factor can be formed, $\tau = 1$ follows from dimensional analysis (Barenblatt, 1996).

The **lower and upper cutoffs**, s_0 and s_{c} respectively, are a measure for the region where scaling applies. Below s_0 simple scaling is not a sensible description of the data and above the characteristic event size s_{c}, or some multiple of it, the PDF drops off dramatically. The scaling function modulates this drop on the scale given by s_{c}.

The independence from L of the metric factors is fundamental to the definition of the two **universal exponents** τ and D, which, in the present case, are called the avalanche size exponent and the avalanche dimension, respectively. If $a^{(\mathrm{s})}$ were allowed

---

[3] Similarly, in ordinary critical phenomena, power law scaling in the reduced temperature is an asymptote even after taking the thermodynamic limit.

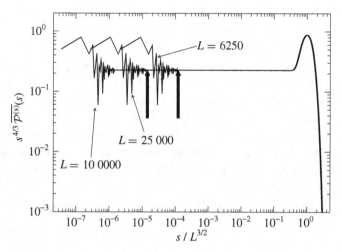

Figure 2.1 Example of a data collapse, based on the data used in Sec. 7.4.2. The *bold, vertical arrows* indicate the value of s_0/s_c above which curves coincide, retracing the scaling function.

to scale arbitrarily in L, say $a^{(s)} = a^{(s)}_0 L^\mu$, then Eq. (2.2) could be rewritten as $a^{(s)}s^{-\tau}\mathcal{G}^{(s)}(s/s_c(L)) = \tilde{a}_s s^{-\tau+\mu/D}\tilde{\mathcal{G}}^{(s)}(s/s_c(L))$ with new, constant metric factor $\tilde{a}_s$ and scaling function $x^{\mu/D}\tilde{\mathcal{G}}^{(s)}(x) = \mathcal{G}^{(s)}(x)$. Similarly, if $b^{(s)}$ could depend on L it would render D ill defined. This is one reason why the notation $\mathcal{P}^{(s)}(s) \propto s^{-\tau}$ can be very misleading, as it is not immediately evident that the factor of proportionality cannot be L dependent. [4] In fact, $\mathcal{P}^{(s)}(s)$ might behave *nowhere* like $s^{-\tau}$, when the scaling function itself follows a power law. If anything, $\mathcal{P}^{(s)}(s)$ follows the **apparent exponent**, $\mathcal{P}^{(s)}(s) \propto s^{-\tilde{\tau}}$, Sec. 2.1.2.1.

In the region $s \gg s_0$ data for different $s_c(L)$ can be collapsed onto a single curve, essentially the scaling function. This can be achieved by plotting $\mathcal{P}^{(s)}(s)s^\tau$ against $s/s_c(L)$, known as a **data collapse** (also Sec. 7.4.2), exemplified in Fig. 2.1. A data collapse is the most direct manifestation of the scaling of the PDF and provides estimates of the exponents, although they are more easily estimated through a moment analysis. By normalising the scaling function using two conditions, which determine the unknown, non-universal metric factors $a^{(s)}$ and $b^{(s)}$, the universal scaling function can be compared across models.

2.1.1 Upper and lower cutoffs

The upper and lower cutoffs deserve further attention. The lower cutoff separates the **non-universal part** $f(s; s_c(L))$ of the PDF from the universal, scaling part

$$\mathcal{P}^{(s)}(s) \approx \begin{cases} a^{(s)}s^{-\tau}\mathcal{G}^{(s)}(s/s_c(L)) & \text{for } s > s_0 & (2.3a) \\ f(s; s_c(L)) & \text{otherwise} & (2.3b) \end{cases}$$

which is an approximation, because of the asymptotic character of the scaling part. Increasing s_0 sufficiently allows Eq. (2.3) to be an arbitrarily good representation of the full

---

[4] This notation alludes to the Pareto distribution, which is a power law PDF without scaling function.

PDF. If the condition $s \gg s_0$ is kept, the right hand side of Eq. (2.3a) becomes again the exact asymptote. Figure 2.4(b) shows a case where s_0 is chosen too small, so that deviations are visible between the supposed universal part (dotted line, Eq. (2.19b)) and the actual (observed) distribution $\mathcal{P}^{(s)}(s)$.

The non-universal part is the part of the PDF that does not scale in s_c, i.e. its asymptote in large s_c is not a power law. It is strongly model and detail dependent and often visibly structured, as can be seen in the ragged appearance of $\mathcal{P}^{(s)}(s)$ for small arguments shown in Fig. 2.1. Generally, the non-universal part converges in the limit of large $s_c(L)$, so that the probability for a particular, small avalanche to occur approaches a certain constant in big systems. This implies that the median avalanche size does not scale (in particular not when $\tau > 1$), i.e. it is a non-universal, typically rather small constant. Only when $s_0 = 0$ does the scaling extend all the way down to the smallest avalanches and, as discussed below, $\tau = 1$ follows. In the following, the limit $s_c(L) \to \infty$ is frequently invoked, motivated by the assumption that the upper cutoff of the event size distribution diverges with the system size, as discussed further below.

Based on the definition (2.1), moments of the PDF are determined by Eq. (2.3),

$$\langle s^n \rangle = f_n^{(s)}(s_0; s_c) + a^{(s)} s_c^{1+n-\tau} g_n^{(s)}(s_0/s_c) + \mathcal{O}(\epsilon) \tag{2.4}$$

where the term $\mathcal{O}(\epsilon)$ is a correction to be added for any finite s_0, s_c; it can be made arbitrarily small by increasing s_0, which in turn constrains s_c, since $s_c \gg s_0$. Its origin is a mismatch at $s = s_0$ of the two cases in Eq. (2.3) for any finite s_0, as scaling is imperfect at any finite s, even in the limit of divergent upper cutoff $s_c \gg s_0$, where the scaling part of Eq. (2.4) dominates (see the mismatch at the dashed line in Fig. 2.4(b)). The two functions $f_n^{(s)}$ and $g_n^{(s)}$ encapsulate features of the non-universal and the universal part of the PDF respectively (Christensen et al., 2008):

$$f_n^{(s)}(s_0; s_c) = \int_0^{s_0} ds\, s^n f(s; s_c), \tag{2.5a}$$

$$g_n^{(s)}(s_0/s_c) = \int_{s_0/s_c}^{\infty} dx\, x^{n-\tau} \mathcal{G}^{(s)}(x). \tag{2.5b}$$

If it were not for this definition of $f_n^{(s)}(s_0; s_c)$ in terms of $f(s; s_c)$ it could absorb the correction $\mathcal{O}(\epsilon)$ of Eq. (2.4). Because $f(s; s_c)$ does not scale in s_c, neither does the non-universal contribution $f_n^{(s)}$. That does not mean, however, that the non-universal part $f_n^{(s)}$ can simply be ignored, because for $1 + n - \tau < 0$ its contribution to $\langle s^n \rangle$ is at least comparable to that of the universal part, $s_c^{1+n-\tau} g_n^{(s)}(s_0/s_c)$, even when the latter does not scale in s_c (Sec. 2.2). The normalisation $\langle s^0 \rangle = 1$ is one such moment, which obviously does not scale. Ignoring $f_0^{(s)}$ and assuming $g_0^{(s)}(x) = \text{constant}$, it has been erroneously concluded many times in the literature that $\tau = 1$. To avoid that conclusion, strictly the correction term $\mathcal{O}(\epsilon)$ is needed as well (Sec. 2.1.2.3). For a more detailed discussion of the normalisation (Sec. 2.1.2.2), it is sensible to introduce the fraction of events governed by the universal part of the PDF,

$$u(s_0; s_c) = a^{(s)} \int_{s_0}^{\infty} ds\, s^{-\tau} \mathcal{G}^{(s)}(s/s_c) = a^{(s)} s_c^{1-\tau} g_0^{(s)}(s_0/s_c). \tag{2.6}$$

Since both $\mathcal{G}^{(s)}(x)$ and $f(s; s_c)$ are non-negative, $u \leq 1$ is a mathematical requirement, with various consequences for example for the structure of the scaling function. In addition, one might impose $u > 0$, i.e. a non-vanishing fraction of 'universal events'. This is not a mathematical necessity and in fact is a matter of interpretation. If u vanishes in the limit of $s_c \to \infty$, then the fraction of universal events vanishes asymptotically, which, however, does not prevent the moments $\langle s^n \rangle$ from scaling. The implications of $u > 0$ are, in a sense, the complement of the implications of $u \leq 1$, discussed further in Sec. 2.1.2.2. As long as u is not fixed to a particular value, the normalisation does not constrain the non-universal metric factors $a^{(s)}$ and $b^{(s)}$, which therefore normally vary across different models even within a universality class.

In ordinary critical phenomena, s_0 is understood to be determined by microscopic details which remain fixed in the thermodynamic limit, or at least converge to a finite value determined by the 'error' ϵ. The same should apply to SOC models, but there are exceptions, most notably the DS-FFM (Sec. 5.2). Here s_0 diverges and as a consequence u vanishes asymptotically. This is clearly visible in the PDF, but not in its moments, as they place the weight at the end of the distribution towards the upper cutoff (Fig. 7.6, p. 226). One can argue that the scaling of the lower cutoff is not a form of simple scaling in the strict sense, a view challenged by the simple scaling of the moments. Unless stated otherwise, it is assumed in the following that s_0 remains constant. The lower cutoff is dimensionful (like the upper cutoff) and often made up from the same details of the model that give rise to the metric factor, most notably $a^{(s)}$.

The divergence of $s_c(L) = b^{(s)}L^D$ with the system size defines the **avalanche dimension** D, which therefore is a **finite size (scaling) exponent**, as are z, D_r, D_A, $\sigma_n^{(s)}$ and every exponent relating an observable to the system size. Unfortunately, $s_c(L)$ is difficult to determine directly from $\mathcal{P}^{(s)}(s)$ and is generally expected to contain corrections, so that $b^{(s)}L^D$ is merely the leading order term,

$$s_c(L) = b^{(s)}L^D(1 + A_1L^{-\omega_1} + A_2L^{-\omega_2} + \cdots) \tag{2.7}$$

with positive exponents ω_i (also known as **confluent singularities**) that characterise the **corrections to scaling** (Wegner, 1972), which are common to all asymptotic properties discussed in this chapter. [5] For example, the simple scaling form Eq. (2.2) itself could be amended by corrections of the leading power $s^{-\tau}$, and so could the scaling function etc. (e.g. Chessa, Stanley, Vespignani, and Zapperi, 1999a).

That the upper cutoff $s_c(L)$ diverges gives rise to FSS; the upper cutoff represents the finite system size as a cluster size with $b^{(s)}$ fixing its dimension. In the presence of a second scale, say Δp, which can dominate the scaling, one might be tempted to incorporate its effect in the upper cutoff, $s_c(L, \Delta p)$. However, in general the scaling function is specific to a particular scaling parameter, such as the finite size, and the presence of a second scale affects not only the upper cutoff but also the scaling function itself. For example, in equilibrium percolation, the deviation Δp from the critical point not only enters into the cutoff of the cluster size distribution, but also changes the scaling function itself, whose behaviour in s/s_c differs significantly for $\Delta p \neq 0$ and $\Delta p = 0$.

---

[5] Some authors use the term 'asymptotic expansion' for expressions of the form Eq. (2.7) in the loose sense of indicating the asymptotic behaviour rather than the strict technical sense.

2.1.2 Scaling function

The scaling function $\mathcal{G}^{(s)}(x)$ in Eq. (2.2) receives remarkably little attention in the SOC community, and the lack of reference data has led to a level of self-sustained neglect. This is compounded by the difficulties in capturing and characterising the many distinctive features as discussed in the following (but see Sec. 2.2.1). The rôle of the scaling function is to cut off the event size distribution at a **characteristic event size** whose scale if given by the upper cutoff. Some authors reserve the term **cutoff function** for a function that serves the same purpose as the scaling function, but converges to a non-zero, finite constant in a suitable limit. This is not necessarily the case for $\mathcal{G}^{(s)}(x)$. In fact, one can show that the limit

$$\lim_{x \to 0} \mathcal{G}^{(s)}(x) = \mathcal{G}_0^{(s)} \tag{2.8}$$

vanishes necessarily [6] if $\tau = 1$. The behaviour of $\mathcal{G}^{(s)}(x)$ in the limit of small arguments $x = s/s_c$ is central to the appearance of the scaling behaviour of the PDF as s_c diverges. Especially in the early days of SOC, $\mathcal{G}^{(s)}(x)$ was often omitted or approximated by an exponential $\exp(-s/s_c)$, which obviously converges to a finite value in large cutoffs. Imposing $\mathcal{G}_0^{(s)} > 0$ leads to a different definition of the avalanche exponent τ, introduced as the **apparent exponent** in Sec. 2.1.2.1.

The normalisation condition on the PDF implies $u \le 1$, Eq. (2.6). Since $u \propto s_c^{1-\tau} g_0^{(s)}(s_0/s_c)$ and $g_0^{(s)}(y) = \int_y^\infty dx\, x^{-\tau} \mathcal{G}^{(s)}(x)$, the boundedness of u translates to a condition on $g_0^{(s)}(y)$ and in turn on $\mathcal{G}^{(s)}(x)$. If $g_0^{(s)}(s_0/s_c) \propto (s_0/s_c)^{1-\tau}$ then the condition $u \le 1$ can be maintained, but $g_0^{(s)}$ cannot diverge faster than that in large s_c. In other words, there must be a constant A such that $g_0^{(s)}(y) - Ay^{1-\tau} \le 0$ for sufficiently small y. If $\tau > 1$, the asymptote can be written as the integral over $A(\tau - 1)x^{-\tau}$, so that for small enough y

$$\int_y^\infty dx\, x^{-\tau} \left(\mathcal{G}^{(s)}(x) - A(\tau - 1) \right) < 0, \tag{2.9}$$

which means that $\mathcal{G}^{(s)}(x)$ is bounded from above for sufficiently small x. If $\tau = 1$, $g_0^{(s)}(y)$ and thus u would be logarithmically divergent, unless $\mathcal{G}^{(s)}(x)$ vanishes in small arguments. Thus, $\mathcal{G}^{(s)}(x)$ is bounded in a finite vicinity around $x = 0$, provided only that it is continuous there, in which case $\mathcal{G}_0^{(s)}$, Eq. (2.8), exists and is finite (not divergent).

In principle, the scaling function itself can be subject to corrections (e.g. Chessa *et al.*, 1999a). Another, fixed scale s^* could be present, which is, however, dominated asymptotically by s_c. One such example is shown in Sec. 2.1.3, where the scaling function is preceded by a factor $1/\operatorname{erfc}(s^*/s_c)$, which converges to unity in large s_c. Such corrections are not normally accounted for.

2.1.2.1 Apparent exponent

One way for $\mathcal{G}_0^{(s)}$ to vanish is when the scaling function itself is a power law in small [7] arguments, $\mathcal{G}^{(s)}(x) \propto x^\alpha$. This is by no means a rarity and has led to considerable confusion and

---

[6] A Pareto distribution which does not incorporate a cutoff function, $\mathcal{G}_0^{(s)} > 0$, therefore has exponent $\tau > 1$.
[7] It can even be a power law throughout, Sec. 6.3.3.

inconsistencies in the literature. The confusion is quite literally a mixing of the exponents τ and α. Defining the **cutoff function** $\tilde{\mathcal{G}}^{(s)}(x)$ and the exponent $\alpha \geq 0$ via

$$\mathcal{G}^{(s)}(x) = x^\alpha \tilde{\mathcal{G}}^{(s)}(x) \tag{2.10}$$

such that $\lim_{x \to 0} \tilde{\mathcal{G}}^{(s)}(x)$ is finite (non-zero), the simple scaling form Eq. (2.2) becomes $\mathcal{P}^{(s)}(s; L) = a^{(s)} s^{-\tau+\alpha} s_c(L)^{-\alpha} \tilde{\mathcal{G}}^{(s)}(s/s_c)$. Because $\tilde{\mathcal{G}}^{(s)}(x)$ is finite in small arguments, α is non-negative. A double logarithmic plot or a simple fit to a power law will yield $\mathcal{P}^{(s)}(s L) \propto s^{-\tau+\alpha}$, where $\tau - \alpha = \tilde{\tau} \leq \tau$ defines the **apparent exponent** $\tilde{\tau}$. It can be derived from the PDF by the scaling symmetry (Christensen *et al.*, 2008)

$$\lim_{s \to \infty} \lim_{s_c \to \infty} \frac{\mathcal{P}^{(s)}(\lambda s; s_c)}{\mathcal{P}^{(s)}(s; s_c)} \lambda^{\tilde{\tau}} = 1 \quad \text{for all } 0 < \lambda \in \mathbb{R}, \tag{2.11}$$

where the first limit, $s_c \to \infty$, reveals the pure power law scaling for sufficiently large s. This region, where the PDF is straight in a double logarithmic plot and not affected by the upper and lower cutoffs, is sometimes (and maybe confusingly so) refered to as the **scaling region**.

A priori, the exponent $\tilde{\tau}$ and its associated scaling function $\tilde{\mathcal{G}}^{(s)}(x)$ provide an equally good definition of simple scaling as τ and $\mathcal{G}^{(s)}(x)$. One might even prefer the former as the more informative definition, because the exponent $\tilde{\tau}$ is what can be measured 'directly' as the slope of the PDF in a double logarithmic plot (a 'straight line fit'). However, a direct measurement of that slope is difficult, because of the upper and lower cutoffs restricting the accessible range and because of the modulation by the scaling function (e.g. Dickman and Campelo, 2003). What is more, *a moment analysis reveals τ rather than the apparent exponent $\tilde{\tau}$*. It is the latter clash of moment analysis and direct inspection that has produced the inconsistencies in the literature.

The power law scaling function can be interpreted as the result of the PDF being so shallow that its normalisation changes with its upper cutoff. A uniform distribution, $\mathcal{P}^{(s)}(s) = s_c^{-1} \theta(s_c - s)$, is an example of this mechanism. The prefactor s_c^{-1} is in fact the normalisation of the Heaviside step function θ. In order to comply with the simple scaling form, the PDF needs to be rewritten as $\mathcal{P}^{(s)}(s) = s^{-1} \mathcal{G}^{(s)}(s/s_c)$ with $\mathcal{G}^{(s)}(x) = x\theta(1 - x)$. When the scaling function vanishes like a power law, its prefactor x^α can be considered as an effect of the normalisation. Such a prefactor leads to a vertical shift $s_c^{-\alpha}$ of the scaling region which is not captured by the apparent exponent $\tau - \alpha$ (Fig. 2.2(b)).

2.1.2.2 Constraints

Based on the requirement $u \leq 1$, a number of constraints can be derived (Christensen *et al.*, 2008) from Eq. (2.6). Since $g_0^{(s)}(s_0/s_c)$ is monotonically increasing in increasing s_c, *the exponent τ cannot be smaller than unity*. If the apparent exponent $\tilde{\tau}$ is found to be less than unity, then this is due to the scaling function being a power law, which entails $\mathcal{G}_0^{(s)} = 0$. However, if $\mathcal{G}_0^{(s)} = 0$, then u vanishes in large s_c unless $\tau = 1$, i.e. *if the fraction of universal events is non-vanishing, then $\mathcal{G}_0^{(s)} = 0$ entails $\tau = 1$*. The converse is a mathematical necessity: *if $\tau = 1$, then $g_0^{(s)}(s_0/s_c)$ and therefore u are logarithmically divergent unless $\mathcal{G}_0^{(s)} = 0$, i.e. if $\tau = 1$ then $\mathcal{G}_0^{(s)} = 0$*.

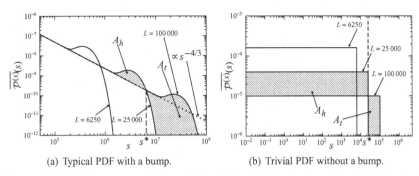

(a) Typical PDF with a bump.   (b) Trivial PDF without a bump.

Figure 2.2 PDFs of the event size in different systems of linear size L as indicated. The *hatched areas* A_h are thought to be the events that are stopped in smaller systems, but would carry on in larger ones, giving rise to the tail A_t. The intersection s^* of two PDFs is marked by a *dashed line*. (a) Area size distribution under a random walker along an absorbing wall, with $\tau = 4/3$ (*dotted line*). The structure is typical for $\tilde{\tau} = \tau > 1$ (also Fig. 2.4(b)). (b) Uniform avalanche size distribution as found in the one-dimensional BTW Model, Eq. (2.14), typical for $\tilde{\tau} < \tau = 1$.

If the apparent exponent $\tilde{\tau}$ is less than unity, then $\mathcal{G}_0^{(s)}$ vanishes and thus $\tau = 1$ for $u > 0$. If the apparent exponent $\tilde{\tau} = \tau - \alpha \leq \tau$ is greater than or equal to unity, it cannot differ from τ, because that would imply $\alpha > 0$, which entails $\mathcal{G}_0^{(s)} = 0$ which implies $\tau = 1$ and thus $\tilde{\tau} < 1$ if $u > 0$. In other words, *if $\tilde{\tau} > 1$ then $\tilde{\tau} = \tau$ provided that $u > 0$.*

Finally, *if the lower cutoff vanishes, $s_0 = 0$, then u is the* entire *normalisation, so that* $1 = a^{(s)} s_c^{1-\tau} g_0^{(s)}(0) + \mathcal{O}(\epsilon)$, *which implies $\tau = 1$.*

One might be tempted to conclude that $\tau = 1$ generally leads to $\tilde{\tau} < 1$, because $\tau = 1$ means $\mathcal{G}_0^{(s)} = 0$ and so $\mathcal{G}^{(s)}(x)$ must have a leading order behaviour like a power law. This is obviously not true, for example

$$\mathcal{P}^{(s)}(s) = s^{-1} \frac{1}{2K_0(2\sqrt{qr})} \exp\left(-q\frac{s_c}{s} - r\frac{s}{s_c}\right) \tag{2.12}$$

has apparent exponent $\tilde{\tau} = 1$, and therefore $\tau = 1$, with dimensionless parameters q and r. Otherwise, this is a rather contrived example, as the scaling region, qs_c to s_c/r, on a logarithmic scale does not actually extend with increasing s_c, but simply translates (Fig. 2.3).

2.1.2.3 The bump

Towards large values of their argument, many scaling functions display a distinctive **bump**, [8] e.g. Fig. 2.2(a). This structure is what is probed in a moment analysis (Sec. 7.3, in particular Fig. 7.6). It is commonly explained as an accumulation of events that would have continued and thus further increased in size, were it not for the finiteness of the system (e.g. Amaral and Lauritsen, 1996b). There is a distinctive size from when it is actually possible for an event to 'feel' the finiteness of the system. The width of the bump increases linearly in the upper cutoff. Events stopped short by the finite system do not all end up at the same size but are increasingly broadly distributed. The process by which the data in Fig. 2.2(a) are generated (Sec. 2.2) is in perfect agreement with this interpretation. Here the area under

---

[8] The naming of this feature is not consistent across the literature; it is also refered to as a hump.

a random walker trajectory is recorded until the walker either hits an absorbing wall or reaches a maximum time, which is considered as the system size. Walkers that reach the maximum typically generate an event size in the bump of the distribution.

The events that are constrained by the finiteness of the system are therefore contained in the area (A_h in Fig. 2.2(a)) where the PDF for one system size exceeds that of a much larger one. The tail of the latter (A_t in Fig. 2.2(a)), where it proceeds beyond the former, is made up of precisely those events that are 'trimmed off' in smaller systems. The two areas can be defined formally using the intersection at s^* of two PDFs for system sizes, L and a multiple of it, αL. Keeping only the universal part,

$$A_h = \int_0^{s^*} ds\, s^{-\tau} \left(\mathcal{G}^{(s)}(s/s_c(L)) - \mathcal{G}^{(s)}(s/s_c(\alpha L)) \right) \tag{2.13a}$$

$$A_t = \int_{s^*}^{\infty} ds\, s^{-\tau} \left(\mathcal{G}^{(s)}(s/s_c(\alpha L)) - \mathcal{G}^{(s)}(s/s_c(L)) \right). \tag{2.13b}$$

Since both PDFs are normalised, it follows that $A_h = A_t$. That such bumps are therefore a necessary feature, is, however, a non sequitur. In fact, Fig. 2.2(a) shows only one of the two typical structures for $\tilde{\tau} = \tau > 1$ (the other one being shown in Fig. 2.4(b)). In the following, it is shown when and why the bump is a necessity.

Firstly, the pronounced peak, although typical, is not necessary. This is shown in Fig. 2.2(b) for the uniform PDF on $s \in [0, \infty)$

$$\mathcal{P}^{(s)}(s) = \theta(L - s)/L = s^{-1} \left(\frac{s}{L} \right) \theta(1 - s/L) \tag{2.14}$$

which is constant where it does not vanish, but formally has $\tau = 1$. The 'hump' A_h in this case is spread out across the entire region of support, which happens whenever the effective normalisation leads to a vertical shift of the PDF with increasing s_c (also Fig. 2.4(a)). Such a regular, uniform vertical shift is the effect of the scaling function being a power law. In other words, Fig. 2.2(b) shows the characteristic structure for $\tilde{\tau} < \tau = 1$.

Secondly, but more importantly, one can show that the areas A_h and A_t scale like $s_c(L)^{1-\tau}$, by noting that s^* as well as $s_c(\alpha L)$ can be written as multiples of $s_c(L)$. For $\tau > 1$ this means that the areas vanish asymptotically and their contributions to the normalisation therefore always drop below the error ϵ introduced in Eq. (2.4). In fact, for $\tau > 1$ a PDF with finite support can simply be *continued as a power law without violating the normalisation condition*. This is illustrated in Fig. 2.4(b), which has a non-vanishing area A_t, but no discernible area A_h. The PDF simply seems to extend further and further, suggesting that its normalisation changes. However, Eq. (2.18) is always perfectly normalised and the whole graph is minutely rescaled downwards (not visible in Fig. 2.4(b)) as it extends further. A finite, discernible bump A_h, which might be 'spread out' is thus a necessity only for $\tau = 1$, even when it is often seen in systems with $\tau > 1$ and is typical for the distributions studied in SOC.

If $\tilde{\tau} = 1$, the scaling of the bump is more subtle. It cannot be spread out as in Fig. 2.2(b), because that implies $\tilde{\tau} < 1$. It cannot be localised as in Fig. 2.2(a), because at $\tilde{\tau} = 1$, the area A_t can be shown to be larger then A_h by a constant amount, regardless of s_c, so that discounting it eventually exceeds any error threshold. Instead, $\tilde{\tau} = \tau = 1$ leads to a shift of the entire scaling region, shown for Eq. (2.12) in Fig. 2.3. This PDF has no lower cutoff, i.e. it scales for all $s \geq 0$ and a collapse would be perfect, even when graphs for different L seem to join in a staggered fashion.

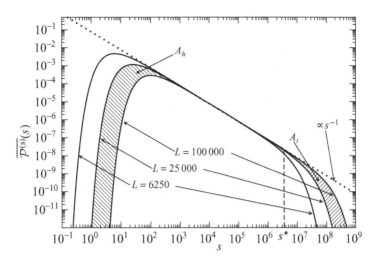

Figure 2.3 Example of a PDF with $\tau = \tilde{\tau} = 1$ and $s_0 = 0$, Eq. (2.12) with $q = r = 10^{-3}$ and $s_c = L$ as indicated. Areas and *other markers* as in Fig. 2.2. Although scaling applies for all $s \geq 0$, i.e. there is no positive lower cutoff s_0 and graphs for different s_c collapse throughout when plotted appropriately, on a logarithmic scale the region parallel to the *dotted line*, $\propto s^{-1}$, does not increase with s_c.

2.1.3 Two examples

Two *analytical* examples for PDFs and in particular their scaling functions are presented in the following, the first for $\tau = 1$, which is a rather special case as seen above, and the second for $\tau = 3/2$, an example taken from Sec. 8.1.3.

An equally instructive as surprising example for a scaling function is a Gaussian, Fig. 2.4(a). To be scale free, its mean must be a multiple of its standard deviation, say

$$\mathcal{P}^{(\mathrm{s})}(s) = \frac{2}{\mathrm{erfc}(-\alpha/\sqrt{2})} \frac{1}{\sqrt{2\pi\sigma^2}} e^{-\frac{(s-\alpha\sigma)^2}{2\sigma^2}}. \tag{2.15}$$

Restricting the support to $s \in [0, \infty)$ means that the mean of this distribution is not $\alpha\sigma$ nor is its variance σ^2. Nevertheless, it is fully parameterised by σ. Interestingly, it can be written in the simple scaling form Eq. (2.2):

$$\mathcal{P}^{(\mathrm{s})}(sL) = s^{-1}\mathcal{G}^{(\mathrm{s})}\left(\frac{s}{\sqrt{2}s_c(L)}\right) \tag{2.16}$$

with $s_c = \sigma$ and

$$\mathcal{G}^{(\mathrm{s})}(x) = \frac{2}{\sqrt{\pi}\,\mathrm{erfc}(-\alpha/\sqrt{2})} x\, e^{-(x-\alpha/\sqrt{2})^2}, \tag{2.17}$$

which has vanishing lower cutoff, $s_0 = 0$, and does not have any corrections, i.e. Eq. (2.16) is identical to Eq. (2.15). No metric factor is needed in front of Eq. (2.16) because $\tau = 1$ produces a PDF with the right dimension. Since $s_0 = 0$, the Gaussian displays scaling throughout, i.e. a collapse will always be perfect. This implies that the scaling function

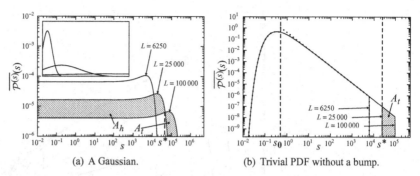

(a) A Gaussian.   (b) Trivial PDF without a bump.

Figure 2.4 Examples of PDFs shown as in Fig. 2.2. (a) A Gaussian distribution of event sizes, Eq. (2.16) with $s_c(L) = L$ (as indicated), $\alpha = 1$, produces an almost constant asymptote, $\tilde{\tau} = 0$, as $\tau = 1$, $\mathcal{G}_0^{(s)} = 0$ and $s_0 = 0$. The inset shows the same data on a linear scale. (b) Return time distribution of a random walker with $\tau = \tilde{\tau} = 3/2$, Eq. (2.18) with $a_0^{(s)} = 1$ and $D = 1/2$. Up to the upper cutoff $s_c(L) = L$, data for different system sizes are indistinguishable. The dotted line is the scaling form Eq. (2.19b), which deviates visibly only around $s_0 = 1/2$ marked by a dashed line; beyond that it is a pure power law with exponent $\tau = 3/2$. The area A_t in the tail is finite, but vanishes asymptotically and there is no visible area A_h, another type of distribution possible for $\tilde{\tau} = \tau > 1$ (also Fig. 2.2(a)).

vanishes asymptotically, which can be verified by inspecting Eq. (2.17). The fact that a Gaussian itself displays scaling seems to contradict the notion that scaling indicates cooperative, i.e. non-Gaussian, behaviour. However, scaling should rather be regarded as a signal for the long-range nature of the cooperative phenomenon; scaling indicates that it does not disappear on some large scale. Clearly, a lack of correlation, as indicated by a Gaussian, possesses that stability as well. *Non-trivial* scaling, in particular $\tau \neq 1$ and non-Gaussian scaling functions, is thus what one should watch out for, indicating correlated behaviour that is stable even on large (spatial) scales. Other PDFs which unexpectedly display perfect scaling are the δ function $\mathcal{P}^{(s)}(s; L) = \delta(s_c(L) - s) = s^{-1}\delta(s/s_c(L) - 1)$ and the exponential $\mathcal{P}^{(s)}(s; L) = s_c \exp(-s/s_c) = s^{-1}s/s_c \exp(-s/s_c)$. The latter example is particularly important as many authors seem to assume that an exponential signals the absence of scaling. This is clearly incorrect (p. 349). This misbelief has its origin probably in the observation that a fixed cutoff independent from the system size, i.e. a lack of scaling, often comes together with an exponential scaling function.

As discussed further in Sec. 2.2.1, Eq. (2.15) is *not* the Gaussian distribution found if the event sizes are compiled from local, independent, positive events. In that case the average scales faster than the width, with both first moment and variance linear in the number of independent contributions N, corresponding to $\alpha \propto \sqrt{N}$.

The second example, Fig. 2.4(b), is the termination probability distribution of a random walker with effective diffusion constant D along an absorbing wall, conditional to termination at or before time L. The walker starts with an initial distance a_0 from the wall and the avalanche size s is given by the time until it runs into the wall. The expression

$$\mathcal{P}^{(s)}(s; L) = \frac{a_0 \, \theta(L - s) \, \exp\left(-\frac{a_0^2}{4Ds}\right)}{\mathrm{erfc}(a_0/\sqrt{4DL})\sqrt{4\pi D}} s^{-3/2} \tag{2.18}$$

is a modified version of Eq. (8.26). It can be *approximated* by

$$
\mathcal{P}^{(s)}(s; L) \approx \begin{cases} \dfrac{a_0\,\theta\,(L-s)\,\mathrm{e}^{-\frac{a_0^2}{4Ds}}}{\mathrm{erfc}\left(\frac{a_0}{\sqrt{4DL}}\right)\sqrt{4\pi D}}s^{-3/2} & \text{for } s < \frac{a_0^2}{4D} = s_0 \qquad (2.19\text{a}) \\[4mm] a^{(s)}s^{-3/2}\mathcal{G}^{(s)}(s/s_c) & \text{otherwise} \qquad\qquad (2.19\text{b}) \end{cases}
$$

with $s_c = L$ and metric factor

$$
a^{(s)} = \frac{a_0}{\sqrt{4\pi D}}, \qquad\qquad (2.20)
$$

which does not contain the $\mathrm{erfc}(a_0/\sqrt{4DL})$ from the normalisation, because it would render $a^{(s)}$ dependent on the upper cutoff $s_c = L$. Because this dependence on s_c does not enter in the form s/s_c, it cannot be absorbed into the scaling function

$$
\mathcal{G}^{(s)}(s/s_c) = \theta(1 - s/s_c) \qquad\qquad (2.21)
$$

either. Omitting it takes nothing away from Eq. (2.19b) being an *asymptote*. The sharp cutoff at $s_c = L$ in this example is a matter of choice, in principle it could be more structured without changing much else.

As s_c increases, this PDF seems to extend magically without violating the normalisation. This impression is a result of the logarithmic scale which dramatically exaggerates the size of A_t – Eq. (2.18) is perfectly normalised. On the other hand, strictly, the scaling form Eq. (2.19b), shown as a dotted line in Fig. 2.4(b), *does* violate the normalisation. Firstly, it assumes that $\mathrm{erfc}(a_0/\sqrt{4DL}) = 1$, even when the right hand side is only the limit of the left in large L. However, this violation is so small that it cannot be spotted in Fig. 2.4(b) as a horizontal shift relative to Eq. (2.18). Secondly, around $s_0 = a_0^2/(4D)$ Eq. (2.19b) differs noticeably from Eq. (2.18). This deviation and violation of the normalisation can be made arbitrarily small by adjusting s_0. It occurs in a region which is irrelevant for the moments discussed in the next section.

Finally, it is worth noting that the dimension of the metric factor Eq. (2.20), which is made up entirely of microscopic details of the model, namely the diffusion constant and the initial gap, is the square root of a time, as needed to cancel the dimension of $1/\sqrt{s}$, such that $\mathcal{P}^{(s)}(s)\,\mathrm{d}s$ is dimensionless.

2.2 Moments

Moments are much more easily determined and characterised than the PDF, even when the former can be derived from the latter much more easily than vice versa (Carleman's theorem in Feller, 1966). The finite size scaling of moments of the PDF is derived in Eq. (2.4). Its analysis is the most common method for determining the exponents characterising the PDF, made popular (again) by De Menech, Stella, and Tebaldi (1998).

According to Eq. (2.4), the asymptote of the nth moment with the upper cutoff is $s_c^{1+n-\tau}g_n^{(s)}(s_0/s_c)$, that is, if it scales. Assuming $\mathcal{G}^{(s)}(x)$ is continuous around $x = 0$, the integral $g_n^{(s)}(y)$ cannot increase faster than $y^{1+n-\tau}$ for $1 + n - \tau < 0$ as $y \to 0$. For

$1 + n - \tau > 0$ it is convergent as $y \to 0$, unless it diverges for all $y > 0$. In finite systems, this is not normally the case, as finite systems usually impose a finite upper bound on the event sizes. However, if this event size is something that does not require a resource or does not cause dissipation, which would eventually limit its size, then some moments may not exist. This can happen, for example, when the event size is a survival time. In that case, a finite probability of indefinite survival is sufficient for divergent moments, but not necessary; a shallow decay is enough, say $\mathcal{P}^{(s)}(s) = (1/2)s^{-3/2}$ for $s \geq 1$ has divergent moments $\langle s^n \rangle$ for all $n \geq 1/2$. Divergence of moments does not preclude a scaling PDF, as illustrated by this example, where the scaling function is a constant and the upper cutoff infinite. Such a scenario is, however, atypical for SOC models, where normally some form of dissipation (at the boundaries) curbs the event size in finite systems.

In the following, it is assumed that all moments exist for finite s_c, so that

$$\lim_{y \to 0} g_n^{(s)}(y) = g_n^{(s)}(0) > 0 \quad \text{for all } n > \tau - 1 \tag{2.22}$$

and given that $f_n^{(s)}(s_0; s_c)$ does not scale at all, $\langle s^n \rangle$ is dominated by $a^{(s)} s_c(L)^{1+n-\tau} g_n^{(s)}(0)$, or more succinctly

$$\lim_{s_c \to \infty} \frac{\langle s^n \rangle}{s_c^{1+n-\tau}} = a^{(s)} g_n^{(s)}(0) \quad \text{for all } n > \tau - 1. \tag{2.23}$$

The integral $g_n^{(s)}(0)$ depends on the non-universal, metric factor $b^{(s)}$ and thus is not (generally) universal. Equation (2.23) describes essentially the asymptote of Eq. (2.4) at $n > \tau - 1$.

Moments with $n < \tau - 1$ cannot diverge as $s_c \to \infty$ and thus are finite in all non-trivial cases and vanish otherwise. The moment $\langle s^{\tau-1} \rangle$ is a special case: it diverges logarithmically if $\mathcal{G}_0^{(s)} > 0$ and remains finite if $\mathcal{G}^{(s)}(x)$ decays as a power law in small arguments. If $\tau = 1$, it is the normalisation, which is bound to be finite, i.e. for $\tau = 1$ the scaling function $\mathcal{G}^{(s)}(x)$ is bound to vanish in small arguments.

The divergence of the first moment as $s_c \to \infty$ is frequently seen as a signature of scaling. It occurs for all $\tau < 2$ and if $\mathcal{G}_0^{(s)} > 0$ also at $\tau = 2$. As τ exceeds 2, the first moment remains finite. This is sometimes interpreted as a finite characteristic avalanche size. However, even when the first moment has the same canonical dimension as the upper cutoff, they are not proportional to each other unless $\tau = 1$. Rather, they are powers of each other, with dimensional consistency being restored by an additional length scale which does not, however, govern the long-range behaviour.

Assuming only that $s_c(L)$ scales in L to leading order like $b^{(s)}L^D$, Eq. (2.2), exponents $\sigma_n^{(s)}$ are defined by producing the finite limit

$$\lim_{L \to \infty} \frac{\langle s^n \rangle}{L^{\sigma_n^{(s)}}} = a^{(s)} \left(b^{(s)} \right)^{1+n-\tau} g_n^{(s)}(0) > 0 \quad \text{for all} \quad n > \tau - 1. \tag{2.24}$$

It thus is a matter of linear regression to determine D and τ from

$$\sigma_n^{(s)} = D(1 + n - \tau), \tag{2.25}$$

see Eq. (2.7) and Eq. (2.23). The gap of the exponents of any two consecutive moments with $n > \tau - 1$ is D, which is therefore often referred to as the **gap exponent** (Pfeuty and Toulouse, 1977), and this particular form of scaling is often referred to as **gap scaling**. If all moments exist, if the moment generating function (2.46) converges and its inverse Laplace transform exists, then gap scaling implies finite size scaling in the form (2.2). If gap scaling applies and for any two distinct $n, m > \tau - 1$ the moments scale to leading order like $\langle s^n \rangle^m \propto \langle s^m \rangle^n$ then $\tau = 1$. This is of particular interest for $n = 1$ and $m = 2$ in the case of FSS in equilibrium critical phenomena, where the Rushbrooke and Josephson scaling laws ensure that the second moment of the order parameter scales like the first moment squared. In equilibrium, the observation $\tau = 1$ can be extended to **critical scaling**, i.e. scaling in the approach to the critical point in the limit of large system sizes.

A more sophisticated ansatz than Eq. (2.24) allows for **corrections to scaling** (Wegner, 1972, or confluent singularities, Eq. (2.7)), which are often crucial in numerical moment analysis (Sec. 7.3, also e.g. Chessa *et al.*, 1999a):

$$\langle s^n \rangle = a^{(s)} g_n^{(s)}(0) \left(b^{(s)} L \right)^{\sigma_n^{(s)}} \left(1 + A_1' L^{-\omega_{n,1}'} + A_2' L^{-\omega_{n,2}} + \cdots \right) + f_n^{(s)}(s_0; s_c). \qquad (2.26)$$

This is a reminder that scaling generally is an *asymptotic* phenomenon whether in FSS or in the approach to the critical point of an equilibrium phase transition. The scaling is *dominated* by the first term only if $\sigma_n^{(s)} > 0$ and so one can summarise

$$\langle s^n \rangle \propto L^{\sigma_n^{(s)}}, \quad \sigma_n^{(s)} = \begin{cases} \mathsf{D}(1 + n - \tau) & \text{for } n > \tau - 1 \qquad (2.27a) \\ 0 & \text{for } n < \tau - 1, \qquad (2.27b) \end{cases}$$

which is illustrated in Fig. 7.5. The same applies, for example, for the moments of the avalanche duration,

$$\langle T^n \rangle \propto L^{\sigma_n^{(T)}}, \quad \sigma_n^{(T)} = \begin{cases} \mathsf{z}(1 + n - \alpha) & \text{for } n > \alpha - 1 \qquad (2.28a) \\ 0 & \text{for } n < \alpha - 1. \qquad (2.28b) \end{cases}$$

A range of numerical techniques to determine exponents on the basis of these moments is discussed in Ch. 7.

2.2.1 Moment ratios

Dimensionless combinations of moments provide a well-defined, unambiguous way to characterise the universal features of the scaling function, without having to make use of any other (estimated) universal quantity, such as the exponents. Moment ratios thus provide an alternative, independent way to characterise the universality class (Privman and Fisher, 1984). Known also as amplitude ratios[9] (e.g. Salas and Sokal, 2000), they are frequently used to characterise ordinary phase transitions, often in the form of a variance normalised higher order cumulant (Dickman and Kamphorst Leal da Silva, 1998). One particularly well known ratio is the so-called **Binder cumulant** (Binder, 1981a,b; Landau and Binder, 2005). Moment ratios are normally designed to converge (Lübeck and Janssen, 2005; Pruessner, 2007), i.e. they can generally be written as a constant plus corrections in powers of the system size L. In this sense, their universal value can be measured more directly, requiring at least one system size, whereas scaling exponents require at least two.

---

[9] This notion originates from the two amplitudes a power law has on the two sides of a phase transition and is also used in the context of corrections to scaling (Privman *et al.*, 1991).

However, as the scaling function in finite size scaling depends on details such as boundary conditions, system shape and aspect ratio, so do moment ratios.

In systems displaying a phase transition, moment ratios usually display trivial values in the two phases, and a non-trivial value at the critical point. With increasing system size, the moment ratios change increasingly sharply across the transition, which can be used to determine the critical point (e.g. Dickman, Tomé, and de Oliveira, 2002).

The key insight when constructing suitable moment ratios is that there are products of moments that scale like $L^{\sigma_n^{(s)}}$ in the denominator of Eq. (2.24). On the basis of Eq. (2.27), however, it is clear that $\langle s^n \rangle$ is generally the only single moment or power thereof that scales like $L^{\sigma_n^{(s)}}$, unless $\tau = 1$. In this case $\langle s^n \rangle$ scales like $\langle s \rangle^n$, so that relevant moment ratios are of the form $\langle s^n \rangle / \langle s \rangle^n$ or $\langle s^{2n} \rangle / \langle s^2 \rangle^n$, as frequently used in ferromagnetic phase transitions (where $\tau = 1$ due to Rushbrooke and Josephson scaling laws).

For $\tau \neq 1$, more complicated moment ratios are necessary. Considering only moments $\langle s^n \rangle$ with $n > \tau - 1$, each moment contributes a factor $L^{\sigma_n^{(s)}}$ with $\sigma_n^{(s)} = D(1 + n - \tau)$. For them to cancel in a ratio, numerator and denominator must both total to the same power of the avalanche size and both must contain the same number of moments. For example

$$\frac{\langle s^{n_1} \rangle \langle s^{n_1} \rangle \ldots \langle s^{n_N} \rangle}{\langle s^{m_1} \rangle \langle s^{m_1} \rangle \ldots \langle s^{m_M} \rangle} \tag{2.29}$$

requires $\sum_i^N \sigma_{n_i}^{(s)} = \sum_i^M \sigma_{m_i}^{(s)}$. For this to hold independently from τ and D one needs $ND(1 - \tau) = MD(1 - \tau)$ as well as $D \sum_i^N n_i = D \sum_i^N m_i$, which implies

$$N = M \quad \text{and} \quad \sum_i^N n_i = \sum_i^N m_i. \tag{2.30}$$

There are obviously infinitely many ways to satisfy these conditions, in non-trivial cases involving at least three distinct moments. One choice is

$$\frac{\langle s^{n-m} \rangle \langle s^{n+m} \rangle}{\langle s^n \rangle^2} = \frac{g_{n-1}^{(s)}(0) g_{n+1}^{(s)}(0)}{g_n^{(s)}(0)^2} + \text{corrections}, \tag{2.31}$$

for $|m| < n + 1 - \tau$ with the corrections vanishing in the limit $L \to \infty$. Since each moment must be of order greater than $\tau - 1$, the lowest integer moment ratio of this type is usually $\langle s \rangle \langle s^3 \rangle / \langle s^2 \rangle^2$. Notably, the limit on the right, derived on the basis of Eq. (2.24), does not depend explicitly on the metric factors. However, the limits $g_n^{(s)}(0)$ *do* depend on the metric factors; as they are arbitrary, any PDF obeying simple scaling in the form Eq. (2.2) can be rewritten using a new scaling function, say $\mathcal{G}_{\text{new}}^{(s)}(x) = \alpha \mathcal{G}^{(s)}(\beta x)$, so that the $g_n^{(s)}(0)$ derived from the two scaling functions are related by powers of α and β. Since the left hand side of Eq. (2.31) is independent of the choice of the scaling function, the right hand side is equally independent when expressed in terms of either the old or the new $g_n^{(s)}(0)$, i.e. the powers of α and β cancel on the right (which can also be shown by direct evaluation).

A particularly convenient combination is

$$\rho^{(s)}_n = \frac{\langle s^n \rangle \langle s \rangle^{n-2}}{\langle s^2 \rangle^{n-1}} = \frac{g_1^{(s)}(0)^{n-2}}{g_2^{(s)}(0)^{n-1}} g_n^{(s)}(0) + \text{corrections}, \tag{2.32}$$

which allows one to use the arbitrariness of the metric factors. Choosing $g_1^{(s)}{}'(0) = g_2^{(s)}{}'(0) = 1$ fixes both of them and results in $\rho^{(s)}{}_n = g_n^{(s)}(0)$ consistent with Eq. (2.32) even for $n = 1, 2$. Moment ratios are thus the most direct way to characterise the scaling function. [2.1]

Moment ratios can also be constructed on the basis of cumulants (Sec. 2.2.3), which can always be written in terms of sums of products of moments. While cumulants are the derivatives of the free energy in equilibrium statistical mechanics and thus have a fundamental physical meaning, they are somewhat problematic in the present context. Firstly, differences of moments might reveal their subleading order which means that their ratios might, unexpectedly, vanish or diverge. For example, $\langle s^3 \rangle \langle s \rangle - \rho_3^{(s)} \langle s^2 \rangle^2$ has a leading order less than $\langle s^3 \rangle \langle s \rangle$ or $\langle s^2 \rangle^2$, according to Eq. (2.32). Secondly, however, if $\tau > 1$ the nth cumulant is dominated by the nth moment and, except for the subleading orders, it thus makes no difference to replace them by these moments. A cancellation *within* a cumulant that reveals its subleading order can only take place when $\tau = 1$.

Thirdly, cumulants are not as indicative of a Gaussian PDF as one might hope. Beyond the variance, they vanish for a Gaussian distribution, but fail to do so if it has only partial support. For example, the cumulants derived from Eq. (2.15) are finite and depend on α, as do ratios like

$$\frac{\langle (s - \langle s \rangle)^3 \rangle \langle s \rangle}{\langle s^2 \rangle^2}. \tag{2.33}$$

One might thus conclude that cumulants cannot be used to test whether the observable in question is the sum of independent local random variables. Luckily, this is not the case, even if the (suspected) local random variables are all positive, so that the resulting distribution necessarily has only partial support. This is caused by the above mentioned subleading orders of moments that can dominate cumulants if $\tau = 1$. While the nth moment of the sum of N random variables scales to leading order with N^n, all its cumulants are only linear in N, so that Eq. (2.33) would vanish like N^{-2}. The example Eq. (2.15) has to be evaluated for $\alpha \propto \sqrt{N}$ to represent a distribution of sums of local, independent contributions to the overall event. Ratios such as Eq. (2.33) from distributions of sums of independent random variables thus always vanish asymptotically, regardless of whether they have only partial support or not.

2.2.2 Joint distributions and conditional moments

In particular during the early days of SOC, event sizes were studied conditional to other event sizes, which effectively makes use of a joint distribution, say of the avalanche size and duration, $\mathcal{P}(s, T)$, so that

$$\langle s^n | T \rangle = \int ds\, s^n \mathcal{P}(s, T) \tag{2.34a}$$

$$\langle T^n | s \rangle = \int dT\, T^n \mathcal{P}(s, T). \tag{2.34b}$$

Assuming that both marginals obey simple finite size scaling, Eq. (1.1) on p. 12 and Eq. (1.2) on p. 14, the first moments are to leading order $\langle s \rangle = a^{(s)}(b^{(s)}L^D)^{2-\tau}g_1^{(s)}$ and

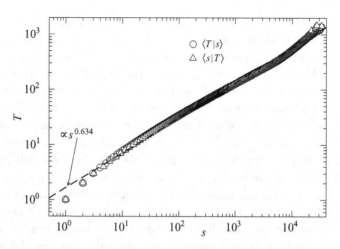

Figure 2.5 Average duration conditional to avalanche size, and average avalanche size conditional to duration in the one-dimensional Oslo Model ($L = 160$, 10^6 and 10^7 avalanches for equilibration and statistics respectively) in a double logarithmic plot. The dashed line shows a power law $s^{0.634}$, since $\langle T|s \rangle \propto s^{\gamma_{Ts}}$, Eq. (2.35b), and $\gamma_{Ts} = z/D \approx 0.634$, Eq. (2.39), using $z \approx 1.42$ and $D \approx 2.24$.

$\langle T \rangle = a^{(T)}(b^{(T)}L^z)^{2-\alpha}g_1^{(T)}$ with suitable definition of $g_1^{(T)}$, provided τ and α are not too big (this condition is relaxed below). Assuming in addition a **narrow joint distribution**, to be discussed in the following, **conditional averages** (also known as **conditional moments** or **conditional expectation values**) of s and T are functions of each other, say (to leading order)

$$\langle s|T \rangle = c_{sT}T^{\gamma_{sT}} \tag{2.35a}$$

$$\langle T|s \rangle = c_{Ts}s^{\gamma_{Ts}}. \tag{2.35b}$$

It is natural to assume $\gamma_{sT} = 1/\gamma_{Ts}$, as if the observable s were bound by T to attain a value proportional to $T^{\gamma_{sT}}$ and correspondingly for T conditional to s. This is shown in Fig. 2.5, where the average duration $\langle T|s \rangle$ plotted against avalanche size s follows the curve of T plotted against conditional avalanche size $\langle s|T \rangle$ very closely, i.e.

$$\langle T|\langle s|T' \rangle \rangle \approx T' \quad \text{and} \quad \langle s|\langle T|s' \rangle \rangle \approx s'. \tag{2.36}$$

The exponents γ_{sT} and γ_{Ts} can thus be interpreted as the (possibly **anomalous**) **dimension** of one observable in terms of the other, which in the following is referred to as the **assumption of narrow distributions** (Jensen et al., 1989; Christensen et al., 1991; Lübeck, 2000). This name alludes to the feature of the joint PDF that the width of its marginals remains so small that the avalanche duration can be regarded a *function* of the size and vice versa, Fig. 2.6.

The scaling Eq. (2.35) and all that follows is immediately extended to other observables, such as the number of distinct sites toppling in an avalanche (area, A), the radius of gyration (characteristic radius, r) and the number of boundary sites (p). As Lübeck (2000) pointed out, the scaling of the expected area covered by an avalanche with its radius determines its fractal dimension, which thus equals γ_{Ar}, bounded from above by the spatial dimension d

of the lattice,[10] in which case avalanches are said to be **compact**. In the early SOC literature 'compact' was used to indicate the absence of holes in the coverage by avalanches, as found in the BTW Model (Christensen and Olami, 1993). More recently, this is often treated as equivalent to $\gamma_{Ar} = d$, as found in the MANNA Model (e.g. Ben-Hur and Biham, 1996; Chessa *et al.*, 1999a), although, strictly, neither implies the other.

If the contribution by **multiple topplings** is negligible, the avalanche area A equals its size s, $\gamma_{As} = 1$. In general, however, $s \geq A$ and thus $\gamma_{sA} \geq 1$. Similarly $s \geq T$ and $\gamma_{sT} \geq 1$, as at least one site topples per time step while an avalanche is running.

It is natural to expect that the characteristic radius of the avalanches scales linearly in the system size (and the correlation length), i.e. the upper cutoff r_c in

$$\mathcal{P}^{(\mathrm{r})}(r) = a_r r^{\tau_r} \mathcal{G}^{(\mathrm{r})}(r/r_c) \tag{2.37}$$

is proportional to L. Clearly, r_c is bounded by a multiple of L. If the ratio r_c/L does not converge to a non-zero value as the lattice spacing vanishes, the geometrical features will collapse to a point in the continuum limit. Anticipating Eq. (2.39), γ_{sr} therefore equals the avalanche dimension D, γ_{Tr} is the dynamical exponent z and similarly for other observables. In any case, linearity of the characteristic radius $r_c \propto L^{D_r}$ is an upper bound, $D_r \leq 1$. The same applies to the characteristic avalanche area $A_c \propto L^{D_A}$, which cannot exceed the system size, the hard cutoff for the avalanche area, so that $D_A \leq d$. Again, compact avalanches suggest $D_A = d$.

Integrating over the nuisance variable with appropriate weight gives, for example

$$\langle s \rangle = \int \mathrm{d}T \, \langle s|T \rangle \, \mathcal{P}^{(\mathrm{T})}(T) = c_{sT} a^{(\mathrm{T})} (b^{(\mathrm{T})}L)^{\mathsf{z}(1+\gamma_{sT}-\alpha)} g^{(\mathrm{T})}_{\gamma_{sT}}. \tag{2.38}$$

Comparison with simple scaling assumed for these moments and using $\gamma_{sT} = 1/\gamma_{Ts}$ gives

$$\gamma_{sT} = \frac{\mathsf{D}}{\mathsf{z}} = \frac{1-\alpha}{1-\tau}, \tag{2.39}$$

which is reproduced by generalising Eq. (2.35) to higher moments such as $\langle s^n|T \rangle \propto T^{n\gamma_{sT}}$ and therefore does not rely on τ and α not being too big. It implies the **fundamental scaling law of narrow joint distributions** (Lübeck, 2000)

$$\mathsf{D}(1 - \tau) = \mathsf{z}(1 - \alpha) =: \Sigma. \tag{2.40}$$

The same line of argument can be followed through for any other observable, each time invoking the narrow distribution property, producing in particular $\Sigma = 1 - \tau_r$ as $D_r = 1$. Because Eq. (2.40) is widely believed to be a mathematical necessity (Sec. 10.1, p. 350), it is not always stated clearly in the literature whether it has been used for numerical estimates of exponents. Table 2.1 exemplifies its validity for the MANNA Model.

Since both τ and α are expected to depend on boundary conditions, so does Σ. In that sense D and z are 'more universal' than τ and α, which are 'enslaved' by the former via scaling laws. The scaling law Eq. (2.40) entails that the moment exponents $\sigma_n^{(\mathrm{s})}$ and

---

[10] As the characteristic radius of gyration can scale sublinearly in the system size, $\gamma_{Ar} \leq d$ does not follow from A being bounded by the volume of the d-dimensional lattice, but from r^d being the maximum volume of any (path-)connected subject with radius of gyration r.

Table 2.1 Illustration of the scaling laws Eq. (2.40) and $2 = \sigma_1^{(s)} = D(2 - \tau)$ using estimates of exponents in the MANNA Model (Table 6.1, most recent data). Some are based on variants of the MANNA Model, but essentially any data from Table 6.1 give consistent results.

d	τ		D		$D(\tau - 1)$ $\stackrel{?}{=} z(\alpha - 1)$	$D(2 - \tau) \stackrel{?}{=} 2$
	α		z			
1	1.11(5)	(a)	2.2(1) ⋄	(b)	0.24(11)	1.96(14)
	1.17(5)	(a)	1.50(4) ♮	(c)	0.25(8)	
2	1.27(3)	(a)	2.764(10)	(d)	0.75(8)	2.02(8)
	1.48(3)	(a)	1.533(24) ♮	(e)	0.74(5)	
3	1.41(2) ♮	(e)	3.36(1)	(f)	1.38(7)	1.98(7)
	1.77(4) ♮	(e)	1.823(23) ♮	(e)	1.46(5)	
MFT	3/2	(g)	4	(g)	2	2
	2	(g)	2	(g)	2	

⋄ Reported for the oMM.

♮ Reported for variants of the MANNA Model (in particular FES, Sec. 9.3.1).

a Bonachela (2008), b Nakanishi and Sneppen (1997), c Dickman (2006), d Lübeck (2000), e Lübeck and Heger (2003a), f Pastor-Satorras and Vespignani (2001), g Lübeck (2004).

$\sigma_n^{(T)}$, for avalanche size Eq. (2.27) and duration Eq. (2.28) respectively, can be written as $\sigma_n^{(s)} = Dn + \Sigma$ and $\sigma_n^{(T)} = zn + \Sigma$.

Very often, the boundary conditions give rise to a certain scaling of the first moment of the avalanche size, fixing $\sigma_1^{(s)}$ typically to values like $\sigma_1^{(s)} = 2$ for **bulk drive** (illustrated in Table 2.1) or $\sigma_1^{(s)} = 1$ for **boundary drive** (Sec. 8.1.1.1, also Nakanishi and Sneppen, 1997) which results in an identity between τ and D. Because the value of $\sigma_1^{(s)}$ is normally a mathematical necessity, the exponents τ and D are often not measured independently. Together with Eq. (2.40), there are then merely two exponents left which determine the scaling of avalanche size and duration. Any further observable and its simple scaling adds only one more independent exponent, as each is expected to obey an equation of the form Eq. (2.40).

If one observable can even be seen as a (deterministic) function of another, say $T = c_{Ts} s^{\gamma_{Ts}}$, then Eq. (2.39) follows from $\mathcal{P}^{(s)}(s) \, ds = \mathcal{P}^{(T)}(T) \, dT$ or, equivalently, by integrating over the nuisance variable of the joint probability density function

$$\mathcal{P}(s, T) = \mathcal{P}^{(s)}(s)\delta(T - c_{Ts} s^{\gamma_{Ts}}). \tag{2.41}$$

In general, avalanche duration and size are not strict functions of each other but random variables, so that $\mathcal{P}(s, T)$ has finite support in s at fixed T and vice versa. Nevertheless, the scaling law Eq. (2.40) holds if to leading order (Chessa et al., 1999a; Pruessner and Jensen, 2004)[11]

$$\mathcal{P}^{(s)}(s) \, ds \simeq \mathcal{P}^{(T)}(\langle T|s \rangle) \, d \langle T|s \rangle. \tag{2.42}$$

[11] Chessa et al. (1999a) discovered some problems in the conditional scaling of the Abelian BTW Model (compared to the AMM) and concluded that symmetric conditional PDFs are needed, which for some observables requires an additional transformation of their distribution.

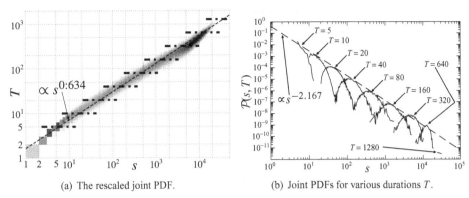

(a) The rescaled joint PDF.   (b) Joint PDFs for various durations T.

Figure 2.6 The joint PDF in the one-dimensional OSLO Model (as Fig. 2.5). (a): Density plot of the joint PDF, with darker greyscales indicating higher values. The ridge has slope $s^{\gamma_{Ts}}$ with $\gamma_{Ts} \approx 0.634$ (dashed line). To maintain roughly constant ridge height and therefore grey level, the data are rescaled to $\mathcal{P}(s, T)s^{1+\alpha\gamma_{Ts}}$ with $\alpha \approx 1.84$, so that $1 + \alpha\gamma_{Ts} \approx 2.167$. This is because the marginal of $\mathcal{P}(s, T)$ across s scales like $\mathcal{P}^{(T)}(T) \propto T^{-\alpha} \propto s^{-\alpha\gamma_{Ts}}$ and this volume is spread out over a width increasing linearly in s. As a result, the maximum of $\mathcal{P}(s, T)$ in s at fixed T drops like $s^{-\alpha\gamma_{Ts}-1}$, as indicated by a dashed line in (b), which also shows samples of $\mathcal{P}(s, T)$ at fixed T as indicated (dash-dotted, *horizontal* lines in (a)). In a double logarithmic plot the width is roughly constant across, i.e. it increases linearly in s. The PDFs are slightly skewed and obey $s \geq T$, by definition of the observables.

This can be further qualified using Bayes' theorem,

$$\mathcal{P}(s, T) = \mathcal{P}^{(s)}(s)\mathcal{P}^{(T)}(T|s) \tag{2.43}$$

so that unconditional moments $\langle T^n \rangle$ can be written in terms of the moments of the conditional PDF $\mathcal{P}^{(T)}(T|s)$. If the variance of the latter scales like $\sigma^2(T|s) \propto s^{2\mu}$ then the (unconditional) second moment of the duration, which is bounded from below by the weighted integral over the conditional variance, scales at least as fast as $L^{D(1+2\mu-\tau)}$. On the other hand, the second moment of the duration is supposed to scale like $L^{z(3-\alpha)}$, i.e. $D(1 + 2\mu - \tau) \leq z(3 - \alpha)$. For Eq. (2.40) to hold, this requires $\mu \leq z/D = \gamma_{Ts}$. In other words, the width of the distribution of the avalanche duration conditional to a certain size should not scale faster than the first conditional moment, Eq. (2.35b). A typical distribution is shown in Fig. 2.6(a) and distributions at different durations are shown in Fig. 2.6(b). Although these, effectively conditional, distributions can be quite skewed, they just about fulfil the condition of $\mu \leq \gamma_{Ts}$. As the width is a characteristic scale, this scaling, $\mu \approx \gamma_{Ts}$, is *typical* and has also been confirmed in the DS-FFM (Pruessner and Jensen, 2004, Fig. 27). Nevertheless, the resulting scaling law, Eq. (2.40), is not a necessity, but a widely expected and accepted assumption.

2.2.3 Moment and cumulant generating functions

The **moment generating function** of the PDF $\mathcal{P}^{(s)}(s)$ is its Laplace transform

$$\mathcal{M}^{(s)}(x) = \int_0^\infty ds\, \mathcal{P}^{(s)}(s)\, e^{sx}, \tag{2.44}$$

also known as its **characteristic function**. Assuming the integral exists, integer moments can be derived by differentiation,

$$\langle s^n \rangle = \left. \frac{d^n}{dx^n} \right|_{x=0} \mathcal{M}^{(s)}(x). \tag{2.45}$$

Its Maclaurin series

$$\mathcal{M}^{(s)}(x) = \sum_{n=0}^{\infty} \frac{x^n \langle s^n \rangle}{n!} \tag{2.46}$$

does not have to exist, even if all moments do.

The logarithm

$$\mathcal{C}^{(s)}(x) = \ln(\mathcal{M}^{(s)}(x)) \tag{2.47}$$

is the cumulant generating function (CGF), i.e. the cumulants of $\mathcal{P}^{(s)}(s)$ are obtained (and defined) by differentiating $\mathcal{C}^{(s)}(x)$ with respect to x and evaluating at $x = 0$. The first cumulant equals the first moment and the second and third cumulants are the second and third central moments, e.g. $\langle (s - \langle s \rangle)^2 \rangle$ and $\langle (s - \langle s \rangle)^3 \rangle$, respectively. In equilibrium critical phenomena the CGF of the order parameter is the free energy. Because the CGF of a Gaussian is a simple quadratic, all its cumulants beyond the second vanish.

Cumulants can be written as sums of products of moments, each term having the same canonical dimension. The nth cumulant contains the nth moment as a term, which is its leading order behaviour for $\tau > 1$. In general, for $\tau > 1$ the variance of the nth moment scales like the $2n$th moment, so that the leading order of the relative error diverges like $L^{\sigma_{2n}^{(s)}/2 - \sigma_n^{(s)}}$, where $\sigma_{2n}^{(s)}/2 - \sigma_n^{(s)} = D(\tau - 1)/2$. Typically, higher moments have larger relative errors, because they draw their weight mainly from increasingly large and thus increasingly rare events (Fig. 7.6, p. 226). The relative error continues to grow with increasing system size, but at least this increase does not depend on the order of the moment. For $\tau = 1$, the leading order of the variance can scale like $L^{\sigma_{2n}^{(s)}}$ or slower and thus the relative error remains either constant with increasing system size or even decreases. Further numerical implications are discussed in Sec. 7.3.

2.3 Algebraic correlations

At the core of a critical phenomenon are **algebraic spatiotemporal correlations**, which can be characterised either directly on the basis of the correlation function of a suitable *local* variable, or indirectly through *global* observables such as avalanche size and duration. The study of SOC models normally concentrates on global, spatially 'averaged' observables, as correlation functions are notoriously noisy and difficult to analyse, not least due to the presence of boundary conditions. In the following, the fundamental importance of correlation functions is discussed, briefly together with their scaling behaviour, which, in the form of block scaling, also provides a means to bridge the gap between local and global observables.

If the state of site $\mathbf{x}$ at time t is given by the local **dynamical variable** $h(\mathbf{x}, t)$, the auto-correlation function to be considered is

$$c(\mathbf{x}, \mathbf{x}' t, t') = \langle h(\mathbf{x}, t) h(\mathbf{x}', t') \rangle_c = \langle h(\mathbf{x}, t) h(\mathbf{x}', t') \rangle - \langle h(\mathbf{x}, t) \rangle \langle h(\mathbf{x}', t') \rangle \qquad (2.48)$$

where $\langle \cdot \rangle$ denotes the ensemble average, often improved by averaging over space and/or time. In general, the **connected correlation function** $c(\mathbf{x}, \mathbf{x}' t, t')$ decays for large spatial or temporal distances, as $\langle h(\mathbf{x}, t) h(\mathbf{x}', t') \rangle$ factorises asymptotically, suggesting that local states eventually become independent, $\langle h(\mathbf{x}, t) h(\mathbf{x}', t') \rangle \rightarrow \langle h(\mathbf{x}, t) \rangle \langle h(\mathbf{x}', t') \rangle$. **Long-range order** is generally associated with non-vanishing local averages $\langle h(\mathbf{x}, t) \rangle$ and so-called **quasi long-range order** with an algebraic decay of $\langle h(\mathbf{x}, t) h(\mathbf{x}', t') \rangle$ (Chaikin and Lubensky, 1995).

Algebraic correlations in isotropic space mean that the equal time correlation function in the stationary state beyond a lower cutoff $r_0 \ll |\mathbf{x} - \mathbf{x}'|$ decays like

$$c(\mathbf{x}, \mathbf{x}'; t, t; L) = a^{(\mathrm{h})} |\mathbf{x} - \mathbf{x}'|^{-(d-2+\eta)} \mathcal{K}\left(\frac{|\mathbf{x} - \mathbf{x}'|}{L} \right), \qquad (2.49)$$

defining the correlation exponent η, a metric factor $a^{(\mathrm{h})}$ and a scaling function $\mathcal{K}(x)$ very similar to the form used in Eq. (2.2). The length L is the system size when finite size scaling applies, as is the case in SOC, and the correlation length, where it dominates the scaling behaviour (see below).

The definition of η in Eq. (2.49) draws on the observation that if the spatiotemporal evolution of $h(\mathbf{x}, t)$ were purely diffusive, or, rather, driven by surface tension ν,

$$\partial_t h(\mathbf{x}, t) = \nu \nabla^2 h(\mathbf{x}, t) - \epsilon h(\mathbf{x}, t) + \xi(\mathbf{x}, t) \qquad (2.50)$$

with $\epsilon h(\mathbf{x}, t)$ representing an effective bulk dissipation (see Sec. 9.2.1) and ξ a Gaussian white noise with amplitude $2D$, then the correlation function would obey, provided it existed,[12] by dimensional analysis

$$c(\mathbf{x}, \mathbf{x}'; L) = \frac{D}{\nu} |\mathbf{x} - \mathbf{x}'|^{-(d-2)} \mathcal{K}\left(|\mathbf{x} - \mathbf{x}'|/L \right), \qquad (2.51)$$

with the scale L now being determined by the (tunable) bulk dissipation, $L = \sqrt{\nu/\epsilon}$. Equation (2.50) describes free particles and a deviation from $\eta = 0$ thus indicates interaction and more specifically non-trivial scale invariance and emergence. It is worth noting that $\eta > 0$ means that correlations decay *faster* than in the free theory.

The key observation in Eq. (2.49) is that *the only relevant length scale is L*, which *dominates* the asymptotic behaviour of the correlation function; there is no other **characteristic length scale** on the long range. By dimensional consistency, other, *competing* length scales and thus additional terms in Eq. (2.50) are needed to produce $\eta \neq 0$.

Similar arguments can be made for temporal correlations as often studied in the context of growth phenomena (Barabási and Stanley, 1995).[13] In SOC, temporal correlations can exist on the microscopic as well as on the macroscopic time scale. The two point spatial

---

[12] Ultraviolet divergences cause problems at and above $d = 4$.
[13] Temporal correlations are often characterised by their **Hurst exponent**.

correlations discussed in the following are taken at equal microscopic and thus macroscopic time.

The scaling form Eq. (2.49) indicates **scale invariance**: the correlations at one length scale are related to that on a different length scale by

$$c(\lambda \mathbf{x}, \lambda \mathbf{x}'; \lambda L) = \lambda^{-(d-2+\eta)} c(\mathbf{x}, \mathbf{x}'; L). \tag{2.52}$$

To illustrate scale invariance, in the following distances $r = |\mathbf{x} - \mathbf{x}'|$ are expressed as multiples of some other arbitrary but fixed distance $r_0 = |\mathbf{x}_0 - \mathbf{x}_0'|$. In the limit of $L \to \infty$ the scaling function is expected to converge,

$$\lim_{L \to \infty} c(\lambda r_0; \lambda L) = \lambda^{-(d-2+\eta)} C_0 \quad \text{with} \quad C_0 = a^{(h)} r_0^{-(d-2+\eta)} \lim_{y \to \infty} \mathcal{K}(y), \tag{2.53}$$

although in principle the same caveats apply as for the avalanche size distribution (Sec. 2.1.2). To avoid such complications, the scale L can be kept finite and r as well as r_0 can be chosen from a range where the scaling function is essentially constant (see footnote 14 on p. 49 about full scale invariance) or close to a power law, which gives rise to an 'apparent' exponent η. Dropping the limit on the left and using $r = \lambda r_0$ produces the desired result:

$$c(r; L) = \left(\frac{r}{r_0}\right)^{-(d-2+\eta)} C_0. \tag{2.54}$$

Full scale invariance (i.e. pure power law scaling) manifests itself by the impossibility of defining a scale from within (see also p. 11, p. 355). One might be tempted, for example, to define the characteristic length $R_{1/2}$ as the length within which correlations decay by a half, $c(r + R_{1/2}; L)/c(r; L) = 1/2$, which determines $R_{1/2}$ according to Eq. (2.54) through $(1 + R_{1/2}/r)^{-(d-2+\eta)} = 1/2$, but only *relative* to the initial scale r. If $c(r; L)$ displayed an exponential decay

$$c(r; L) = e^{-r/r_0} C_0, \tag{2.55}$$

$\exp(-R_{1/2}/r_0) = 1/2$ would reveal the value of $R_{1/2}$ as a multiple of the fixed distance r_0.

It is worth noting that scaling does not mean the complete absence of any scales, as exemplified by r_0 in Eq. (2.54). Without r_0 on the right of Eq. (2.54) no dimensionless quantity could be formed and $d - 2 + \eta$ is bound to vanish as C_0 has the same dimension as $c(r; L)$. Full scale invariance means the absence of a fixed **dominant scale** and in non-trivial cases the presence of **competing scales**, which allow the formation of dimensionless quantities and thus power laws with non-trivial exponents. In turn, the observation of non-trivial exponents, i.e. those deviating from the engineering dimension, generally implies the presence of other scales, which, however, do not dominate. These scales are often microscopic features of the system. For example, the effective diffusion constant D and the initial position a_0 of a random walker enter in the return time distribution of a random walker, Eq. (8.26), to form the characteristic time a_0^2/D it takes to explore the initial gap a_0 diffusively.

Correlation functions are generally difficult to measure directly. Fortunately, algebraic correlations leave their signature in other observables, such as the avalanche size and the coarse-grained variables discussed below.

2.3.1 Coarse graining and block scaling

That scale invariance is reflected in correlations is somewhat counter-intuitive, as 'invariance under rescaling' seems to suggest that there is no structure at all that could spoil the system looking 'the same' on different scales. However, a fully scale invariant [14] system does not lack any structure, rather it lacks a **characteristic structure size**.

The coarse grained dynamical variable

$$H(\mathbf{x}; l) = l^{-d} \int_{l^d(\mathbf{x})} d^d x' \, h \qquad (2.56)$$

is the local average over a block with volume l^d centred at $\mathbf{x}$. The spatial correlations of $H(\mathbf{x}; l)$ can be derived from Eq. (2.49) assuming that the distances involved are large compared to the lower cutoff r_0, which entails $l \gg r_0$:

$$\langle H(\mathbf{x}; l) H(\mathbf{x}'; l) \rangle_c = a^{(h)} |\mathbf{x} - \mathbf{x}'|^{-(d-2+\eta)} \mathcal{K}\left(\frac{|\mathbf{x} - \mathbf{x}'|}{L}\right) + \mathcal{O}\left(\frac{l}{|\mathbf{x} - \mathbf{x}'|}\right). \qquad (2.57)$$

The correlations of h are thus equally visible in $H(\mathbf{x}; l)$ irrespective of the coarse graining scale l, Eq. (2.49). Since $|\mathbf{x} - \mathbf{x}'| \gg l$ is required to suppress corrections, the correlations eventually fizzle out as $|\mathbf{x} - \mathbf{x}'|$ approaches the large scale L (not necessarily the system size). The fact that Eq. (2.57) is independent of l is the key-signature of scale invariance: the dynamical variable 'looks the same' on different levels of coarse graining.

Evaluating $\langle H(\mathbf{x}; l) H(\mathbf{x}'; l) \rangle_c$ for $\mathbf{x} = \mathbf{x}'$ gives the local variance $\sigma^2(H; l)$. A priori, this is a *local, patch-wise* variance taken over an ensemble of independent realisations. Yet, assuming translational invariance, any instantaneous realisation of a system can be decomposed into patches of size l^d, and the ensemble average improves by averaging over space as well. In fact, the patches do not need to be strictly distinct. This strategy is in effect a form of sub-sampling. Repeating the calculation above, the variance can be shown to scale like

$$\sigma^2(H; l) = a^{(h)} l^{-(d-2+\eta)} \tilde{\mathcal{K}}\left(\frac{l}{L}\right) \qquad (2.58)$$

with $\tilde{\mathcal{K}}(x)$ asymptotically scaling like $x^{-(2-\eta)}$, i.e. for $l \gg L$ the variance vanishes like l^{-d}. This is the behaviour expected from the central limit theorem for independent random variables. Unless anti-correlations prevail, $\sigma^2(H; l)$ cannot drop off faster in l than it does in the complete absence of any correlation, which gives $\sigma^2(H; l) \propto l^{-d}$, with a Gaussian distribution of $H(\mathbf{x}; l)$ according to the central limit theorem. [15] In that sense, an exponent $\eta > 2$ can be considered as indicating the absence of all correlations. Equation (2.58) is known as **block scaling** or **box scaling** (Binder, 1981a) and provides direct access to the correlation exponent η without the need to consult the correlation function directly (Pruessner, 2008).

---

[14] Strictly, a fully scale invariant system has no characteristic scale at all. In particular, the thermodynamic limit must have been taken already. Above, that has been approximated at an intermediate length scale in a large albeit not infinite system.

[15] However, a Gaussian distribution does not in turn imply absence of scale invariance, Sec. 2.1.3.

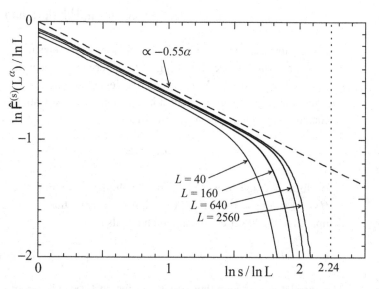

Figure 2.7 Example of the transformation according to Eq. (2.59), based on the avalanche size distribution of the simplified OsLo Model (system sizes as indicated, 10^7 avalanches). With increasing system size, the resulting graph approaches $f(\alpha)$ where $\alpha = \ln s / \ln L$. The dashed line shows the expected asymptote $f(\alpha) = \alpha(1 - \tau)$, where $\tau \approx 1.55$ in the OsLo Model. The dotted line indicates $D \approx 2.24$, the value of α where $f(\alpha)$ ceases to exist.

2.4 Multiscaling

Multiscaling, also known as **multifractal scaling**, is a form of scaling in the absence of gap scaling, i.e. $\sigma_{n+1}^{(s)} - \sigma_n^{(s)}$ as defined via Eq. (2.24) on p. 38 varies in n even for $n > \tau - 1$. Unfortunately, the term has frequently been applied to systems that display broad distributions but strictly no scaling.

Based on the complementary cumulative distribution function, $\hat{F}^{(s)}(z) = \int_z^\infty ds\, \mathcal{P}^{(s)}(s)$, the multifractal spectrum $f(\alpha)$ is defined as (Tebaldi, De Menech, and Stella, 1999)

$$f(\alpha) = \lim_{L \to \infty} \frac{\ln \hat{F}^{(s)}(L^\alpha)}{\ln L}. \tag{2.59}$$

If simple scaling applies, Eq. (2.2), then $f(\alpha) = \alpha(1 - \tau)$ for $\alpha \le D$, for other values of α the spectrum $f(\alpha)$ is undefined, as the integral $\hat{F}^{(s)}(z)$ vanishes asymptotically.

Equation (2.59) describes the scaling of the complementary cumulative distribution function expressed in a form suitable for calculating moments. They can be shown to scale like

$$\langle s^q \rangle \propto L^{\sigma_q^{(s)}} \propto \int_0^D d\alpha\, \ln(L) L^{\alpha q + f(\alpha)}, \tag{2.60}$$

whose scaling is dominated by the maximum, or more accurately by the supremum over $\alpha q + f(\alpha)$, if $f(\alpha)$ is continuous,

$$\sigma_q^{(s)} = \sup_\alpha (\alpha q + f(\alpha)). \tag{2.61}$$

Again, if simple scaling applies, then $\alpha q + f(\alpha) = \alpha(q + 1 - \tau)$ becomes maximal for the largest value α for which $f(\alpha)$ exists, i.e.

$$D(1 + q - \tau) = \sigma_q^{(s)}, \tag{2.62}$$

reproducing Eq. (2.27).

Based on $\alpha = d\sigma_q^{(s)}/dq$, the spectrum $f(\alpha)$ can also be interpreted as the Legendre transform of the scaling exponent of the avalanche size moments (De Menech *et al.*, 1998). Equation (2.59) suggests a collapse, as plotting $\ln \hat{F}^{(s)}(L^\alpha) / \ln L$ versus $\alpha = \ln s/\ln L$ produces the spectrum $f(\alpha)$. Up to the normalisation with respect to $\ln L$, this amounts to a simple double logarithmic plot of the cumulative distribution function, the slope of which is $\tau - 1$ in the case of simple scaling.[16] Strictly, Eq. (2.59) does not describe a *collapse*, but a limit.

An example of a multiscaling analysis of a system displaying gap scaling is shown in Fig. 2.7. Although the present technique was introduced early and used successfully to characterise the scaling of models related to the original BTW Model (Kadanoff *et al.*, 1989), the main emphasis in SOC remains on gap scaling. In fact, it is difficult to entertain the concepts of scale invariance and universality in the presence of multiscaling (but Kadanoff, 1990).

---

[16] It does *not*, however, show the apparent exponent, $\tilde{\tau} - 1$, as illustrated by $\mathcal{P}^{(s)}(s) = \theta(L - s)/L$ which has $\tilde{\tau} = 0$, $\tau = 1$ and $f(\alpha) = 0$.

3　Experiments and observations

A number of experiments and observations have been undertaken to test for SOC in the 'real world'. Ultimately, these observations motivate the research based on analytical and numerical tools, although the latter provide the clearest evidence for SOC whereas experimental evidence is comparatively ambiguous. What evidence suffices to call a system self-organised critical? One might be inclined to say scale invariance without tuning, but as discussed in Sec. 9.4, the class of such systems might be too large and comprise phenomena that traditionally are regarded as distinct from criticality, such as, for example, diffusion.

In most cases, systems suspected to be self-organised critical display a form of scaling and a form of avalanching, suggesting a separation of time scales. Because of the early link to $1/f$ noise (Sec. 1.3.2), some early publications regard this as sufficient evidence for SOC. At the other end of the spectrum are systems that closely resemble those that are studied numerically and whose scaling behaviour is not too far from that observed in numerical studies. Yet it remains debatable whether any numerical model is a faithful representation of any experiment or at least incorporates the relevant interactions.

At first sight, solid experimental evidence for scaling or even universality is sparse among the many publications that suggest links to SOC. This result is even more sobering as evidence for SOC is heavily biased – there are very few publications (e.g. Jaeger, Liu, and Nagel, 1989; Kirchner and Weil, 1998) on *failed* attempts to identify SOC where it was suspected. One might therefore conclude that the empirical basis of SOC is much thinner than originally thought. However, the enormous number of reports on *noisy, skewed and non-universal* power laws in natural systems can hardly be dismissed as spurious. In some interpretations, SOC can account for that (e.g. Pruessner and Peters, 2006). Even when the (sometimes rather faint) evidence for scaling is not taken as a sign for a particular type of interaction that governs across length and time scales, some authors (e.g. Linkenkaer-Hansen, Nikouline, Palva, and Ilmoniemi, 2001) still regard it as a constraint for models, which need to reproduce at least some sort of scaling.

Most of the experiments have spun off a bulk of literature on models which manage to reproduce certain experimental features, which cannot be found in the models originally intended to apply. Some early variations of sandpile models are vivid examples of such attempts to recuperate lost territory, many of which were ultimately in vain. When the BTW Model (Sec. 4.1) did not seem to capture the experimental reality, inertia was added or randomness, friction etc. The following sections do not discuss these models, rather they focus on the primary literature about the original experiments and observations, which referred to SOC either as the suspected underlying principle or at least as a potential 'explanation'. Some large areas of experimental work can only be mentioned in passing, in particular the many sociological studies, often heavily influenced by **Zipf's law**

(1949), and even the main results and their enormous literature can only be discussed very briefly.

About one third of the following is dedicated to biological systems, where, however, SOC faces a fundamental problem. If they are not governed by deterministic, mechanical laws which naturally give rise to driving and relaxation rules, SOC has little to add beyond the language for describing the events in terms of avalanches. SOC cannot provide a basis to derive such rules, which are necessarily beyond its realm. Only once these rules are established can SOC attempt to explain the overall behaviour, help to characterise the scaling and identify the relevant underlying interaction.

Malcai, Lidar, Biham, and Avnir (1997, also Avnir, Biham, Lidar, and Malcai, 1998) reviewed the literature available on empirical studies of fractals to determine the range of scales over which fractals have been identified. They found 'hardly any' publication on data exceeding two decades, not least because it is difficult to undertake experiments and observations over larger spatiotemporal scales. Their finding seems to translate naturally to the empirical basis of SOC. In fact, Avnir *et al.* (1998) concluded that labelling something as a fractal was, most of the time, merely a matter of taste of the experimentalist, because most observations were not backed by a theory suggesting a particular scale invariant behaviour. It is difficult to untie this Gordian knot. If experimental data are truly reliable only if supported by a suitable theory, they cannot serve independently as support for the latter. This, however, is often the intended purpose of empirical studies, namely to show that certain theories and models are justified and supported by nature.

A number of reviews of the experimental work in SOC are available. The review by Jaeger, Nagel, and Behringer (1996) is a very useful resource for the physics of granular matter, and Kakalios (2005) gives a more up-to-date comprehensive literature overview of the same subject. Altshuler and Johansen (2004) reviewed avalanching in superconductors and Stanley, Amaral, Buldyrev, *et al.* (1996) as well as Brown and West (2000) and Gisiger (2001) provide detailed overviews on scaling and SOC in biological systems. General reviews of experimental evidence for SOC can be found in most books and review papers discussed in Sec. 1.1, in particular Turcotte (1999) and Dickman *et al.* (2000).

3.1 Granular media

Probably the earliest experiment on $1/f$ noise in sandpiles is due to Schick and Verveen (1974) who observed an hourglass with a particularly long neck and found a $1/f$ power spectrum in the flow measured by a laser beam that is blocked out whenever particles pass. The earliest sandpile experiments in response to the BTW Model are due to Jaeger *et al.* (1989, but note Evesque and Rajchenbach, 1989), sometimes refered to as the **Chicago group**, who used three different geometries, such as Fig. 3.1, and two different types of granular material, namely essentially monodisperse glass beads or rough aluminium oxide particles. These systems were driven either by slowly tilting a pile of the material in a drum (radius 5 cm) until avalanches occurred (tilt loading) or by slowly and randomly adding particles to the surface (top loading). The slope of granular material varies between the

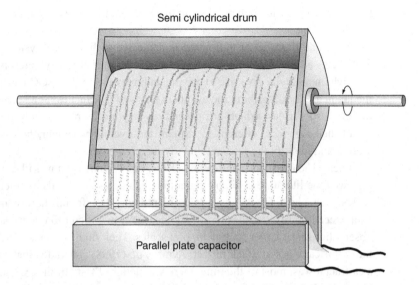

Semi cylindrical drum

Parallel plate capacitor

Figure 3.1   The semicylindrical drum used by Jaeger *et al.* (1989). Avalanches form when it is slowly rotated with particles spilling over the rim. The amount of particles can be measured using a parallel plate capacitor, where they function as a dielectric.

angle of repose θ_r, through a region where avalanching is imminent up to the **maximum angle of stability** θ_m above which avalanching must occur and frequently spreads through the entire system. The pile could be vibrated (Evesque and Rajchenbach, 1989) to help particles overcome static friction, reducing $\theta_m - \theta_r$.

The avalanche sizes were measured by their **drop number** (a terminology common in the early 1990s, introduced by Kadanoff *et al.*, 1989), i.e. the amount of particles falling off the rim of the container. Their time series displayed oscillatory behaviour rather than scale invariance. In particular, the power spectrum in the low frequency region levelled off, consistent with the existence of a characteristic frequency. It was concluded that the experiment showed little evidence for SOC and that therefore Bak *et al.*'s (1987) sandpile metaphor was 'not well founded'.

This result was further qualified by Held, Solina, Solina, *et al.* (1990), the **IBM group**, who implemented FSS as well as a perfect separation of time scales, as discussed by Jensen *et al.* (1989) in response to the experiments discussed above. The avalanche sizes were measured by monitoring the mass of the pile sitting on a circular base plate, Fig. 3.2, again with particles falling off the rim of the base. Avalanches were triggered by depositing particles whenever the pile was quiescent, i.e. the weight remained constant. Making the usual FSS ansatz,

$$\mathcal{P}^{(s)}(s) = as^{-\tau}\mathcal{G}(s/(bL^D)), \tag{3.1}$$

the authors concluded $\tau = 2$ (in their notation β/ν, see Kadanoff *et al.*, 1989) on the basis that the mass added per avalanche is constant (a single grain) and equal to the average avalanche size in the stationary state (assuming that avalanches of vanishing size had *not* been discounted). However, incorporating a lower cutoff, which is more than likely to exist

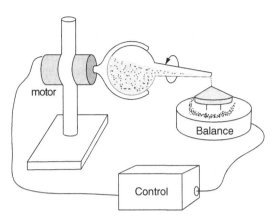

Figure 3.2 The setup used by Held *et al.* (1990). The pile rests on an electronic balance and is fed very slowly and intermittently by grains falling out of a rotating funnel that stops as soon as the weight of the pile changes, i.e. an avalanche is running. Particles added and those falling off the pile are detected by a change in weight.

in an experimental setup, and allowing a power law asymptote for the scaling function, [1] the constraint of finite first moment leads more generally to $D(2 - \tau) = \sigma_1^{(s)} \leq 0$ and thus $\tau \geq 2$. In a direct measurement, the authors found $\tau = 2.5$, but used $\tau = 2$ and $D = 0.9$ in a data collapse.

Most interestingly, the power spectrum on the macroscopic time scale displayed $1/f^2$ noise, quite consistently so, even in the low frequency region, and was detected in the BTW Model by Jensen *et al.* (1989). Given that the sandpile is hardly more than a metaphor of the BTW Model, this is a rather surprising coincidence and might be regarded as a sign of universality. However, the largest system size used, $L = 3$ in (7.62 cm), displayed a breakdown of the scaling behaviour.

Jensen *et al.* (1989, also Dendy and Helander, 1998) emphasised the difference between the avalanche size measured as the number of particles falling off the pile (**drop number**) and that measured as the total number of particles moving (**flip number**). They found that the statistics of the drop numbers in model and experiment were not so different after all and advocated comparing like with like in model and experiment. The drop numbers measure only dissipative avalanches (Drossel, 2000, also Sec. 4.2.4), and only the amount dissipated, which might not even be proportional to their total volume. Measurement of the flip number was first implemented by Bretz, Cunningham, Kurczynski, and Nori (1992) using a CCD camera, which allowed a measurement of the number of particles that had moved during an avalanche. They found $\tau = 2.134$ in the avalanche size distribution over one decade. Their experimental setup was similar to that used by Jaeger *et al.* (1989) and they also found periodicity in the 'large sliding events'.

---

[1] Strictly, finite $g_1^{(s)}(0)$ is needed, Eq. (2.4) on p. 29, which requires either $\tau > 2$ with any scaling function, or $\tau = 2$ provided the scaling function decays fast enough, for example like a power law. This discussion is identical to that about the normalisation, Sec. 2.2. Unfortunately, many of the early experiments are riddled with omissions and misunderstandings with regard to scaling (e.g. Feder 1995; Frette, Christensen, Malthe-Sørenssen, *et al.*, 1996).

Bretz *et al.* (1992) also introduced the experimental 'trick' of glueing a few layers of granular material to the base plate of the tilting box. The impact of this measure was later studied in detail by Altshuler, Ramos, Martínez, *et al.* (2001), who found that an irregular configuration of the base improved the scaling behaviour of avalanches in top loaded piles of steel beads significantly. They determined exponents $D = 1.35$ and $\tau = 1$. Although the authors measured avalanche sizes as drop numbers, $\tau = 1 < 2$ does not violate conservation, as they explicitly discounted avalanches of size zero. Costello, Cruz, Egnatuk, *et al.* (2003) performed a similar study finding $\tau \approx 1.5$ in many different settings.

Cohesion and thus **friction** was identified by all experimentalists as a problem to be avoided or controlled (Jaeger and Nagel, 1992; Nagel, 1992; Albert, Albert, Hornbaker, *et al.*, 1997), focusing mostly on reducing humidity (but Somfai, Czirok, and Vicsek, 1994). Clearly, static friction together with geometrical effects gives rise to a non-zero angle of repose as well as the crucial threshold above which an avalanche is triggered. Evesque (1991) was the first who raised **inertia and gravity** as a dominant scale in avalanches (Krug, Socolar, and Grinstein, 1992), introducing a characteristic size and an effective cutoff in the avalanche size distribution. He characterised the flow of glass beads in a slowly rotating drum by the sound they emit and detected neither $1/f$ nor periodicity nor power law distributed avalanche durations. Prado and Olami (1992, also Mehta, 1992; Puhl, 1992; Bouchaud, Cates, Prakash, and Edwards, 1995; Herrmann and Luding, 1998; Nerone and Gabbanelli, 2001) studied the effect of inertia numerically and (qualitatively) reproduced many of the experimental results. It is now widely accepted that inertia plays a crucial rôle as it allows grains to gain momentum thus changing the microscopic dynamics depending on the distance travelled (Grinstein, 1995; Jensen, 1998; Dickman *et al.*, 2000). In small avalanches, where friction dominates over momentum, it is generally less important than in big avalanches. It seems therefore sensible to *increase* friction, rather than suppress it. Mehta and Barker (1991, also Mehta, 1992, 2007) placed more emphasis on the surface dynamics of a sandpile as particles are transported. According to them, the correlations between particles can affect subsequent avalanches and suppress scale invariant behaviour (Mehta and Barker, 1994; Barker and Mehta, 1996), which fits with earlier findings that disorder and 'dirt' can change the scaling of sandpiles (Toner, 1991; Ding, Lu, and Ouyang, 1992; Puhl, 1993; Barker and Mehta, 2000).

Grumbacher, McEwen, Halverson, *et al.* (1993) took great care to deliver a single bead at a time to the top of the pile and studied the ensuing avalanches in an otherwise similar setup to that used by Held *et al.* (1990), see Fig. 3.2. The avalanche size distribution based on the drop number was compatible with $\tau \approx 2.5$, independent of many details, such as bead material, diameter of the base and the covering of the base. Based again on the setup used by Held *et al.* (1990), Rosendahl, Vekić, and Kelley (1993) tried to unravel the cause of the apparent difference between small and large avalanches, the former following an approximate power law, the latter occurring periodically. The power law distribution of small avalanches was found to persist even in large piles (8 cm, comparable to the maximum size used by Held *et al.* (1990), but using smaller grains [2]), but was supplemented by an approximately uniform

---

[2] If the same arguments apply as in the usual SOC sandpile models, the scale is set by the ratio between system size and grain size.

distribution of large avalanches (Rosendahl, Vekić, and Rutledge, 1994) which accounted for an increasing fraction 'of the total avalanched mass' (Rosendahl *et al.*, 1993).

Jánosi and Horváth (1989) performed a different kind of avalanching experiment around the same time when Jaeger *et al.* (1989) published the first results on an experimental sandpile. Using a camera and image processing software, they recorded the time series of the water coverage of a glass pane that was sprinkled with small water droplets. On the glass, the droplets merged and ran down the tilted pane. The power spectrum of the coverage showed a clear power law decay $\approx 1/f^{2.15}$ over about one order of magnitude. In principle, Jánosi and Horváth (1989) had access to the flip number, which, however, they did not analyse, possibly because of the tedious image analysis that would have been required. Plourde, Nori, and Bretz (1993) analysed the drop numbers in a similar system, namely that of merging droplets which formed inside a Perspex®dome as it was continuously sprayed with a fine mist of water. Once small streams formed and reached the rim of the dome, water drops fell on a detector from which the signal was processed to determine the size and duration of such avalanches. Their distribution was compatible with a power law stretching roughly over one order of magnitude with an exponent between $\tau \approx 1.93$ and $\tau \approx 2.66$, depending on the viscosity of the water, which was controlled by its temperature.

The length to which some authors went in order to find SOC in experimental granular systems is strangely at odds with the claim that SOC is ubiquituous. Observational data yielded mixed results. The Himalayan avalanches analysed by Noever (1993) seemed to provide little support for SOC, even when the data stretched over twelve orders of magnitude and produced a broad, smooth distribution. On the other hand, the distribution of snow avalanches analysed by Faillettaz, Louchet, and Grasso (2004) seemed to follow a power law over about two orders of magnitude, $\tau = 1.2(1)$. Feder (1995) reviewed many experimental findings and concluded that none displayed SOC as originally envisaged. While the overall conclusion stands unquestioned, the analysis is based on the finding that many experimental results are compatible with a stretched exponential (also Laherrère and Sornette, 1998). Retaining most of the original notation, it can be rewritten as standard FSS

$$\mathcal{P}^{(\mathrm{s})}(s) = as^{-\beta/\nu}\frac{\gamma}{\xi^{1/\gamma}\Gamma(1/\gamma)}\underbrace{\left(\frac{s}{L^{\nu}}\right)^{\beta/\nu}e^{-\left(\frac{1}{\xi}\left(\frac{s}{L^{\nu}}\right)^{\gamma}\right)}}_{\mathcal{G}(bs/L^{\nu})}, \tag{3.2}$$

which contrary to the original claim, *supports* SOC in these experiments, with avalanche exponent $\tau = \beta/\nu$. Since the scaling function $\mathcal{G}(x)$ is bound to vanish for small arguments, $\tau = \beta/\nu = 1$ follows (Sec. 2.1.2.2 and Christensen *et al.*, 2008), provided that an asymptotically finite fraction of events is governed by this scaling form (but Rosendahl *et al.*, 1993), which clashes with $\tau \geq 2$ required for a finite first moment. [3]

Frette *et al.* (1996) used rice as the granular material in an effectively one-dimensional pile between two sheets of Perspex®as shown in Fig. 6.7, p. 191, on which avalanches were triggered by grains being added at a very low rate (top loading). Like Bretz *et al.* (1992) they used a CCD camera to record changes in the pile and derived from that the change

---

[3] Originally, it was shown that $\beta/\nu = 2$, but similar arguments as on p. 55 apply, i.e. $\beta/\nu \geq 2$.

in its potential energy. The data analysis has been partly revised (Christensen *et al.*, 2008), suggesting $\tau = 1$, and remains somewhat inconclusive. However, the distribution of avalanche sizes indicated power law scaling and even universality much more clearly than in any of the previous experiments. As discussed in further detail in Sec. 6.3.2, this has been traced back to the rice grains' prevalence for sliding rather than rolling, thereby suppressing inertial effects. The ricepile experiment was extended to two dimensions by Ahlgren, Avlund, Klewe, *et al.* (2002) focusing, however, on the residence time of particles in an approach similar to that used by Christensen *et al.* (1996) for a one-dimensional pile. Aegerter, Günther, and Wijngaarden (2003, also Aegerter, Lőrincz, Welling, and Wijngaarden, 2004a) measured the avalanche sizes by means of a projected pattern of coloured lines whose changing shape was recorded by a CCD camera. The resulting data allow a data collapse with exponents $D = 1.99(2)$ and $\tau = 1.21(2)$. Lőrincz and Wijngaarden (2007) investigated the effect of different boundary conditions, finding again periodic large avalanches for one type of boundary. The avalanche size distribution followed a power law very closely over almost four orders of magnitude with slopes $\tau = 1.12(2)$ and $\tau = 1.15(3)$ depending on the boundary condition. While none of the ricepile experiments match the exponent of the numerical ricepile model, Table 6.2 on p. 189, they certainly represent the best experimental results so far for SOC in granular media. A promising new route has been taken recently by Corté, Chaikin, Gollub, and Pine (2008, also Corté, Gerbode, Man, and Pine, 2009) considering granular systems with conserved particle numbers under shear, subject, however, to tuning.

3.2 Superconductors

The link between superconductors and 'sand hills' that display avalanching was made very early by de Gennes (1966, p. 83, according to some authors also by Bean, 1962, 1964), to illustrate the movement of **flux lines** penetrating a **type II superconductor**. [4] Exposing the superconductor to an external magnetic field, above a certain strength the magnetic field starts to penetrate the superconductor, without destroying the superconducting state. Once inside the material, the magnetic field is encapsulated in **vortex lines** consisting of superconducting electrons revolving around it. These vortex lines, each effectively carrying a quantum of magnetic flux $\phi_0 = h/(2e)$, enter from the surface and become **pinned at structural defects**. Due to an effective repulsion between the lines, only a finite number can reside at a defect at any given time. As their number increases, they start migrating deeper into the bulk, becoming trapped at defects until their capacity is exhausted as well. The state where (nearly) all defects are fully loaded is known as the **Bean critical state** (Bean, 1962). A further increase in the magnetic field leads to avalanches of vortex lines moving through the system, somewhat reminiscent of the **Barkhausen effect** and **charge density waves** (Fisher, 1998), making the subject of vortex pinning closely related to that of interface propagation in random media (e.g. Frette, 1993). One of the earliest measurements

---

[4] Type I superconductors completely exclude flux lines (complete Meisner effect) in the superconducting phase.

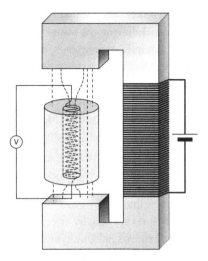

Figure 3.3   Experimental setup used by Field *et al.* (1995). The hollow tube consists of superconducting material which is penetrated by the external field above a certain strength. The vortex lines formed inside the superconductor are pinned at structural defects, which can only accommodate a finite number of them. Eventually, avalanches of vortex lines form and 'spill over' into the coil in the centre of the tube. Their arrival is detected by a change of voltage across the coil.

of the sudden changes in flux through a lead alloy superconductor with slowly increasing external field was conducted by Heiden and Rochlin (1968, but see M. R. Beasley's 1967 Ph.D. thesis cited by Ling, Shi, and Budnick, 1991), with a pickup coil wound around the specimen. They saw broad, exponential distributions of avalanches entering and leaving the sample at low fields.

Probably the first to adopt the SOC framework to 'avalanche flux motion' (Ling *et al.*, 1991) in type II superconductors were Pla and Nori (1991) and Ling *et al.* (1991), the former in a computer model of the **stick-slip motion of the flux lattice**, the latter in an experiment studying the current-voltage characteristic, inspired by the analogy made with 'traditional' critical phenomena by Tang and Bak (1988a). Both were consistent with avalanche-like dynamics, with Pla and Nori (1991) finding the displacement distribution followed a power law. Wang and Shi (1993) continued the experimental work focusing on the magnetic relaxation of the superconductor when exposed to an external field. Depending on the driving current, flux motion is thermally activated at first but then dominated by avalanches (Tang, 1993).

Placing a pickup coil *inside* a tube of superconducting material, Fig. 3.3, allows a precise measurement of such avalanches propagating *through* the material and arriving at the centre – again avalanche sizes are measured by the drop number, i.e. by the number of vortex lines leaving the superconductor. Motivated by sandpile experiments, Field, Witt, Nori, and Ling (1995) derived the avalanche size distribution in a niobium-titan superconductor from the voltage spikes induced in the central coil as the external field is slowly ramped up starting from three different initial field strengths. This setup was a significant improvement over previous attempts to characterise flux avalanches in superconductors. The resulting histograms followed power laws for up to 1.5 decades, characterised by exponents $\tau \approx 1.4$, $\tau \approx 1.8$, $\tau \approx 2.2$, depending on the initial field strength.

Compared with sandpile experiments, superconductivity experiments have significant advantages: **inertia effects** can be assumed to be small (but Zieve, Rosenbaum, Jaeger, *et al.*, 1996) and the system size is much larger, accommodating about 9000 vortex lines in the tube wall. However, the measurements suffer from some systematic problems. Firstly, not all flux moves in 'bundles'; some of it sweeps through the superconductor without significant interaction with pinning centres and other vortex lines (Yeh and Kao, 1984), possibly because of thermal activation. The situation is somewhat similar to grains rolling down the slope of a sandpile without triggering the movement of other grains. [5] Secondly, some flux bundles might intersect only parts of the pickup coil, which thus would fail to detect the whole vortex line, leading to a reduced signal which suggests smaller avalanches. As long as all vortex lines are equally affected, the signal measured remains useful. At high ramping rates of the external field, individual avalanches cannot be resolved any longer, i.e. a breakdown of time scale separation is visible in the data.

Zieve *et al.* (1996) performed a similar experiment on a different material, measuring the magnetic field *inside* the superconducting material by means of a Hall probe. This corresponds to measuring the total mass of a sandpile, rather than the grains falling off, but still amounts to a measurement of the drop number. Zieve *et al.* (1996) found no indication of power law distributed avalanche sizes, which Pla, Wilkin, and Jensen (1996) explained by the different nature of the pinning centres: 'Sharp and dense pins [as in niobium-titan] lead to a broad distribution of vortex avalanches. Pinning centres of longer range and lower density [as in the material used by Zieve *et al.* (1996)] produce a narrow distribution of events.'

Nowak, Taylor, Liu, *et al.* (1997) used two microscopically small Hall probes to measure the magnetic field locally in niobium and found very little evidence of SOC. In fact, the correlations of the two probes suggested that the flux matrix moved uniformly rather than in avalanches. Similar methods employed by Ooi, Shibaushi, and Tamegai (2000) yielded some evidence for scaling in the noise spectrum. Like Wang and Shi (1993), Aegerter (1998) used a SQUID (superconducting quantum interference device) to determine the magnetisation inside a superconductor when exposed to a strong external field. Avalanches were measured as changes in magnetisation over fairly long time spans (20 s per measurement) and showed a power law with exponent $\tau = 2$ over about one order of magnitude.

Behnia, Capan, Mailly, and Etienne (2000, also 2001) were the first to measure avalanches inside the material, i.e. the flip numbers rather than the drop numbers, using an array of 64 Hall probes on a square sample of size $0.8 \times 0.8\,\text{mm}^2$. The resulting avalanche size distribution recorded in niobium, a superconductor similar to the sample used by Field *et al.* (1995), was fitted against a power law with exponents $\tau = 2.05$ as well as a stretched exponential, suggesting that a length scale much smaller than the system size dominated the avalanche size distribution. At lower temperatures, they observed narrowly distributed 'catastrophic avalanches'.

Vlasko-Vlasov, Welp, Metlushko, and Crabtree (2004), Altshuler, Johansen, Paltiel, *et al.* (2004) and Aegerter, Welling, and Wijngaarden (2004b) used a magneto-optical technique

---

[5] Thermal activation here corresponds to vibrating the sandpile.

to image the flux pattern inside the superconductor. Vlasko-Vlasov *et al.* (2004) focused on the flux profile across the sample, which is analysed as an evolving interface for its roughness (Surdeanu, Wijngaarden, Visser, *et al.*, 1999), and found it to belong to the qKPZ universality class (e.g. Alava and Muñoz, 2002). Altshuler *et al.* (2004) measured an avalanche size exponent of $\tau = 3.0(2)$ over about two and a half orders of magnitude of the distribution of flip numbers in niobium. Based on the same observable but in a different material, Aegerter *et al.* (2004b) obtained $D = 1.89(3)$ and $\tau = 1.29(2)$ over about two orders of magnitude, in a data collapse for block finite size scaling (Sec. 2.3.1). The finite size here is the size of the patch examined rather than the actual sample size. Later experiments in niobium and control of the disorder (and thus the density of pinning centres) yielded $D = 2.25(5)$ and $\tau = 1.07(2)$ (Welling, Aegerter, and Wijngaarden, 2005) at high disorder, in line with the results by Pla *et al.* (1996). The values are close to those found for the *one-dimensional* MANNA universality class (see Table 6.1 on p. 168 and Table 6.2 on p. 189), but it remains unclear why this is not a mere coincidence. Computer simulations of vortex pinning by Bassler and Paczuski (1998, also Olson, Reichhardt, and Nori, 1997) were compatible with the two-dimensional MANNA universality class.

Invoking arguments similar to those discussed in Sec. 8.5.4.1 (p. 308), Aegerter *et al.* (2004b) extracted from their experimental data a roughness exponent for 'flux avalanches' and found it to be compatible with the value determined by Vlasko-Vlasov *et al.* (2004) and with later measurements (Welling, Aegerter, and Wijngaarden, 2004; Aegerter, Welling, and Wijngaarden, 2005). Their comparison (Aegerter *et al.*, 2005) to the qEW equation, however, seems to draw on a scaling relation, $z = \sigma_1^{(s)}$, that is not necessarily valid in the present case. [6] While the experimental avalanche size exponent $\tau = 1.29(2)$ (Aegerter *et al.*, 2004b) matches that of the *two-dimensional* MANNA Model, Table 6.1 on p. 168, the avalanche dimension does not seem to match. Nevertheless, the body of work on SOC in superconductors, reviewed briefly by Wijngaarden, Welling, Aegerter, and Menghini (2006) and more comprehensively by Altshuler and Johansen (2004), represents by far the most compelling evidence for SOC in experimental systems.

3.2.1 Superfluid helium

Self-organised criticality in **superfluid** ^{4}He is a somewhat unusual topic to mention in the present chapter, because it is difficult to see to what extent it represents an instance of SOC at all. However, some authors, for example Corté *et al.* (2009) regard it as '[p]erhaps the clearest previous observation of SOC'. The dissemination of its early theory (Onuki, 1987) almost coincides with that of SOC itself and describes a phenomenon around the superfluid (lambda) transition, whereby the heat flux in a sample of ^{4}He establishes a temperature profile matching that of the transition temperature T_λ, which changes across the sample under the influence of the gravitional field. As a result, the reduced temperature is constant throughout. Various theories have been developed to understand and describe the phenomenon (e.g. Onuki, 1996; Haussmann, 1999; Weichman and Miller, 2000). While fascinating in its own right, it is clearly distinct from all other SOC phenomena. The system

---

[6] In directed models, Sec. 8.4.3, $z = \sigma_1^{(s)} = 1$.

is carefully *tuned* to its critical point, it does not display avalanching or time scale separation, the central observable is not subject to any noteworthy fluctuations and does not display scaling. In summary, it might better be described as 'self-organisation at criticality'.

It is not entirely clear at what stage this phenomenon was first classified as SOC; the first experimental work by Moeur, Day, Liu, *et al.* (1997) uses this term and refers to the original work by Bak *et al.* (1988) without elaborating on the parallels between the two phenomena. The work by Chatto, Lee, Duncan, and Goodstein (2007) contains a brief review of important experimental and theoretical results in this area.

3.3 Barkhausen effect

The **Barkhausen effect** (Barkhausen, 1919) is the noise in the magnetisation produced by a ferromagnet when exposed to a changing external magnetic field. As magnetic domains rearrange in response to the field the magnetisation of the material changes, and this can be picked up by a coil. The resulting voltage signal can be analysed to determine the duration of a domain wall avalanche as well as its size, which is the area in the plot of voltage versus time. Because magnetic domain walls have negligible (effective) inertia and are subject to pinning by defects, the Barkhausen effect is related to avalanching in sandpiles (Feynman, Leighton, and Sands, 1964, Sec. 37–9), sandpile models, superconductors (Bean, 1964) and generally to interfaces in random media (Fisher, 1998; Sethna, Dahmen, and Myers, 2001).

Long before SOC came into existence, the distribution of magnetisation discontinuities (jumps) had captured the interest of researchers. Stierstadt and Boeckh (1965) were the first to measure their distribution. Babcock and Westervelt (1990, also 1989; Babcock, Seshadri, and Westervelt, 1990; Che and Suhl, 1990) probably were the first to study magnetic domain growth and destruction, 'topological avalanches', as an instance of SOC, using magneto-optical techniques (Gopal and Durian, 1995, performed a similar study in foam). After such avalanches, which can last for 45 s, the resulting pattern is 'barely stable', ready for the next avalanche to occur. The avalanche size distribution they found, based on a count of the domains eliminated in an avalanche, was broad and could be approximated by a power law with exponent $\tau \approx 2.5$. According to Urbach, Madison, and Markert (1995) topological avalanching is fundamentally different from that dominated by pinning effects, which operates on a much faster time scale (75 s for an entire field cycle). Topological avalanching has been further analysed by Bak and Flyvbjerg (1992) who concluded that such (experimental) systems organised themselves into a *sub*critical state.

The first to study Barkhausen noise more specifically in the light of SOC were probably Geoffroy and Porteseil (1991a,b, 1994), who showed that its power spectrum displayed strong correlations, $1/f$ behaviour and avalanching (also Alessandro, Beatrice, Bertotti, and Montorsi, 1990a,b). Based on very similar experiments and around the same time, Cote and Meisel (1991) concluded that Barkhausen noise exhibited all the features of SOC as mentioned by Bak *et al.* (1987), in particular power law distributed event sizes and durations ($\alpha = 2.26(35)$) as well as a power law relation of the two, as suggested by Jensen *et al.* (1989). As in superconductors, the experimental exponents are subject to significant variations within and across different studies, even when Meisel and Cote (1992) identified some signs of universality with $\tau \approx 1.74-1.88$ (but $1.44-1.60$ in Cote and Meisel, 1993) and $\alpha \approx 1.64-2.1$, at the same time, however, revising their earlier results as experimental artefacts (aliasing).

O'Brien and Weissman (1994) strongly criticised the conclusion that the Barkhausen effect was an expression of SOC. They listed various experimental problems as well as a counter-example (Hardner, Weissman, Salamon, and Parkin, 1993) of a system in perfect thermodynamic equilibrium that displayed $1/f$ noise, at that time taken as a hallmark of SOC. In their experiments, O'Brien and Weissman found no evidence for SOC in the power spectrum of the noise and pointed at the simple (mean-field) model put forward by Alessandro *et al.* (1990a) to explain the observed phenomena.

Urbach *et al.* (1995) reconciled these *dynamical* results with previous findings, by emphasising and clarifying the rôle of the demagnetisation field, whose effect is eliminated in the annular geometry used by O'Brien and Weissman (1994). They also addressed the apparent clash between interfaces in random media generally requiring tuning of the driving force and the apparent scale invariance in crackling noise without the need of tuning, which lies, of course, at the heart of SOC. If SOC is not present, alternative explanations for the Barkhausen effect (and, for that matter, for avalanching in superconductors, p. 61) are an unusually broad critical region of this phenomenon ('plain old criticality', discussed below) or correlations of the pinning centres as suggested by O'Brien and Weissman (1994). The question whether power law distributed observables are just an expression of the fractal structure of an underlying disorder was raised earlier by Cote and Meisel (1993). The origin of the fractal disorder might have an explanation very different from SOC.

Urbach *et al.* (1995) drew a picture very much in the spirit of the AS mechanism (Sec. 9.3), whose proponents regard Barkhausen noise as one of its experimental manifestations (Dickman *et al.*, 2000). The exponent $\tau \approx 1.33$ they measured is compatible with that found in the **random field Ising Model** at the critical point (Ji and Robbins, 1992). They corroborated their findings by numerical simulations of a model seemingly closely related to models in the MANNA universality class (Sec. 6.2.1.2) which, depending on the presence of a uniform or a locally varying demagnetisation field, required tuning or not.

The connection to the random field Ising Model (RFIM) was made earlier by Sethna, Dahmen, Kartha, *et al.* (1993, also Dahmen and Sethna, 1993, 1996; Dahmen, Kartha, Krumhansl, *et al.*, 1994; Carrillo, Mañosa, Ortín, *et al.*, 1998), culmulating in the explanation of Barkhausen noise by Perković, Dahmen, and Sethna (1995) as an effect of 'plain old criticality', for '[s]everal decades of scaling without tuning a parameter need not be self-organized criticality: it can be vague proximity to a plain old critical point'. This statement emphasises the importance of probing for finite size scaling. A finite distance to the critical point can be revealed by a system-size independent, finite scale, which ultimately limits finite size scaling.[7]

Perković *et al.* (1995) also performed numerical simulations of the RFIM at zero temperature (Sethna *et al.*, 1993), which compare well with experimental results for Barkhausen noise, and found $\tau = 1.60(6)$. The numerical and experimental study by Spasojević,

---

[7] Experimentally and numerically, only a lower bound for such a scale is given by the largest system size probed at which FSS is fully intact.

Bukvić, Milošević, and Stanley (1996) concluded that both explanations of Barkhausen noise, as scaling close to criticality or as an instance of SOC, bear some validity and thus they remained undecided about the issue. Zapperi, Cizeau, Durin, and Stanley (1998) argued that the RFIM approach to Barkhausen noise neglected long-range dipole interactions (Narayan, 1996) as well as the demagnetisation effects mentioned above. They favoured a more traditional perspective of Barkhausen noise as being the result of interfaces pulled through a random medium and the scaling being the result of the vicinity to the critical point, characterised by $\tau = 3/2$. Their model recovers earlier (mean-field-type) results by Alessandro et al. (1990a) and traces the cutoff in the distributions back to the demagnetisation field, rather than the variance of the random field (Perković et al., 1995). The demagnetisation field could be interpreted as an effective dissipation mechanism in the spirit of the AS mechanism (Sec. 9.3), giving the latter further experimental support. The AS mechanism could also explain the rôle of the driving rate, whose effect on the exponents measured was first studied by Alessandro et al. (1990a) and confirmed by Bertotti, Durin, and Magni (1994, also Durin, Bertotti, and Magni, 1995), who also found $\tau = 3/2$ in the limit of vanishing driving rate.

Over the years, the interest in SOC as an 'explanation' for Barkhausen noise has slowly declined. It seems generally accepted that the scaling behaviour is the expression of a critical point and that the distance to that critical point can be controlled by parameters such as the driving rate, the demagnetisation factor or the strength disorder. Whether it belongs to the universality class of the RFIM or is the result of an interface moving through a random medium and whether it displays universality at all remain subjects of ongoing research. Most recently, Durin and Zapperi (2000) identified two distinct universality classes and Mehta, Mills, Dahmen, and Sethna (2002, also Zapperi, Castellano, Colaiori, and Durin, 2005) found a universal pulse shape in Barkhausen noise.

3.3.1 Mechanical instabilities

The Barkhausen effect is closely related to the wider class of **mechanical instabilities** (Dickman et al., 2000), together with crackling noise, fracture, rupture and stick-slip motion. A number of experiments have been performed in the context of SOC. Garcimartín, Guarino, Bellon, and Ciliberto (1997) measured the acoustic energy of cracking of inhomogeneous materials such as plaster, wood and fibreglass, by recording their **acoustic emission**. The data displayed scaling over about $2\frac{1}{2}$ orders of magnitude, which led the authors to conclude that cracking dynamics is a critical phenomenon. The authors were cautious with regard to classifying it as SOC, as there was some tuning involved by controlling the stress rather than the strain (Guarino, Garcimartín, and Ciliberto, 1998).

Maes, Van Moffaert, Frederix, and Strauven (1998) took a similar stance in their work on microfracturing in cellular glasses, which produced an exponent $\tau = 1.5(1)$ compatible with results for the Burridge–Knopoff Model, Sec. 5.3, in particular the mean-field theory (also Zapperi, Ray, Stanley, and Vespignani, 1999). Other, earlier SOC-inspired fracturing experiments such as those on volcanic rocks (Diodati, Marchesoni, and Piazza, 1991), on hydrogen precipitation in niobium (Cannelli, Cantelli, and Cordero, 1993), on synthetic plaster (Petri, Paparo, Vespignani, et al., 1994) and on creep deformation in ice (Weiss and Grasso, 1997), all based on acoustic emission (but Bons and van Milligen, 2001), had produced similar results and all were concluded to be either well compatible with SOC or even experimental confirmation. In a related study on paper crumpling, on the other hand, Houle and Sethna (1996) concluded that this process was not 'explain[ed]' by SOC, due to a lack of prolonged noise bursts characteristic of avalanches. What is more, there is no (obvious) underlying dynamical mechanism that would suggest a self-tuning to a critical state (but Caldarelli, Di Tolla, and Petri, 1996a). This latter argument is somewhat problematic given that some authors regard extremal dynamics, which does not exhibit any form of self-tuning either, as SOC (Sec. 5.4.3 and Sec. 9.4.3). There is no general consensus as to whether (micro-)fracturing displays SOC or not, although attempts to reconcile the different perspectives have been made (Zapperi, Vespignani, and Stanley, 1997, but Sornette and Andersen, 1998).

The large group of **stick-slip experiments** (also known as **dry friction** experiments) display mechanical instabilities as well. The first SOC-inspired experiment in this class was probably Feder and Feder's (1991a) analysis of jerky motion of sandpaper being pulled over a carpet, which they placed in the context of the well-studied BURRIDGE–KNOPOFF Model (Sec. 5.3; for earlier experiments see Tullis and Weeks, 1986). Feder and Feder's (1991a) finding of $\tau = 1.79(5)$ however does not compare well with previous modelling attempts. Vallette and Gollub (1993) studied the slipping of latex on a glass rod but did not find power law behaviour, pointing out, however, that their experiment was affected by resolution problems and wear. Johansen, Dimon, Ellegaard, *et al.* (1993) found $\tau = 1.6(3)$ for the slip force distribution, measured as sudden drops in force when a work hardened metal block slipped while being dragged over a steel surface. This distribution displayed signs of a cutoff and generally depended on the dragging speed, with the power law distribution appearing at low speed.

Ciliberto and Laroche (1994) performed experiments on elastic rough surfaces in an annular and a linear geometry, with constant distance between the dragged surfaces and constant force, respectively. The rough surface was created by embedding 2 mm steel balls in rubber which (elastically) impede relative motion when two such surfaces are dragged across each other. Avalanches were defined by the force difference at sudden drops and their distribution was well approximated by power laws, whose slope approached a value close to $\tau \approx 2$ as the loading speed was decreased. More recently, Buldyrev, Ferrante, and Zypman (2006) found much steeper distributions of force drops with scaling over about 11/2 orders of magnitude in dry sliding friction experiments (metal on metal).

It is difficult to draw a clear line between experiments and observations, because at the core of an experiment always lies an observation. The key distinction drawn here is that most of the laboratory setups described above can be manipulated arbitrarily, i.e. media can be changed, measuring equipment positioned differently, viscosity tuned. Even when the relevant process occurs on a microscopic level and is hidden from the naked eye, many details are still a matter of choice. This is different when observational and empirical data are collected, possibly over long periods, and analysed long after the events, with limited means to interfere and to probe. Data selection and processing has the most immediate effect on the outcome of such studies and direct interference with the system in question is impossible or significantly reduced. The empirical evidence for SOC discussed in the following is found in systems that generally cannot be tested in 'artificial' laboratory experiments, because they are too vast or too complex to be (re-)built on a lab bench.

3.4 Earthquakes

Although closely related to the mechanical instabilities discussed in Sec. 3.3.1, earthquakes need to be discussed in their own right. From early on and for a long time, they were regarded as the prime example of SOC (e.g. Sornette, 1991). Many if not all of SOC's key features (Sec. 1.2) are present: scaling, slow driving, thresholds, avalanching, correlations. The importance of this phenomenon, the wide availability of very detailed observational data and the general interest in the subject has led to an enormous amount of literature, reviewed by Turcotte (1993) as well as Carlson, Langer, and Shaw (1994) and more recently extensively by Hergarten (2002). One particularly successful model of earthquakes is the OFC Model, which is introduced in Sec. 5.3, where some of the literature specific to that model is reviewed in further detail. In the following, only some of the major contributions of SOC to the current research into earthquakes are discussed.

As early as 1988, Bak, Tang, and Wiesenfeld (1989b) discussed at a conference in Kyoto 'the self-organized critical behavior of the earth dynamics', an idea further developed by Bak and Tang (1989a). Independently, Sornette and Sornette (1989) and Ito and Matsuzaki (1990) proposed earthquakes as an instance of SOC at work in nature, which found wide acceptance even in the geophysical community (e.g. Kagan, 1991a; Scholz, 1991; Sornette, 1991). The observation of a power law relation between the energy release during an earthquake and its frequency goes back to Gutenberg and Richter (1954) and is firmly established in the literature. The **Gutenberg–Richter law** characterises the cumulative probability distribution $F^{(E)}(E)$ of earthquakes which release energy E or more by an exponent b, often referred to as the 'b value',

$$F^{(E)}(E) = aE^{-b} \tag{3.3}$$

with a dimensionful amplitude a. In terms of the probability density, which is more commonly used in SOC, that means $\tau = b + 1$, Eq. (1.1). Although detailed analyses (e.g. Pacheco, Scholz, and Sykes, 1992) reveal more structure and inconsistencies (Kagan, 2003), there is general consensus in the literature that the probability density function of earthquakes is well approximated by a power law of its energy release (Kagan, 1991a). Yet, the value of b varies across regions, earthquake depth and size. Evernden (1970) found values between 0.65 (Alaska) and 1.46 (Iran) and even higher values in some more ambiguous cases; Main and Burton (1986) found $b = 0.89$ for frequent earthquakes of intermediate size and $b = 0.51$ for rare large ones; Kagan (1991b), who fitted against a gamma function, found values decreasing from $b = 0.74(4)$ to $b = 0.55(8)$ with increasing depth; finally, Pacheco *et al.* (1992) found values ranging from $b = 0.88$ to $b = 1.51$ depending on the earthquake catalogue and the depth of the event (also Carlson, 1991). Interestingly, Vere-Jones (1976, 1977) found $b = 3/4$ in a model with uncorrelated sites (although Chen, Bak, and Obukhov, 1991, quote $b = 1/2$), close to the most frequently reported values around $b = 0.85$.

The question whether earthquakes exhibit SOC is of great importance for a number of reasons. If SOC is a critical phenomenon in the traditional sense, then universality is one of its essential features. If earthquakes exhibit universal behaviour, then very basic microscopic interactions might be identified, governing earthquake dynamics on virtually all scales, which in turn would not only provide greater insight into the mechanism leading to earthquakes, but might also pave the way to better experiments, modelling and forecasting. Even without universality, SOC might still help to understand the processes underlying earthquakes. Finally, SOC might just illustrate how thresholds and slow driving can give rise to sudden outbursts of energy as observed in earthquakes.

Originally, Bak and Tang (1989a) concluded from the irregularity of the activity in the earthquake model they studied that earthquakes are **inherently unpredictable**. This claim can be supported by the observation of scaling, namely that events that fall under the same scaling behaviour are governed essentially by the same underlying interaction, producing, however, an event of a particular size at one moment and, for no apparent reason, of a size orders of magnitude different at the next moment. At closer inspection, this statement requires further qualification, which is often ignored: scaling is an asymptotic behaviour in large event sizes in systems with large upper cutoffs. In fact, Pacheco *et al.* (1992) argued that different scaling regimes are set by the size of the moving fault. Generally, a power law asymptote might be governed by physics very different from that on the most relevant scale. In fact, the very existence of a lower cutoff implies a scale below which other mechanisms govern the behaviour of the system.

Predictability is at the heart of earthquake research and dismissing it is almost equivalent to a dismissal of the discipline as a whole. Some explain the unpredictability claim mentioned above therefore by the amount of attention it generates. In fact, SOC, i.e. non-trivial scaling, suggests the opposite, namely **strong, algebraic spatial and temporal correlations**, quite the opposite of unpredictability (Sornette and Werner, 2009). Such correlations are in fact found in the non-conservative OFC Model (Christensen and Olami, 1992b, also Ramos, Altshuler, and Måløy, 2009) and can be exploited for predictions (Pepke and Carlson, 1994) using techniques similar to those used in earthquake forecasting (e.g. Helmstetter, Kagan, and Jackson, 2006). The **Omori law** (sometimes the **Omori–Utsu law**, Omori, 1894; Utsu, 1961; Utsu, Ogata, and Matsu'ura, 1995) is a further indicator of correlations, stating how the frequency of aftershocks decays. Christensen and Olami (1992b) investigated such correlations which were later reconciled (Olami and Christensen, 1992) with the Omori law with exponent 1, consistent with empirical data. In the conservative OFC Model no significant correlations were found (Christensen and Olami, 1992b; Olami and Christensen, 1992). Ten years later, Hergarten and Neugebauer (2002) identified the Omori law in the two-dimensional OFC Model not only for aftershocks, but also for foreshocks, and found a significant dependence of the exponents on the level of conservation. [8] However, even their largest exponent for the Omori law is still far away from unity, which, as they argue, might be due to some restrictions in the OFC Model.

---

[8] The authors, both apparently with a background in geophysics, unwittingly illustrate the difference in attitude between seismology and statistical mechanics: '. . . conclude that the OFC Model exhibits SOC . . . However, . . . the question [whether or not the OFC Model displays SOC] may be of academic interest rather than important in seismology.'

Bak, Christensen, Danon, and Scanlon (2002a,b) demonstrated how to collapse the waiting time (also known as interoccurence time) between earthquakes (Christensen and Olami, 1992b) in Southern California onto a single, *universal* curve, using data obtained from the Californian earthquake catalogue, by spatial subsampling (Ito, 1995) – in sharp contrast to the earlier claim of a lack of universality (p. 135, Christensen and Olami, 1992a), culminating in the appeal: 'One should not look for universal values of [*b*] in nature.'

In the finding by Bak *et al.* (2002a), the power law prefactor corresponds to the Omori law (but Davidsen and Goltz, 2004), whereas the scaling of the cutoff represents the Gutenberg–Richter law, so that *all* data are incorporated into a single, unified scaling law, with $\tau - 1 = b \approx 1$. Capturing all events in a single framework challenges the very notion of fore and aftershocks. This point, as well as the universality of seismic events and its meaning, in particular with respect to correlations and thus the wider implications with regard to predictability, are discussed widely and controversially in the literature (Corral, 2003, 2004b,c; Davidsen and Paczuski, 2005; Lindman, Jónsdóttir, Roberts, *et al.*, 2005, 2006; Corral and Christensen, 2006; Davidsen and Paczuski, 2007; Werner and Sornette, 2007; Sornette and Werner, 2009).

3.5 Evolution

Like earthquakes, **evolution** (reviewed in Gisiger, 2001) is very difficult to observe in the laboratory, because of the long temporal and spatial scales involved. When Bak and Sneppen (1993) proposed their model of evolution (BS Model, Sec. 5.4) with its essential feature of punctuated equilibrium (Eldredge and Gould, 1972), the response from evolutionary biologists was not unanimous for several reasons. Firstly, the statistics of biological extinction events is not studied as widely and in a similar fashion in biology as earthquake statistics are studied in geology, so the question addressed by the BS Model in particular and by SOC in general was not one that was widely asked. Secondly, punctuated equilibrium was (and still is) a matter of controversy (Gould and Eldredge, 1993). Its very existence as a distinctive feature is questioned by some, who regard it as accelerated gradualism. Thirdly, the proponents of SOC appeared to be forcing their insight on the biologists' community without paying much attention to established knowledge and understanding: 'That physicists are itching to take over biology is now well attested. [. . .] But surely only a brave physicist would take on Darwin on his home ground . . . ' (Maddox, 1994). Bak (1998) showed little appreciation for the achievements of biologists when he asked: 'Is biology too difficult for biologists?' In summary, the relation between SOC and evolution has never been an easy one. In the following, some of the evidence for evolution being self-organised critical is reviewed. Sepkoski (1982) famously collected data on the extinction of marine families, which he extended and revised in 1993 and 2002 (published posthumously). Raup's (1986) statistical analysis of the data (also Raup, 1976) favoured new extinction mechanisms, concluding that '[e]xtinction may be episodic at all scales. . . ' and that 'dinosaurs may not have done anything "wrong" in a Darwinian sense.' Global events which are not selective on physiology or habitat were proposed by Raup and Boyajian (1988). Raup and Sepkoski (1984) had initially suggested periodic extinction events, which was challenged by Benton's

(1995) analysis of the fossil records, drawing on the database he had created earlier (Benton, 1993). At the same time, Sneppen, Bak, Flyvbjerg, and Jensen (1995) identified scaling in the statistics presented by Raup (1986) (although 'too scanty to allow for a real quantitative test'), promoting, at the same time, the BS Model as a suitable toy model, as it produced punctuated equilibrium and displayed scaling in the event size distribution. They found that the number of genera as a function of the time span they existed scaled over two orders of magnitude roughly with exponent 2 (survivorship curves in Raup, 1991). This was further supported by a power spectrum close to $1/f$ noise (Solé, Manrubia, Benton, and Bak, 1997, also Halley, 1996) in time series of the number of families derived from Benton's (1993) 'Fossil Record 2'. The interpolation procedures used have been strongly criticised by Kirchner and Weil (1998) who argued that the power laws are artefacts of the interpolation techniques. Newman and Eble (1999) showed that there was a degree of ambiguity in the data, its analysis and the interpretation, concluding that there are 'clear deviations from' $1/f$. Dimri and Prakash (2001b) applied several different techniques to analyse Sepkoski's (1982) data and found the power law scaling of the power spectrum confirmed, which was disputed in a comment by Kirchner (2001, also Dimri and Prakash, 2001a).

In their review, Solé and Bascompte (1996) re-analysed Raup's (1986) data, fitting it to both a power law ($\tau \approx 1.95$) and an exponential. In a series of papers, Newman (1996, 1997a,b) argued that the power laws found in empirical data do not conclusively link evolution to SOC, favouring (global) environmental stress instead. He tested a host of models which lack the characteristics of SOC models, yet displayed power laws close to the observed one.

Burlando (1990) presented evidence that the division of taxa into sub-taxa (such as a taxonomic family into sub-families or a class into sub-classes) generally follows a power law, hinting at scaling in evolutionary processes. He later extended the study from living to extinct organisms (Burlando, 1993) using the fossil records such as those mentioned above. Burlando's method of analysing the 'tree of life' on many levels simultaneously is a form of sub-sampling, which allowed him to overcome a predicament any statistical analysis of evolution faces: as of now, experiments on the extinction of species would require prohibitively large time and length scales. One has to resort to observations of a single instance of the phenomenon available, which can be divided into sub-systems (similar to block-scaling, e.g. Sec. 2.3.1, Binder, 1981a) and examined on different levels of coarse graining.

3.6 Neural networks

Hopfield (1994, crediting A. Herz and J. Rundle) was the first to link **neural networks** to SOC in direct reference to the OFC Model. These networks are made up of individual elements (**neurons**) (e.g. Koch, 1997) that can interact via an **integrate-and-fire mechanism**. They act like capacitors whose potential increases when they receive a current from neighbouring neurons or an external source (integrate). If the voltage exceeds a threshold, they fire by passing their charge to neighbouring neurons, reminiscent of sites in sandpiles receiving and shedding sand grains. In general, the charge transferred is regarded as a signal (action potential), whose effect on the receiving cell can be either excitatory or inhibitory, depending on the type of **synapse** involved.

Shortly after Hopfield's (1994) overview article, a number of authors studied the occurrence of avalanching and SOC in models of neural networks (Chen, Wu, Guo, and Yang, 1995; Corral, Pérez, Díaz-Guilera, and Arenas, 1995; Herz and Hopfield, 1995), identifying many of the well-known features of the OFC Model. Teich, Heneghan, Lowen, *et al.* (1997) found $1/f$ noise in the time series of the neural signal of the cat's visual system, but did not link this result to SOC and neither did Lowen, Cash, Poo, and Teich (1997), who identified self-similar behaviour in the neurotransmitter secretion in neurons and non-neuronal cells. Toib, Lyakhov, and Marom (1998) found a scaling relation between the duration and recovery time for events in mammalian Na^+ channels which stretched from milliseconds to minutes. Papa and da Silva (1997, also da Silva, Papa, and de Souza, 1998, using the BS Model) were probably the first to make the link to SOC by re-analysing the data obtained by Gattass and Desimone (1996) for the **waiting time distribution** between firings of **cortex neurons** of macaques. Linkenkaer-Hansen *et al.* (2001, also Linkenkaer-Hansen, Nikouline, and Ilmoniemi, 2000) recorded very long time series from magnetoencephalography and electroencephalography measurements and identified SOC, considering $1/f^\alpha$ noise as the hallmark of a (self-organised) critical state (cf. Bédard, Kröger, and Destexhe, 2006; de Arcangelis, Perrone-Capano, and Herrmann, 2006). Using the squared potential difference as the avalanche size, Worrell, Cranstoun, Echauz, and Litt (2002) derived an avalanche size exponent of $\tau \approx 1.9$ for the size distribution of the energy released in epileptic human brains. They obtained the time series from electroencephalography recordings during operations.

The first to analyse spatially extended signals with many electrodes in neural networks of various sizes were probably Segev, Benveniste, Hulata, *et al.* (2002), who identified scale invariant behaviour of the spatiotemporal signal of the firing pattern, but did not relate it to avalanching and SOC. Shortly afterwards, Beggs and Plenz (2003) identified 'neural avalanches' as spatially extended, correlated events, characteristic of SOC. They placed slices of the cortex of rats on an array of electrodes, which recorded the local voltage as the tissue displayed spontaneous activity induced by a suitable drug. Integrating over short time windows, an avalanche was defined by a number of electrodes above a certain threshold potential, the statistics of which displayed scaling close to the mean-field value of $\tau = 3/2$ (also found by Mazzoni, Broccard, Garcia-Perez, *et al.*, 2007). Beggs and Plenz (2003, reviewed by Beggs, 2008) argued that this exponent was the result of optimised information transmission, while at the same time suppressing 'runaway' of excitation (but Worrell *et al.*, 2002). They extended their work later to include mechanisms of memory (Beggs and Plenz, 2004).

Some authors are rather critical when assessing what SOC can add to the understanding of neuronal processes (e.g. Gisiger, 2001). Surely, SOC has little to offer when it comes to explaining the *biological* origin and functional rôle of avalanches (Mazzoni *et al.*, 2007), yet provides a detailed understanding of avalanching itself. Given that exponents are close to their mean-field values, one might conclude that avalanching in brains is not affected by long-range correlation and therefore simple, random walker based theories should suffice to explain many of the features observed. However, if one accepts the evidence for 'SOC in the brain', one puzzling question is how and why the brain tunes itself to a critical state. [9] A number of authors have suggested that 'plastic' or 'dynamic' synapses are responsible for the tuning of neural networks to a critical state (e.g. Stassinopoulos and Bak, 1995; de Arcangelis *et al.*, 2006; Levina, Herrmann, and Geisel, 2006, 2007). To curb their destabilising effect Royer and Pare (2003) introduced an additional homeostatic mechanism, thereby resolving the apparent dichotomy of stability and plasticity.

The wider context of complexity and brain function has been reviewed by Tononi and Edelman (1998). Gisiger's (2001, before Beggs and Plenz's 2003 results) more specific review on scale invariance in biology contains a large number of references to further experiments into the present subject with an emphasis on the relation to established results in neurophysiology. Chialvo's (2004) brief review, on the other hand, focuses more on the input from physics and the results obtained for neural network models from a complexity point of view.

3.7 Other systems

The following is a collection of publications on SOC in a diverse range of systems, which have a smaller literature base compared to those discussed above.

3.7.1 Meteorology

Various **meteorological time series** have been investigated for evidence of SOC. The earliest work is probably that of to Vattay and Harnos (1994) who probed air **humidity** fluctuations for $1/f$ behaviour and found scaling over about one and a half orders of magnitude. Kardar (1996) referred to **rainfall** as being 'suggested to show SOC', but the first publication dedicated to that question seems to be the work by Andrade, Schellnhuber, and Claussen (1998, also Koscielny-Bunde, Bunde, Havlin, *et al.*, 1998), who found drought time distribution characterised by exponents between 1.5 and 2.5, depending on the region. The data used by Peters, Hertlein, and Christensen (2002) had a significantly finer resolution, spanning four orders of magnitude, and approximated a power law with exponent 1.42 well. These authors also compiled an event size distribution from the time series and found $\tau \approx$ 1.36 over about three orders of magnitude (also Peters and Christensen, 2002, but Dickman, 2004a). Sarkar and Barat (2006) performed a similar analysis for the rainfall in India and found exponents between $\tau = 1.00(1)$ and $\tau = 1.67(3)$ (for SOC in air pollution, see Zhu, Zeng, Zhao, *et al.*, 2005, and more locally Shi and Liu, 2009). Peters and Neelin (2006) found the signature of an ordinary second order phase transition meteorological systems might be sweeping over, very much in the sense of Sornette (1994). While illustrating the existence of an ordinary critical point underlying a supposed SOC phenomenon, these observations are also compatible with a complete absence of self-organisation; rather they could be the product of a control parameter moving (occasionally) across the transition, being subject to an equation of motion which drives it around the transition but does not (in any limit) pin it there, in contrast to the AS mechanism (Sec. 9.3). Yet, there is evidence

---

[9] The 'how' is, of course, one of the central questions that drives the field of SOC.

that the system is *attracted* by the instability (Neelin, Peters, and Hales, 2009), which sits well with the traditional view of a **quasi-equilibrium** in the troposphere (Arakawa and Schubert, 1974). Due to the large amount and fine resolution of the available data, some authors regard meteorology as the most promising hunting ground for SOC in nature.

3.7.2 High energy physics

Various phenomena in **high energy physics** have been considered as exhibiting SOC, all of which are some form of **plasma** or are closely related to it. Meng, Rittel, and Zhang (1999, also Jinghua, Ta-chung, Rittel, and Tabelow, 2001) presented theoretical arguments for the existence of SOC on the level of quarks, which was well supported by experimental evidence. As particle physics is closely related to statistical field theory, one could question whether the scaling observed is actually the expression of a traditional scale invariant phenomenon. As for some other phenomena in this section, it is difficult to see how avalanching and SOC could be reconciled with existing, successful theories and explanations in the respective field.

Tokamaks are large devices that confine a hot plasma in a magnetic field. One important question is whether certain transitions in the behaviour of the confined plasma are catastrophic (akin to first order phase transitions) or gradual (akin to a continuous phase transition; cf. Chapman, Dendy, and Hnat, 2001). **Turbulent transport** is the main transport mechanism in the confinement zone of a tokamak and is subject to large and small scale noise (Diamond and Hahm, 1995). The plasma is only marginally stable and can respond to small perturbation with sudden bursts of transport. SOC was suggested as a paradigm for turbulent transport in plasmas probably first by Diamond and Hahm (1995) and has since been widely used for modelling (Carreras, Newman, Lynch, and Diamond, 1996; Newman, Carreras, Diamond, and Hahm, 1996; Chapman *et al.*, 2001; Sanchez, Newman, and Carreras, 2001; Graves, Dendy, Hopcraft, and Jakeman, 2002). The first detailed experimental evidence was probably presented by Rhodes, Moyer, Groebner, *et al.* (1999, also Politzer, 2000), who found scaling in the power spectra of density, potential fluctuations and particle transport. Other authors studied **astrophysical plasmas**, such as **accretion disks**, first discussed by Mineshige, Ouchi, and Nishimori (1994a, also Mineshige, Takeuchi, and Nishimori, 1994b) and briefly reviewed by Dendy *et al.* (1999) as well as Chapman, Dendy, and Rowlands (1999). Another subject area is that of **solar flares**, first analysed in the light of SOC by Lu and Hamilton (1991), who also introduced a model of the magnetic reconnection process. The research into solar flares proved particularly fruitful, in terms of both data analysis and modelling (e.g. Hughes, Paczuski, Dendy, *et al.*, 2003; Paczuski, Boettcher, and Baiesi, 2005; Uritsky, Paczuski, Davila, and Jones, 2007). Even **geomagnetic storms**, which are caused by solar flares, have been discussed as an instance of SOC at an early stage (Chang, 1992). Whether SOC and **intermittent turbulence**, which was very early on studied as an SOC phenomenon by Bak, Chen, and Tang (1990), are actually the same phenomena (Chang, 1999; Paczuski *et al.*, 2005) or not (Boffetta, Carbone, Giuliani, *et al.*, 1999; Carbone, Cavazzana, Antoni, *et al.*, 2002, also Solomon, Weeks, and Swinney, 1994) is disputed in the literature.

3.7.3 Ecology, epidemiology and population dynamics

Malamud, Morein, and Turcotte (1998) analysed the statistics of forest fires, where they are expected to be a key mechanism in shaping the population pattern. The term **Yellowstone effect** has been coined for the percolating effect that dense, contiguous growth has, which can be the result of regularly extinguishing small forest fires, but which promotes the occurrence of large fires that spread over large distances. '... [V]ery large fires might have been prevented or reduced if ... there had not been a policy of [small] fire suppression. Many ... recognize that the best way to prevent the largest forest fires is to allow the small and medium fires to burn.' (Malamud *et al.*, 1998). This obviously assumes that the total extent of a fire can be determined accurately to allow those in charge to decide whether the fire should be extinguished or not. Also, the conclusion seems to rest on the assumption that many small fires are less detrimental to the population than a few large ones, which can be prevented by letting the former burn, i.e. small and large fires are mutually exclusive. The problem of prioritisation and aims becomes more obvious in epidemics, say in human populations, where any outbreak is suppressed even at the cost of making larger outbreaks possible by providing contiguous population coverage.

Malamud *et al.* (1998) identified some power law behaviour in the frequency-area distribution of forest fires in various regions across the world. The exponents extracted, ranging from $\tau \approx 1.31$ to $\tau \approx 1.49$, vary between locations and do not seem to coincide with those found in the Forest Fire Model (FFM) and its variants discussed in Sec. 5.1. One difficulty in establishing the link between model and observation is the fact that the shape of patches burning down in forest fire models is usually determined by previous forest fires, i.e. the dynamics is dominated by the fires. This might not be the case in most forest fires considered in natural systems, where geography and policy (i.e. fire fighting) play a crucial rôle.

The non-universality in forest fires was confirmed in more detailed studies, such as those by Reed and McKelvey (2002) and Malamud, Millington, and Perry (2004) who even produced a map of exponents, $\tau \in [1.1, 1.8]$, for the United States. Studies for other countries, such as China (Song, Weicheng, Binghong, and Jianjun, 2001) and Italy (Ricotta, Avena, and Marchetti, 1999; Ricotta, Arianoutsou, Díaz-Delgado, *et al.*, 2001), reached similar conclusions. It is difficult to tell what these exponents actually mean or imply. Corral, Telesca, and Lasaponara (2008), on the other hand, did not find a power law in the size distribution. Yet, they found one in the **waiting time** for forest fires reported in Italy over the course of five years.

Epidemic spreading in human populations has been studied against the background of the FFM as well. The applicability of the FFM might be similarly controversial here, as for real forest fires, since environmental factors might dominate the spread of diseases. Yet, one key ingredient of the FFM can be recognised much more easily than in the case of real forest fires: patches 'consumed' by an epidemic are shaped by the epidemic itself, because humans contracting the disease (fire) become immune and therefore cannot host it again (ash). Correspondingly, 'tree (re-)growth' maps to birth in the epidemics studied (Rhodes, Jensen, and Anderson, 1997). Rhodes and Anderson (1996) found exponents compatible with the two-dimensional DS-FFM (Sec. 5.2), across three different populations for the spreading of a type III epidemic (measles) before the advent of vaccination. Over three orders of magnitude the data for the Faroe Islands were found to be very well approximated by $\tau = 1.28$, Bornholm gave $\tau \approx 1.28$ (although noisier) whereas Reykjavik produced $\tau \approx 1.21$. Rhodes *et al.* (1997) extended the study to different epidemics (measles, whooping cough and mumps) and found exponents consistent with those in DS-FFM at higher dimensions. It remains unclear how to interpret this latter finding. If it is not coincidental, it might be the expression of the presence of an effective interaction that is, in fact, higher dimensional than flat geography suggests.

There is a range of smaller studies on **population dynamics** and **ecology**, some of which are closely related to studies on SOC in evolution (Halley, 1996). Solé and Manrubia (1995) characterised the (multi-)fractal nature of **rain forests**, loosely relating it to SOC, which might tie in with the studies on forest fires. Keitt and Marquet (1996) analysed the **extinction pattern of bird species** introduced mostly by settlers to Hawaiian islands over the course of about 150 years. This study is remarkably ambitious, given the sparsity of the statistics and the difficulties of identifying driving and relaxation mechanisms in analogy with sandpiles. Finally, Miramontes and Rohani (1998) re-analysed data on *laboratory insect populations* to determine the extinction avalanche exponents τ ranging from 1.70 to 2.79.

3.7.4 Physiology

Some other **physiological systems** have been tested for SOC behaviour. Suki, Barabasi, Hantos, *et al.* (1994) investigated the inflation of dog **lungs**, as the airway resistance changes in jumps when initially closed up parts of the lung open up and contribute to the inbreathing. During the first few seconds the jumps are large, as large sections of the lung join the process, while at later stages increasingly smaller jumps occur. The jump size distribution follows a power law with $\tau = 1.8(2)$ for a bit more than one order of magnitude. The observation ties in with the fractal nature of the lung (Mandelbrot, 1983; Shlesinger and West, 1991). Peng, Buldyrev, Goldberger, *et al.* (1992) as well as Voss (1992) reported power laws in the power spectrum of DNA sequences, the former using a random walker mapping, the latter using a special 'equal-symbol multiplication'. Although the authors referred to SOC it did not enter much into the discussion about the origin of the observed scaling, except for the hint to non-equilibrium phenomena and the general background (also Buldyrev, Goldberger, Havlin, *et al.*, 1993; Voss, 1993, reviewed by Buldyrev, Dokholyan, Goldberger, *et al.*, 1998).

The variability of the **waiting time** between **heart beats** on time scales above 200 beats also displays scaling, depending on the state of health of the heart (Peng, Mietus, Hausdorff, *et al.*, 1993), which has been linked to SOC (Peng, Havlin, Stanley, and Goldberger, 1995). **Allometric scaling laws** in biology (West, Brown, and Enquist, 1997) are sometimes cited as instances of SOC, but were later explained solely on the basis of geometry rather than the hydrodynamics of fractals (Banavar, Maritan, and Rinaldo, 1999; West, Brown, and Enquist, 1999).

3.7.5 Financial markets, sociology and psychology

Various forms of **(collective) human behaviour** have been analysed for signals of SOC. Bak, Chen, Scheinkman, and Woodford (1993, also Scheinkman and Woodford, 1994) attempted to model fluctuations in **economic systems** by means of a directed sandpile, where nodes represent participants in the economy, purchasing goods from suppliers and selling them on to customers. The model arrived at a time when the application of concepts from complexity to economics questions gained momentum and the notion of 'econophysics' became established. Mantegna and Stanley (1995) were probably the first to identify various power law relationships in economics data from an SOC perspective, however considering **turbulence** as an alternative scenario that leads to scale invariant behaviour (Ghashghaie, Breymann, Peinke, *et al.*, 1996; Mantegna and Stanley, 1997). Gopikrishnan, Plerou, Amaral, *et al.*'s (1999) detailed analysis of the distribution of returns as a function of time revealed a clear dependence of the scaling exponents on the time scale. Bartolozzi, Leinweber, and Thomas (2005) analysed smoothened financial time series and found indications of a lack of temporal correlations, yet convincing power laws for avalanche size distributions, whose exponents, however, varied from $\tau = 1.7(2)$ to $\tau = 2.39(4)$ depending on the time series considered (also Lux and Marchesi, 1999, but Stanley, Amaral, Gopikrishnan, *et al.*, 2002; Gabaix, Gopikrishnan, Plerou, and Stanley, 2003). Such results give little hope that universality might be at work, but they might nevertheless provide useful tools and give insight into the mechanism operating at various scales. Canning, Amaral, Lee, *et al.* (1998) related the variance of GDP growth rates to the GDP and found an inverse power law with exponent 0.15 spanning about three orders of magnitude.

Smethurst and Williams (2001) published some remarkable results on SOC in **hospital waiting times**, which were widely met with bewilderment. The study was based on the histogram of the monthly relative changes in the length of waiting lists of four dermatology consultants working for the British NHS, which displayed noisy, skewed power law behaviour over about one and a half orders of magnitude. The authors concluded, rather contentiously, that '... healthcare systems probably organize themselves in such a way as to reduce the impact of any attempted intervention. They self-regulate to buffer against differing levels of demand, thereby creating bottomless pits that absorb all resources made available. [...] If self-organization is occurring, then, the system must be operating at an efficient equilibrium for the level of input.'

There is hardly another study on SOC that is similarly politically charged, as documented by the comments, for example by The British Library Science Technology and Business (STB) (2001, also Ball, 2001) quoting the authors of the study when summarising that '[t]he NHS system should be judged, "by measuring the overall quality of medical care, rather than by the length of hospital waiting lists". The self-regulation of healthcare systems reduces the impact of any intervention.' In their study under the title 'Is the National Health Service at the edge of chaos?', Papadopoulos, Hadjitheodossiou, Chrysostomou, *et al.* (2001) concluded that certain pay schemes and policy decision will have little bearing on the waiting lists, because '[s]elf-organized criticality is a law of nature that enables complex systems to optimize their efficiency, and [...] NHS surgical waiting lists are no exception.'

Although waiting time distributions have some significance in SOC (e.g. Christensen and Olami, 1992b, but Sánchez, Newman, and Carreras, 2002), the empirical basis in the present case is rather thin and there is little evidence for critical behaviour in the dynamics of changes in hospital waiting lists. As Freckleton and Sutherland (2001) pointed out, **negative feedback** might be present, as more patients might be willing to join a short queue, which might lead to a form of avalanching. The authors concluded that the data of the original study would be consistent with a non-interacting, random walker based null model (also Sornette, 2002).

As for other areas of human behaviour, SOC has been identified in **traffic flow models** (Nagel and Herrmann, 1993), **urban development** (Batty and Xie, 1999, also Batty and Longley, 1994; Portugali, 2000), **popular album charts** (Bentley and Maschner, 1999), **evolution of artefact styles** (Bentley and Maschner, 2001), and, leaving the realm of good taste in the far distance, wars (Roberts and Turcotte, 1998, $\tau \approx 1.27$). $1/f$ noise was identified in **human coordination** (Chen, Ding, and Kelso, 1997) as well as **human cognition** (Gilden, Thornton, and Mallon, 1995) and SOC was proposed as possible framework to account for it. The distribution of the sizes of US firms follows a power law (Axtell, 2001, $\tau = 2.059(54)$), which reaches into the domain of Zipf's (1949) law, according to which ranked histograms follow power laws.

3.7.6 Virtual, electrical and data networks

Human built networks have also been subject to the search for scale invariance, $1/f$ noise and SOC. Huberman and Adamic (1999) analysed the size of sites in the **World Wide Web** and found a distribution well approximated by a power law with an exponent ranging from 1.647 to 1.909 (also Crovella and Bestavros, 1997; Huberman, Pirolli, Pitkow, and Lukose, 1998; Willinger, Govindan, Jamin, *et al.*, 2002). Carreras, Newman, Dobson, and Poole (2004) analysed the time series of **power blackouts** in North America and identified systematic scaling for a number of observables. Defining the avalanche size as the number of customers losing power, the PDF is characterised by an exponent $\tau \approx 1.7$. Applying the same analysis to a time series from a sandpile model showed remarkably good quantitative agreement. Carreras *et al.* (2004) argued in favour of an analogy between sandpile models and electric networks: loads in electrical networks are slowly ramped up by consumers until they might trigger equipment failure. About one half of the events are caused by weather, which was analysed separately but did not affect scaling. Once a part of the network fails, similar to an avalanche, the redistribution and shedding of load may lead, over the course of minutes to hours, to failures in other components in the network.

3.8 A very brief conclusion

It is difficult to draw firm conclusions from results as diverse as those reviewed above. There has been an enormous range of studies of natural and human-built systems in an attempt to identify SOC in nature. These attempts are riddled with difficulties, not least because it is very demanding to find convincing evidence for a particular asymptotic behaviour, as long spatial and temporal scales are hard to reach and control experimentally. The exponents found are often close to their typical mean-field values (Sec. 8.1.3), in particular $\tau = 3/2$. Nevertheless, some experimental results, for example those on granular media (Sec. 3.1) and even more so those on superconductors (Sec. 3.2), but also observational data of earthquakes (Sec. 3.4) and precipitation (Sec. 3.7.1) provide compelling evidence for SOC in natural systems.

PART II

MODELS AND NUMERICS

Many different models have been developed in order to study particular features of SOC, such as $1/f$ noise, non-conservation and anisotropy. In Part II, some of the more important models are introduced and their general properties discussed. At the beginning of each section, the definition and characteristics of each model are catalogued in a box and the exponents listed in a table. Each section is essentially independent, discussing a model in its own right for its particular qualities. Nevertheless, relations to other models are emphasised and a minimal set of common observables (see Sec. 1.3), in particular exponents (Ch. 2), is discussed for each of them.

The attempt to tabulate exhaustively all numerical estimates for exponents of the various models is futile; it is practically impossible to find all published exponents for a model. It is similarly fruitless to draw a clear line between genuine estimates and exponents derived from others using (assumed) scaling relations. Wherever possible, exponents are only listed if the sole underlying assumption is simple scaling as stated in the caption. The tables of exponents therefore serve only to illustrate the variety and sometimes the disparity of results. The exponents are listed in historical order, which often means that the data towards the bottom of the tables are based on more extensive numerics and are thus more reliable. The tables should enable the reader to judge whether a given model displays systematic, robust scaling behaviour or not. The exponents are reported in a form that facilitates easy comparison between models, which means that concessions have been made regarding the symbols and historic form in which the exponents were published. Again, it is not easy to draw a line. Throughout this book, τ is reported in its standard form (Sec. 1.3), even for models where the overwhelming majority of authors publish $\tau + 1$ (e.g. for the DS-FFM, Sec. 5.2). In the case of model-specific exponents, results are listed in the established form, even when other models use a different form, as for example in the case of σ for the BS Model (Sec. 5.4) which is *not* converted to $1/\sigma$, corresponding to the exponent λ of the DS-FFM.

Along the same lines, avoiding clashes in the terminology has received priority over the model-specific use of certain terms. Wherever possible, a consistent nomenclature is applied to describe similar models, even when the original definition of such models (or other defining publications) uses a different language. For example the traditional terms 'grains' and 'heights' are avoided in the description of the MANNA Model (Sec. 6.1), as they clash with the BTW Model (Sec. 4.1), where they have a slightly different meaning. Nevertheless, using the established terminology, the MANNA Model is presented in terms of 'particles', which correspond to slope units in the BTW Model.

Finally, only a restricted set of exponents is reported. All models have exponents τ and D to describe avalanche sizes, most have α and z characterising avalanche durations, but other exponents are listed only when deemed essential to the model.

The models are divided into three chapters: the classic sandpile-like models, dissipative models and models with intrinsic stochasticity. Their connections are often deep-running and many have been developed with reference to the original BTW Model. The models presented are amongst those with the greatest impact on the development of the entire field. Many of them are still being actively investigated, though sometimes with an emphasis very different from that initially intended. For example, the MANNA Model was intended as a simplified version of the BTW Model, but is currently intensely studied for its links to non-equilibrium critical phenomena. The models *not* discussed but with significant impact on the field are: The Game of Life (Bak, Chen, and Creutz, 1989a, also Bak, 1992), Jensen's (1990) lattice gas, the FEDER–FEDER Model (Feder and Feder, 1991a), the ZAITSEV Model (Zaitsev, 1992), Eulerian walkers (Priezzhev, Dhar, Dhar, and Krishnamurthy, 1996a), the MASLOV–ZHANG Model (Maslov and Zhang, 1996) and the SNEPPEN–NEWMAN Model (Sneppen and Newman, 1997). A list of models mentioned in this book, including references, can be found under *models* in the index.

The models listed on the following two pages were initially thought to obey simple scaling and most were therefore expected to display universal behaviour. For the majority of the models, these assumptions had to be revised. To a certain extent the lack of robust scaling in some models can be formalised using the concept of multiscaling, where individual moments scale systematically, while the overall PDF fails to do so, undermining the traditional understanding of universality.

To this day, there is not a single exactly solved SOC model that displays non-trivial spatiotemporal correlations. Nevertheless, exactly solvable models play an important rôle in developing a theory of SOC. As discussed in Sec. 8.4, most of them derive from one of the models presented below.

The following table provides an overview of the models discussed. The first column contains the name of the model, the authors and its key reference as well as the section where it is introduced in the present book. The second column lists some of the model's key features, i.e. whether states are continuous or discrete (in the latter case q refers to the coordination number), the lattice the model is most frequently studied on, whether the model is Abelian and finally whether its relaxation is conservative. The third column contains details of the dynamics of the model, in particular whether initialisation, driving and relaxation are stochastic or deterministic. The second table is a (sometimes subjective) summary of the scaling behaviour of the various models, whether they scale robustly against changing details of the setup, whether exponents are easily reproducible and whether any tuning is necessary. The second column lists values of exponents in the typical setting (as repeated in the third column) for comparison.

Overview of models – general characteristics		
Model	States	Initialisation
Authors	Lattice	Driving
Section	Abelianess	Relaxation
	conservation	
Bak–Tang–Wiesenfeld (BTW) Model	q states	arbitrary
Bak *et al.* (1987)	2d square	stochastic
Sec. 4.1, p. 85	Non-Abelian	deterministic
	conservative	
Abelian Sandpile Model (ASM)	q states	arbitrary
Dhar (1990a)	2d square	stochastic
Sec. 4.2, p. 91	Abelian	deterministic
	conservative	
Zhang Model	continuous	arbitrary
Zhang (1989)	2d square	stochastic
Sec. 4.3, p. 104	non-Abelian	deterministic
	conservative	
Forest Fire Model (FFM)	3 states	arbitrary
Bak *et al.* (1990)	2d square	stochastic
Sec. 5.1, p. 112	n/a	deterministic
	non-conservative	
Drossel–Schwabl Forest Fire Model (DS-FFM)	2 states	arbitrary
Drossel and Schwabl (1992a)	2d square	stochastic
Sec. 5.2, p. 115	n/a	stochastic
	non-conservative	
Olami–Feder–Christensen (OFC) Model	continuous	stochastic
Olami, Feder, and Christensen (1992)	2d square	deterministic
Sec. 5.3, p. 125	non-Abelian	deterministic
	(non)-conservative	
Bak–Sneppen (BS) Model	continuous	arbitrary
Bak and Sneppen (1993)	1d	deterministic
Sec. 5.4, p. 141	non-Abelian	stochastic
	non-conservative	
original Manna Model (oMM)	2 states	arbitrary
Manna (1991b)	2d square	stochastic
Sec. 6.1, p. 162	non-Abelian	stochastic
	conservative	
Abelian Manna Model (AMM)	2 states	arbitrary
Dhar (1999a)	2d square	stochastic
Sec. 6.2, p. 166	Abelian	stochastic
	conservative	
Oslo Model	$q + 1$ states	arbitrary
Christensen *et al.* (1996)	1d	deterministic
Sec. 6.3, p. 183	(non)-Abelian	stochastic
	conservative	

Overview of models – scaling			
Robustness Reproducible exponents Tuning Upper critical dimension	Exponents	Setup	Model
not robust normally studied as ASM no tuning			BTW Model
not robust some (waves) no tuning $d_c = 4$	$\tau = 1.2$–1.3 D $= 2.5$–2.7 $\alpha = 1.3$–1.5 z $= 1.0$–1.5	2d square	ASM
not robust unclear no tuning $d_c = 4$	$\tau = 1.2$–1.3 D $= 1.4$–3.4 $\alpha \approx 1.5$ z $= 1.2$–1.7	2d square	Zhang Model
not robust(?) some(?) implicit tuning			FFM
not robust(?) some(?) tuning $(\theta \to \infty)$ $d_c = 6$	$\tau = 1.0$–1.2 $\lambda = 1.0$–1.2 $\alpha = 1.2$–1.3 $\nu' \approx 0.6$	2d square	DS-FFM
not robust(?) some(?) no tuning	values depend on degree of non-conservation; conservative case (2d square): $\tau = 1.2$–1.3 D $= 3.0$–3.4		OFC Model
not robust fairly some tuning (f_0) $d_c = 4$ or 8	$\tau = 1.0$–1.1 $\sigma \approx 0.35$	1d	BS Model
robust solid no tuning $d_c = 4$	$\tau = 1.25$–1.30 D ≈ 2.75 $\alpha \approx 1.5$ z ≈ 1.4	2d square bulk driven	oMM
robust solid no tuning $d_c = 4$	as oMM		AMM
robust solid no tuning	$\tau \approx 1.55$ D ≈ 2.25 $\alpha = 1.7$–1.9 z ≈ 1.42	1d boundary driven	Oslo Model

4 Deterministic sandpiles

The models discussed in this chapter resemble some of the phenomenology of (naïve) sandpiles. As discussed earlier (Sec. 3.1), it is clear that the physics behind a real relaxing sandpile is much richer than can be captured by the following 'sandpile models'. Yet, it would be unjust to count that as a shortcoming, because these models were never intended to describe all the physics of a sandpile. The situation is similar to that of the 'Forest Fire Model' which is only vaguely reminiscent of forest fires and was, explicitly, not intended to model them. The names of these models should not be taken literally, they merely serve as a sometimes humorous aide-memoire for their setup, similar to Thomson's Plum Pudding Model which is certainly not a model of a plum pudding.

In the following section, the iconic BAK–TANG–WIESENFELD Model and its hugely important derivative, the Abelian Sandpile Model, are discussed in detail. This is followed by the ZHANG Model, which was intended as a continuous version of the BAK–TANG-WIESENFELD Model. Their common feature is a deterministic, rather than stochastic, relaxation rule. Although a lot of analytical and numerical progress has been made for all three models, their status quo, in particular to what extent they display true scale invariance, remains inconclusive.

4.1 The BAK–TANG–WIESENFELD Model

The publication of the BAK–TANG–WIESENFELD (BTW) Model (see Box 4.1) (Bak *et al.*, 1987) marks the beginning of the entire field. The BTW Model is often referred to as *the* Sandpile Model, but there is some confusion with the *Abelian* Sandpile Model (see Box 4.2, p. 91) which was introduced later by Dhar (1990a) as a mathematically more tractable, generalised version of the BTW Model. Where a clear distinction is required, the former is called the 'original BTW Model' in the following, otherwise it is referred to as just the 'BTW Model'.

The BTW Model was introduced by Bak, Tang and Wiesenfeld as a model for the avalanching behaviour reminiscent of that observed at the sandpile in an hourglass (see Schick and Verveen, 1974, for $1/f$ noise in an hourglass). However, it was not the original objective of the BTW Model to capture all the physics of a real sandpile, which would require the incorporation of friction and inertia (e.g. Puhl, 1992; Mehta, 1992, review by Kakalios, 2005).

Box 4.1   **The Bak–Tang–Wiesenfeld Sandpile**

General features

- First published by Bak, Tang, and Wiesenfeld (1987).
- Motivated by avalanching behaviour of a real sandpile.
- In one dimension rules represent downward movement of sand grains.
- Defined in any dimension, exactly solved (trivial) in one.
- Stochastic (bulk) drive, deterministic relaxation.
- Non-Abelian in its original definition.
- Many results actually refer to Dhar's (1990a) Abelian sandpile, Sec. 4.2.
- Simple scaling behaviour disputed, multiscaling proposed.
- Exponents listed in Table 4.1, p. 92, are for the Abelian BTW Model.

Rules

- d dimensional (usually) hyper-cubic lattice and q the coordination number (on cubic lattices $q = 2d$).
- Choose (arbitrary) critical slope $z^c = q - 1$.
- Each site $\mathbf{n} \in \{1, \ldots, L\}^d$ has slope $z_\mathbf{n}$.
- *Initialisation*: irrelevant, model studied in the stationary state.
- *Driving*: add a grain at $\mathbf{n}_0$ chosen at random and update all uphill nearest neighbours $\mathbf{n}_0'$ of $\mathbf{n}_0$:
$$z_{\mathbf{n}_0} \longrightarrow z_{\mathbf{n}_0} + q/2$$
$$z_{\mathbf{n}_0'} \longrightarrow z_{\mathbf{n}_0'} - 1.$$
- *Toppling*: for each site $\mathbf{n}$ with $z_\mathbf{n} > z^c$ distribute q grains among its nearest neighbours $\mathbf{n}'$:
$$z_\mathbf{n} \longrightarrow z_\mathbf{n} - q$$
$$\forall_{\mathbf{n}'.\text{nn}.\mathbf{n}} \, z_{\mathbf{n}'} \longrightarrow z_{\mathbf{n}'} + 1.$$
In one dimension site $\mathbf{n} = L$ relaxes according to $\qquad z_L \longrightarrow z_L - 1$
$$z_{L-1} \longrightarrow z_{L-1} + 1.$$
- *Dissipation*: grains are lost at open boundaries.
- *Parallel update*: discrete microscopic time, sites exceeding z^c at time t topple at $t + 1$ (updates in sweeps).
- *Separation of time scales*: drive only once all sites are stable, i.e. $z_\mathbf{n} \leq z^c$ (quiescence).
- *Key observables* (see Sec. 1.3):
  avalanche size s, the total number of topplings until quiescence;
  avalanche duration T, the total number of parallel updates until quiescence.

On a d dimensional square lattice of linear size L, each site $\mathbf{n} \in \{1, \ldots, L\}^d$ has an integer $h_\mathbf{n}$ associated with it, which measures the height of a column of sand grains resting on that site. The slope $z_\mathbf{n}$ derived from the height generally (i.e. in $d > 1$) would be vector valued, but for simplicity is represented by an integer. In one dimension, this poses no restriction and one defines

$$z_\mathbf{n} = h_\mathbf{n} - h_{\mathbf{n}+1} \tag{4.1}$$

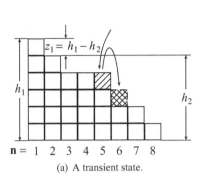 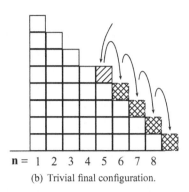

(a) A transient state.   (b) Trivial final configuration.

Figure 4.1 The original BTW Model with $z^c = 1$. In (a) a transient configuration is shown together with the definition of height $h_{\mathbf{n}}$ and slope $z_{\mathbf{n}}$. The hatched box represents a newly arrived height unit about to topple to the neighbouring site (cross-hatched). In (b) the trivial final configuration is displayed together with the intermediate configurations (distributed grains shown as dashed, cross-hatched boxes), as a single grain topples through the system starting from $\mathbf{n}_0 = 5$. This configuration, $z_{\mathbf{n}} = 9 - \mathbf{n}$, is the only recurrent one and is reached again after every avalanche (Frette, 1993).

where $h_{L+1} = 0$, so that the height configuration can be derived from the slope configuration, using

$$h_{\mathbf{n}} = \sum_{\mathbf{n}'=\mathbf{n}}^{L} z_{\mathbf{n}'}. \tag{4.2}$$

Sample configurations are shown in Fig. 4.1.

A material constant of granular media is the angle of repose, which is the steepest angle of a stable pile, i.e. above this angle avalanches will form. Motivated by this observation, a critical slope z^c is introduced, and a **toppling rule**: each site $\mathbf{n}$ with slope $z_{\mathbf{n}}$ exceeding the threshold z^c distributes one sand grain towards the downhill sites.[1] In one dimension that means $h_{\mathbf{n}} \to h_{\mathbf{n}} - 1$ and $h_{\mathbf{n}+1} \to h_{\mathbf{n}+1} + 1$, unless $\mathbf{n} = L$, in which case the grain leaves the system, so that $h_{\mathbf{n}} \to h_{\mathbf{n}} - 1$ without addition anywhere. Given Eq. (4.1) and the **boundary condition** $h_{L+1} = 0$ the toppling is equivalently described by updates of the slope. For bulk sites $z_{\mathbf{n}} \to z_{\mathbf{n}} - 2$ and $z_{\mathbf{n}'} \to z_{\mathbf{n}'} + 1$ for all nearest neighbours $\mathbf{n}'$ (two in the bulk), apart from the rightmost site $\mathbf{n} = L$ which topples according to $z_L \to z_L - 1$ and $z_{L-1} \to z_{L-1} + 1$. The relation between the toppling rules in terms of height units $h_{\mathbf{n}}$ and in terms of slope units $z_{\mathbf{n}}$ is illustrated in Fig. 4.1(a).

In this original definition, height units (grains) are dissipated in one dimension whenever site $\mathbf{n} = L$ topples, while the other boundary site $\mathbf{n} = 1$ is exceptional only in the sense that it does not receive any height units from an uphill site $\mathbf{n} = 0$. In that sense, the slope at site $\mathbf{n} = 0$ is fixed below z^c (Neumann boundary conditions) while the value h_{L+1} remains unchanged after charge (Dirichlet boundary conditions). It might appear more suitable to

---

[1] Originally, $z_{\mathbf{n}} = z^c$ was stable (Bak *et al.*, 1987, 1988; Dhar, 1990a). As this is a matter of definition of z^c, some authors regard $z_{\mathbf{n}} = z^c$ as unstable (Ruelle and Sen, 1992; Dhar, Ruelle, Sen, and Verma, 1995).

call the boundary at $\mathbf{n} = 1$ closed and that at $\mathbf{n} = L$ open.[2] Reformulating the rules in terms of slope units interchanges the character of both boundaries (see p. 94 and Sec. 4.2.2), as slope units are dissipated only at $\mathbf{n} = 1$, whereas $\mathbf{n} = L$ is closed for slope units.

The toppling rules in the bulk are fully **conservative**, with respect to both height and slope. Hwa and Kardar (1989a, also Sec. 9.2) identified bulk conservation and open, i.e. dissipative, boundaries as crucial ingredients for scale invariance, which was tested by Manna, Kiss, and Kertész (1990) for various non-conservative versions of the BTW Model (also Ghaffari, Lise, and Jensen, 1997; Tsuchiya and Katori, 2000; Vázquez, 2000; Lin, Chen, Chen, *et al.*, 2006). In particular, they found that the avalanche size exponent changed from $\tau = 1.22$ (Manna, 1990) to $\tau = 1.515(20)$ for local non-conservation, but global conservation on average, suggesting a change of universality class.[3] Local non-conservation was implemented as annealed noise, which dilutes correlations (Pérez, Corral, Díaz-Guilera, *et al.*, 1996, for quenched noise Bak *et al.*, 1988; Černák, 2002). Global non-conservation generated a characteristic avalanche size independent of the system size and was therefore identified as a relevant variable. The rôle of conservation, Sec. 9.2.2, became more prominent with the advent of models such as the OFC Model, Sec. 5.3.

Having translated the toppling rules, the **driving** can be expressed in terms of slope units as well. Originally, a single grain of sand is added to site $\mathbf{n}_0$ (chosen at random or fixed[4]), so that $h_{\mathbf{n}_0} \to h_{\mathbf{n}_0} + 1$, while expressing the same process in slope units means $z_{\mathbf{n}_0} \to z_{\mathbf{n}_0} + 1$ and $z_{\mathbf{n}_0-1} \to z_{\mathbf{n}_0-1} - 1$, unless $\mathbf{n}_0 = 1$, where $z_{\mathbf{n}_0} \to z_{\mathbf{n}_0} + 1$ only. Thus, while driving in terms of height units is non-conservative at every site, only site $\mathbf{n}_0 = 1$ is non-conservative with respect to the driving in terms of slope units, where units can be *gained* by addition. This is also reflected by Eq. (4.2), implying that height h_1 is the total of all slope units in the system which decreases only when site $\mathbf{n} = 1$ topples. To summarise the rules in one dimension:

$$
\begin{array}{llll}
\text{Driving at } \mathbf{n}_0 & z_{\mathbf{n}_0} \to z_{\mathbf{n}_0} + 1 & & \\
& z_{\mathbf{n}_0-1} \to z_{\mathbf{n}_0-1} - 1 & \text{if } \mathbf{n}_0 > 1 & \\
\text{Toppling at } \mathbf{n} & z_{\mathbf{n}} \to z_{\mathbf{n}} - 2 & \text{if } \mathbf{n} \neq L & \\
& z_{\mathbf{n}} \to z_{\mathbf{n}} - 1 & \text{if } \mathbf{n} = L & (4.3) \\
& z_{\mathbf{n}+1} \to z_{\mathbf{n}+1} + 1 & \text{if } \mathbf{n} < L & \\
& z_{\mathbf{n}-1} \to z_{\mathbf{n}-1} + 1 & \text{if } \mathbf{n} > 1. &
\end{array}
$$

The driving reveals a surprising anisotropy in the definition of the model. Grains always only move to the right, instead of in the direction of maximum slope. In one dimension this anisotropy is induced by the definition of slope in terms of height, Eq. (4.1) and the condition $z_{\mathbf{n}} > z^c$ rather than $|z_{\mathbf{n}}| > z^c$, which means that by driving the BTW Model only at sites $\mathbf{n} > \mathbf{n}_0$, all sites $\mathbf{n} = 1, 2, \ldots, \mathbf{n}_0$ will remain empty and a sharp drop develops towards

---

[2] Unfortunately, these attributes are not used consistently throughout the literature. For example, open boundaries are sometimes called free, and closed boundaries reflecting.

[3] The local non-conservation was implemented as a randomisation of the neighbouring slope, which Christensen (1992) interpreted, quite intuitively, as a de facto randomisation of the neighbour itself, so that the model produces mean-field behaviour.

[4] In the following, random $\mathbf{n}_0$ is assumed unless stated otherwise.

the left of $n_0 + 1$ rather than a pile similar to that covering sites $\mathbf{n} = \mathbf{n}_0 + 1, \mathbf{n}_0 + 2, \ldots, L$. Yet, contrary to the formulation in terms of height units, the bulk relaxation in terms of slope units is fully isotropic.

The avalanche size is measured as the number of topplings between inserting a particle and all sites being stable again, i.e. $z_\mathbf{n} \leq z^c$ for all $\mathbf{n}$. The height picture provides the easiest way of calculating the average number of topplings per particle added, i.e. the **average avalanche size** (also Sec. 8.1.1.1). In the stationary state, the one-dimensional BTW Model evolves into a trivial configuration where all $z_\mathbf{n} = z^c = 1$ (Bak *et al.*, 1987), see Fig. 4.1(b). In that state, a grain added at $\mathbf{n}_0$ will perform exactly $L + 1 - \mathbf{n}_0$ topplings involving all sites downhill from $\mathbf{n}_0$ until it leaves the system, moving to the right in each toppling event. After the avalanche the pile remains unchanged, as illustrated in Fig. 4.1(b) for a system driven at $\mathbf{n}_0 = 5$. The average avalanche size therefore depends only on the distribution of **driving positions**. If $\mathbf{n}_0$ is distributed uniformly, the resulting PDF of the avalanche size and duration is uniform, $\mathcal{P}^{(\mathrm{BTW})}(s) = L^{-1}\theta(L - s) = s^{-1}(s/L)\theta(1 - s/L)$, while driving at a fixed site $\mathbf{n}_0$ produces an avalanche of size $L - \mathbf{n}_0$ only, i.e. $\mathcal{P}^{(\mathrm{BTW})}(s) = \delta\left(s - (L - \mathbf{n}_0)\right) = s^{-1}(s/L)\delta\left((s/L) - (1 - \mathbf{n}_0/L)\right)$, so that $\tau = \alpha = 1$ and $D = z = 1$ in both cases. While this is formally correct and the exponents are unique, it looks contrived and the exponents hide the fact that there are no spatiotemporal correlations.

Kadanoff *et al.* (1989) studied a number of variations of the one-dimensional BTW Model, which differ from the original version only in very few details, and found non-trivial behaviour. The scaling of the avalanche sizes or 'flip numbers' seem to follow **multifractal scaling** much better than (simple) FSS.

4.1.1　Higher dimensions

Above one dimension the local slope $z_\mathbf{n}$ is defined without reference to a height, instead the rules are generalised from one dimension in a natural way. In two dimensions the original rules (Bak *et al.*, 1988) read:

$$
\begin{aligned}
\text{Driving at } (x_0, y_0) \quad & z_{(x_0,y_0)} \to z_{(x_0,y_0)} + 2 \\
& z_{(x_0-1,y_0)} \to z_{(x_0-1,y_0)} - 1 \quad \text{if } x_0 > 1 \\
& z_{(x_0,y_0-1)} \to z_{(x_0,y_0-1)} - 1 \quad \text{if } y_0 > 1 \\
\text{Toppling at } (x, y) \quad & z_{(x,y)} \to z_{(x,y)} - 4 \\
& z_{(x\pm1,y)} \to z_{(x\pm1,y)} + 1 \quad \text{if } x \begin{cases} < L \\ > 0 \end{cases} \\
& z_{(x,y\pm1)} \to z_{(x,y\pm1)} + 1 \quad \text{if } y \begin{cases} < L \\ > 0 \end{cases}
\end{aligned}
\tag{4.4}
$$

where the two conditions on the right refer to the two directions ± 1 on the left. Again, a net gain of slope units occurs only when charging (i.e. driving) the system at the 'upper' boundary sites, $x_0 = 0$ or $y_0 = 0$. In contrast to one dimension, the toppling of any boundary site leads to a loss of slope units.

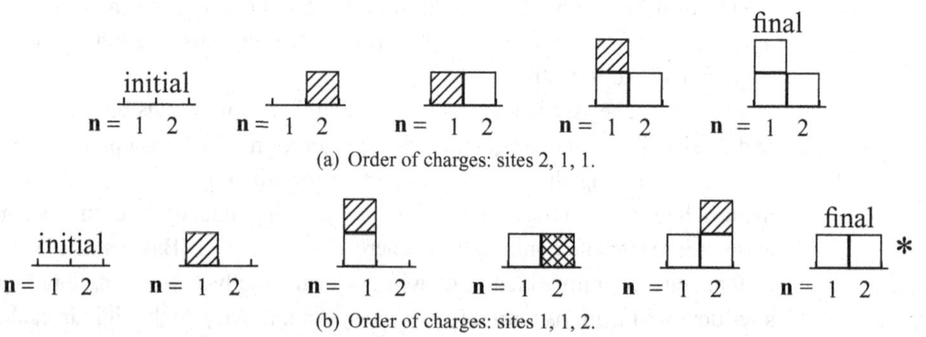

Figure 4.2 Example of the non-Abelian nature of the BTW Model ($L = 2$), resulting in different final configurations if the order of charges is changed. Time advances from left to right, newly added grains (height units) are shown hatched. If a new charge induces toppling, the resulting state is shown to the right, redistributed grains are shown cross-hatched. The order of charges in (a) is such that all sites remain stable, while in (b) two topplings (after the second and the third charge) lead to a different final configuration with one grain dissipated (shown as a star).

The generalisation to higher dimensions is fairly straightforward, see Box 4.1, p. 86. The in-built **anisotropy** in the driving rule retains some of the character of a real sandpile. However, the BTW Model was originally not intended to allow for negative slopes which would be needed to model a peaked pile as found in an hourglass.

4.1.2 The Abelian symmetry

Contrary to common belief, the BTW Model in its original definition is *not* Abelian, i.e. the final state of the model depends on the order in which different sites are charged by the external drive and generally on the order of updates of unstable sites (Sec. 1.2, p. 10, Sec. 8.3). This is due to its specific driving rule, which is motivated by the height picture (Dhar, 1990a).[5] Figure 4.2 illustrates how changing the sequence of charges by the external drive leads to a different final configuration. At the heart of the problem is the removal of slope from the site upstream from the one subject to the driving. Charging the neighbouring site before or after a given site is charged might alter the final outcome, because charging the neighbour first might prevent the local slope from exceeding the critical value.

In the following section, an Abelian version of the BTW Model is introduced as one specific realisation of a large class of models often referred to as *the* Abelian Sandpile Model. The vast majority of numerical studies (but not all) of the BTW Model are based on the Abelian Sandpile Model, which might account for some of the numerical inconsistencies reported for the BTW Model, see Table 4.1. However, the difference between the definitions of the two models is so minute that one might not expect it to affect the scaling behaviour. The situation is similar in the MANNA Model, see Sec. 6.1.

---

[5] It is possible but untested that, in the stationary state, the original, non-Abelian BTW Model does not differ from the ASM discussed below.

Box 4.2	The Abelian Sandpile Model

General features

- First published by Dhar (1990a), generalised, Abelian variant of the BTW Model, Box 4.1.
- Defined in any dimension.
- Stochastic (bulk) drive, deterministic relaxation.
- Abelian.
- Simple scaling behaviour disputed, multiscaling proposed.
- Large number of exact results.
- Exponents listed in Table 4.1, p. 92.
- Upper critical dimension $d_c = 4$ (Grassberger and Manna, 1990).

Rules

- Finite set $\{1, 2, 3, \ldots, N\}$ of sites, usually on an arbitrary lattice in arbitrary dimension.
- Introduce a $N \times N$ toppling matrix Δ with integer elements obeying for all $\mathbf{n}, \mathbf{n}' \in \{1, \ldots, N\}$ $\Delta_{\mathbf{nn}} > 0$ and $\Delta_{\mathbf{nn}'} \leq 0$ for $\mathbf{n}' \neq \mathbf{n}$.
- No creation of particles by toppling: for all $\mathbf{n} \in \{1, \ldots, N\}$ the toppling matrix obeys $\sum_{\mathbf{n}'=1}^{N} \Delta_{\mathbf{nn}'} \geq 0$.
- Each site $\mathbf{n} \in \{1, \ldots, N\}$ has a slope $z_{\mathbf{n}} \in \mathbb{Z}$ subject to the dynamics and a fixed critical slope $z_{\mathbf{n}}^c$.
- Set $z_{\mathbf{n}}^c = \Delta_{\mathbf{nn}}$ without loss of generality.
- *Initialisation*: irrelevant, model studied in the stationary state.
- *Driving*: add a particle at randomly chosen $\mathbf{n}_0$, resulting in a change of slope: $z_{\mathbf{n}_0} \to z_{\mathbf{n}_0} + 1$.
- *Toppling*: for each site $\mathbf{n}$ with $z_{\mathbf{n}} > z_{\mathbf{n}}^c$ ('active' or 'unstable' site) distribute slope units among other sites, $\forall_{\mathbf{n}' \in \{1, \ldots, N\}} z_{\mathbf{n}'} \to z_{\mathbf{n}'} - \Delta_{\mathbf{nn}'}$, with $\sum_{\mathbf{n}'=1}^{N} \Delta_{\mathbf{nn}'} \geq 0$ units dissipated.
- *Dissipation*: typically, slope units are lost at boundaries.
- *Parallel update*: discrete microscopic time, sites becoming active at time t topple at $t + 1$ (updates in sweeps).
- *Separation of time scales*: drive only once all sites are stable, i.e. $z_{\mathbf{n}} \leq z_{\mathbf{n}}^c$ (quiescence).
- *Key observables* as in the BTW Model, see Box 4.1.

4.2 Dhar's Abelian Sandpile Model

Dhar (1990a) generalised the BTW Model in a way that rendered it (fully) Abelian and opened the door to a detailed and very elegant analysis (see also Sec. 8.3, p. 269). This variant of the BTW Model, known as Abelian Sandpile Model(s) (ASM(s)) and defined in Box 4.2, is probably the best studied model in SOC, in particular in two dimensions, in the form of the Abelian BTW Model. Dhar (1999a) wrote a succinct review of the ASM and models derived from it.

Practically all that is known analytically about the Sandpile Model is based on this variant of the original BTW Model. The key steps from the BTW Model to the ASM were

Table 4.1 Exponents of the Abelian BTW Model. $\mathcal{P}^{(s)}(s) = a^{(s)}s^{-\tau}\mathcal{G}^{(s)}(s/(b^{(s)}L^{D}))$ is the avalanche size distribution and $\mathcal{P}^{(T)}(T) = a^{(T)}T^{-\alpha}\mathcal{G}^{(T)}(T/(b^{(T)}L^{z}))$ is the avalanche duration distribution, Sec. 1.3. Scaling relations: $D(2-\tau) = \sigma_1^{(s)} = 2$ (bulk drive) and $(1-\tau)D = (1-\alpha)z$ (sharply peaked joint distribution, widely assumed) in all dimensions (Sec. 2.2.2).

d	τ		D		α		z	
1	1	(a)	2	(a)	1	(a)	1	(a)
2	0.98 *	(b)	2.56	(o)	0.97 *	(b)	1.234	(f)
	1.22 ▷	(c)	2.50(5)	(p)	1.38	(c)	1.168	(g)
	1.21 ▷*	(d)	2.73(2)	(k)	1.316(30)	(e)	1.52(2)	(k)
	1 ▷	(d)			1.480(11)	(i)	1.02(5)	(q)
	1.2007(50)	(e)			1.16(3)	(n)	1.082−1.284	(m)
	1.253	(f)						
	1.253	(g)						
	6/5	(h)						
	1.293(9)	(i)						
	1.27(1)	(k)						
	1.122−1.367	(m)						
	1.13(3)	(n)						
3	1.35 *	(b)	2.96	(r)	1.59 *	(b)	1.6	(r)
	1.33 ▷	(o)	3.004	(t)	1.625	(d)	1.618	(t)
	1.47 ▷*	(d)			1.597(12)	(s)		
	1.37 ▷	(d)						
	1.35 ▷	(r)						
	1.333(7)	(s)						
4	1.61 ▷*	(d)	4	(s)	2	(s)	2	(s)
	1.5 ▷	(d)						
	3/2	(s)						
MFT	3/2 ▷	(u)	4	(w)	1	(u)	2	(w)
	3/2 ▷	(v)			2	(v)		

∗ Based on the (original) non-Abelian BTW Model.
▷ Reported as $\tau + 1$.
a Ruelle and Sen (1992), *b* Bak *et al.* (1987), *c* Manna (1990), *d* Christensen *et al.* (1991), *e* Manna (1991a), *f* Pietronero, Vespignani, and Zapperi (1994), *g* Vespignani and Zapperi (1995), *h* Priezzhev, Ktitarev, and Ivashkevich (1996b), *i* Lübeck and Usadel (1997b), *k* Chessa *et al.* (1999a), *m* Lin and Hu (2002), *n* Bonachela (2008), *o* Grassberger and Manna (1990), *p* De Menech *et al.* (1998), *q* De Menech and Stella (2000), *r* Ben-Hur and Biham (1996), *s* Lübeck and Usadel (1997a), *t* Lübeck (2000), *u* Tang and Bak (1988a), *v* Vergeles, Maritan, and Banavar (1997), *w* Tang and Bak (1988b).

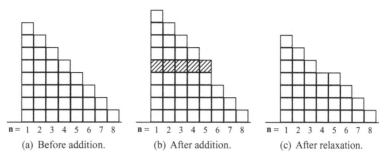

n = 1 2 3 4 5 6 7 8   n = 1 2 3 4 5 6 7 8   n = 1 2 3 4 5 6 7 8
      (a) Before addition.   (b) After addition.   (c) After relaxation.

Figure 4.3   In the ASM, driving and dissipation are unphysical. (a) The single recurrent state of the BTW Model and one of the $L + 1$ states recurrent in the ASM. Adding a slope unit at $\mathbf{n} = 5$, introduces 5 height units, shown hatched in (b). The resulting avalanche of size $s = 20$ (not shown) results in a configuration similar to (a) but with $z_4 = 0$. During the course of the avalanche 2 slope units are lost, with every toppling at $\mathbf{n} = L = 8$ ($\mathbf{n} = 1$) dissipating L (1) height units, here $8 + 1 = 9$ in total. The topplings can be decomposed in five sequences: the first grain starts at $\mathbf{n} = 5$ and, after 4 intermediate steps, falls off the edge at $\mathbf{n} = 8$, dissipating $L + 1$ grains, so that, for example, $h_1 = 8$. This is followed by four more sequences of 4 topplings, starting at $\mathbf{n} = 4, 3, 2, 1$ and reaching $\mathbf{n} = 8, 7, 6, 5$ respectively. The final state is shown in (c).

to generalise the toppling by means of a **toppling matrix**, which is essentially the adjacency matrix of the lattice, and to change the driving mechanism, which in the ASM consists in adding a single *slope unit*,[6] rather than a grain. The effect of this change on equilibration, stationarity and scaling was discussed by Christensen *et al.* (1991). As illustrated in Fig. 4.3, the driving in the ASM has a dramatic (un)physical effect when translating the slope units to height, Eq. (4.2), but simplifies the analysis in terms of slope units considerably. Most importantly, the resulting model is Abelian. The ASM normally dissipates slope units on all boundaries, a further simplification compared to the BTW Model. As a result, in the one-dimensional Abelian BTW Model all $L + 1$ states with no more than one site with $z_{\mathbf{n}} = 0$ are recurrent (Ruelle and Sen, 1992), compared to only one recurrent state in the BTW Model. It is characteristic for the one-dimensional ASM that a single 'hole', $z_{\mathbf{n}} = 0$, either disappears or moves through the system depending on where the external drive takes place.

The similarity of the slope units in the ASM and the grains in the BTW Model has led to considerable confusion in the literature, as $z_{\mathbf{n}}$ is now often referred to as 'height'. Kadanoff *et al.* (1989) studied a large number of variants of the BTW Model and the ASM and reported a failure of simple scaling and a surprising plethora of universality classes. They coined the terms **critical height model** and **critical slope model** to motivate the distinction of different toppling rules in two dimensions. Yet, what reaches a critical slope is the height and what reaches a critical height is best interpreted as the slope (also Manna, 1991a). To complete the muddle, based on a conversion of slope to height along the lines of Eq. (4.2) and Fig. 4.1, boundary conditions are reversed in the two pictures (Corral,

---

[6] Strictly, there is not (necessarily) a height picture of the ASM, which calls into question the suitability of the term 'slope unit'. It is used in the following to keep the ASM in the context of the BTW Model; the corresponding entity in the MANNA Model is a 'particle'.

2004a), which makes it very important to specify what a boundary condition applies to (height or slope unit): the boundary that is lossy with respect to toppling grains (the right boundary in Fig. 4.1) is conservative with respect to toppling slopes, and the boundary where no grains are lost (the left boundary in Fig. 4.1) is lossy with respect to slope units, see p. 88. Finally, using open (dissipative) boundary conditions (for slope units) throughout the ASM means that downward boundaries dissipate the entire lowest layers of height units whenever a slope unit is dissipated, similar to the one added in Fig. 4.3(b), i.e. the right hand boundary being lossy with respect to slope makes it even more lossy with respect to height.

The most important feature of the ASM is its Abelian symmetry, which, strictly, means that its final state is independent of the order in which external charges are applied. This is normally extended to the final configuration being independent of the order in which sites topple during a single avalanche (see end of Sec. 4.2.1, p. 98). To see that the ASM is Abelian, Dhar (1990a) invoked an argument on the microscopic time scale. Any site $\mathbf{n}_i$ is charged $-(\Delta_{\alpha i} + \Delta_{\beta i})$ slope units when sites $\mathbf{n}_\alpha$ and $\mathbf{n}_\beta$ topple regardless of the order in which they topple. 'By a repeated use of this argument, we see that in an avalanche, the same final stable configuration is reached irrespective of the sequence in which unstable sites are toppled' (Dhar, 1990a). This argument is immediately extended to toppling sites being charged at the same time, supported by the important observation that any discharge or external charge anywhere else in the system can only *add* slope units to any given site. As long as a given unstable site has not yet toppled, it cannot turn from unstable to stable by a relaxation of another site or by an external charge. This condition is not met in the original BTW Model, because the external drive at a site reduces the slope at some of its neighbours, see Fig. 4.2.

Based on this derivation of the Abelian symmetry the final state of an ASM does not depend on the order of updates. Every active site can topple at any time, all that matters is that it is active. This is often used in numerical simulations, where all active sites can be held on a stack, often a LIFO (see App. A.1), and processed in arbitrary order. In contrast to the ZHANG Model (Sec. 4.3), the amount of slope units redistributed from a toppling site $\mathbf{n}$ is independent of its slope $z_{\mathbf{n}}$.

Ignoring the difference in the drive, the original BTW Model on a hyper-cubic lattice corresponds to the ASM with toppling matrix $\Delta_{\mathbf{nn}} = z_{\mathbf{n}}^c + 1 = 2d$ and $\Delta_{\mathbf{nn}'} = -1$ for all nearest neighbours $\mathbf{n}'$ of $\mathbf{n}$ with open boundaries providing dissipation (for bulk dissipation see Lin *et al.*, 2006). In that case $\Delta_{\mathbf{nn}'}$ is the lattice Laplacian, which allows easy calculation of the average avalanche size, Sec. 8.1.1.1 (also Sec. 4.2.2). All other entries in the toppling matrix vanish. This setup is generally known as the **Abelian BTW Model** [7], although, as a matter of convenience, the critical slope is often chosen as $z_{\mathbf{n}}^c = \Delta_{\mathbf{nn}} = q$ (Dhar, 1999c). In one dimension special boundary conditions might be applied for the relaxation of the rightmost site, where $\Delta_{LL} = 1$. Ruelle and Sen (1992) have studied this one-dimensional Abelian Sandpile Model in great detail. In an effort to investigate universality in the ASM, Ali and Dhar (1995a,b) studied more complicated variants of the one-dimensional ASM and found that generally the scaling form by Ruelle and Sen (1992) is replaced by a linear

---

[7] It is common practice to refer to the Abelian BTW Model as just the BTW Model.

combination of two simple scaling forms,

$$\mathcal{P}^{(1\mathrm{DASM})}(s) = s^\tau \left(\mathcal{G}_1(s/L) + \mathcal{G}_2(s/L^2) \right) \tag{4.5}$$

with *non-universal* scaling functions $\mathcal{G}_1$ and $\mathcal{G}_2$. This expression can be tested for a multi-scaling, Sec. 2.4.

4.2.1 Operator approach to the ASM

The ASM has been analysed in great detail on the basis of an operator algebra (Dhar *et al.*, 1995; Dhar, 1999a,c, see also Sec. 8.3 in particular Sec. 8.3.2.2 and Sec. 8.3.2.3). **Operators** $a_\mathbf{n}$ represent the addition of a slope unit at site $\mathbf{n}$ of a stable configuration $\mathcal{C}$ together with the entire subsequent relaxation, such that all sites are stable again in $a_\mathbf{n}\mathcal{C}$, i.e. $z_{\mathbf{n}'} \le z_{\mathbf{n}'}^c$ for all sites $\mathbf{n}'$. The Abelian symmetry means that the operators **commute** – the Abelian symmetry is the invariance of the final state under a change of order in external charge with subsequent relaxation, i.e. it is a property of the *macroscopic* time scale. Under the action of the $a_\mathbf{n}$, with $\mathbf{n}$ indicating the 'driven' site, where a slope unit has been dropped (usually chosen at random), the ASM passes through **transient configurations**, which are accessible only from certain initial conditions, until it reaches a **recurrent state**,[8] i.e. one that can be obtained again by repeated application of the operators $a_\mathbf{n}$. Dhar *et al.* (1995) (also Dhar, 1999c, 2004) suggested that the length of the transient, i.e. the number of external charges needed to reach the stationary state, is of the order L^d.

There are even local *sub-configurations* which cannot possibly be recurrent, known as **forbidden sub-configurations** (FSCs) (Dhar, 1990a, 1999c). For example, every site that ever receives a slope unit eventually maintains $z_\mathbf{n} > z_\mathbf{n}^c - \Delta_{\mathbf{nn}}$ forever, because it loses only $\Delta_{\mathbf{nn}}$ units whenever $z_\mathbf{n} > z_\mathbf{n}^c$. Another, slightly less trivial example is that of two sites that charge each other by toppling. If the toppling matrix is symmetric, every pair of sites either charges each other (equally) or not at all. Two sites that charge each other cannot both have $z_\mathbf{n} = 1 + z_\mathbf{n}^c - \Delta_{\mathbf{nn}}$, which is their slope after they have toppled. Such a slope implies that any site charging $\mathbf{n}$ has not toppled since $\mathbf{n}$ toppled last time. So, if $z_\mathbf{n} = 1 + z_\mathbf{n}^c - \Delta_{\mathbf{nn}}$ all sites $\mathbf{n}'$ charged by $\mathbf{n}$ obey $z_{\mathbf{n}'} > 1 + z_{\mathbf{n}'}^c - \Delta_{\mathbf{n}'\mathbf{n}'}$.

More formal and sophisticated methods have been developed to determine whether a configuration is recurrent (Dhar, 1999c), most notably the **burning test** (Majumdar and Dhar, 1992), applicable to the ASM with symmetric toppling matrix.[9] The so-called 'script-test' can be applied in more complicated situations (Speer, 1993; Dhar, 1999a). A configuration $\mathcal{C} = \{z_\mathbf{n}\}$ is recurrent under the operator $a_\mathbf{n}$ if there is a (smallest) **period** $p_\mathbf{n} \ge 1$ such that $a_\mathbf{n}^{p_\mathbf{n}}\mathcal{C} = \mathcal{C}$, so that the powers of $a_\mathbf{n}$ operating on the recurrent configurations form a cyclic group.[10]

---

[8] Below, many results are derived considering only the set of recurrent states. The Abelianness of the operators applies to transient states as well.

[9] Asymmetric, randomised toppling matrices have been investigated by Karmakar *et al.* (2005) and found to belong to the MANNA universality class, see Sec. 6.2.1.2.

[10] The restriction of a property of an operator $a_\mathbf{n}$ to its recurrent states means formally that a new operator $\tilde{a}_\mathbf{n}$ is to be defined as $a_\mathbf{n}$ for its operation on the recurrent states of $a_\mathbf{n}$ and undefined otherwise. Strictly, $\tilde{a}_\mathbf{n}^{p_\mathbf{n}} = \mathbf{1}$ as $a_\mathbf{n}^{p_\mathbf{n}}\mathcal{C} = \mathcal{C}$ for all its recurrent states, but $a_\mathbf{n}^{p_\mathbf{n}} \ne \mathbf{1}$ as $a_\mathbf{n}^{p_\mathbf{n}}\mathcal{C}' \ne \mathcal{C}'$, if $\mathcal{C}'$ is transient.

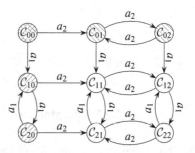

States that are transient with respect to one operator can be recurrent under another. The figure shows the evolution of the two-site system described in the text. Operators $a_\mathbf{n}$ change only site $\mathbf{n}$, implementing the toppling matrix $\Delta_\mathbf{nn} = z_\mathbf{n}^c = 2$ for $\mathbf{n} = \{1, 2\}$ and $\Delta_\mathbf{nn'} = 0$ for $\mathbf{n} \neq \mathbf{n'}$. Starting from $z_1 = z_2 = 0$, the 9 stable states (circles) are $C_{z_1 z_2}$ connected by arrows indicating the action of an operator. Hatched circles indicate transient states (north-west for transients of a_2 and north-east for a_1).

Under the action of $a_\mathbf{n}$ a repeating cycle of configurations (limit cycle) is visited deterministically, each with equal frequency. As a consequence, the total number of recurrent states of all operators is given by the determinant of the toppling matrix (Dhar, 1990a), in two dimensions $\det(\Delta) \propto \exp(4.67L^2)$ (Wiesenfeld, Theiler, and McNamara, 1990; Markošová and Markoš, 1992; Dhar *et al.*, 1995).

If $a_\mathbf{n}$ has period $p_\mathbf{n}$, $a_\mathbf{n}^{p_\mathbf{n}-1}$ is the **inverse (operator)** of $a_\mathbf{n}$ with respect to its action on any recurrent configuration $C_n = a_\mathbf{n}^{p_\mathbf{n}} C$. The same is not true for $a_\mathbf{n}$ acting on a transient configuration, i.e. generally no power of $a_\mathbf{n}$ can undo the action of $a_\mathbf{n}$ on a transient state.

In general, a state that is transient under one operator is not necessarily transient under another operator (but see Dhar, 1999c). A system consisting of two sites which do not charge each other at toppling, for example, initialised with $z_1 = 0$, $z_2 = 0$, will display periodically occurring states under the action of, say, a_1, but as long as $z_2 = 0$ is transient under a_2 each such state is a transient with respect to a_2, see Fig. 4.4. However, if state C' is transient under $a_\mathbf{n}$, then there is no operator $a_\mathbf{n'}$ such that $C' = a_\mathbf{n'}C$ for any recurrent state $C = a_\mathbf{n}^{p_\mathbf{n}} C$, i.e. once $a_\mathbf{n}$ enters its limit cycle, *there is no operator (or, generally, composite thereof) that can generate a configuration that is a transient state of $a_\mathbf{n}$; **transients are inaccessible from recurrent states**, even using another operator.* [4.1] This is a consequence of the Abelian symmetry, since for any $C' = a_\mathbf{n'}C$ with $C = a_\mathbf{n}^{p_\mathbf{n}} C$

$$a_\mathbf{n}^{p_\mathbf{n}} C' = a_\mathbf{n}^{p_\mathbf{n}} a_\mathbf{n'} C = a_\mathbf{n'} a_\mathbf{n}^{p_\mathbf{n}} C = a_\mathbf{n'} C = C' \tag{4.6}$$

so that C' is recurrent under $a_\mathbf{n}$ (but not necessarily also recurrent under $a_\mathbf{n'}$ as illustrated above). If all sites are charged with a finite probability, i.e. all operators $a_\mathbf{n}$ appear with a non-vanishing frequency, then all operators will eventually reach their space of recurrent states, i.e. all states that appear are recurrent with respect to all operators (jointly recurrent states). In general, the entire space of recurrent states of one operator cannot be accessed through that operator alone, i.e. there is generally no q such that $C' = a_\mathbf{n'}C$ can be accessed from C using $a_\mathbf{n}$ only, $C' = a_\mathbf{n}^q C$, even if C' is recurrent under $a_\mathbf{n}$, so that $a_\mathbf{n}^{p_\mathbf{n}} C' = C'$.

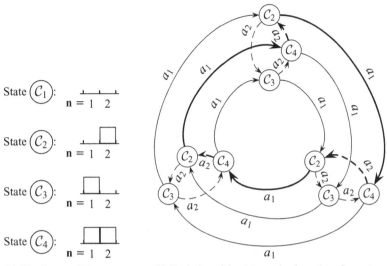

(a) The four possible states.   (b) Evolution of the states under the action of operators.

Figure 4.5 The one-dimensional, Abelian BTW Model with $L = 2, z^c = 1$ and open boundaries. The 4 stable states of this system are shown in (a). The effect of the two operators a_1 (solid lines) and a_2 (dashed lines) on the recurrent states is illustrated in (b). The circles indicate the states shown in (a) and the arrows between them the action of the operators. State C_1 is transient with respect to both operators. Operator a_1 visits $\ldots, C_4, C_3, C_2, C_4, C_3, C_2, \ldots$, whereas a_2 takes the reverse order $\ldots, C_2, C_3, C_4, C_2, C_3, C_4, \ldots$. Various identities can be read off, such as $a_1^2 = a_2$ and $a_2^2 = a_1$, as well as $a_1 a_2 = 1$. Applying alternately $a_1 a_2 a_1 a_2 \ldots$ (bold arrows) explores only 2 of 3 recurrent states.

If two states C and C' are related by actions of some a_{n_i}, say $C' = a_{n_1} a_{n_2} \ldots a_{n_m} C$, then the Abelian property guarantees the same periodicity for C' as for C, i.e. if $a_n^{p_n} C = C$ then $a_n^{p_n} C' = C'$. The existence of an inverse (in the space of recurrent configurations) means that C can in turn be written in terms of some a_{n_i} acting on C', hence a unique periodicity p_n of a given operator a_n. Yet, periods p_n generally differ for different $\mathbf{n}$.

In the ASM the set of states which are recurrent with respect to all operators can be arranged into an N dimensional torus (N being the number of sites), where each dimension represents the action of an operator a_n, see Fig. 4.5. Repeatedly applying a_n to a state on the torus will produce a circular trajectory, of a radius equal to the periodicity of a_n. If driven sites and therefore the a_n are picked at random and independently, each with probability $c_n > 0$, then eventually the entire torus will be covered homogeneously (Dhar, 1999c). The situation is more complicated if some c_n vanish, or if the system is driven deterministically.[11] Alternating, for example, between two driven sites will produce a set of states that depends crucially on whether the periodicities of the two operators are relative prime or not, resulting in partial or complete coverage of the two-dimensional torus of

---

[11] The fully deterministic Abelian BTW Model has been studied for example by Wiesenfeld *et al.* (1990) and more recently by Lübeck, Rajewsky, and Wolf (2000) and Creutz (2004).

configurations accessible to the two operators. For example, applying a_1 and a_2 alternately in Fig. 4.5 leads to only 2 states out of 3 being visited at all.

A state $\mathcal{C}$ generally appears more than once on the torus, i.e. there are integers q and q' different from the respective periods $p_{\mathbf{n}}$ and $p_{\mathbf{n'}}$, such that $a_{\mathbf{n}}^q a_{\mathbf{n'}}^{q'} \mathcal{C} = \mathcal{C}$. This can be seen in Fig. 4.5, where $a_1 a_2 \mathcal{C} = \mathcal{C}$ and $a_1^2 a_2^2 \mathcal{C} = \mathcal{C}$ for all recurrent configurations $\mathcal{C}$.

The Abelian symmetry and the existence of an inverse means that if this holds for one particular $\mathcal{C}$, it must hold for all states accessible through actions of other operators, i.e. in particular for all jointly recurrent states. Each state therefore appears equally often on the torus and at homogeneous coverage is therefore equally likely to occur. The outline of a more formal proof is presented in Sec. 8.3.1.

Regarding the avalanche size, two operators commuting does not imply that the set of avalanche sizes produced is independent of the order in which the operators are applied. The avalanche sizes generally will depend on the order if there is at least one site that topples during the avalanches and is charged, directly or indirectly, by both operators. Yet, the sum of the avalanches, i.e. the total number of topplings, is invariant under a change of the order. [4.2]

The proof of the Abelian symmetry is based on the dynamics of the ASM on the microscopic time scale, i.e. within an individual avalanche, the order in which unstable sites, $z_{\mathbf{n}} > z_{\mathbf{n}}^c$, are relaxed does not affect the final state either. This is consistent with representing a configuration containing a certain set of unstable sites $\{\mathbf{n}_1, \mathbf{n}_2, \ldots, \mathbf{n}_n\}$ by a set of operators $\{a_{\mathbf{n}_1}^{q_1}, a_{\mathbf{n}_2}^{q_2}, \ldots, a_{\mathbf{n}_m}^{q_m}\}$ with $q_i = z_{\mathbf{n}_i} - z_{\mathbf{n}_i}^c > 0$ acting on a stable configuration. The operators in this set can be applied in any order because they commute, thereby greatly simplifying the numerical implementation of the ASM. At any point in microscopic time, the (sub-)avalanche caused by a particular toppling can be followed through until the system is stable again and other charges are processed. The sum of the sizes of the sub-avalanches is independent of the order in which the operators $a_{\mathbf{n}_i}$ are applied.

4.2.2 Analytical results for the two-dimensional Abelian BTW Model

The Abelian BTW Model on a two-dimensional square lattice is the ASM with the particular toppling matrix

$$\Delta_{\mathbf{nn'}} = \begin{cases} 4 & \text{for } \mathbf{n} = \mathbf{n'} & \text{(4.7a)} \\ -1 & \text{for } \mathbf{n'} \text{ nearest neighbour of } \mathbf{n} & \text{(4.7b)} \\ 0 & \text{otherwise} & \text{(4.7c)} \end{cases}$$

which, in the bulk, is in fact a **lattice Laplacian**. As mentioned above, if a boundary site $\mathbf{n}$ topples, it sheds 4 particles, while only its nearest neighbours receive one, thereby implementing open boundaries throughout. The Abelian BTW Model differs from the BTW Model therefore not only with respect to the driving but also regarding the dissipation mechanism. [12] A large number of analytical results are available for this special case. In particular, Majumdar and Dhar (1991) derived the asymptotic fraction of sites with $z_{\mathbf{n}} = 1$

---

[12] With all boundaries being equally lossy, the translation of particles (as slope units) to height units is plagued by physical inconsistencies, such as the one shown in Fig. 4.3.

in the thermodynamic limit, as well as the probabilities of certain local configurations, such as $z_{\mathbf{n}} = 1$ neighbouring $z_{\mathbf{n}} = 2$ and more complicated ones. Most interestingly, they showed that the Abelian BTW Model displays **algebraic correlations**: if $I_1(\mathbf{n})$ is an indicator function,

$$I_1(\mathbf{n}) = \begin{cases} 1 & \text{if } z_{\mathbf{n}} = 1 \\ 0 & \text{otherwise} \end{cases} \qquad\qquad (4.8a) \\ (4.8b)$$

then

$$\langle I_1(\mathbf{n}) I_1(\mathbf{n}') \rangle - \langle I_1(\mathbf{n}) \rangle \langle I_1(\mathbf{n}') \rangle = -\frac{\langle I_1(\mathbf{n}) \rangle^2}{2|\mathbf{n} - \mathbf{n}'|^4} + \mathcal{O}(|\mathbf{n} - \mathbf{n}'|^{-6}). \qquad (4.9)$$

The average $\langle \cdot \rangle$ is taken over all recurrent states. The probability for a site to be in state 1 is derived as $\langle I_1(\mathbf{n}) \rangle = 2/\pi^2 - 4/\pi^3$. In higher dimensions it is shown that the anti-correlations (see sign in Eq. (4.9)) decay like r^{-2d} for large distances $|\mathbf{n} - \mathbf{n}'| = r$. Although the decay is algebraic, the exponent is so large that the state of sites can be regarded as uncorrelated on large length scales, see Sec. 2.3.1 (but Christensen, 1992). Numerical results (Grassberger and Manna, 1990; Manna, 1990) are consistent with the conclusion that there are essentially no long-range spatial correlations in the ASM.

Priezzhev (1994) hugely extended the work by Majumdar and Dhar by deriving the fractions of sites with slope $z_{\mathbf{n}} = 2, 3, 4$ in closed form using very elaborate combinatorial techniques.

The ASM was put into the context of **conformal field theory** (CFT) very early on by Majumdar and Dhar (1992). Based on the burning test, they showed that allowed configurations in the ASM are spanning tree configurations, which in turn is a Potts Model in the limit $q \to 0$ (Dhar, 1999a), described by a CFT with central charge $c = -2$. Assuming power law behaviour of observables such as the avalanche size and duration, they derived various scaling relations. However, they could not derive the exponents or show that the statistics actually is characterised by power laws at all. The CFT approach was exploited further by Jeng (2005b) in the most impressive way, calculating a vast number of correlation functions and correcting some important details in previous derivations (Ivashkevich, 1994, also Mahieu and Ruelle, 2001; Ruelle, 2002; Jeng, 2005a; Jeng, Piroux, and Ruelle, 2006; Saberi, Moghimi-Araghi, Dashti-Naserabadi, and Rouhani, 2009; Azimi-Tafreshi et al., 2010).

An important step towards an analytical characterisation of the scaling of the avalanche size distribution was the introduction of a so-called **inverse avalanche** by Dhar and Manna (1994). In a more direct approach, Ivashkevich, Ktitarev, and Priezzhev (1994b, also Priez-zhev et al., 1996b; Ivashkevich and Priezzhev, 1998) showed that individual avalanches in the Abelian BTW Model can be decomposed into **waves**, the scaling of which can be determined exactly. Every site that topples at all during an avalanche topples at most once, as long as the first toppling site topples only once. The sequence of topplings of the first, driven site naturally gives rise to a decomposition of an avalanche into waves. In two dimensions, Dhar and Manna (1994) derived $\tau = 11/8$ for the exponent describing the size distribution of the last wave (more results are reviewed briefly by Hu and Lin, 2003). Based on exact results for the waves and *assuming* various scaling behaviour (but not the actual value of the exponents), Priezzhev *et al.* (1996b) derived the avalanche size exponent $\tau = 5/4$ for individual waves and $\tau = 6/5$ for the total number of topplings in two dimensions. There is only a single wave for avalanches started at corners of angle θ, which are therefore *exactly* characterised by the avalanche size exponent $\tau = 1 + \pi/(2\theta)$ and $D = 2$ (Ivashkevich, Ktitarev, and Priezzhev, 1994a, illustrated by Stapleton, 2007, pp. 149–157). *For boundary drive, the two-dimensional Abelian BTW Model is thus exactly solvable, displays power law scaling, is fully characterised, but lacks multiple topplings (Sec. 8.4.4) and therefore interaction.* In higher spatial dimensions d, boundary drive leads to $D = d$, which can only hold up to $d = 4$, when D reaches the mean-field value of the ASM (Dhar and Majumdar, 1990; Janowsky and Laberge, 1993; Nakanishi and Sneppen, 1997). [13]

The angle $\theta = 2\pi$ corresponds to a toppling triggered at the end of a half-line, which might be expected to give rise to fewer topplings than in the bulk and thus might imply the area exponent being bounded from above by $5/4$ (Dhar, 1999a). De Menech and Stella (2000) found evidence that unlike in the two-dimensional BTW Model, in the three-dimensional BTW Model the overall avalanche exponent should coincide with that characterising waves, which had been discussed in great detail earlier (Ktitarev, Lübeck, Grassberger, and Priezzhev, 2000).

Waves are compact, i.e. have no holes (Ivashkevich *et al.*, 1994b; Priezzhev *et al.*, 1996b; Karmakar *et al.*, 2005). A site surrounded by neighbours, all of which have toppled at least n times, must have toppled at least n times as well, after having received their charges. In turn, a site all of whose neighbours have toppled not more than n times, cannot have received more charges to have toppled more than n times. As a result, the sites that have toppled $n = 1, 2, 3, \ldots$ times can be ordered in strata and then form terraced, compact layers resembling a wedding cake structure (Milshtein *et al.*, 1998; Biham *et al.*, 2001); the number of distinct sites toppling, the avalanche area, thus is the size of the first wave. As avalanches are formed from waves, they are compact as well, but compactness does not imply that the dimension is the same as that of the embedding space (Nakanishi and Sneppen, 1997).

Interestingly, the numerics (Manna, 1991a) quoted by Priezzhev *et al.* (1996b) in support of $\tau = 6/5$ from the results on waves was considered a 'one-off' by Majumdar and Dhar (1992). Based on a numerical study, Paczuski and Boettcher (1997) criticised some of the assumptions made by Priezzhev *et al.* (1996b). Dhar (1999a), discussing other lines of critique as well, pointed out that the scaling suggested by Paczuski and Boettcher (1997) leads to $\tau = 1$. One might thus summarise that the scaling behaviour of the BTW Model remains disputed.

Further exact results for the two-dimensional Abelian BTW Model were derived by Jeng *et al.* (2006) who conjectured [14] the average slope, $N^{-1} \sum_n z_n$ in a system with N sites, to be $25/8$ for $z^c = 4$ being stable (also Priezzhev, 1994; Vespignani, Dickman,

---

[13] The mean-field theory developed by Janowsky and Laberge (1993) also establishes a link to a zero range process.

[14] The reason why the result is not exact is that it relies on a numerical integration which gives the conjectured value within 10^{-12}.

Muñoz, and Zapperi, 2000; Dhar, 2006). Many more properties of the ASM are known on trees, i.e. Bethe lattices (Dhar and Majumdar, 1990), ring-structures (Fey, Levine, and Wilson, 2010b, on the basis of local toppling invariants) and other (especially one but also higher dimensional) lattices (Fey-den Boer and Redig, 2005; Meester and Quant, 2005; Fey, Meester, and Redig, 2009; Fey, Levine, and Peres, 2010a; Fey, Levine, and Wilson, 2010c). Most interestingly, for certain lattices Fey *et al.* (2010b) were able to *prove* that the fixed-energy sandpile (Sec. 9.3.1 and Sec. 9.3.4) does *not* have the same average slope as the corresponding ASM (but Jo and Jeong, 2010).

Most of the analytical results so far are, accepting certain assumptions, exact. Pietronero *et al.* (1994, also Díaz-Guilera, 1994; Vespignani and Zapperi, 1995; Ben-Hur and Biham, 1996; Corral and Díaz-Guilera, 1997; Lin and Hu, 2002) have developed an approximate scheme to determine the scaling of the BTW Model, sometimes refered to as a **dynamically driven renormalisation group** (DDRG). The results are included in Table 4.1. The method was inspired by **real space renormalisation group** (RSRG) methods as used in equilibrium critical phenomena. [15] Similar attempts have been made to analyse the ASM on fractal lattices (e.g. Daerden and Vanderzande, 1998, also Kutnjak-Urbanc *et al.*, 1996).

SOC models suffer from being defined solely through a set of rules, so that standard methods, for example via a Langevin equation or a Fokker–Planck equation, are not readily available. The RSRG approach by Pietronero *et al.* (1994) therefore is somewhat ad hoc, ignoring multiple topplings (Grassberger and Manna, 1990; Ben-Hur and Biham, 1996; Lübeck and Usadel, 1997a; Lübeck, 2000; Paczuski and Bassler, 2000) and assuming, for example, that the dynamics of the Abelian BTW Model on a coarse grained level is fully captured by a five-component vector-field. This vector describes, locally, how many sites are 'critical' (topple at the next charge) and how many neighbouring sites are hit by a toppling. To this extent the scheme ignores spatial correlations, which can be justified, see Sec. 4.2.2, p. 99. Even if spatial correlations can safely be ignored, however, temporal correlations between sites during an avalanche might not. It is remarkable that non-trivial exponents are recovered seemingly without accounting for any spatiotemporal correlations.

Many of the analytical results above are equally applicable to the so-called Abelian Distributed Processor Model (Dhar, 1999a), which is a generalisation of the ASM. Other generalisations have been suggested by Chau and Cheng (1993, also Chau, 1994).

4.2.3 A directed sandpile

Dhar and Ramaswamy (1989) introduced a **directed version** of the BTW Model, which is exactly solvable. In this variant of the ASM (reviewed in Dhar, 1999a,c, 2006), toppling sites on a hyper-cubic lattice redistribute their slope units in one direction only. For example, if site $\mathbf{n} = (x_1, x_2, \ldots, x_d)$ topples, then the receiving nearest neighbours are $\mathbf{n}' = (x_1 + 1, x_2, \ldots, x_d), \mathbf{n}' = (x_1, x_2 + 1, \ldots, x_d), \ldots, \mathbf{n}' = (x_1, x_2, \ldots, x_d + 1)$. The directed ASM is in a different universality class than directed percolation, but can be obtained in the limit of vanishing 'stickiness' in a generalised model which generically belongs to the DP

---

[15] They are sometimes considered a 'black art', because it is very difficult to improve results in a controlled fashion (Burkhardt and van Leeuwen, 1982).

universality class for a finite degree of stickiness (Dhar, 1999a). In Sec. 8.4 various directed models are discussed and compared.

The directed ASM is equivalent to a model by Takayasu (1989) and in two dimensions to a much older model by Scheidegger (1967). In both models, there is no obvious separation of time scales, so that Dhar (1999a) concluded that a separation of time scales should not be regarded a necessary condition for SOC. This conclusion can be questioned. In fact, solved models, which are directed or develop into a product state, can usually be mapped to some properties of random walkers, which develop under dynamics that do not make (explicit) use of a separation of time scales. For example, random neighbour models often map to random walkers along an absorbing wall (Sec. 8.1.4) which simply do not possess any temporal or spatial scales to be separated. Rather, a separation of time scales is a *prerequisite* for the very mapping of the SOC model – the mapping is applicable only in the critical state of the SOC model which generally requires a separation of time scales. Whether such random walker models themselves are instances of SOC is best left as a matter of definition.

From this perspective, for the separation of time scales to be regarded as irrelevant, it is not enough that it should be absent in a mapped model, it also needs to be irrelevant in the mapping itself. In the present case that does not seem to apply.

4.2.4 Numerical results

The Abelian version of the BTW Model is undoubtedly the best studied model of SOC. Because it is trivial in one dimension, most attention has focused on the two-dimensional model. Most publications on numerical results for the BTW Model model contain estimates for exponents describing the scaling of PDFs for many different observables, such as the avalanche size and duration, but also the **avalanche area**, that is the number of distinct sites toppling, as well as the **radius of gyration**, both of which show a much cleaner scaling behaviour than avalanche size and duration (e.g. Lübeck, 2000). Nevertheless, the following discussion focuses on the avalanche size and duration. The results are summarised in Table 4.1, p. 92.

Unfortunately, the two-dimensional Abelian BTW Model is notorious for producing inconsistent numerical results, possibly due to strong logarithmic corrections (Manna, 1990; Lübeck and Usadel, 1997b; Lübeck, 2000). The first avalanche size exponent reported for it (Bak *et al.*, 1987) was $\tau \approx 0.98$ (however, this was for the non-Abelian, original version), which is impossible as an asymptote (Sec. 2.1.2.2). This estimate was, like those reported by Bak *et al.* (1988), based on a naïve straight line fit (Sec. 2.1.2.1 and Sec. 7.4.2) in a double-logarithmic plot of the avalanche size distribution, which was in common usage (e.g. Christensen *et al.*, 1991) until the **moment analysis** was made popular by De Menech *et al.* (1998, also Sec. 7.3). Ironically, Jensen *et al.* (1989, also Christensen *et al.*, 1991) found that the power spectrum of the activity (based on the distribution of weighted avalanche durations) in the two-dimensional Abelian BTW Model is much better described by $1/f^2$ than by $1/f$ (see Sec. 1.3.2), which was the original conclusion by Bak *et al.* (1987, also Bak *et al.*, 1988). Interestingly, a similar result was found experimentally by Jánosi and Horváth (1989, but also Jaeger *et al.*, 1989; Held *et al.*, 1990). Kertész and Kiss (1990) confirmed these results and introduced a criterion for the occurrence of $1/f^2$ noise, based on a relation involving other exponents; temporal scaling does not guarantee a non-trivial power spectrum, just like power law correlations do not guarantee a deviation from Gaussian long-range behaviour, see Sec. 4.2.2.

In particular during the early days of SOC, some authors considered the cluster size distribution measured as being weighted by the cluster size, which is common practice in percolation (Christensen and Olami, 1993; Stauffer and Aharony, 1994). There, $n(s)$ is the site normalised cluster size distribution $n(s)$, which is the average number of clusters of size s per site, and scales like $s^{-\tau}$. A randomly chosen occupied site belongs to a cluster of size s with probability $sn(s) \propto s^{1-\tau}$. If the avalanche size distribution $\mathcal{P}^{(s)}(s)$ is assumed to follow $s^{1-\tau}$, rather than $s^{-\tau}$, the exponent quoted is off by one. All exponents in Table 4.1 (and all other tables) are converted to the same convention where necessary, which is indicated by a '$\triangleright$'. The situation can become very confusing when authors use the term 'distribution' without saying whether it refers to 'cumulative distribution' as commonly used in statistics, or probability density function. If $\mathcal{P}^{(s)}(s)$ scales like $s^{-\tau}$, its cumulative distribution scales like $s^{1-\tau}$, so that the exponent τ quoted is *not* off by one, even when $s^{1-\tau}$ looks like the 'old style convention'.

Manna (1990) analysed the second moment of the avalanche size distribution and found $\sigma_2^{(s)} = 4.79$, so that together with $\sigma_1^{(s)} = 2$ (for random drive, Sec. 8.1.1.1 and Dhar, 1990a) and $\sigma_n^{(s)} = D(1 + n - \tau)$ (Sec. 2.2) this entails $D = 2.79$ and $\tau = 1.28$, whereas the slope suggests $\tau = 1.22$. Similarly $\alpha = 1.38$ for the avalanche duration (see Sec. 1.3), for which no exact result is known, so that z can be determined only by making additional assumptions, such as $(1 - \tau)D = (1 - \alpha)z$ (Sec. 2.2.2), which gives $z = 2.06$. Chessa et al. (1999a) published a critique and revision of earlier numerical results, which they rejected as biased, and concluded that the BTW Model and the MANNA Model are in the same universality class. In addition, they identified a lack of FSS for the avalanche duration (also Lübeck, 2000).

Dorn, Hughes, and Christensen (2001) used much bigger lattices and more sophisticated methods to measure τ as the slope of the avalanche size distribution in a double-logarithmic plot and found $\tau = 1.22$, consistent with the value found by Manna (1991a), where a more accurate estimate of $\tau = 1.2008$ is reported as well. Lübeck and Usadel (1997b) found $\tau = 1.293(9)$ for the avalanche size and $\alpha = 1.48(1)$ for the duration, again using a more sophisticated method for determining the slope in a double-logarithmic plot. They suggested some exact values for the exponents, some of which clash with the earlier exact result by Priezzhev et al. (1996b), notably the exponent characterising the number of distinct sites toppling (4/3 according to Lübeck and Usadel, 1997b, compared to 5/4 by Priezzhev et al., 1996b).

Based on a moment analysis, De Menech et al. (1998) found $\tau = 1.2$ and $D = 2.5$, but also evidence for the BTW Model displaying multiscaling. This observation is based on the FSS of the nth moment in the range $0 < n < 1$, which, however, is affected by the lack of scaling for $n < \tau - 1$ and the strong corrections around $n = \tau - 1$ (Sec. 7.3, Fig. 7.5 and Sec. 2.2, p. 37). Tebaldi et al. (1999) confirmed this result based on a more sophisticated analysis (also Stella and De Menech, 2001). Drossel (2000) found further evidence for the failure of FSS and the hypothesis of simple scaling in the ASM, based on numerical results and scaling assumptions, studying dissipative and non-dissipative avalanches separately, with the former displaying somewhat 'cleaner' power laws than the latter. [16] Lübeck (2000) found in a similar analysis, that the avalanche dimension D and the dynamical exponent z are not well defined for the two-dimensional Abelian BTW Model at all. Karmakar *et al.* (2005) identified a multiscaling universality class of the Abelian BTW Model, i.e. they found another model which displays equally complicated scaling behaviour as the BTW Model.

In higher dimensions $d > 2$, the BTW Model seems to be better behaved as there are no claims of a lack of simple scaling. Bak *et al.* (1987) reported exponents $\tau = 1.35$ and $\alpha \approx 1.59$ [17] (again for the non-Abelian version) for $d = 3$, while Christensen *et al.* (1991) found $\tau = 1.37$ and $\alpha = 1.625$. In a moment analysis Lübeck (2000) estimated $\tau = 1.33$, $D = 3.004$, $\alpha = 1.63$ and $z = 1.62$ using $(1-\tau)D = (1-\alpha)z$. The upper critical dimension is reported as $d_c = 4$ (Grassberger and Manna, 1990; Lübeck and Usadel, 1997a; Lübeck, 1998; Priezzhev, 2000), but as $d_c = 5$ by Christensen (1992). At this dimension the model displays the expected mean-field behaviour (Sec. 8.1.3 and Vergeles *et al.*, 1997).

4.3 The ZHANG Model

The ZHANG Model (Zhang, 1989, Box 4.3) is one of the oldest models of SOC. Similar to the BTW Model, its dynamics is deterministic. Unlike the BTW Model, however, it has a continuous local degree of freedom associated with each site. The ZHANG Model was the first such model, making it accessible to a wide range of analytical tools. A degree of discreteness is left in the form of the finite **energy increment** δ (cf. the OFC Model, Sec. 5.3), which, however, is broken down by the dynamics very efficiently. The ZHANG Model is usually studied in the stationary state, which is, as in the ASM, reached within a time proportional to the number of sites (Giacometti and Díaz-Guilera, 1998).

The central issue of the ZHANG Model is its **non-Abelian** nature, which is a consequence of the mechanism by which the energy increments are broken down. The ZHANG Model was designed to be in the same universality class as the BTW Model, but eventually turned out to be characterised by different exponents, which could mean that the Abelian symmetry is one of the determining factors of the universality class. As a consequence of the lack of Abelian symmetry, the dynamics of the ZHANG Model must be regarded as an integral part of its definition. Different 'choices' for the updating mechanism, such as parallel or random sequential, and the problems arising on lattices where neighbouring sites can topple simultaneously, can and have led to different results, not only for its dynamical, but also for its static properties (Díaz-Guilera, 1992; Pastor-Satorras and Vespignani, 2000a).

---

[16] The preprint (Drossel, 1999b) is less cautious and much more strongly worded than the published version (Drossel, 2000) and the first version of this preprint, Drossel (1999a), is almost dismissive about the ASM: 'It is the purpose of this letter to show that these [avalanche] exponents do not exist, ... '

[17] The exponent α needs to be derived from the weighted distribution of avalanche durations, which gives $\alpha + \gamma$ and $\gamma = (\tau - \alpha)/(2 - \tau)$ in their notation.

Box 4.3   The ZHANG Model

General features

- First published by Zhang (1989).
- Motivated by the BTW Model (Sec. 4.1), a *continuous* model supposedly in the same universality class.
- Stochastic (bulk) drive and deterministic relaxation.
- Non-Abelian.
- Simple scaling behaviour disputed.
- In the limit of vanishing energy increment equivalent to the conservative OFC Model, Sec. 5.3.
- Upper critical dimension $d_c = 4$ (Jánosi, 1990).
- Exponents listed in Table 4.2, p. 106 are for the original definition as well as for random energy increment.

Rules

- Lattice with coordination number q ($q = 2d$ in the hyper-cubic case).
- Each site $\mathbf{n} \in \{1, \ldots, L\}^d$ (on a hyper-cube) has local energy $E_\mathbf{n}$.
- Choose critical energy $E_{\max} = 1$ without loss of generality.
- Choose fixed-energy increment $\delta \in [0, E_{\max}]$ (in a variant this is chosen at random in every driving step).
- *Initialisation*: irrelevant, model studied in the stationary state.
- *Driving*: pick site $\mathbf{n}_0$ at random and add energy increment δ, so that $E_{\mathbf{n}_0} \to E_{\mathbf{n}_0} + \delta$.
- *Toppling*: for each site $\mathbf{n}$ where $E_\mathbf{n} \geq E_{\max}$ (active site), add $E_\mathbf{n}/q$ to every nearest neighbour $\mathbf{n}'$, $E_{\mathbf{n}'} \to E_{\mathbf{n}'} + E_\mathbf{n}/q$, then remove all energy from the toppling site, $E_\mathbf{n} \to 0$.
- *Dissipation*: energy is lost at open boundaries.
- *Parallel update*: discrete microscopic time, sites becoming active at time t topple at time $t + 1$ (updates in sweeps).
- *Separation of time scales*: drive only once all sites are stable, i.e. $E_\mathbf{n} \leq E_{\max}$ (quiescence).
- *Key observables* (see Sec. 1.3):
  avalanche size s, the total number of topplings until quiescence;
  avalanche duration T, the total number of parallel updates until quiescence.

   With the dynamics being an integral part of the definition of the model, the definition of the **microscopic time** becomes part of the model as well. In the ZHANG Model, an avalanche starts with driving the model at microscopic time t by adding an energy increment δ. If the resulting energy exceeds the threshold $E_{\max}$, the driven site topples at $t + 1$, possibly triggering new topplings at $t + 2$ and so on.

   Giacometti and Díaz-Guilera (1998) suggested that a random, uniformly distributed energy increment $\delta \in [0, 1/q]$ is statistically better behaved, yet produces the same critical behaviour as fixed δ. This finding is surprising, because random energy increments cause a breakdown of the correspondence to the BTW Model in one dimension and a breakdown of the correspondence to the OFC Model in the limit $\delta \to 0$. Lübeck (1997) identified a more complex dependence of the critical exponents on δ, which disappears for sufficiently

Table 4.2 Exponents of the ZHANG Model. $\mathcal{P}^{(s)}(s) = a^{(s)}s^{-\tau}\mathcal{G}^{(s)}(s/(b^{(s)}L^{D}))$ is the avalanche size distribution and $\mathcal{P}^{(T)}(T) = a^{(T)}T^{-\alpha}\mathcal{G}^{(T)}(T/(b^{(T)}L^{z}))$ is the avalanche duration distribution, Sec. 1.3. Scaling relation: $(1-\tau)D = (1-\alpha)z$ (sharply peaked joint distribution, widely assumed) in all dimensions (Sec. 2.2.2). Analytical arguments suggest $z = (d+2)/3$ and $\tau = 2 - 2/d$ (Zhang, 1989). MFT is that of the BTW Model.

d	τ		D		α		z	
2	1.2	(a)	1.46	(c)	1.512(14)	(b)	1.36(3)	(g)
	1.282(10)	(b)	1.84(6)	(d)			≈1.365	(h)
	1.25(1) ▷	(c)	3.39	(f)			1.2	(i)
	1.288(19)	(d)					1.19(4)	(d)
	1.26(2) ▷	(e)					1.34(2)	(d)
							1.74	(f)
3	1.55	(a)	2.54(9)	(d)			1.34(4)	(d)
	1.454(41)	(d)					1.65(2)	(d)
4	1.9	(a)						
MFT	3/2 ▷	(k)			2	(k)		

▷ Reported as $\tau + 1$.

a Jánosi (1990), b Lübeck (1997), c Milshtein *et al.* (1998), d Giacometti and Díaz-Guilera (1998), e Biham *et al.* (2001), f Pastor-Satorras and Vespignani (2000a), g Díaz-Guilera (1992), h Díaz-Guilera (1994), i Ghaffari and Jensen (1996), k Vergeles *et al.* (1997).

small δ. Biham *et al.* (2001) considered stochastic variants of the ZHANG Model, one of which was shown to be Abelian in a restricted sense, and found that they belonged either to the universality class of the ASM (Abelian case) or to the universality class of the MANNA Model (stochastic case).

4.3.1 Relation to the BTW Model

One reason for the development of the ZHANG Model was to find an analytically tractable system which is in the same universality class as the BTW Model. A number of numerical studies address the question whether that is actually the case, with far from unambigious outcomes. Zhang (1989) initially found that his model's exponents are compatible with those of the BTW Model in two and three dimensions. Based on more extensive simulations, some authors (Lübeck, 1997; Biham *et al.*, 1998; Giacometti and Díaz-Guilera, 1998) confirmed that both models share the same universality class in two dimensions, while Giacometti and Díaz-Guilera (1998) found disagreement in three dimensions. They also found that the scaling of the characteristic time scale of the spatial variance of the energy is different from that of the characteristic time scale of the avalanches. That is slightly worrying, because one would normally expect a single characteristic time scale; its absence suggests that the ZHANG Model does not follow what is commonly expected from simple scaling.

One of the most interesting questions that arises for the ZHANG Model is whether it is its non-Abelian nature which causes its scaling behaviour to differ from the BTW Model, despite the arguably strong similarities between them. This question (see also Sec. 8.3) has also been addressed for certain directed sandpile models (Paczuski and Bassler, 2000; Kloster *et al.*, 2001; Hughes and Paczuski, 2002; Zhang, Pan, Sun, *et al.*, 2005a). That Zhang (1989) did not discuss this important issue when he proposed the model is hardly surprising, since the whole notion of Abelian versus non-Abelian was introduced with the ASM about a year later (see Sec. 4.2).

Pastor-Satorras and Vespignani (2000a) found in a moment analysis that the relevant observables in the ZHANG Model do not display simple scaling, which renders a comparison with the BTW Model almost impossible. Given the ambigious numerical evidence for the scaling of the BTW Model, one might wonder whether both models, the BTW as well as the ZHANG Model, suffer from the same lack of scaling.

4.3.2 Non-Abelian dynamics

The ZHANG Model is non-Abelian because the amount of energy redistributed during a toppling event depends on its current value and it is therefore crucial whether a neighbouring site topples before or after a given site. If two nearest neighbours of site $\mathbf{n}$ are active, i.e. have a local energy greater than or equal to $E_{\max}$, then the charge due to a toppling of $\mathbf{n}$ depends on which of the active neighbours had toppled by the time $\mathbf{n}$ topples. This property makes the ZHANG Model slightly more difficult to handle numerically, because sites toppling at t and at $t + 1$ have to be kept separate.

In its original definition, the ZHANG Model is defined only for hyper-cubic lattices. Prima facie this is a minor detail, however hyper-cubic lattices [18] have the useful property that they naturally decompose into even and odd **sublattices**: no pair of nearest neighbours of a given site are themselves nearest neighbours of each other. Consequently, in the ZHANG Model on hyper-cubic lattices, toppling sites can trigger activity only in nearest neighbouring sites, which are themselves never nearest neighbours to each other, so that toppling sites never topple onto each other. This avoids any ambiguity in the order of updates within a time step t.

A priori it is not obvious how to redefine the ZHANG Model to render it Abelian. If the amount of energy redistributed among the neighbours is fixed, one effectively recovers the BTW Model. Destroying this correspondence by randomising the charges [19] spoils the deterministic character of the model.

4.3.3 Energy histogram

While the slope variable in the BTW Model can obtain only discrete values, in the ZHANG Model the energy $E_{\mathbf{n}}$ at a site is continuous and is a priori not bounded from above while an avalanche is running. For sufficiently large lattices, the energy increment δ is broken down during an avalanche to arbitrarily small fractions. Nevertheless, charges are bounded

---

[18] And equally vertices of a hexagonal lattice, as opposed to other lattices such as the triangular one.
[19] The continuous OSLO Model (Pruessner, 2003b) is an example of a *stochastic* model that is both continuous and Abelian.

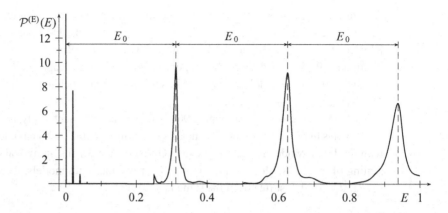

Figure 4.6 The histogram $\mathcal{P}^{(\mathrm{E})}(E)$ of the energy E in a two-dimensional Zhang Model of linear size $L = 128$ and with energy threshold $E_{\max} = 1$ and increment $\delta = 0.02$. The peak at $E = 0$ is cut off. The vertical dashed lines mark the peaks with constant gap $E_0 \approx 0.312 > E_{\max}/4$. The peaks become increasingly blurred towards higher energies. Close to $E = 0$ the energy increment δ has produced small 'echoes' of the peak at $E = 0$. Their height in the figure depends on the resolution of the histogram.

from below, because at least $E_{\max}/q$ is deposited by a toppling site. This is reflected in the **histogram** of the energies $E_{\mathbf{n}}$ in the system, see Fig. 4.6. Sites that have not received any charge since their last toppling have energy $E_{\mathbf{n}} = 0$, no site has an energy in the open interval $(0, \min(\delta, E_{\max}/q))$ and sites with at least one charge from a neighbour have energy $E_{\mathbf{n}} \geq E_{\max}/q$. For not too small δ some 'echoes' of $E_{\mathbf{n}} = 0$ are visible in the histogram of energies (Jánosi, 1990; Lübeck, 1997) from sites that have received $\delta, 2\delta, 3\delta \ldots$ external driving charges. Their existence and distribution depends on the lifetime of an empty site. A sharp rise in the histogram can be seen at $E_{\max}/q$ due to sites that have received a single charge from a toppling neighbour. The locations of the peaks are *not* (quite) multiples of $E_{\max}/q$, as discussed below. They are blurred by the external drive and by charges larger than $E_{\max}/q$. They disappear completely when modifying the Zhang Model by a randomised energy redistribution at toppling (Pietronero, Tartaglia, and Zhang, 1991). It has been claimed that the finite width of the peaks is a finite size effect (Díaz-Guilera, 1992), which would imply that the Zhang Model on large scales is governed by discrete dynamics, very similar to the BTW Model. The peak repeats several times but becomes less and less structured with increasing energy. In total, q peaks can be identified, which become increasingly blurred with increasing anisotropy (Jánosi, 1990, also Azimi-Tafreshi, Dashti-Naserabadi, and Moghimi-Araghi, 2008).

The peaks, originally introduced as *quasi-**unit*** (emphasis in the original, Zhang, 1989), are a discretisation effect, due to the presence of a lattice and a finite drive δ. A continuum description lacks such peaks, because of the absence of a lattice, but also because transport in the continuum is usually diffusive and takes place in continuous time, i.e. not in 'chunks' of approximate size $E_{\max}/q$ (see discussion below). An understanding of the energy histogram, therefore, is highly important for the understanding of the implicit assumptions in a continuum description of the model.

At first sight, it might be surprising that the positions of the peaks are not multiples of E_{max}/q, corresponding to the charge of a site toppling at energy E_{max}. However, the majority of the toppling sites topple at a higher energy, therefore delivering a higher charge to nearest neighbours. In addition to this, at small external drives δ almost all sites will receive multiple external charges before receiving a charge by a nearest neighbour. These two contributions, the fraction $1/q$ of the energy of a toppling site and the total of the external drive, constitute the gap between the peaks. Based on a number of heuristic arguments, Lübeck (1997) derived the approximate gap size $E_{max}(q+1)/q^2$. The 'emergence of quasiunits' has been analysed in the one-dimensional Zhang Model by Sadhu and Dhar (2008).

4.3.4 Analytical approaches

The Zhang Model has been studied extensively by analytical methods. In the original letter (Zhang, 1989) a scaling analysis is traced out based on the energy flux through the system, which is essentially diffusive. The exponents found are $\tau = 2(1 - 1/d)$ and $z = (d+2)/3$, the latter in agreement with some numerical results, notably by Giacometti and Díaz-Guilera (1998).

The renormalisation group (RG) approach by Díaz-Guilera (1992, 1994, non-conservative version by Ghaffari et al., 1997), is one of the first attempts to employ methods used in dynamical critical phenomena, in particular growth equations, to an SOC model. Because of its continuum nature, the Zhang Model is well suited for this approach. Díaz-Guilera (1994, also Corral and Díaz-Guilera, 1997) unified and compared the Zhang Model with the BTW Model within the same continuum description. The first step towards an analytical treatment is to find an equation of motion for the energy field $E_{\mathbf{n}}$. This equation must incorporate the toppling, that is the **diffusive transport** of the energy through the system, triggered by the local energy exceeding a local threshold. One way of implementing this threshold-driven dynamics is via a **Heaviside step function**, $\theta(x)$; Díaz-Guilera found for the Zhang Model

$$\partial_t E(\mathbf{r}, t) = \alpha \nabla^2 \Big(\theta \left(E(\mathbf{r}, t) - E_{max} \right) E(\mathbf{r}, t) \Big) + \xi(\mathbf{r}, t) \tag{4.10}$$

after taking the continuum limit. The additive noise $\xi(\mathbf{r}, t)$ has vanishing mean and is δ-correlated. The fundamental difficulty is the non-analyticity of the Heaviside step function which needs to be regularised before applying the renormalisation group procedure (Pérez et al., 1996; Corral and Díaz-Guilera, 1997). It remains somewhat questionable whether such a **regularisation** is actually permitted. The importance of the threshold and the step function has been discussed (in broader terms) by Cafiero, Loreto, Pietronero, et al. (1995), who introduced the notion of 'local rigidity' and argued that the threshold has to remain finite even when taking the continuum limit, otherwise the model eventually behaves purely diffusively.

Even though some bold assumptions are necessary to start the RG procedure, such as the regularisation of the step function as well as the continuum description of the model itself, the exponents produced are impressively close to Zhang's (1989) original result. Moreover, they vary systematically, suggesting that by taking into account a sufficiently large number of couplings, results for the dynamical exponent are in good agreement with the suggested analytical result $z = (d + 2)/3$ even at one-loop level. Although Zhang (1989) and Díaz-Guilera (1994) used complementary methods, the former even neglecting some important spatiotemporal correlations, the two approaches are related in that they effectively consider the diffusive flux of energy through the system.

Ghaffari and Jensen (1996) argued that the extrapolation technique used by Díaz-Guilera to derive the exponent z which incorporates infinitely many couplings to one-loop order can be improved by considering the exponent z as a function of the deviation of the Heaviside step function from its regularisation. The regularisation itself is rather arbitrary, because

$$\lim_{\beta \to \infty} \frac{f(\beta x) - f_-}{f_+ - f_-} = \theta(x) \tag{4.11}$$

for any $f(x)$ which has two different limits $\lim_{x \to \pm \infty} f(x) = f_{\pm}$ with $f_+ \neq f_-$ (and provided $\theta(0)$ is matched as required). Using their extrapolation scheme, Ghaffari and Jensen found $z = 1.2$ in two dimensions which is closer to the theoretical finding of $z = 1.234$ by Pietronero *et al.* (1994, for the BTW Model) than $z \approx 1.365$ found by Díaz-Guilera (1994).

Because of its analytical tractability (a natural continuum description in conjunction with deterministic dynamics), the ZHANG Model has been analysed from many different points of view, for example its Lyapunov exponents (Cessac, Blanchard, and Krüger, 2001), its transformation properties on the macroscopic time scale (Blanchard, Cessac, and Krüger, 1997) or its 'limiting shapes' (Fey-den Boer and Redig, 2008).

4.3.5 Numerical results

In the light of the numerical findings, which are summarised in Table 4.2, one might question some of the above analytical results – or question that the model displays scale invariant behaviour at all. The numerical situation for the ZHANG Model is inconclusive, similar to the situation for the BTW Model. The fact that some authors change the dynamical rules for quicker convergence is a contributing factor to this confusion. In their favour one might argue that the original definition is not fully consistent, for example, on the precise choice of the energy increment, suggesting a fixed value as well as a continuous range.

Based on a moment analysis, Pastor-Satorras and Vespignani (2000a) came to the rather pessimistic conclusion that the ZHANG Model, at least in its original definition, violates simple scaling. The authors also studied a number of dynamical variants of the model, using random sequential updates as well as a stochastic version, where the energy is 'stochastically redistributed', similar to a continuous version of the MANNA Model. The resulting exponents confirm that the model is in the **MANNA universality class**. This is a remarkable result, because it places an Abelian model (namely the MANNA Model) and a non-Abelian model (namely the stochastic, parallel ZHANG Model) in the same universality class, which is known to be very robust, Sec. 6.1.

5 Dissipative models

Dissipation is a major theme in SOC for several reasons. Like every relaxation process, avalanching in a sandpile generally can be seen as a form of dissipation, quite literally so for the sand grains that dissipate potential energy in the BTW Model. In that sense, sandpile models are inherently dissipative. Yet, their dynamics can be expressed in terms of variables, which are conserved under the ***bulk* dynamics** (also '*local*' dynamics), such as the number of slope units in the Abelian BTW Model.

The models described in the present chapter, however, go a step further by obeying dynamical rules without local **bulk conservation** (also 'local' conservation), although all models develop towards a stationary state even in the non-conserved variable, i.e. overall there is asymptotic conservation on average. The observation of scale-free dissipation in turbulence triggered the development of the Forest Fire Model (Sec. 5.1), i.e. it was explicitly designed in a dissipative fashion. The situation is somewhat similar for the OFC Model (Sec. 5.3) which incorporates a dissipation parameter α, whereas in the BS Model (Sec. 5.4) dissipation is a necessary by-product. Both the OFC Model and the BS Model are examples of models driven by extremal dynamics, which consists of identifying the 'weakest link' among all sites and starting relaxation from there.

In the light of Hwa and Kardar's (1989a) work (Sec. 9.2.2), which suggested that scale invariant phenomena arise naturally, even generically in the presence of bulk conservation, the existence of non-conservative SOC models is particularly important. A mechanism explaining SOC, which necessarily requires conservation in the bulk, fails at these models and thus might be deemed either incomplete or wrong. The rôle of conservation in SOC continues to cause concern (e.g. Dickman *et al.*, 2000) and attract attention (e.g. Bonachela and Muñoz, 2009).

Because the particle density is often seen as a temperature-like control parameter (Sec. 1.2), which is tuned to the critical value by self-organisation, dissipation seems to spoil that very process and therefore needs to be compensated by another mechanism, such as particle influx (Vespignani and Zapperi, 1998). On a phenomenological level, dissipation reduces the activity in an avalanche and stops it from spreading throughout the system, thereby effectively imposing a cutoff (in particular Sec. 5.3), as illustrated by a mass term in a Gaussian Langevin equation (Bonachela and Muñoz, 2009). If dissipation takes place only at the boundaries and avalanches are triggered in the bulk, avalanches are bound to reach there (Paczuski and Bassler, 2000). This enforced (particle) flux through the lattice is absent or restricted in models with bulk dissipation. Similar to the models discussed in the previous chapter, there is limited solid numerical and no analytical evidence that any of the following models exhibits scale invariance.

5.1 The BAK–CHEN–TANG Forest Fire Model

The **Forest Fire Model** (FFM) is a completely different type of lattice model compared with the sandpile-like models discussed in Ch. 4, as there is **no flux of any locally conserved quantity** through the system (Sec. 9.2.2, p. 327). Moreover, as opposed to previous models, its driving is stochastic by definition. Finally, the notion of Abelian versus non-Abelian does not reasonably apply.

The dynamics of the FFM is governed by percolating clusters of burning trees, so that classic **percolation** (Stauffer and Aharony, 1994) plays an important rôle in its analysis. There are two main versions of the FFM, and from the literature it is not immediately obvious who actually invented the model. The earliest version was presented in a talk by Henley (1989) as 'self-organised percolation' in an attempt to create an SOC model that is even simpler than the BTW Model. The later version by Bak *et al.* (1990) published as 'a forest-fire model' (in the following called only 'FFM', see Box 5.1) made it popular. This original model will be presented first. The focus has shifted to the second model (see Box 5.2, p. 116) presented in this chapter, which is commonly known as the DS-FFM (Drossel and Schwabl, 1992a). It is now generally agreed that the FFM by Bak *et al.* is governed by very simple and well-understood dynamics similar to Turing patterns (e.g. Murray, 2003). Curiously, the DS-FFM is precisely Henley's original version and therefore should rather be called Henley's FFM. His original authorship went almost completely unnoticed in the literature and only to maintain the standard nomenclature, it will be called DS-FFM in the following.

One could argue that the development of the FFM opened Pandora's box, in that it made, for the first time, a direct reference from SOC to a natural phenomenon, the control and understanding of which is of crucial importance at least in some parts of the world. Turcotte's (1999) review gives a very colourful, at times somewhat misconstrued, account of the lessons to be learnt from the FFM.

The motivating question for the original model was 'How can a uniform energy injection result in a fractal dissipation?' (Bak *et al.*, 1990). The original FFM seemed to illustrate the problem very well, reproducing some features observed in turbulence. While the BTW Model and ZHANG Model are conservative in the bulk, the FFM has no such conservation and is purely dissipative. This simple computer model would therefore facilitate the search for a mechanism producing fractal dissipation, as is well understood in the case of the $-5/3$ law in the Kolmogorov cascade (Frisch, 1995).

Bak *et al.* (1990) realised that the two time scales present in their model (Box 5.1), namely the burning which is fast (rate 1) compared to the slow growth [5.1] (rate p), would produce an effective length and time scale, with p^{-1} time steps between growing attempts on every site. Scale-free behaviour can only be expected when the time scales of fire propagation and growing are fully separated, $p \to 0$. This limit requires, correspondingly, larger and larger lattices, otherwise there are no densely occupied regions available to feed the fire. From a different perspective, the correlation length ξ is bounded by the linear system size $L, \xi \propto L$, and is expected to scale with the **temperature-like control parameter** p (also Sec. 1.2) like $\xi(p) \propto p^{-\nu}$.

| Box 5.1 | Forest Fire Model (original) |

General features

- First published by Bak, Chen, and Tang (1990).
- Motivated by fractal dissipation events in turbulence.
- Defined on any regular lattice, mostly studied on hyper-cubic lattices in two and three dimensions.
- In one dimension, quickly develops into a state without fire.
- Deterministic relaxation, stochastic drive.
- Non-conservative (bulk) dynamics.
- Lightning mechanism present as well, but deemed 'statistically insignificant' (Bak *et al.*, 1990).
- Widely accepted as not being scale invariant (Grassberger and Kantz, 1991; Moßner, Drossel, and Schwabl, 1992).
- Link to percolation.
- Critical point reached by tuning p to its trivial critical value $p_c = 0$ (complete separation of time scales of growing and burning).
- Superseded by the DS-FFM (see Box 5.2, p. 116 and Table 5.1, p. 120), exponents reported in the original article are the fractal dimension of the set of burning sites (Bak *et al.*, 1990; Grassberger and Kantz, 1991).

Rules

- d dimensional regular lattice, usually periodic boundary conditions.
- Every site $\mathbf{n} \in \{1, \ldots, L\}^d$ (hyper-cubic case) is in one of three states $s_{\mathbf{n}} \in \{A, T, B\}$, **A**sh, **T**ree or **B**urning.
- *Initialisation*: irrelevant, model studied in the stationary state.
- *Discrete time t* (updates in sweeps).
- **A**→**T**: a site containing **A**sh at time t becomes occupied by a **T**ree at time $t + 1$ with (small) probability p.
- **T**→**B**: a site occupied by a **T**ree at time t and with a nearest neighbour burning at time t **B**urns itself at time $t + 1$.
- **B**→**A**: a site **B**urning at time t contains **A**sh at time $t + 1$.
- *Key observables* (see Sec. 1.3):
  number of sites on fire, $N_f(t)$, as a function of t and its histogram;
  correlation function of the state of sites.

One might argue that with the introduction of the control parameter p, SOC is back in the realm of classical critical phenomena which require the tuning of a parameter, in order to bring about non-trivial scaling. Yet, in the present model, the critical value of the tuning variable is trivial and corresponds merely to a separation of time scales. Asymptotically vanishing, it cannot provide a scale in its own right in the same way a critical temperature does (Sec. 1.3.1).

Bak *et al.* (1990) measured the exponent ν in a remarkably elegant fashion by means of Monte Carlo (real space) renormalisation group techniques, which is a form of numerical coarse graining (e.g. Stanley, Reynolds, Redner, and Family, 1982; Gawlinski and Redner,

1983), namely by looking at pairs p_1, p_2 of the growth probability, such that the coarse grained number of occupied sites in a small system equals that in a larger system with more coarse graining iterations. Assuming that this equality implies an equality of the coarse grained correlation length, one can derive the ratio of the correlations length in the 'bare' (non-coarse grained) lattices. With $\xi \propto p^{-\nu}$ one has $\xi(p_1)/\xi(p_2) = (p_1/p_2)^{-\nu}$.

As for its Abelian or non-Abelian nature, the FFM is very different from sandpile models, where the entire evolution can be expressed by a single type of operator which 'kicks' and subsequently relaxes the whole system. In contrast, in the FFM two very different processes take place simultaneously, one feeding it and one relaxing it. Clearly, operators creating trees on empty sites and the time-evolution operator responsible for the spreading of the fire do not commute, so in that sense the model is non-Abelian. However, the model is defined with stochastic growth and constant deterministic burning, so it looks somewhat contrived to express these two entirely different processes as operators on an equal footing. Focusing solely on the burning, the model is Abelian in the sense that the order of updates during the spreading of the fire, which is effectively done in parallel anyway, does not change the behaviour of the model.

Comparing with some of the discussions in the context of the BTW Model, for example regarding the effect of annealed noise (p. 88), it might come as a surprise that the stochasticity of the growth in the FFM does not ultimately destroy correlations. Although weak, they have been observed to scale similarly to those in a fully deterministic variant (Bak *et al.*, 1990, probably referring to Chen, Bak, and Jensen, 1990).

5.1.1 Critique

It was noted very early [1] (Grassberger and Kantz, 1991) that in the original definition, the FFM develops into an essentially deterministic state, producing **spiralling patterns** that become more sharply defined with increasing system size, or, more precisely, the width of a fire front on a spiralling arm grows with system size $1/p$ more slowly than the distance between fronts.

Figure 5.1 shows realisations of the FFM for the same values of $p = 0.04$ but on lattices of different size L. The smaller (or rather coarser) lattice, Fig. 5.1(a), displays a less correlated pattern of fires, while the large one, Fig. 5.1(b), has more clearly defined fire fronts, some of which curl into spirals. Grassberger and Kantz (1991) confirmed the fractal dimension D of the set of burning sites ($D = 1$ in $d = 2$), but found that this one-dimensional object is simply the burning front in a spiralling pattern, similar to a Turing pattern as studied frequently in mathematical biology (Tuszyński, Otwinowski, and Dixon, 1991; Maini, Paintera, and Chaub, 1997; Murray, 2003). They concluded that on the relevant length scales, the FFM is 'not governed by stochastic fluctuations, but is essentially deterministic.' They argued that the nearest neighbour interaction of the fire propagation produces some finite correlations between burning sites and as their density vanishes with p, there must be some (length) scale that diverges.

---

[1] In fact, Grassberger and Kantz (1991) submitted their critique, citing a 1988 preprint by Bak, Chen and Tang, three months *before* the original article (Bak *et al.*, 1990) was published.

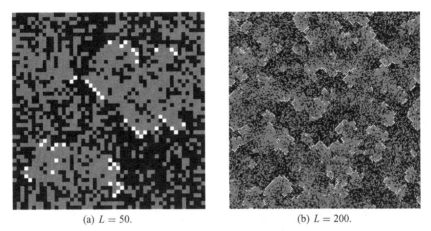

(a) $L = 50$.   (b) $L = 200$.

Figure 5.1   Realisations of the FFM for the same $p = 0.04$ and the same number of iterations (290 000 sweeps) on differently sized lattices. Fires are white, trees are black and empty sites grey. The smaller lattice has fewer and less clearly defined fire fronts.

Grassberger and Kantz (1991) also made the connection to epidemic models (e.g. SIR, Hethcote, 2000; Murray, 2003) and investigated a potential link to uncorrelated percolation. However, this was not corroborated by their numerical results, which might have been a finite size effect. In two dimensions the percolation scenario is supported by the analytic result that the critical behaviour of percolation mixes with the critical behaviour of the phase transition in an underlying lattice model (Coniglio and Klein, 1980). However Moßner *et al.* (1992) found plausible arguments that even in the limit $p \to 0$, no non-trivial scaling is expected. Their argument is based on an analysis of the speed of the fire fronts, which they found approaches a constant value for small p.

5.2 The DROSSEL–SCHWABL Forest Fire Model

The original Forest Fire Model was revised by Drossel and Schwabl (1992a),[2] but the resulting model was first proposed by Henley as early as 1989, *before* the publication of the original model by Bak *et al.* (1990). The crucial step was to do away with the burning time scale altogether and make it instantaneous, thereby realising the separation of time scales completely, i.e. $p = 0$ if that rate is measured on the microscopic time scale of fire propagation.

Consequently, there are no fires left on the time scale of growing. In the original model, the absence of any fire in the system eventually leads to the **absorbing state** of all sites being occupied. To avoid that, Bak *et al.* (1990) had introduced an 'external' **lightning mechanism** (footnote 2 in Bak *et al.* (1990)), which was claimed to be statistically

[2] Some authors cite Drossel and Schwabl (1992b) which is essentially the same article.

Box 5.2   **DROSSEL–SCHWABL Forest Fire Model**

General features

- First published by Henley (1989), but commonly attributed to Drossel and Schwabl (1992a) as a revised version of the original model by Bak *et al.* (1990), Box 5.1.
- Originally defined on any regular lattice (periodic boundary conditions), but mostly studied in two dimensions on a square lattice.
- Stochastic, explicit lightning, burning time scale fully separated from lightning and growing.
- Exact results known in one dimension (Drossel, Clar, and Schwabl, 1993; Paczuski and Bak, 1993).
- Stochastic drive and stochastic relaxation.
- Non-conservative (bulk) dynamics.
- Critical point reached by tuning $\theta = p/f$ to its trivial critical value $\theta_c^{-1} = 0$ (complete separation of time scales of growing and lightning).
- Two distinct length scales found (Honecker and Peschel, 1997; Schenk, Drossel, Clar, and Schwabl, 2000; Schenk, Drossel, and Schwabl, 2002).
- Scale invariance disputed (Grassberger, 2002; Pruessner and Jensen, 2002a).
- Avalanches correspond to clusters burnt.
- Exponents listed in Table 5.1, p. 120.
- Upper critical dimension $d_c = 6$ (Grassberger, 1993).

Rules

- d dimensional hypercubic lattice.
- Every site $\mathbf{n} \in \{1, \ldots, L\}^d$ is in one of two states $s_\mathbf{n} \in \{A, T\}$, Ash or Tree.
- *Initialisation*: irrelevant, model studied in the stationary state.
- *Discrete time t* (updates in sweeps).
- **A→T**: every site containing Ash at time t becomes occupied by a Tree at time $t + 1$ with (small) probability p.
- **T→A**: every site containing a Tree at time t is ignited with (small) probability $f \ll p$. If that happens, the entire cluster of Trees connected to that site, defined through nearest neighbour interactions, is turned to Ash. Ignitions at multiple sites are normally ignored.
- *Key observables* (see Sec. 1.3):
  size (mass, radius) of clusters removed by fire;
  burning time of clusters, assuming that fire spreads on the burning time scale by nearest neighbour interaction (see Box 5.1).

---

insignificant and was considered unnecessary for sufficiently large lattices by Grassberger and Kantz (1991).

Making the fires instantaneous, Drossel and Schwabl (1992a) had to introduce an explicit lightning mechanism.[3] In their model, trees ignite spontaneously with rate f which is small

---

[3] An OFC-like mechanism was introduced by Chen *et al.* (1990) in a fully deterministic model.

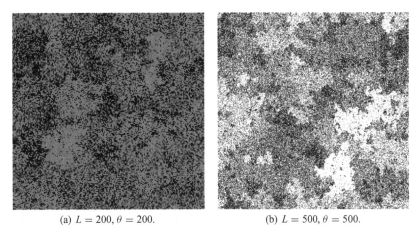

(a) $L = 200, \theta = 200.$ (b) $L = 500, \theta = 500.$

Figure 5.2   The patchy structure of the tree distribution in the D<small>ROSSEL</small>–S<small>CHWABL</small> FFM. For better comparison in (a) trees are shown in black and empty sites in grey as in Fig. 5.1(b). In (b) a larger lattice is shown at higher contrast.

compared to growing, while burning occurs instantaneously with respect to both growing and lightning. The resulting model never gets trapped in the absorbing state for $f > 0$, but only at the expense of a new, tunable parameter f. Drossel and Schwabl therefore make use of the full potential of the original model by making the lightning mechanism explicit, which allows them to realise the separation of time scales between burning and growing completely. With this adjustment the model is also slightly more reminiscent of real forest fires.

The resulting model was intended as a non-trivial, *critical* version of the FFM and is now commonly known as the D<small>ROSSEL</small>–S<small>CHWABL</small> Forest Fire Model (DS-FFM). It was introduced together with a set of (mostly) integer exponents derived from various (scaling) assumptions. That the cutoff s_c must scale like its first moment $\langle s \rangle$, 'since there is only one diverging length scale in a critical system' (Drossel and Schwabl, 1992a) is a popular fallacy ('first moment' on p. 349). As shown in Ch. 2, $s_c \propto \langle s \rangle$ implies $\tau = 1$, assuming simple scaling, so if the argument about the uniqueness of the (length) scale applied, then all critical systems would have $\tau = 1$. This is not the case due to the presence of dimensionful metric factors, so that $\langle s \rangle$ might diverge slower (but never faster) than s_c.

The analytical results by Drossel and Schwabl (1992a) were shown to be 'very wrong' (Grassberger, 1993) on the basis of numerical results first by Grassberger (1992) who immediately questioned the proper scaling behaviour of the model. Henley (1993) confirmed these findings by analysing, among other observables, the correlation function.

Without the removal of clusters inherent to the DS-FFM, site occupation would be completely uncorrelated, because tree growth is uncorrelated. The removal process leads to a patchy appearance (see Fig. 5.2), which is another way of saying that densities are locally (namely patch-wise) correlated – where a cluster is removed it leaves behind a region which over time acquires a homogeneous tree density.[4] Henley (1993) remarked that the

---

[4] Yet, as Grassberger (1993) pointed out, these patches might contain small unconnected (isolated) islands which are left behind after a fire has swept through.

arguments presented by Drossel and Schwabl (1992a) to derive $\tau = 1$ cannot be correct and/or complete, because they would apply equally to percolation, which is characterised by a non-trivial exponent τ. [5.2]

As in the original FFM (see the remark on p. 114), it makes little sense to state whether or not the DS-FFM is Abelian.

5.2.1 Double separation of time scales

The original FFM (see Box 5.1) has, like most SOC models, two time scales, but unlike them, in the original FFM their complete separation is impossible, because this would imply growing new trees only after all burning has stopped, which would trap the model in a state without fire. In the DS-FFM, that cannot happen, because forest patches are constantly set alight with a small rate f. The DS-FFM has three time scales: the first one for burning (rate 1), if that is allowed to operate on its own time scale, for example in order to define a burning time; the second for growing (rate p); the third for lightning (rate f). The condition for the removal of all explicit time scales is $1 \gg p \gg f$. By growing and setting alight only once all burning is done, the first separation, $1 \gg p$, can be realised completely, because then growing and lightning is arbitrarily (read infinitely) slow compared to burning. The second separation of time scales requires tuning as discussed below.

The condition $1 \gg p \gg f$, known as a '**double separation of time scales**' (Drossel and Schwabl, 1992a), can be understood on the level of the microscopic processes. In the stationary state, the tree density ρ is expected to fluctuate around a mean value $\langle \rho \rangle$. If the growth rate is p and the tree density essentially constant at $\langle \rho \rangle$, then the average number of trees grown per time unit is $p(1 - \langle \rho \rangle)V$, where V is the volume of the system, on a hyper-cubic lattice $V = L^d$. The total lightning rate in the system being $f \langle \rho \rangle V$ the average time between two lightnings is $(f \langle \rho \rangle V)^{-1}$, so that the number of trees grown between two lightnings on average is $(1 - \langle \rho \rangle)p/(f \langle \rho \rangle)$. In the stationary state, this is exactly the average size of a patch removed by lightning, and therefore

$$\langle s \rangle = \frac{1 - \langle \rho \rangle}{\langle \rho \rangle} \theta, \tag{5.1}$$

where $\theta = p/f$, although some authors use θ to denote the inverse (cf. Vespignani and Zapperi, 1998 versus Grassberger, 1993). On the burning time scale, it is often (Clar, Drossel, and Schwabl, 1996) assumed that the average (or characteristic) time for the removal of the patch is a power law of the the average (or characteristic) size of the patch, so that the typical burning time T_c can be written as $T_c \propto \theta^{\nu'}$. It is *that* time which needs to be so short that no new trees grow and no trees are set alight during the process of burning a patch,

$$\theta^{\nu'} = \left(\frac{p}{f} \right)^{\nu'} \ll p^{-1}, f^{-1}. \tag{5.2}$$

By making the burning instantaneous compared to all other processes, $p, f \to 0$ with non-zero p/f, this separation of the burning time scale is perfectly realised.

To remove the remaining scale set by p and f, a second separation of time scales is needed for the limit $p^{-1} \ll f^{-1}$, meaning that the growth rate is arbitrarily (infinitely) greater than the lightning rate, or $\theta = p/f \to \infty$. However, taking this limit in a finite system leads to a dense forest, $\langle \rho \rangle = 1$. Once the system is completely filled with trees, all trees would burn down at the next lightning. This is a finite size effect and the scales to be compared are p/f (as a measure of the average cluster size) and the volume V, so θ is to be tuned to infinity but kept small compared to V. The strength of finite size effects is given by the value of the ratio θ/V. As in classical critical phenomena, one might be tempted to use these finite size effects in finite size scaling, however, as explained above, too large θ renders the model trivial, i.e. studying the effect of finiteness of V at the 'critical point' $\theta \to \infty$ makes little sense. Unfortunately, the model does not behave systematically at intermediate values of θ either (Pruessner and Jensen, 2004). That is the reason why the exponent λ for the scaling of the characteristic cluster size with the control parameter, $s_c \propto \theta^\lambda$, and ν' for the characteristic fire duration are reported in Table 5.1 rather than the finite size scaling exponents as, for example, for the Abelian BTW Model, Table 4.1 on p. 92.

The incomplete separation of time scales has attracted some critique in the literature (e.g. Drossel and Schwabl, 1992a; Loreto, Pietronero, Vespignani, and Zapperi, 1995, 1997; Drossel, 1997), because θ can be considered as a relevant control parameter, i.e. a temperature-like variable. This makes the DS-FFM different from other SOC models, where no parameter requires tuning, often because the separation of time scales is fully realised, as discussed in the previous chapter. From that point of view, the DS-FFM appears like an instance of ordinary critical phenomena. However, while an ordinary temperature-like variable requires intricate tuning, very large θ could be quite naturally realised in many physical systems which are slowly 'loaded' and relax in sudden outbursts.[5]

5.2.2 Lack of scaling in the two-dimensional DS-FFM

Very early on, some unusual behaviour of the scaling in the DS-FFM was reported for the two-dimensional DS-FFM (Grassberger, 1993; Honecker and Peschel, 1997) even for system sizes of up to $17\,408 \times 17\,408$. Later, the DS-FFM was realised on even larger lattices of up to $65\,536 \times 65\,536$ and θ up to $256\,000$ in order to probe its critical properties (Grassberger, 2002; Pruessner and Jensen, 2002a, 2004). Grassberger (2002) investigated the avalanche size distribution (size distribution of the burnt patches) and found that simple scaling (see Sec. 2.1) is violated. Similarly, the expected power law scaling for the **radius of gyration** was not confirmed and Grassberger concluded that he did 'not see many indications for *any* power laws in this model.' Pruessner and Jensen (2002a) analysed the avalanche size distribution in detail and came to the same conclusion, namely that the model does not follow simple scaling. A form of partial data collapse *is* possible, provided one allows for an ever increasing lower cutoff or a constant upper cutoff (Figs. 17 and 16 respectively in Pruessner and Jensen, 2004), precisely the opposite of what is expected in simple scaling (Sec. 2.1). Interestingly, a moment analysis draws a very different picture, with moments following power laws robustly (Pruessner and Jensen, 2004). This is caused

---

[5] See also the remark on the temperature-like control parameter on p. 112.

Table 5.1 Exponents of the DS-FFM. $\mathcal{P}^{(s)}(s) = a^{(s)}s^{-\tau}\mathcal{G}^{(s)}(s/(b^{(s)}\theta^{\lambda}))$ is the (burnt) cluster size distribution and $\mathcal{P}^{(T)}(T) = a^{(T)}T^{-\alpha}\mathcal{G}^{(T)}(T/(b^{(T)}\theta^{\nu'}))$ is the burning time distribution (Sec. 1.3 and Clar *et al.*, 1994). Scaling relations: $\lambda(2-\tau) = \sigma_1^{(s)} = 1$ (stationarity) and $\lambda(1-\tau) = \nu'(1-\alpha)$ (sharply peaked joint distribution, widely assumed), in all dimensions (Sec. 2.2.2 and Pruessner and Jensen, 2004). For MFT results see Sec. 8.1.3.

d	τ		λ		α		ν'	
1	1 ▷	(*a*)	1	(*c*)	1	(*d*)	1	(*d*)
	0.95(5) ▷	(*b*)	1	(*d*)				
	1 ▷	(*c*)						
	1 ▷	(*d*)						
2	1 ▷	(*a*)	1	(*a*)	1.27(7)	(*d*)	0.58	(*d*)
	1.150(5) ▷	(*e*)	1.08(2)	(*f*)	1.27(1)	(*k*)	0.59(1)	(*k*)
	1.15(2) ▷	(*f*)	1.17(2)	(*e*)	1.24	(*q*)	0.6	(*q*)
	1.16(5) ▷	(*b*)	1.15(3)	(*d*)				
	1.14(3) ▷	(*d*)	1.17(2)	(*h*)				
	1.48 ▷	(*g*)	1.09(1)	(*k*)				
	1.159(6) ▷	(*h*)	1.10	(*q*)				
	1.02–1.16	(*i*)						
	1.08(1)	(*k*)						
	1.11 ▷	(*m*)						
	1.16	(*n*)						
	1.45	(*o*)						
	1.10 ▷	(*p*)						
	1.19 ▷	(*p*)						
	1.22 ▷	(*p*)						
	1.09 ▷	(*q*)						
3	1.24(5) ▷	(*b*)	1.30(6)	(*d*)	1.47(9)	(*d*)	0.64(6)	(*d*)
	1.23(3) ▷	(*d*)						
4	1.33(5) ▷	(*b*)	1.56(8)	(*d*)	1.73(10)	(*d*)	0.78(8)	(*d*)
	1.36(3) ▷	(*d*)						
5	1.39(5) ▷	(*b*)	1.82(10)	(*d*)	1.89(11)	(*d*)	0.92(10)	(*d*)
	1.45(3) ▷	(*d*)						
6	1.47(5) ▷	(*b*)	2.01(12)	(*d*)	1.97(11)	(*d*)	1.04(11)	(*d*)
	1.50(3) ▷	(*d*)						
MFT	3/2 ▷	(*b*)			2	(*b*)		

▷ Reported as $\tau + 1$.

a Drossel and Schwabl (1992a), *b* Christensen *et al.* (1993), *c* Drossel *et al.* (1993), *d* Clar *et al.* (1994), *e* Henley (1993), *f* Grassberger (1993), *g* Patzlaff and Trimper (1994), *h* Honecker and Peschel (1997), *i* Malamud *et al.* (1998), *k* Pastor-Satorras and Vespignani (2000b), *m* Grassberger (2002), *n* Loreto *et al.* (1995), *o* Schenk *et al.* (2002), *p* Pruessner and Jensen (2002a), *q* Pruessner and Jensen (2004).

by the moment analysis not being sensitive to a lack of scaling (scaling of the *lower* rather than the upper cutoff), Sec. 2.1.1 (also Sec. 7.3, Fig. 7.6).

5.2.3 Analytical approaches

The analytical results for the one-dimensional DS-FFM by Drossel *et al.* (1993) and Paczuski and Bak (1993) mentioned earlier rely on the fact that a one-dimensional lattice is loop free, so that the path connecting any two sites is unique. Growth is independent throughout the lattice and densities are therefore correlated only in regions that formerly were covered by a patch. In one dimension, a patch is a continuous strip of occupied sites.

Drossel *et al.* (1993) derived some analytical properties of the one-dimensional DS-FFM, mostly asymptotes, as 'the results become more and more exact when the critical point is approached.' Some of the underlying assumptions, such as ignored correlations, are not easy to test or to prove, but the results are compatible with independent derivations by Paczuski and Bak (1993).

Loreto *et al.* (1995, also Loreto, Vespignani, and Zapperi, 1996) have formulated a real space renormalisation group (RSRG) approach of the one- and the two-dimensional DS-FFM, similar to that devised for sandpiles (Pietronero *et al.*, 1994; Vespignani and Zapperi, 1995). Consequently, it suffers from similar shortcomings, taking place in real space (Burkhardt and van Leeuwen, 1982) and ignoring certain correlations. However, in one dimension, the findings coincide with the asymptotes mentioned above and in two dimensions the results are in perfect agreement with older numerical results. The latter were cast into doubt by more recent numerical findings (Grassberger, 2002; Pruessner and Jensen, 2002a), which question the very scale invariance of the DS-FFM (Sec. 5.2.2).

A mean-field theory (MFT) for the DS-FFM was suggested by Christensen, Flyvberg, and Olami (1993), essentially mapping it to a branching process (Sec. 8.1.4). Clar, Drossel, and Schwabl (1994) later clarified that the MFT of the DS-FFM does not coincide with the DS-FFM on a Bethe lattice as expected from standard percolation. The MFT was instrumental in substantiating the link to percolation (see below) by showing that percolation is the correct description of the DS-FFM at $d = 6$, which is the upper critical dimension of percolation. The derivation of the MFT draws heavily on percolation ideas, lending further support to the existence of this link.

A wide range of analytical considerations are based on the model's relation to standard **percolation** (Stauffer and Aharony, 1994). The original publication of the model made use of that link already (Drossel and Schwabl, 1992a), to estimate the average tree density, which was thought to be a result of a maximisation principle, whereby the fires remove the largest possible amounts of trees. The comparison of non-universal features of the DS-FFM and percolation is controversial (Grassberger, 1993, 2002), because of the density correlations in the DS-FFM. According to Henley (1989), these correlations decay so slowly in space that the DS-FFM should display an asymptotic behaviour that deviates from that of uncorrelated percolation (Weinrib, 1984), i.e. even universal quantities, in particular critical exponents, should be very different in the DS-FFM compared to uncorrelated (standard) percolation.

Schenk *et al.* (2002) derived $\tau = 1.45$ assuming uncorrelated percolation, but taking into account the temporal evolution of the occupation density. This value was backed up by numerical simulations of some effective models, yet it remains to be shown that this is the true asymptote of the avalanche size distribution in the DS-FFM. Grassberger (2002) estimated that the asymptotic regime of the DS-FFM in its original form is situated above $\theta \approx 10^{40}$, where the tree density equals the occupation density found in standard percolation at the critical point. This value of θ is off the chart for current numerical methods which reach only up to $\theta \approx 10^5$. Interestingly, using an improved MFT approach that incorporates some of the correlations, Patzlaff and Trimper (1994) calculated the exponent $\tau = 1.48$, not much different from $\tau = 1.45$ found by Schenk *et al.* (2002). It is unclear how their results at $d = 6$ of $\tau = 5/2$ can be reconciled with the numerical finding of $\tau \approx 1.5$ and the general notion that the DS-FFM coincides with standard percolation, $\tau = 3/2$, in $d = 6$ and beyond. [6]

5.2.4 Physical relevance

As summarised in Sec. 3.7.3, Malamud *et al.* (1998) and others (e.g. Reed and McKelvey, 2002; Malamud *et al.*, 2004) analysed forest fires occurring in the wild from the point of view of the FFM and found some agreements. This might come as a surprise. The basic ingredients of both models described above are removal of randomly selected patches by fires and random regrowth up to a certain maximum density, effectively the (local) carrying capacity. In nature, many other factors come into play, such as human control, geography, climate, season, spreading of regrowing species and their numerous interactions, many of which might be relevant to the supposed scaling behaviour. One might therefore conclude that only in very rare cases is the spatial population pattern of real forests controlled by fires sweeping through. Even where fires occur frequently, the shape and size of burning patches are, in contrast to the FFM, rarely determined by previous fires in the neighbourhood. In the real world, many other environmental factors determine the shape and size of forest patches, not least the geography and the climate of a region, as well as the weather during the fire. In fact, in most of the world inhabited by humans, forest fires are shaped by fire brigades, whose effect is unlikely to be captured by any of the models described in this book.

5.2.5 Numerical methods

The DS-FFM has been studied extensively in computer simulations using various approximations and simplifications, which are briefly discussed in the following.

In the original definition of the DS-FFM, p and f are rates, [7] so that every finite time interval Δt can, in principle, contain any number of growing (empty site turns occupied) and lightning events (occupied sites turn empty) on every site on the lattice. Implementing

---

[6] Note that τ is shifted by 1 compared to the notation used in percolation.
[7] They are probabilities 'during one time step', but that time step is made to vanish.

this process in the most naïve way means choosing a small, finite time interval δt and performing the task described in the following pseudocode:

```
while (1) {                                              /* Run forever. */
    t=t+δt;                                         /* Time ticks in units of δt. */
    for all sites s do {                            /* Scan the entire lattice. */
        if (s empty) {                      /* If empty, occupy with probability δt p: */
            with probability δt p: s=occupied;
        } else {                       /* If occupied, start a fire with probability δt f: */
            with probability δt f:
                burn(entire cluster connected to s);
        }
    }
}
```

To implement correctly the concurrent **Poisson processes** of growing and lightning one has to take the limit $\delta t \rightarrow 0$, which is of course highly inefficient, as no event takes place during most time steps. The most accurate way of implementing the time t correctly, while avoiding this inefficiency, is to draw waiting times between events. This technique is not very economical if that level of accuracy of the representation of time is not needed. Given the (normally) large number of empty and occupied sites in the system, a good compromise is to choose a time slice δt such that the larger of the two rates, normally p, gives rise to one event per site and interval δt, i.e. $p\delta t = 1$. To avoid any spatial bias, the sites have to be visited randomly. This method has been used by Henley (1993) and Honecker and Peschel (1997). Assuming a constant tree density $\langle \rho \rangle$ the average number of trees grown between two fires is $(1 - \langle \rho \rangle)p/(\langle \rho \rangle f)$, but there are significant fluctuations around this value, with the variance equal to the average (Poisson process).

However, by far the most popular technique (Grassberger, 1993; Clar *et al.*, 1994; Malamud *et al.*, 1998; Schenk *et al.*, 2000) is to ignore the fluctuations altogether and attempt p/f times to grow a tree at randomly chosen sites, followed by a lightning attempt, producing an average growth of $(1 - \langle \rho \rangle)p/(\langle \rho \rangle f)$ trees between two lightnings (assuming constant tree density). As a result, one finds spurious peaks in the time series of the tree density. This effect has been investigated by Schenk *et al.* (2000); one might wonder to what degree the spatial correlations in the DS-FFM would be reduced if lightning were implemented in a truly Poissonian fashion.

To burn an entire cluster, one normally uses a stack (last in, first out, LIFO, Sec. A.1). At first, the site hit by lightning is put on the stack (PUSH) and its state is turned to empty. As long as the stack is non-empty, a site is taken from the stack (POP) and all occupied nearest neighbours visited. Occupied neighbours then have their state turned to empty and are put on the stack as well. By only considering occupied sites and turning them empty as soon as they are put on the stack, no site is put on the stack more than once.

The algorithm described above is known as **depth-first search** (DFS) (Cormen, Leiserson, and Rivest, 1996), because the last site put on the stack is the first one whose neighbours are visited when sites are taken from the stack. That way, the fire will very quickly reach the perimeter of the cluster, while the bulk of the cluster remains intact. This type of burning is popular because the stack needs very little maintenance: it requires a small amount of memory and grows only in one direction. However, if one wants to measure the burning time as defined in the original FFM, where fire spreads from site to site via nearest neighbour interaction at a rate of one step per time, a LIFO stack is unsuitable. There are two obvious solutions: either one employs two LIFOs, one containing all sites to be burnt at time t, one containing all sites to be burnt in the next time step (Pruessner and Jensen, 2004), or one performs a **breadth-first search** (BFS) using a first in, first out stack (FIFO, Sec. A.1) so that all sites to be burnt at t are turned to empty before sites to be burnt at $t+1$ (Grassberger, 2002).

The **Hoshen–Kopelman algorithm** (Hoshen and Kopelman, 1976) is a more sophisticated technique to identify and, to some extent, maintain *all* clusters on a lattice, as it allows simple bookkeeping as clusters merge when trees bridge gaps between them. At the centre of the technique is a particular data structure and the facilities to maintain them can be extended almost arbitrarily (Hoshen, Berry, and Minser, 1976; Hoshen, 1997; Pruessner and Jensen, 2002a, 2004; Moloney and Pruessner, 2003; Pruessner and Moloney, 2006). The Hoshen–Kopelman algorithm is famed for its minimal memory requirements in percolation, where clusters are generated on the fly while the lattice is being filled. This feature plays no rôle in the FFM, whose configurations evolve in time and cannot be generated *ex nihilo* without time evolution. The feature most relevant for the FFM is the algorithm's ability to characterise all clusters in a single sweep. It has typical computational complexity $\mathcal{O}(V \log V)$ for a lattice with V sites. To traverse all clusters on a lattice, there are more naïve methods such as a DFS, which visits each site several times: once when probing from all surrounding occupied sites and once when trawling for untouched sites. While the computational complexity of the latter is only of order $\mathcal{O}(V)$, in practice the Hoshen–Kopelman algorithm, in particular for large lattices (10^6 sites and more) and complex cluster structures, is often faster and easier to maintain. In particular, it offers the option to grow and merge clusters dynamically, i.e. adding a few trees results in few changes to the data tables of the Hoshen–Kopelman algorithm, while the DFS would require exploration of the entire cluster affected, making the former orders of magnitude faster than the latter. Comparing the computational complexity of the two approaches, it is clear that multiple sweeping results in an amplitude in front of V that can easily exceed $\log V$ for any reasonable V.

5.2.5.1 Numerical results

This section summarises the numerical results, many of which have already been discussed above. Table 5.1 provides an overview of the many studies available. Working with very small values of $\theta \approx 70$ and probably equally small lattices, in their original publication Drossel and Schwabl (1992a) found $\tau = 1$ in $d = 1, 2, 3$ confirming their analytical arguments (see the discussion on p. 117). While the scaling of the one-dimensional model can to some extent be determined analytically (Drossel *et al.*, 1993; Paczuski and Bak, 1993), in higher dimensions most of the results have been obtained numerically. Christensen *et al.* (1993) investigated the model in up to 6 dimensions and Clar *et al.* (1994) up to 8 dimensions, confirming the upper critical dimension as $d_c = 6$ (Christensen *et al.*, 1993; Grassberger, 1993) as in standard percolation (Stauffer and Aharony, 1994). In such high dimensions, the system size is very limited, for example $L = 20$ in 6 dimensions in the study by Clar *et al.* (1994), suppressing the effect of large fluctuations and thereby pushing the behaviour of the system more towards the mean-field limit. [8]

As discussed in Sec. 5.2.1, the volume L^d of the lattices represents a constraint to the maximum value θ. In Clar *et al.* (1994) values of up to $\theta = 32\,000$ are considered in two dimensions, whereas Christensen *et al.* (1993) are more cautious and simulate up to $\theta = 4000$, similar to Grassberger (1993). Pruessner and Jensen (2004) suggested that the lattice size required for $\theta = 4000$ is at least $L = 8000$. Pastor-Satorras and Vespignani (2000b) applied a moment analysis to numerical results for systems up to $L = 19\,000$ and $\theta = 32\,768$. The resulting exponents for the cluster size distribution show clear deviations from previous results. Pruessner and Jensen (2002a) studied systems of a size up to $L = 32\,000$ (in two dimensions) with reliable results up to $\theta = 32\,000$, whereas Grassberger (2002) used $L = 65\,536$ and $\theta = 256\,000$. He also pointed out some inconsistencies in the studies by Pastor-Satorras and Vespignani (2000b).

Some of the exponents reported in Table 5.1 are derived from scaling relations and are therefore not independent. Although sometimes suggested differently (e.g. Paczuski and Boettcher, 1996), these relations are not identities in the sense of a mathematical necessity. For the DS-FFM, their validity is discussed in further detail by Pruessner and Jensen (2004).

5.3 The OLAMI–FEDER–CHRISTENSEN Model

The OLAMI–FEDER–CHRISTENSEN earthquake model, more commonly known as the OFC Model, went through several stages of development and has quite an intertwined history (see also the brief summary of the history of the OFC Model by Leung, Müller, and Andersen, 1997). Originally conceived by Burridge and Knopoff (1967) in one dimension, see Fig. 5.3(a), it remains somewhat unclear whether it was later extended to two dimensions (see Fig. 5.3(b)) by Otsuka (1971, 1972a,b) or invented and investigated independently. Takayasu and Matsuzaki (1988) studied its (ordinary) critical behaviour using a (tunable) control parameter Δ. In the BURRIDGE–KNOPOFF Model, the interface at an earthquake fault is represented by a set of blocks that are pulled over a surface where they experience static friction. The pulling force is exerted by a second surface, to which the blocks are attached elastically, which is modelled by flat (leaf) springs with spring constant K_L. As soon as the pulling force overcomes the static friction, the blocks start to slip. The blocks are interconnected by springs to form a chain in one dimension or a grid in two dimensions. The movement of one block to its new position where it experiences no net force changes the forces acting upon its neighbours and its slipping might therefore cause neighbouring blocks to slip in turn, giving rise to an avalanche, or rather, an earthquake.

---

[8] This is an unfortunate bias characteristic for the hunt for mean-field exponents in higher dimensions.

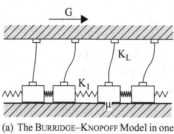

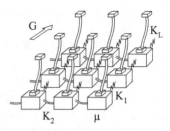

(a) The BURRIDGE–KNOPOFF Model in one dimension.

(b) The BURRIDGE–KNOPOFF Model in two dimensions.

Figure 5.3 The BURRIDGE–KNOPOFF Model. (a) In its one-dimensional version, blocks subject to static friction (indicated by μ) are connected by springs (spring constant K_1) and pulled by a uniform driving force (arrow, G) applied through leaf springs with spring constant K_L. (b) In two dimensions, blocks are coupled by additional leaf springs (K_2), in the direction transverse to the driving.

Carlson and Langer (1989) studied the one-dimensional **BURRIDGE–KNOPOFF Model** as an instance of SOC, focusing on its deterministic, continuum nature. The closely related **train model** which differs in the external driving, was later introduced by de Sousa Vieira (1992) and conjectured to be in the MANNA universality class by Paczuski and Boettcher (1996, also de Sousa Vieira, 2000). Nakanishi (1990) reformulated the BURRIDGE–KNOPOFF Model as a **coupled map lattice** (Kaneko, 1983, 1984, 1989) and studied the limit of slow drive, at the same time identifying the ratio of the spring constants as the fundamental model parameter. Brown, Scholz, and Rundle (1991) took a similar approach to the two-dimensional version (Fig. 5.3(b)) using discretised local states and effectively taking the conservative limit, yet with a built-in randomness. Feder and Feder (1991a,b), conducted experiments on **stick-slip behaviour** by pulling sandpaper over a carpet and introduced a non-conservative, continuous version of the BTW Model. Inspired and guided by this work Olami, Feder, and Christensen (1992) finally generalised the BURRIDGE–KNOPOFF Model comprehensively and formulated it most succinctly along the lines of a continuous sandpile model, see Box 5.3. The conservative OFC Model is related to the ZHANG Model, which, however, has a randomised driving rule. The OFC Model is recovered in the limit of vanishing energy increment in the ZHANG Model, $\delta \to 0$, effectively implementing a continuous, global, uniform drive (Pietronero et al., 1991; Pérez et al., 1996).

The BURRIDGE–KNOPOFF Model remains the mechanical equivalent of the OFC Model, so in that sense the OFC Model is not a new model. The achievement of Olami et al. (1992) is the identification of an explicit, tunable **level of conservation** (also 'conservation parameter') as the ratio of the spring constants, in two dimensions (Fig. 5.3(b) and Brown et al., 1991)

$$\alpha_1 = \frac{K_1}{2K_1 + 2K_2 + K_L} \tag{5.3a}$$

$$\alpha_2 = \frac{K_2}{2K_1 + 2K_2 + K_L}, \tag{5.3b}$$

| Box 5.3 | OLAMI–FEDER–CHRISTENSEN Model |

General features

- First published by Olami, Feder, and Christensen (1992).
- Coupled map lattice version of the BURRIDGE–KNOPOFF Model (Burridge and Knopoff, 1967), a **stick-slip model** of earthquake fault dynamics.
- Defined in any dimension, extensively studied in two dimensions.
- Deterministic, uniform driving, deterministic relaxation, random initialisation.
- Non-Abelian.
- Separation of time scales for driving and relaxation.
- Simple scaling widely accepted only in the conservative limit.
- Exponents listed in Table 5.2 and Table 5.3, p. 129.

Rules

- Lattice with coordination number q ($q = 2d$ in the hyper-cubic case).
- A force $F_{\mathbf{n}}$ acts on each site $\mathbf{n} \in \{1, 2, \ldots, L\}^d$ of a d dimensional hyper-cubic lattice (extension to other lattices obvious) with open boundaries.
- Conservation level $\alpha \in [0, 1/q]$ (anisotropic extension possible).
- Global threshold F_{th} usually unity.
- *Initialisation*: $F_{\mathbf{n}}$ taken randomly and independently from a uniform distribution in $[0, F_{\mathrm{th}}]$.
- *Driving*: find $\mathbf{n}_{\mathrm{max}}$ with the (globally) largest force F_{max} and increase all forces by $1 - F_{\mathbf{n}_{\mathrm{max}}}$.
- *Toppling*: for each site $\mathbf{n}$ where $F_{\mathbf{n}} \geq F_{\mathrm{th}}$, move $\alpha F_{\mathbf{n}}$ to the nearest neighbours $\mathbf{n}'$ of $\mathbf{n}$,

$$F_{\mathbf{n}'} \to F_{\mathbf{n}'} + \alpha F_{\mathbf{n}}$$

and reset $\mathbf{n}$,

$$F_{\mathbf{n}} \to 0.$$

- *Dissipation*: energy is lost at toppling for $\alpha < 1/q$ and at open boundaries.
- *Parallel update*: discrete microscopic time, sites that become active at time t topple at time $t + 1$ (updates in sweeps).
- *Separation of time scales*: drive only once all sites are stable, i.e. $F_{\mathbf{n}} \leq F_{\mathrm{th}}$ (quiescence).
- *Key observable* (see Sec. 1.3):
  avalanche size s, the total number of topplings until quiescence.

the reformulation as a coupled map lattice [9] as well as recognising **quasi-stationarity** as the appropriate limit to study the model. Quasi-stationarity means driving the model infinitely slowly which amounts to a separation of time scales between avalanching and loading.

The formulation of the BURRIDGE–KNOPOFF Model in terms of a coupled map lattice, whose relevance to SOC was already remarked on by Kaneko (1989) in the context of

---

[9] Originally it was called a **continuous cellular automaton** (Olami *et al.*, 1992), an oxymoron that has, to the displeasure of some (e.g. Grassberger, 1994), well caught on. Strictly, in a cellular automaton the *discrete* state of a site in the next *discrete* time step is a (traditionally deterministic) function of a finite neighbourhood on the lattice.

Table 5.2 Exponents of the non-conservative OFC Model, $\alpha < 1/q = 1/4$, open boundary conditions, in two dimensions. $\mathcal{P}^{(s)}(s) = a^{(s)}s^{-\tau}\mathcal{G}^{(s)}(s/(b^{(s)}L^D))$ is the avalanche size distribution, Sec. 1.3. Scaling relation: $D \leq d$ (Klein and Rundle, 1993).

d	τ			D		
2	1.89(10)	$(\alpha = 0.20)$	(a)	2.2(1)	$(\alpha = 0.20)$	(a)
	2.72(10)	$(\alpha = 0.10)$	(a)	2.3(1)	$(\alpha = 0.10)$	(a)
	1.8 $\triangleright$	$(0.18 < \alpha < 1/4)$	(b)	2.15(10)	$(\alpha = 0.20)$	(c)
	1.9(1) $\triangleright$	$(\alpha = 0.20)$	(c)	2	$(\alpha = 0.15, 0.18, 0.21)$	(e)
	1.63	$(\alpha < 1/4)$	(d)	2	$(\alpha < 1/4)$	(f)
	1.8	$(\alpha = 0.15, 0.18, 0.21)$	(e)			
	1.8	$(\alpha < 1/4)$	(f)			

$\triangleright$ Reported as $\tau + 1$.
a Christensen and Olami (1992a), b Grassberger (1994), c Pérez *et al.* (1996), d Ceva (1998), e Lise and Paczuski (2001b), f Lise and Paczuski (2001a).

conservation laws, means that the redistribution of forces can be done using simple local rules as described in Box 5.3, instead of simultaneously solving and numerically integrating a set of difference equations which are highly non-linear due to the presence of static friction (Burridge and Knopoff, 1967). The parameters introduced in Eq. (5.3), which characterise the redistribution of forces by slipping blocks, arise naturally in the set of local rules (Christensen, 1992; Christensen and Olami, 1992a,b; Olami *et al.*, 1992; Christensen and Moloney, 2005). The two dimensionless parameters α_1 and α_2 are usually chosen to be equal (but see Christensen and Olami, 1992a), despite the very different nature of the two types of springs they characterise, see Fig. 5.3(b): one accounts for the forces on the system due to (normal) strain, the other for the forces due to local shear (strain).

On a square lattice, force is conserved only in the conservative limit $2\alpha_1 + 2\alpha_2 = 1$ (and similarly for other lattices). In addition, dissipation occurs on the boundaries, which are usually chosen to be open, i.e. the force that a toppling site deposits at each neighbour is lost where that neighbour is missing compared to a bulk site.[10] In the stationary state, the dissipation is met by an external drive, which according to the mechanical origin is global and uniform. For a complete separation of time scales, the driving is effectively infinitely slow compared to the avalanches produced by toppling sites, i.e. no driving takes place while sites are toppling.

Notwithstanding claims to the contrary (Jánosi and Kertész, 1993), the driving speed in the OFC Model does not amount to an additional time scale compared to, say, the ASM. Both models have an external driving speed which *defines* the **macroscopic time scale**.

The force needed to trigger the next avalanche is given by the site that experiences the largest pulling force already. In a numerical simulation, after an avalanche, the site **n** with the maximum force $F_\mathbf{n}$ is determined and all forces increased[11] by $F_{\mathbf{th}} - F_\mathbf{n}$. The dynamics therefore is governed, quite naturally and in tune with stick-slip motion (but

---

[10] Without bulk dissipation boundaries have to be open, otherwise no stationary state exists.
[11] Numerically, this can be done much more efficiently by changing $F_{\mathbf{th}}$, see Appendix A.

Table 5.3 Exponents of the conservative OFC Model, $\alpha = 1/q$, open boundary conditions, in two dimensions. $\mathcal{P}^{(s)}(s) = a^{(s)}s^{-\tau}\mathcal{G}^{(s)}(s/(b^{(s)}L^D))$ is the avalanche size distribution, Sec. 1.3. Scaling relation: $D(2 - \tau) = \sigma_1^{(s)} = 2$ (Christensen, 1992; Jánosi and Kertész, 1993) in all dimensions (Sec. 2.2.2).

d	τ		D	
2	1.22(5)	(a)	3.3(1)	(a)
	1.253	(b)	3.01	(b)
MFT	3/2	(c)		

a Christensen and Olami (1992a), b Christensen and Moloney (2005), c Bröker and Grassberger (1997).

note the critique in Chen *et al.*, 1991), by an extremum principle, which places the model in the realm of extremal dynamics, see Sec. 5.4.3, p. 149. Randomness, sometimes regarded as a crucial ingredient in SOC (Pérez *et al.*, 1996), enters the OFC Model only through the initial condition, in contrast to most other models in SOC, where randomness is at least part of the driving (as in the ZHANG Model or the BTW Model). Some authors explicitly average over different initial conditions (e.g. Torvund and Frøyland, 1995; Ceva, 1998).

As long as the model is studied on hyper-cubic lattices and only one site reaches the threshold by external drive, sites are updated alternately on even and odd sublattices and no ambiguities arise as to the order of updates, see p. 369 and the corresponding discussion of the ZHANG Model, p. 107.

In the following, some of the most interesting, but also the most controversial, questions surrounding the OFC Model are discussed, concerning the effect of dissipation in this model, the relevance of the OFC Model to real earthquakes and some interesting suggestions about the true nature of the (apparently critical) behaviour of the OFC Model.

5.3.1 Non-conservation

The advent of the FEDER–FEDER Model (Feder and Feder, 1991a), as the precursor of the OFC Model, coincided with the heyday of the concept of generic scale invariance put forward by Hwa and Kardar (1989a) and others (Sec. 9.2), which had conservation as a central theme. Dhar (1990a) as well as Manna *et al.* (1990) had also highlighted conservation in the bulk dynamics and anisotropy induced by boundary conditions as the precondition for scale invariance (Sec. 9.2.2, p. 327). The FEDER–FEDER Model and much more explicitly so the OFC Model both led to the belief that an alternative mechanism for generic scale invariance (for that matter synonymous with SOC) exists, which does not require conservation. As opposed to the other models discussed in this chapter (see also the review by Pérez *et al.*, 1996), the OFC Model[12] is controlled by a (locally stored)

[12] The underlying mechanical model, the BURRIDGE–KNOPOFF Model, is not designed to make the level of conservation of the local force tunable.

quantity which performs a diffusion-like motion through the system and could in principle be conserved. In that sense the OFC Model is **explicitly non-conservative** depending on the level of conservation, α.

Pruessner and Jensen (2002b) introduced a non-conservative random neighbour model that displays scaling and is analytically solvable, in direct response to the questions about the relevance of bulk conservation first raised by Hwa and Kardar (1989a). However, resembling closely the DS-FFM, the model is constantly externally 're-charged' to allow for divergent avalanche sizes. In a finite system, that charge must be finely adjusted to its size, otherwise either it runs empty of **susceptible sites** (sites that topple after a charge) or they take over the entire system, similar to a DS-FFM with complete tree coverage. In the analytical treatment, this problem is circumvented by taking the thermodynamic limit first and considering the limit of divergent total external drive. The question whether the OFC Model is in fact scale invariant, i.e. displaying SOC, in the non-conservative regime, $\alpha < 1/q$, remains disputed in the literature.

Olami *et al.* (1992) introduced the term 'non-localised behaviour' to contrast system-wide avalanching with avalanches confined to an area of typical (small) size, which constitutes a characteristic scale. They suggested a transition from localised to non-localised behaviour at $\alpha_c \approx 0.05$. The transition manifests as a change in the avalanche size distribution from algebraic to exponential behaviour. An exponential generally precluding algebraic behaviour is a widespread misconception – a priori an exponential is simply a particular scaling function (Sec. 2.1.2 and comment on p. 349).

Yet, the use of the term 'exponential' alludes to the presence of a scale s^* independent of the system size, which enters the exponential in the form $\exp(s/s^*)$, as, for example, in a Poisson process. The existence of a characteristic avalanche scale s^* independent of the system size rules out FSS. One might conclude that 'any amount of dissipation introduces a characteristic length scale and thus destroys criticality' (Lise and Jensen, 1996).

Lise and Jensen (1996) used an infinite average avalanche size as an indicator for the presence of 'non-localised behaviour'. Ignoring correlations and making some ruthless, but perfectly legitimate approximations, they derived $\alpha_c = 2/9$ on a two-dimensional square lattice in a random neighbour approximation, such that average avalanche size diverges [13] for all $\alpha \geq \alpha_c$. Their derivation can be retraced in three steps. (1) Assuming that the forces of all sites are *uniformly distributed* in the interval $[0, F_{\text{th}}]$ (see Fig. 5.4), the average number of sites made unstable by a toppling site is $\sigma = q\alpha \langle F^+ \rangle / F_{\text{th}}$, if $\langle F^+ \rangle$ is the average force present in a toppling site. (2) The average force present at a (stable) site which topples after a charge by a toppling site, i.e. after a charge by $\alpha \langle F^+ \rangle$ is $\langle F^- \rangle$, assuming *independence* and *uniform* distributions of such sites means $\langle F^- \rangle = F_{\text{th}} - (1/2)\alpha \langle F^+ \rangle$. (3) By self-consistency, the force of a toppling site is the average force of a stable site that will topple plus the average charge $\langle F^+ \rangle = \langle F^- \rangle + \alpha \langle F^+ \rangle$. Imposing $\sigma = 1$ (see below), Lise and Jensen (1996) derived $\alpha_c = 2/(2q + 1)$.

The **branching process** (see Sec. 8.1.4), parameterised by the **branching ratio** σ, has always played a prominent rôle in the analysis of SOC models (e.g. Alstrøm, 1988;

---

[13] Divergent average avalanche size is not a condition for a power law avalanche size distribution, neither necessary ($\tau > 2$), nor sufficient (see for example the supercritical branching process, Sec. 8.1.4).

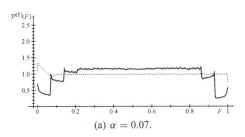

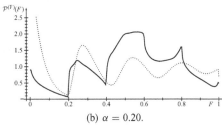

(a) $\alpha = 0.07$.   (b) $\alpha = 0.20$.

Figure 5.4 Histograms $\mathcal{P}^{(F)}(F)$ of the force in an OFC Model with open boundary conditions on a square lattice (solid line) or with random neighbours (dotted line). (a) The histogram for a low level of conservation, $\alpha = 0.07$, so that most structure is washed out compared to $\alpha = 0.20$ in (b). If the uniform ramping up of all forces by the external drive were suppressed in the histogram, it would resemble the energy histogram in the Zhang Model, Fig. 4.6.

Christensen and Olami, 1993; García-Pelayo, 1994a; Zapperi, Lauritsen, and Stanley, 1995). In a mapping of the random neighbour OFC Model on a branching process, σ is the ratio of newly created unstable sites (offspring) and the number of toppling sites (parents). On the lattice, where individual topplings are correlated, the concept of the branching ratio is less useful (also Sec. 5.4.5.6) and in the past has led to a number of erroneous conclusions (de Carvalho and Prado, 2000; Christensen, Hamon, Jensen, and Lise, 2001). The total number of unstable sites generated in a finite avalanche of size s is $s - 1$, whereas the number of toppling sites is s, so that the branching ratio for that avalanche is $\sigma = (s-1)/s$. The avalanche size weighted average is (Christensen $et\ al.$, 2001)

$$\langle \sigma \rangle = 1 - \frac{1}{\langle s \rangle}, \tag{5.4}$$

which converges to unity for $\langle s \rangle \to \infty$. In fact, the average avalanche size diverges for any $\sigma \geq 1$.

The derivation of $\alpha_c = 2/(2q+1)$ by Lise and Jensen (1996), as outlined above, ignores certain correlations, not least by casting the problem as a branching process. Apart from that, the line of argument seems quite strong and one might expect that it remains essentially intact even after incorporating correlations. At first sight, the possible inconsistency of the various distributions [5.3] of forces entering into the derivation looks like a mere technicality. In fact, Lise and Jensen (1996) found numerical evidence for a lack of FSS around $\alpha_c = 0.22$.

It turns out, however, that the distributions envisaged to give rise to the above derivation are not only inconsistent, but also unstable in that they cannot be sustained in a finite system. The result for α_c is highly sensitive to the choice of the distribution (Kinouchi, Pinho, and Prado, 1998; Pinho, Prado, and Kinouchi, 1998). Using the correct distributions, one would find that the minimal value of α required to produce a divergent average avalanche size in the random neighbour model is (asymptotically in N, the number of sites) $1/q$ as shown by Bröker and Grassberger (1997) as well as Chabanol and Hakim (1997), who determined the limiting distribution numerically by iterating over the underlying convolutions. They found that the average avalanche size diverges only in the conservative limit, $\alpha \to 1/q$, and confirmed its rapid increase as α approaches $1/q$, which renders numerical estimates by direct simulation rather unreliable. *The mean-field theory of the OFC Model is critical only in the conservative case.*

5.3.1.1 Average avalanche size

One might wonder whether the average avalanche size can be derived from random walker arguments similar to those presented in Sec. 8.1.1.1. However, in the present case, due to the global, uniform driving, the situation is slightly more complicated, because the total influx (of force) per avalanche is unknown. Nevertheless, a slightly weaker result can be derived.

The value of α for the average avalanche size to diverge can be derived promptly using the stationarity condition and *assuming an independent, random force distribution* $\mathcal{P}^{(F)}(F)$ even in a finite system of N sites, which is non-zero at F_{th}. Figure 5.4 shows the force distribution for the random neighbour as well as the two-dimensional OFC Model with open boundary conditions and different values of α. An independent, random force distribution means that the average maximum force is

$$\langle F_{max} \rangle = F_{th} - \int_0^{F_{th}} \mathrm{d}G \left(\mathsf{F}^{(F)}(G) \right)^N , \tag{5.5}$$

where $\mathsf{F}^{(F)}(G) = \int_0^G \mathrm{d}F\, \mathcal{P}^{(F)}(F)$ is the cumulative distribution function (CDF). On average, the total force influx by the uniform drive across N sites therefore is $N(F_{th} - F_{max}) = N \int_0^{F_{th}} \mathrm{d}G\, \mathsf{F}^{(F)}(G)^N$. As toppling sites have a force of at least F_{th}, the average dissipation is bounded from below by $(1 - q\alpha) \langle s \rangle F_{th}$ which is compensated by the influx at stationarity, so that

$$\langle s \rangle \leq \frac{N \int_0^{F_{th}} \mathrm{d}G \left(\mathsf{F}^{(F)}(G) \right)^N}{(1 - q\alpha)F_{th}}. \tag{5.6}$$

Using $\mathcal{P}^{(F)}(F_{th}) \neq 0$ and assuming that the PDF is analytic in the vicinity of F_{th}, the CDF can be approximated by $\mathsf{F}^{(F)}(G) \approx 1 - (F_{th} - G)\mathcal{P}^{(F)}(F_{th})$ and the integral in Eq. (5.6), for large N dominated by the behaviour of $\mathsf{F}^{(F)}(G)$ for G around F_{th}, becomes, asymptotically in large N,

$$\int_0^{F_{th}} \mathrm{d}G \left(\mathsf{F}^{(F)}(G) \right)^N \approx \frac{1}{(N+1)\mathcal{P}^{(F)}(F_{th})}. \tag{5.7}$$

In large N, the average avalanche size therefore is bounded by

$$\langle s \rangle \leq \frac{1}{F_{th}(1 - q\alpha)\mathcal{P}^{(F)}(F_{th})}, \tag{5.8}$$

which for $\alpha < 1/q$ to leading order is independent of N and diverges only in the conservative limit, $\alpha \to 1/q$, even under more general conditions than those originally set out by Lise and Jensen (1996) – under the assumption of finite $\mathcal{P}^{(F)}(F_{th})$ and independent forces [14] the average avalanche size is finite and asymptotically independent of the system size in the presence of dissipation. The above argument can be generalised and refined using general results from extreme value statistics (Gumbel, 1958, also Sec. 5.4.3).

---

[14] If correlations in F_n decay sufficiently fast, the derivation above remains valid even without independence.

5.3.1.2 Marginal phase locking

Bröker and Grassberger's (1997) analysis mentioned above was guided by the belief 'that spatial structures are crucial for the emergence of scaling' which is put into perspective by the scaling of the conservative random neighbour OFC Model. Indeed, **spatial inhomogeneities** in the form of **boundary conditions**, or even as quenched disorder, to be discussed below, see for example Fig. 5.6(c), are of fundamental importance in the spatial OFC Model. In the conservative limit, dissipation can only take place at the boundaries, which, for the sake of a stationary state, cannot be periodic. Even in the non-conservative regime, periodic boundary conditions destroy scaling and lead to periodic states (Socolar, Grinstein, and Jayaprakash, 1993; Grassberger, 1994; Middleton and Tang, 1995). This observation has motivated several numerical studies of the rôle of the boundary, as discussed in the following.

Grassberger (1994) found that two-dimensional systems with *periodic boundary conditions* and $\alpha \lesssim 0.18$ display almost exclusively avalanches of size 1, i.e. the external drive triggers a single toppling which leads to no subsequent topplings.[15] In this regime, the system visits N different states sequentially with period $1 - q\alpha$ (influx of force per site). At the same time, the distribution of forces across the system is structured but very broad. With increasing α avalanches tend to become larger and to spread further, so that the strict periodicity disappears (but see Socolar *et al.*, 1993), whereas the distribution of forces becomes sharper, somewhat reminiscent of the ZHANG Model, see Fig. 4.6, p. 108. Randomising the energy distribution at toppling is known to make the spiky structure disappear (Pietronero *et al.*, 1991). At *open boundaries* the number of *charging* nearest neighbours is reduced. Sites at the boundary redistribute the same fraction of their force as bulk sites, but the number of sites that can charge them is reduced. In the mechanical picture of the BURRIDGE–KNOPOFF Model, open boundaries correspond to a fixed, rigid frame around the bulk sites. This has been criticised as unphysical (Leung *et al.*, 1997), however, remedying the situation seems to destroy scale invariance. This observation is somewhat at odds with the work by Christensen and Olami (1992a), who derived critical exponents for a whole range of different boundary conditions. Since sites at open boundary are exposed to a different local environment compared to bulk sites, they break the periodic behaviour the bulk otherwise would slip into. For example, if every site topples once on a square lattice, namely as soon as it reaches F_{th} due to the external drive, all bulk sites lose in total force $F_{th}(1 - 4\alpha)$, while boundary sites lose more, $F_{th}(1 - 3\alpha)$ or even $F_{th}(1 - 2\alpha)$ at corners.

Instead of generating *more* disorder, however, open boundaries lead to **marginal (phase) locking**, a term coined by Middleton and Tang (1995). For any small α, periodicity breaking avalanches invade their neighbourhood from the boundaries and, as a result, correlations start spreading through the bulk. As shown in Fig. 5.5 and Fig. 5.6, small synchronised (i.e. phase locked) patches, similar to those found in the DS-FFM, Fig. 5.2, and somewhat reminiscent of ferromagnetic domains, appear at the open edge of a lattice and slowly creep

---

[15] Correspondingly, Olami *et al.* (1992) and Bottani and Delamotte (1997) showed that at $\alpha \approx 0.18$ the average avalanche size remains finite in the thermodynamic limit, because $\tau > 2$ below that value.

towards its centre, producing increasingly large synchronised patches on their way. [16] The boundary sites with their different period prevent full synchronisation, leaving the system marginally locked in patches (Socolar *et al.*, 1993). Lise (2002) captured these patch-wise correlations by examining the PDF of the block-wise averaged force (Sec. 2.3.1, also Binder, 1981a; Pruessner, 2008). The clear deviation from a Gaussian and an obvious skewness signals the presence of long-range correlations, yet one might dispute the suggested presence of a BHP PDF (Bramwell, Holdsworth, and Pinton, 1998; Bramwell, Christensen, Fortin, *et al.*, 2000).

The OFC Model, which otherwise would be trapped in a periodic loop of states, even with very few inhomogeneities explores a much larger phase space volume (see Fig. 5.6(c), Grassberger, 1994; Ceva, 1995, 1998; Middleton and Tang, 1995; Torvund and Frøyland, 1995; Mousseau, 1996). A similar effect has been observed in the fully deterministic (point-driven) Abelian BTW Model, which required some form of disorder to display scaling features (Lübeck *et al.*, 2000, also Wiesenfeld *et al.*, 1990), and also experimentally (Altshuler *et al.*, 2001; Ramos *et al.*, 2009, but see Costello *et al.*, 2003 for disparate results). The inhomogeneity, however, has a rather counter-intuitive effect, because *while destroying temporal periodicity, it causes phase locking, i.e. the formation of patches*. These patches in turn prevent the system from periodically cycling through the same states producing a chain of avalanches of size $s = 1$, which are often explicitly excluded from a scaling analysis because of their dominating statistical weight (Grassberger, 1994; Ceva, 1998; Lise and Paczuski, 2001a; Drossel, 2002; Wissel and Drossel, 2006).

As even a very low level of inhomogeneity or quenched noise changes its behaviour completely (see Fig. 5.6(c); Torvund and Frøyland, 1995; Mousseau, 1996), the OFC Model with periodic boundary conditions is highly unstable (Ceva, 1998), whereas systems with open boundaries have been reported to remain unaffected by noise (Olami *et al.*, 1992). Randomising the threshold F_{th} locally, however, destroys the critical behaviour in the non-conservative regime altogether (Jánosi and Kertész, 1993, also Mehta and Barker, 1994). Without dissipation, this does not seem to happen (also Manna, 1990; Christensen, 1992, also Sec. 4.1, p. 88).

The slow creep of phase locked patches implies very long transients, which for the OFC Model have been reported to be as long as 10^{10} avalanches (Middleton and Tang, 1995). The long transients, compared to models like the ASM and the ZHANG Model, are even more surprising given that the OFC Model is driven *globally*, i.e. the macroscopic time scale is set by an energy influx *per site*. The macroscopic time in units of F_{th} indicates the minimal number of times each site must have toppled, but does not determine the total number of topplings throughout the system, as dissipation differs for boundary sites. With such a rescaled time, the ASM and the ZHANG Model have a constant transient time (see Sec. 4.2.1, p. 95 and Sec. 4.3, p. 104), whereas Lise (2002) found the (macroscopic) transient time in the OFC Model to be a universal power law of the system size.

The marginal locking mechanism mentioned above suggests that the OFC Model is somehow constrained in the phase space it can explore, so that averaging over several

---

[16] Sites on the same sublattice acquire the same force, $F_n = 0$, when relaxing simultaneously, which subsequently leads to a clearly visible checkerboard type pattern.

random initialisations might be required (Ceva, 1998). This apparent constraint might be due to the deterministic nature of the model. There are, however, no obvious, rigorous arguments why the OFC Model displays (temporal) periodicity with periodic boundary conditions and some non-trivial behaviour with open boundary conditions. The ASM, for example, is periodic with open boundaries (Dhar, 1999c), when driven deterministically (on the same site). This is a necessity given its deterministic, discrete nature which means that there is a finite set of states the system explores deterministically. The continuum nature of the OFC Model allows it to escape from that necessity.

Figure 5.5 shows the evolution of the OFC Model as first presented by Middleton and Tang (1995), who analysed the scaling of the invasion process in further detail. The boundary conditions are periodic at the top and the bottom and open on the left and right sides, so that the spatial homogeneity can be attributed to the open boundaries. As illustrated in Fig. 5.6(a) and Fig. 5.6(b), one might conclude that the OFC Model displays some form of scale invariance in its patchiness. Lise (2002) found in a numerical study of the approach to stationarity that patches adjacent to the boundary have a transient that depends on their size, whereas patches placed at the centre of the system have a size-independent transient time, of the order of the longest time observed at the boundary. This suggests a 'self-organisation mechanism', which gains a foothold at the boundary and gradually invades the whole system until the 'self-organised region spans the whole system' (Lise, 2002). The important rôle of synchronisation in the OFC Model placed it in the somewhat unexpected context of synchronisation as observed in the living world (Peskin, 1975; Winfree, 1980; Grassberger, 1994).

5.3.1.3 Universality in the non-conservative regime

Universality and robustness (Bak and Tang, 1989a) is key to the relevance of SOC as a whole and became the subject of a major debate in the context of the OFC Model. Originally, Olami et al. (1992) suggested that the exponents found in the OFC Model depend very strongly on the level of conservation, which, in fact, is consistent with the variation of the exponents found when fitting empirical data to the **Gutenberg–Richter law** (Olami et al., 1992). This observation of non-universal, i.e. parameter-dependent scaling has been extended to the power spectrum (Christensen et al., 1992) and confirmed in a number of numerical studies (Christensen and Olami, 1992a; Jánosi and Kertész, 1993) as well as in a range of other models (e.g. Lübeck and Usadel, 1993; Lübeck, Tadić, and Usadel, 1996; Newman, 1996; Tadić, 1999; Černák, 2002; Khfifi and Loulidi, 2008). The OFC Model is remarkably rich in its features and has been studied systematically with respect to a wide range of variations. In addition to the 'localisation transition' (p. 130), Christensen and Olami (1992a) identified a transition controlled by the level of anisotropy and a discontinuous transition in the conservative limit, $\alpha \rightarrow 1/q$.

This latter transition has been confirmed by Jánosi and Kertész (1993), whereas Middleton and Tang (1995) found in their effective model only a smooth dependence of the relevant exponent on α. Grassberger (1994), who was the first to study reasonably large system sizes (see Sec. 5.3.4 and the Appendix A), however found that the apparent avalanche size exponent is $\tau \approx 1.8$ independent of α and that FSS is violated. He reiterated a point made earlier

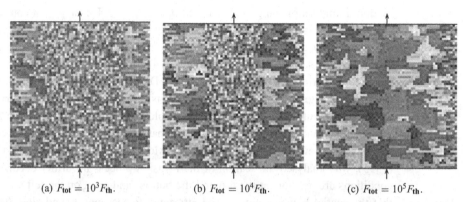

(a) $F_{tot} = 10^3 F_{th}$.   (b) $F_{tot} = 10^4 F_{th}$.   (c) $F_{tot} = 10^5 F_{th}$.

Figure 5.5   The OFC Model at $\alpha = 0.07$ with linear size $L = 64$, open boundaries at the sides and periodic boundary conditions in the vertical direction as suggested by the arrows (Middleton and Tang, 1995). Darker shades of grey indicate lower force. F_{tot} measures the total force influx per site by the external drive. Initialised with uniformly distributed random forces, after a short period small patches appear along the open boundaries ((a) at $F_{tot} = 10^3 F_{th}$ about $5.5 \cdot 10^6$ avalanches have been triggered) which grow towards the disordered centre ((b) at $F_{tot} = 10^4 F_{th}$ after about $53 \cdot 10^6$ avalanches) until they cover the entire system ((c) at $F_{tot} = 10^5 F_{th}$ after about $483 \cdot 10^6$ avalanches) in patches of increasing size. The open boundaries break the periodicity of single site toppling, as obtained with periodic boundaries, and induce phase locked patches.

by Klein and Rundle (1993), that the avalanche dimension must fulfil the relation $D \leq d$ (but Table 5.2) for $\alpha < 1/q$ because every avalanche can at most dissipate $L^d F_{th}$ in at most $L^d F_{th}/(1 - q\alpha)$ topplings. There are strictly no avalanches bigger than $L^d F_{th}/(1 - q\alpha)$. The cutoff s_c is proportional to an avalanche size in the tail of the distribution, i.e. an avalanche size with finite probability, so that $s_c \mu \leq L^d F_{th}/(1 - q\alpha)$ for some fixed proportionality factor μ and therefore $s_c \propto L^D \in \mathcal{O}(L^d)$ which means $D \leq d$. For any system size L the cutoff never clashes with the constraint $s_c \in \mathcal{O}(L^d)$, and as long as L is small enough, an estimated $D > d$ will not clash with $L^D \in \mathcal{O}(L^d)$ in any apparent way. For sufficiently small system sizes, one might therefore measure $D > d$, as reported by Christensen and Olami (1992a) and confirmed by Grassberger (1994), who was aware of the asymptotic constraint $D \leq d$. Christensen (1993) determined conservative estimates ($L \approx 760\,000$)[17] of the smallest system sizes required to render the scaling of $s_c \propto L^D$ with $D > 2$ visibly wrong.

Ceva (1998) found an avalanche size exponent independent of the level of conservation, α, of about $\tau \approx 1.63$ illustrating at the same time the dramatic deviations from that value caused by too small lattices. The book by Christensen and Moloney (2005) contains a brief discussion of the OFC Model, concluding that, based on a moment analysis, the model might not display simple scaling in the non-conservative regime. Their results are compatible with those by Lise and Paczuski (2001b) who found instead some evidence for multiscaling (Sec. 2.4, p. 50). Nevertheless, they also obtained a universal (i.e. α

---

[17] Conservative because the true asymptotic behaviour, if there is any, might set in at significantly smaller system sizes, since the avalanche sizes are likely to be cut off before the theoretically largest possible avalanche size.

(a) $F_{tot} = 3.8 \cdot 10^5 F_{th}, L = 64$.   (b) $F_{tot} = 5 \cdot 10^7 F_{th}, L = 256$.   (c) $F_{tot} = 1.3 \cdot 10^8 F_{th}, L = 64$, single site inhomogeneity at the centre, PBC.

Figure 5.6   Similar to Fig. 5.5, three snapshots of the non-conservative OFC Model ($\alpha = 0.07$) in different setups. (a) A typical configuration for a fairly small system ($L = 64$), just after the end of the transient (about $2 \cdot 10^8$ avalanches), low level of conservation, $\alpha = 0.07$, and (the typical) open boundary conditions. (b) Same as (a), but for a larger system size (after $4.2 \cdot 10^{12}$ avalanches). (c) Same as (a), but periodic boundary conditions (PBC), so that the system would remain disordered if it were not for the single site inhomogeneity with $\alpha = 0.17$. A small island of homogeneity appears around it after very many iterations (about $7.1 \cdot 10^{11}$ avalanches). Stationarity has apparently not been reached.

independent) apparent exponent $\tau \approx 1.8$, a result that has since been questioned by Boulter and Miller (2003). Similarly, Wissel and Drossel (2006) found 'dirty power laws', 'apparent power laws [that do] not reflect true scale invariance of the system.'

The numerical situation improves significantly when the scaling analysis is applied to avalanches fully contained within a given, smaller patch of the whole system. Lise and Paczuski (2001a) recovered simple scaling for such patches. Their location with respect to the boundary plays an important rôle (Lise, 2002), as one would expect according to the mechanism discussed above. These studies are somewhat reminiscent of the finding by Drossel (2000) in the ASM, where avalanches behave differently depending on whether or not they reach the boundary, i.e. are dissipative or not.

The results for the OFC Model are remarkably inconsistent. Drossel (2002) argued that this might be caused partly by numerical problems such as limited floating point precision (Sec. A.4.5, p. 379). In fact, changing the type used to store the forces F_n has a visible impact on the avalanche size distribution, with a limited precision contributing spuriously to the phase-locking.

5.3.2 Conservative limit

From a physical point of view, the conservative, isotropic limit is a rather peculiar case, where the driving spring constant, K_L, happens to vanish, Eq. (5.3) on p. 126. Vanishing K_L means that there is no restoring force caused by the displacement of a site relative to the driving plate, Fig. 5.3(b). This, however, contradicts the driving mechanism which allows the force on all sites to be increased uniformly everywhere by pulling; $K_L = 0$ implies that

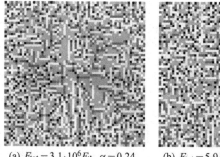

(a) $F_{tot} = 3.1 \cdot 10^6 F_{th}$, $\alpha = 0.24$.   (b) $F_{tot} = 5.9 \cdot 10^5 F_{th}$, $\alpha = 0.25$.

Figure 5.7   Similar to Fig. 5.5, two snapshots of the OFC Model with $L = 64$ and open boundary conditions, (a) close to ($\alpha = 0.24$, $5 \cdot 10^9$ avalanches) and (b) at conservation, respectively, after $2 \cdot 10^9$ avalanches. The slightly non-conservative system shows visibly more structure than the conservative one, which nevertheless contains small synchronised islands.

there is no spring attached to the driving plate (Christensen and Moloney, 2005). Until the advent of the OFC Model, the BURRIDGE–KNOPOFF Model was primarily investigated in the conservative limit (Olami *et al.*, 1992).

In the (locally) conservative limit some form of open boundaries become a necessity, because the continuous influx of external force needs to be dissipated somewhere to prevent the system from running indefinitely. This feature is shared with many other SOC models, such as the BTW Model, and adds to the perception that boundaries and bulk conservation laws are the key ingredients of any SOC model. Christensen and Olami (1992a) investigated a wide range of boundary conditions, including 'free', reflecting and 'open' boundaries, the choice of which changes the scaling behaviour. Figure 5.7 shows two examples of the configurations of relatively small systems on a two-dimensional square lattice, $L = 64$, close to ($\alpha = 0.24$) and at the conservative limit ($\alpha = 0.25$).

Because of its physical appeal and its greater relevance both to (geo-)physics and SOC, the non-conservative regime has received much more attention than the conservative limit. This is regrettable, as the latter behaves much more consistently compared to the former (Christensen and Moloney, 2005). Bottani and Delamotte (1997) identified $\alpha = 0.24$ as a threshold above which the OFC Model displays critical behaviour without synchronisation. Jánosi and Kertész (1993) confirmed the findings by Christensen and Olami (1992a) that the avalanche dimension D changes discontinuously with the degree of non-conservation, jumping from $D \approx 1.9$ at around $\alpha = 0.22$ to $D \approx 3.25$ at $\alpha = 0.25$. They invoked the usual random walker argument (Sec. 8.1) to show $D(2 - \tau) = 2$ in the conservative limit, whereas $D(2 - \tau) < 2$ is strongly suggested from the numerics in the non-conservative regime. In fact, $D(2 - \tau)$ shows the strongest sign for a discontinuity in the exponents. The singular behaviour of the OFC Model as a function of α is also found in three dimensions (Peixoto and Prado, 2004).

Christensen and Olami (1992a) ascribe the change in the exponent to a change of 'shape' of the avalanche, where the height is given by the number of topplings of a site during a single avalanche. In the conservative limit the height of the cone-shaped avalanche is proportional to its radius, whereas any level of non-conservation prescribes some maximum height of that cone. These results are corroborated by a similar discussion by Christensen and Moloney (2005), suggesting $D \approx 3.01$ and $\tau \approx 1.253$ from a moment analysis.

The claim by Middleton and Tang (1995) that the conservative OFC Model belongs to the BTW Model universality class does not seem to be supported by numerics.

5.3.3 Geophysical relevance

Within statistical mechanics, the OFC Model is a unique SOC model, incorporating, to the great surprise of many, non-conservation and all features that are quintessential ingredients for SOC models: slow drive and avalanching (separation of time scales), thresholds, local dynamics. The credibility of the OFC Model benefits greatly from its geophysical origin as the Burridge–Knopoff Model. To date, earthquakes are a showcase natural phenomenon for SOC (Sec. 3.4).

When Bak and Tang (1989a) entered the stage of geophysics, they dismissed essentially all research on earthquakes and the mechanism underlying the occurrence of a power law relation between frequency and earthquake size. The ensuing rift between statistical mechanics and geophysics, which was partly fuelled by this disregard, was briefly retraced by Sornette and Werner (2009). In contrast, Sornette (1991) wrote a much more enthusiastic review, which ties together the basic ideas of SOC and the more technical concepts of seismology. Recently, Hergarten (2002) published an extensive overview of the different areas in earth sciences, which could benefit from the 'SOC concept'.

As Bak and Tang (1989a) pointed out, a model of earthquakes should be robust with respect to local details. For example, introducing some quenched randomness, as one expects to be present in the fault features, should not spoil the scaling of the model. Yet, the statistical characteristics of the OFC Model depend sensitively on such details: noise, boundary conditions, anisotropies, all affect the scaling behaviour. This behaviour is more pronounced in the OFC Model, but is not entirely new in SOC – very early on, Kadanoff *et al.* (1989) found what they thought were a number of different universality classes and broken FSS in a range of variants of the BTW Model.

The bemoaned (e.g. Torvund and Frøyland, 1995) lack of robustness in the OFC Model becomes a welcome feature in the eyes of anyone who attempts to fit it against empirical data. In fact, a whole range of exponents has been reported for earthquakes. The exponent found in the Gutenberg–Richter law is in the range $[1.8, 2.05]$ (Olami *et al.*, 1992; Christensen and Olami, 1992a, but Pacheco *et al.*, 1992, also Sec. 3.4), corresponding to $\alpha \approx 0.2$ (on a square lattice), which is of particular significance, as $\alpha_1 = \alpha_2 = \alpha = 1/5$ for $K_1 = K_2 = K_L$ in Eq. (5.3). In the light of the different nature of the different spring constants, see Fig. 5.3(b), this latter choice, however, appears difficult to rationalise. In a brief review, Hergarten (2003) concluded that existing SOC models, including the OFC Model, in their current setup do not model earthquakes appropriately, either for (geo-)physical reasons or because they 'strongly overestimate the number of large events.'

5.3.4 Numerical methods

There is a striking difference in the parameter range of the older simulations and the more recent ones, especially in terms of system size and the number of iterations, which is mainly caused by the dramatic improvement in the implementations of the OFC Model. Grassberger (1994) listed a number of very effective measures to improve the performance of implementations of models like the OFC Model. Firstly, a list or **stack** of active sites should be kept, which by now is commonplace in many models. Secondly, **sites** can be **addressed** in different ways, the efficiency of which might depend on the programming language used and the details of the implementation. Thirdly, the **threshold** F_{th} is to be **lowered**, rather than **sweeping** over the entire lattice and increasing all forces by the same amount. [18] This improvement has by far the most drastic effect on the efficiency. Finally, Grassberger proposed a very simple algorithm using '**boxes**' that greatly improves the speed with which the largest force in the system is located. The algorithm differs significantly from the naïve approach of constantly scanning the lattice for the maximum force and from the more conventional method of using a **binary search tree**. Further details of Grassberger's scheme are deferred to the Appendix A, which illustrates some general numerical techniques using a sample implementation of the OFC Model. Efficient algorithms are of particular importance in the OFC Model, where enormous transients have been found (Grassberger, 1994; Middleton and Tang, 1995).

Drossel (2002) identified limited **floating point precision** as one cause of marginal phase locking, concluding that only 'single-flip avalanches' would occur if only the precision were high enough. Similarly, noise in natural systems could bring about phase locking, even when an analytically exact model without any noise eventually slips into a strictly periodic sequence of states. If that is correct, all attempts to perfect the numerical implementation of the OFC Model counteract the aim of modeling a *natural* system. Here, once more, one sees the almost paradoxical nature of the OFC Model, where phase locking is caused by disorder (Sec. 5.3.1.2, in particular p. 134).

5.3.5 Numerical results

In this section the numerical results as collected in Table 5.2 and Table 5.3 are discussed. Given the enormous amount of literature on the model and its close links to experiments and observations, it is striking how few solid numerical results are reported beyond rough estimates. This is partly to blame on the continuous parameter α with very few choices coinciding across publications. The continuity has also led some authors to present their data in a viewgraph illustrating the dependence of the relevant exponent as a function of α (e.g. Jánosi and Kertész, 1993; Bottani and Delamotte, 1997; Ceva, 1998) rather than listing the results in a table. [19] Where exponents are reported, they focus on the avalanche size (the total energy release), whereas hardly any results are reported for the avalanche

---

[18] This 'trick' has some rather unexpected consequences for the numerical precision, see Sec. A.4.4, p. 376.
[19] Such results are not listed in Table 5.2 or Table 5.3.

duration (on the microscopic time scale), which could, quite naturally, be measured as the number of parallel updates.

Despite being the simplest case, the literature for one dimension is remarkably sparse. Wissel and Drossel (2005) could rule out power law behaviour in sufficiently large, one-dimensional systems (trivial periodic behaviour found by Zhang, Huang, and Ding, 1996), contrasting that to other, closely related, one-dimensional models (Crisanti, Jensen, Vulpiani, and Paladin, 1992; Blanchard, Cessac, and Krüger, 2000).

In two dimensions different authors focus on different values of α. Christensen and Olami (1992a) argue in favour of $\alpha = 0.20$ as the value for the level of conservation most relevant to natural phenomena which are characterised by $\tau \in [1.8, 2.05]$, compatible with the values reported in Table 5.2. This is contrasted by the much larger exponent of $\tau = 2.7(1)$ at $\alpha = 0.10$. Grassberger (1994) questions critical behaviour at such small values of α but confirms an approximate power law with $\tau \approx 1.8$. Lise and Paczuski's (2001a) results, as tabulated in Table 5.2, are based on box-scaling (Sec. 2.3.1, p. 49). They found $\tau = 1.8$ for all values of α in the non-conservative regime. Lise and Paczuski (2001b) concluded that the OFC Model lacks simple scaling behaviour for the overall avalanche size distribution, whereas events conditional to their confinement to small subsystems were found to scale robustly with universal critical exponents (Lise and Paczuski, 2001a).

Originally, Bak and Tang (1989a) argued that the earth's crust is a *three*-dimensional critical system rather than a two-dimensional fault, as assumed later (e.g. Christensen and Olami, 1992a). Yet, the literature on the three-dimensional or any higher dimensional version of the OFC Model is virtually non-existent. Unfortunately, Peixoto and Prado (2004) did not determine the avalanche size exponents in their study, but focused on the statistics of epicentres.

5.4 The BAK–SNEPPEN Model

The BAK–SNEPPEN Model (BS Model; Bak and Sneppen, 1993), as defined in Box 5.4, was intended to model so-called **punctuated equilibrium** as introduced by Eldredge and Gould (1972, reviewed by Gould and Eldredge, 1993) and backed by empirical evidence by Kellog (1975). Punctuated equilibrium is widely and controversially discussed in evolutionary biology as the underlying mechanism that leads to **speciation**, i.e. drastic changes in the genotype (Gould, 2002). In the aftermath of the BS Model, sandpiles themselves and the whole of SOC was perceived as an instance of punctuated equilibrium (Bak and Paczuski, 1995; Yang and Cai, 2001; Gould, 2002).

As opposed to all other models introduced in this book, the BAK–SNEPPEN Model does *not* require an explicit separation of time scales (Box 5.4). The model evolves at a constant rate exhibiting a form of **extremal dynamics** (Sec. 5.4.3). Avalanches are defined a posteriori, by imposing thresholds on the key observable, which is the **minimal fitness** in the system. This fitness is updated stochastically.

The model is easily implemented and displays non-trivial behaviour in one dimension already, as reported in extensive numerical and analytical studies. Because only a single site

General features

- First published by Bak and Sneppen (1993).
- Intended to capture some key features of evolution.
- Originally defined in one dimension, easily extended to higher dimensions.
- Deterministically chosen site and nearest neighbours are stochastically updated.
- Single update rule; no separation of time scales for driving and relaxation.
- Non-conservative dynamics.
- Not robust with respect to the microscopic dynamics.
- Extensively studied numerically and analytically.
- Exponents listed in Table 5.4, p. 143.
- Upper critical dimension disputed, $d_c = 4$ or $d_c = 8$.

Rules

- Each site $\mathbf{n} \in \{1, 2, \ldots, L\}^d$ of $N = L^d$ sites on a d dimensional hyper-cubic lattice (extension to other lattices obvious) with periodic boundaries has a fitness $f_\mathbf{n} \in [0, 1]$.
- *Initialisation*: $f_\mathbf{n}$ randomly and independently taken from a uniform distribution on $[0, 1]$.
- *Update*: find $\mathbf{n}_{\min}$ with the (globally) smallest fitness $f_{\min} = f_{\mathbf{n}_{\min}}$. Assign new fitnesses $f_\mathbf{n}$, $f_{\mathbf{n}'}$ picked uniformly and independently at random from $[0, 1]$ to $\mathbf{n} = \mathbf{n}_{\min}$ and all its nearest neighbours $\mathbf{n}'$. One such update constitutes a time step.
- Key observables:
  minimum fitness $f_{\min}(t)$ in the system as a function of time t;
  gap function, $G(t) = \max\{f_{\min}(t') | 0 \leq t' \leq t\}$;
  fitness distribution;
  non-stationary avalanche duration, i.e. time between two increases in $G(t)$;
  f_0-avalanches, that is duration of $f_{\min}(t)$ below threshold f_0;
  duration and size are used synonymously.

and its neighbourhood are updated at a time, the question whether updates commute does not actually arise, i.e. the attribute Abelian or non-Abelian does not apply.

The BS Model was inspired by a related model invented by Sneppen (1992, also Zaitsev, 1992; Sneppen and Jensen, 1993; Tang and Leschhorn, 1993) in the context of growth processes. In their extensive review, Paczuski *et al.* (1996) compared it in particular to the linear interface model (LIM), which is in the same universality class as the MANNA Model (Sec. 6.1, 6.2) and the OSLO Model (Sec. 6.3). The Reggeon field theory (Gribov, 1967; Brower, Furman, and Moshe, 1978; Liggett, 2000) proposed for the BS Model (Paczuski, Maslov, and Bak, 1994b) has been (partly) withdrawn (Paczuski, Maslov, and Bak, 1994a) and a tractable stochastic equation of motion, i.e. a Langevin equation, is still lacking.

The BS Model was designed to capture some key mechanisms of evolution. Species live in a **fitness landscape** and their genetic code is subject to constant mutations. Some of

Table 5.4 Exponents of the BS Model.
$\mathcal{P}^{(s)}(s) = a^{(s)} s^{-\tau} \mathcal{G}^{(s)}(s/(b^{(s)}(f_c - f_0)^{-1/\sigma}))$ is the f_0-avalanche size distribution. The exponent $1/\sigma$ plays a rôle similar to λ in the DS-FFM, see Table 5.1, p. 120. Observables and exponents are defined in Sec. 1.3 (scaling relations and further exponents in Maslov, 1996; Paczuski *et al.*, 1996; Marsili, De Los Rios, and Maslov, 1998).

d	τ		σ	
1	0.9(1)	(a)	0.35(2)	(b)
	1.08(5)	(b)	0.346(5)	(e)
	1.08	(c)	0.343(4) ◇	(f)
	1.1204	(d)		
	1.073(3)	(e)		
	1.07(1)	(f)		
	1.07	(g)		
2	1.1	(a)	0.444(6) ◇	(f)
	1.245(10)	(f)		
	1.25	(g)		
3	1.35	(g)		
4	1.41	(g)		
MFT	3/2	(h)	1/2	(h)

◇ Derived via $\sigma = (2 - \tau)/\gamma$ (Paczuski *et al.*, 1996, e.g. p. 423). *a* Bak and Sneppen (1993), *b* Jovanović, Buldyrev, Havlin, and Stanley (1994), *c* Paczuski *et al.* (1994b), *d* Marsili (1994a), *e* Grassberger (1995), *f* Paczuski *et al.* (1996), *g* De Los Rios, Marsili, and Vendruscolo (1998), *h* de Boer, Derrida, Flyvbjerg, *et al.* (1994).

these mutations might lead to improved **fitness**, while others reduce it. A **barrier height** represents the amount of genetic change required for a given species to arrive at a new (local) fitness maximum, not necessarily higher than the current. The lower the barrier, the sooner a genetic change occurs. A species that crosses the barrier can be considered as mutating as a whole or giving way to another species invading and eventually displacing it from its ecological niche. The new fitness maximum has associated with it a new barrier height. In fact, the barrier height is encoded in the fitness, as a low fitness might suggest that the species is prone to mutation and, conversely, a low barrier indicates easy access to improvement and therefore low fitness. In turn, high fitness in a given environment leads to stability and thus a high barrier. Interaction between species means that the mutation of one changes the fitness landscape around another. The model implements this over-simplified mechanism by assigning in every (discrete) time step to the species with the globally lowest fitness and its nearest neighbours, new fitnesses taken independently, uniformly and randomly from the interval [0, 1].

Because the notion of barriers in evolution is open to debate, as is the nature of the model as a model of evolution, many authors have abandoned the initial terminology and refer to the barriers as local fitness (e.g. Paczuski *et al.*, 1996), or simply as random numbers (e.g. Boettcher and Paczuski, 2000). The terminology of local fitnesses is adopted in the following.

The BS Model is **non-conservative** in the sense that the (bulk) dynamics does not conserve the total fitness of the sites involved (Sec. 9.2.2). Yet, the notion of non-conservation in the BS Model might appear slightly contrived, since the bulk dynamics does not involve the fitness as some transported quantity, but merely as a means to determine the order of updates.

Prima facie the model does not seem to incorporate a separation of time scales between driving and relaxation, simply because there are not two distinct mechanisms for these two processes. The model requires global control, in that the overall minimum of all fitnesses has to be found. As suggested by Bak and Sneppen (1993, also Sneppen, 1995a), such a global update can be replaced by a local mechanism, whereby low fitnesses are updated exponentially faster than high fitnesses, say with rate $\exp(-\alpha f_{\mathbf{n}})$ in the limit of $\alpha \to \infty$, which amounts to a separation of infinitely many (local) time scales. Observing a fixed subset of sites over time, the model displays **punctuated equilibria** as bursts of mutation within this subset. Similarly, observing the model on the exponential time scale, it displays times of calm (**stasis**) with intermittent eruptions of activity.

5.4.1 Histogram and gap function

Over time, the global fitness histogram converges to a step function, with very few fitnesses at any given time below a certain f_c (in one dimension $f_c = 0.667\,02(3)$, Paczuski *et al.*, 1996), where the histogram rises sharply, see Fig. 5.8(a). The distribution of the global minimum is shown in Fig. 5.8(b), indicating an approximately [20] linear decline from 0 to f_c, above which there is almost a continuum of sites available, so that the global minimum will very rarely ever be higher than f_c. In the thermodynamic limit, the probability of the smallest fitness $f_{\min}(t)$ being higher than f_c is expected to vanish (Paczuski *et al.*, 1996).

The maximum of $f_{\min}(t')$ across all $t' \in \{0, \ldots, t\}$ is known as the **gap function** $G(t)$. By definition, $G(t)$ is a monotonically increasing function of time t, which converges to 1 in any finite system. However, even for moderately large systems, with $N \geq 1000$ sites, within computationally accessible time, $G(t)$ will never exceed the asymptotic value f_c by more than a few per cent (Tabelow, 2001).

The self-organisation of the histogram into the distinctive shape shown in Fig. 5.8(a) is well understood. The evolution of a finite system developing these histograms is illustrated in Fig. 5.9. Starting from a uniform, random initialisation, $f_{\mathbf{n}} \in [0, 1]$, most updates lead to fitnesses above the current $f_{\min}$, so that the gap function $G(t)$ increases gradually but quickly. At these early stages, $G(t)$ represents the apparent dividing line between the majority of the sites above and the few fitnesses below. However, as opposed to $G(t)$, this

---

[20] The slight deviation from a perfectly straight line suggests that there is no simple argument for that behaviour. This is corroborated by the corresponding graph of the random neighbour model, see Fig. 5.12(b).

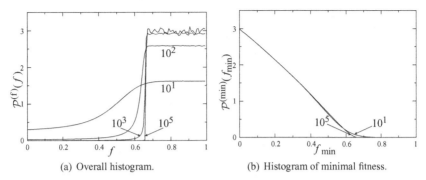

(a) Overall histogram.   (b) Histogram of minimal fitness.

Figure 5.8 Numerical results for the one-dimensional BS Model in the stationary state. (a) The histograms of all fitnesses $\mathcal{P}^{(f)}(f)$ for different system sizes, $N = 10^1$, 10^2, 10^3, 10^5. Asymptotically the distribution is a step function. Larger systems are noisier because an increasingly large fraction of sites remains untouched for increasingly long times. (b) The histograms of minimal fitness $\mathcal{P}^{(min)}(f_{min})$. The asymptotic shape is reached for very small system sizes and there is virtually no difference between $N = 10^2$ and $N = 10^5$.

apparent dividing line will stop rising as soon as the update of the minimum site produces typically one fitness beneath it.

In the stationary state, the number of sites moved across *any* given f is the same for sites moved from below f and from above, so the apparent dividing line in Fig. 5.9 is difficult to define in terms of a flux, in particular in a finite system. At stationarity, almost all f_{min} are found below the dividing line, apart from a few rare events, when they are taken from its immediate vicinity above. Although there is no net flux across any fitness f, fluctuations populate the region underneath the dividing line. They also allow $G(t)$ to venture into the region above. [5.4] In the thermodynamic limit the dividing line is located at f_c.

The asymptotic behaviour of a finite system is independent of its initial setup. Starting from a configuration with all fitnesses far above f_c, the system will very quickly relax to the same histogram as shown in Fig. 5.8(a) and will look indistinguishable from the final stage Fig. 5.9(c). The fitness distribution above f_c reflects the distribution the new fitnesses are drawn from (Paczuski *et al.*, 1994b; Ray and Jan, 1994; Tabelow, 2001), which, by definition, is uniform.

Considering further the approach to stationarity, for a given realisation of the BS Model, the gap function $G(t)$ will display plateaus while the minimal fitness $f_{min}(t)$ does not increase. The times between sudden jumps in $G(t)$, i.e. whenever $f_{min}(t)$ increases above the current $G(t)$, define a type of avalanche [21] with duration $s_G(t)$, during which some fitnesses are below $G(t)$. Because the fitness distribution above $G(t)$ is uniform, its average increase ΔG at the end of an avalanche is $\Delta G = (1 - G(t))/N$ (Paczuski *et al.*, 1996). [5.5] The average avalanche duration $\langle s_G(t) \rangle$ at a given time t is independent of the jump in the gap function, because the latter is determined only by the fitness distribution above a certain threshold. The avalanche is not affected by the actual values of these fitnesses, provided

---

[21] So-called 'backward avalanches' have been introduced as well (Maslov, 1995).

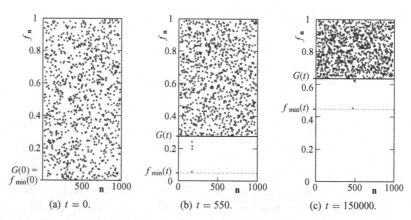

(a) $t = 0$.   (b) $t = 550$.   (c) $t = 150000$.

Figure 5.9 Three snapshots of a one-dimensional BS Model with $N = 1000$ sites at different times t. Each dot represents the fitness $f_\mathbf{n}$ at site $\mathbf{n}$. At the beginning of each time step t, the minimal fitness $f_{\min}(t)$ is recorded (dotted line). The gap $G(t)$ (solid line) is the maximum value the minimal fitness has attained in the time up to t.

they are above $G(t)$. The average rate of change of $G(t)$ therefore [22] is given by the **gap equation** (Paczuski *et al.*, 1994b)

$$\frac{\mathrm{d}}{\mathrm{d}t}\langle G(t)\rangle \cong \frac{1 - \langle G(t)\rangle}{N\langle s_G(t)\rangle}. \tag{5.9}$$

In a finite system the average avalanche duration $\langle s_G(t)\rangle$ does not diverge (Tabelow, 2001), because at most N fitnesses can be below $G(t)$ and there is a finite probability ϵ, which depends only on N and $G(t)$, that they are all above $G(t) < 1$ after N or fewer updates, even when starting from all N fitnesses below $G(t)$. In that sense, ϵ describes the worst case scenario. For large N and $G(t)$ close to unity, ϵ is extremely small, yet, it cannot vanish [23] for any $G(t) < 1$ and finite N. Similarly, the probability that all fitnesses are above $G(t)$ within $N + 1$ to $2N$ updates is underestimated by $(1 - \epsilon)\epsilon$. Continuing this argument for j periods of N updates, gives rise to an upper bound for the number of updates required,

$$\langle s_G(t)\rangle < \sum_{j=1}^{\infty}(jN)(1 - \epsilon)^{j-1}\epsilon = \frac{N}{\epsilon}, \tag{5.10}$$

so that Eq. (5.9) has the bound $\frac{\mathrm{d}}{\mathrm{d}t}\langle G(t)\rangle > (1 - \langle G(t)\rangle)\epsilon/N^2$ for all t, which means that

$$\lim_{t\to\infty} G(t) = 1, \tag{5.11}$$

for all finite N. Numerical simulations seem to tell a different story, as $\langle s_G(t)\rangle$ displays a power law divergence with $G(t)$ approaching f_c (Paczuski *et al.*, 1996), suggesting that $\frac{\mathrm{d}}{\mathrm{d}t}\langle G(t)\rangle$ vanishes there and therefore $\lim_{t\to\infty} G(t) = f_c$. This discrepancy between

---

[22] In a more careful derivation one would need to analyse the increments of $\langle G(t)\rangle$, which occur across the discretised time interval $[t, t + \langle s_G(t)\rangle]$.

[23] One possible realisation is that an updated fitness and all its neighbours arrive above $G(t)$, so that $\epsilon > (1 - G(t))^{(2d+1)N}$, assuming (overestimating) N such updates.

numerics and analytical derivation is not unusual, caused by $\langle s_G(t) \rangle$ becoming intractably large in numerical simulations. One cannot take the thermodynamic limit $N \to \infty$ first, as $\frac{d}{dt} \langle G(t) \rangle$ vanishes in that case, Eq. (5.9). Equation (5.11), however, does not imply that the fitness histogram asymptotically has all weight moved to ever increasing values of f. On the contrary, the fitness histogram has a non-trivial stationary state. Equation (5.11) means only that there are *some (rare) instances* when the overall fitness histogram has all weight above an ever increasing threshold $G(t)$, whereas *in the stationary state* it has *on average a shape similar to that shown in Fig. 5.8(a).*

5.4.2 f_0-avalanches and the γ-equation

While the avalanches described above are sensibly defined only in the approach of $G(t)$ to f_c, the object of interest in the **stationary state** is a different type of avalanche. In fact, the time trace of the minimal fitness $f_{\min}(t)$ defines an entire hierarchy of avalanches, given by the time spans between successive occurrences of $f_0 > f_{\min}(t)$, see Fig. 5.10. These are known as f_0-**avalanches**. Their statistical features depend strongly on the value of f_0 and it is generally assumed that scaling is obtained at and in the approach to $f_0 = f_c$. In the thermodynamic limit, the choice $f_0 > f_c$ leads to a single, infinite avalanche, so that the following derivations focus on $f_0 \leq f_c$.

The first moment of the f_0-avalanche durations is $\langle s \rangle_{f_0}$, the first moment of the area covered by such avalanches, i.e. the number of distinct sites whose value is changed during an f_0 avalanche, is correspondingly denoted by $\langle n_{\text{cov}} \rangle_{f_0}$. The notation n_{cov} as opposed to the slightly more common A for the avalanche area is used here for historic reasons.

Another asymptotic equation for $\langle s_G(t) \rangle$ introduced by Paczuski *et al.* (1996) is the γ-**equation**. It is based on the observation (Paczuski *et al.*, 1994a) that all sites must have fitness values above f_0 at the beginning and at the end of an f_0-avalanche. During such an avalanche of average duration $\langle s \rangle_{f_0}$ on average $\langle n_{\text{cov}} \rangle_{f_0}$ sites receive a new fitness, many of them multiple times (not least by multiple topplings). Given that $f_{\mathbf{n}} > f_0$ for all sites before and after the avalanche, the probability that the smallest fitness afterwards is in the interval $[f_0, f_0 + df_0]$ is $\langle n_{\text{cov}} \rangle_{f_0} df_0/(1 - f_0)$ for $f_0 < f_c$ and infinitesimal df_0, again using independence and uniform distribution of all new fitnesses above the threshold f_0 and assuming that the density of fitnesses around f_0 vanished at the beginning of the f_0-avalanche, i.e. merging (see Fig. 5.11) is caused only by updated sites. [5.6] This, at the same time, is the probability (or the fraction of times) that by lifting f_0 by df_0 the avalanche continues, as illustrated in Fig. 5.11. Given that the size of the thereby added avalanche is independent of the merging happening at all, the expected increase in $\langle s \rangle_{f_0}$ by lifting f_0 by df_0 is, to leading order [24] in df_0, the probability of an additional contribution, $\langle n_{\text{cov}} \rangle_{f_0} df_0/(1 - f_0)$ times the potential gain, which is $\langle s \rangle_{f_0}$. The total $\langle s \rangle_{f_0 + df_0}$ can therefore be written as the γ-equation (Paczuski *et al.*, 1994a),

$$\langle s \rangle_{f_0 + df_0} = \langle s \rangle_{f_0} + \langle s \rangle_{f_0} \frac{\langle n_{\text{cov}} \rangle_{f_0} \, df_0}{1 - f_0} \tag{5.12}$$

---

[24] Apart from those contributions mentioned in endnote [5.6], p. 392, higher orders would also enter when considering $f_{\min}$ at the termination of $s_2(f_0)$ in Fig. 5.11 to be in $[f_0, f_0 + df_0]$ as well.

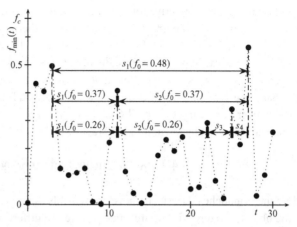

Figure 5.10 The minimal fitness $f_{\min}(t)$ as a function of time t and the different f_0-avalanche sizes derived from it by changing the value of f_0. For the largest value of $f_0 = 0.48$ the extent of a single avalanche ($s_1(f_0 = 0.48)$, top arrow) is marked, which splits into four avalanches for the smallest value of $f_0 = 0.26$. The function $f_{\min}(t)$ is discrete and the dotted lines connecting the points are only to guide the eye.

or

$$\frac{\mathrm{d}\ln\langle s\rangle_{f_0}}{\mathrm{d}f_0} = \frac{\langle n_{\mathrm{cov}}\rangle_{f_0}}{1 - f_0}. \tag{5.13}$$

In the limit $f_0 \to f_c$ the first moment $\langle s\rangle_{f_0}$ is expected to diverge like $(f_c - f_0)^{-\gamma}$, so that

$$\gamma = \lim_{f_0 \to f_c} \frac{\langle n_{\mathrm{cov}}\rangle_{f_0}(f_c - f_0)}{1 - f_0}. \tag{5.14}$$

This is a very unusual result, in that it relates an *exponent* on the left to a simple expectation value on the right.

Because after an f_0-avalanche all fitnesses that have changed during the course of the avalanche have a value above f_0, the fraction $(f_c - f_0)/(1 - f_0)$ is the expected fraction of new fitnesses between f_0 and f_c, again using the uniform distribution of new fitnesses. The (universal) critical exponent γ therefore represents the average number of distinct sites whose new fitness after an f_0-avalanche is below the critical value f_c. Because of the linear term $f_c - f_0$ in Eq. (5.14), the coverage $\langle n_{\mathrm{cov}}\rangle_{f_0}$ must diverge like $(f_c - f_0)^{-1}$ in all dimensions, provided that $\langle s\rangle_{f_0}$ diverges like a power law.

The rôle of the tunable control parameter f_0 and its critical value f_c in the BS Model is markedly different from that of the usual temperature-like control parameter. The model itself runs without any tuning, unlike, say, the DS-FFM which requires trivial tuning of $\theta \to \infty$. On the other hand, to extract the (arguably) most relevant observable, the threshold f_0 has to be tuned very accurately to its critical value f_c. In that sense, the tuning of f_0 in the BS Model is non-trivial, because it is not merely achieved by a separation of time scales, yet, this is not a requirement for reaching the critical state.

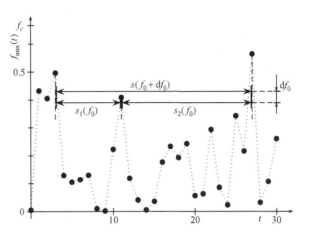

Figure 5.11 Illustration of the γ-equation, Eq. (5.12), based on the same data as Fig. 5.10. For a particular value of f_0 an avalanche $s_1(f_0)$ is obtained (lower left interval), which happens to terminate at a value of $f_{\min}$ in a small interval between f_0 and $f_0 + \mathrm{d}f_0$. If f_0 is increased by $\mathrm{d}f_0$, it is lifted over the terminating $f_{\min}$ and (what would have been) the subsequent f_0-avalanche (lower right interval) of size $s_2(f_0)$ merges with the preceding one to form the avalanche $s(f_0 + \mathrm{d}f_0) = s_1(f_0) + s_2(f_0)$ (upper interval).

5.4.3 Extremal dynamics

A number of SOC models, such as the BS Model and the OFC Model,[25] are governed by extremal dynamics (Paczuski, Bak, and Maslov, 1995; Sneppen, 1995a, also 'marginal dynamics', Sneppen, 1995b) where sites at the global minimum or maximum are subject to the external driving or updating. Extremal dynamics is a familiar concept in various areas of statistical mechanics (Paczuski *et al.*, 1996), such as interface depinning (Sneppen, 1995b), invasion percolation and, closely related, fluid imbibition (e.g. Buldyrev, Barabási, Caserta, *et al.*, 1992a), and plays a central rôle in other biological models (e.g. Papa and da Silva, 1997; da Silva *et al.*, 1998). The statistics of extremes is a well-studied field in mathematics (Gumbel, 1958) and has received renewed attention (e.g. Dahlstedt and Jensen, 2001; Majumdar and Comtet, 2004; Guclu, Korniss, and Toroczkai, 2007) in particular after the hypothesis of universal fluctuations (Bramwell *et al.*, 2000) and further, under the title extremal dynamics, **record dynamics** or **extreme value statistics** (Krug, 2007), in the context of biological evolution (Anderson, Jensen, Oliviera, and Sibani, 2004), where the **fitness landscape** has been compared to the energy of frustrated, glassy systems.[26]

Paczuski *et al.* (1996) compared the BS Model as a self-organised process to models of interface depinning, which require tuning (also Sec. 6.3). In these models, the system evolves in a **quenched random environment**, which means that noise enters as a function of the system's configuration rather than as a function of time. Quenched noise (Marsili, 1994b; Marsili, Caldarelli, and Vendruscolo, 1996) and interface depinning (Sneppen,

---

[25] See also the ZAITSEV or 'Robin Hood' Model (Zaitsev, 1992), which is not discussed in this book.

[26] In fact, the key feature of glassy systems, ageing, has been observed in the BS Model (Boettcher and Paczuski, 1997a).

1992; Alava, 2002) is discussed widely in the literature as a mechanism producing SOC (Ch. 9).

The rôle of extremal dynamics in the BS Model has been tested by various modifications of its dynamics. In an extension of the original exponential updating (p. 144), Cafiero, De Los Rios, Dittes, *et al.* (1998) introduced a stochastic updating rule, where the probability for a site to be updated is proportional to a power law with exponent α of the fitness. They found that the scaling exponents of this modified BS Model depended directly on α. Similarly, Datta, Christensen, and Jensen (2000) tested the robustness of the BS Model by updating the *two* lowest fitnesses simultaneously in every time step and found that the scaling behaviour changes dramatically, even in the random neighbour version of the model. This finding is rather surprising and could be tested analytically. Assuming that more than one site is normally updated in real physical systems it challenges the relevance of the original BS Model to natural systems.

5.4.4 Correlations

Because the values of the fitnesses are chosen at random and independently, at first sight there are no correlations at all between the forces f_n at different sites. On the other hand, there are clearly some global correlations, because the site with the *globally* minimal fitness $f_{\min}$ will change its fitness in the next time step. Although the model is so simple, it is not easy to pin down what observables are correlated and how. Neglecting correlations has led, according to Sornette and Dornic (1996), to the 'erroneous' conclusion that the BS Model is in the DP universality class (Paczuski *et al.*, 1994b).

The smaller the value the less likely a site's fitness is to survive many updates. It remains unchanged as long as the site itself and its neighbours are 'screened' by other, even lower fitnesses. Because of this screening, the probability of finding a number of small fitnesses just below f_0 is different from that same probability conditional on the presence of a number of even smaller fitnesses. In other words, the probabilities do not factorise and the fitnesses are correlated (Paczuski *et al.*, 1994b), even in a random neighbour version of the BS Model, see Sec. 5.4.5.4.

Sites that have been updated recently are more likely to be updated again, with a power law relation between their age and the probability of an update (Marsili, 1994b; Cafiero, Gabrielli, Marsili, and Pietronero, 1996; De Los Rios *et al.*, 1998). The probability distribution of all $f_n < f_0$ during an f_0-avalanche therefore depends on the history of $f_{\min}(t)$. Moreover, *consecutive* minimal fitnesses are correlated in value and position (Gabrielli, Marsili, Cafiero, and Pietronero, 1996).

On the other hand, sites that acquire a new fitness during the course of an f_0-avalanche have mutually independent fitnesses at the end of it. Provided that $f_0 < G(t)$, on average they have smaller values than untouched fitnesses, because their new f_n has been taken from a uniform distribution (effectively) conditional on $f_n > f_0$ rather than $f_n > G(t) > f_0$, as applies to the other untouched sites. Even though updated sites have a fitness effectively taken from a different distribution compared to untouched sites, their values are chosen independently and randomly. Their respective histograms are those of independent, uniformly distributed random numbers subject to the constraint of different minimal values.

At the end of an $f_0 = G(t)$ avalanche, all sites have a fitness taken independently from that distribution above f_0 – a dense 'curtain' of fitnesses just above f_0 screens the equally dense bulk. The smaller a fitness, the sooner it will be 'weeded out' and the more likely it is that a small fitness is nearby, as they are likely to have been generated within the same $f_0 = G(t)$ avalanche. The correlations of the positions of the minimal site $\mathbf{n}_{\min}(t)$ were studied by Bak and Sneppen (1993), Sneppen and Jensen (1993) as well as Paczuski *et al.* (1996). In fact, Bak and Sneppen (1993) used the power law found in the distance distribution of consecutively updated sites to conclude that 'the system is critical'.

The effect and the origin of correlations becomes most evident when comparing the BS Model from the point of view of the so-called **BS branching process** (Sec. 5.4.5.6). When comparing with an uncorrelated variant, introduced as 'Model 3' by Jovanović *et al.* (1994, also Sec. 5.4.5.7, in particular Fig. 5.13), one observes that during an f_0-avalanche in the BS Model, fitnesses below but close to f_0 tend to 'linger' around, while most of the updates occur somewhere else in the system, until it 'remembers' that the site is still to be updated. This happens when an updated fitness has a value just below f_0, so that other fitnesses are updated first, and coincidentally the position of sites where updates occur, the 'centre of activity', starts to move away in a random fashion. The high fitness will remain untouched because none of its neighbours are updated and its own update has to wait until updates somewhere else in the system have ceased. If the order of updates is randomised, as in Model 3, then the updating of any site below f_0 is never postponed long enough for the centre of activity to get far away. So, in this model, by the time a site is updated, the avalanche of sites $f_{\mathbf{n}} < f_0$ it produces is likely to join other such sites, Fig. 5.13.

5.4.5 Analytical approaches

The BS Model has attracted considerable attention from theoreticians, because of its minimalistic yet undeniably non-trivial nature and its lack of certain spatial correlations, Sec. 5.4.4, which makes it a good candidate for an exact solution. In the following, a number of analytical approaches, tackling the BS Model from various different angles, are presented.

5.4.5.1 Determining f_c

The estimate of $f_c = 0.67(1)$ for the one-dimensional BS Model in (Bak and Sneppen, 1993) suggested an exact value of $2/3$, later conjectured explicitly by Sneppen (1995b), which would have supported the supposition that the model is exactly solvable, or at least that f_c can be derived analytically. Grassberger (1995) addressed this question very early and found $f_c = 0.667\,02(8)$, concluding $f_c \neq 2/3$, which was confirmed by Paczuski *et al.* (1996) who found $f_c = 0.667\,02(3)$. To determine f_c, they extrapolated $(1 - f_0)/\langle n_{\mathrm{cov}}\rangle_{f_0}$ as a function of f_0 to 0, whereas Grassberger inspected a double-logarithmic plot of the avalanche size distribution. Garcia and Dickman (2004) used the uniformity of the distribution of fitnesses above $G(t)$ and found $f_c = 0.6672(2)$ based on finite size scaling. By now, it is generally accepted that $f_c \neq 2/3$ and Meester and Znamenski (2003) were able to find rigorous mathematical arguments for $f_c < 1$. In two dimensions, Paczuski *et al.* (1996) determined the critical fitness to be $f_c = 0.328\,855(4)$, clearly distinct from any simple rational number.

5.4.5.2 Renormalisation group approach

Marsili (1994a) developed a real space renormalisation group (RSRG) approach for the BS Model, using a transformation (Marsili, 1994b) of the quenched noise, represented by the updated fitnesses, into an annealed noise, which is much easier to handle analytically. This transformation was later corrected by Gabrielli et al. (1996) who also commented on the assumption of independence of fitnesses, which is implicit to this transformation. Marsili found fair agreement with directed percolation (Sec. 5.4.5.7) of the coarse grained model.

5.4.5.3 Upper critical dimension

Significant effort has been spent on determining the upper critical dimension d_c of the BS Model. Identifying d_c serves two purposes, firstly regarding a possible agreement with directed percolation and secondly to assess the validity of mean-field theory above d_c. De Los Rios et al. (1998) found $d_c = 8$ using a method introduced by Marsili et al. (1998), but they also found different statistical behaviour below and above $d = 4$. In the thermodynamic limit at $d \geq 4$ the probability for a site to be updated again is less than unity, while for $d \leq 3$ all sites ever updated will be updated again almost surely. Nevertheless, the critical exponents, measured numerically, coincide with mean-field theory only at and above $d_c = 8$.

This result cannot be reconciled with the earlier results by Boettcher and Paczuski (1996), who found $d_c = 4$ in a certain limit of a generalised version of the BS Model. Later, Boettcher and Paczuski (2000) used a re-formulation of the original BS Model in terms of the BS branching process (Sec. 5.4.5.6) to verify their earlier results and found them confirmed. The apparent disagreement of the numerical estimates of the exponents by Boettcher and Paczuski (2000) and De Los Rios et al. (1998) might be explained by the significantly larger system sizes used by the former authors.

5.4.5.4 Mean-field theory

Although mean-field theory (MFT, Sec. 8.1.3) generally fails to capture the effect of fluctuations in the large time, large length scale limit, it often provides important insight into the microscopic details of the dynamics. Flyvbjerg, Sneppen, and Bak (1993), in a letter published back-to-back with the original paper (Bak and Sneppen, 1993), derived many of the key results for the generalised random neighbour version of the BS Model. Their results have been further refined by de Boer et al. (1994). Marsili et al. (1998) used MFT as a starting point for a perturbation theory.

The random neighbour version of the BS Model evolves into the critical state in the sense that its overall fitness histogram evolves into a step function (Fig. 5.12(a)) similar to that of the BS Model on a lattice discussed earlier (Fig. 5.8(a)). In the random neighbour version, the step develops precisely at the critical value $f = 1/K$ (de Boer et al., 1994; de Boer, Jackson, and Wettig, 1995; Daerden and Vanderzande, 1996). The distribution of minimal fitnesses, Fig. 5.12(b), on the other hand, is markedly different from the nearest neighbour version (Fig. 5.8(b)).

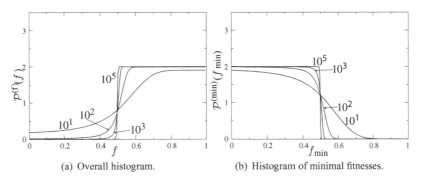

(a) Overall histogram.   (b) Histogram of minimal fitnesses.

Figure 5.12 Numerical results for the random neighbour BS Model with $K = 2$. (a) The histograms of all fitnesses $\mathcal{P}^{(\mathrm{f})}(f)$ for different system sizes, $N = 10^1$, 10^2, 10^3, 10^5. As in the nearest neighbour version, Fig. 5.8(a), the distribution approaches a step function with a jump at $f_0 = 1/K$. (b) The histograms of minimal fitnesses $\mathcal{P}^{(\mathrm{min})}(f_{\mathrm{min}})$. In contrast to the nearest neighbour version, Fig. 5.8(b), the distribution approaches a step function as well.

While correlations cannot be established in an infinite system, where sites are never visited twice at random, this is not true for a system with a finite number of sites. In this case correlations are relevant, even in a random neighbour version. In other words, the joint probability densities of fitnesses at different sites generally do not factorise in a finite system, not even in the random neighbour model.

The mean-field theory of the BS Model is presented in Sec. 8.1.3 in further detail as a generic example. It ignores any correlations (random neighbours version in the thermodynamic limit) and shows the expected phase transition in f_0, provided the system is prepared in the correct initial state. The approach to that state can be studied only in finite systems (de Boer *et al.*, 1995), as *self-organisation requires correlations and MFT will therefore generally struggle to display self-organisation*. Contrary to the popular belief that such correlations build up in the system as spatial correlations, which is suggested by the patchy structure found in other models such as the OFC Model and the DS-FFM, they can equally well be of a temporal nature – the random neighbour BS Model organises into the critical state by storing information in the system [27] without the need of any spatial relation between sites.

The size distribution exponent for f_0-avalanches of $\tau = 3/2$ in the MFT (Sec. 8.1.3) is to be compared to the numerical results for the spatial BS Model, Table 5.4, p. 143. The avalanche duration exponent $\alpha = 2$, also derived in Sec. 8.1.3, is only of limited significance, because it relies on a generational time scale not normally adopted in the BS Model.

5.4.5.5 $1/f$ noise in the BS Model

Testing the BS Model for the presence of $1/f$ noise is a natural line of enquiry, given that the 'ubiquity of $1/f$ noise' was the motivation for SOC as a whole, Sec. 1.3.2. Paczuski *et al.*

---

[27] 'Storing information' in the sense of building up (temporal) correlations; small fitnesses are destroyed quickly while large ones are likely to survive for much longer depending on the presence of smaller ones.

(1996) investigated the temporal auto-correlation of the activity, based on the distribution of waiting times between any two updates, consecutive or not, at a given site. The power spectrum derived for this observable is constant in the mean-field limit and $1/f^{0.58}$ in one dimension. Daerden and Vanderzande (1996) similarly analysed the power spectrum of the number of fitnesses below a certain threshold in the system and presented analytical arguments that the random neighbour BS Model displays $1/f$ noise. In addition, they investigated the one-dimensional BS Model and found that the power spectrum scales with an exponent close to 1. Their findings have been revised by Davidsen and Lüthje (2001) who pointed out that the master equation governing the activity distribution in the random neighbour BS Model is, up to boundary conditions, an Ornstein–Uhlenbeck process, which has Ornstein–Zernike-type correlations and a Lorentzian power spectrum. In small frequencies, the power spectrum converges to a finite value, i.e. this asymptote is characterised by the exponent 0.

5.4.5.6 The BS branching process

The **BS branching process** has been introduced as an equivalent formulation of the BS Model in terms of a **reaction-diffusion process** which makes it accessible to standard techniques (for example those discussed in the review by Täuber, Howard, and Vollmayr-Lee, 2005). The term 'BS branching process' introduced by Paczuski *et al.* (1994b) is a slight misnomer, suggesting that the reformulation is related to the Galton–Watson branching process (Sec. 5.4.5.6; Harris, 1963) which is concerned with non-interacting, independent objects, best pictured in tree-like structures, as illustrated in Fig. 8.4, p. 258 (also Grassberger, 1995). Similar concerns apply in the case of the OFC Model (p. 131).

In the BS branching process (Paczuski *et al.*, 1994b; Boettcher and Paczuski, 2000), the system is initialised only at the origin $\mathbf{n} = \mathbf{0}$ with $f_\mathbf{n} = 0$ (Grassberger, 1995), all other sites have a fitness $f_\mathbf{n} > f_0$, the precise value of which is irrelevant (assuming without proof that the overall fitness distribution converges to a step function, see Sec. 5.4.1). A site with $f_\mathbf{n} < f_0$ is called a particle. The particle $\mathbf{n}$ with the smallest $f_\mathbf{n}$ is updated by drawing new fitness values from [0, 1] for $\mathbf{n}$ and its nearest neighbours. At each update, if present, a particle is destroyed and a new one produced with probability f_0. The system runs until all particles have vanished or until a cutoff time T^* has been reached. The underlying lattice allows existing particles to be removed ($f_\mathbf{n} > f_0$ after the update) and their values to change whenever a neighbour is updated as the minimal site, which introduces interaction and correlations that do not conform to the traditional understanding of branching processes (Harris, 1963). The term 'branching' is probably meant to stress the ability of particles to create new ones, similar to a reaction-diffusion process.

The above procedure is obviously the description of the BS Model initially prepared such that all fitnesses are distributed above f_0, except for the fitness at the origin which vanishes. A technical advantage of this reformulation of the BS Model is that the approach of the fitness histogram to a particular shape can be skipped. Rather, the system is studied directly for a particular value of f_0. Moreover, keeping track only of particles instead of the entire lattice, computer memory constrains only the maximum number of particles that can be simulated rather than the system size, so that in practice the model will not experience the finiteness of the underlying lattice. For that matter, the numerical implementations can run in the thermodynamic limit, thereby avoiding finite size effects (Yang and Cai, 2001). The cutoff T^* poses the new constraint on the avalanche size.

Particles are spatially correlated by the underlying lattice so that an update of one particle changes or even destroys another, which means that the particles cannot be managed as unrelated entities and the algorithm will therefore need to map efficiently from the lattice position to the particle number (Boettcher and Paczuski, 2000, see also Sec. 5.4.7). As soon as an f_0-avalanche is over, the entire system is reset and a new avalanche is started, thereby producing independent samples, even for different f_0 simultaneously as suggested by Grassberger (1995).

The major theoretical advantage of this reformulation of the BS Model is closely related. The system is no longer described in terms of all fitnesses, but only by a set of particle coordinates and their fitnesses $f_n < f_0$. Once the system is devoid of particles, i.e. a vacuum is reached, it can be thought as having hit the *unique* absorbing state. In this description, the purpose of the threshold f_0 is extended beyond its initial rôle of defining avalanches to a temperature-like control parameter that determines whether or how the system reaches that absorbing state.

Having cast it in terms of reaction-diffusion processes one may hope that a field-theoretic description of the model can be formulated. As discussed in the next section, the reformulation as the BS branching process suggests that the BS Model belongs to the directed percolation universality class and is therefore correctly described by a Reggeon field theory (Paczuski *et al.*, 1994b).

5.4.5.7 Relation to directed percolation

The question whether the BS Model belongs to the **directed percolation** (DP) universality class is the subject of a major debate in the literature, not least because such models usually require careful tuning of their parameters. Janssen (1981) and Grassberger (1982) famously conjectured that systems displaying a continuous phase transition into a unique absorbing state belong to the DP universality class (Marro and Dickman, 1999; Hinrichsen, 2000; Lübeck, 2004), provided that the order parameter has a single, non-negative component, interaction is short range and no additional symmetries (such as a conserved parity, Ódor and Menyhárd, 2008), anisotropies, quenched disorder etc. are present. The conjecture is somewhat open to interpretation[28] but strong enough to suggest that the BS Model is in the DP class. Analytical and numerical arguments in support of that view have been put forward by Paczuski *et al.* (1994b)[29] as well as by Ray and Jan (1994) and Jovanović *et al.* (1994). The latter authors cast the BS Model somewhat confusingly in the terminology of fluid imbition (Alava, Dubé, and Rost, 2004) in order to draw a parallel with **invasion percolation**, which itself is considered an instance of SOC by some authors (Sec. 9.1,

---

[28] From a physicist's point of view, the lack of additional symmetries and other 'special attributes' (Hinrichsen, 2000) might be the most contentious or hazy property, while a mathematician might dispute the uniqueness and/or the very existence of an order parameter, not to mention its usefulness.

[29] Revised in the erratum Paczuski *et al.* (1994a).

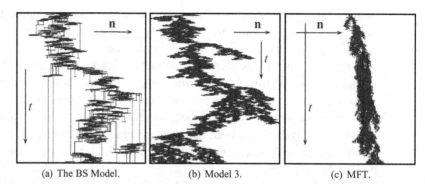

(a) The BS Model.   (b) Model 3.   (c) MFT.

Figure 5.13 Realisations of the BS Model and two of its variants. The length and time (1000 updates per row) scales are the same in all three graphs. A black point at position $\mathbf{n}$ indicates the presence of a particle, i.e. $f_{\mathbf{n}} < f_0$. (a) The original BS Model with $f_0 = 0.667\,02$ defining a particle. Particles 'linger' for long times without being updated. (b) Model 3 as introduced by Jovanović et al. (1994), where particles are updated in a random order, so that the local value $f_{\mathbf{n}}$ serves only the purpose of defining a particle via $f_{\mathbf{n}} < f_0$, here $f_0 = 0.635$. Particles cannot linger long before being (randomly) updated. (c) The visualisation of the MFT as a 'branching process' involving non-interacting particles (a site is black if occupied by at least one particle). Particles are updated in a random order and can split in up to three new particles, namely on the same site and its two neighbours, in a Bernoulli process with probability $f_0 = 1/3$.

Grassberger and Manna, 1990). Most interestingly, their 'Model 3' is a version of the BS Model where the particle to be updated is chosen at random, i.e. every $f_{\mathbf{n}} < f_0$ has the same probability of being updated, instead of always updating the site with the smallest fitness.

In fact, the whole notion of fitness is lost in Model 3 when considered as a reaction-diffusion model, whereas it survives in the description in terms of fluid imbibition. In the reaction-diffusion picture, all that matters is whether a site is (occupied by) a particle, i.e. has any value of $f_{\mathbf{n}} < f_0$. As Grassberger (1995) pointed out, Model 3 is a variant of the contact process and therefore belongs naturally to the DP universality class. Generally, DP is recovered as soon as the extremal updating rule of the BS Model is replaced by a global mechanism that treats all particles as equal, as done by Sornette and Dornic (1996) as well as Cafiero, Valleriani, and Vega (1999).

As for the original BS Model, based on an RSRG approach, Marsili (1994a) stated somewhat cautiously that '[t]he possibility that the BS model is not in the universality class of DP should not be ruled out'. Grassberger (1995), on the other hand, provided numerical evidence that it does not belong to the DP universality class, which was discussed further by Paczuski et al. (1996).

Figure 5.13 shows realisations of the BS Model and two variants, where particles are shown as black dots. The time scale is the same in all three versions, so that the same number of particles has been updated at the same vertical position in different graphs. In the BS Model, Fig. 5.13(a), a few particles might remain behind (see also the discussion in Sec. 5.4.4), while the system is updated most intensely somewhere else, whereas in Model 3, Fig. 5.13(b), particles stay close to each other. The markedly different visual appearance suggests the existence of long-range correlations in the BS Model. Figure 5.13(c) is a realisation of the mean-field theory, where sites are allowed to be multiply occupied by non-interacting particles, each being subject to reproduction attempts individually. A black dot corresponds to a site with *at least one* particle, so that the width of the black (i.e. active) region translates into a much greater number of particles.

Judging solely from these snapshots, one might wonder whether it is the existence of effective long-range interactions which causes the BS Model not to fulfil the above mentioned criteria required in the DP universality class.

One problem in the BS Model which complicates a direct comparison with DP is the definition of time. In every time step, the site with the smallest fitness is identified and updated (together with its nearest neighbours) irrespective of the number of particles in the system. In DP models, such as the contact process (Hinrichsen, 2000), on the other hand, updating is normally Poissonian, random sequential or parallel, such that every particle is updated with the same (average) rate. Implementing this in the BS Model by incrementing the time in units of the inverse of the number of particles does not resolve the apparent mismatch between the BS Model and DP (Paczuski *et al.*, 1994b, 1996; Grassberger, 1995). Various temporal observables and definitions of time (Jovanović *et al.*, 1994; Grassberger, 1995; Marsili *et al.*, 1996; Cafiero *et al.*, 1999) have been proposed in order to remedy the discrepancy. Grassberger (1995) found that a redefinition of time does not render the BS Model DP. Generally, the BS Model suffers from strong corrections to scaling and has a more complex scaling behaviour than is normally observed in DP systems.

5.4.5.8 Exact results

The BS Model was designed for analytical tractability, and it has kept that promise to a large extent. Paczuski *et al.* (1996) reviewed a large number of exact results, such as the gap equation (Paczuski *et al.*, 1994b), Eq. (5.9), and the γ-equation (Sec. 5.4.2; Paczuski *et al.*, 1994a). The mathematicians Meester and Znamenski (2003) showed rigorously that in the thermodynamic limit of the one-dimensional model, the stationary state has an average fitness which is bounded away from 1, consistent with a fitness distribution of the form shown in Fig. 5.8(a). Maslov (1996) introduced a hierarchy of master equations, which are the starting point for the numerical and analytical results presented by Marsili *et al.* (1998). Yet, as for every non-trivial SOC model,[30] scaling in the BS Model remains an *assumption* without which none of the scaling results hold, i.e. no scaling is found without assuming it first.

Boettcher and Paczuski (1996) constructed a variant of the BS Model, the **multitrait version**, where each site is characterised by $M > 1$ fitnesses instead of only one (also Head and Rodgers, 1998). Reminiscent of the spherical model, in the limit $M \to \infty$ the new model becomes exactly solvable, corresponding essentially to the MFT.

Barbay and Kenyon (2001) introduced a variant of the BS Model with discrete fitnesses $f_{\mathbf{n}} \in \{0, 1\}$ chosen at random, say zero with probability f_0, with the aim of proving rigorously some of the features found in the original version. They showed that there is a positive, critical f_0 below which the system in the stationary state contains an asymptotically

---

[30] Non-trivial in the sense of displaying spatial *and* temporal correlations beyond mean-field.

vanishing fraction of zeros. Their results were extended by Meester and Znamenski (2002), who showed that for sufficiently large f_0 the mean fitness is bounded away from 1, i.e. must contain at least some zeros. Recently a rigorous link was established (Bandt, 2005) between the discrete BS Model and DP. The original fitnesses do not enter in the discrete BS Model, so the model is reduced once more to a lattice with sites being empty or containing a particle, as discussed in Sec. 5.4.5.7. Like Model 3 it fulfills the criteria of the DP conjecture and DP behaviour therefore is expected. Bandt (2005) constructed the relation between DP and the discrete BS Model explicitly. Such mathematically rigorous results are highly desirable, not least as they lend further credibility to the DP conjecture.

5.4.6 Biological relevance

Some authors (de Boer *et al.*, 1994; Grassberger, 1995) are almost apologetic or even openly critical when it comes to the oversimplifications built into the BS Model as a model of evolution.[31] Maddox (1994), as editor of *Nature*, offered a fierce critique of the BS Model, noting that '[t]he sandpile, of course, is not a good model for evolution. [. . .] Instead, they use points along the length of a line to represent the species of an entire ecosystem. [. . .] Representing an entire species by a single entity and ascribing to it [. . .] a scalar number [. . .] will drive many evolutionists to despair.' Later, he concluded in passing (Maddox, 1995): 'So the question remains as to what the model really applies.'

One and a half years after its publication in *Phys. Rev. Lett.* the BS Model was introduced to a broader audience of biologists in an article[32] (Sneppen *et al.*, 1995) in *Proc. Natl. Acad. Sci. USA*, communicated by Stephen Jay Gould. The authors described in rather bold terms their understanding of **intermittency in biological evolution** as it arises from the BS Model, displaying little appreciation for 150 years of research by biologists: 'However, there is no theory deriving the consequences of Darwin's principles for macroevolution. This is a challenge to which we are responding.'

There can be little doubt, however, that the biological data (Sepkoski, 1993) the authors presented are consistent with an underlying mechanism similar to that captured in the BS Model (also Sec. 3.5). As Gisiger (2001, p. 189, also Newman, 1996) pointed out, however, ' . . . the results reviewed . . . have been computed using an extremely small sample, [because] . . . species alive today, which run into the millions, largely outnumber the 250 000 or so fossil specimens uncovered to date.' In fact, Kirchner and Weil's (1998) analysis concluded that scaling and self-similarity were mere artefacts of the interpolation techniques applied to the data set (Solé and Bascompte, 1996). In a critical comparison of the BS Model and an alternative model with respect to their relevance to the theory of biological evolution, Solé and Bascompte (1996) concluded that the BS Model is capable of capturing a number of important features of evolution, even when they call the model an 'oversimplification'.

---

[31] Surprisingly, the biologists Kauffman and Johnsen (1991) 'first introduced the idea of criticality in the modelling of ecosystems' (Gisiger, 2001, p. 191).

[32] Curiously, for legal reasons, the article had to be marked as an advertisement because page charges had been paid.

The BS Model makes the case for mass extinction without the need for an external cause. Solé and Bascompte (1996) pointed out that every major event of mass extinction has been linked convincingly to changes on a global scale. One might argue, however, that some of these scenarios draw their explanatory power from the expectation of such an external cause, which makes one inclined to accept readily even more unlikely causes (also Gisiger, 2001). Constructing explicitly a trivial, interaction-free model cast in the language of evolution, Newman (1996, also Newman and Sneppen, 1996; Newman, 1997a,b) argued against the discriminatory power of power law distributions. Every dynamics tested produced power law distributions, with an exponent, depending on parameters, close to that found in the fossil record and similar models of evolution (Newman and Roberts, 1995; Solé and Bascompte, 1996).

The BS Model cannot capture the complexity of life, for example the fact that life spreads across many orders of magnitude in time and space. Closely related is the fact that different species have considerably different importance within an ecosystem. The extinction of a so-called 'key-stone species' has repercussions markedly different from that of a species whose rôle is easily taken over by others. Also, the drastic effect that the extinction of one species has on the fitness level of others in the BS Model is a rather dramatic oversimplification of possible processes in nature.

Studying the BS Model on networks is appealing for more than one reason. First of all, many types of network make the model analytically tractable at least in some limit, often rendering MFT exact. Secondly, they play a prominent rôle in biological evolution, lending actual relevance to MFT. Christensen, Donangelo, Koiller, and Sneppen (1998) studied the BS Model on random networks, in an early attempt to make the BS Model more realistic. Since then, a host of numerical studies have been produced, for example by Kulkarni, Almaas, and Stroud (1999) and Moreno and Vazquez (2002) as well as analytical results, for example by Masuda, Goh, and Kahng (2005).

5.4.7 Numerical methods

At every update in the BS Model, first the site **n** with the smallest fitness[33] $f_\mathbf{n}$ has to be found. In a naïve implementation, before every update, all sites would need to be visited in order to determine the one with minimal fitness. An algorithm that sweeps over all sites before an update is computationally disastrously inefficient and the bandwidth to memory will essentially determine the speed of the simulation. Sweeps are acceptable only where significant computation is done on every site, so that the relative costs of the sweeps themselves are small compared to the overall computational costs.

Locating the minimum site efficiently, therefore, is the best route to optimising the algorithm. The obvious method is to organise the sites in a tree-like structure, so that accessing and updating the tree is typically logarithmically fast in the system size.[34] One

---

[33] On many platforms operations on integers are much faster than on floating point numbers. Given that most random number generators return by default an integer anyway or at least return only a certain number of random bits, the fitnesses are often best implemented as integers.

[34] The data shown in Fig. 5.8 are based on $2 \cdot 10^9$ updates, which on a moderately fast PC in 2008 took 499.80 s, 495.71 s, 733.40 s and 666.01 s for $N = 10^1$, 10^2, 10^3 and $N = 10^5$ respectively.

implementation (Grassberger, 1995) uses a binary rooted tree for a constantly updated ranking of all fitnesses. Each site is fixedly assigned a leaf of the tree, the internal nodes of which store the minimum of their children. A change in fitness at a leaf might require an update of a sequence of nodes on the path to the root. Another, maybe more standard choice is a **(binary) search tree** (Cormen *et al.*, 1996) with f_n as the key. It often pays to search first for the minimum among the sites most recently updated. In principle, the tree is updated every time the fitness f_n of a site changes, but in the BS Model one can introduce (and even dynamically update) a threshold above which sites are not maintained in the tree, because they are too unlikely to become the minimum site. [5.7] In general, four basic operations on the tree are needed.

(1) `minimum`: Finding the globally minimal fitness, i.e. its value $f_{n_{min}}$ as well as the site n_{min} with that value.

(2a) `locate_site`: Locating a site **n** in the tree (may be implemented by referencing the tree from the lattice and vice versa), when updating neighbours of the minimal site.

(2b) `delete`: Deleting a site from a tree, its position **n** as well as its fitness f_n, when a fitness value changes.

(2c) `insert`: Inserting a site in the tree, i.e. adding an entry for **n** and f_n, whenever a fitness changes.

The two operations `delete` and `insert` can be implemented as a single operation `replace` where appropriate. Along the same lines, `locate_site`, can be avoided completely by identifying nodes of the tree and sites, and merge with `replace` to form `update_site`.

The data structures used to implement the tree and the algorithms, such as `find_site` and `insert_site`, operating on the data go hand in hand. The design of such **trees and heaps** is subject to ongoing research and algorithms of varying degrees of sophistication exist (reviewed by Cormen *et al.*, 1996). Deciding for a particular data structure is a matter of balancing the programming costs (including the debugging time) and the computational costs. When comparing computational costs, care must be taken as the tree is not accessed randomly, which often means that the true costs are closer to the worst case rather than the amortised costs reported for a given algorithm.

5.4.8 Numerical results

The table of exponents, Table 5.4, p. 143, looks surprisingly sparse given that the model is so easily implemented. One reason for this is the wide variety of observables and exponents reported for the model, which are not all listed in the table. This is illustrated for instance by Paczuski *et al.* (1996) who gave a comprehensive overview of numerical results and scaling relations, yet, did not report σ, which can be derived from scaling relations in several ways (reported in Table 5.4 in this form).

Most authors do not distinguish avalanche size and duration unless they introduce a different time scale, such as parallel time (Paczuski *et al.*, 1996) the validity of which was discussed by Grassberger (1995). Both terms are used synonymously above. While the avalanche size exponent τ can be compared immediately with the corresponding avalanche size exponent in other tables, the exponent σ has a slightly different nature compared to all the others. It characterises the divergence of the characteristic avalanche size as f_0 is tuned towards f_c. As discussed earlier, this process is clearly distinct from $\theta \to \infty$ in the DS-FFM and that is the main reason why σ was not converted to $1/\lambda$ in Table 5.4.

A wide range of results is available in one dimension, such as the earliest numerical estimate of $\tau = 0.9(1)$ (Bak and Sneppen, 1993, to be compared to $\tau \geq 1$, Sec. 2.1.2.2), or analytical results, for example based on a RSRG (Marsili, 1994a). The numerical data and the robustness of the scaling behaviour are not as good as for other models, for example the Oslo Model, as can be assessed in the data collapses by Datta *et al.* (2000, also Paczuski *et al.*, 1996). Grassberger (1995) discussed in great detail corrections to scaling in various observables different from the avalanche size. Comparatively little has been published for two dimensions and beyond; unfortunately, De Los Rios *et al.* (1998) did not publish results for σ. In a competing work, Boettcher and Paczuski (2000) studied the avalanche dimension D, rather than τ or σ, so that their results are omitted from Table 5.4. One would normally assume that this avalanche dimension D is the finite size scaling exponent D, for example as listed for the Abelian BTW Model, Table 4.1 on p. 92. Yet, because of the particular way it has been measured, this equivalence cannot be derived from the simple scaling ansatz as stated in the caption of Table 5.4 (Paczuski *et al.*, 1996) without making further assumptions, so that these data are omitted as well. The only derived exponents listed in Table 5.4 are based on the scaling exponents of the first moment of the avalanche size, $\sigma_1^{(s)} = (2 - \tau)/\sigma$, reported as γ in Paczuski *et al.* (1996).

6 Stochastic sandpiles

Both of the following models incorporate a form of **stochasticity** in the relaxation mechanism. In the MANNA Model particles topple to sites randomly chosen among nearest neighbours and in the OSLO Model the local critical slopes are chosen at random. They both display robust scaling and belong to the same *enormous* universality class, which contains two large classes of ordinary (tuned), non-equilibrium critical phenomena: **directed percolation with conserved field** (C-DP) and the **quenched Edwards–Wilkinson equation** (qEW equation). The former is paradigmatically represented by the MANNA Model, the latter by the OSLO Model.

Both models are generally considered to be Abelian, even when they strictly are not. In their original versions, the relaxation of the MANNA Model is non-Abelian and so is the driving in the OSLO Model. This can be perceived as a shortcoming, not only because the BTW Model has been understood in much greater detail by studying its Abelian variant, but also because of the simplification of their implementation, as the final configurations become independent of the order of updates. Nowadays, the MANNA Model and, where the issue arises, the OSLO Model are studied in their Abelian variant. The MANNA Model is currently probably the most intensely studied model of SOC.

In their Abelian form, both models can be described in terms of stochastic equations of motion (Sec. 6.2.1.3 and Sec. 6.3.4.2). These look very different for the two models. The MANNA Model is generally studied using an effective theory, C-DP (Vespignani, Dickman, Muñoz, and Zapperi, 1998; Pastor-Satorras and Vespignani, 2000d; Vespignani *et al.*, 2000), which encapsulates the main features of its dynamics, whereas the OSLO Model translates directly to the qEW equation (Paczuski and Boettcher, 1996; Pruessner, 2003a). The noise term in the two equations has a fundamentally different nature, but (numerical) evidence indicates that they are equivalent and different from DP (see Sec. 6.2.1.2). Showing their equivalence analytically remains a formidable task.

6.1 The MANNA Model

The MANNA Model, also known as the 'MANNA Sandpile Model' and the 'Stochastic Sandpile Model' was introduced relatively early (Manna, 1991b) and since then has been studied in great detail, not least because it is akin to the original sandpile models, yet shows interesting behaviour even in one dimension. In its original definition, the MANNA Model is not Abelian, due to its relaxational rules. Rendering it Abelian gives rise to the Abelian MANNA Model (AMM), a prototypical model of the so-called MANNA, OSLO or

C-DP universality class. The models in this class are at the centre of what can unreservedly be called 'self-organised criticality'.

Four main lines of investigation are associated with the Manna Model: its relation to the BTW Model, its Abelian versus its non-Abelian character, its relation to absorbing state phase transitions and directed percolation (DP, discussed separately in Sec. 8.4.5), and finally its relation to other models in the Manna universality class. This is roughly reflected with varying weighting, in the following. First, the original Manna Model is introduced; similar to the BTW Model, this original version is often confused with its Abelian version, which was proposed only much later by Dhar (1999a), yet a similar amount of numerical work is published on both models. The exponents of the Manna Model are summarised in Table 6.1. To avoid confusion, wherever relevant, the *original* version of the Manna Model is referred to as the oMM and the *Abelian* version as the AMM.

6.1.1 Relation to the BTW Model

Similar to the Zhang Model, the Manna Model was invented as 'a two-state version of the sandpile model' (Manna, 1991b) in an effort to simplify the original BTW Model [1] and explore its universality class. A major part of the early literature on the model focuses on this topic. The only difference between the Abelian BTW Model and the AMM is the toppling mechanism, which is randomised in the Manna Model and restricted to two **particles**, which correspond to **slope units** in the BTW Model. There is an unfortunate muddle in the literature: particles in the Manna Model reach a maximum *absolute* height, whereas grains in the BTW Model reach a maximum slope, i.e. a maximum *relative* height. Nevertheless, in the following the usual nomenclature of particles is used, rather than slope units or 'grain' and 'height' (originally the latter two were used synonymously for 'particle' in the Manna Model literature).

The fact that the oMM (see the definition in Box 6.1) is non-Abelian, because *all* particles are redistributed from a site when it topples, is frequently overlooked. Strictly speaking, the number of particles shed when a site topples (possibly multiple times at once) always depends on the order of updates, regardless of whether or not the model is Abelian, but in the Abelian version the particles are redistributed in pairs, with a single particle potentially staying behind, while in the original, non-Abelian version *all* particles are redistributed. [2] Interestingly, Manna (1991b) pointed out a feature in the non-Abelian model that is normally attributed to the Abelian Oslo Model (Dhar, 2004, also p. 201), namely that the stationary state is immediately entered by setting $h_n = h^c$ throughout the system and charging any site.

Because of the close link, the numerics for the oMM follows closely that of the BTW Model (Grassberger and Manna, 1990). In the original study, Manna (1991b) concluded on

---

[1] Where not indicated differently, 'BTW Model' refers to its Abelian version.

[2] This allows much more freedom in the choice of the details of the dynamics in the Abelian version, because geometrical features of the avalanches, such as size and area, are invariant under a change of order of updates, Sec. 8.3.2.3.

| Box 6.1 | The original MANNA Model |

General features

- First published by Manna (1991b).
- A stochastic, 'two-state version' of the BTW Model, initially thought to be in the same universality class.
- Defined in any dimension, non-trivial even in one.
- Stochastic (bulk) drive and relaxation.
- Non-Abelian in its original definition (Dhar, 1999a).
- Exponents listed in Table 6.1, p. 168.
- Upper critical dimension probably $d_c = 4$ (Lübeck and Heger, 2003b).

Rules

- d dimensional (usually) hyper-cubic lattice.
- Each site $\mathbf{n} \in \{1, \ldots, L\}^d$ contains $h_\mathbf{n} \in \mathbb{N}_0$ particles.
- *Initialisation*: irrelevant, model studied in the stationary state.
- *Driving*: add a particle at $\mathbf{n}_0$ (chosen at random), $h_{\mathbf{n}_0} \rightarrow h_{\mathbf{n}_0} + 1$.
- *Toppling*: for each site $\mathbf{n}$ with $h_\mathbf{n} > h^c = 1$ distribute *all* particles randomly and independently among its nearest neighbours $\mathbf{n}'$, $h_{\mathbf{n}'} \rightarrow h_{\mathbf{n}'} + 1$ ($h_\mathbf{n}$ times) and reset $h_\mathbf{n} \rightarrow 0$.
- *Dissipation*: particles are lost at open boundaries.
- *Parallel update*: discrete microscopic time, sites exceeding h^c at time t topple at time $t + 1$ (updates in sweeps). Sometimes random sequential.
- *Separation of time scales*: drive only once all sites are stable, i.e. $h_\mathbf{n} \leq h^c$ (quiescence).
- *Key observables* (see Sec. 1.3):
  avalanche size s, the total number of topplings until quiescence;
  avalanche duration T, the total number of parallel updates until quiescence.

the basis of numerics for the two-dimensional case 'that our model should be in the same universality class [as] the sandpile model'. The exponents were found to be $\tau = 1.28(2)$, $D = 2.75$, $\alpha \approx 1.50$ and $z = 1.55$, very close to the values in the Abelian BTW Model, Table 4.1 on p. 92. In one dimension, the MANNA Model is still non-trivial, whereas the ASM develops into a trivial state, i.e. they are apparently different.

Ben-Hur and Biham (1996) grouped various sandpile-like models into different universality classes and suggested that the MANNA Model and the (Abelian) BTW Model are in different universality classes. This result is in line with a similar proposition by Milshtein *et al.* (1998), who, however, assumed that their two-state model, which is a variant of the AMM, was non-Abelian. Chessa *et al.* (1999a) revised these results and advocated the idea of a single universality class comprising the BTW Model, the ZHANG Model and the MANNA Model. This conclusion was founded on a (numerical) moment-analysis for the overall avalanche size, area and duration distributions. While they clearly aimed to revise the work by Ben-Hur and Biham (1996), their numerics is however based on the

Abelian variant of the Manna Model (see below). Interestingly, they presented numerical and analytical arguments why the method of conditional averages (see Sec. 2.2.2) employed by Ben-Hur and Biham (1996) was biased.

Pradhan and Chakrabarti (2001) measured the response of the Abelian BTW Model and the AMM to external perturbations as a function of the average particle density in the system and detected no significant difference between the two models. In fact, that the Manna Model and BTW Model are in the *same* universality class, is supported by various heuristic and analytical arguments (Pietronero *et al.*, 1994; Vespignani and Zapperi, 1995; Dickman *et al.*, 1998; Vespignani *et al.*, 1998). Nevertheless, more recent numerical simulations focusing on an extended range of observables place the BTW Model and the Manna Model in distinct universality classes (Lübeck, 2000; Alava and Lauritsen, 2001; Biham *et al.*, 2001; Stella and De Menech, 2001; Dickman and Campelo, 2003; Karmakar *et al.*, 2005). Assuming that this is the final word on the issue implies that theoretical efforts suggesting a correspondence of BTW and Manna Model's miss an important feature, such as multiple topplings (Ben-Hur and Biham, 1996; Paczuski and Bassler, 2000). Some of these results can be questioned on the basis that the very existence of scaling in the BTW Model remains disputed.

As opposed to the BTW Model, in the Manna Model particles move independently and randomly (Nakanishi and Sneppen, 1997) and need, in one dimension, on average exactly $(L+1-\mathbf{n}_0)\mathbf{n}_0$ moves to leave the system (Sec. 8.1.1.1). The number of topplings is exactly half the number of moves and each $\mathbf{n}_0$ is chosen with probability $1/L$, so that the average avalanche size is (Sec. 8.1.1.1)

$$\langle s \rangle = \frac{1}{2L} \sum_{\mathbf{n}_0=1}^{L} (L+1-\mathbf{n}_0)\mathbf{n}_0 = \frac{1}{12}(L+2)(L+1), \qquad (6.1)$$

and thus $\langle s \rangle \propto L^{\sigma_1^{(s)}}$ with $\sigma_1^{(s)} = 2$. Similar arguments can be used in higher dimensions, and apply equally to the oMM and the AMM .

6.1.2 The lack of Abelian symmetry

The oMM defined in Box 6.1 is not Abelian because of the toppling procedure, where the number of particles redistributed among randomly chosen neighbours depends on the number stored at the toppling site. This number is in fact constant, namely $h^c + 1$, if sites are updated instantly (in parallel), so that sites never get charged again before they discharge, and, more importantly, at most one charge arrives at a time. If, however, multiple charges arrive at a site simultaneously from one or more neighbours, *all* particles are removed if the site exceeds the threshold h^c. The final state will therefore change, for example, if two toppling sites charge a third simultaneously or sequentialy. Strictly, the oMM does not leave any ambiguity here, so changing the sequence of updates means investigating a different model (Sec. 8.3). Also, strictly, Abelianness is the invariance of the final, stable state under a change in the order of external charges and subsequent relaxation, rather than a change in the microscopic dynamics, such as the order of updates *during* an avalanche. One therefore

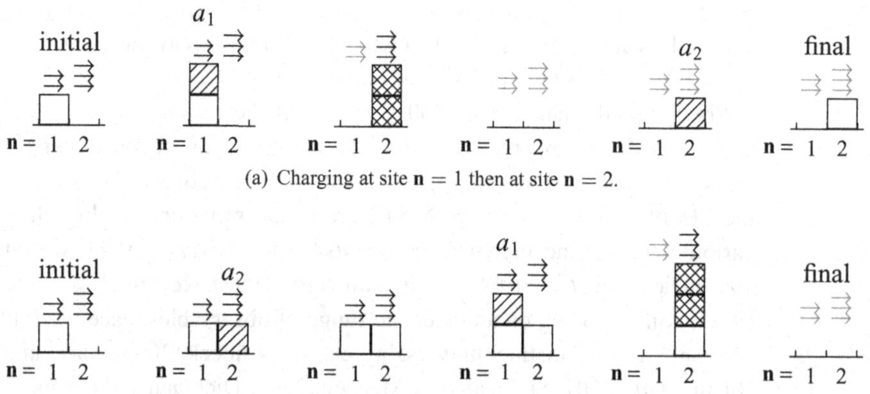

(a) Charging at site $\mathbf{n} = 1$ then at site $\mathbf{n} = 2$.

(b) Charging at site $\mathbf{n} = 2$ then at site $\mathbf{n} = 1$.

Figure 6.1 The one-dimensional oMM of size $L = 2$ charged at sites $\mathbf{n} = 1$ and $\mathbf{n} = 2$ in two different orders. The sequence of randomly chosen directions of redistributed particles (boxes) is fixed (Sec. 8.3.2.2) and indicated by the arrows. For example, the triple arrow indicates that all three particles are moved in the same direction (only two particles are available in (a)). Grey arrows have been used at past topplings. Hatched boxes are added by the external drive, cross-hatched boxes are added by neighbours toppling. In (a) the system is charged at site $\mathbf{n} = 1$ first and then at site $\mathbf{n} = 2$. In (b) the order is reversed. The difference in the final state means that the model is not Abelian.

has to probe the Abelianness of a model by considering the final states after two external charges are applied in two different orders. The lack of Abelian symmetry in the oMM is shown by way of (counter) example in Fig. 6.1.

Initially, Abelianness was assumed to be incompatible with stochasticity (Milshtein *et al.*, 1998), because it implies that charging and relaxing the model in different orders will generally produce different (namely random) results. While this is true when considering the charge-and-relax operators $a_{\mathbf{n}}$ as used for the ASM, which operate on individual states of the system, using operators $\hat{a}_{\mathbf{n}}$ acting on the *distribution* of states resolves the problem (Sec. 8.3.1 and Sec. 8.3.2). In Fig. 6.1, the stochasticity is suppressed by fixing the directions of the redistributed particles. The difference from the Abelian version where the number of units shed in a toppling is constant (the next model, Sec. 6.2), is so minute that some authors do not distinguish between them (e.g. Chessa *et al.*, 1999a).

6.2 Dhar's Abelian MANNA Model

In contrast to the BTW Model, which is non-Abelian only due to its driving rule (Sec. 4.1.2), the MANNA Model in its original definition is non-Abelian because of its relaxation rule. Either because the rules were not laid out clearly enough in the original publication (Manna, 1991b), or maybe because there was little sensitivity towards the relevance of Abelian symmetry, a rich variety of variants of the MANNA Model is used throughout the literature, among them early Abelian variants of the MANNA Model, without any emphasis on this

The Abelian Manna Model

General features

- First published by Dhar (1999a), directly derived from the original Manna Model, Box 6.1.
- Prototype of a universality class, probably including the oMM.
- Strong link to interfaces and absorbing state phase transitions (Vespignani and Zapperi, 1997).
- Defined in any dimension, non-trivial even in one.
- Stochastic (bulk) drive and relaxation.
- Investigated analytically as well as numerically.
- Exponents listed in Table 6.1, p. 168.
- Upper critical dimension probably $d_c = 4$ (Lübeck and Heger, 2003b).

Rules

- d dimensional (usually) hyper-cubic lattice.
- Each site $\mathbf{n} \in \{1, \ldots, L\}^d$ contains $h_{\mathbf{n}} \in \mathbb{N}_0$ particles.
- *Initialisation*: irrelevant, model studied in the stationary state.
- *Driving*: add a particle at $\mathbf{n}_0$ (chosen at random), $h_{\mathbf{n}_0} \rightarrow h_{\mathbf{n}_0} + 1$.
- *Toppling*: for each site $\mathbf{n}$ with $h_{\mathbf{n}} > h^c = 1$ distribute *two* particles randomly and independently among its nearest neighbours $\mathbf{n}'$, $h_{\mathbf{n}'} \rightarrow h_{\mathbf{n}'} + 1$ and $h_{\mathbf{n}} \rightarrow h_{\mathbf{n}} - 2$.
- *Dissipation*: particles are lost at open boundaries.
- *Parallel update*: discrete microscopic time, sites exceeding h^c at time t topple at time $t + 1$ (updates in sweeps). Alternatively, often random sequential.
- *Separation of time scales*: drive only once all sites are stable, i.e. $h_{\mathbf{n}} \leq h^c$ (quiescence).
- *Key observables* (see Sec. 1.3):
  avalanche size s, the total number of topplings until quiescence;
  avalanche duration T, the total number of parallel updates until quiescence.

point (e.g. Ben-Hur and Biham, 1996). Some authors adopt the view that the Abelian Manna Model (AMM) is the model originally formulated by Manna (1991b) and cite Dhar (1999a,b) for the proof for its Abelianness.

Almost in passing, Dhar (1999a) introduced formally the Abelian version of the Manna Model discussed in the following. It was presented in greater detail in a dedicated article (Dhar, 1999b). Dhar introduced two variants of the original Manna Model (oMM), which differ in the way the receiving sites are chosen at a toppling. In the first variant, the target sites always lie opposite each other,[3] while in the second variant, the targets are chosen independently. The constraint in the first version leads to various invariants, similar to the **forbidden sub-configurations** in the ASM (Sec. 4.2.1), so that only a subset of stable states is recurrent (Dhar, 1999b). Furthermore, these **toppling invariants** give rise to ladder operators, which generate eigenvectors of the evolution operator.

---

[3] So particles do not perform independent random walks, Sec. 8.1.1.1.

Table 6.1 Exponents of the AMM (and the oMM where marked by ◇, supposedly in the same universality class). $\mathcal{P}^{(s)}(s) = a^{(s)} s^{-\tau} \mathcal{G}^{(s)}(s/(b^{(s)} L^D))$ is the avalanche size distribution and $\mathcal{P}^{(T)}(T) = a^{(T)} T^{-\alpha} \mathcal{G}^{(T)}(T/(b^{(T)} L^z))$ is the avalanche duration distribution, Sec. 1.3. Scaling relations: $D(2-\tau) = \sigma_1^{(s)} = 2$ (bulk drive, Sec. 8.1.1.1) and $(1-\tau)D = (1-\alpha)z$ (sharply peaked joint distribution, widely assumed) in all dimensions (Sec. 2.2.2).

d	τ		D		α		z	
1	1.09(3)◇	(a)	2.2(1)◇	(a)	1.18(2)★	(b)	1.47(7)◇	(a)
	1.11(2)★	(b)			1.17(3)♮	(c)	1.66(7)♮	(e)
	1.11(2)♮	(c)			1.17(5)	(d)	1.45(3)♮	(f)
	1.11(5)	(d)					1.54(5)♮	(f)
							1.393(37)♮	(c)
							1.50(4)♮	(g)
2	1.28(2)◇	(h)	2.75◇	(h)	1.47(10)◇	(h)	1.55◇	(h)
	1.253	(i)	2.54♮	(q)	1.50(1)	(m)	1.234	(i)
	1.253	(k)	2.7(1)◇	(a)	1.55(4)★	(b)	1.168	(k)
	1.26(3)◇	(a)	2.73(2)	(m)	1.50(3)♮	(c)	1.49♮	(q)
	1.27(1)	(m)	2.74(2)	(n)	1.48(3)	(d)	1.50(2)	(m)
	1.27(1)	(n)	2.764(10)	(o)			1.50(2)	(n)
	1.275(11)	(o)					1.540(10)	(o)
	1.25(2)▷♮	(p)					1.572(7)♮	(f)
	1.30(1)★	(b)					1.57(4)♮	(r)
	1.28(14)♮	(c)					1.533(24)♮	(c)
	1.27(3)	(d)						
3	1.43▷♮	(q)	3.33♮	(q)	1.78(2)	(s)	1.8♮	(q)
	1.41(1)	(s)	3.302(10)	(o)	1.77(4)♮	(c)	1.713(10)	(o)
	1.41(2)♮	(c)	3.36(1)	(s)			1.76(1)	(s)
							1.80(5)♮	(r)
							1.823(23)♮	(c)
MFT	3/2	(t)	4	(t)	2	(t)	2	(t)

◇ Reported for the oMM.

★ Reported for non-dissipative avalanches.

♮ Reported for variants of the MANNA Model (in particular FES, Sec. 9.3.1).

▷ Reported as $\tau + 1$.

a Nakanishi and Sneppen (1997), *b* Dickman and Campelo (2003), *c* Lübeck and Heger (2003a), *d* Bonachela (2008), *e* Dickman, Alava, Muñoz, *et al.* (2001), *f* Dickman *et al.* (2002), *g* Dickman (2006), *h* Manna (1991b), *i* Pietronero *et al.* (1994), *k* Vespignani and Zapperi (1995), *m* Chessa, Vespignani, and Zapperi (1999b), *n* Chessa *et al.* (1999a), *o* Lübeck (2000), *p* Biham *et al.* (2001), *q* Ben-Hur and Biham (1996), *r* Alava and Muñoz (2002), *s* Pastor-Satorras and Vespignani (2001), *t* Lübeck (2004).

In the second version, by construction, the empty lattice is accessible from any state within a finite number of charges. Starting from the empty lattice, every stable state can be constructed by adding at most one particle at every site. Thus, the number of recurrent stable configurations which, of course, do not all occur with the same frequency, is 2^{L^d} where $N = L^d$ is the number of sites (Dhar, 1999b). This second model is discussed in the following as the AMM. It differs from the oMM solely by the number of redistributed grains being constant (two).

Dhar (1999a) invoked arguments similar to those used in the ASM to demonstrate the Abelian symmetry in his version of the Manna Model. The key step is to fix the sequence of random numbers, as discussed in Sec. 8.3.2 (cf. Fig. 6.1). Some arguments, in particular regarding the nature of the fully stochastic operators $\hat{a}_{\mathbf{n}}$ versus the fully deterministic operators $a_{\mathbf{n}}$, were clarified elsewhere (Dhar, 1999b). Most of this material, however, is based on Dhar's first variant of the AMM.

Similar to the Abelian variant of the BTW Model, p. 90, it is fair to say that almost all theoretical and numerical progress in the Manna Model is based on the AMM. Casting it in terms of activated random walkers (Sec. 9.3), relating it to interface models (Sec. 8.5), generalising it as an operator algebra (Sec. 6.2.1.1), drawing parallels with the Oslo Model (Sec. 6.3) or introducing anisotropy (Sec. 8.4), is all based on the AMM rather than the oMM.

6.2.1 Logarithmic corrections and variants of the Manna Model

What makes the numerics of the Manna Model particularly difficult is the apparent presence of **strong logarithmic corrections** (Dickman and Campelo, 2003), the significance of which depends on the dimensionality and on the type of avalanche considered. In addition, the effective exponents characterising these corrections display a strong dependence on the system size. Logarithmic corrections are an essential feature of the system behaviour at the upper critical dimension (Lübeck and Heger, 2003b, but also Dhar, 1999a), but are otherwise rather undesirable and difficult to probe and account for. Dickman and Campelo's (2003) study suffers somewhat from ignoring the scaling function. For example, collapsing $L^D \mathcal{P}^{(s)}(s)$ by plotting it versus s/L^D implies $\tau = 1$, consistent with their observation of the apparent exponent $\tilde{\tau} < 1$ (see Sec. 2.1.2.2). Dickman and Campelo distinguished dissipative and non-dissipative avalanches, similarly to Drossel (2000) or Lise and Paczuski (2001a) in the ASM and the DS-FFM respectively, and found no logarithmic corrections for dissipative avalanches in one dimension, but strong corrections for non-dissipative avalanches. In the original study, Manna (1991b) recognised a 'considerable curvature' in the distribution of avalanche duration. One can only speculate whether or rather how much the logarithmic corrections contribute to the apparent splitting of the Manna universality class in one dimension, as observed by Lübeck and Heger (2003a).

Malcai, Shilo, and Biham (2006) replaced the **boundary dissipation** in the two-dimensional AMM by **bulk dissipation**, where every redistributed particle has probability ϵ of being removed from the system (similarly in the ASM, Lin et al., 2006, also Chen and Lin, 2009). The scaling of ϵ was chosen such that the average avalanche size of the ordinary AMM was reproduced, i.e. $\epsilon \propto L^{-2}$ since $\langle s \rangle \propto L^2$ and $\epsilon \langle s \rangle = 1$ from conservation. The resulting scaling coincided with that of the ordinary AMM (with boundary dissipation). This finding is important for the understanding of SOC in terms of absorbing states (Sec. 9.3), where dissipation usually takes place in the bulk. The avalanche dimension D seems to be directly affected by the choice of the scaling for ϵ (Alava, Laurson, Vespignani, and Zapperi, 2008), which is surprising given the robustnes of D observed otherwise (Nakanishi and Sneppen, 1997; Pruessner and Peters, 2008).

De Menech and Stella (2000) introduced a **wave decomposition** of avalanches in the AMM along the lines of waves in the BTW Model (Ivashkevich *et al.*, 1994b). Contrary to the BTW Model, the same exponent was found to characterise the overall distribution of avalanches and waves, which did not display significant correlations. This might be partly caused by the definition of waves in the MANNA Model being somewhat ad hoc, in particular as individual sites in a wave can topple multiple times. A wave appears to be an almost random subset of topplings in an avalanche.

A host of variants of the MANNA Model have been introduced, some of which display more robust scaling behaviour than others. In the MANNA Model with **height restriction** [4] (Dickman *et al.*, 2002; Dickman, 2006), one specific version of which is also known as the conserved threshold transfer process (CTTP, Mendes, Dickman, Henkel, and Marques, 1994; Rossi, Pastor-Satorras, and Vespignani, 2000; Lübeck, 2002a; Lübeck and Heger, 2003a), the local particle number h_n cannot exceed two. This has several major implications. First of all, the number of stable and unstable states is finite, which simplifies the time evolution operator on the microscopic time scale considerably (p. 271) and is a crucial ingredient in a cluster approximation (ben-Avraham and Köhler, 1992; Dickman *et al.*, 1998, 2000). Secondly, **multiple discharges** [5] are avoided and so specifying that all sites with $h_n = 2$ topple with the same Poissonian frequency provides an unambigious definition of the microscopic dynamics. If h_n can take arbitrary values, the dynamics has to specify whether *all pairs* on a toppling site topple at once or sequentially, [6] i.e. whether a single pair or all pairs are moved in an update (see also below). With parallel updates of all sites active at the beginning of a time step, $h_n \leq 2$ ensures that updates can take place alternately on even and odd sublattices as time progresses.

The height restrictions are implemented by suppressing any individual particle movement that violates them, so that in some topplings less than two particles are redistributed. As the number of redistributed particles depends on the local configuration, it is straightforward to construct a non-Abelian scenario such as that of Fig. 6.2, which raises the question whether the **Abelian symmetry** is **relevant** in the field theoretic sense, i.e. whether its presence or absence changes the scaling behaviour of the system and therefore its universality class (see Sec. 8.3.3, p. 283). Multiple discharges can in principle be avoided even in the AMM

---

[4] Also known as the 'restricted sandpile'.

[5] Multiple discharges are redistributions of more than one pair of particles during an update of an unstable site until it is stable again. **Multiple topplings** (Paczuski and Bassler, 2000) on the other hand are generally multiple redistributions during the same avalanche but (typically) involving separate updates of a site that is charged again after having toppled (Sec. 8.4.4).

[6] Sequential toppling and thus independent movement of pairs is more in line with the activated random walker (Dickman *et al.*, 2000) picture.

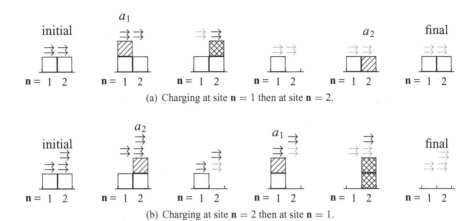

(a) Charging at site **n** = 1 then at site **n** = 2.

(b) Charging at site **n** = 2 then at site **n** = 1.

Figure 6.2 The one-dimensional height restricted MANNA Model of size $L = 2$ charged at sites **n** $= 1$ and **n** $= 2$ in two different orders (symbols as in Fig. 6.1). In (a) the height restriction constrains the relaxation of the first site, so that only one particle is passed to the right. In (b) none of the relaxations is constrained (as a result the right site topples twice). The difference in the final state means that the model is not Abelian.

by carefully planning the order of updates, and generally do not occur very frequently. In that sense the Abelian symmetry is only weakly broken (but see the note by Ben-Hur and Biham, 1996). That the height restriction has little impact on the overall behaviour of the model is corroborated by the small deviations observed in the critical particle density (Dickman *et al.*, 2002). These additional tests are of great importance, as it is somewhat problematic to use the height restricted MANNA Model to pin down the exponents of the MANNA universality class more accurately and determine its universality class at the same time. One might end up characterising a *different* universality class very accurately.

Whenever time-dependent or time-averaged quantities are measured, a clear definition of the microscopic time is vital (Sec. 8.3.2.3). Dickman and Campelo (2003) tested the effect of different implementations of time, parallel and random sequential, and found no systematic effect in the AMM.

Other variants of the AMM are the *n*-state models used by Ben-Hur and Biham (1996), where *n* particles are distributed in a random fashion at each toppling, or, more popular, so-called **fixed-energy sandpiles** (FES, Sec. 9.3.1; Chessa, Marinari, and Vespignani, 1998; Dickman *et al.*, 1998, 2002; Vespignani *et al.*, 1998, 2000; Alava and Lauritsen, 2001), where the boundaries of a regular AMM are periodically closed. This type of model displays an **absorbing state phase transition** (Vespignani and Zapperi, 1997, 1998; Tang and Bak, 1988a; Dickman *et al.*, 2000; Rossi *et al.*, 2000; Dickman, 2002b, also Sec. 9.3).

A range of directed stochastic models exists, some of which are closely related to or derived from the AMM. They are presented separately, in Sec. 8.4, as they share a common theme and are often developed as models in their own right. The link to DP initiated by Tadić and Dhar (1997) is concerned with 'directed sticky sandpiles' (Sec. 8.4.5.1) which are much more closely related to the directed version of the OSLO Model.

A very elegant, entirely equivalent formulation of the AMM is that of **activated random walkers** (ARW, Dickman *et al.*, 2000). In this description, each particle is considered as a random walker which can be in one of two states. On sites with $h_\mathbf{n} = 1$ it is dormant and does not move. Once it is paired up with a second particle, each particle constituting the pair moves independently to one of the neighbouring sites. They move together as a pair with probability $1/2$, otherwise they split, possibly joining with isolated random walkers at neighbouring sites.

Stapleton (2007, but see Woodard, Newman, Sánchez, and Carreras, 2007) challenged the established understanding of SOC and suggested investigating the AMM from a different perspective. After a short transient, the AMM develops into a stationary state. Each stable state encountered could be regarded as an element in an **SOC ensemble**. Picking from that ensemble repeatedly elements $\mathcal{C}$ with the appropriate (stationary) weight $w_\mathcal{C}$ and triggering an avalanche produces the well known statistical results and scaling properties of the AMM along with a set of stable states $\mathcal{C}'$ occurring with frequencies $w_{\mathcal{C}'}$, which are in fact simply a new incarnation of the stationary ensemble.

Spatial correlations are clearly visible within this ensemble and a widely accepted interpretation of SOC (e.g. Grinstein, 1995; Dickman *et al.*, 1998; Lise, 2002) tells the narrative of spatial correlations spreading slowly throughout a system until they reach its boundaries at which stage SOC is fully developed, displaying the known (finite size) scaling features. Stapleton (2007) asked what happened if these correlations were suppressed. This can be achieved by generating initial stable configurations $\mathcal{C}$ where each site is independently occupied by a particle with a probability equal to the occupation density found in the AMM under normal SOC conditions. An avalanche is triggered by adding a particle at a fixed site (see Sec. 6.3). Neither the initial set $\{\mathcal{C}, \dots\}$ nor the resulting set $\{\mathcal{C}', \dots\}$ corresponds to the SOC ensemble.

The scaling found in this setting is clearly distinct from that of the MANNA Model, see Fig. 6.3 (Stapleton, 2007, finds $\tau = 1.31$ and $D = 2.25$). Nevertheless, it is systematic, fully compatible with gap scaling and clearly non-trivial, which suggests that spatial correlations in the initial states are indeed largely irrelevant. If that is true, the non-trivial scaling is due to the correlations developed by the microscopic dynamics *during an avalanche*, rather than the spatial correlation found in the ensemble. Correlations would need to spread within a single avalanche, rather than by repeated driving. Although surprising in its interpretation, Stapleton's proposition amounts, in effect, to measuring SOC observables in an absorbing states (AS)-type setting, see Sec. 9.3. It differs from the typical AS setup only by keeping the boundaries open rather than closing them periodically. Taking the original perspective, one might assume that many of the randomly generated states are at least very similar to a recurrent one and probing the random ensemble amounts to probing the recurrent set with different weights.

6.2.1.1 A generalised AMM in terms of operators

The directed variants of the AMM (Sec. 8.4, e.g. Paczuski and Bassler, 2000; Kloster *et al.*, 2001), which have exact solutions, are included in a generalisation using an operator

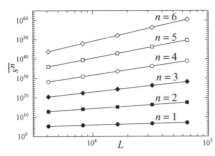 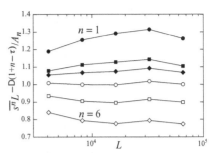

(a) Moments of the randomised MANNA model.   (b) Rescaled moments of the randomised MANNA
                                                model.

Figure 6.3 Numerical estimates of the avalanche size moments $\overline{s^n}$ for $n = 1, 2, \ldots, 6$ of the one-dimensional MANNA Model with size $L = 4096, 8192, \ldots, 65536$ initialised repeatedly with a random configuration with site occupation probability $\zeta = 0.9488$ (cf. Fig. 7.4). The data shown double-logarithmically in (a) suggest robust gap scaling, which is confirmed by extracting estimates of the exponents $\sigma_n^{(s)}$ and amplitudes A_n from the leading order behaviour $\overline{s^n} = A_n L^{\sigma_n^{(s)}} + \cdots$ of the various moments. A linear regression of $\sigma_n^{(s)}$ versus n gives $\tau = 1.39(2)$ and $D = 2.47(1)$ for $\sigma_n^{(s)} = D(1 + n - \tau)$, incompatible with the exponents in Table 6.1. The results in (b) (same symbols as in (a)) confirm that the moments $\overline{s^n}$ are well approximated by $A_n L^{-D(1+n-\tau)}$ based on D and τ extracted in (a).

representation of the AMM (Pruessner, 2004c).[7] Because the stable states in the AMM are binary, it can be regarded as an Ising lattice with creation and annihilation operators (Chua, 2003, see also Eq. (6.18)), depositing and removing a particle respectively, when the local state allows this operation:

$$C = \begin{pmatrix} 0 & 1 \\ 0 & 0 \end{pmatrix} \quad \text{and} \quad C^\dagger = \begin{pmatrix} 0 & 0 \\ 1 & 0 \end{pmatrix}. \tag{6.2}$$

These two 2×2 matrices, which operate on a single site, constitute the Markov matrices which operate on the distribution vector. For example, the AMM in one dimension can be written as

$$\hat{a}_{\mathbf{n}} = C_{\mathbf{n}}^\dagger + \frac{1}{4}(\hat{a}_{\mathbf{n}+1} + \hat{a}_{\mathbf{n}-1})^2 C_{\mathbf{n}} \tag{6.3}$$

with suitably defined $C_{\mathbf{n}}^{(\dagger)} = \mathbf{1}_2^{\otimes(\mathbf{n}-1)} \otimes C^{(\dagger)} \otimes \mathbf{1}_2^{\otimes(L-\mathbf{n})}$, where $\mathbf{1}_2$ is a 2×2 identity matrix and $\otimes$ denotes the direct product (Kronecker product). Boundary conditions are met by defining $\hat{a}_0 = \hat{a}_{L+1} = \mathbf{1}$, the latter being a $2^L \times 2^L$ identity matrix. If $\hat{a}_{\mathbf{n}}$ and $C_{\mathbf{n}}^{(\dagger)}$ commute, it would be trivial to recover Eq. (8.63) on p. 278,

$$\hat{a}_{\mathbf{n}}^2 = \frac{1}{4}(\hat{a}_{\mathbf{n}+1} + \hat{a}_{\mathbf{n}-1})^2 = \frac{1}{4}(\hat{a}_{\mathbf{n}+1}^2 + 2\hat{a}_{\mathbf{n}+1}\hat{a}_{\mathbf{n}-1} + \hat{a}_{\mathbf{n}-1}^2) \tag{6.4}$$

by squaring Eq. (6.3), using $\{C, C^\dagger\} = CC^\dagger + C^\dagger C = \mathbf{1}_2$, the Abelianness $\hat{a}_{\mathbf{n}+1}\hat{a}_{\mathbf{n}-1} = \hat{a}_{\mathbf{n}-1}\hat{a}_{\mathbf{n}+1}$ and the nilpotency $(C^{(\dagger)})^2 = 0$, which follows immediately from Eq. (6.2), but is of course a general feature of the algebra (after changing state, the system is in a different state) and thus independent of the particular matrix representation. However, one can show

---

[7] A different operator representation can be found in Stilck, Dickman, and Vidigal (2004). An explicit calculation for system sizes up to 12 was done by Sadhu and Dhar (2009).

explicitly that $\hat{a}_n$ and $C_n^{(\dagger)}$ do *not* generally commute, using, for example, a system of size $L = 1$, where $\hat{a}_n = C_n^\dagger + C_n$. Spelled out, Eq. (6.4) means that charging a site twice amounts to the same as charging one of its neighbours twice with probability $1/4$ each or each one once with probability $1/2$.

With the short-hand $W = (\frac{1}{4})(\hat{a}_{n+1} + \hat{a}_{n-1})^2$ in Eq. (6.3) one finds

$$\hat{a}_n^2 = WC_nC_n^\dagger + C_n^\dagger WC_n + WC_nWC_n, \tag{6.5}$$

as $(C^\dagger)^2 = 0$. In the same notation, one has $\hat{a}_nC_n^\dagger = WC_nC_n^\dagger$ and $\hat{a}_nC_n = C_n^\dagger C_n$. Because W is defined on the basis of $\hat{a}_{n\pm 1}$, the Abelian symmetry $\hat{a}_n\hat{a}_m = \hat{a}_m\hat{a}_n$ implies commutation of $\hat{a}_n$ and W, $\hat{a}_nW = W\hat{a}_n$ so that

$$\hat{a}_n^2C_n^\dagger = \hat{a}_nWC_nC_n^\dagger = WC_n^\dagger C_nC_n^\dagger \tag{6.6a}$$
$$\hat{a}_n^2C_n = WC_nC_n^\dagger C_n, \tag{6.6b}$$

using the nilpotency of $C^{(\dagger)}$ again. After multiplication from the right with C_n and $C_n^\dagger$ respectively, the sum of the left hand sides is $\hat{a}_n^2$, using $\{C_n^\dagger, C_n\} = \mathbf{1}_2^{\otimes L} = \mathbf{1}_{2^L} = \mathbf{1}$. Similarly, the square of the identity is $\mathbf{1}^2 = C_n^\dagger C_nC_n^\dagger C_n + C_nC_n^\dagger C_nC_n^\dagger$, which simplifies the sum of the right hand sides of Eq. (6.6) to $\hat{a}_n^2 = W$, Eq. (6.4).

Along the lines of the n-state model (Ben-Hur and Biham, 1996) mentioned above, which assigns probabilities to certain particle redistributions individually, one can generalise the model by assigning individual probabilities to the different processes in Eq. (6.3),

$$\hat{a}_n = C_n^\dagger + \left(p_r\hat{a}_{n+1}^2 + p_l\hat{a}_{n-1}^2 + p_0\hat{a}_{n+1}\hat{a}_{n-1}\right)C_n \tag{6.7}$$

where $0 \le p_{r,l,0} \le 1$ are probabilities with $p_r + p_l + p_0 = 1$. The model is fully defined by fixing the boundary conditions and specifying the external drive, such as $\mathcal{O} = \frac{1}{L}\sum_n^L \hat{a}_n$ for uniform influx, see Eq. (8.58) on p. 274. For Eq. (6.7) the same identity Eq. (6.4) applies as for Eq. (6.3),

$$\hat{a}_n^2 = p_r\hat{a}_{n+1}^2 + p_l\hat{a}_{n-1}^2 + p_0\hat{a}_{n+1}\hat{a}_{n-1}. \tag{6.8}$$

For $p_r = p_l$ this model is isotropic on average. [8] For $p_r = p_l = 1/4$, the AMM is recovered exactly, but the same universal behaviour is found for all $0 < p_r = p_l < 1/2$ (Pruessner, 2004c). If one of the three probabilities is unity, the system is deterministic, where $p_0 = 1$ represents the one-dimensional Abelian BTW Model, Eq. (6.19). If $p_0 = 0$, the equation describes diffusion (with convection if $p_l \ne p_r$), with pairs of particles moving through the system until they reach the boundaries. As they never split, $p_0 = 0$, they effectively never interact and their movements leave the system unchanged so that only the external drive leads to an evolution of the stable state, not dissimilar to the one-dimensional BTW model (Ruelle and Sen, 1992). Two charges at the same site will therefore bring the system back to its original state, $\hat{a}_n^2 = \mathbf{1}$ for all $p_l + p_r = 1$.

For $p_l \ne p_r$ the boundary conditions are crucial for the exact (scaling) behaviour. If they are both open, then any amount of convection $p_r - p_l$ will eventually dominate, producing pure convection for $p_0 = 0$ or the scaling of the TAOM for $p_0 > 0$ (Pruessner and Jensen, 2003; Pruessner, 2004b, also Sec. 6.3.5). The situation is more complicated

---

[8] Ben-Hur and Biham (1996) use the term 'non-directed on average'.

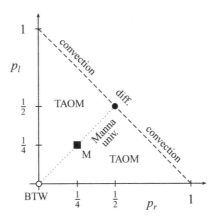

Figure 6.4 The phase diagram of the generalised AMM, Eq. (6.3). The BTW Model (open circle) is recovered at $p_r = p_l = 0$, the AMM at $p_r = p_l = 1/4$ (filled square). The entire dotted line $0 < p_r = p_l < 1$ (excluding the endpoints) belongs to the Manna universality class. It terminates in pure diffusion (filled circle). All models on the dashed line $p_r + p_l = 1$ ($p_0 = 0$) have only pairs of particles moving through the system, with convection asymptotically dominating. All $p_0 > 0$ and $p_r \neq p_l$ are expected to belong to the TAOM universality class, subject to appropriate boundary conditions.

with particles possibly 'piling up' if one of the boundaries is closed. The phase diagram is shown in Fig. 6.4. Dissipation can be added by allowing discharge $C_{\mathbf{n}}$ without charging the neighbours, i.e. by adding $\epsilon \mathbf{1} C_{\mathbf{n}}$ to Eq. (6.7), and reweighting the other processes, so that $p_r + p_l + p_0 = 1 - \epsilon$.

In principle, Eq. (6.7) can be used to expand recursively any $\hat{a}_{\mathbf{n}}$ in terms of $C_{\mathbf{n}}^{(\dagger)}$ and weights. However, such an expansion will always produce infinitely many terms, even in a finite system, unless the model is **totally asymmetric**, $p_0 = p_l = 0$ or $p_0 = p_r = 0$. Mathematically, this problem originates from squares, $\hat{a}_{\mathbf{n}\pm 1}^2$, in $\hat{a}_{\mathbf{n}}$. These squares, in turn, contain squares, in particular $\hat{a}_{\mathbf{n}}^2$, Eq. (6.8), so that the expansion of every $\hat{a}_{\mathbf{n}}$ contains $\hat{a}_{\mathbf{n}}^2$, each of which is expanded to $\hat{a}_{\mathbf{n}}^2$ and so on. This problem can be traced back to a finite weight w of the identity in the expansion,[6.1] $\hat{a}_{\mathbf{n}} = \cdots + w\hat{a}_{\mathbf{n}}\mathbf{1}$, so that the model can go through the same state multiple times, i.e. there is a finite probability for every avalanche size $s \in \mathbb{N}$ in the stationary state for any $L \geq 2$.

The Oslo Model, which is widely accepted to be in the same universality class as the AMM (Nakanishi and Sneppen, 1997), is somewhat better behaved in this respect, see Sec. 6.3.4.1. Physically, the origin of this inconvenience is the possibility of an arbitrarily long sequence of pure diffusion in the Manna Model, where a pair of particles moves through the system until it breaks up after n moves with probability $(1 - p_0)^{n-1}p_0$. One might therefore replace the quadratic terms by pairs of terms of the form $\hat{a}_{\mathbf{n}'+1}\hat{a}_{\mathbf{n}'-1}$ over a range of sites $\mathbf{n}'$ and appropriate weights.

An interesting path to explore the above formalism is a complete **fermionisation** of the operators, which is a standard technique in many-body physics (Mahan, 1990, pp. 46–53), possibly along the lines of a **Jordan–Wigner transform** in the Ising Model (Wannier, 1987, pp. 356–365). Other techniques which might be applicable to the present problem are the **density matrix renormalisation group** (Kaulke, 1999; Peschel, Wang, Kaulke, and Hallberg, 1999) and the **matrix product ansatz** (Derrida, Evans, Hakim, and Pasquier, 1993; Hinrichsen, 2000). However, most of these methods have been applied to the actual *microscopic* dynamics, whereas the Markov matrices developed above operate on a macroscopic time scale, in that they map from one stable state to another. The dynamics is fully wrapped up in the expansion Eq. (6.7).

An equation of motion for the AMM is easily devised and follows the same procedure exemplified in Sec. 8.3.2.1 for the Abelian BTW Model. The number of charges $g(\mathbf{n}, t)$ received by site $\mathbf{n}$ has a deterministic and a stochastic component. The deterministic component originates from the number of topplings $G(\mathbf{n}', t)$ performed by the neighbours $\mathbf{n}'$ of $\mathbf{n}$, the stochastic component from the randomness of the particle distribution. The latter can be encoded in the noise $\xi(\mathbf{n}', G)$, which is 1 if both particles went to the right at the Gth toppling on site $\mathbf{n}'$. It is -1 if both particles went to the left and vanishes otherwise.

In one dimension, the number of charges arriving at $\mathbf{n}$ from the right neighbour $\mathbf{n} + 1$ therefore is $G(\mathbf{n} + 1, t) - \sum_{G'=1}^{G(\mathbf{n}+1,t)} \xi(\mathbf{n} + 1, G')$, so that

$$g(\mathbf{n}, t) = G(\mathbf{n}+1, t) - \sum_{G'=1}^{G(\mathbf{n}+1,t)} \xi(\mathbf{n}+1, G') + G(\mathbf{n}-1, t) + \sum_{G'=1}^{G(\mathbf{n}-1,t)} \xi(\mathbf{n}-1, G'), \quad (6.9)$$

with appropriate initial and boundary conditions, say $g(\mathbf{n}, 0) = 0 = G(\mathbf{n}, 0)$, $g(0, t) = 0 = g(L+1, t)$ and $G(0, t) = 0 = G(L+1, t)$. In the following discussion, these boundary conditions are not considered further.

The number of topplings at a site is a simple function of the charges

$$G(\mathbf{n}, t+1) = \left\lfloor \frac{g(\mathbf{n}, t)}{2} \right\rfloor \quad (6.10)$$

so that one can in principle derive an equation of motion,

$$g(\mathbf{n}, t+1) = \left\lfloor \frac{g(\mathbf{n}+1, t)}{2} \right\rfloor - \sum_{G'=1}^{\lfloor g(\mathbf{n}+1,t)/2 \rfloor} \xi(\mathbf{n}+1, G') + \left\lfloor \frac{g(\mathbf{n}-1, t)}{2} \right\rfloor$$

$$+ \sum_{G'=1}^{\lfloor g(\mathbf{n}-1,t)/2 \rfloor} \xi(\mathbf{n}-1, G'), \quad (6.11)$$

which, in this form, is practically intractable. The fundamental problem is the summation over the noise (Bonachela, Chaté, Dornic, and Muñoz, 2007), which means that $g(\mathbf{n}, t)$ does not reach a stationary state even in a co-moving frame. For example, if the ensemble average $\langle \cdot \rangle$ over the noise vanishes, then

$$\langle g(\mathbf{n}, t+1) \rangle = \left\langle \left\lfloor \frac{g(\mathbf{n}+1, t)}{2} \right\rfloor \right\rangle + \left\langle \left\lfloor \frac{g(\mathbf{n}-1, t)}{2} \right\rfloor \right\rangle, \quad (6.12)$$

closely related to the equation of motion of the Abelian BTW Model, Eq. (8.66) on p. 278. Subtracting Eq. (6.12) from Eq. (6.11) leaves the sums over the noise on the right, which are divergent as $t \to \infty$. The 'interface' $g(\mathbf{n}, t)$ therefore cannot be expected to exhibit asymptotically finite fluctuations about its average profile. On the other hand, the configuration $h(\mathbf{n}, t)$ of the system reaches a stationary state, but is given by the net influx $h(\mathbf{n}, t) = g(\mathbf{n}, t) - 2G(\mathbf{n}, t)$, which is the difference between two non-stationary quantities.

This unfortunate complication in the Manna Model does not arise in the Oslo Model, see Sec. 6.3.4.2.

6.2.1.2 The Manna universality class

Universality is a central theme in SOC – without it, there is little justification for studying apparently over-simplified models on small scales in order to understand more complicated systems, which may operate on much wider length and time scales. The **Manna universality class**, first identified and distinguished as such by Ben-Hur and Biham (1996), is arguably the only universality class relating the *robust* scaling of models ranging from SOC systems, such as the Manna Model and the Oslo Model (Nakanishi and Sneppen, 1997), to established non-equilibrium critical phenomena, such as the qEW equation (Paczuski and Boettcher, 1996, also Sec. 6.3.4.2).

Depending on the emphasis, the Manna universality class is known by many names: the **Oslo universality class**, the **conserved lattice gas** (CLG) universality class or the **C-DP universality class**, more specifically the **universality class of absorbing phase transitions with (non-diffusive [9]) conserved field**. The last description underlines its distinctive feature compared to the DP universality class, i.e. the existence of an additional symmetry (conservation of particles in local relaxations, e.g. Dickman *et al.*, 2001) which justifies the non-applicability of the DP conjecture (Janssen, 1981; Grassberger, 1982; Rossi *et al.*, 2000; Muñoz *et al.*, 2001). [10] Rossi *et al.* (2000) proposed C-DP as a universality class distinct from DP, which has been confirmed numerically many times (e.g. Dickman *et al.*, 2001, 2002; Ramasco, Muñoz, and da Silva Santos, 2004; Dornic, Chaté, and Muñoz, 2005; Dickman, 2006; Bonachela, 2008; Bonachela and Muñoz, 2008, also Table 8.4, p. 306). However, the numerical values of some exponents are so similar (Muñoz, Dickman, Vespignani, and Zapperi, 1999; Dickman *et al.*, 2002, also Table 8.2, p. 294) that it has been suggested to approach C-DP as a pertubation of DP (Bonachela and Muñoz, 2007). The name C-DP may also be understood in the style of the nomenclature of dynamical critical phenomena (Dickman, 2006) introduced by Hohenberg and Halperin (1977), where model C has a conserved energy density to which the order parameter couples quadratically (see also the more recent review by Täuber, 2005). The C-DP universality class is defined by Rossi *et al.*'s (2000, also Pastor-Satorras and Vespignani, 2000d) conjecture that

> in the absence of additional symmetries, absorbing phase transitions in stochastic models with infinite absorbing states and activity coupled to a non-diffusive conserved field define a unique and per se universality class.

Nakanishi and Sneppen (1997, confirmed for example by Pan, Zhang, Sun, and Yin, 2005b and Malthe-Sørenssen, 1999, who uses a variant of the Oslo Model and interprets

---

[9] The term 'non-diffusive' (sometimes also 'static') is used to stress that particles remain trapped locally and do not diffuse unless activated. This field therefore does not contain a diffusion term in its Langevin equation. Coupling to a diffusive conserved field leads to a different universality class (van Wijland, Oerding, and Hilhorst, 1998).

[10] The existence of **multiple absorbing states** does not generally preclude DP behaviour after all (Mendes, 2000; Dickman, 2002b; Ódor, 2004). Park, Kang, and Kim (2005) presented a model with particle conservation and two absorbing states displaying neither DP nor C-DP scaling.

his findings very differently) showed numerically that the AMM and the OSLO Model belong to the same universality class, if driven in the same way, i.e. both in the bulk, typical for the MANNA Model, or both at the boundary, typical for the OSLO Model. In both models, the avalanche dimension D does not change with the type of driving (see also Paczuski and Boettcher, 1996), while the avalanche exponent τ does according to $D(2-\tau) = \sigma_1^{(s)}$ with $\sigma_1^{(s)}$ being fixed by conservation and diffusion arguments, see Sec. 8.1.1.1. One might interpret this finding as an expression of the **universality of the underlying *micro*-dynamics**, which is linked to the qEW equation in a Langevin description, whereas τ is determined by *global* **conservation laws**. The robustness of D distinguishes the MANNA Model from the ASM where both exponents, D as well as τ, depend on the driving ($D = 2$ and $\tau = 1 + \pi/(2\theta)$ for boundary drive with opening angle θ, Ivashkevich *et al.*, 1994a, whereas $D \neq 2$ for bulk drive).

The range of a priori unrelated models in the **MANNA universality class** is remarkable:

- The MANNA Model, the OSLO Model and many of their variants (Amaral and Lauritsen, 1996b; Paczuski and Boettcher, 1996; Malthe-Sørenssen, 1996, 1999; Nakanishi and Sneppen, 1997; Zhang, 1997; Alava and Lauritsen, 2001; Pan *et al.*, 2005b; Stapleton, 2007).
- The stochastic ZHANG Model (Nakanishi and Sneppen, 1997; Biham *et al.*, 2001), the MASLOV–ZHANG Model (Maslov and Zhang, 1996; Bonachela and Muñoz, 2008), the stochastic parallel ZHANG Model (Pastor-Satorras and Vespignani, 2000a), and possibly the ZAITSEV Model (Zaitsev, 1992; Roux and Hansen, 1994; Paczuski *et al.*, 1996; Nakanishi and Sneppen, 1997), which is governed by extremal dynamics, as is the continuous version of the OSLO Model by Malthe-Sørenssen (1999).
- The sandpile model by Maslov and Zhang (1996, see Bonachela and Muñoz, 2008), the isotropic version of the sandpile with sticky grains (Mohanty and Dhar, 2002; Bonachela, Ramasco, Chaté, *et al.*, 2006, but see Mohanty and Dhar, 2007).
- The conserved lattice gas (CLG, Rossi *et al.*, 2000, derived from Jensen's 1990 model), which displays $1/f$ noise, and the conserved threshold transfer process (CTTP, Mendes *et al.*, 1994; Rossi *et al.*, 2000; Lübeck and Heger, 2003a), or generally the restricted sandpile (Dickman *et al.*, 2002; Dickman, 2006).
- The conserved reaction-diffusion model (CRD, Pastor-Satorras and Vespignani, 2000d; Rossi *et al.*, 2000; Muñoz *et al.*, 2001; Alava and Muñoz, 2002).
- A two-species, conserved reaction-diffusion system (Pastor-Satorras and Vespignani, 2001) as well as the linear interface model (LIM, Sec. 8.5.4.1, Barabási and Stanley, 1995; Paczuski *et al.*, 1996; Paczuski and Boettcher, 1996; Nakanishi and Sneppen, 1997; Vespignani *et al.*, 2000; Alava and Lauritsen, 2001) also known as the quenched Edwards–Wilkinson equation (qEW equation, Pruessner, 2003a).
- The *deterministic* [11] train model (de Sousa Vieira, 1992; Paczuski and Boettcher, 1996; Malthe-Sørenssen, 1999; de Sousa Vieira, 2000), the ASM with quenched random toppling matrix (Karmakar *et al.*, 2005) and a Burridge–Knopoff-type model on a random substrate (Cule and Hwa, 1996; Chianca, Sá Martins, and de Oliveira, 2009).

---

[11] In this model, the noise seems to be substituted by its chaotic dynamics (Paczuski and Boettcher, 1996).

Some inconsistencies remain, however. Exponents in one dimension, in particular the dynamical exponent z, show a significant scatter, see Table 6.1. Lübeck and Heger (2003a) reported a **splitting of the Manna universality class** at $d = 1$ where some of the exponents of the Manna Model and the CTTP (a version of the height restricted Manna Model) differ (also Dickman, 2006). This might be due to strong corrections (Sec. 6.2.1) or seemingly irrelevant interactions becoming relevant in one dimension (Muñoz *et al.*, 2001) or a matter of long correlation times (Alava and Muñoz, 2002). It has also been argued (Lübeck and Heger, 2003a) that this discrepancy is due to the CTTP becoming predominantly deterministic and trivial in one dimension.[12] This argument is supported by the CLG, in the Manna universality class for $d = 2$, becoming fully deterministic in one dimension (Rossi *et al.*, 2000; Lübeck and Heger, 2003a). The failure of such models to display the same scaling behaviour as the Manna Model in one dimension might therefore be rooted in their dynamics being pathological in one dimension, where a single relaxation mechanism dominates which is screened by others in higher dimensions. There is also some discord regarding the upper critical dimension of the entire universality class. While van Wijland (2002) identified $d_c = 6$ in a field theoretic approach, numerical evidence suggests $d_c = 4$ (Lübeck and Heger, 2003b; Janssen, 2005), consistent with results for the qEW equation (Nattermann, Stepanow, Tang, and Leschhorn, 1992; Le Doussal, Wiese, and Chauve, 2002).

A clear **distinction between DP and C-DP** has been subject to some debate, not least because exponents are numerically very close, for example in one dimension $\tau = 1.11(2)$ for C-DP and $\tau = 1.108(1)$ for DP; the values become increasingly close in higher dimensions and eventually coincide in $d = d_c = 4$ (Bonachela and Muñoz, 2008).[13] There are at least two models, which have been claimed to belong to the DP universality class, as opposed to C-DP (Bonachela and Muñoz, 2008), which they should belong to according to the conjecture by Rossi *et al.* (2000) cited above: the sandpile with **sticky grains** by Mohanty and Dhar (2002, also Sec. 8.4.5.1) and the (continuous) sandpile model by Maslov and Zhang (1996), with a random redistribution of energy similar to the Manna Model. These discrepancies have been addressed and resolved (also Bonachela *et al.*, 2006; Bonachela and Muñoz, 2007, 2008, but see Mohanty and Dhar, 2007), by probing the response of the models under scrutiny to the introduction of anisotropy or different types of boundaries.

Dickman *et al.* (2001) and Kockelkoren and Chaté (2003) found evidence that the Manna Model (as a representative of the C-DP universality class) and the LIM (the qEW equation) in one dimension do not share the same universality class, while Vespignani *et al.* (2000) discovered some 'anomalies' in the dynamical critical properties in two dimensions. These discrepancies were later partly resolved by Bonachela *et al.* (2007) by identifying the correct

---

[12] One could argue that the AMM suffers from similar 'trivial', i.e. diffusion-like events. The deterministic event of a doubly occupied site being trapped in a sea of singly occupied ones (Lübeck, 2004, Fig. 40) is probably very rare. It is remarkable that about 40% of all relaxations are deterministic in one dimension (Lübeck and Heger, 2003a; Lübeck, 2004). This might refer to the frequency with which the second of two redistributed particles can be placed only at one neighbouring site. Some of these problems may be avoided by implementing the external field in a different way (Lübeck, 2002b).

[13] The situation seems to be slightly clearer for the exponents D and z listed in Table 8.2 (also Table 8.4).

observable (namely g rather than G, Sec. 6.2.1.1; Pruessner, 2003a; Bonachela *et al.*, 2007). Dickman *et al.* (2002) confirmed the anomalous behaviour, since they failed to see standard FSS in the FES version of the MANNA Model and had to resort to scaling at the 'effective critical point'. Again, some of these results might be due to the strong corrections to scaling and pronounced finite size corrections present in the MANNA Model (Sec. 6.2.1, Nakanishi and Sneppen, 1997; Dickman and Campelo, 2003; Dickman, 2006).

At closer inspection, the OSLO Model being in the MANNA universality class is a surprise, because of the fundamentally different rôle of noise in the OSLO Model and the MANNA Model. Anticipating some of the results discussed in Sec. 6.3, the **stochasticity** in the OSLO Model was initially thought to be 'conceptually different from transport in media with quenched disorder' (Christensen *et al.*, 1996), because it is part of the dynamics, influencing the motion of grains and at the same time its product, as the 'random' rice grain configurations in the experiment give rise to it. Paczuski and Boettcher (1996), however, immediately saw the connection to well-understood **quenched noise**, at this stage still employing a Heaviside step function. This was later confirmed in a direct mapping of the sOM to the **quenched Edwards–Wilkinson equation** (qEW equation, Pruessner, 2003a), the scope of which was further extended by Alava (2002). In the MANNA Model, the noise enters its equation of motion as an unusual, unbounded sum, Eq. (6.11), whereas in the OSLO Model the equation of motion has the standard form of the qEW equation, Eq. (6.34) on p. 205. In fact, a Langevin formalism, in particular one with quenched noise, is applied much more naturally to the OSLO Model than to the MANNA Model. Similarly, the algebraic properties of the AMM and the OSLO Model seem to be fundamentally different, producing recursive matrix representations with infinitely many terms in the case of the AMM, see p. 175, and with a finite number of terms in the case of the OSLO Model, p. 200. In turn, the **absorbing state** formalism and the perspective of C-DP is much more suited for the MANNA Model than for the OSLO Model. The continuum version of the OSLO Model suggested by Pruessner (2004c, p. 210), which is capable of encapsulating both the OSLO Model and MANNA Model, could provide a way of understanding which feature of the MANNA Model takes on the rôle of the quenched noise in the OSLO Model.

The observation that the qEW equation and C-DP share the same scaling behaviour (compare Table 8.4, p. 306, and Table 8.5, p. 315), whether considered as SOC models, as interfaces or as AS models, is subject to ongoing research. The apparent mismatch of the models can be aligned with the difficulties mentioned by Alava and Lauritsen (2001) as well as Alava and Muñoz (2002) of mapping Langevin equations with *quenched* noise, such as the qEW equation, and those with *RFT-type, **annealed** noise* and non-diffusive conserved field, representing C-DP (Rossi *et al.*, 2000), onto each other.[14] It seems that significant progress in the understanding of both types of equations individually has to be made before they can be reconciled in a common field theory, which would explain the shared universal behaviour by generating the same set of relevant vertices. At the same time, one might hope that the field theory of C-DP will eventually enlighten field theories with quenched noise generally.

---

[14] This type of 'activated' noise, rather than the quenched noise in the qEW equation, was probably what Christensen *et al.* (1996) as quoted above had in mind.

With the MANNA Model entering the realm of absorbing states and **interfaces**, most of the studies mentioned above were carried out comparing the fixed energy sandpile (FES, Sec. 9.3.1) version of the MANNA Model to another absorbing state model. Correspondingly, the key observables were those of absorbing state phase transitions (activity density and its moments, survival probability, spreading etc.) rather than avalanches as characteristic of SOC (but Pruessner, 2007). This makes it particularly difficult to determine common features of SOC models in the MANNA universality class. Given that oMM and AMM scale identically, Abelianness does not seem to be a necessary condition (Sec. 8.3.3) and neither does quenched noise, given that the OSLO Model is in the same universality class.

Karmakar *et al.* (2005) investigated the ASM (Sec. 4.2), averaging over randomised, conservative, asymmetric toppling matrices, and found that the resulting model belongs to the MANNA universality class. As the randomness was quenched, the ASM was still deterministic and averages were taken over different realisations of the toppling matrix. The characteristic multiscaling of the Abelian BTW Model was recovered only for symmetric toppling matrices, where the charge Δ_{nm} of site **m** at a toppling of site **n** equals Δ_{mn} for the toppling of site **m**. Like other deterministic models listed on p. 178, this result shows that *stochasticity is not a key ingredient of the* MANNA *universality class either.*[15] It also suggests that local anisotropy in the toppling dynamics might be one. Yet, the OSLO Model is locally and globally isotropic and belongs to the same universality class. The necessary and sufficient conditions for a model to belong to the MANNA universality class remain an open issue (see also Ódor, 2004).

6.2.1.3 Analytical approaches

Apart from Dhar's (1999b) work on the Abelian operator algebra of the AMM, there has been limited exact work on it. One of the major obstacles is the infinite number of terms that would need to be considered in a direct expansion of the operators and, equivalently, in the infinite number of 'decay channels', see Sec. 6.2.1.1. Even a system of size $L = 2$ has a finite (but exponentially decaying) probability of generating an avalanche of any (finite) size.

Dickman (2002a) characterised the one-dimensional MANNA Model with height restriction and periodic boundary conditions (i.e. an FES version of the MANNA Model) by means of a **truncated correlation function** (ben-Avraham and Köhler, 1992; Ferreira and Mendiratta, 1993; Dickman *et al.*, 2002). The height restriction is required so that the number of unstable states remains finite,[16] the latter because the correlation function represents the microscopic dynamics of the model. One might wonder whether this can be avoided by employing effective dynamics or making the particle positions rather than the local particle numbers the subject of the evolution.

In the truncation scheme, correlations beyond n sites are assumed to factorise, whereas the full joint probabilities are retained for up to n sites. The truncation is, again, a necessity

---

[15] One might speculate that stochastic variants of deterministic models, such as the MANNA Model as the stochastic version of the Abelian BTW Model (in one dimension), generally fare better with respect to scaling, because constraints in deterministic models prevent them from exploring the phase space more fully, see also p. 134 for the beneficial effect of impurities in the OFC Model.

[16] There are 2^L stable states, not all of which are necessarily accessible, Sec. 6.2.

to keep the number of coupled differential equations finite and affects in principle *all* correlations even those in a range less than n. However, systems of size $L = n$ are represented (almost) exactly, so that finite size scaling or any variation thereof can be used to determine the scaling behaviour. One such scheme is known under the curious name of 'coherent anomaly method' (Suzuki and Katori, 1986; Suzuki, Katori, and Hu, 1987). The resulting estimates of absorbing state exponents are in line with numerics (Dickman *et al.*, 2002).

A range of **effective (field) theories** has been developed (e.g. Pastor-Satorras and Vespignani, 2000d; Dornic *et al.*, 2005) and analysed using real space renormalisation group techniques (e.g. Pietronero *et al.*, 1994), numerical integration (Dickman, 1994; López and Muñoz, 1997; Ramasco *et al.*, 2004) or more general arguments based on interfacial phenomena (e.g. Vespignani *et al.*, 2000; Alava and Lauritsen, 2001). Because of the link between the MANNA Model and OSLO Model and from the OSLO Model to the quenched Edwards–Wilkinson (qEW) equation (Pruessner, 2003a), it is clear that any analytical progress in the MANNA Model implies analytical progress in the qEW, whose quenched noise is notoriously difficult to handle (Nattermann *et al.*, 1992; Leschhorn, Nattermann, Stepanow, and Tang, 1997). There is notable exchange between the two fields, most remarkably the determination of qEW equation exponents by means of an SOC model (Leschhorn, 1994, for the quenched KPZ equation see Leschhorn and Tang, 1994; Sneppen, 1992), long before the extent of the link between the two was fully appreciated (Paczuski and Boettcher, 1996). Table 8.5 on p. 316 shows a comparison of avalanche size exponents of the MANNA Model (Table 6.1 on p. 168) and those derived from computer simulations of qEW-type interfaces.

A number of real space renormalisation group studies have been performed on the MANNA Model (Pietronero *et al.*, 1994; Vespignani and Zapperi, 1995), which were modified by Hasty and Wiesenfeld (1997, also Hasty and Wiesenfeld, 1998b), who considered a one-dimensional model very reminiscent of the one-dimensional AMM. A diagrammatic perturbation theory (but without subsequent renormalisation group analysis) was performed by Dickman and Vidigal (2002) again on an FES version of the AMM, which they further refined later (Vidigal and Dickman, 2005).

6.2.2 Numerical results

Conceived in $d = 2$, the MANNA Model is most frequently studied in two dimensions, which is reflected in the table of exponents, Table 6.1. Nevertheless, extensive studies have also been published in the entire range up to and even beyond the (supposed) upper critical dimension.

Initially, properties of the avalanches were probed directly. In one and to some extent in two dimensions, however, corrections to scaling make it difficult to determine exponents accurately and cast some doubt over the very scaling behaviour of the system. As the link to absorbing state phase transition grew stronger, many authors changed their focus to dynamical features of the FES version of the MANNA Model or its interface interpretation. These dynamical properties can be linked to SOC observables (Sec. 9.3), but have a somewhat limited meaning in Abelian models (Sec. 8.3.2.3). Many exponents listed in Table 6.1 have been derived that way. All of those listed, however, are based on the MANNA

Model or one of its variants, rather than other less directly related models in the same universality class.

It took a remarkably long time for the first extensive numerics to be published on the **one-dimensional version** of the Manna Model (Nakanishi and Sneppen, 1997). This is even more surprising, as the one-dimensional Manna Model is non-trivial, which makes it clearly distinct from the BTW Model, lending strong support to the hypothesis of a separate universality class, which is one of the central questions raised initially. One might speculate that this clear distinction between the Manna Model and BTW Model in one dimension is related to the splitting of the Manna universality class due to deterministic behaviour taking over in some models (p. 179 and Lübeck and Heger, 2003a).

Ben-Hur and Biham (1996) considered a generalised form of the Manna Model, where the particle redistribution does not occur independently, similar to the first model introduced by Dhar (1999a). They also used an extended range of observables and exponents, including various geometrical features of avalanches. A similar approach taken in later studies (Biham *et al.*, 2001) proved useful in discriminating different universality classes. Chessa *et al.* (1999a) questioned some of the methods used and concluded, as mentioned above, that the BTW Model (Sec. 6.1.1), Zhang Model and Manna Model belong to the same universality class.

Despite these differences, the presence of simple FSS in the two-dimensional Manna Model has been reported quite consistently and exponents generally agree (Table 6.1 on p. 168). The wider spread for temporal exponents might be due to the problem of defining the microscopic time scale in Abelian models, Sec. 8.3.2.3. More recent numerical studies, such as those by Chessa *et al.* (1999b) and Lübeck (2000), mostly rely on a moment analysis. Relatively little work has been done above two dimensions, with the notable exception of Lübeck and Heger (2003b), who investigated the Manna Model in up to five dimensions, confirming its coincidence with mean-field behaviour there and determining some features with striking accuracy. A random neighbour version of the Manna Model was introduced by Zapperi *et al.* (1995) in the form of a self-organised branching process (Sec. 8.1.4.2). On the whole, the Manna Model is undoubtedly one of the most robust models displaying SOC.

Accepting the link between the Manna Model and the Reggeon Field Theory of the C-DP universality class, numerical work such as that by Ramasco *et al.* (2004) could be added to the list of numerical results for the Manna Model. In recent years, the focus has generally shifted towards understanding the link between SOC models and more traditional phase transitions into absorbing states (e.g. Bonachela and Muñoz, 2008) as well as the link between the latter and the qEW equation (Bonachela *et al.*, 2007).

6.3 The Oslo Model

The Oslo Model is the youngest model to be presented here in detail. The name 'Oslo Model' was coined by Paczuski and Boettcher (1996) who published their findings on its universal behaviour back to back with the definition of the model by Christensen *et al.*

(1996). As for most other models, for historic reasons, there has been some confusion regarding the name of the model and its precise definition. Nakanishi and Sneppen (1997), who found that the OSLO Model is in the MANNA universality class (Sec. 6.2.1.2, but see Fig. 6.5), called the model the 'OSLO ricepile model', as the model was inspired and informed by the **ricepile experiment** by Frette *et al.* (1996, see also Sec. 6.3.2). However, Amaral and Lauritsen (1996b) introduced a very closely related 'ricepile model' almost simultaneously, which is, in fact, a generalisation of the OSLO Model. In the current literature, it is the OSLO Model that is often referred to as the 'ricepile model', perhaps in order to contrast it with the 'sandpile', which it very much resembles. In fact, the algebraic structure of the OSLO Model, Eq. (6.19), is *identical* to that of the ASM, which might be why Jettestuen and Malthe-Sørenssen (2005) referred to it as the 'Oslo sandpile model'.

Leaving the naming muddle aside, there are two slightly different definitions of the model: on the one hand the original, widely used definition, called the 'original OSLO Model' (oOM) in the following, which is mainly derived from the one-dimensional BTW Model in the spirit of sand grains tumbling down a descent as defined in Box 6.3; on the other hand a numerically and sometimes analytically more convenient definition based on slope units (particles), in the following called the 'simplified OSLO Model' (sOM), defined in Box 6.4. Most results apply equally to both versions, which are, to a large extent, discussed in parallel below. Where no distinction is needed, 'OSLO Model' is used. Despite the close link to the original BTW Model, strictly speaking, the oOM is Abelian and so is the sOM.

Because of its simplicity, its robustness and proven relation to established models of non-equilibrium phase transitions (Paczuski and Boettcher, 1996; Pruessner, 2003a), it has received increased attention over the last few years. Since it is so well behaved even in one dimension, it would probably be a better representative of the MANNA universality class than the MANNA Model itself, if it were not for the particular way it is driven. Because of its link to an experiment and in order to exclude **external randomness** (Frette, 1993) due to the driving, the oOM is **driven at the boundary** and originally derives its dynamical rules from a height picture, which makes it somewhat ambiguous in higher dimensions and sets it apart from models with (random) **bulk drive**, such as the MANNA Model. The different type of driving affects only the avalanche size exponent τ, not the avalanche dimension D (Nakanishi and Sneppen, 1997; Christensen, 2004). The driving might affect similarly α and z, which in addition suffer from some ambiguity in Abelian models (Sec. 8.3.2.3) so that the mismatch of α between the OSLO Model and MANNA Model is no surprise. Given that z has a wide spread in the MANNA Model, the only immediate point of contact between the OSLO Model and MANNA Model is therefore D. To corroborate their shared universal behaviour, Stapleton (2007, in particular Fig. 1.11 on p. 47) compared the scaling function of the avalanche size distribution of the two models and found some, yet surprisingly imperfect agreement, probably caused, however, by a mismatch of boundary conditions, see Fig. 6.5, and Bonachela (2008) implemented the OSLO Model with bulk drive, reproducing the avalanche exponents of the MANNA Model.

The first version of the OSLO Model presented in this chapter is its original definition, Box 6.3. It is closely linked to models developed by Ding *et al.* (1992), who intended to incorporate inertia, and in particular to those developed by Frette (1993), who made the case for a model that differs from the BTW Model in that it is non-trivial even in one dimension, does not develop into a trivial state in the absence of external randomness and reflects

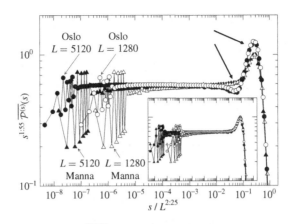

Figure 6.5 Attempt to collapse the binned histograms $\overline{\mathcal{P}^{(s)}}(s)$ of the avalanche size for the one-dimensional oOM (circles) and AMM (triangles) for system sizes L as indicated (10^7 avalanches for equilibration, $5 \cdot 10^8$ for statistics), both driven at the first site only. The data of the MANNA Model are shifted horizontally by a factor $1/0.3$. While the collapse works very well within each model, deviations between the MANNA Model and original OSLO Model are visible (pointed out by arrows). They disappear once the same boundary conditions are applied in both models, as shown in the *inset* where the data of the sOM collapses with that of the AAM (no shifts are applied).

the randomness and anisotropy expected to govern the relaxation mechanism in granular media. In fact, a number of experiments, which later bore out the ricepile experiment, preceded Frette's model (Christensen, 2008; Frette, 2009, see reference 12 in Frette, 1993 and Frette, Feder, Jøssang, and Meakin, 1993). Frette (1993) suggested randomising the **local critical slope**, which is in fact the *only* difference between the BTW Model and the OSLO Model defined three years later, but Frette (1993) implemented the local critical slope in a very different way compared to the OSLO Model. Consequently, Frette's (1993) model does not seem [17] to display the same scaling as the OSLO Model. At the time Frette published his model, only very few other models featured internal stochasticity: the MANNA Model (Manna, 1991b) which relaxes stochastically, the lattice gas model by Jensen (1990, random pinning centres in addition to random drive, also Chau and Cheng, 1992) and the ZAITSEV Model (Zaitsev, 1992, Poissonian relaxation). It is not immediately obvious which physical mechanism actually gives rise to the noise in the present models. As O'Brien and Weissman (1994, also Ch. 3, p. 61 and p. 63) pointed out, internal noise and disorder might not be so desirable after all, because broad event size distributions caused by the internal noise rather than underlying cooperative phenomena might be mistaken for genuine scale invariance, or the latter might be (solely) caused by disorder.

The ricepile model presented by Amaral and Lauritsen (1996b), which is very closely related to the OSLO Model, was developed as a response to the ricepile experiment (Frette *et al.*, 1996), just like the OSLO Model itself. The critical slopes are random variables as well, but can have values over a wider range. Sites with slope exceeding a lower threshold, $z_\mathbf{n} > S_1$, topple with a certain probability p. Not to topple effectively amounts to a larger

---

[17] The numerical results and their presentation in the study (Frette, 1993) are somewhat unclear.

Box 6.3   **The original Oslo Model**

General features

- First published by Christensen, Corral, Frette, *et al.* (1996).
- Motivated by the ricepile experiment (Frette *et al.*, 1996).
- Defined in one dimension.
- Deterministic (boundary) drive, stochastic relaxation.
- Abelian when not driven in bulk (Dhar, 2004).
- Closely related to the BTW Model.
- In the Manna universality class (Nakanishi and Sneppen, 1997).
- Mapped to the qEW equation (Pruessner, 2003a).
- Investigated analytically as well as numerically.
- Exponents listed in Table 6.2, p. 189.
- For simplified version see Box 6.4.

Rules

- One-dimensional grid.
- Each site $\mathbf{n} \in \{1, \ldots, L\}$ has local threshold $z_{\mathbf{n}}^c \in \{1, 2\}$ and contains $h_{\mathbf{n}} \in \mathbb{N}_0$ grains. Define $h_{L+1} = 0$ for simplicity.
- *Initialisation*: irrelevant, model studied in the stationary state.
- *Driving*: add a grain at $\mathbf{n} = 1, h_{\mathbf{n}_0} \rightarrow h_{\mathbf{n}_0} + 1$.
- *Toppling*: for each site $\mathbf{n}$ with $z_{\mathbf{n}} = h_{\mathbf{n}} - h_{\mathbf{n}+1} > z_{\mathbf{n}}^c$ move 1 grain downhill, $h_{\mathbf{n}+1} \rightarrow h_{\mathbf{n}+1} + 1$ and $h_{\mathbf{n}} \rightarrow h_{\mathbf{n}} - 1$. Then choose a new threshold $z_{\mathbf{n}}^c \in \{1, 2\}$ randomly and with equal probabilities.
- *Dissipation*: grains are lost at open boundaries.
- *Parallel update*: discrete microscopic time, sites exceeding $z_{\mathbf{n}}^c$ at time t topple at time $t + 1$ (updates in sweeps). Alternatively, sometimes random sequential.
- *Separation of time scales*: drive only once all sites are stable, i.e. $z_{\mathbf{n}} \leq z_{\mathbf{n}}^c$ (quiescence).
- *Key observables* (see Sec. 1.3):
  avalanche size s, the total number of topplings until quiescence;
  avalanche duration T, the total number of parallel updates until quiescence;
  residence time R, the number of particles added until a certain grain added, leaves the system.

local critical slope. Whenever a site's slope changes, toppling takes place with probability p. Only if the slope exceeds another threshold, $z_{\mathbf{n}} > S_2$, toppling *must* take place. For $S_1 = 1$, $S_2 = 2$ and $p = 1/2$, Amaral and Lauritsen's (1996b) ricepile is *identical* to the oOM.

The Oslo Model and its immediate relatives share two features: BTW-like local relaxations and stochastic, local, critical slopes. By randomly changing critical slopes, grains can become buried and thus trapped in the system, which effectively introduces a degree of non-conservation of grains and slope units. Zhang (2000) found numerically that the fluctuation of the average slope, or equivalently the height of the pile at $\mathbf{n} = 1$, displays varying types of $1/f^\alpha$ noise.

Chua and Christensen (2002) used some heuristic but very persuasive arguments to show that *recurrent*, stable configurations in the oOM, i.e. configurations that appear with non-vanishing frequency when repeatedly triggering avalanches, obey two simple rules. Their result was later confirmed by Dhar (2004, also Corral, 2004a) using an operator approach. The rules preclude the *majority* of stable states, so that the fraction of recurrent states is only about $((3 + \sqrt{5})/6)^L$, to be compared to a similar result for the Abelian BTW Model (Dhar, 1990a, also Sec. 4.2.1, p. 96). In contrast to the Abelian BTW Model, not all recurrent states are equally likely, a feature which Chua and Christensen call '**non-ergodic**'. Although the phase space is so severely restricted, this is not an integral feature of the Oslo Model, as widening it by generalising the rules and adding additional randomness in the redistribution of particles at toppling (Sec. 6.2.1.1) does not change its scaling behaviour (Pruessner and Jensen, 2003).

6.3.1 A simplified version of the Oslo Model

In its original version, Box 6.3, the Oslo Model is not easily extended to higher dimensions (Aegerter, 2003; Pradhan and Dhar, 2006; Bonachela, 2008) and suffers generally from the same problems as the BTW Model, namely a scalar, local 'slope' $z_{\mathbf{n}}$ even when $d > 1$, which should rather be the vector-valued [18] gradient $\nabla h_{\mathbf{n}}$. Also, the model is more elegantly formulated in terms of slope units (particles), not least to allow **bulk drive** (i.e. adding particles at non-boundary sites, possibly chosen at random) without violating the Abelian symmetry. Finally, to avoid anisotropies, boundary conditions are best chosen to be identical, i.e. open everywhere. Such a simplified version of the Oslo Model is defined in Box 6.4. Similar to the Abelian formulation of the BTW Model, the differences between the oOM and its simplified version are minute. Firstly, instead of adding grains, slope units are added. While this makes no difference for boundary drive at $\mathbf{n} = 1$, it is crucial to maintain the Abelian symmetry in the presence of bulk drive. Secondly, all boundaries are equivalent and open with respect to slope units [19], introducing the unphysical toppling illustrated in Fig. 4.3, p. 93. In this simplified version, the only difference between the Oslo Model and the Abelian BTW Model is the randomness of the critical slope $z_{\mathbf{n}}^c$. With bulk drive, the Oslo Model displays essentially the same critical behaviour as the Manna Model (Nakanishi and Sneppen, 1997; Stapleton, 2007; Bonachela, 2008, also Pan *et al.*, 2005b).

It is not easy to trace back the origin of the simplified Oslo Model (sOM). Malthe-Sørenssen (1999) introduced a bulk driven variant of the Oslo Model, which differs, however, from the sOM in several other respects. Finding $\tau = 1.11(2)$ in one dimension (and $D = 2.25(3)$), he concluded that his model is in a different universality class than the Oslo Model, while the difference in τ can easily be explained by the difference in driving which leads to $\sigma_1^{(s)} = 1 = D(2 - \tau)$, Sec. 8.1.1.1. The exponent $\tau = 1.13(2)$ was reported earlier for the bulk driven linear interface model (Paczuski *et al.*, 1996; Paczuski and Boettcher, 1996) and is similarly obtained when applying boundary drive next to a

---

[18] Welling, Aegerter, and Wijngaarden (2003) interpreted $z_{\mathbf{n}}$ as the *magnitude* of the gradient.

[19] In contrast, the right boundary of the oOM is closed with respect to the slope units and open with respect to rice grains, see p. 94.

Box 6.4   The simplified Oslo Model

General features

- Fully Abelian version of the oOM with simplified boundary conditions.
- Defined in any dimension.
- Deterministic (boundary) drive, stochastic relaxation.
- Stochastic (bulk) drive possible.
- Exponents listed in Table 6.2, p. 189.
- In the Manna universality class.
- Investigated analytically as well as numerically.
- Closely related to the Abelian BTW Model.

Rules

- d dimensional (usually) hyper-cubic lattice with coordination number q.
- Each site $\mathbf{n} \in \{1, \ldots, L\}^d$ contains $z_\mathbf{n} \in \mathbb{N}_0$ particles.
- *Initialisation*: irrelevant, model studied in the stationary state.
- *Driving*: add a particle at $\mathbf{n}_0$ (usually fixed, or chosen at random, possibly along a boundary), $z_{\mathbf{n}_0} \longrightarrow z_{\mathbf{n}_0} + 1$.
- *Toppling*: for each site $\mathbf{n}$ with $z_\mathbf{n} > z_\mathbf{n}^c$ distribute one particle to each nearest neighbour $\mathbf{n}'$, $z_{\mathbf{n}'} \longrightarrow z_{\mathbf{n}'} + 1$ and $z_\mathbf{n} \longrightarrow z_\mathbf{n} - q$. Then choose a new threshold $z_\mathbf{n}^c \in \{q - 1, 2q - 2\}$ with equal probability (see text).
- *Dissipation*: particles are lost at open boundaries.
- *Parallel update*: discrete microscopic time, sites exceeding $z_\mathbf{n}^c$ at time t topple at time $t + 1$ (updates in sweeps). Alternatively, often random sequential.
- *Separation of time scales*: drive only once all sites are stable, i.e. $z_\mathbf{n} \leq z_\mathbf{n}^c$ (quiescence).
- *Key observables* (see Sec. 1.3):
  avalanche size s, the total number of topplings until quiescence;
  avalanche duration T, the total number of parallel updates until quiescence.

reflecting wall (Bonachela and Muñoz, 2008). Dhar (2004) introduced a generalisation of the Oslo Model, which contains the sOM as a special case, but did not discuss bulk drive specifically. Christensen (2004), published back-to-back with Dhar's article, mentioned the sOM in passing, referring to a private communication with Stapleton (see Stapleton, 2007). Pruessner and Jensen (2003) probably introduced the simplification of the boundary conditions, moving the language from 'rice grains' to 'slope (units)'. Bonachela (2008) derived the Oslo Model and thus the sOM as a stochastic variant of the Abelian sandpile model.

There remains some ambiguity about what range to choose the random $z_\mathbf{n}^c$ from in higher dimensions. As for the BTW Model, the dynamics remains unchanged if all $z_\mathbf{n}^c$ are shifted by the same offset. In one dimension, the possible choices for $z_\mathbf{n}^c$ differ by one, which has been generalised in the form of the ricepile model by Amaral and Lauritsen (1996b) and directly for the Oslo Model by Zhang (1997), who found no indication for a change of scaling

Table 6.2 Exponents of the oOM (and the sOM where marked by ⋆ or †).
$\mathcal{P}^{(\mathrm{s})}(s) = a^{(\mathrm{s})}s^{-\tau}\mathcal{G}^{(\mathrm{s})}(s/(b^{(\mathrm{s})}L^{\mathrm{D}}))$ is the avalanche size distribution and
$\mathcal{P}^{(\mathrm{T})}(T) = a^{(\mathrm{T})}T^{-\alpha}\mathcal{G}^{(\mathrm{T})}(T/(b^{(\mathrm{T})}L^{z}))$ is the avalanche duration distribution, Sec. 1.3. Scaling
relations: $\mathrm{D}(2-\tau) = \sigma_1^{(\mathrm{s})} = 1$ (boundary drive), $\mathrm{D}(2-\tau) = \sigma_1^{(\mathrm{s})} = 2$ (random bulk drive)
and $(1-\tau)\mathrm{D} = (1-\alpha)z$ (sharply peaked joint distribution, widely assumed) in all dimensions
(Sec. 2.2.2 and Sec. 8.1.1.1).

d	τ		D		α		z	
1	1.55(10)	(a)	2.25(10)	(a)	1.83(4)♮	(d)	1.42(3)	(p)
	1.53(5)♮	(b)	2.20(5)♮	(b)	1.84(5)♮	(h)	1.42(3)♮	(d)
	1.53(3)♮	(c)	2.2(2)♮	(c)	1.74(1)	(m)	1.40(5)♮	(h)
	1.55(2)♮	(d)	2.24(3)♮	(d)	1.16(3)†	(n)	1.45	(q)
	1.53(5)♮	(e)	2.20(5)♮	(e)			1.45(1)	(m)
	1.53♮	(f)	2.20(5)♮	(h)				
	1.57(5)♮	(g)	2.25(2)⋆	(i)				
	1.53(5)♮	(h)	2.2496(12)	(k)				
	1.556(4)⋆	(i)	2.2509(6)	(o)				
	1.5555(2)	(k)	2.2498(3)⋆	(o)				
	1.53(1)	(m)	2.233(2)†	(o)				
	1.10(3)†	(n)	2.12(1)	(m)				
2	1.40(5)	(r)			1.48(3)†	(n)		
	1.26(3)†	(n)						

♮ Reported for variants of the oOM.
⋆ Reported for the sOM with boundary drive.
† Reported for the sOM with bulk drive.
a Christensen *et al.* (1996), *b* Amaral and Lauritsen (1996b), *c* Zhang (1997), *d* Amaral and
Lauritsen (1997), *e* Bengrine, Benyoussef, Kenz, *et al.* (1999a), *f* Bengrine, Benyoussef, Mhirech,
and Zhang (1999b), *g* Markošová (2000b), *h* Markošová (2000a), *i* Pruessner and Jensen (2003),
k Christensen (2004), *m* Zhang, Sun, Pan, *et al.* (2005c), *n* Bonachela (2008), *o* Pruessner (2004c),
p Paczuski and Boettcher (1996), *q* Zhang, Sun, Li, *et al.* (2005b), *r* Aegerter (2003).

behaviour. In higher dimensions, a similar choice suggests $z_{\mathbf{n}}^c \in \{q-1, q\}$, with q being
the coordination number (Aegerter, 2003; Welling *et al.*, 2003; Lőrincz and Wijngaarden,
2008). However, in one dimension the choice $z_{\mathbf{n}}^c \in \{1, 2\}$ allows two topplings to occur
consecutively after only two charges, i.e. a site that topples because $z_{\mathbf{n}} = 3 > z_{\mathbf{n}}^c = 2$,
so that $z_{\mathbf{n}} = 1$ afterwards, topples again when charged, $z_{\mathbf{n}} = 2$, if $z_{\mathbf{n}}^c = 1$ after the first
toppling. This is reproduced in higher dimensions by including a largest and smallest $z_{\mathbf{n}}^c$ of
difference $q-1$; with this choice a site is guaranteed to be stable after toppling, but might
become unstable after the next charge.

Because the sOM and oOM differ in the right boundary condition, some of their universal
features disagree (allowing the sOM to collapse with the AAM, Fig. 6.5), which is expected
from ordinary critical phenomena (Privman *et al.*, 1991). Unsurprisingly, amplitudes change
as well, for example $\langle s \rangle = L$ exactly in the one-dimensional oOM, whereas $\langle s \rangle = L/2$ in
the sOM when driven at $\mathbf{n} = 1$ ($\sigma_1^{(\mathrm{s})} = 1$ in both cases), identical to the one-dimensional

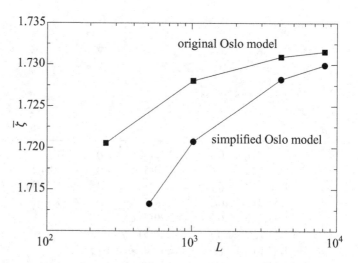

Figure 6.6 The numerical estimate $\overline{\zeta}$ of the average density of particles in the oOM and the sOM. With increasing system size, the densities in the two models seem to converge to the same value.

BTW Model and the Abelian BTW Model respectively (Sec. 8.1.1.1, Eq. (8.2) and Eq. (8.3) evaluated for $\mathbf{n} = 1$; for bulk drive see Eq. (6.1) on p. 165). [6.2] Similarly, moment ratios depend on boundary conditions, but, on the other hand, apparently not exponents like D, τ, α and z (but see below). Unexpectedly, the (stationary) average density of particles in the system, $\langle \zeta \rangle$, seems to converge to the same value, see Fig. 6.6, suggesting that in the bulk the effect of the boundary conditions vanishes asymptotically, even in the presence of algebraic correlations.

Changing the driving in the OSLO Model from boundary drive to bulk drive changes the first moment of the avalanche size from $\langle s \rangle \propto L$ for boundary drive to $\langle s \rangle \propto L^2$, see Sec. 8.1.1.1. Consequently $D(2 - \tau) = 1$ or $D(2 - \tau) = 2$ depending on the boundary condition, so that at least one of the two exponents, D or τ, changes depending on the boundary condition. Stapleton (2007, also Christensen, 2004) argued that D is the fractal dimension of the avalanches and an expression of the microscopic dynamics, whereas τ is determined by external constraints such as the boundary conditions (cf. Sec. 4.2.2). This is confirmed by the link from the OSLO Model to the qEW equation, which implies $D = d + \chi^{(\mathrm{int})}$, Sec. 8.5.4.1, regardless of such external constraints, with $\chi^{(\mathrm{int})}$ the roughness exponent of the interface model of the equation of motion (Paczuski and Boettcher, 1996, see also Sec. 6.3.3 and Sec. 8.5). However, Alava $et\ al.$ (2008) found that the exponent D is subject to changes in the presence of bulk dissipation.

6.3.2 The ricepile experiment

The **ricepile experiment** (for the wider experimental context see Sec. 3.1) in one dimension by Frette $et\ al.$ (1996), shown in Fig. 6.7, informed the OSLO Model in two important aspects. Firstly, stochasticity became part of the microscopic dynamics (Frette, 1993) and secondly, observables in the model were designed to resemble those accessible to experimentalists.

Figure 6.7 Photo of the ricepile experiment, courtesy of V. Frette, K. Christensen, A. Malthe-Sørenssen, J. Feder, T. Jøssang and P. Meakin. The pile is contained between two Perspex® sheets. An additional small pile can be seen to the right. The avalanche sizes were derived from the structural changes of the pile captured by a CCD camera. The avalanche size distribution of the large pile could not be measured well enough because of resolution limitations.

There remain some inconsistencies in the experimental paper (Frette *et al.*, 1996, also Amaral and Lauritsen, 1996b). The authors suggest that the normalisation of the avalanche size (dissipated energy E) distribution

$$\mathcal{P}^{(\mathrm{E})}(E\,L) = L^{-\beta_E} f(E/L_E^\nu) \tag{6.13}$$

implies $1 = \int_0^\infty \mathrm{d}E\, L^{-\beta_E} f(E/L_E^\nu) = L^{\nu_E - \beta_E} \int_0^\infty \mathrm{d}x\, f(x)$ for every system size L and therefore $\nu_E = \beta_E$. Strictly, the statement is correct and rewriting $\mathcal{P}^{(\mathrm{E})}(E\,L) = E^{-\beta_E/\nu_E} \tilde{f}(E/L_E^\nu)$ with $x^{-\beta_E/\nu_E} \tilde{f}(x) = f(x)$, one recognises $1 = \beta_E/\nu_E \hat{=} \tau$. In its generality, it applies equally to every avalanche size distribution, which should all therefore be characterised by the avalanche size exponent $\tau = 1$. This rather surprising result is correct *provided there is no lower cutoff* and $f(x)$ behaves favourably, see Sec. 2.1.2.2. That means, in turn, that $\nu_E = \beta_E$ is a consequence of the authors' ignoring the possible (and very likely) existence of a lower cutoff – taking it into account gives the bound $\beta_E \geq \nu_E$, namely the usual $\tau \geq 1$.

Furthermore, for two types of rice (Geisha®naturris and Geisha®middagsris), the scaling function $f(x)$, Eq. (6.13), was found to be characterised *itself* by a power law, $f(x) \propto x^{-\alpha_E}$ with $\alpha_E \approx 2.06$ for one of them and similar for the other. This, however, is clearly impossible as the energy dissipation is (algebraically) bounded in a finite experimental system, i.e. for every L there must be a (algebraically increasing) cutoff beyond which $\mathcal{P}^{(E)}(E\,L)$ vanishes strictly. A third type of rice (Geisha®grøtris) displayed scaling behaviour with a stretched exponential scaling function and $\tau = 1$. Yet, its upper cutoff scaling linear in the system size was deemed 'inconsistent with the idea of SOC' (Frette *et al.*, 1996). In the light of standard scaling arguments, one might rather conclude quite the opposite, namely that the latter displayed standard scaling whereas the scaling of the former was incompatible with some strict physical bounds.

That avalanche histograms of elongated types of rice (Geisha®naturris and Geisha®middagsris) were found to be an approximate power law, whereas those of the other, rounder type were not (Geisha®grøtris), led Frette *et al.* (1996) to suggest that the former was an instance of SOC, whereas the latter was not and to conclude that '... SOC is not as "universal" and as insensitive to the details of a system as was initially supposed...' (also Amaral and Lauritsen, 1997; Altshuler, Ramos, Martínez, *et al.*, 2003, for similar findings in sand). Being subject to inertia and friction, rather than to slide (and thereby dragging other grains), the rounder rice grains roll down the slope, which generally displayed less fluctuations (the rôle of the shape of the grains has inspired many studies, e.g. Amaral and Lauritsen, 1996b; Luding, 1996; Newman and Sneppen, 1996; Head and Rodgers, 1997, 1999; Quartier, Andreotti, Douady, and Daerr, 2000; Gleiser, 2001; Gleiser, Cannas, Tamarit, and Zheng, 2001; de Sousa Vieira, 2002; Pruessner and Jensen, 2003; Khfifi and Loulidi, 2008). Similarly, the interpretation of the experiment by Christensen *et al.* (1996) strongly alludes to the presence of non-universality, which questions the entire approach and goal (Kardar, 1996) of SOC (also Sec. 5.3, p. 135). In passing, this conclusion was later partly revised (Christensen *et al.*, 2008); the finding of a power law scaling function for a strictly bounded observable suggests it might be worthwhile to revisit the original data.

While the experiments focus on the loss of potential energy in the ricepile, the avalanche size in the OSLO Model is usually measured as the number of topplings. Amaral and Lauritsen (1997, also Markošová, 2000a,b) showed for a range of models that the distribution of the energy dissipation of avalanches is very similar to that of their size, although significant deviations can occur when the dissipated energy is estimated as in the experiments (Amaral and Lauritsen, 1996a). Most numerical studies of the OSLO Model in recent years have used the usual observables in sandpile-like systems, such as avalanche size and duration.

The ricepile experiments on a **two-dimensional** substrate by Ahlgren *et al.* (2002) did not display systematic FSS, even when the residence times of particles in the pile stretched over five orders of magnitude. Focusing on transport properties, the authors concluded that inertial effects were responsible for the lack of FSS (Krug *et al.*, 1992) and that therefore friction has to be large enough to prevent 'catastrophic avalanches' in larger systems. Their findings are consistent with the sandpile experiments by Held *et al.* (1990). Aegerter *et al.* (2003, also Aegerter *et al.*, 2004a) performed experiments with rice in a two-dimensional substrate of size 1×1 m^2 which developed into a wedge shaped pile, being driven at random positions along the edge (Aegerter, 2003; Lőrincz and Wijngaarden, 2007). The authors made use of a sophisticated technique to scan the changes on the surface by measuring the distortion they cause to lines of projected light, which was incident at oblique angles. The exponents of $\tau = 1.21(2)$ and $D = 1.99(2)$ they found are significantly different from those of the OSLO universality class listed in Tables 6.1 and 6.2. Using various scaling arguments,

Aegerter *et al.* (2003) found that the roughness and growth behaviour of the ricepile's surface is consistent with that of the Kardar–Parisi–Zhang equation, which is somewhat surprising as the dynamics of the former is to good approximation conservative (barring problems such as holes), while the latter is not. In a different experiment, some of the authors found striking similarity of the roughness and growth exponents in superconductors and a version of the two-dimensional Oslo Model (Welling *et al.*, 2003).

6.3.3 Residence time distribution

Avalanches can be seen as a superdiffusive, collective, activated and effectively ballistic **transport phenomenon**. Transport properties are easier to determine in experimental systems than avalanche sizes, and are often of much greater significance there (e.g. Hwa and Kardar, 1992; Newman *et al.*, 1996; Dendy *et al.*, 1999; Bons and van Milligen, 2001; Ahlgren *et al.*, 2002; van Milligen and Pons, 2002; Kakalios, 2005). Nevertheless, they generally feature much less prominently in the SOC literature than standard observables like avalanche size and duration. Yet, in the Oslo Model they have been studied in great detail (e.g. Christensen *et al.*, 1996; Boguñá and Corral, 1997; Zhang, 1997; Bengrine *et al.*, 1999a; Carreras, Lynch, Newman, and Zaslavsky, 1999; Corral and Paczuski, 1999; Head and Rodgers, 1999; Dhar, 2006), not least because there is actual quantitative agreement between the ricepile experiment and the Oslo Model (Christensen *et al.*, 1996; Malthe-Sørenssen, Feder, Christensen, *et al.*, 1999, similar for the surface roughness, Malthe-Sørenssen, 1999, but see Newman and Sneppen, 1996). Variants of the Oslo Model (Chapman, 2000) [20] have also been used to model transport in plasmas (Chapman *et al.*, 2001; Graves *et al.*, 2002, also Carreras *et al.*, 1996; Newman *et al.*, 1996; Sanchez *et al.*, 2001). Dendy and Helander (1998) challenged the significance of the transport properties, as they found discrepancies between a model's particle flow and its 'internal dynamics' with the latter displaying SOC while the former does not.

Christensen *et al.* (1996) investigated the transport properties of the ricepile experiment, which are less prone to experimental limitations such as the resolution of the camera mentioned by Frette *et al.* (1996). Individual, marked, i.e. distinguishable, rice grains used as **tracer particles** were fed into the system and the **residence time** (called 'transit time' by Christensen *et al.*, 1996) until they left the system determined. This time was measured as the number of rice grains added in the meantime, i.e. the number of avalanches triggered. The resulting histogram can be analysed along the same lines as the avalanche duration distribution; in particular, FSS of the characteristic residence time should produce an estimate of the dynamical exponent z. This interpretation, however, is disputable: the dynamical exponent associated with avalanche duration characterises the system on a microscopic time scale only, whereas the residence time is measured on the macroscopic time scale which is arbitrarily separated from the microscopic one. Corral and Paczuski (1999, in particular p. 574) seem to suggest that the scaling exponent ν_R of Eq. (6.14) equals the dynamical exponent, $\nu_R = z$, maybe because they measured the residence time on a microscopic time scale. They also confirmed $\nu_R = \beta_R$, see Eq. (6.14).

---

[20] Apparently, this variant of the Oslo Model has been developed independently of the Oslo Model.

Corral and Paczuski (1999, also Hwa and Kardar, 1992; Woodard *et al.*, 2007; Lőrincz and Wijngaarden, 2008) studied the breakdown of complete separation of time scales from the point of view of the particle current in the OSLO Model as the external driving rate is increased (first considered in Tang and Bak, 1988a), using observables such as **dissipative avalanches**, **residence times** and the **surface roughness** of the ricepile. They obtained data collapses plotting these observables as a function of effective parameters involving the driving rate, which thereby provides a quantitative criterion to characterise 'sufficiently slow drive', where SOC can be found. In a slow drive regime the average slope of the pile scales with the driving rate with a much smaller exponent than in the fast drive regime, where the average slope increases almost with the fourth power of the driving rate; such a sharp division might suggest that the transport mechanism underlying the continuous flow in the fast drive regime is fundamentally different from that in the slow drive regime and the continuous flow is therefore not a 'superposition of finite avalanches' (Corral and Paczuski, 1999).[21] The sensitivity of the system to the driving rate might also explain some of the numerical inconsistencies reported in the literature (but Birkeland and Landry, 2002).

Even when the OSLO Model is Abelian (Sec. 8.3, p. 270), the residence time and generally the transport properties of the OSLO Model depend on the microscopic dynamics, i.e. on the order of updates. For example, random sequential and parallel updates generally give different results (also p. 171).

Parallel updates do not even define the microscopic dynamics of the grain movement unambiguously, because neighbouring sites can be active simultaneously, certainly with the boundary conditions of the oOM, when the last site might remain active after toppling once, see Fig. 6.8. In principle, multiple discharges can occur in the bulk as well, although this is very rare in one dimension. As long as topplings move only one grain it can happen when two nearest neighbours topple simultaneously. Only if active sites are fully relaxed in every time step do the updating rules induce a pattern of the even and the odd sublattice toppling alternatingly. However, if neighbouring sites topple simultaneously, updating those further to the right *after* updating those to the left leads generally to shorter residence times, see Fig. 6.8. This illustrates how certain observables are affected by the order of updates within an avalanche, even in Abelian models.

In the OSLO Model, the avalanche size as well as the duration has a strict upper bound imposed by the system size, because each grain added can only topple L times until it leaves the system at the boundary, in line with the expectation of *finite events in finite systems*. Consequently, all moments of avalanche size and duration in the OSLO Model are finite. However, this argument does not apply to the **residence time** R, higher momenta of which seem to diverge in a finite system, although that cannot be probed in a numerical simulation, which has to terminate after a finite time. The residence time distribution $\mathcal{P}^{(R)}(R)$ of the oOM displays regular FSS (Christensen *et al.*, 1996; Zhang, 1997)

$$\mathcal{P}^{(R)}(R) = R^{-\beta_R/\nu_R}\tilde{F}(R/L^{\nu_R}), \qquad (6.14)$$

---

[21] Surprisingly, the crossover time scale between the two regimes diverges *much* slower than L^z.

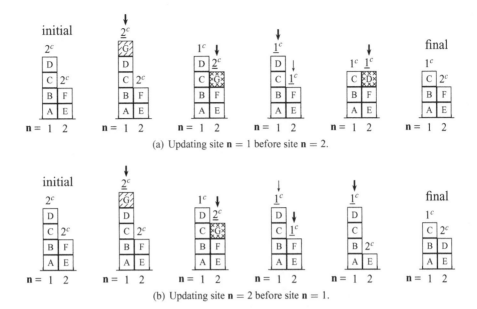

(a) Updating site $n = 1$ before site $n = 2$.

(b) Updating site $n = 2$ before site $n = 1$.

Figure 6.8 The one-dimensional oOM of size $L = 2$ in the height picture updated in two different orders. Individual grains are labelled by letters and critical slopes indicated as numbers with superscript c, which are underlined if exceeded by the local slope. The sites due to topple are marked by an arrow, which is bold where the next toppling takes place. Grains added by the external drive are hatched, grains moved to neighbouring sites are cross-hatched. Sequences (a) and (b) are identical up to the fourth configuration, when both sites exceed the local critical slope. Up to that point, only grain G has left the system, but depending on the order of updates, grain D is lost in (a), while grain F is lost in (b). Both sequences arrive at the same final configuration.

with the unusual feature that for sufficiently large arguments the scaling function $\tilde{F}(x)$ appears *itself* to be a power law,[22]

$$x^{-\beta_R/\nu_R} \tilde{F}(x) \propto x^{-\alpha_R}, \tag{6.15}$$

so that $\mathcal{P}^{(R)}(R) \propto L^{\nu_R \alpha_R - \beta_R} R^{-\alpha_R}$ for large R, Fig. 6.9, which has also been observed experimentally.[23] On the same (controversial) grounds as in the energy distribution, see Eq. (6.13), it was concluded (Christensen *et al.*, 1996) that $\beta_R = \nu_R$ which has been confirmed numerically several times.

Boguñá and Corral (1997) determined the exponent α_R in Eq. (6.15) to be equal to the avalanche dimension D in an analysis based on modelling the transport in the ricepile as a series of *independent* Lévy flights of grains (Hopcraft, Jakeman, and Tanner, 1999, also Pastor-Satorras, 1997; Carreras *et al.*, 1999; Hopcraft, Jakeman, and Tanner, 2001a; Hopcraft, Tanner, Jakeman, and Graves, 2001b) with intermittent periods of trapping.

---

[22] As known, for example, from scaling functions in the Family–Vicsek scaling ansatz, which characterises the crossover from a rougheneing surface with increasing width to a stationary state (Barabási and Stanley, 1995).

[23] As mentioned above, the avalanche size distributions obtained in the ricepile experiment (Frette *et al.*, 1996) seem to possess a power law scaling function as well.

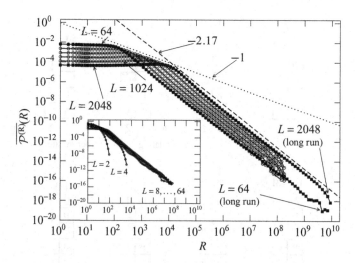

The binned histogram $\overline{\mathcal{P}^{(R)}}(R)$ of the residence time R of grains passing through the oOM (dynamics: update right to left, see Fig. 6.8b, one toppling at a time) for different system sizes $L = 64, 128, \ldots, 2048$ as indicated, see Fig. 4 in Christensen *et al.* (1996). All runs have the same transient of $5 \cdot 10^7$ iterations. The histograms shown as *hollow* circles are collected over $2 \cdot 10^8$ iterations, which provide the same cutoff towards long times for all system sizes L. The histograms shown as filled squares are based on $1 \cdot 10^{10}$ iterations and confirm the origin of the sharp drop towards long times as imposed by the simulation time. Up to the intermediate drop, which itself traces out a power law with slope -1 (dotted line, 'landmark' in Sec. 7.4.2) residence times are roughly constant, and their frequencies differ by a constant factor for different L. The apparent exponent is 0 (constant), which implies $\beta_R/\nu_R = 1$, Sec. 2.1.2.1, while the power law decay has a slope of about $-2.17 \approx \alpha_R$ (dashed line). The inset shows a range of small systems, $L = 2, 4, \ldots, 32, 64$.

Numerical estimates for the exponents are for example $\beta_R = 1.4(2)$, $\nu_R = 1.5(2)$, $\alpha_R = 2.4(2)$ (Christensen *et al.*, 1996), $\beta_R = \nu_R = 1.25(2)$, $\alpha_R = 2.4(1)$ (Zhang, 1997) and $\beta_R = \nu_R = 1.25(1)$, $\alpha_R = 2.21(5)$ (Boguñá and Corral, 1997), see Figs. 6.9 and 6.10.

Since $\alpha_R > 2$, the first moment of the residence time distribution scales with the system size like $\langle R \rangle \propto L^{\nu_R(2-\beta_R/\nu_R)}$ which equals L^{ν_R} as $\beta_R = \nu_R$. Moments of order $\alpha_R - 1$ and higher, on the other hand, are divergent even in finite systems. It should be noted, however, that the power law behaviour of the scaling function F in Eq. (6.14) is only an asymptotic behaviour. As shown in the inset of Fig. 6.9, a sharp cutoff in the distribution can be probed in sufficiently small systems, $L = 2, 4$. This might be caused by the lower cutoff or by a correction to scaling. If it is a genuine feature of the leading order, it seems to diverge faster than any power law, suggesting that the scaling function actually depends on the system size explicitly (rather than only through a ratio R/L_R^ν) for example like

$$\tilde{F}(R/L^{\nu_R}; L) = \begin{cases} A_1 \dfrac{R}{L_R^\nu} & \text{for } R < BL_R^\nu & (6.16a) \\[3mm] A_2 \left(\dfrac{R}{L_R^\nu}\right)^{-1.17} e^{-\frac{CR}{\exp(C'L^3)}} & \text{for } R \geq BL_R^\nu & (6.16b) \end{cases}$$

with amplitudes A_1, A_2, B, C and C', violating the FSS hypothesis because of the term $\exp(C'L^3)$, which plays the rôle of a second cutoff. Dhar and Pradhan (2004) argued,

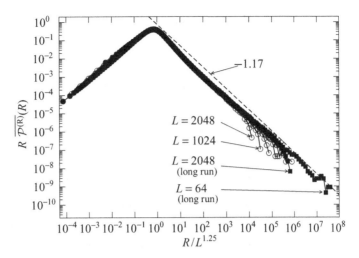

Plotting $R\mathcal{P}^{(R)}(R)$ versus $R/L^{1.25}$ collapses the data shown in Fig. 6.9 and confirms $\beta_R/\nu_R = 1$ and $\nu_R \approx 1.25$, see Eq. (6.14). The asymptote of the scaling function $\tilde{F}(x)$ can be approximated by a power law shown as a dashed line with exponent -1.17, so that $\alpha_R \approx 2.17$.

analytically, that $\langle R \rangle \propto L^2$ in one dimension, while $\langle R \rangle \propto L^{\nu_R}$ with $\nu_R < 2$ numerically, which they attributed to a numerical bias towards short residence times (also Ahlgren et al., 2002). The same authors identified an exponential upper cutoff as illustrated in Eq. (6.16b) (Dhar, 2006; Pradhan and Dhar, 2006, 2007) and, based on a range of numerical and analytical evidence, concluded that, strictly, the residence time distribution does not follow simple finite size scaling.

Numerically, however, the simple FSS hypothesis Eq. (6.14) seems to be confirmed up to the maximum R probed, which leads to a sudden drop in the histogram around the same value ($R = 2 \cdot 10^8$) in Fig. 6.9, and an apparent (or rather spurious) mismatch in the collapse shown in Fig. 6.10. To confirm the origin of the mismatch as the maximum simulation time, $L = 64$ and $L = 2048$ have been run for much longer times (10^{10} avalanches). These two runs reach much further and again drop off around the same maximum time.

Since $\nu_R > 1$, the average grain velocity $L/\langle R \rangle \propto L^{1-\nu_R}$ vanishes with increasing system size; the tracer particles seem to 'have information on the system size' (Christensen et al., 1996). The experiment hinted at the existence of an **active zone**, a layer close to the surface of the ricepile where grain transport takes place (Jaeger et al., 1996). Its depth scales with the system size and can be related to the **surface roughness** of the pile (Krug et al., 1992; Paczuski and Boettcher, 1996; Malthe-Sørenssen et al., 1999, also Lübeck and Usadel, 1993), the scaling of which suggests that it is a self-affine structure (Welling et al., 2003). As discussed towards the end of Sec. 6.3.4.2, the surface roughness exponent $\chi^{(pile)}$ can be derived from the avalanche dimension D and in one dimension $\chi^{(pile)} = D - 2$. The current through the ricepile on the macroscopic time scale is one particle per time step, which is sustained by an active layer of thickness $w \propto L^{\chi^{(pile)}}$ if all grains in that layer move on average $1/w$ lattice spacings per particle added, so that one particle drops out at $\mathbf{n} = L$ for every particle thrown in at $\mathbf{n} = 1$. To travel the distance of L lattice spacings, every particle in the active layer therefore on average lingers in the pile for $L^{\nu_R} \propto \langle R \rangle \propto L/(1/w) \propto L^{1+\chi^{(\text{pile})}} = L^{D-1}$ particle additions (Christensen *et al.*, 1996; Paczuski and Boettcher, 1996), so that $\nu_R = D - 1$, which is fully compatible with the numerics.

6.3.4 Relation to the BTW Model

The oOM and the BTW Model are closely related and differ in fact only in the details of the driving (normally the oOM is driven at site $\mathbf{n} = 1$, the BTW Model randomly in the bulk) and in the varying, **local critical slope** $z_{\mathbf{n}}^c$. Bengrine *et al.* (1999b) effectively changed the probabilities with which the different values of $z_{\mathbf{n}}^c$ were chosen and found oOM scaling for sufficiently large system sizes as soon as $z_{\mathbf{n}}^c$ is allowed to vary. Because of the link to the BTW Model, one might expect the oOM to be non-Abelian, for the same reason as the original BTW Model is non-Abelian, namely because the local slope can be reduced by adding a grain (Sec. 4.1.2). In the oOM, however, this situation does not arise, as long as it is driven only at the boundary at the top of the pile (Dhar, 2004).

The boundary conditions themselves are somewhat complicated when translated into the slope picture, and identical to those originally applied in the BTW Model (see Sec. 4.2, p. 94 for the mapping between height and slope). In the bulk, $1 < \mathbf{n} < L$, moving a grain downhill changes the slopes by $z_{\mathbf{n}\pm 1} \to z_{\mathbf{n}\pm 1} + 1$ and $z_{\mathbf{n}} \to z_{\mathbf{n}} - 2$. One slope unit is always lost when $\mathbf{n} = 1$ topples, $z_2 \to z_2 + 1$, but $z_1 \to z_1 - 2$, whereas no slope is lost on the right boundary $\mathbf{n} = L$, where $z_{L-1} \to z_{L-1} + 1$ and $z_L \to z_L - 1$. Although messy, based on these boundary conditions, the slope picture can be translated back into the height picture, where every grain added leaves the system after *exactly* L topplings downhill, which means that the average avalanche size, including avalanches of size 0 in the statistics, is $\langle s \rangle = L$ (Paczuski and Boettcher, 1996). With non-Abelian uniform bulk drive $\langle s \rangle = L^{-1} \sum_{\mathbf{n}=1}^{L} (L + 1 - \mathbf{n}) = (L+1)/2$, so that $\sigma_1^{(\text{s})} = 1$ in both cases. This is a somewhat unusual result, as bulk drive of *slope units* implies $\sigma_1^{(\text{s})} = 2$ (Sec. 8.1.1.1), which is the correct exponent for bulk drive in the sOM (Box 6.4). The situation parallels that of the BTW Model: in its original definition bulk drive implies $\sigma_1^{(\text{s})} = 1$, whereas the Abelian variant has $\sigma_1^{(\text{s})} = 2$.

Because slope units, or particles as the corresponding entity is called in the MANNA Model, in the OSLO Model are not redistributed independently (Nakanishi and Sneppen, 1997), a priori average avalanche sizes cannot be derived by mapping their motion to random walks. It turns out, however, that their trajectories nevertheless (asymptotically) trace out *all* paths with the correct weight (Shilo and Biham, 2003, also Sec. 8.1.1). Stapleton (2007) used a similar approach (Dhar, 1990a) to calculate average avalanche sizes for different driving locations, which are identical in the MANNA Model and the BTW Model, e.g. Eq. (6.1) on p. 165.

For the non-Abelian, bulk driven version of the oOM, Zhang (1997, also Bengrine *et al.*, 1999a) found significant deviations ($\tau = 1.20(6)$ and $D = 1.25(10)$) from established values, Table 6.2, and, independently, Zhang *et al.* (2005b) concluded even that the bulk driven oOM does not obey simple scaling. The latter authors, however, also failed to reproduce the established scaling exponents for the oOM and, as opposed to Zhang (1997), their results seem to be at odds with $\sigma_1^{(\text{s})} = 1$.

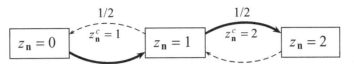

The transitions between the three possible states of a site in the one-dimensional Oslo Model. Randomness enters only when a site with slope $z_\mathbf{n} = 1$ is charged, at which point the critical slope is decided to be $z_\mathbf{n}^c = 1$ or $z_\mathbf{n}^c = 2$ with equal probabilities (1/2, as indicated). A change of state occurs by charge, a bold arrow indicates a change of state without toppling, a dashed arrow indicates change of state with toppling, i.e. by charging neighbouring sites (Pruessner, 2003a).

The **algebra** of the sOM and, after adjusting for different boundary conditions, the oOM, is identical to that of the Abelian BTW Model, see Eq. (6.19) and Eq. (8.52), and thus many results can be transferred between them (e.g. Dhar, 2004). The algebraic structure of a model is apparently insufficient to determine its scaling behaviour, as an Oslo Model and Abelian BTW Model clearly do not share the same universality class (Dhar, 1999a, for a similar observation on Eulerian random walkers). It remains an open question which properties of the matrices $S_\mathbf{n}$ and $T_\mathbf{n}$ defined below are relevant.

The algebra of the Abelian BTW Model and sOM is the same, because the redistribution of grains or, for that matter, slope units is deterministic whenever a site topples. Randomness enters the Oslo Model only when a new $z_\mathbf{n}^c$ is drawn after a site topples. If a site $\mathbf{n}$ has slope $z_\mathbf{n} = 0$, a single charge cannot trigger a new toppling (because $z_\mathbf{n}^c > 0$), so that drawing a new $z_\mathbf{n}^c$ can be postponed to the time when the site is charged a second time. If, on the other hand, the slope is $z_\mathbf{n} = 2$ but stable, the site's slope is apparently $z_\mathbf{n}^c = 2$ and a further charge makes it topple. The randomness of $z_\mathbf{n}^c$ therefore enters only when a site with $z_\mathbf{n} = 1$ is charged (Slanina, 2002): if $z_\mathbf{n}^c = 1$ it topples, otherwise $z_\mathbf{n}^c = 2$ and it remains stable. A charge increases the slope by one unit and a toppling reduces it by two, so that the net effect of a charge is $z_\mathbf{n} \to z_\mathbf{n} \pm 1$. This mechanism is summarised in Fig. 6.11.

6.3.4.1 The Oslo algebra

Charging a site results either in its toppling, or in an increase of the local slope. Apart from the stochastic element entering whenever $z_\mathbf{n} = 1$ (see Fig. 6.11), the dynamics is identical to that of the BTW Model. Fixing all $z_\mathbf{n}^c$ throughout the system at any value would, in fact, reproduce precisely the BTW Model.

A stable state of the system specifies only the local slopes $z_\mathbf{n}$, because information about the $z_\mathbf{n}^c$ is superfluous, as described above. In one dimension, such a stable state can be written as one vector $(z_1, z_2, \ldots, z_L) \in \{0, 1, 2\}^L$ out of 3^L. The evolution of their distribution is, accordingly, described by a 3^L-dimensional distribution vector $|P\rangle$, which changes under the action of the Markovian operator $\hat{a}_\mathbf{n}$, corresponding to charges at site $\mathbf{n}$ (for further details see Sec. 8.3). An sOM with a single site evolves under the Markov matrix

$$\begin{pmatrix} 0 & 1/2 & 0 \\ 1 & 0 & 1 \\ 0 & 1/2 & 0 \end{pmatrix} \tag{6.17}$$

where the final state 0 is now fixed to the top row (contribution $\frac{1}{2}$ from former state 1), state 1 to the middle and 2 to the bottom. The stationary state is therefore $(\frac{1}{4}, \frac{1}{2}, \frac{1}{4})^{\mathsf{T}}$. The Markov operators in larger systems are defined recursively, using two matrices which account for the non-toppling (S) and the toppling (T) change of state of a single site

$$S_{\mathbf{n}} = \mathbf{1}_3^{\otimes(\mathbf{n}-1)} \otimes \begin{pmatrix} 0 & 0 & 0 \\ 1 & 0 & 0 \\ 0 & 1/2 & 0 \end{pmatrix} \otimes \mathbf{1}_3^{\otimes(L-\mathbf{n})} \tag{6.18a}$$

$$T_{\mathbf{n}} = \mathbf{1}_3^{\otimes(\mathbf{n}-1)} \otimes \begin{pmatrix} 0 & 1/2 & 0 \\ 0 & 0 & 1 \\ 0 & 0 & 0 \end{pmatrix} \otimes \mathbf{1}_3^{\otimes(L-\mathbf{n})} \tag{6.18b}$$

where $\otimes$ denotes the direct product and $\mathbf{1}_3$ is a 3×3 identity matrix. The single site system Eq. (6.17) is recovered as $S_1 + T_1$ with $L = 1$. The Markov matrix for a system of size L can be written as

$$\hat{a}_{\mathbf{n}} = S_{\mathbf{n}} + \hat{a}_{\mathbf{n}+1}\hat{a}_{\mathbf{n}-1}T_{\mathbf{n}}. \tag{6.19}$$

Suitable boundary conditions, such as $\hat{a}_0 = \mathbf{1}_3^{\otimes L}$ (**Dirichlet**, i.e. fixed or absorbing) and $\hat{a}_{L+1} = \hat{a}_L$ (**Neumann**, i.e. reflecting, see Dhar, 2004) in the oOM terminate the recursion. In the height picture of the oOM, every toppling moves a grain towards the right hand boundary and as there is a finite number of such grains in the system, the strict maximum number of topplings [24] is $L + L(L+1)(2L+1)/3$, which is also an upper bound for the number of topplings in the boundary driven sOM. [25] Consequently, the expansion of $\hat{a}_{\mathbf{n}}$ contains only a finite number of terms in both versions of the Oslo Model. For $L = 2$ and Dirichlet boundary conditions one finds $\hat{a}_1 = (S_1 + S_2 T_1)(1 - T_2 T_1)^{-1}$ where the inverse can be interpreted as a geometric series, which, however, contains only a finite number of non-vanishing terms, as $T_2 T_1$ is nilpotent.

The boundary drive is realised by repeated application of $\hat{a}_1$. The simplified version of the Oslo Model defined in Box 6.4, which corresponds to $\hat{a}_0 = \hat{a}_{L+1} = \mathbf{1}_3^{\otimes L}$, allows bulk drive as well, which corresponds to applying $\hat{a}_{\mathbf{n}}$ for any $\mathbf{n} = 1, 2, \ldots, L$.

The algebra Eq. (6.19) is *identical* to that of the one-dimensional Abelian BTW Model (Eq. (8.52) in Sec. 8.3), but the matrices $S_{\mathbf{n}}$ and $T_{\mathbf{n}}$ differ. The Oslo Model has been generalised by allowing the two entries of $1/2$ to vary (Bengrine *et al.*, 1999a, also Corral, 2004a), i.e. $z_{\mathbf{n}}^c = 1$ and $z_{\mathbf{n}}^c = 2$ are not picked with equal probabilities. In the limiting case that one of the two values is not picked at all, certain configurations become inaccessible and BTW Model behaviour is recovered. As long as both entries appear with non-vanishing frequency, the scaling behaviour of the oOM is sustained, provided the system is large enough. From this point of view, the Abelian BTW Model appears like a limiting case of the Oslo Model, reminiscent of the phase diagram of the generalised Manna Model, Fig. 6.4, p. 175.

---

[24] This is a pile with slope $z_{\mathbf{n}} = 2$ throughout emptying, $\sum_{\mathbf{n}=1}^{L}(L + 1 - \mathbf{n})2(L + 1 - \mathbf{n})$ after being charged at L, giving rise to an additional L topplings.

[25] Whenever the last site topples in the sOM, it removes an entire row of grains, see Fig. 4.3, p. 93.

Dhar (2004) identified the Oslo Model as a 'special case of the **[A]belian distributed processor model**' (Dhar, 1999c) and showed that the model is indeed Abelian. Based on that property and using $\hat{a}_n^3 = \hat{a}_{n-1}\hat{a}_n\hat{a}_{n+1}$ (see Eq. (8.63) on p. 278), Dhar showed that the distribution of states produced by charging the one-dimensional oOM $L(L+1)$ times is stationary, since

$$\hat{a}_1^{L(L+1)+1} = \hat{a}_1^{L(L+1)}. \tag{6.20}$$

In other words, starting from an arbitrary initial state, after $L(L+1)$ charges, the distribution vector of states remains invariant under further charges and therefore is stationary. The system is immediately in the stationary state, i.e. a final state is picked from the stationary distribution if the most unstable configuration $(2, 2, \ldots, 2)$ is charged once (Corral, 2004a).[26]

Restricting the action of $\hat{a}_1$ to recurrent configurations, the number of charges required to reach stationarity can be shown to be as low as $L(L+1)/2$ (Dhar, 2004). Equation (6.20) implies

$$\lambda_i^{L(L+1)+1} |e_i\rangle = \lambda_i^{L(L+1)} |e_i\rangle \tag{6.21}$$

for all eigenvectors $|e_i\rangle$ of $\hat{a}_1$ which have eigenvalue λ_i and therefore $\lambda_i \in \{0, 1\}$. Because all recurrent configurations are accessible from each other, the matrix $\hat{a}_1$ is not reducible (van Kampen, 1992, Ch. V.3) so that *the stationary state is unique*, i.e. there is exactly one eigenvalue 1, say $\lambda_1 = 1$ and $|e_1\rangle = |P_{\text{stat}}\rangle = |0\rangle$ is the unique stationary state (Corral, 2004a; Dhar, 2004). Dhar (2004) presented a line of argument that relates the number of non-zero eigenvalues in the Oslo Model to the number of recurrent states in the ASM. In one dimension and with the boundary conditions of the oOM, the ASM has a single, recurrent state, Fig. 4.1(b), p. 87.

This remarkable property, Eq. (6.20), seems to challenge the typical approach of the stationary state in Markov chains, which is exponentially slow, with **correlations** dying off according to the eigenvalues. In the oOM, all eigenvalues except one vanish, which seems to suggest that a single charge is required to produce the stationary state from any initial distribution. In fact, if the initial distribution $|P\rangle$, which might be a pure state, i.e. $|P\rangle = (0, 0, \ldots, 0, 1, 0, \ldots, 0)$ can be written as $|P\rangle = \sum_i A_i |e_i\rangle$ then $\hat{a}_1 |P\rangle = \sum_i A_i \lambda_i |e_i\rangle = A_1 \lambda_1 |e_1\rangle$. Assuming that $|P\rangle$ is normalised, $(1, 1, \ldots, 1) |P\rangle = 1$ and given that $(1, 1, \ldots, 1) = \langle 0|$ is a left eigenvector of $\hat{a}_1$ with eigenvalue 1 due to the Markov property, it follows that $A_1 = 1$ and therefore $\hat{a}_1 |P\rangle = |P_{\text{stat}}\rangle$. This result suggests that there are no correlations and a single charge is always sufficient to get the oOM into the stationary state, as if all columns of $\hat{a}_1$ were identical to $|P_{\text{stat}}\rangle$ and its rank one. This conclusion is wrong, because the spectral theorem does not apply to Markovian matrices[27] and so its eigenvectors $|e_i\rangle$ do not span the entire vector space of distribution vectors $|P\rangle$ regardless of whether or not it is restricted to recurrent configurations. As discussed in Sec. 6.3.5, Pruessner (2004b) added bulk dissipation to the totally asymmetric Oslo Model in order to make the eigenvectors of the relevant Markov matrix span the entire

---

[26] Manna (1991b) seems to have suggested the same for the Manna Model.

[27] If the spectral theorem applies, the Markov process is often trivial. If, for example, $\hat{a}_1$ is symmetric, $(1, 1, \ldots, 1)^{\mathsf{T}}$ is necessarily a right eigenvector, i.e. the uniform distribution is a stationary state.

vector space. Generally, $|P\rangle$ cannot be written in the form $\sum_i A_i |e_i\rangle$ and therefore $\hat{a}_1 |P\rangle$ is generally different from $|P_{\text{stat}}\rangle$. Nevertheless, Eq. (6.20) implies that no correlations on the macroscopic time scale exist in the oOM beyond $L(L+1)$ charges, as long as the observable is based on the configuration of the system in terms of slopes (which is not the case for its transport properties, see p. 194).

Corral (2004a, also Sadhu and Dhar, 2009) calculated some of the Markov matrices explicitly for small, one-dimensional systems, $L \leq 8$. The results rely crucially on the boundary conditions. In fact, the sOM possesses multiple non-zero eigenvalues, in particular $\exp(2\pi \imath i/(L+1))$ for $i = 0, 1, \ldots, L$ (Chua and Pruessner, 2003), which again corresponds to the $L + 1$ recurrent configurations of the corresponding Abelian BTW Model (Ruelle and Sen, 1992, see p. 93).

The boundary condition of the oOM has the additional advantage that the avalanche size is given exactly by the difference in potential topplings Φ

$$\Phi = \sum_{\mathbf{n}=1}^{L} h_{\mathbf{n}}(L - \mathbf{n} + 1) \tag{6.22}$$

before and after the avalanche (Corral, 2004a).[28] The avalanche size is thus solely given by initial and final states, even when the sequence of topplings between them is not unique. In the moment generating function $\mathcal{M}^{(s)}(x)$ of the avalanche size distribution, every term contributing to the transition from one particular initial state to a particular final state therefore comes with the same prefactor in x. The potential topplings Φ can be derived easily for a given slope configuration $(z_1, z_2, \ldots, z_L)$ and using direct products, a diagonal matrix $\mathcal{U}(x)$ can be constructed without difficulty, consisting of a power x^Φ for each individual state, $\mathcal{U}(1/x)\mathcal{U}(x) = \mathbf{1}$. The matrix $\mathcal{U}(1/x)\hat{a}_1\mathcal{U}(x)$ contains a term $x^{\Phi-\Phi'}$ for each transition from a state with potential topplings Φ to one with potential topplings Φ'. The driving adds L potential topplings, so that

$$\mathcal{M}^{(s)}(x) = x^L \langle 0| \mathcal{U}(1/x)\hat{a}_1\mathcal{U}(x) |0\rangle, \tag{6.23}$$

which can be advantageous compared to the more direct approach used in Sec. 6.3.5. Since $\hat{a}_{\mathbf{n}}\hat{a}_{\mathbf{m}} = \hat{a}_{\mathbf{m}}\hat{a}_{\mathbf{n}}$, the moment generating function for the joint avalanche size of charges at two sites $\mathbf{n}$ and $\mathbf{m}$ does not depend on the order of updates,

$$\begin{aligned}
\mathcal{M}^{(s_{\mathbf{n}} + s_{\mathbf{m}})}(x) &= x^{2L+2-\mathbf{n}-\mathbf{m}} \langle 0| \mathcal{U}(1/x)\hat{a}_{\mathbf{n}}\hat{a}_{\mathbf{m}}\mathcal{U}(x) |0\rangle \\
&= x^{2L+2-\mathbf{m}-\mathbf{n}} \langle 0| \mathcal{U}(1/x)\hat{a}_{\mathbf{m}}\hat{a}_{\mathbf{n}}\mathcal{U}(x) |0\rangle = \mathcal{M}^{(s_{\mathbf{m}} + s_{\mathbf{n}})}(x)
\end{aligned} \tag{6.24}$$

with $|0\rangle$ now being the stationary state of the double charge.

The effect of the boundary conditions and the small number of eigenvectors complicate the usage of more sophisticated techniques such as a **matrix product ansatz** (Derrida and Evans, 1997; Hinrichsen, 2000) or the **density matrix renormalisation group** (Kaulke, 1999; Peschel *et al.*, 1999). The applicability of such techniques is also hindered by the lack of a simple iteration scheme to extend an existing system. For example, the unique eigenvector $|0\rangle_{L+1}$ for a system of size $L + 1$ is not related in any obvious way to the

---

[28] This is not the total potential energy of the ricepile, $\sum_{\mathbf{n}=1}^{L} h_{\mathbf{n}}(h_{\mathbf{n}} + 1)/2$.

eigenvector $|0\rangle_L$ for a system of size L, even when the former can be constructed from the latter by adding a single site to the left of $\mathbf{n} = 1$. That single site interacts with the remaining system by charging the first site in a string of L sites and one might therefore conclude that the stationary state of that subsystem is equal to that of a system of size L. If that assumption were right, the stationary state would be a **product measure**, i.e. local states are uncorrelated, even independent. This is incorrect because of **backward avalanches** (Kadanoff *et al.*, 1989; Frette *et al.*, 1996; Head and Rodgers, 1997; Corral, 2004a), as any transition in the subsystem implies a certain number of charges of the added site, which in turn charges the subsystem. Some of these problems are potentially alleviated by enabling bulk dissipation in the sOM, similar to the totally asymmetric Oslo Model, Sec. 6.3.5.

6.3.4.2 An equation of motion

Paczuski and Boettcher (1996) suggested very early that the Oslo Model is, in its **interface representation**, governed by a quenched Edwards–Wilkinson (qEW) equation and, based on the assumption of compact avalanches, concluded that the roughness exponent (Barabási and Stanley, 1995, also Sec. 8.5) of the qEW equation $\chi^{(\mathrm{int})}$ is related to the avalanche dimension D in the Oslo Model by (Paczuski *et al.*, 1996, also Alava, 2002 and Sec. 8.5.4.1)

$$\mathsf{D} = \chi^{(\mathrm{int})} + d, \tag{6.25}$$

which has further numerical support in other interface models (Buldyrev *et al.*, 1992a; Buldyrev, Havlin, and Stanley, 1992b; Amaral, Barabási, Buldyrev, *et al.*, 1995). The suggested equation of motion for the interface contains effectively a Heaviside step function, which induces motion in the interface at positive net forces (Díaz-Guilera, 1994; Alava, 2002). The step function is not easy to handle analytically (Pérez *et al.*, 1996; Corral and Díaz-Guilera, 1997), but can be avoided by changing the dynamical variable from the number of topplings $G(\mathbf{n}, t)$ at site $\mathbf{n}$ and after microscopic time t, to the number of charges $g(\mathbf{n}, t)$. Along the lines of Sec. 8.3.2.1 (Pruessner, 2003a), in one dimension the equation of motion of the Oslo Model can be derived as follows (extensions to higher dimensions are straightforward). Assuming an initial state with $z_\mathbf{n}(t = 0) = 1$ everywhere, the number of topplings of a bulk site fulfils

$$2G(\mathbf{n}, t) + z_\mathbf{n}(t) - 1 = g(\mathbf{n}, t), \tag{6.26}$$

which a priori is only a matter of particle conservation and does not fix the microscopic dynamics. The right boundary site, $\mathbf{n} = L$, obeys $G(L, t) + z_L(t) - 1 = g(L, t)$ in the oOM, and in the sOM the same equation as all bulk sites. Imposing that a site is fully relaxed at time $t + 1$ with respect to all charges up to time t (but ignoring those arriving at $t + 1$), implements a particular dynamics, namely $2G(\mathbf{n}, t + 1) + z_\mathbf{n}(t + 1) - 1 = g(\mathbf{n}, t)$ and $z_\mathbf{n}(t + 1) \in \{0, 1, 2\}$. In this expression, G and $z_\mathbf{n}$ account only for charges arriving up to and including t, not for those arriving at $t + 1$ (or later). For bulk sites, $z_\mathbf{n}(t + 1)$ has the opposite parity to $g(\mathbf{n}, t)$ and only if it is odd can there be any ambiguity in the state, see

Fig. 6.11. Introducing the quenched noise [29]

$$\xi(\mathbf{n}, g) = \begin{cases} 1 & \text{if } g \text{ odd, with probability } 1/2 & (6.27a) \\ & \text{corresponding to } z_{\mathbf{n}} = 0 \\ 0 & \text{if } g \text{ even, corresponding to } z_{\mathbf{n}} = 1 & (6.27b) \\ -1 & \text{if } g \text{ odd, with probability } 1/2 & (6.27c) \\ & \text{corresponding to } z_{\mathbf{n}} = 2, \end{cases}$$

the number of topplings G obeys

$$G(\mathbf{n}, t+1) = \frac{1}{2} \left(g(\mathbf{n}, t) + \xi(\mathbf{n}, g(\mathbf{n}, t)) \right). \tag{6.28}$$

The number of charges of a site is given (in the bulk) by the number of topplings of the neighbours,

$$g(\mathbf{n}, t) = G(\mathbf{n}+1, t) + G(\mathbf{n}-1, t), \tag{6.29}$$

so that in the bulk

$$z_{\mathbf{n}}(t) = G(\mathbf{n}+1, t) - 2G(\mathbf{n}, t) + G(\mathbf{n}-1, t) + 1, \tag{6.30}$$

without referring to the dynamics. The 'slope' $z_{\mathbf{n}}(t)$ thus is essentially the 'curvature' in the number of topplings. Integrating $z_{\mathbf{n}}$, Eq. (4.2) on p. 87, and ignoring further complications by boundary conditions (see footnote 31 on p. 205), the height $h_{\mathbf{n}}(t)$ of the ricepile at $\mathbf{n}$ is

$$(G(\mathbf{n}-1, t) - G(\mathbf{n}, t)) - (G(L, t) - G(L+1, t)) + (L+1-\mathbf{n}), \tag{6.31}$$

where $L + 1 - \mathbf{n}$ reflects the initial condition $z_{\mathbf{n}} = 1$.

Using Eq. (6.28) in Eq. (6.29) gives rise to the exact equation of motion for the number of charges $g(\mathbf{n}, t)$ at site $\mathbf{n}$ up to time t:

$$g(\mathbf{n}, t+1) - g(\mathbf{n}, t) = \frac{1}{2} \Big(g(\mathbf{n}-1, t) - 2g(\mathbf{n}, t) + g(\mathbf{n}+1, t)$$
$$+ \xi(\mathbf{n}+1, g(\mathbf{n}+1, t)) + \xi(\mathbf{n}-1, g(\mathbf{n}-1, t)) \Big). \tag{6.32}$$

This expression differs from the equation of motion of the Abelian BTW Model, Eq. (8.66) on p. 278, by the noise terms and the absence of any rounding, [30] and from that of the MANNA Model, Eq. (6.11) on p. 176, by the additional absence of a summation over the noise.

The two noise terms effectively encapsulate the stochasticity and the Heaviside step function (Pruessner, 2003a), which is somewhat surprising as **thresholds** were traditionally considered a crucial ingredient in SOC (Bak *et al.*, 1987; Cafiero *et al.*, 1995; Jensen, 1998). Translating from the lattice directly to the continuum gives

$$\partial_t g(\mathbf{n}, t) = \frac{1}{2} \partial_x^2 g(\mathbf{n}, t) + \left(1 + \frac{1}{2} \frac{d^2}{dx^2} \right) \xi(\mathbf{x}, g(\mathbf{n}, t)) \tag{6.33}$$

---

[29] There are degrees of 'quenchedness'. The noise in Eq. (6.27) is quenched in that it does not change in time, but only with $z_{\mathbf{n}}$. But the quenched random numbers entering, say, the toppling matrix of the model studied by Karmakar *et al.* (2005) *never* change throughout the lifetime of its realisation.

[30] The rounding $\lfloor\ \rfloor$ in Eq. (8.66) on p. 278 encapsulates the Heaviside step function.

where the complicated noise terms can be simplified, using Galilean invariance (Meakin, 1998) and Middleton's (1992) no-passing. The latter guarantees the uniqueness of the interface $g(\mathbf{n}, t)$, which in the Oslo Model is a consequence of the Abelian symmetry of the bulk dynamics on the micoscopic time scale (Pruessner, 2003a). Collecting the noise into a new, effective term reproduces the qEW equation,

$$\partial_t g = D\partial_{\mathbf{x}}^2 g + \tilde{\xi}(\mathbf{x}, g). \qquad (6.34)$$

It can be shown (Pruessner, 2003a) that the noise term $\tilde{\xi}(\mathbf{x}, g)$ has correlator

$$\left\langle \tilde{\xi}(\mathbf{x}, g)\tilde{\xi}(\mathbf{x}', g') \right\rangle = 2\Gamma^2 \delta(\mathbf{x} - \mathbf{x}')\Delta(g - g') \qquad (6.35)$$

with the correlator $\Delta(g - g')$ developing a cusp singularity under renormalisation (Leschhorn *et al.*, 1997).

With the effective noise, Eq. (6.34) is much easier to handle than the continuum version (6.33) of Eq. (6.32) and has been analysed with great interest as a model of **interface depinning** using a range of methods (Bruinsma and Aeppli, 1984; Koplik and Levine, 1985; Nattermann *et al.*, 1992; Leschhorn, 1994; Le Doussal *et al.*, 2002). The exponent $\chi^{(\text{int})}$ in Eq. (6.25) is the roughness of the interface given by g. Models of interface depinning display ordinary critical behaviour in that a (tunable) control parameter, the pulling force, is applied homogeneously throughout the system, i.e. Eq. (6.34) acquires an additional, constant source term f on the right. In the Oslo Model, however, the driving is implemented in the form of a boundary condition, namely by adding a site at $\mathbf{n} = 0$ with $g(0, t) = 2E(t)$. This site is otherwise not subject to any dynamics but is fixed to that value, so that $G(0, t + 1) = E(t)$, providing the charging of the first site, $\mathbf{n} = 1$, by the **external drive** $E(t)$. This boundary condition gives rise to a current of charges $-\partial_{\mathbf{x}}g$ through the system to the boundary at $\mathbf{n} = L$. In the sOM, a Dirichlet boundary condition on g, $g(L, t) = 0$, induces an ever increasing current, while for the Neumann boundary conditions as in the oOM, it eventually levels off, see Fig. 8.13, p. 303, and there is no net transport of slope units through the system. The current is closely related to the height of the ricepile, see Eq. (6.31), and is thus clearly space dependent. [31] Due to boundary pinning and the noise $\tilde{\xi}$, which act as (local) sinks and sources, spatially changing current in g does not necessarily mean that its profile constantly evolves in time. Instead, the ricepile can be in a ***dynamic equilibrium*** or **non-equilibrium steady state** (rather than at equilibrium, where detailed balance holds), where the shape of the profile is maintained by suitable sinks and sources.

In the discrete Oslo Model, the external drive increases only very slowly in the microscopic time t and remains constant while an avalanche is running. If the external drive is meant to be ignorant about the presence or absence of avalanching, SOC is achieved in the limit $\dot{E} \rightarrow 0$, i.e. by separation of time scales, with $\dot{E}$ being the conversion parameter. [6.3]

The interfacial properties of the field $g(\mathbf{n}, t)$, characterised by roughness and growth exponents, are to be distinguished from the interfacial properties of the height profile $h_{\mathbf{n}}(t)$ of the ricepile (Hwa and Kardar, 1989a), 'consider[ing] the profile dynamics as

---

[31] The translation from g to the height of the ricepile strictly breaks down in the case of a Dirichlet boundary condition (sOM) on $g(L, t)$, as the number of charges received at the virtual site $\mathbf{n} = L + 1$ is unknown.

the wandering of a front with one end fixed' (Frette, 1993). As mentioned above, in one dimension,[32] Paczuski and Boettcher (1996) argued that the roughness exponent $\chi^{(\text{pile})}$ of the height profile obeys $\chi^{(\text{pile})} = \chi^{(\text{int})} - 1$, which confirmed above, as the height profile is determined by the gradient of G, Eq. (6.31), which differs from g essentially only by $z_{\mathbf{n}}$, Eq. (6.26). With Eq. (6.25) this roughness exponent is related to the avalanche dimension, $\chi^{(\text{pile})} = D - 2$. Numerically, Zhang (2000) found the relation confirmed in the oOM only when driving the model at $\mathbf{n} = 1$, whereas random bulk drive (which is non-Abelian) produced a smoother surface.

The qEW equation is often illustrated by a rubber string pulled over a rough surface. This analogy is somewhat problematic in both versions of the OSLO Model: in the oOM the slope at the right boundary would somehow need to be kept fixed, in the sOM the string would be anchored at $g(L, t) = 0$ and extended indefinitely. Interestingly, this analogy was used very early by Burridge and Knopoff (1967) to motivate their model of earthquakes, modified versions of which display the same (roughness) scaling as the OSLO Model (Cule and Hwa, 1996) or can be mapped to a variant of it (Chianca et al., 2009).

6.3.5 The totally asymmetric OSLO Model

Introducing any small amount of **anisotropy** in the toppling rules of the OSLO Model changes, in the thermodynamic limit, its scaling behaviour to that of the totally asymmetric OSLO Model (TAOM), introduced in one dimension by Pruessner and Jensen (2003). The TAOM is a particularly convenient representative of this universality class as it can be solved exactly (Pruessner, 2004b; Stapleton and Christensen, 2005, 2006; Welinder et al., 2007), yielding $\tau = 4/3$ and $D = 3/2$. The exact solution of the model relies, like most directed models (Sec. 8.4), on the absence of spatial correlations, which allows them to be mapped to a random walker. In fact, the exponents of other (one-dimensional, stochastic, directed) models, which are in the same universality class as the TAOM, had already been derived by Paczuski and Bassler (2000, transverse dimension $d_\perp = 0$) and for a similar model by Kloster et al. (2001, in one dimension, also Maslov and Zhang, 1995; Priezzhev, Ivashkevich, Povolotsky, and Hu, 2001). The exponents are identical to those of the *two-dimensional* (transverse dimension $d_\perp = 1$) deterministic, directed sandpile model by Dhar and Ramaswamy (1989, also Pastor-Satorras and Vespignani, 2000e).

In the TAOM, only the downstream neighbours (the right neighbour in one dimension) are charged when site $\mathbf{n}$ topples, i.e. $z_{\mathbf{n}} \to z_{\mathbf{n}} - 1$ and $z_{\mathbf{n}+1} \to z_{\mathbf{n}+1} + 1$. It is somewhat unusual in that the co-dimension $d_\perp$ vanishes, which, in the case of the directed MANNA Model led to ambiguities in its definition (Hughes and Paczuski, 2002). The equation of motion of the TAOM is a quenched Edwards–Wilkinson equation with drift v (Pruessner and Jensen, 2003, also Maslov and Zhang, 1995),

$$\partial_t g = D\partial_x^2 g - v\partial_x g + \tilde{\xi}(\mathbf{n}, g), \tag{6.36}$$

any small amount $v > 0$ of which leads to a change in universal behaviour. The TAOM therefore affords an exact solution of a Langevin equation with quenched noise. The result

---

[32] In higher dimensions, the argument is complicated by the difficulties in mapping g to a height profile.

D = 3/2 and $\chi^{(\text{int})} = 1/2$ (Pruessner, 2004c, also Tang, Kardar, and Dhar, 1995) confirms Eq. (6.25).

An analysis of the rôle of a drift term in the presence of *thermal* noise hinted at a very powerful mechanism (Pruessner, 2004a), rendering the roughness exponent $\chi^{(\text{int})}_d$ of the model *with* drift as well as its growth exponent β_d equal to the growth exponent β of the model *without* drift (Barabási and Stanley, 1995). However, the drift affects models with quenched noise differently, so that β of the qEW equation (in one dimension $\beta = 0.88(2)$, Leschhorn, 1993) cannot be found from $\chi^{(\text{int})}_d = 1/2$ of the qEW equation with drift (as is the case for the thermal Edwards–Wilkinson equation).

It remains somewhat unclear whether the original ricepile experiment suffered from any form of anisotropy. Because of the boundary conditions, preferential orientation of elongated rice grains at input and output amounts to bulk dissipation of slope, whereas elasticity in the stack might give rise to a slope current. In a sandpile-type model, Mehta and Barker (1994) found that elongated grains that are subject to random reorientation do *not* display scale-free behaviour (also Toner, 1991). Jettestuen and Malthe-Sørenssen (2005) implemented grain dissipation as a negative drift of slope units which introduces a finite, system size independent upper cutoff in the avalanche size distribution without changing the exponent τ.

The TAOM is solved exactly by defining the set of eigenvectors of its Markov matrix $\hat{a}_L$ recursively for different system sizes L (Pruessner, 2004b)

$$\hat{a}_L = S \otimes \mathbf{1}_2^{\otimes(L-1)} + T \otimes \hat{a}_{L-1} + U \otimes \hat{a}_{L-1}^2 \tag{6.37}$$

with matrices S, T and U corresponding to adding a particle at the site, leaving it unchanged and removing one respectively (similar to Eq. (8.54) on p. 272, also Sec. 6.3.4.1). For the set of eigenvectors to form a basis of the vector space, bulk dissipation ϵ

$$\hat{a}_L = \epsilon \mathbf{1}_2^{\otimes L} + (1 - \epsilon) \left(S \otimes \mathbf{1}_2^{\otimes(L-1)} + T \otimes \hat{a}_{L-1} + U \otimes \hat{a}_{L-1}^2 \right) \tag{6.38}$$

has to be added and the limit $\epsilon \to 0$ taken afterwards. [33] As is typical for directed models (Paczuski and Bassler, 2000), the stationary state of the TAOM has **product measure**, i.e. the probability of a site to be in a particular state is independent of the state of any other site. The exact avalanche size distribution can be derived by 'decorating' each toppling with a factor x

$$\hat{a}_L(x) = \epsilon \mathbf{1}_2^{\otimes L} + (1 - \epsilon) \left(S \otimes \mathbf{1}_2^{\otimes(L-1)} + T \otimes x\hat{a}_{L-1}(x) + U \otimes x^2 \hat{a}_{L-1}^2(x) \right). \tag{6.39}$$

If $|P_{\text{stat}}\rangle = |0\rangle$ is the resulting stationary state and $(1, 1, \ldots, 1) = \langle 0|$ the corresponding left eigenvector, the moment generating function of the avalanche size distribution is given by $\mathcal{M}^{(L)}(x) = \langle 0| \hat{a}_L(x) |0\rangle$. After generalising $\mathcal{M}^{(L)}(x)$ for multiple charges, the moments can be derived as solutions of recurrence relations. Taking the continuum limit, the avalanche size distribution of the TAOM is identified as the distribution of the areas enclosed by the trajectory of a random walker along an absorbing wall, see Fig. 7.3, p. 223, where the absorbing wall is in fact generated by the trajectory of the previous walker. If the two

---

[33] An alternative route might be the introduction of generalised eigenvectors.

values of the local critical slope, $z_{\mathbf{n}}^c \in \{1, 2\}$, are not chosen with equal probabilities, the trajectories display an additional drift, while their relative distance (which is a random walk in its own right) does not. Only if one of the two critical slopes does not appear at all, does the TAOM become deterministic and the trajectories become ballistic, so that all avalanches have the same size and $\tau = D = 1$, similar to the situation of the OSLO Model degenerating to a BTW Model (Sec. 6.3.4).

The mapping to the area under a random walker trajectory links the TAOM to extreme value statistics (Majumdar and Comtet, 2004, 2005) and allows the derivation of a number of exact and asymptotic results (Welinder *et al.*, 2007) in particular for generalisations (Stapleton and Christensen, 2005, 2006), which include systems that evolve under (internally) *correlated* avalanches, while developing into a product state. These models display a crossover, controlled by the system size, from uncorrelated, mean-field-like behaviour ($\tau = 3/2$ and $D = 2$, corresponding to $d_\perp = 2$, i.e. upper critical dimension, in Paczuski and Bassler, 2000) to a state with temporal correlations ($\tau = 4/3$ and $D = 3/2$, corresponding to $d_\perp = 0$). The generalised directed Abelian algebras studied by Alcaraz and Rittenberg (2008) contain the TAOM. While they reproduced $\alpha = D = 3/2$ in one dimension, Eq. (8.72), they could not confirm $\alpha = D = 7/4$ in two dimensions (Paczuski and Bassler, 2000; Kloster *et al.*, 2001).

What makes the TAOM so easily solvable is the complete absence of any **backward avalanches** (Sec. 8.4.4), which are retained in other members of this universality class of anisotropic models. In terms of a diffusion-convection equation with diffusion constant D and drift v, the anisotropy v imposes a cutoff D/v on the spatial (but not the temporal) correlations in the system, which are entirely trivial in the TAOM, making it solvable. In contrast, there are strong spatial correlations present in the OSLO Model, which is illustrated by the discrepancy between the probabilities with which the different values of $z_{\mathbf{n}}^c$ are picked and the frequency with which they appear in the stationary state. In its original definition, only about 15% of the sites have $z_{\mathbf{n}}^c = 1$, because that choice frequently implies that a site topples, in turn triggering a toppling at neighbouring sites, which can lead to another toppling [34] at $\mathbf{n}$. In the TAOM, $z_{\mathbf{n}}^c = 1$ and $z_{\mathbf{n}}^c = 2$ appear with the prescribed frequencies, testifying to the absence of spatial correlations.

6.3.6 Numerical results

The natural formulation of the OSLO Model in terms of a local height, its strong link to an experiment and the fact that it displays robust, non-trivial scaling behaviour in one dimension meant that it has been 'at home in one dimension' since its conception. Most of the numerical studies published therefore focus on the one-dimensional oOM along with a plethora of variants, in particular Amaral and Lauritsen's (1996b) ricepile model. These results are included in Table 6.2, suggesting that the variants are in the same universality class as the oOM. For reference purposes, a few exponents for the bulk driven, Abelian sOM are included as well (marked by †). As discussed above, p. 198, the scaling behaviour of the bulk driven (non-Abelian) oOM deviates significantly from that of the boundary driven oOM.

---

[34] In effect, $z_{\mathbf{n}}^c = 1$ implies that the choice is revised rather sooner than later.

Even with the link established between the Oslo Model and Manna Model, very few results are available in dimensions above one (Aegerter, 2003; Bonachela, 2008), and none above two, extensively studied for the Manna Model, see Table 6.1. Implementations and interpretations of the Oslo Model in two dimensions differ considerably, owing to the original formulation in terms of grains (compare, for example, Aegerter, 2003 and Pradhan and Dhar, 2006). Because the Oslo Model is comparatively young, many of the results are based on a moment analysis (e.g. Zhang *et al.*, 2005b) rather than a direct analysis of the relevant PDFs (e.g. Zhang, 1997). Christensen (2004) suggested that the exact exponents in one dimension are $D = 9/4$ and $\tau = 14/9$, consistent with $\sigma_1^{(s)} = D(2 - \tau) = 1$, which is a consequence of particle conservation and driving.

In the more recent past, the focus has shifted towards the relation of the Oslo Model to the Manna universality class and the fact that the quenched noise, which is an integral part of the Oslo Model, does not seem reflected well by most other models in the Manna universality class, not least because quenched noise is normally associated with disorder which has a fundamentally different nature compared to annealed noise (e.g. Alava and Muñoz, 2002).

The Oslo Model has received considerable attention for its transport behaviour, in particular where it has been linked to experimental situations for example by Chapman *et al.* (2001) and Welling *et al.* (2003). Numerically, keeping track of tracer particles is a challenge very different from the usual avalanching behaviour, where particles are indistinguishable slope units. The most straightforward implementation of a ricepile with tracer particles uses a $d + 1$ dimensional array of cells, each being able to hold all relevant information of a grain, such as the time it entered the stack. A stack of grains over the d-dimensional substrate consists of corresponding entries along the transverse dimension of the array. Because the maximum average slope is 2, the array must have about $2L^{d+1}$ cells. To avoid searches through this array, a second array keeps track of local heights and a third array keeps track of local slopes. Apart from being somewhat clumsy, this implementation suffers from frequent copying of cells as the grains move down the slope.

A more elegant and extensible implementation uses a vector of pointers `site[i]` which point to the top grain at each lattice site `i`. Each stack of grains is maintained as a NULL terminated singly linked list. The entries represent grains and contain a time stamp and any other relevant observable (e.g. average depth buried in, time stamps for entering and departing individual sites etc.) as well as a pointer to the grain directly underneath if present, i.e. they could be implemented using a structure like

```
struct grain_struct {
    int t_in; /* time entered */
    struct grain_struct *below;
};.
```

Local slopes are maintained in a separate array, while grains are moved by changing pointers (rather than copying content). Adding grain g to site r is implemented as `g->below=site[r]; site[r]=g;` and (cautious) removal of the top grain as `if (site[r]) site[r]= site[r]->below;`. The results shown in Fig. 6.9 were obtained using this scheme. With maximum slope 2 a wedge shaped pile requires around L^{d+1} grain structures.

Numerical methods and data analysis

This chapter describes a number of (numerical) techniques used to estimate primarily universal quantities, such as exponents, moment ratios and scaling functions. The methods are applied during post-processing, i.e. after a numerical simulation, [1] such as the OFC Model, Appendix A, has terminated. Many methods are linked directly to the scaling arguments presented in Ch. 2, i.e. they either probe for the presence of scaling or derive properties *assuming* scaling.

A time series is the most useful representation of the result of a numerical simulation, because it gives insight into the temporal evolution of the model and provides a natural way to determine the variance of the various observables reliably. Most of the analysis focuses on the **stationary state** of the model, where the statistics of one instance of the model with one particular initial state is virtually indistinguishable from that with another initial state. The end of the **transient** can be determined by comparing two or more independent runs, or by comparing one run to exactly known results (such as the average avalanche size) or to results of much later times. The transient can be regarded as past as soon as the observables are within one standard deviation. It pays to be generous with the transient, in particular when higher moments or complex observables are considered.

In the stationary state, the ensemble average (taking, at equal times, a sample across a large number of realisations of the model) is strictly time independent. A sequence of instantaneous measurements of an observable will usually produce some fluctuations about this stationary value. To some extent, the definition of a stationary state depends on the definition of the ensemble. Strictly, a fully deterministic model cannot reach stationarity in a variable that remains subject to (deterministic) changes forever. In such a model, additional averaging is needed, for example about the initial configuration (see Sec. 5.3), or some further qualification, such as averaging over recurrent states.

Very often the transient is visible as an exponential relaxation of an ensemble averaged mean to its stationary value, see Fig. 7.1. If the duration τ_t of the transient on the microscopic time scale can be quantified reliably, it can be an interesting observable in its own right, sometimes associated with the **dynamical (critical) exponent** z by $\tau_t \propto L^z$. In the following, it is assumed that the sample has been stripped of any transient effects, while it almost certainly suffers from temporal correlations to be discussed next.

---

[1] 'Simulation' is strictly a misnomer commonly applied to the numerical implementation of stochastic processes. Frequently, they are represented exactly in code and thus are not intended to 'imitate', i.e. simulate something. The same applies to many 'computer models' which are nothing but perfect models of themselves.

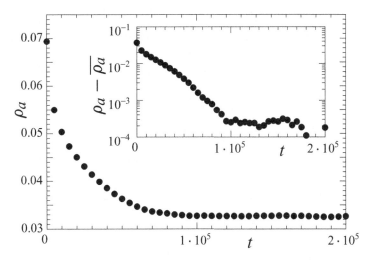

Figure 7.1 Relaxation of the activity ρ_a in the absorbing state variant of the AMM. Starting at a fraction of about 0.07 active sites the system relaxes to some (quasi-) stationary value, $\langle \rho_a \rangle \approx 0.032$. The inset shows the same data with the estimate $\overline{\rho_a}$ of $\langle \rho_a \rangle$ subtracted, revealing an exponential relaxation in a linear-log plot.

7.1 Independence and correlations

In order to measure the average of an observable, say s, in a numerical simulation, one has to resort to a finite sample.[2] In a **Markov Chain Monte Carlo**[3] simulation, the sample will be generated as a *sequence* of N observations $s_1, s_2, \ldots, s_N$. This sequence itself is subject to a joint probability density $\mathcal{P}^{(N)}(s_1, s_2, \ldots, s_N)$, which may or may not factorise into a product of probability density functions (PDFs) of individual observations: $\mathcal{P}^{(1)}(s_1)\mathcal{P}^{(1)}(s_2)\ldots\mathcal{P}^{(1)}(s_N)$. Only if the joint PDF factorises, are the individual observations said to be **independent**; in particular, two observables are (mutually) independent if the conditional PDF of one of them is identical to the unconditional PDF: $\mathcal{P}^{(1|1)}(s_1|s_2) = \mathcal{P}^{(2)}(s_1, s_2)/\mathcal{P}^{(1)}(s_2) = \mathcal{P}^{(1)}(s_1)$.

If the PDF factorises, so do the moments:

$$
\begin{aligned}
\langle s_1^a s_2^b \ldots s_N^z \rangle &= \int ds_1 \, ds_2 \, \ldots ds_N \, \mathcal{P}^{(s)}(s_1, s_2, \ldots, s_N) s_1^a s_2^b \ldots s_N^z \\
&= \int ds_1 s_1^a \mathcal{P}^{(s)}(s_1) \int ds_2 s_2^b \mathcal{P}^{(s)}(s_2) \ldots \int ds_N s_N^z \mathcal{P}^{(s)}(s_N) \\
&= \langle s_1^a \rangle \langle s_2^b \rangle \ldots \langle s_N^z \rangle .
\end{aligned}
\tag{7.1}
$$

---

[2] In the following, the avalanche size s is used to illustrate the techniques, but they generally apply to all observables obtained in a similar fashion.

[3] A Monte Carlo algorithm generally performs an integration by random sampling. The sample can be generated in a Markov chain, i.e. by letting the system evolve in time, which normally leads to a correlated sample.

The converse, however, is generally not true. Just because some moments factorise, the observables are not necessarily independent (Mari and Kotz, 2001). However, if the moments factorise such that the covariance

$$\text{cov}(s_1, s_2) = \langle s_1 s_2 \rangle - \langle s_1 \rangle \langle s_2 \rangle \tag{7.2}$$

vanishes and if, in addition, the joint PDF of s_1 and s_2 is Gaussian, then they are **independent** (van Kampen, 1992).

The **auto-correlation function**

$$c_i = \langle s_1 s_{i+1} \rangle - \langle s \rangle^2 = \langle (s_1 - \langle s \rangle)(s_{i+1} - \langle s \rangle) \rangle \tag{7.3}$$

is the standard observable used to test for correlations in an observable. In Eq. (7.3) stationarity has been assumed, so that $\langle s_1 \rangle = \langle s_i \rangle = \langle s \rangle$. At $i = 0$ the correlation function coincides with the variance. While, in general, $c_0 > 0$, independence implies $c_i = 0$ for $i > 0$. One naturally expects asymptotically vanishing correlations, i.e. c_i can be made arbitrarily small by increasing i, in particular in the case of a continuous time Markov process. In that case, one can prove that the correlations generally decay as a sum of exponentials. Again, even if the s_i are uncorrelated, $c_i = 0$, they are not necessarily independent (unless the s_i are Gaussian distributed); only the converse holds generally.

If the sample displays correlations, one often derives a **correlation time** τ by fitting c_i to $c_0 \exp(-|i|/\tau)$, even though strictly there is no necessity for the correlations to decay exponentially for a general observable, for example if the correlation function shows damped oscillations. If the exponential decay applies exactly, the parameter c_0 is the variance of the s_i, i.e. $c_0 = \sigma^2(s)$. The **correlation time** helps to quantify the number of 'effectively independent measurements'. For example, calculating the average of s from a sample of size N, the estimate $S_N/N = N^{-1} \sum_i^N s_i$ has variance $\langle S_N^2/N^2 \rangle - \langle S_N/N \rangle^2 = N^{-2} \sum_{ij} c_{i-j}$. Assuming that c_i is an exponential, the right hand side behaves, in the limit of large N, like $N^{-1} c_0 (1 + \exp(-1/\tau))/(1 - \exp(-1/\tau))$. The fraction behaves like $2\tau + 1/(6\tau) + \cdots$ for large τ and converges to 1 for small τ. It is therefore well approximated by $2\tau + 1$ in both regimes. [4] The variance of the estimate is thus approximated by $\langle S_N^2/N^2 \rangle - \langle S_N/N \rangle^2 \approx N^{-1}(2\tau + 1)c_0$, whereas uncorrelated events s_i produce an estimate with expected variance $N^{-1} c_0$, which means that the variance on the basis of N events s_i with correlation time τ is approximately equal to the variance on the basis of $N/(2\tau + 1)$ uncorrelated events. This can be regarded as the *definition* of the correlation time. If the correlation time is found to be divergent, i.e. there is no way to perform effectively independent measurements, there is little hope of proceeding with any numerical estimate.

In principle, different observables have different correlation times, but it is common to assume the correlation time to be the characteristic time scale of the entire system rather than that of individual observables or 'projections'. The scaling of the correlation time with system size is often assumed to follow $\tau \propto L^z$, where z denotes the dynamical critical exponent. However, to coincide with the scaling of the avalanche duration, L^z should give a microscopic time scale, whereas the correlation time τ is a macroscopic time in the

---

[4] For all $\tau \geq 0$ the approximation slightly overestimates, $(2\tau+1) \cdot 0.65 \leq (1+\exp(-1/\tau))/(1-\exp(-1/\tau)) \leq (2\tau+1)$.

preceding discussion. The two are not easily converted into each other, and the conversion depends on more than just the average avalanche duration $\langle T \rangle$. A couple of short, small avalanches triggered in different regions of the lattice are less correlated than the same small number triggered at the same spot. Because $\langle T \rangle$ is dominated by long durations, assuming $\tau \propto L^z / \langle T \rangle$ drastically underestimates the macroscopic correlation time. A more pessimistic conversion time scale is the median avalanche duration (Sec. 2.1.1, p. 29), which suggests a *much* longer correlation time τ.

The divergence of the (microscopic) correlation time with the systems size, $\tau \propto L^z$, has led to the notion of **critical slowing down**, in particular in ferromagnetic phase transitions (e.g. Binder and Heermann, 1997).

An estimate of the standard deviation, square root of the estimated variance of the estimator (rather than the variance of the actual observable), is the numerical error usually stated for a numerical estimate of an ensemble averaged observable. If the correlation time is finite, the statistical error will therefore generally vanish with increasing sample size N like $N^{-1/2}$.

7.1.1 Efficient, consistent and unbiased estimators

As Monte Carlo simulations generally do not allow a particular observable to be determined *exactly*, [5] as done, for example, in exact enumeration, an estimator is needed to derive an approximation from the simulation data. An estimator is a technique, usually a formula, devised to estimate a quantity that is typically much more easily described in terms of a population average. [6] For example, the arithmetic mean of the avalanche sizes $\langle s \rangle$ (i.e. the average avalanche size) is well estimated by

$$\bar{s} := \frac{1}{N} \sum_{i=1}^{N} N s_i \tag{7.4}$$

where s_i for $i = 1, 2, \ldots, N$ are, again, consecutive measurements, which normally include vanishing event sizes, for example when the external drive did not result in an avalanche. What characterises a good estimator?

Efficient Given that CPU time is the currency of numerics, efficiency is of great importance for an estimator, especially if the estimation is done on the fly, i.e. in the simulation code. Because the standard deviation scales like $N^{-1/2}$ in the sample size,

---

[5] A randomised algorithm that determines the exact result is sometimes called a **Las Vegas algorithm**.

[6] In principle, one should thoroughly distinguish numerical estimates of observables (such as the average avalanche size) and what they are estimating, i.e. the exact population averages. This distinction should be carried through to the notation, but, for readability this is done in the following only where absolutely necessary; a numerical estimate is then denoted by an overline, e.g. $\bar{a}$, whereas the full population average is marked by the usual angular brackets, e.g. $\langle a \rangle$. Where necessary, the former is defined explicitly using ':=' to underline the origin of the estimator as a construct. For readability the symbol a used in $\bar{a}$ normally refers to the observable whose population average it attempts to estimate, i.e. $\bar{a}$ estimates $\langle a \rangle$. This, however, makes it difficult to denote an estimator of $\langle a \rangle^2$, since $\overline{a^2}$ estimates $\langle a^2 \rangle$ along the lines above, whereas $\bar{a}^2$ suggests the square of $\bar{a}$, for clarity written as $(\bar{a})^2$ in the follwing, which is a rather poor estimator of $\langle a \rangle^2$. Such ambiguities are avoided below as much as possible.

quadrupling the CPU time cuts the numerical error only in half. The power of an algorithm is thus best characterised by the product of CPU time and variance, smaller numbers indicating a better algorithm.

Consistent If an estimator converges to the population average with increasing sample size, it is said to be consistent (Brandt, 1998). Often, it is 'apparent' that an estimator is consistent, but it can be very difficult to prove.

Unbiased If the population average of an estimator depends on the sample size, it is said to be **biased** (Brandt, 1998). Bias is often difficult to avoid, but at the same time it is probably the most common and widely underrated problem in numerics. Unsuspectingly, it can produce spurious results, especially where differences and ratios of observables are concerned and (effective) sample sizes are small. The prime example of a biased observable is the naïve estimator for the variance, which is biased due to the naïve estimator of the square of the average: $(\bar{s})^2$, the square of Eq. (7.4), is a biased estimator of the squared arithmetic mean $\langle s \rangle^2$. This can be seen by taking the population average of $(\bar{s})^2$, even assuming mutual independence of the s_i, i.e. $\langle s_i s_j \rangle = \langle s \rangle^2$ for $i \neq j$, while $\langle s_i s_i \rangle = \langle s^2 \rangle$. Using Eq. (7.4), one has

$$\left\langle (\bar{s})^2 \right\rangle = N^{-2} \sum_{ij} \langle s_i s_j \rangle = \left(1 - \frac{1}{N}\right) \langle s \rangle^2 + \frac{\langle s^2 \rangle}{N} = \langle s \rangle^2 + \frac{\sigma^2(s)}{N} \tag{7.5}$$

using the population variance $\sigma^2(s)$ of s. This expression holds for higher moments as well,

$$\left\langle \left(\overline{s^n}\right)^2 \right\rangle = \langle s^n \rangle^2 + \frac{\sigma^2(s^n)}{N} \quad \text{with} \quad \sigma^2(s^n) = \left\langle s^{2n} \right\rangle - \langle s^n \rangle . \tag{7.6}$$

Given that $\overline{s^2} := N^{-1} \sum_i s_i^2$ is an unbiased estimator of $\langle s^2 \rangle$, the unbiased estimator of $\langle s \rangle^2$ reads

$$\frac{N}{N-1} \left((\bar{s})^2 - N^{-1}\overline{s^2} \right), \tag{7.7}$$

with N replaced by (approximately, depending on the definition of the correlation time) $N/(2\tau + 1)$ in the presence of correlations.

The numerical error of an estimator stated, relies itself on an estimate of the square root of the variance of the estimator, which often becomes extremely large if it is dominated by a few, **rare events**, i.e. there are a few measurements s_i which make up most of the numerical estimate. **Importance sampling** is a way to bias the sampling towards those measurements which dominate the observables. In equilibrium statistical mechanics, importance sampling in the form of **Markov Chain Monte Carlo** is de facto the norm to sample according to the Boltzmann distribution, but no general techniques are available in complex systems. The problem of a poor sampling technique can be detected by investigating the distribution of estimates from **subsamples** of the original sample, which are produced naturally, if multiple instances of the same simulation run in parallel ('**poor man's parallel computing**'), or when a single instance writes data out in regular intervals (Sec. A.4.3), generating '**chunks**' of data. Its deviation from Gaussian is a vague indicator of the quality of the data. Having the data divided into chunks can also be exploited in a quantitative fashion, giving rise to reliable and easily implemented estimators of errors of estimators. This technique is discussed in the following.

7.2 Subsampling

Subsampling allows a more direct characterisation of the distribution of an estimate, rather than assuming a certain, effective number of independent measurements and deriving the error of an estimator from the variance of the original observable, as suggested in the discussion about the correlation time, p. 212. Given a single sample $\{s_1, s_2, \ldots, s_N\}$ containing N independent measurements, the naïve estimate of the average mentioned earlier, Eq. (7.4), $\bar{s} := \sum_i^N s_i/N$, is an unbiased estimator of the arithmetic mean of the population, $\langle s \rangle$, because $\langle \bar{s} \rangle = \langle s \rangle$, independent of N. The variance of that estimator, i.e. its error squared, is $\sigma^2(\bar{s}) = \langle (\bar{s})^2 \rangle - \langle \bar{s} \rangle^2$. From Eq. (7.5) it follows that $\sigma^2(\bar{s}) = \sigma^2(s)/N$ for N independent measurements. In other words, the variance of the estimator of the arithmetic mean is the variance of the observable, $\sigma^2(s)$, over the sample size N. To state the error of the estimator $\bar{s}$, the square root of the variance of the estimator is determined, which requires, the variance $\sigma^2(s)$, estimated in the following. [7]

The variance consists of two terms to be estimated: $\langle s^2 \rangle$ and $\langle s \rangle^2$. The first one is trivial. Defining $\overline{s^2} := \frac{1}{N}\sum_i^N s_i^2$ immediately gives $\langle \overline{s^2} \rangle = \langle s^2 \rangle$. The unbiased estimator of $\langle s^2 \rangle$ is stated in Eq. (7.7), so that $\sigma^2(s)$ is estimated well by

$$\bar{\sigma}^2(s) := \frac{N}{N-1}\left(\overline{s^2} - (\bar{s})^2\right) \tag{7.8}$$

and

$$\bar{\sigma}^2(\bar{s}) := \frac{\overline{s^2} - (\bar{s})^2}{N-1} \tag{7.9}$$

is in fact an unbiased estimator of the variance of the estimator $\bar{s}$. If N is large enough, the divisor might be replaced by N, but as soon as differences, ratios and sums of observables are considered, this can be a dangerous thing to do. Similarly one finds that the covariance of estimates of averages of two observables a and b with $\langle \bar{a} \rangle = \langle a \rangle$ and $\langle \bar{b} \rangle = \langle b \rangle$, which is defined as $\mathrm{cov}(\bar{a}\bar{b}) = \langle \bar{a}\bar{b} \rangle - \langle \bar{a} \rangle \langle \bar{b} \rangle$, has the unbiased estimator

$$\overline{\mathrm{cov}}\left(\bar{a}, \bar{b}\right) := \frac{\overline{ab} - \bar{a}\bar{b}}{N-1}. \tag{7.10}$$

This expression is used below for $a = s^m$ and $b = s^n$.

The derivation above is based on the assumption of mutual independence of (consecutive) measurements, or at least a lack of correlations. If the sample is correlated, the situation is not fundamentally different. As introduced above, see p. 212, the correlation time τ enters in the form $\sigma^2(\bar{s}) = \frac{2\tau+1}{N}\sigma^2(s)$, which one might regard as the *definition* of the correlation

---

[7] Strictly, the square root of the estimate of the variance is only a biased estimator of the square root of the variance.

time,[8] which is generally specific to a particular observable. Since $\langle (\overline{s})^2 \rangle = \sigma^2(\overline{s}) + \langle \overline{s} \rangle^2$ by definition and because $\langle \overline{s^2} \rangle = \langle s^2 \rangle$ even in the presence of correlations, one has

$$\left\langle \frac{1}{\frac{N}{2\tau+1} - 1} \left(\overline{s^2} - (\overline{s})^2 \right) \right\rangle = \sigma^2(\overline{s}) = \frac{2\tau + 1}{N} \sigma^2(s), \tag{7.11}$$

i.e. $\frac{2\tau+1}{N-2\tau-1} (\overline{s^2} - (\overline{s})^2)$ is a good estimator of the variance of the estimator $\overline{s}$, if measurements are correlated. It follows that $\overline{s^2} - (\overline{s})^2$ is a biased estimator of the variance,

$$\left\langle \overline{s^2} - (\overline{s})^2 \right\rangle = \frac{N - 2\tau - 1}{N} \sigma^2(s). \tag{7.12}$$

A sample is practically always correlated and estimating the error involves, according to Eq. (7.11), the correlation time, which needs to be estimated as well. Extracting that by fitting the correlation function against an exponential (Sec. 7.1), introduces another source of error and can be very time consuming and tedious. An alternative that avoids estimation of the correlation time altogether is therefore welcome. One such alternative is subsampling, closely related to so-called **jackknife** and **bootstrap** techniques (Efron, 1982; Berg, 1992). The jackknife is probably the most straightforward way of estimating the error of any estimator and functions thereof. It requires very little computational and programming effort, but usually needs subsamples of equal size. Naïvely, these are based on *all* measurements bar a single one, which, in principle, requires storing all measurements generated (but Pruessner, Loison, and Schotte, 2001). Given a sample size of order $N = 10^{10}$ this would generate prohibitively large files, written using vast amounts of CPU time. However, averages over subsamples can be regarded as the principal measurement to be fed into the jackknife procedure. These can be realised easily in a simulation by writing out averages over subsamples periodically.

There is another, flexible and fairly natural method to determine error bars which is discussed in the following. As for other 'subsampling plans', the variance of the estimator of the average is estimated without measuring the correlation time. This scheme works even if subsamples have different sizes, for example, when they are generated by multiple instances of the same code running for the same amount of real-time on differently powered computers. Errors of functions of estimators are determined through error propagation, e.g. Eq. (7.39) and Eq. (7.40), and thus through covariances, which can make the method somewhat tedious to use for this type of observable.

Given T subsamples $i = 1, 2, \ldots, T$ in the form $\{s_{i1}, s_{i2}, \ldots, s_{iN_i}\}$ with individual (sub)sample size N_i, one naturally defines estimators based on individual subsamples, such as

$$m_i := \frac{1}{N_i} \sum_{j}^{N_i} s_{ij} \tag{7.13}$$

and the overall estimator of the mean

$$\widetilde{m} := \frac{1}{\mathcal{N}} \sum_{i}^{T} N_i m_i \tag{7.14}$$

---

[8] With this definition (see p. 212), the correlation time does not need to be calculated explicitly, as the subsampling discussed below incorporates it automatically.

where $\mathcal{N} = N_1 + \cdots + N_T$ is the total sample size. Here and in the following, the tilde as used in Eq. (7.14) denotes a weighted average. The m_i are produced in the simulation, while $\tilde{m}$ is generated in post-processing. The following derivations are based on the first moments m_i of the subsamples, higher moments of the subsamples enter only when higher moments such as $\langle s^2 \rangle$ are estimated.

Obviously $\langle m_1 \rangle = \langle m_2 \rangle = \ldots = \langle m_T \rangle = \langle \tilde{m} \rangle = \langle s \rangle$, i.e. $\tilde{m}$ is an unbiased estimator of $\langle s \rangle$. The variance of this estimator is simply $\sigma^2(\tilde{m}) = \sigma^2(s)/\mathcal{N}$, provided each individual subsample consists of uncorrelated measurements and subsamples are **mutually uncorrelated**. That is what is assumed for the time being, but the assumption of uncorrelated individual measurements will be relaxed below. The only condition upheld throughout is that the correlation time is small [9] compared to the size of every subsample, $\tau \ll N_i$, which implies that subsamples are mutually uncorrelated, $\langle m_i m_j \rangle = \langle m_i \rangle \langle m_j \rangle$ for $i \neq j$. In practice, subsamples can be merged until this condition is met (monitored by the convergence of estimates, as used in the closely related scheme by Flyvbjerg and Petersen, 1989), at the expense of a decreasing number of (supposedly independent) subsamples. When done *during* the simulation, subsamples usually cannot be recreated after being merged, so they are best generated afterwards from chunks of data written out at regular intervals (Sec. A.4.3) and chosen as small as diskspace allows.

Defining

$$\widetilde{m^2} := \frac{1}{\mathcal{N}} \sum_i^T N_i m_i^2 \tag{7.15}$$

the expectation value of $\widetilde{m^2}$ is easily calculated, because all measurements are mutually uncorrelated, $\langle s_{ij} s_{kl} \rangle - \langle s \rangle^2 = \sigma^2(s) \delta_{ik} \delta_{jl}$. One finds

$$\left\langle \frac{1}{T-1} \left(\widetilde{m^2} - (\tilde{m})^2 \right) \right\rangle = \sigma^2(\tilde{m}) = \langle (\tilde{m})^2 \rangle - \langle \tilde{m} \rangle^2 = \frac{1}{\mathcal{N}} \sigma^2(s) \tag{7.16}$$

which means that the expression averaged over on the left is an unbiased estimator of the variance of the estimator $\tilde{m}$.

The situation is more complicated if one assumes, realistically, that subsamples are uncorrelated but measurements within a subsample are correlated with correlation time τ, which can be considered as being defined via the effective sample size, see footnote 8 on p. 216. If the number of effectively uncorrelated measurements, $N_i/(2\tau + 1)$, is sufficiently large [10] compared to the number of subsamples T, the variance of the estimator $\tilde{m}$, Eq. (7.14), is

$$\sigma^2(\tilde{m}) = \frac{2\tau + 1}{\mathcal{N}} \sigma^2(s) \tag{7.17}$$

and from Eq. (7.11) one has

$$\langle m_i^2 \rangle = \sigma^2(m_i) + \langle m_i \rangle^2 = \frac{2\tau + 1}{N_i} \sigma^2(s) + \langle s \rangle^2 \tag{7.18}$$

---

[9] The jackknife applied to the sample $\{m_1, m_2, \ldots, m_T\}$ is the method of choice to relax even that condition.

[10] If this condition is violated, then the number of effectively independent subsamples exceeds $\mathcal{N}/(2\tau + 1)$. In that case Eq. (7.17) overestimates the variance of $\tilde{m}$ and thus its error, because more measurements than just $\mathcal{N}/(2\tau + 1)$ are uncorrelated. In the extreme case that $(2\tau + 1) > N_i$, the number of (supposedly truly independent) subsamples, T, exceeds $\mathcal{N}/(2\tau + 1)$. The effective number $\mathcal{N}/(2\tau + 1)$ used in the following is a conservative estimate, also accounting, in principle, for correlation across subsamples.

so that

$$\left\langle \widetilde{m^2} \right\rangle = \frac{2\tau + 1}{\mathcal{N}} T \sigma^2(s) + \langle s \rangle^2 \tag{7.19}$$

using the definition (7.15). Because subsamples are (mutually) uncorrelated and using Eq. (7.18) one finds

$$\left\langle (\widetilde{m})^2 \right\rangle = \langle s \rangle^2 + \frac{2\tau + 1}{\mathcal{N}} \sigma^2(s) \tag{7.20}$$

(see Eq. (7.5)) so that finally

$$\left\langle \frac{1}{T-1} \left(\widetilde{m^2} - (\widetilde{m})^2 \right) \right\rangle = \sigma^2(\widetilde{m}) = \frac{2\tau + 1}{\mathcal{N}} \sigma^2(s) \tag{7.21}$$

identical to Eq. (7.16), in other words

$$\overline{\sigma}^2(\widetilde{m}) := \frac{1}{T-1} \left(\widetilde{m^2} - (\widetilde{m})^2 \right) \tag{7.22}$$

is an unbiased estimator of the variance $\sigma^2(\widetilde{m})$ of the unbiased estimator $\widetilde{m}$ for $\langle s \rangle$, under the assumption that $N_i/(2\tau + 1) \gg T$. Ignoring for a moment that Eq. (7.22) applies generally, even for varying N_i, it can be understood as the estimator of the variance of the estimator of the mean $\widetilde{m}$ like Eq. (7.9), as if there were T equivalent independent measurements, which is the case when all subsamples have equal size $N_i = \mathcal{N}/T$. If the N_i vary, they enter in Eq. (7.15) as the correct weight to produce Eq. (7.19) via Eq. (7.18). The square root of Eq. (7.22) is to be quoted as the estimated error of the estimator Eq. (7.14).

At first sight the identity of Eq. (7.16) and Eq. (7.21) is rather surprising, because the correlation time does not enter explicitly. However, the subsampling effectively reflects the correlation time directly in the variance of the mean based on individual subsamples. If the correlation time is large, the estimates drawn from individual subsamples display a large variance. Due to the weighting by N_i employed in Eq. (7.14) and Eq. (7.15) the technique applies even if the subsample sizes N_i vary. It is therefore the natural choice if numerical estimates are based on multiple, independent runs.

The variance of s (rather than its arithmetic mean) can be estimated using the simple unbiased estimator of the second moment $\overline{s^2} := \frac{1}{\mathcal{N}} \sum_i^T \sum_j^{N_i} s_{ij}^2$, which draws on the *second* moments of the subsamples for the first time (see the remark on p. 217). Subtracting the first moment squared, Eq. (7.20), the variance is found up to a prefactor

$$\left\langle \overline{s^2} \right\rangle - \left\langle (\widetilde{m})^2 \right\rangle = \frac{\mathcal{N} - 2\tau - 1}{\mathcal{N}} \sigma^2(s) \tag{7.23}$$

to be compared to Eq. (7.12). The estimator $\overline{s^2} - (\widetilde{m})^2$ in Eq. (7.23) becomes unbiased by moving $(\mathcal{N} - 2\tau - 1)/\mathcal{N}$ to the left. In practice τ is difficult to estimate, but since $(\mathcal{N} - 2\tau - 1)/\mathcal{N}$ is usually very close to 1, the factor can often be ignored. The ratio of the expectation values of Eq. (7.22) and Eq. (7.23) gives $(2\tau+1)/(\mathcal{N}-(2\tau+1)) \approx (2\tau+1)/\mathcal{N}$, and therefore provides a (possibly quite noisy) path to estimating the correlation time. If the correlation time is known, the estimate of $\frac{2\tau+1}{\mathcal{N}} \sigma^2(s) = \sigma^2(\widetilde{m})$, Eq. (7.22), can be used to estimate $\sigma^2(s)$ in an unbiased fashion. In turn, the knowledge of $\sigma^2(s)$ provides an estimator of τ using Eq. (7.23) or rather Eq. (7.22).

More generally, one defines subsampled **higher moments** [11]

$$m_i^{(n)} = \frac{1}{N_i} \sum_j^{N_i} s_{ij}^n \qquad (7.24)$$

so that

$$\widetilde{m^{(n)}} = \frac{1}{\mathcal{N}} \sum_i^T N_i m_i^{(n)} \qquad (7.25)$$

is an unbiased estimator [12] of $\langle s^n \rangle$. For example, $\widetilde{s^2}$ used in Eq. (7.23) is identical to $\widetilde{m^{(2)}}$. Given this unbiased estimator for all moments, the variance of the nth moment, $\langle s^{2n} \rangle - \langle s^n \rangle^2$ is easily estimated as well (up to a prefactor as in Eq. (7.23)). The unbiased [13] estimator of the variance of the estimator $\widetilde{m^{(n)}}$ or generally covariances of such estimators are given by

$$\overline{\mathrm{cov}}\left(\widetilde{m^{(n)}}, \widetilde{m^{(m)}}\right) = \frac{1}{T-1}\left(\widetilde{m^{(n)} m^{(m)}} - \widetilde{m^{(n)}}\,\widetilde{m^{(m)}}\right), \qquad (7.26)$$

where

$$\widetilde{m^{(n)} m^{(m)}} = \frac{1}{\mathcal{N}} \sum_i^T N_i m_i^{(n)} m_i^{(m)}, \qquad (7.27)$$

using

$$\left\langle m_i^{(n)} m_i^{(m)} \right\rangle - \left\langle m_i^{(n)} \right\rangle \left\langle m_i^{(m)} \right\rangle = \frac{2\tau+1}{N_i}\left(\langle s^n s^m \rangle - \langle s^n \rangle \langle s^m \rangle\right) = \frac{2\tau+1}{N_i}\,\mathrm{cov}\left(s^n, s^m\right), \qquad (7.28)$$

which can, again, be regarded as the very definition of the correlation time. The measurements s_{ij}^n and s_{ij}^m can be completely different observables, say s_{ij}^n the avalanche size and s_{ij}^m the avalanche duration, and Eq. (7.26) can be used to estimate their covariance.

Finally, the assumption of uncorrelated subsamples m_i is reconsidered. It has been assumed that the subsample sizes N_i are large compared to $2\tau+1$, which means that any correlations between two consecutive subsamples m_i are guaranteed to have died off within a time which is small compared to the size of the subsample. The bulk of the measurements constituting m_i can therefore be assumed to be independent of all other subsamples m_j. Alternatively, the totality of all measurements can be seen as one big sample of size $\mathcal{N}$, which is reduced to effective size $\mathcal{N}/(2\tau+1)$ due to correlations, regardless of whether or not they stretch over the entire sample and across subsamples. This is reflected in Eq. (7.17); Eq. (7.21) is then only a way of expressing efficiently the factor $2\tau+1$, provided it is small compared to all N_i.

7.2.1 Functions of moments

Above, an unbiased estimator for moments and their variance was devised. In general, more complicated (numerical) objects are considered, in particular moment ratios (see Sec. 2.2.1

---

[11] Higher moments often refer to moments beyond the fourth, which are not also captured by variance, skewness or kurtosis. Here they refer to any moment beyond the first.

[12] This notation distinguishes $\widetilde{m^{(2)}}$, Eq. (7.25), from $\widetilde{m^2}$, Eq. (7.15).

[13] Provided again that $N_i/(2\tau+1) \gg T$.

and Sec. 7.3.4) and higher order cumulants (see Sec. 2.2.3). Furthermore, the moments for different system parameters are fitted against a certain expected scaling behaviour, most notably in finite size scaling (FSS), Sec. 2.1, to derive exponents.

Error estimation of these more complicated objects can be very intricate and is a subject of the literature on numerical methods and Monte Carlo methods (e.g. Binder and Heermann, 1997; Newman and Barkema, 1999; Berg, 2004; Landau and Binder, 2005). Subsampling almost always provides a fairly easy way out of the thicket of error estimation. It is regrettable when systematic error calculation is replaced by rough visual estimates.

7.3 Moment analysis

The easiest route to estimate exponents from numerical data is via moments, which was made popular as a standard tool in SOC by De Menech *et al.* (1998, also Tebaldi *et al.*, 1999; Pastor-Satorras and Vespignani, 2000e; Christensen and Moloney, 2005).[14] *To leading order* the moments behave *asymptotically* like a power law in the **upper cutoff** s_c (also called the **characteristic scale**),

$$\langle s^n \rangle = A_n s_c^{\mu_n^{(s)}} \tag{7.29}$$

for $\mu_n^{(s)} \geq 0$, as introduced in Ch. 2 (in particular Eq. (2.24) and Eq. (2.26)). If the system displays **gap scaling**, then $\mu_n^{(s)} = 1 - \tau + n$, where τ is the **avalanche size exponent** to be estimated in the following. This is a three stage process. Firstly, moments $\langle s^n \rangle$ are estimated together with their errors (Sec. 7.2) for a range of (normally integer-valued, small) n and different upper cutoffs s_c or different system sizes L (see below). Secondly, exponents $\mu_n^{(s)}$ or $\sigma_n^{(s)}$ are determined in a numerical fit, using, for example, the Levenberg–Marquardt algorithm (mrqmin in Press, Teukolsky, Vetterling, and Flannery, 2007) to fit data against Eq. (7.32). The theoretical aspects, the pitfalls and dangers of this step are the main focus of the following discussion. Thirdly, the exponents τ and D are found in a simple linear regression of $\mu_n^{(s)} = 1 + n - \tau$ and $\sigma_n^{(s)} = D(1 + n - \tau)$ respectively.

Because the (dimensionful and non-universal) coefficients A_n in Eq. (7.29) are generally unknown and depend on n, considering $\langle s^n \rangle$ as a function merely of n does not lead to an estimate of τ. The usual procedure therefore is to change s_c, which might itself be a function or, in fact, be a power law of another system parameter. Strictly, one cannot estimate the exponents of a PDF with cutoff function without varying the cutoff (but Clauset, Shalizi, and Newman, 2009). In the case of FSS the relevant parameter is the system size, i.e. $s_c = CL^D$ to leading order, so that

$$\langle s^n \rangle = B_n L^{\sigma_n^{(s)}} = B_n L^{D(1-\tau+n)}. \tag{7.30}$$

Even when neither the relevant system parameter nor the upper cutoff is known, moments can still be used to estimate τ, using any other moment $\langle s^m \rangle$ in the form

$$\langle s^n \rangle = \frac{A_n}{A_m} \langle s^m \rangle^{\mu_n^{(s)}/\mu_m^{(s)}} \tag{7.31}$$

---

[14] It is remarkable that this approach, used in equilibrium critical phenomena for at least 50 years, has not always been the modus operandi in SOC.

assuming that Eq. (7.29) holds (or similarly Eq. (7.30)) and that neither $\langle s^n \rangle$ nor $\langle s^m \rangle$ is constant in the system parameter that is changed across different realisations, i.e. $\mu^{(s)}_{n,m} > 0$. Using Eq. (7.31) relaxes the requirement that the system parameter has to be known which $\langle s^n \rangle$ is a power law of. As $L^D \propto \langle s^n \rangle^{1/\mu^{(s)}_n}$ it can equally be used for collapsing the PDF ($\mu^{(s)}_n = n$ for $\tau = 1$, as normally found in equilibrium, e.g. Koba, Nielsen, and Olesen, 1972).

7.3.1 Corrections to scaling

In all three cases, Eq. (7.29), Eq. (7.30) and Eq. (7.31), the moments are fitted against a power law of the relevant system parameter for fixed n. This procedure generates estimates of $\mu^{(s)}_n, \sigma^{(s)}_n$ and $\mu^{(s)}_n / \mu^{(s)}_m$ respectively. Because any scaling is only asymptotic, the fit should account for corrections to scaling (see Sec. 2.2, in particular Eq. (2.26) on p. 39), for example

$$\langle s^n \rangle = B_n L^{\sigma^{(s)}_n} \left(1 + C_1 L^{-\omega_{n,1}} \right) \tag{7.32}$$

which can change the estimate of $\sigma^{(s)}_n$ significantly. Typically, corrections become stronger with increasing moment n and generally become weaker with increasing cutoff s_c. Moreover, the estimated error of $\sigma^{(s)}_n$ increases dramatically with the number of additional unknown parameters, such as C_1 and $\omega_{n,1}$ above. Some fitting routines do not converge or report an error for an expression like Eq. (7.32) because of the ambiguity of the two exponents $\sigma^{(s)}_n$ and $\sigma^{(s)}_n - \omega_{n,1}$. This problem can be controlled by demanding that $\omega_{n,1} > 0$, or an even stronger lower bound or, depending on the exact fitting procedure, setting the initial estimates of $\sigma^{(s)}_n$ and $\omega_{n,1}$ to some reasonable, graphically determined values. Often, very good fits can be achieved by setting $\omega_{n,i} = i$, e.g.

$$\langle s^n \rangle = B_n L^{\sigma^{(s)}_n} \left(1 + C_1 L^{-1} + C_2 L^{-2} \right). \tag{7.33}$$

Which corrections can be taken into account is, of course, also a matter of the sample size, for example the number of different system sizes L tested, versus the number of fitting parameters. It can pay to determine the leading order in Eq. (7.32) by fitting against $B_n L^{\sigma^{(s)}_n}$ only and then to fit $\langle s^n \rangle$ divided by $B_n L^{\sigma^{(s)}_n}$ based on these estimates against $1 + C_1 L^{-\omega_{n,1}}$. In the example based on the TAOM (see below), the corrections are not systematic or hardly detectable, so that $\overline{s^n}$ rescaled by the $L^{\sigma^{(s)}}$, with the exponent taken from a power law fit, produces the amplitudes listed in Eq. (7.34), see Fig. 7.4(b). It is generally very difficult to determine the precise nature of a deviation of a numerical estimate from the expected value. Finite sample sizes and small system sizes, missing the critical point, corrections to scaling, combined with incomplete fitting functions (lack of corrections) might suggest statistical errors, even when they are systematic and vice versa. The systematic error most frequently overlooked is the fitting function being a *hypothesis* forced upon the data, whose compatibility is often merely a matter of enough parameters to tweak.

While some of these systematic errors can be reflected in the goodness-of-fit (Press *et al.*, 2007), they are difficult to include appropriately in the error of the quantity estimated. This gives rise to another caveat: the more accurate the data and the larger the sample size, the more correction terms are needed in the function to fit against in order to accommodate all the tiny details visible in the data. Of course, the number of free variables in the fit is bounded from above by the number of data points, and they can be determined reliably only on the basis of a sufficiently large sample – the quality of a fit says little about the validity of a hypothesised scaling behaviour, if the degrees of freedom are sufficient to match *any* data set (which is typically the case if the degrees of freedom equals the sample size). For example, it makes little sense to determine avalanche size moments with unprecedented accuracy if the number of free parameters needed to account for the fine details is not significantly smaller than the number of different system sizes probed. The order which an analytical expression (assuming it exists) would need to be expanded to in order to match a numerical result within its error is, of course, unbounded.

The potential problems can be illustrated by fitting the fourth order polynomial $x^4 + x^3 + x^2 + x + 1$ evaluated at $x = 2^1, 2^2, \ldots, 2^n$ and with (fake) relative errors of, say, 10^{-4} against different functions and for varying sample size n. Using the Levenberg–Marquardt algorithm (`mrqmin` in Press *et al.*, 2007) a fit against Ax^μ for $n = 6$ gives $\mu = 3.84143(4)$ and a fit against $Ax^\mu(1 + C_1 x^{-1})$ for $n = 6$ gives $\mu = 4.0809(2)$, which is quite disillusioning. In both cases the vanishing goodness-of-fit correctly indicates the disagreement of data and fitting function. The correction term is more realistically chosen to be $x^{-\omega_1}$, or more reliably $x^{-|\omega_1|}$, which for $n = 4$ gives $\mu = 3.9529(6)$, about 82 standard deviations off the correct result. Unsurprisingly the goodness-of-fit is unity, as the number of fitting parameters matches the sample size – the hypothesis of the fitting function *determines* its coefficients. Sample size $n = 6$ gives $\mu = 3.97757(14)$ and $n = 8$ gives $\mu = 3.98809(6)$, i.e. about 192 standard deviations off. At the same time, the goodness-of-fit is practically 0, so the errors quoted are in fact meaningless. Only when including another correction, $Ax^\mu(1 + C_1 x^{-|\omega_1|} + C_2 x^{-|\omega_2|})$, which may require careful setting of initial values (found in preliminary fits), does the result eventually make sense with a goodness-of-fit close to 1. A sample of size $n = 6$, which is the smallest possible for 6 free parameters, gives $\mu = 3.999(2)$ and $n = 8$ gives $\mu = 3.9995(4)$. Forcing $Ax^\mu(1 + C_1 x^{-1/2})$ upon the data for $n = 6$ gives $\mu = 4.077(17)$. The 'benefit' of larger errors is illustrated by fitting a sample of size $n = 4$ against $Ax^\mu(1 + C_1 x^{-|\omega_1|})$ with relative error 10^{-2}, which produces $\mu = 3.95(6)$, namely 0.82 standard deviations off.

It is advisable to rescale moments of observables, e.g. $\langle s^n \rangle / L^{\sigma_n^{(s)}}$, to isolate systematic corrections, which should level out with increasing L, as shown in Fig. 7.2(a) (also Fig. 7.4(b)). This is often missed in χ^2 fits of data to a simple power law, which capture corrections by adjusting an effective exponent (usually at the cost of reduced goodness-of-fit). This is illustrated in Fig. 7.2(a), where the rescaled data suggest that a further increase in L would lead to even larger deviations, even when a double logarithmic plot indicates good agreement with a power law. In turn, including larger L into a fit would reduce the estimate of the leading order power. However, fitting against a power law with correction suggests a smaller leading order and corrections, which eventually level out, see Fig. 7.2(b).

7.3.2 Exponents

In the following, the numerics of a well-understood process is used to illustrate the procedures involved in estimating exponents. This process is essentially the **totally asymmetric OSLO Model** (TAOM, Pruessner and Jensen, 2003; Pruessner, 2004b, see also Sec. 6.3.5) which maps directly to a fair random walker (without drift) on a lattice with effective

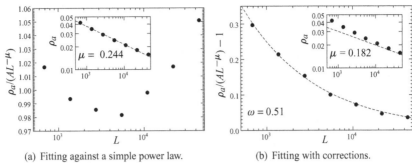

(a) Fitting against a simple power law.   (b) Fitting with corrections.

Figure 7.2 Fitting the activity ρ_a in a variant of the MANNA Model against different functions of the system size L. (a) The double logarithmic plot (inset) suggests that a single power law, $AL^{-\mu}$ (shown as a dashed line), might be a suitable fitting function. Performing that fit, $\mu = 0.244$, and dividing the result out (main panel) shows corrections *increasing* in L. (b) The same data, as the deviation from $AL^{-\mu}$. These can be approximated by a power law (dashed line) with exponent $\omega = 0.51$, determined by fitting the data against $AL^{-\mu}(1 + A'L^{-\omega})$. The inset shows that a power law with exponent $\mu = 0.182$ by itself is far too shallow, yet the corrections are captured properly, as illustrated in the main panel.

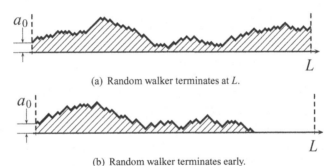

(a) Random walker terminates at L.

(b) Random walker terminates early.

Figure 7.3 Two trajectories of a random walker, both with the same (effective) diffusion constant and initial distance a_0 to an absorbing wall. The hatched area under the trajectory maps to the avalanche size in the TAOM. The walker in (a) reaches the end of the system at L and terminates there, whereas in (b) it terminates as it touches the absorbing wall.

diffusion constant D along an absorbing wall. The area under the trajectory of the random walker, which starts at distance a_0 away from the wall (in the TAOM an even number of lattice spacings), corresponds to the avalanche size in the TAOM. The path of the walker terminates if it collides with the wall or when it reaches a maximum time, corresponding to the system size L, see Fig. 7.3.

To leading order, integer moments of the area size distribution are known exactly (Pruessner, 2004b; Stapleton and Christensen, 2006; Welinder *et al.*, 2007):

$$\langle s \rangle = a_0 L \tag{7.34a}$$

$$\langle s^2 \rangle = \frac{32}{15\sqrt{\pi}} a_0 \sqrt{D} L^{5/2} \tag{7.34b}$$

$$\langle s^3 \rangle = \frac{15}{8} a_0 D L^4 \tag{7.34c}$$

$$\langle s^4 \rangle = \frac{4064}{693\sqrt{\pi}} a_0 D^{3/2} L^{11/2}. \tag{7.34d}$$

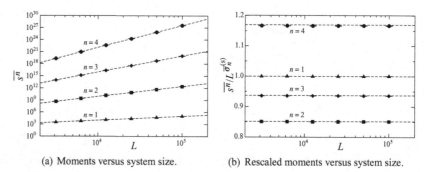

(a) Moments versus system size.   (b) Rescaled moments versus system size.

Figure 7.4 Avalanche size moments in the TAOM estimated from numerical simulations. (a) The moments $n = 1, 2, 3, 4$ as a function of the system size L. The dashed lines represent the fit to a power law. (b) The same data as before, but divided by the estimated exponent $\overline{\sigma}_n^{(s)}$. The resulting amplitudes follow those in Eq. (7.34), which are shown as dashed lines.

These power laws as well as the amplitudes are reproduced very clearly in the numerical data, see Fig. 7.4.

While the exponents for individual moments are interesting in their own right, it is much more common to derive τ assuming gap scaling and, in addition, D. From Eq. (7.34) one derives $\sigma_n^{(s)} = D(1 - \tau + n)$ for $n \geq \tau - 1$, where $D = 3/2$ is the avalanche dimension and $\tau = 4/3$ is the avalanche size exponent. For $n < \tau - 1$ the moments approach a constant. [15] Figure 7.4(a) shows the scaling and the numerical fits for some moments $\overline{s^n}$ obtained in numerical simulations of the process described above. Plotting the resulting estimated exponents $\overline{\sigma}_n^{(s)}$ as a function of n reveals D as the slope and $\tau - 1$ as the intercept of the sloped, linear part of the graph and the abscissa, see Fig. 7.5(b). In a self-consistent estimation of τ, moments $n < \tau - 1$ are excluded.

Around $n \approx \tau - 1$ the graph in Fig. 7.5(b) shows some rounding, which is an artefact caused by the finiteness of the upper cutoff and the presence of a lower cutoff. Including increasingly large systems in FSS usually suppresses the rounding and leads to better estimates of exponents. Standing well clear of the rounded region, a linear fit to the data shown in Fig. 7.5(a) gives $(\tau - 1)D = 0.50019(2)$ and $D = 1.500134(6)$, whereas the exact values are $(\tau - 1)D = 1/2$ and $D = 3/2$. While the relative difference from the exact values is extremely small, the estimated statistical error *excludes* the exact exponents. The situation improves significantly [16] by excluding small system sizes and moments beyond $n = 3$, which gives $(\tau - 1)D = 0.4999(1)$ and $D = 1.50011(7)$. In some systems, the lower cutoff itself is divergent in some (tunable) control parameter such as the system size (Pruessner and Jensen, 2002a, 2004), which can lead to moments $1 - \tau + n < 0$ that are divergent. Such behaviour leads to a more pronounced rounding around $n \approx \tau - 1$, but is

---

[15] If the event size s never vanishes, which is a matter of definition, then moments $n < 0$ exist and do not scale either. In order to maintain the validity of some exact results (Sec. 8.1.1.1), vanishing event sizes should be included when calculating moments.

[16] The statistical errors derived in the present example are, of course, ridiculously small; in fact they have been derived *without* considering correlations of the exponents $\sigma_n^{(s)}$ across different n, which are very densely spread in the present case (see p. 226).

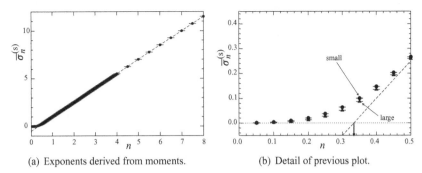

(a) Exponents derived from moments.   (b) Detail of previous plot.

Figure 7.5 (a) The exponent $\overline{\sigma}_n^{(s)}$ derived from moments, as illustrated in Fig. 7.4(a). Data for $n > 4$ are pruned for illustration purposes. The dashed line shows the result of the fit to $\overline{\sigma}^{(s)} = D(1 - \tau + n)$. (b) A detail of (a), showing the deviation from linear behaviour around $1 - \tau + n \approx 0$. The large system sizes (triangles, $L = 25\,000$, $50\,000$, $100\,000$) produce a slightly sharper pick-up than the small system sizes (circles, $L = 3125$, 6250, $12\,500$). The dotted line shows the abscissa, and the intersection with the linear fit of the exponents (dashed line) marks $\tau - 1$.

best detected by direct inspection of the PDF. As a result, some moments $1 - \tau + n < 0$ will display scaling but draw their characteristics from the non-universal part of the PDF.

While Fig. 7.5(a) contains, for illustration purposes, hundreds of moments, in particular non-integer ones, such data are, realistically, not normally available. Firstly, non-integer moments can be computationally expensive to generate, when they require the repeated use of math-library functions. This problem can be contained to some extent, see Sec. A.2, p. 363. Secondly, large n moments have a disproportionately large statistical error if $\tau > 1$. In that case the variance of the nth moment is dominated by $\langle s^{2n} \rangle \propto s_c^{1 - \tau + 2n}$, the square root of which is to be compared to $\langle s^n \rangle \propto s_c^{1 - \tau + n}$. To leading order this gives a relative error

$$\frac{\sqrt{\sigma^2 (s^n)}}{\langle s^n \rangle} \propto s_c^{(\tau - 1)/2}. \tag{7.35}$$

Normally, higher moments, subject to steeper scaling and stronger corrections, are more seriously affected by this. As a result, frequently, data for $n > 4$ are too noisy to be included in the analysis. If $\tau = 1$, the variance of the nth moment can scale *slower* than $\langle s^{2n} \rangle$, Eq. (7.35), i.e. the relative error might vanish asymptotically, which is known as **self-averaging** (Ferrenberg, Landau, and Binder, 1991). If correlations are finite, for example, the variance of a density scales like L^{-d}, known as **strong self-averaging**, whereas slower scaling is known as **weak self-averaging**. Depending on the amount of CPU time needed to generate independent configurations and to determine the observables, it can be advantageous to consider larger systems. Only in the rare presence of **anti-correlations** can the scaling of the variance exceed strong self-averaging and decay even faster than L^{-d} (Welinder *et al.*, 2007, also Sec. 8.5.4.1). Typically, SOC models display at best weak self-averaging, which, together with a positive dynamical exponent z (i.e. divergent correlation times) and increased computational effort, means that large system sizes carry disproportionally increased statistical error, which is often outweighed by the reduced corrections to scaling (e.g. Manna *et al.*, 1990).

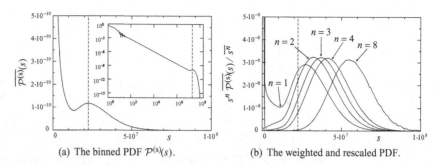

(a) The binned PDF $\mathcal{P}^{(s)}(s)$.   (b) The weighted and rescaled PDF.

Figure 7.6   The (estimated) PDF of the process described on p. 222 (see Fig. 7.3) for $L = 100\,000$. The dashed line marks a measure of the cutoff (local maximum) in both graphs. (a) The original PDF on a linear scale (linearly binned) and on a logarithmic scale in inset (logarithmically binned). (b) Linear plot of the rescaled integrand of $\int d^d s\, s^n \mathcal{P}^{(s)}(s) = \langle s^n \rangle$ (PDF binned linearly), i.e. $s^n \mathcal{P}^{(s)}(s)\, \langle s^n \rangle^{-1}$. The rescaling by $\langle s^n \rangle^{-1}$ ensures equal (unit) areas under the graph to make them comparable. The original PDF is shown dotted, but is hardly visible on a linear scale, as it is close to 0 for almost all $s > 0$. The higher the order of a moment, the more weight is drawn from the tail of the PDF, visible in the maximum being shifted to the right. Other system sizes give the same graph up to a change of scale.

The two problems mentioned above (computational costs of non-integer moments and large statistical errors at large n) can be avoided if moments can be estimated reliably from the PDF collected during the simulation, i.e. if the PDF is collected in the simulation rather than only its moments. This, however, is not always possible, because most histograms collected in simulations require some form of binning at the time of collection and the resolution of the resulting estimate of the PDF might be insufficient to allow the generation of moments in post-processing. The error, which is easily estimated by comparing moments taken during the simulation with those based on the histogram, is due to an increased weight of large events for large moments. The resolution towards large event sizes drops necessarily, but this is where large n moments $\langle s^n \rangle$ draw most of their weight from. Figure 7.6 illustrates the relative weight of different regions of the histogram, for different moments. In Fig. 7.6(b) the $n = 8$th moment is shown to have a significant contribution beyond $s = 6 \cdot 10^7$, where $\mathcal{P}^{(s)}(s)$, according to Fig. 7.6(a) has practically vanished and would probably need to be implemented with a much coarser resolution.

The relative weights shown in Fig. 7.6(b) illustrate a source of error that has not been mentioned so far. Different moments n are correlated and so are the scaling exponents $\sigma_n^{(s)}$ derived from them. Considering them *simultaneously* as done in a fit like Fig. 7.5 to derive D and τ should in principle take these correlations into account when quantifying the error. This can be done systematically on the basis of the distribution of exponent estimates based on subsamples, for example using the jackknife. The problem might be alleviated by spacing the different n that enter into the fit more widely than in Fig. 7.5. An upper bound of the error can be determined very pragmatically by multiplying the error of each moment by the square root of the number of moments considered simultaneously, as if each was the result of an independent, distinct sample (e.g. Pruessner and Moloney, 2003).

7.3.3 Cumulants

Finally, **cumulants**, as introduced in Sec. 2.2.3, provide an alternative path to data analysis. If $\tau > 1$, then all cumulants $\langle s^n \rangle_c$ are dominated (asymptotically) by their highest order moment $\langle s^n \rangle$, so that analysis of the cumulants does not add much to the analysis of the moments, although it might suffer from different (possibly improved) corrections to scaling.

In the case of $\tau = 1$, cumulants provide a comfortable path to test for a Gaussian PDF, [17] the presence of which suggests that the correlations in the system ultimately decay so fast that the observable can be considered as a sum of independent random variables. A big avalanche in an SOC system might then be seen as the sum of randomly and independently drawn smaller avalanches. Rescaling the resulting PDF appropriately (see Sec. 7.4.2, p. 238) then reveals the Gaussian, produced by the **central limit theorem**. Since all cumulants of third and higher order of a Gaussian vanish, one might therefore expect that the cumulants of such a Gaussian avalanche signal would vanish as well.

This is generally not the case without rescaling the signal first, because the cumulants of the *sums* of N independent random variables are all proportional to N, despite the nth moments being proportional to N^n *to leading order*. In other words, the cumulants of the sums of random variables generally *diverge* with increasing N, only they do so very slowly. The unknown effective value of N can be taken from any cumulant, so that for large N the ratio $\langle s^n \rangle_c / \langle s^m \rangle_c$ converges for any $n, m \geq 1$, if s is the (effective) sum of N independent random variables, whereas $\langle s^n \rangle_c / \langle s^n \rangle$ vanishes in large N for all $n \geq 2$.

In statistics known as the kurtosis, $\langle s^4 \rangle_c / \langle s^2 \rangle_c^2$ vanishes if the underlying distribution is a Gaussian; in statistical mechanics the **Binder cumulant** (Binder, 1981a,b; Landau and Binder, 2005), $1 - \langle s^4 \rangle / (3 \langle s^2 \rangle^2)$ vanishes if the underlying distribution is a Gaussian with vanishing mean. [18] It vanishes asymptotically in large N, if $\langle s^4 \rangle - 3 \langle s^2 \rangle^2$ diverges in N slower than $\langle s^2 \rangle^2$, which is the case for s being the sum of independent random variables with vanishing mean.

7.3.4 Moment ratios

A set of universal quantities that is often overlooked are **moment ratios**, which are essentially features of the scaling function, but in a form that excludes its ambiguous metric factors. However, the dependence of the scaling function on the boundary conditions, the drive and the topology of the system, as well as geometrical features such as the aspect ratio (Privman *et al.*, 1991) is inherited by the moment ratios. On the other hand, exponents suffer from similar effects, for example τ in the OSLO Model depending on the external drive (Nakanishi and Sneppen, 1997; Malthe-Sørenssen, 1999) and D depending on the dissipation (Alava *et al.*, 2008), which can be understood in models where $\sigma_1^{(s)} = D(2 - \tau)$ is determined by the conservative, diffusive movement of particles, see Sec. 8.1.

---

[17] If $\tau > 1$ then a Gaussian PDF is not possible.

[18] If the mean vanishes, the kurtosis in fact simplifies to $\langle s^4 \rangle_c / \langle s^2 \rangle_c^2 = \langle s^4 \rangle / \langle s^2 \rangle^2 - 3$.

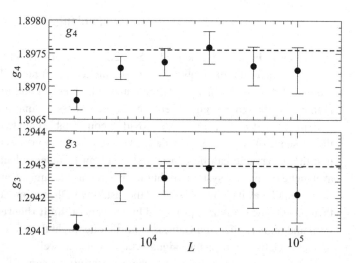

Moment ratios $g_3^{(s)}$ and $g_4^{(s)}$ for the random walker example introduced above, using Eq. (7.36) and Eq. (7.40) for the error. The dashed lines are the exact values derived from Eq. (7.34), $g_3^{(s)} = 3375\pi/8192 = 1.294\,296\ldots$ and $g_4^{(s)} = 47\,625\pi/78\,848 = 1.897\,554\ldots$ (Stapleton and Christensen, 2006).

The following discussion is restricted to the numerics of the class of moment ratios defined in Eq. (2.32) (see Fig. 7.7)

$$g_n^{(s)}(0) = \frac{\langle s^n \rangle \langle s \rangle^{n-2}}{\langle s^2 \rangle^{n-1}} \qquad n \geq 1 \tag{7.36}$$

which are a compromise between the number of different terms and the order of moments involved. To ease notation, in the following the argument is dropped from $g_n^{(s)}(0)$. Another useful moment ratio is $\langle s^{n-m} \rangle \langle s^{n+m} \rangle / \langle s^n \rangle^2$.

Since the first two moment ratios are $g_1^{(s)} = g_2^{(s)} = 1$ by definition, the first non-trivial moment ratio is $g_3^{(s)} = \langle s^3 \rangle \langle s \rangle / \langle s^2 \rangle^2$. The numerical challenges are twofold. Firstly, the product or ratio of estimates is generally a biased estimator of the product or ratio of the population averages, just like $(\bar{s})^2$ is a biased estimator of $\langle s \rangle^2$, see Sec. 7.1.1. **Jackknife** and **bootstrap** schemes provide a way out in cases where this bias is intolerable. In the following, the numerical estimate

$$\overline{g_n^{(s)}} = \frac{\overline{s^n}\,(\bar{s})^{n-2}}{\left(\overline{s^2}\right)^{n-1}} \tag{7.37}$$

is used, where $\overline{s^n}$ denotes the unbiased numerical estimator of $\langle s^n \rangle$ as introduced in Sec. 7.1.1, so that $\langle \overline{s^n} \rangle = \langle s^n \rangle$.

The second challenge is of a more general nature and is frequently overlooked. It is caused by the different moments in Eq. (7.37) being correlated. Ignoring covariances gives for the variance (i.e. the square of the statistical error of the estimate $\overline{g_n^{(s)}}$ of $g_n^{(s)}$),

$$\sigma^2\left(\overline{g_n^{(s)}}\right) = \left(\overline{g_n^{(s)}}\right)^2 \left(\frac{\sigma^2\left(\overline{s^n}\right)}{\left(\overline{s^n}\right)^2} + (n-2)^2 \frac{\sigma^2(\bar{s})}{(\bar{s})^2} + (n-1)^2 \frac{\sigma^2\left(\overline{s^2}\right)}{\left(\overline{s^2}\right)^2} \right), \tag{7.38}$$

where each of the variances, $\sigma^2\left(\overline{s^n}\right)$, $\sigma^2(\overline{s})$ and $\sigma^2\left(\overline{s^2}\right)$, is the variance of a numerical estimate, which typically scales with the inverse of the sample size. If the moments follow gap scaling, Eq. (7.29), the leading order of Eq. (7.38) is $s_c{}^{\tau-1}$, as $\sigma^2\left(\overline{s^n}\right) \propto \left\langle s^{2n}\right\rangle$ and $\left\langle \overline{s^n}\right\rangle = \langle s^n\rangle$. However, this expression obviously overestimates the errors of $g_1^{(s)}$ and $g_2^{(s)}$ which actually vanish exactly, while both are non-zero according to Eq. (7.38).

Including the covariances in standard error propagation gives

$$\sigma^2\left(f(x,y)\right) = \sigma^2\left(x\right) f_x^2 + \sigma^2\left(y\right) f_y^2 + 2\mathrm{cov}\left(x,y\right) f_x f_y \tag{7.39}$$

and thus for the variance of the estimator $\overline{g_n^{(s)}}$ of $g_n^{(s)}$

$$\sigma^2\left(\overline{g_n^{(s)}}\right) = (\overline{g_n^{(s)}})^2 \Bigg(\frac{\sigma^2\left(\overline{s^n}\right)}{\left(\overline{s^n}\right)^2} + (n-2)^2 \frac{\sigma^2(\overline{s})}{(\overline{s})^2} + (n-1)^2 \frac{\sigma^2\left(\overline{s^2}\right)}{\left(\overline{s^2}\right)^2}$$
$$- 2(n-2)(n-1)\frac{\mathrm{cov}\left(\overline{s^2},\overline{s}\right)}{\overline{s}\,\overline{s^2}} - 2(n-1)\frac{\mathrm{cov}\left(\overline{s^2},\overline{s^n}\right)}{\overline{s^2}\,\overline{s^n}}$$
$$+ 2(n-2)\frac{\mathrm{cov}\left(\overline{s},\overline{s^n}\right)}{\overline{s}\,\overline{s^n}} \Bigg), \tag{7.40}$$

where $\mathrm{cov}\left(\overline{s^m},\overline{s^n}\right) = \left\langle \overline{s^m}\,\overline{s^n}\right\rangle - \left\langle \overline{s^m}\right\rangle\left\langle \overline{s^n}\right\rangle$, e.g. Eq. (7.26), is the covariance of the numerical estimates $\overline{s^m}$ and $\overline{s^n}$, so that $\sigma^2\left(\overline{s^n}\right) = \mathrm{cov}\left(\overline{s^n},\overline{s^n}\right)$. To leading order in the sample size, Eq. (7.37) and Eq. (7.40) are recovered using more sophisticated techniques, such as the **jackknife** (Pruessner *et al.*, 2001; Pruessner, 2003b; Stapleton, 2007). In fact, if subsamples of equal size are available (see Sec. 7.2), the variance of the chunk-wise average (divided by that number minus one) provides the most convenient way of determining the variance of the estimator $\overline{g_n^{(s)}}$.

Similar to the variance, the covariances of the numerical estimates decrease with increasing sample size. For $n = 1, 2$, Eq. (7.40) vanishes as it should. Unbiased estimators for the variance and the covariance have been introduced in Sec. 7.2, in particular Eq. (7.9), Eq. (7.10), Eq. (7.22) and Eq. (7.26). The first three terms of Eq. (7.40) are identical to Eq. (7.38), but the former takes correlations into account that are ignored in the latter. It is worth noting that generally neither Eq. (7.37) nor Eq. (7.40) (nor, for that matter, Eq. (7.38)) is unbiased, because of the fractions they involve, even when the estimators of $\sigma^2\left(\overline{s^n}\right)$ they use are unbiased.

7.4 PDF estimation

Traditionally, one of the central observables in SOC models was the **probability density function** (PDF) of the avalanche sizes, because it seems to provide the most immediate access to scaling, via a data collapse or by measuring the apparent exponent (Clauset *et al.*, 2009). Over the years, however, direct analysis of the PDF has been replaced by the far more quantitative moment analysis discussed above, which gained further popularity due to De Menech *et al.* (1998). Nevertheless, an analysis of the PDF remains important, for example because a moment analysis is rather insensitive to the lower cutoff, see Sec. 2.1.1, which might be divergent itself and thus incompatible with simple scaling. More importantly, a data collapse of the PDF measured for different values of the cutoff confirms the simple scaling hypothesis, which underpins all scaling arguments. Some practical aspects of PDF estimation are discussed further by Pruessner (2009) and from a very applied point of view by White, Enquist, and Green (2008).

The main purpose of a direct inspection of the PDF is a qualitative assessment rather than a quantitative one. The latter suffers from two key problems. To start with, it is not clear what a PDF is to be compared with or fitted against. In the few, rare cases when the scaling function can be guessed or is known, it is difficult to quantify the quality of the match. Secondly, estimating the PDF $\mathcal{P}^{(s)}(s)$ with an error is not a simple task. The unbiased estimator of $\mathcal{P}^{(s)}(s)$ on the basis of a sample $\{s_1, s_2, \ldots, s_N\}$ is the mean value of δ_{s,s_i} if s is discrete. Its variance is given by $\langle \delta_{s,s_i}^2 \rangle - \langle \delta_{s,s_i} \rangle^2$, which can be estimated by $\mathcal{P}^{(s)}(s)(1 - \mathcal{P}^{(s)}(s))$, yet the variance of the estimator of $\mathcal{P}^{(s)}(s)$ has to account for correlations in the sample. What is more, considering the estimated PDF $\mathcal{P}^{(s)}(s)$ for different values of s simultaneously, the error of $\mathcal{P}^{(s)}(s)$ would need to account for correlations between them. The most promising route here is a subsampling scheme, which is, however, beyond the scope of the following.

For a qualitative, visual inspection, such as a data collapse, a histogram gathered in a numerical simulation is normally too fine grained, e.g. Fig. 7.9(a), and thus needs to be smoothed. The first step therefore is **binning**, i.e. averaging over suitably defined ranges along the histogram. The resulting data should be displayed in rescaled form which takes the expected behaviour into account, so that the resulting graph can be thought of as the deviation from the expected behaviour (Sec. 7.4.2). Finally, the PDF can undergo a data collapse, thereby testing for **simple (finite size) scaling**, i.e. verifying whether the PDF follows

$$\mathcal{P}^{(s)}(s) = a^{(s)} s^{-\tau} \mathcal{G}^{(s)} \left(\frac{s}{b^{(s)} L^D} \right) \quad \text{for } s > s_0 \tag{7.41}$$

with **metric factors** $a^{(s)}$ and $b^{(s)}$, **lower cutoff** s_0, **scaling function** $\mathcal{G}^{(s)}$ and **scaling exponents** τ and D, see Sec. 2.1. Equation (7.41) is sometimes amended by corrections, but this is more common on the level of moments, see Secs. 7.3.1 and 2.2. The algebraic decay $s^{-\tau}$ in Eq. (7.41), sometimes referred to as **fat tails**, is best investigated on a double logarithmic scale. [19] The three stages, binning, displaying, collapsing, are discussed in the following.

7.4.1 Binning

It is assumed in the following that the numerical simulation produced a normalised histogram of the form h_i, so that $h_i r_i$ avalanches have occurred in the interval $\{t_i, t_i+1, \ldots, t_i + r_i - 1\}$ for discrete observables. The extension to continuous observables is obvious and normally consists in changing the bin ranges to half open intervals $[t_i, t_{i+1})$. The ranges r_i parameterise a form of binning which has taken place on the fly during the simulation (e.g. Sec. A.6.2), while binning schemes discussed in the following are applied during

---

[19] Plotting the logarithm of observables on linear scales gives the same graph as a double logarithmic plot, but with axes that are difficult to interpret. Nevertheless, the former practice remains widespread.

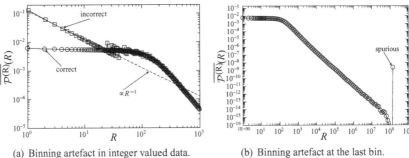

(a) Binning artefact in integer valued data.   (b) Binning artefact at the last bin.

Figure 7.8 (a) Residence time distribution in the Oslo Model binned exponentially in two different ways. The correct procedure (circles) makes use of integer valued bins, matching the integer valued residence times. The incorrect procedure (squares) uses real valued bins, most of which are empty at small values of the observables, producing a spurious slope -1 as indicated by the dashed line. (b) Data as for (a), but binned such that the last bin size is much smaller than all other bins. As a result a spurious data point appears as a terminal peak. As a guide for the eye, data points are connected by a line here and in most of the following graphs.

post-processing. [7.1] If enough memory is available and the observables are discrete, then binning might not take place at all during the simulation, i.e. $r_i = 1$ and $t_i = i$. Using a sophisticated binning scheme during the simulation usually amounts to a significant waste of CPU time. Every binning procedure blurs results, which is acceptable only if the underlying PDF varies sufficiently slowly.

The binning schemes presented in the following average the histogram determined in the simulation over windows or bins j ranging from b_{j-1} inclusive to $b_j - 1$ inclusive. Bin j thus has size $b_j - b_{j-1}$ and contains the data for all t_i in the interval $\mathbb{I}_j = \{b_{j-1}, b_{j-1} + 1, \ldots, b_j - 1\}$:

$$\overline{\mathcal{P}^{(s)}}(\mathbb{I}_j) = \frac{1}{b_j - b_{j-1}} \sum_{i \mid t_i \in \mathbb{I}_j} h_i r_i, \tag{7.42}$$

where the sum runs overall entries i of the original histogram whose intervals start within bin j.

Binning generally can produce a number of (related) artefacts.

Too small bins Bin sizes should not be smaller than the resolution r_i in the original data, as that leads to spuriously empty bins. This problem occurs typically when the observable is discrete, but the bins are calculated as real numbers. When the bin sizes $b_j - b_{j-1}$ are much smaller than the resolutions r_i in their range, most of them will be empty. The few that are not empty are rescaled by the size of the bin, which in the case of exponential binning (see below) is linear in b_j itself. As a result a spurious slope -1 can occur in a double logarithmic plot, illustrated in Fig. 7.8(a). This artefact is prevented by using integer valued bins when the data are discrete, and generally by keeping the resolution of the binning at or below the resolution of the data.

Incompatible resolutions The resolution of the bin ranges b_j should not be smaller than the resolution r_i of the data collected. For example, pronounced artefacts occur at small bin sizes (small values of the observable), when the bins are calculated as real numbers, but the data is discrete. In that case, bins of seemingly similar size might contain different numbers of entries t_i. For example, the interval from $b_1 = 0.9$ to $b_2 = 2.1$ contains the *two* integers 1 and 2, whereas the interval from $b_2 = 2.1$ to $b_3 = 3.3$ contains only one integer, 3. Normalising by the bin width $b_2 - b_1 = b_3 - b_2 = 1.2$ then produces a misleading result. The correct normalisation counts the number of entries falling in the bin.

Double binning Binning the data in the simulation as well as during post-processing, can produce rather strong artefacts due to coarse resolution of the data produced in the simulation and (slight) mismatches of the bin boundaries of the two different binning procedures.

Maximum bin Aiming to place the end of the last bin, say b_M, at the maximum entry of the histogram, say t_M, can turn out to be a bad choice. Dramatic artefacts occur around t_M, when rounding leads to b_M ending up just below t_M, creating a new bin reaching to $b_{M+1} > t_M$. The sparse data at large values of the observable need to be averaged over a sufficient range to be representative. If t_M is slightly larger than b_M, then only very few entries make it into the new bin, which is thus very noisy. What is worse, its value is essentially determined by the normalisation $(b_{M+1} - b_M)^{-1}$: as there are no further data points beyond t_M one might wrongly choose b_{M+1} to be just above t_M. The final bin size then ends up being much smaller than the preceding bins, so that its normalisation $(b_{M+1} - b_M)^{-1}$ in Eq. (7.42) is significantly larger, suggesting a much higher probability density than in the preceding bins, see Fig. 7.8(b). To prevent these problems, bin boundaries should be chosen independently from t_M (see also next item).

Mismatching bins Comparing PDFs for different values of the upper cutoff generally benefits from using the same bin boundaries b_j for all datasets, to allow direct comparison of the PDF $\mathcal{P}^{(s)}(s)$ for the same effective values of s. For example, the pattern in the small s region of Fig. 7.12(a) where the PDF is very rugged[20] would not repeat perfectly if different bins were used for different data sets.

Sparse data Sparse data and high resolution of bins for large entries can also produce a spurious power law with slope -1 when using exponential binning, Fig. 7.9(b). This is a common, avoidable error, discussed below in further detail.

Figure 7.9 shows the histogram for the area s under the trajectory of a random walker along an absorbing wall as introduced in Sec. 7.3 for $L = 10^5$. The PDF of this process, parameterised by the initial distance a_0, system size L (see Fig. 7.3) and effective diffusion constant D, is known to leading order (Pruessner, 2004b; Majumdar and Comtet, 2004, 2005; Stapleton and Christensen, 2006; Welinder *et al.*, 2007):

$$\mathcal{P}^{(s)}(s) = \left(\frac{a_0^3}{D}\right)^{1/3} s^{-4/3} \mathcal{G}^{(s)}\left(\frac{s}{\sqrt{D}L^{3/2}}\right), \tag{7.43}$$

so that a comparison with Eq. (7.41) gives $\tau = 4/3$ and $\mathsf{D} = 3/2$.

---

[20] It is worth stressing that this ruggedness is 'real', i.e. not a statistical artefact as in noisy data (e.g. Fig. 7.9(b)).

Figure 7.9(a) shows the raw histogram based on a sample size of only 10^4 walks. Visual inspection of the PDF is greatly facilitated by binning, see the upper graph in Fig. 7.9(b). Three different schemes are in common use, which differ by the choice of the bin boundaries. This choice depends strongly on the format in which the data are to be displayed.

Coarse graining means that the ranges have equal size $R \in \mathbb{N}$ on a linear scale, which are large compared to any of the ranges r_i, so that bin j starts at $b_j = jR$. The numerical estimate of the PDF then is

$$\overline{\mathcal{P}^{(s)}}(jR) = \frac{1}{R} \sum_{i|t_i \in \{(j-1)R,...,jR-1\}} h_i r_i \tag{7.44}$$

for integers $j = 1, 2, \ldots$ where the sum runs over all distinct entries i of the original histogram with $(j-1)R \leq t_i \leq jR - 1$. The method is best suited for presenting data in a linear plot, [21] which is rarely relevant to PDFs displaying scaling. An example is shown in Fig. 7.10(a) for $R = 5000$. Because this plot is double logarithmic, bins seem to be very sparse for small s and very dense for large s, which means that a lot of information is lost compared to the other two binning schemes.

All binning schemes possess some ambiguity regarding the position of a bin along the abscissa. For example, one might wonder whether the value of $\overline{\mathcal{P}^{(s)}}(jR)$ should be shown at $s = jR$ or, say, $s = (j - 1/2)R$. The pragmatic view is: if the position of the bins makes a significant difference to any estimate derived from it (by fitting or based on a data collapse), then they are probably too widely spaced.

Exponential binning is the most common binning scheme, which uses bin sizes proportional to their position, which means that bins are evenly distributed on a logarithmic abscissa. [22] The beginning of each bin is $b_j = \lfloor cR^j \rfloor$, exponentially increasing in j, with constants c and $R > 1$, so that bins have size $b_{j+1} - b_j \approx b_j(R - 1)$, which is approximate only because the b_j are discrete. The data in bin $j = 1, 2, \ldots$ are an average according to

$$\overline{\mathcal{P}^{(s)}}(cR^j) = \frac{1}{b_j - b_{j-1}} \sum_{i|t_i \in \{b_{j-1},...,\max\{b_{j-1},b_{j-1}\}\}} h_i r_i, \tag{7.45}$$

where the sum allows for repeated values of the bin boundaries b_j, which are discrete in order to avoid the artefact of too small bins described on p. 231. In practice, repeated bin boundaries are dropped. The same applies to power law binning below. The natural position of an exponential bin along the abscissa is $cR^{j-1/2}$.

Figure 7.10(a) contains some exponentially binned data with about 200 bins. It is technically straightforward to determine c and R for a given number of bins – in the example cR^{200} reaches just beyond the maximum entry in the original data and c is the lowest entry arriving in the first bin. For exponential binning on the fly, it can be sufficient to set $c = 1$ and $R = 2$ so that the bin is the most significant bit in the integer representation of s in the simulation (Grassberger and Manna, 1990).

While exponential binning is best suited for double logarithmic plots, it can produce very misleading artefacts, as illustrated in Fig. 7.9(b). If the number of bins is so large that many

---

[21] Coarse graining is therefore sometimes called linear binning.
[22] Exponential binning is therefore sometimes called logarithmic or (confusingly) linear binning.

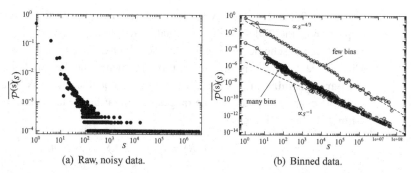

(a) Raw, noisy data.   (b) Binned data.

Figure 7.9 The size distribution of the area under the trajectory of a random walker along an absorbing wall with cutoff after $L = 100\,000$ steps in a double logarithmic plot. (a) The raw data of a rather sparse histogram. The majority of the data in the tail consists of single events. (b) The same data in binned form. The upper graph consists of about 50 bins, which reveals the main feature, a slope of $-4/3$ shown as a dashed line. The lower graph (shifted downwards by a factor 10^{-3}) contains too many bins (about 500), which leads to a spurious slope of -1. Due to the sparsity of the data, the characteristic bump (see Fig. 7.10(a)) is barely discernible in the upper graph and absent in the lower.

of them, in particular towards large event sizes, are either empty or contain only a single entry, then the resulting densities in the bins either vanish or are anti-proportional to their size, i.e. $\propto (b_{j+1} - b_j)^{-1} \propto b_j^{-1}$, which suggests a spurious, noisy exponent of 1 (see the lower graph in Fig. 7.9(b)). The few bins that contain a few more entries lie just above and follow the same slope, seemingly confirming the finding. Because the exponent of 1 or a value close to it is not uncommon for τ, this artefact can be very dangerous. Reducing the number of bins recovers the correct result (see the upper graph in Fig. 7.9(b)). It remains to point out that sparse data generally hide the detailed structure of the scaling function, the bump (Sec. 2.1.2.3), which is crucial for a data collapse, see Sec. 7.4.2.

One major drawback of logarithmic binning is the decreasing number of events per bin with increasing event size. Assuming equal correlation times across all event sizes, the statistical error therefore increases in particular towards the significant bump of the PDF.

Bins with increasing size even on a logarithmic scale are used for **power law binning**, with bins starting at $b_j = \lfloor c(M - j)^{-1/(\tau'-1)} \rfloor$ for $\tau' > 1$, $j = 0, 1, \ldots, M - 1$ and some constant c. The last bin is open, $b_M \to \infty$, and the parameters are thus best chosen so that the last finite bin boundary b_{M-1} lies beyond the maximum histogram entry t_M, $t_M < b_{M-1}$. Normally, τ' is chosen to match (roughly) the exponent τ of the PDF, Eq. (7.41). Assuming for simplicity a pure power law $\mathcal{P}^{(s)}(s) \propto s^{-\tau}$ and real valued bin boundaries, the fraction of events in bin $j > 0$ is

$$\int_{b_{j-1}}^{b_j} ds\, s^{-\tau} = \frac{c^{1-\tau}}{\tau - 1} \tag{7.46}$$

for $\tau' = \tau$ and therefore constant in j. For $\tau = 1$ this is the same for exponential binning. The lowest graph in Fig. 7.10(a) uses power law binning with exponent 4/3. Because of the increasing size of the bins towards large s, the bump is very poorly resolved. Another major downside of this scheme is its reliance on the very same quantity that it should help estimate, namely the exponent τ. Because of possible artefacts, it is generally unwise to feed into an estimation scheme even a rough estimate of the quantity to be gained from it.

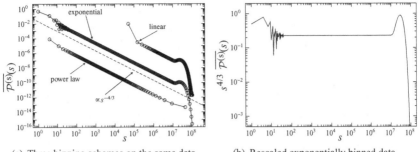

(a) Three binning schemes on the same data.   (b) Rescaled exponentially binned data.

Figure 7.10 (a) Illustration of the three different binning schemes discussed in the text, based on the same large data set of the process described in Fig. 7.9. Linear and power law schemes are shifted by a factor 10^3 and 10^{-3} respectively. Circles mark actual data, in particular the horizontal position resulting from the choice of bin position, the line connecting them is a visual guide. The dashed line follows the intermediate power law according to Eq. (7.43). (b) Rescaling the data according to their expected behaviour reveals their main features.

7.4.2 Data collapse

There is a remarkable disparity in the way PDFs are displayed in the literature. In particular, many early publications show the raw data, such as Fig. 7.9(a). While binning the data and plotting $\overline{\mathcal{P}^{(s)}}(s)$ directly in a double logarithmic plot, Fig. 7.9(b), helps appreciate its power law behaviour for intermediate values of s, it is difficult to tell the deviation from a power law and to make out the structure of the crucial bump towards large s (Sec. 2.1.2.3). The ordinate spans about 14 orders of magnitude, so that any detail of the PDF will drown in the steep decline of the function.

The hypothesis that $\overline{\mathcal{P}^{(s)}}(s)$ is well described by simple scaling (see Sec. 2.1), $\mathcal{P}^{(s)}(s) = a^{(s)}s^{-\tau}\mathcal{G}^{(s)}(s/s_c(L))$, with $s_c(L) = b^{(s)}L^D$ is best tested by plotting its deviation from that behaviour. To this end, the histogram is re-weighted by s^τ using some estimate of the exponent τ, which reveals the scaling function up to a prefactor. A good guess of the **apparent exponent** $\tilde{\tau}$ (Sec. 2.1.2.1) can be gained in a 'straight line fit' or 'ruler fit' of a suitable intermediate section of the PDF. Using this value to rescale the PDF by $s^{\tilde{\tau}}$ will result in a graph that is almost perfectly straight and horizontal in the mid-section, as illustrated in Fig. 7.10(b). Suitably defined (Sec. 2.1.2.1), the apparent exponent $\tilde{\tau}$ equals the scaling exponent τ provided that the scaling function $\mathcal{G}^{(s)}(x)$ does not vanish in the limit $x \to 0$, Sec. 2.1.2.2.

If further details of the scaling function are known or expected, the PDF can be plotted correspondingly, moving more and more terms from the right hand side of Eq. (7.41) to the left hand side. This procedure is often hampered by unknown constants, such as the metric factors. The rescaled PDF $\overline{\mathcal{P}^{(s)}}(s)s^\tau$ might be best analysed on a linear scale, which gives more space to large values on the abscissa, see Fig. 7.13(a).

In a **data collapse**, the scaling hypothesis Eq. (7.41) is tested by extracting essentially the scaling function from numerical estimates of the PDF for a range of system sizes. Provided that simple scaling applies, for some $s > s_0$ different system sizes produce the same graph $a^{(s)}\mathcal{G}^{(s)}(s/s_c)$ when plotting $\overline{\mathcal{P}^{(s)}}(s)s^\tau$ against s/s_c. At the same time, the upper cutoff s_c

should follow $b^{(s)}L^D$, but it can help to impose these two demands independently, i.e. perfecting the collapse without constraining s_c to scale like L^D, which can be tested separately.

A data collapse is performed in a few steps. Firstly, the binned data, such as those shown in Fig. 7.10(a), for a range of system sizes are plotted in rescaled form $\overline{\mathcal{P}^{(s)}}(s)\,s^{\tau'}$ using some rough estimate of the slope in the intermediate range, say $\tau' = 1.5$ as shown in Fig. 7.11(a). In this plot, some '**landmarks**' can be identified, for example the maxima of the bumps, as indicated by arrows in Fig. 7.11(a). The estimate of the scaling exponent τ is then adjusted according to the slope of the line connecting the landmarks (Christensen, 2004; Pruessner, 2009), which in the present example gives $\tau' = 1.5 - 0.166 = 1.334$, so that the landmarks all have roughly the same vertical position, Fig. 7.11(b). Dividing s by the position of the landmarks $\overline{s_c}$ on the the abscissa for each graph, thereby determining the **upper cutoffs** $s_c \propto L^D$, the landmarks all arrive at the same point. If simple scaling applies, increasingly large sections of the graphs overlap, see Fig. 7.12(a). For small arguments s, they deviate in the plot with rescaled abscissa, but collapse without rescaling, see Fig. 7.11(b). The point $\overline{s_0}$, below which graphs deviate from each other in the rescaled plot (see arrows from below in Fig. 7.12(a)), i.e. below which the graph cannot be approximated by a straight line in a double logarithmic plot, is the **lower cutoff**. Above the lower cutoff, the PDF is (approximately) universal up to the metric factors; below the lower cutoff, even systems with different *upper* cutoff behave very similarly. The characteristics in that region are determined by the details of the microscopic physics and the definition of the observable, i.e. the behaviour of the PDF in this region is **non-universal**. This is clearly visible in the spiky pattern at small s, cf. Fig. 7.11(b).

In the present example, the lower cutoff is directly related to the lattice spacing and the initial distance of the random walker from the absorbing wall, a_0. The discreteness of the lattice causes the spiky structure at small s. It is absent in a continuum description, [7.2] which approximates the discrete one well provided a_0 is large compared to that lattice spacing. [23] On the lattice as well as in the continuum, systems with different size but the same a_0 all display the same behaviour at small s. The $s^{-4/3}$ power law takes over for s large compared to a_0^3/D, which is therefore identified as the lower cutoff. A lower cutoff thus exists in the continuum and on the lattice, yet the latter suffers in addition from discretisation effects, which often dominate the appearance of the PDF at small s.

After identifying the lower cutoff in the rescaled plot as s_0/s_c, below which curves for different s_c deviate, it is scaled back to the original units s and compared for different system sizes. Simple scaling means that it has the same value for all system sizes or at least is bounded from above.

In addition to using landmarks, there is a second, popular method of extracting the upper cutoff from a PDF (e.g. Jánosi and Kertész, 1993), namely by identifying it with the **maximum event** size in the distribution, i.e. the upper cutoff is simply regarded as the largest event encountered during the simulation. This approach, however, has some serious shortcomings. First of all, the maximum event size is obviously dependent on the sample size. Increasing the sample size indefinitely will push the record towards the theoretical

---

[23] See, for example, Eq. (8.26) on p. 256, which is the asymptotic termination probability density of random walkers along an absorbing wall in an infinite system.

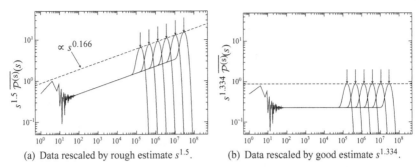

(a) Data rescaled by rough estimate $s^{1.5}$.   (b) Data rescaled by good estimate $s^{1.334}$.

Figure 7.11 (a) In a first step towards a data collapse, the binned data for different system sizes, $L = 3125, 6250, \ldots,$ $100\,000$, are rescaled using a rough estimate of the exponent (here $\tau' = 1.5$) characterising the intermediate range, see Fig. 7.10(a). The arrows mark a characteristic scale (bump) in each graph. The slope of the dashed line connecting the landmarks determines the correct scaling exponent used in (b), which shows the same data rescaled by $s^{1.334}$. The horizontal positions of the maxima have changed only very slightly compared to (a) and are marked by arrows. The peaks of the landmarks are roughly equally high (dashed line). The vertical scale of both graphs is the same for illustration purposes only.

maximum, which typically scales with a higher power than the correct cutoff. In the example used here, Fig. 7.3, the theoretical maximum event size is $a_0 L + L(L + 1)$ (on the lattice), which scales quadratically and thus with the wrong power compared to the correct upper cutoff, $s_c \propto L^{3/2}$, and which occurs with a ridiculously small probability, $(D/4)^L \leq (1/4)^L$. In any case, this estimator of the upper cutoff depends on the sample size. Even when it is not an average, it can be considered as biased and might even be inconsistent, if the PDF to be characterised has infinite support, so that the maximum typically found in a finite sample increases indefinitely with the sample size.

This points to a general problem when estimating any observable that depends on the upper cutoff, such as most moments. In some cases, the upper cutoff is not determined by the system itself, but by some technical constraint, such as a maximum value beyond which the PDF under consideration cannot be probed. For example, in the residence time distribution of the OSLO Model, Sec. 6.3.3, the upper cutoff is detected only if the simulation is run long enough. Even when it is detected, all moments derived are bound to be finite due to a finite simulation time, while the scaling suggests that no upper cutoff exists, so that all moments $\langle R^n \rangle$ with $n > \alpha - 1 \approx 1.2$ diverge.

Realistically, the maximum event size is not too far off the cutoff, as illustrated in Fig. 7.14. Fixing the sample size N, one might crudely approximate [7.3] the maximum by assuming that, in a system with discrete event sizes, it is produced once in the entire sample, so $\mathcal{P}^{(s)}(s^*) = 1/N$. Equating s^* and s_c overestimates small cutoffs and underestimates large ones. If the terminating bump of the PDF, i.e. the scaling function, is not properly resolved, $s_c = s^*$ might seriously underestimate the cutoff. In this case it might actually converge to a finite value even for increasing (actual) upper cutoff and not generally increase monotonically with it.

So far, the procedures described above have produced an estimate only of the avalanche exponent τ in the form τ'. As shown in Fig. 7.12(b), plotting the position of the landmarks

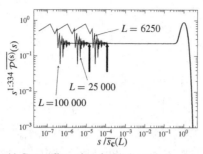
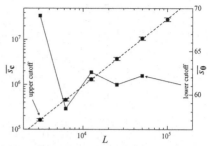

(a) Data collapse in a double logarithmic plot.   (b) Scaling of upper and lower cutoffs.

Figure 7.12 (a) Data collapse for three different system sizes, by rescaling the data and presenting them in a double logarithmic plot to assist inspection of the power law region and the scaling function. Because small arguments are resolved in detail, the different system sizes, $L = 6250, 25\,000, 100\,000$, are clearly distinguishable in that region. The estimated position of the lower cutoff is marked by a bold arrow from below (for clearer visibility three system sizes have been removed with respect to Fig. 7.11(a)). (b) Double logarithmic plot of the estimated upper and lower cutoffs, $\overline{s_c}$ and $\overline{s_0}$, versus the system size L estimated from Fig. 7.11(b) and Fig. 7.12(a) respectively. The right hand side ordinate shows the estimated value of the lower cutoffs $\overline{s_0}$ (squares connected by straight lines), the left hand side ordinate shows the estimated upper cutoffs $\overline{s_c}$ (circles). The dashed line shows a fit to a power law, $s_c \propto L^{1.5}$, i.e. $D \approx 1.5$.

$\overline{s_c}(L)$ versus the system size L (or any suspected control parameter) in a double logarithmic plot reveals, to leading order, the exponent D, since $\overline{s_c} \propto L^D$. Following the same procedure as for moments (see Sec. 7.3), the cutoff can be analysed in further detail, in particular by allowing for corrections. However, the uncertainty in the position of the upper cutoff is given by the bin size (shown as the error bar in Fig. 7.12(b)), which is large towards large s, and reducing their size increases the noise in the data, which in turn increases the (curvature dependent) uncertainty in the location of the landmark, a local maximum. In general, a data collapse has limited quantitative power, not least because the error in the collapse itself is difficult to quantify (Bhattacharjee and Seno, 2001). Figure 7.12(b) also shows the absence of scaling in the lower cutoffs (estimated from Fig. 7.12(a)), which remains at around the same small value $\overline{s_0} \approx 65$.

The data collapse described above is not unique. Plotting $\overline{\mathcal{P}^{(s)}}(s)\, s^\tau f(s/L^D)$ as a function of $x = s/L^D$ or versus $g(s/L^D)$ produces a collapse for any two functions $f(x)$ and $g(x)$, namely to $\mathcal{G}^{(s)}(x) f(x)$ versus $g(x)$. For example, an easy way to probe for the presence of a Gaussian is to plot $\mathcal{P}^{(s)}(s)\sqrt{\sigma^2(s)}$ against $(s - \langle s \rangle)/\sqrt{\sigma^2(s)}$ (e.g. Lise, 2002), which produces a collapse for any PDF with $\tau = 1$ and $\sqrt{\sigma^2(s)} \propto L^D$. In this case $f(x) = (c_1 x)^{-1}$ and $g(x) = c_1 x - c_2$ with $x = s/L^D$ and constants $c_1 = L^D/\sqrt{\sigma^2(s)}$ and $c_2 = \langle s \rangle/\sqrt{\sigma^2(s)}$. That a Gaussian has $\tau = 1$ (also Sec. 2.1.3) can be seen by writing it in the form[24]

$$\mathcal{P}^{(s)}(s) = s^{-1}\frac{s}{\sqrt{2\pi\sigma^2}}e^{-\frac{1}{2}\left(\frac{s}{\sqrt{\sigma^2}}-\alpha\right)^2} \tag{7.47}$$

---

[24] Equation (7.47) allows for negative event sizes s, as opposed to Eq. (2.15) on p. 35. For large α the difference between the two is insignificant.

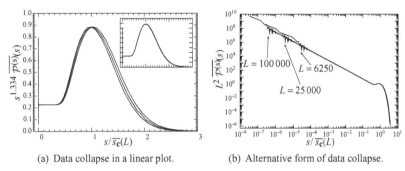

(a) Data collapse in a linear plot.   (b) Alternative form of data collapse.

Figure 7.13 (a) The same data as shown in Fig. 7.12(a) in a linear plot. The bump towards large arguments is better resolved, whereas all details for small arguments as well as most of the intermediate power law region disappear in a single vertical line. In this view, the collapse can be further improved by tweaking the estimates for the upper cutoff $\overline{s_c}$, as shown in the inset. (b) The same data as shown in (a), but rescaled only by powers of L. Many details of the scaling function drown in the large range the graph covers on the ordinate.

with constant dimensionless α, average $\langle s \rangle = \alpha \sqrt{\sigma^2}$ and variance $\sigma^2(s) = \sigma^2$, so that $\sqrt{\sigma^2}$ effectively corresponds to s_c, the only scale characterising the PDF. If the average $\langle s \rangle$ does not scale like $\sqrt{\sigma^2(s)}$, i.e. in the presence of a second, independent scale, the Gaussian does not follow simple scaling, Eq. (7.41), yet plotting $\mathcal{P}^{(s)}(s)\sqrt{\sigma^2(s)}$ versus $(s - \langle s \rangle)/\sqrt{\sigma^2(s)}$ still produces a collapse. In a linear-log plot, the Gaussian becomes an upside-down parabola. Multiplying by $\sqrt{2\pi}\exp(x^2/2)$ should give 1, but with very noisy tails that are best investigated in a double logarithmic plot; subtracting $\exp(-x^2/2)/\sqrt{2\pi}$ should give 0, best scrutinised on a linear scale.

Some authors choose $f(x) = x^{-\tau}$ and plot $\overline{\mathcal{P}^{(s)}}(s)\,L^{D\tau}$ versus $g(x) = x = s/L^D$, which, however, typically blurs the details of the collapse in a steep decline of the graph, see Fig. 7.13(b). The resulting collapse shows $\mathcal{G}^{(s)}(x)x^{-\tau}$, reducing the weight in the tail of the distribution. A power law $f(x)$ is useful where the scaling function is itself a power law, in particular where $\tau = 1$, see Sec. 2.1.2.1 and Sec. 6.3.3. The smaller the spread of the resulting data across the ordinate, the more details of the collapse are visible. In general, to assess the validity of the assumption, data are best plotted so that perfect agreement leads to no spread, so that deviations from the assumption become deviations from a straight line. A central part of a data collapse is the power law behaviour in the intermediate region, which can be inspected in great detail in a plot like Fig. 7.12(a). If, however, the focus is the structure of the scaling function, the stretch of the abscissa at small arguments in a double logarithmic plot is not helpful and a linear plot as shown in Fig. 7.13(a) is more appropriate, offering a more direct way to approximate the scaling function by visual inspection. Using the improved resolution of the scaling function, the collapse can be further refined by changing the estimates for the upper cutoff.

Similarly, derivatives and, much more frequently, integrals of the PDF are studied, in particular the complementary cumulative distribution $\hat{F}^{(s)}(s)$, the probability that an avalanche

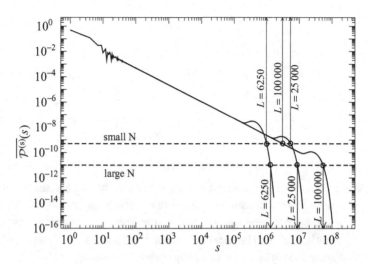

Figure 7.14 The maximum event size (area under a random walker trajectory) observed is approximated by s^* with $\mathcal{P}^{(s)}(s^*) = 1/N$, the inverse sample size. The dashed lines show two different choices for N and the intersection (circles) of the PDF with the dashed lines indicate the resulting s^* (indicated by arrows). The lower dashed line (large N) produces good estimates for the upper cutoff, whereas the upper dashed line produces distorted results, with the larger systems ($L = 100\,000$) even producing an upper cutoff smaller than that of a smaller system ($L = 25\,000$). All systems larger than $L = 100\,000$ would produce similar cutoffs.

will be larger than s:

$$
\hat{F}^{(s)}(s) = \int_s^\infty \mathrm{d}s'\, \mathcal{P}^{(s)}(s')
$$

$$
= \int_s^\infty \mathrm{d}s'\, a^{(s)} s'^{-\tau} \mathcal{G}^{(s)}(s'/s_c)
$$

$$
= a^{(s)} s^{1-\tau} \int_1^\infty \mathrm{d}x\, x^{-\tau} \mathcal{G}(xs/s_c) \tag{7.48}
$$

using Eq. (7.41) and assuming, for simplicity, $s > s_0$. The resulting cumulative distribution can be written as $\hat{F}^{(s)}(s) = as^{1-\tau}\mathcal{F}(s/(s_c))$, similar to the original PDF. While corrections can in principle be smaller for the cumulative PDF, its main advantage is its smaller spread across the ordinate for direct inspection, in particular when $\tau \approx 1$. This is probably achieved in a more controlled fashion by rescaling as done in Fig. 7.11(b).

Finally, it can be very helpful, and is popular in particular among experimentalists, to analyse the data of a 'de-dimensionalised' histogram (e.g. Planet, Santucci, and Ortín, 2009). This, however, can forestall the hunt for exponents, if it is done by dividing the observable s by its average or standard deviation, say $z = s/\langle s \rangle$ (or $z = (s - \langle s \rangle)/\sqrt{\sigma^2(s)}$). Any data collapse of $\mathcal{P}^{(z)}(z)$ versus z implies $\tau = 1$, which can be shown considering $\mathcal{P}^{(z)}(z)\,\mathrm{d}z = \mathcal{P}^{(s)}(s)\,\mathrm{d}s$ and assuming simple scaling for $\mathcal{P}^{(s)}(s)$ (Planet, Santucci, and Ortín, 2010; Pruessner, 2010).

PART III

THEORY

Analytical results

In this chapter, some important analytical techniques and results are discussed. The first two sections are concerned with mean-field theory, which is routinely applied in SOC, and renormalisation which has had a number of celebrated successes in SOC. As discussed in Sec. 8.3, Dhar (1990a) famously translated the set of rules governing an SOC model into operators, which provides a completely different, namely algebraic perspective. Directed models, discussed in Sec. 8.4, offer a rich basis of exactly solvable models for the analytical methods discussed in this chapter. In the final section, Sec. 8.5, SOC is translated into the language of the theory of interfaces.

It is interesting to review the variety of theoretical languages that SOC models have been cast in. Mean-field theories express SOC models (almost) at the level of updating **rules** and thus more or less explicitly in terms of a **master equation**. The same applies for some of the renormalisation group procedures (Vespignani, Zapperi, and Loreto, 1997), although Díaz-Guilera (1992) suggested very early an **equation of motion** of the local particle density in the form of a Langevin equation. The language of **interfaces** overlaps with this perspective in the case of Hwa and Kardar's (1989a) **surface evolution** equations, whereas the **absorbing state (AS) approach** as well as **depinning** use a similar formalism but a different physical interpretation – what evolves in the former case is the configuration of the system, while it is the number of charges in the latter. Dhar's truly original operator approach has little in common with any other technique, except perhaps the depinning representation and its boundary conditions, Sec. 8.5.3.

8.1 Random walkers and mean-field approaches

Random walkers (reviews for example by Fisher, 1984; ben-Avraham and Havlin, 2000; Redner, 2001; Rudnick and Gaspari, 2004) play an important rôle in SOC, firstly because in some models, such as the MANNA Model model, 'particles'[1] actually perform **random walks** (Kadanoff *et al.*, 1989; Dhar, 1990a; Grassberger and Manna, 1990), for which *exact* results are easily available. Secondly, in many models which lack spatial correlations, such as models on the Bethe lattice (Dhar and Majumdar, 1990), the directed sandpile model (Sec. 8.4) or the TAOM, **spatial structures**, which can be considered as interfaces, can be mapped to trajectories of random walkers. Thirdly, on a more abstract level, taking a

---

[1] In the following any randomly moving constituent of a model is called a 'particle', even when they are termed differently in the context of the specific model.

Figure 8.1 The local mean square displacement per microscopic time step, i.e. the local (effective) diffusion constant, in the MANNA Model. Since a particle moves by one lattice spacing in every toppling, the diffusion constant is solely determined by the toppling frequency, larger values of which are indicated a lighter greyscale. Four trajectories of particles injected at random positions are shown as black curves.

mean-field approach and ignoring correlations, **observables** such as the number of toppling sites can often be cast in the language of random walks.

Due to the lack of correlations, random walkers are particularly simple to handle analytically from various angles; they are a stochastic process, their equation of motion is a Langevin equation with additive noise and their Fokker–Planck equation is the heat equation. Normally, the main difficulty is implementing the initial and boundary conditions, the latter, in the form of absorbing and reflecting walls, are usually most easily applied to a Fokker–Planck formalism.

In the following the three perspectives are briefly put in the context of the relevant models. Some key results for random walkers are presented and the closely related branching process discussed in some detail.

8.1.1 Particles performing random walks

The MANNA Model is the only major model where all particles *independently* perform a random walk. Marking any particle in the system as a **tracer** (see Fig. 8.1), it will depart from a site to a nearest neighbouring site irrespective of the movement of any other particle, i.e. the target site is chosen at random and independently. Unless there is a deliberate anisotropy, target sites are chosen with uniform probability. While the resulting trajectory is that of a random walker, particles might become stuck at individual sites for varying lengths of time. There is no unique time scale such that all particles in the system move around with the same effective diffusion constant, [2] see Fig. 8.1. In fact, their momentary

---

[2] Random walks are discrete and strictly a diffusion constant characterises only continuum Brownian motion. Taking the continuum limit, the variance per time step σ^2 of the displacement of a random walker in d dimensions is related to the effective diffusion constant D by $2dD = \sigma^2$.

mean square displacement per microscopic time step is correlated in time and space due to the instantaneous location of the avalanche.

8.1.1.1 Average avalanche size

The number of moves a particle performs from entering to leaving the system, its **escape or residence time**, is *exactly* the total number of times this particle has been *removed* from a site. The avalanche size is the **number of topplings** and assuming that the ratio $\tilde{q}$ of removals per toppling is fixed because of particle conservation in the stationary state,[3] the average number of topplings per particle added, i.e. the average avalanche size $\langle s \rangle$ (including avalanches of vanishing size), is exactly proportional to the average number of moves $\langle m \rangle$ by each particle added, $\langle s \rangle = \langle m \rangle / \tilde{q}$. As long as the fraction of non-zero avalanches does not vanish (like a power law) the scaling of the average over all avalanches considered in the following equals that over non-zero avalanches. Introducing a local time for each particle, which ticks whenever the particle moves (Nakanishi and Sneppen, 1997), the average avalanche size is the average residence time divided by $\tilde{q}$. Higher moments of the avalanche size distribution cannot be derived, because no information is available about the *collective* movements of random walkers that make up an avalanche. This is also the reason why the random walks can be independent when the avalanches generally are not. Both sample completely different ensembles and only the average avalanche size and the average escape time happen to be proportional.

In models where particles do not move independently, individual particles might still perform a random walk, for example in the Abelian BTW Model,[4] where the toppling procedure can be implemented in a way that guarantees random and history-independent movement for any individual particle. For example, at toppling, particles could be reshuffled randomly before being distributed amongst the neighbouring sites. Regardless of the origin of an individual particle, at each toppling it takes part in, it will arrive at a randomly chosen site. In that way, individual trajectories of particles, even in the BTW Model or the OSLO Model, are fair random walks, with independent, unbiased moves. However, in these models, where particles are distributed evenly among neighbours, any two trajectories are not independent; if two tracer particles happen to be removed from the same site at the same time, they cannot move to the same site (Nakanishi and Sneppen, 1997).

In principle, such a constraint precludes the random walker mapping. Yet, models such as the OSLO Model and the Abelian BTW Model indeed generate all possible trajectories with the correct weight (Shilo and Biham, 2003). In each toppling, one of the trajectories passing through the toppling site can be thought of as being extended in all possible directions,

---

[3] The number of removals per toppling is the coordination number if one particle is transferred to each nearest neighbour at a toppling and generally equals the number of particles discharged at a toppling. Up to boundary effects (or bulk dissipation), the total number of removals (the number of times particles have been (re)moved or the sum of the number of particles removed) equals the total number of (internal) charges. As particles move simultaneously during an avalanche, not much can be said about the avalanche duration except that each is at most as large as the avalanche size.

[4] ... but not the original BTW Model or any other model with non-Abelian driving, such as the oOM when driven in the bulk, where slope units can disappear when neighbouring sites are charged and the movement of *grains* (rather than slope units) is downhill biased.

with trajectories terminating only once they cross the boundary of the lattice. In this way, whenever a site topples, a trajectory branches into as many new ones as there are sites being charged. Trajectories start with a particle being deposited at a site by the external drive.

On a more formal level, the average number of moves $\langle m \rangle (\mathbf{n})$ a particle makes between arriving at site $\mathbf{n}$ and leaving the system fulfils the (self-consistency) equation

$$\langle m \rangle (\mathbf{n}) = 1 - \sum_{\mathbf{n}' \neq \mathbf{n}} \frac{\Delta_{\mathbf{n}\mathbf{n}'}}{\Delta_{\mathbf{n}\mathbf{n}}} \langle m \rangle (\mathbf{n}'), \tag{8.1}$$

where $\Delta_{\mathbf{n}\mathbf{n}} \geq -\sum_{\mathbf{n}' \neq \mathbf{n}} \Delta_{\mathbf{n}\mathbf{n}'}$ is the total number of particles removed from toppling site $\mathbf{n}$ (corresponding to $\tilde{q}$ above) and $-\Delta_{\mathbf{n}\mathbf{n}'}$ is the number of particles received at site $\mathbf{n}'$, in the notation of the ASM (Sec. 4.2, p. 91). For conservative relaxations $\Delta_{\mathbf{n}\mathbf{n}} = -\sum_{\mathbf{n}'} \Delta_{\mathbf{n}\mathbf{n}'}$, but $\Delta_{\mathbf{n}\mathbf{n}} > -\sum_{\mathbf{n}'} \Delta_{\mathbf{n}\mathbf{n}'}$ where $\mathbf{n}$ is a boundary site, as if there were a receiving site $\mathbf{m}$ so that $-\sum_{\mathbf{n}'} \Delta_{\mathbf{n}\mathbf{n}'} - \Delta_{\mathbf{n}\mathbf{m}} = \Delta_{\mathbf{n}\mathbf{n}}$ but with $\langle m \rangle (\mathbf{m}) = 0$. The sum in Eq. (8.1) accounts for the moves made away from the receiving sites which are reached by making exactly one move. Equation (8.1) can be simplified further to $\sum_{\mathbf{n}'} \Delta_{\mathbf{n}\mathbf{n}'} \langle m \rangle (\mathbf{n}') = \Delta_{\mathbf{n}\mathbf{n}}$, similar to Eq. (8.8).

The precise meaning of the averaging procedure that gives rise to $\langle m \rangle (\mathbf{n})$ is of fundamental importance. Firstly, the average gives equal weight to all particles, which is why each (relaxing) site contributes with weight $-\Delta_{\mathbf{n}\mathbf{n}'}/\Delta_{\mathbf{n}\mathbf{n}}$. Secondly, and more subtly, the averaging accounts for the driving regardless of whether it is stochastic or deterministic; in principle, $\langle m \rangle (\mathbf{n})$ could depend on how the system is driven (on which site or how a stochastic driving is distributed), even when Eq. (8.1), given $\mathbf{n}$, applies to any drive. Generally, however, $\langle m \rangle (\mathbf{n})$ is a matter of a mere counting exercise (see above) and the origin of particles leading to topplings that eventually wash out a particle after a number of moves is irrelevant.

In one dimension it is easy to determine $\langle m \rangle (\mathbf{n})$ (Shilo and Biham, 2003; Malcai et al., 2006, for two dimensions). It is the average number of moves required to escape after entering at $\mathbf{n}$, which takes one move to a neighbour and subsequent escape from there, $\langle m \rangle (\mathbf{n}) = 1 + (\langle m \rangle (\mathbf{n}+1) + \langle m \rangle (\mathbf{n}-1))/2$, with boundary conditions $\langle m \rangle (0) = \langle m \rangle (L+1) = 0$ (particles are lost there, known as open, fixed, absorbing or Dirichlet boundary conditions), which is indeed Eq. (8.1) in one dimension. The solution is $\langle m \rangle (\mathbf{n}) = (L + 1 - \mathbf{n})\mathbf{n}$, which applies to random walkers (as in the MANNA Model) as well as to models with deterministic 'Laplacian' moves, such as the sOM and the Abelian BTW Model, which is governed by a toppling matrix $\Delta_{\mathbf{n}\mathbf{n}'}$, Box 4.2, p. 91, that is essentially a **lattice Laplacian** (Dhar, 1990a; Majumdar and Dhar, 1991). In these models, where $\tilde{q} = 2$, the expected size for an avalanche started at $\mathbf{n}$ is therefore

$$\langle s \rangle (\mathbf{n}) = \frac{1}{2} \langle m \rangle (\mathbf{n}) = \frac{1}{2}(L + 1 - \mathbf{n})\mathbf{n} \tag{8.2}$$

regardless of where and how the system is driven otherwise.[5] Solutions for other lattices can be found in the literature (e.g. Spitzer, 2001). When the number of removals per toppling $\tilde{q}$ is not a constant throughout the lattice, converting the number of moves to the number of topplings, i.e. the avalanche size, can be difficult. However, provided that the

---

[5] At least in stochastic models, Eq. (8.2) is derived without any reference to the driving as such, which might take place at locations different from most $\mathbf{n}$ for which it can be evaluated.

ratio converges to a non-vanishing value in the thermodynamic limit, the exponent $\sigma_1^{(s)}$, $\langle s \rangle \propto L^{\sigma_1^{(s)}}$, can be derived easily. If gap scaling applies, $D(2 - \tau) = \sigma_1^{(s)}$ amounts to a **scaling law** and $\tau \geq 1$ implies $\sigma_1^{(s)} \leq D$ (Sec. 2.1.2.2).

Exact expressions for the average avalanche size generally depend on the driving, the dimensionality and the boundary conditions (further details in Stapleton, 2007). The above result for $\langle m \rangle$ in one dimension and thus for $\langle s \rangle$ is easily extended to a right reflecting (closed, Neumann) boundary condition (BC), $\langle m \rangle (L + 1) = \langle m \rangle (L)$, which means that slope units cannot escape from the system when $\mathbf{n} = L$ topples. In that case

$$\langle s \rangle (\mathbf{n}) = \frac{1}{2}(2L + 1 - \mathbf{n})\mathbf{n} \tag{8.3}$$

by extending the system by L sites to the right of $\mathbf{n} = L$, so that $\langle m \rangle (L-\mathbf{n}) = \langle m \rangle (L+1+\mathbf{n})$, mirroring the actual system of size L. With its particular boundary condition, Eq. (8.3) can also be understood on the basis of Fig. 4.3, p. 93, where adding a unit of slope in the bulk adds one unit of height at each $\mathbf{n}' = 1, 2, \ldots, \mathbf{n}$, which each topple $L + 1 - \mathbf{n}'$ times, i.e. $\sum_{\mathbf{n}'=1}^{\mathbf{n}} (L + 1 - \mathbf{n}') = (2L + 1 - \mathbf{n})\mathbf{n}/2$ (see also endnote [6.2], p. 393).

In particular, for various models in one dimension driven at site $\mathbf{n}_0$

$$\langle s \rangle = \begin{cases} \dfrac{1}{2}L & \mathbf{n}_0 = 1, \text{ open BC (Abelian BTW} \tag{8.4a} \\ & \text{Model, boundary driven sOM)} \\[6pt] L & \mathbf{n}_0 = 1, \text{ left BC open, right BC} \tag{8.4b} \\ & \text{closed (original BTW Model, oOM)} \\[6pt] \dfrac{1}{12}(L + 2)(L + 1) & \mathbf{n}_0 \text{ random, open BC (\textsc{Manna}} \tag{8.4c} \\ & \text{Model, bulk driven sOM),} \end{cases}$$

where Eq. (8.4c) is the average over $\mathbf{n}$ of Eq. (8.2). The scaling of the first moment can be read off:

$$\sigma_1^{(s)} = \begin{cases} 1 & \mathbf{n}_0 = 1, \text{ open BC (Abelian BTW Model, boundary} \tag{8.5a} \\ & \text{driven sOM)} \\[6pt] 1 & \mathbf{n}_0 = 1, \text{ left BC open, right BC closed (original} \tag{8.5b} \\ & \text{BTW Model, oOM)} \\[6pt] 2 & \mathbf{n}_0 \text{ random, open BC (\textsc{Manna} Model, bulk driven} \tag{8.5c} \\ & \text{sOM).} \end{cases}$$

The mapping of the various boundary conditions when translating between slope units (particles) and height units (grains) is crucial for the correct interpretation of the above results in the various models (see p. 94). Equation (8.4a) is derived on p. 189 for the sOM, Eq. (8.4b) on p. 198 for the oOM and Eq. (8.4c) for the \textsc{Manna} Model, Eq. (6.1) on p. 165. Results for the one-dimensional Abelian BTW Model are somewhat trivial, as the model is fully deterministic, see p. 89. If $\mathbf{n}_0 \neq 1$, the results above are not applicable to models with non-Abelian drive, because the bulk removal of slope units in the latter terminates trajectories in the bulk. In the one-dimensional original BTW Model $\sigma_1^{(s)} = 1$ results even for bulk drive (see p. 198 for the oOM), as if the model were directed and the particles were subject to a drift (Sec. 8.4.3, Christensen and Olami, 1993; Bonachela, 2008, p. 99).

If one is only interested in the scaling exponent of the first moment, $\sigma_1^{(s)}$, the above results can easily be generalised to higher dimensions (Christensen and Olami, 1993). If the driving position has d_0 coordinates whose (average) distance to an open boundary is a constant proportional to x_0 in a system of size L, the escape time of the random walker scales in L like $L^{2-d_0} x_0^{d_0}$ for $d_0 = 0, 1$, logarithmically for $d_0 = 2$ and converges to a constant if $d_0 > 2$ (Nakanishi and Sneppen, 1997). In two dimensions d_0 is more generally π/θ, where θ is the angle of aperture at the driving position (Ivashkevich et al., 1994a). In summary [6]

$$
\sigma_1^{(s)} = \begin{cases} 2 - d_0 & \text{for } d_0 < 2 & (8.6a) \\ 0 \text{ (logarithmically)} & \text{for } d_0 = 2 & (8.6b) \\ 0 \text{ (constant)} & \text{for } d_0 > 2 . & (8.6c) \end{cases}
$$

This is in line with the findings above. For example, according to Eq. (8.2), in one dimension, if the driving position $\mathbf{n}_0$ is fixed, then to leading order $\langle s \rangle \approx L\mathbf{n}_0/2$. If, on the other hand, $\mathbf{n}_0 = \alpha L$ for some fixed $1 \geq \alpha > 0$, then $\langle s \rangle \approx L^2 (1 - \alpha)\alpha/2$. If $\mathbf{n}_0$ is chosen at random with uniform probability or is proportional to L, then α is constant on average and thus $\sigma_1^{(s)} = 2$. If, however, its probability is biased and scales in L, or if $\mathbf{n}_0$ scales in L, any exponent $1 \leq \sigma_1^{(s)} \leq 2$ is attainable (Christensen, Moloney, Peters, and Pruessner, 2004) and assuming that D in $\sigma_1^{(s)} = D(2 - \tau)$ is unaffected (Nakanishi and Sneppen, 1997), τ becomes a matter of choice.

The boundary condition next to the driven site is crucial. While $\mathbf{n}_0 = L$ in Eq. (8.2) gives $\sigma_1^{(s)} = 1$, Eq. (8.3) gives $\sigma_1^{(s)} = 2$, as this system is effectively still driven in the bulk. In this case the term 'boundary drive' can be very misleading.

An equation similar to Eq. (8.1), but based on the expected number of topplings rather than the expected number of moves, can be derived using a slightly different reasoning (Dhar, 1990a; Stapleton, 2007). If $G_{\mathbf{nm}}$ is the expected number [7] of topplings of site $\mathbf{m}$ due to a particle added at $\mathbf{n}$, the average number of charges received by site $\mathbf{m}$ after charging site $\mathbf{n}$ is $-\sum_{\mathbf{n}' \neq \mathbf{m}} G_{\mathbf{nn}'} \Delta_{\mathbf{n}'\mathbf{m}}$. Whenever that site topples, it loses $\Delta_{\mathbf{mm}}$ particles. By particle conservation in the stationary state, at every site the particle outflow must balance the inflow including the driving at site $\mathbf{m}$, so that

$$
G_{\mathbf{nm}} \Delta_{\mathbf{mm}} = - \sum_{\mathbf{n}' \neq \mathbf{m}} G_{\mathbf{nn}'} \Delta_{\mathbf{n}'\mathbf{m}} + \delta_{\mathbf{n},\mathbf{m}} \tag{8.7}
$$

or using matrix notation

$$
G\Delta = \mathbf{1}. \tag{8.8}
$$

The average avalanche size due to a charge at $\mathbf{n}$ is given by $\langle s \rangle (\mathbf{n}) = \sum_{\mathbf{m}} G_{\mathbf{nm}}$. More generally, if $|1\rangle$ is a column vector of ones and $\langle p|$ is a row of probabilities of charging each site, then

$$
\langle s \rangle = \langle p| G |1\rangle. \tag{8.9}
$$

---

[6] In a fractal, the escape time of a random walker from the bulk no longer scales like L^2 but rather like L^{d_w}, so for bulk drive $\sigma_1^{(s)} = d_w$ with $d_w \neq 2$ (ben-Avraham and Havlin, 2000).

[7] Even when $G_{\mathbf{nm}}$ is the average response at $\mathbf{m}$ after driving the system (once) at $\mathbf{n}$, as pointed out after Eq. (8.2), the following result is a matter of self-consistency and does not depend on the details of the ensemble, in particular it does not depend on how and where the system is driven otherwise.

In the Abelian BTW Model, Δ is the lattice Laplacian (see p. 94) with prefactor $-2d$, so that the solution of Eq. (8.8) is essentially its inverse, the **lattice Green function** (Spitzer, 2001, in the continuum Butkovskiy, 1982). Because Δ, the lattice Laplacian, is little more than the adjacency matrix, the average avalanche size can easily be calculated for many different lattices and networks via Eq. (8.8) and Eq. (8.9), even for stochastic models, because essentially only particle conservation has been used above.

8.1.2 Random walker structures

The first SOC model whose internal structure bore features of a random walker trajectory was Dhar and Ramaswamy's (1989) exactly solvable directed sandpile model. Based on the uniqueness of the succeeding state (see Sec. 8.3.2.2, p. 281) the authors showed that every recurrent state is equally likely, which means that at any point in time, new avalanches form in a 'perfectly random' environment. Each (compact) avalanche is outlined by two **annihilating or vicious random walkers** (Fisher, 1984; ben-Avraham and Havlin, 2000) which depart from the same point until their trajectories intersect, so that the avalanche size is given by the area traced out. The relative motion of two annihilating random walkers is a random walk along an absorbing wall with twice the variance (see Fig. 7.3, p. 223), which is easily shown using generating functions.

As is typical in **directed sandpile models** no spatial correlations develop (Sec. 8.4), but temporal correlations occur in the sequence of avalanche sizes, i.e. on the macroscopic time scale (Welinder *et al.*, 2007). [8] Without spatial correlations, some features of the model can invariably be mapped to a random walk. In fact, the annihilating random walk can also be identified in the TAOM (Pruessner, 2004b) and its variants (Stapleton and Christensen, 2005, 2006), the directed model considered by (Kloster *et al.*, 2001) and a similar class of models discussed by Paczuski and Bassler (2000).

Solving a random walker problem on the lattice (e.g. Eq. (8.14)) can be very tedious (Dhar, 1990a; Shilo and Biham, 2003; Stapleton, 2007, numerically Walsh and Kozak, 1981, 1982; Soler, 1982) and as far as only asymptotic long time and large scale behaviour is concerned, a continuum approach is often more suitable. Some of the pitfalls in taking the continuum limit are discussed below, Sec. 8.1.3. Rescaling time and space naïvely, a random walker with non-vanishing drift will move ballistically with divergent velocity. Alternatively, the effective diffusion constant vanishes and the drift remains finite. The continuum limit is obtained in a more consistent manner by considering a Taylor series expansion of the original difference equation on the lattice. This implies, however, that the continuum limit is generally *not* the *exact* long time and large space limit of the difference equation, as illustrated in Fig. 8.2(a).

Luckily, the relative motion of two equivalent annihilating random walkers contains no drift and therefore does not pose any of these problems. Many features of non-intersecting and annihilating random walks have been studied as models of polymers (Gillet, 2003; Tracy and Widom, 2007), as reaction-diffusion processes (ben-Avraham and Havlin, 2000)

---

[8] The microscopic time in such models usually coincides with a particular spatial direction, as the avalanche advances through a lattice with a well-defined front in only one direction.

and more recently in particular due to their link to extreme value statistics (Majumdar and Comtet, 2004, 2005).

In the following section, random walker trajectories are identified in observables other than spatial structures, for example the number of active sites. In some systems both pictures are present simultaneously. For example, in the TAOM the evolution of the configuration can be mapped to random walk trajectories just as well as the number of charges received by a site. This is not a coincidence: the independence of the local degree of freedom manifested by the mapping to a random walk trajectory suggests that other observables might equally well be described by a random walk. The duality of the random walk mapping, spatial structure on the one hand and less directly accessible observables on the other hand, is also present in descriptions of SOC models in terms of interfaces. Originally their height profile was described by a stochastic equation of motion, i.e. a growth equation, whereas later the local activity, the number of charges received by a site, or the number of times it has toppled was described in this way.

8.1.3 Mean-field theory

Mean-field theories come in many disguises, [8.1] but generically involve ignoring correlations at some level, including spatial fluctuations, where they are shaped by correlations and feedback of a local degree of freedom to itself. In systems, which are supposedly (fully) scale invariant, i.e. which possess algebraically decaying correlations, this has a drastic effect and mean-field theories therefore usually provide rather crude approximations. In SOC, the 'typical' mean-field theory (MFT) produces $\tau = 3/2$ and $\alpha = 2$. In many mean-field theories, finite size scaling is not accessible and where it is, typical values of the exponents are $D = 4$ and $z = 2$ with $D = 2z$ from $D(1 - \tau) = z(1 - \alpha)$, see Sec. 2.2.2. Because finite size is the one fundamental scaling variable in (non-trivial) SOC models, mean-field theories have far less explanatory power in SOC than in ordinary critical phenomena. Furthermore, due to a lack of spatial structure, boundary conditions normally do not enter mean-field theories at all.

If the PDF of a model's state factorises into a product of PDFs of local states, it is said to be in a **product state**. In this case, all connected n-point correlation functions involving at least 2 distinct points in space vanish strictly, because then the n-point cumulant generating function is simply a sum $\sum_i \mathcal{C}(z_i)$, which vanishes when differentiated with respect to any two distinct variables $z_i, z_j, i \neq j$. The same applies, possibly only approximately or asymptotically, where correlations are destroyed explicitly by a mechanism like **perfect mixing** (or similar, Bröker and Grassberger, 1995). In such models, an MFT is often an exact description of the dynamics. Where this is not the case, MFTs can often be improved systematically by inluding some correlations (ben-Avraham and Köhler, 1992; Ferreira and Mendiratta, 1993).

For similar reasons, MFT is generally expected to be the correct asymptotic description at sufficiently high dimensions $d \geq d_c$ (Gaveau and Schulman, 1991; Dhar, 1999a), as spatial correlations decay increasingly quickly with increasing dimensions, so that they can be ignored at sufficiently large distances (see Sec. 2.3.1). Generally, universal logarithmic corrections are expected at the upper critical dimension $d = d_c$ (Lübeck, 1998), while the scaling at dimensions $d > d_c$ can be dominated by the short length scales, i.e. the *lower* cutoff (e.g. Krug, 1997, p. 175) or altered significantly due to the presence of dangerously irrelevant variables (Ma, 1976, for example Pruessner, 2004a).

Even when all spatial correlations can be ignored, an MFT might still be hard to come by. The two main caveats both relate to temporal correlations, which might play an important rôle. Firstly, on the macroscopic time scale, **successive avalanches** can be correlated, as for example in the TAOM (Welinder *et al.*, 2007). Secondly, on the microscopic time scale, **finite size** frequently means that the PDF for finding a constituent in a particular state changes during the course of an avalanche, as the system evolves and constituents are visited multiple times. The latter is often neglected, i.e. the same (stationary) PDF is assumed throughout an avalanche, even when the effect of multiple visits is not exponentially weak. [8.2]

Tang and Bak (1988b) were the first to carry out an MFT-type calculation for the BTW Model, focusing on the distribution of local slopes in the stationary state, and various MFT schemes for virtually all SOC models have been developed since then (Alstrøm, 1988; Dhar and Majumdar, 1990; Gaveau and Schulman, 1991; Christensen *et al.*, 1993; Flyvbjerg *et al.*, 1993; Janowsky and Laberge, 1993; Zapperi *et al.*, 1995; Katori and Kobayashi, 1996; Bröker and Grassberger, 1997; Vergeles *et al.*, 1997; Vespignani and Zapperi, 1997, 1998; Lübeck, 2003; Yang, 2004), some being apparently blissfully unaware of each other, but many with fundamentally different perspectives and levels of approximations. [9] The 'dynamical MFT' of Vespignani and Zapperi (1997, 1998), operating on the level of the **master equation**, deserves particular mention for its breadth and the scope it provided for understanding the mechanism underlying SOC.

The MFT of the BS Model (Flyvbjerg *et al.*, 1993, for the BS Model itself see Sec. 5.4) can be used as a template for the MFT of most other SOC models, and is discussed in detail in the following. Suppressing *all* spatial correlations in the BS Model means that when a site topples, it charges a 'neighbouring' site which is chosen at random. Assuming in addition infinite system size, sites are never revisited. [10] Never returning activity and the absence of **backward avalanches** are the signature of a lack of (spatial) correlations.

In the infinite, random neighbour BS Model the distribution of fitnesses cannot change after a finite avalanche so that the fitness of a neighbour is effectively drawn from the same stationary distribution each time it is considered. Flyvbjerg *et al.* (1993) managed to avoid this additional assumption to some extent and similarly de Boer *et al.* (1994, also de Boer *et al.*, 1995) studied finite systems, illustrating that an MFT can have multiple topplings and returning activity (in contrast to Paczuski and Bassler, 2000).

The mapping of the avalanche size to the **survival time of a random walker** (Paczuski *et al.*, 1996) and of its duration to the (generational) **extinction time of a branching process** (Harris, 1963; Alstrøm, 1988), is a generic feature of mean-field theories of SOC models, or, more specifically, of infinite random neighbour SOC models. In the following, the f_0-avalanche size distribution in an infinite BS Model is derived along the lines of the

---

[9] Still, 'a mean-field theory is a mean-field theory is a mean-field theory' (Schotte, 1999, but it surely can be non-Gaussian, Brézin and Zinn-Justin, 1985; Luijten and Blöte, 1995).

[10] Paczuski and Bassler (2000) attributed this to the absence of **multiple or recurrent topplings**, but showed that multiple topplings (due to multiple discharges) do not imply sites being revisited and are thus compatible with the absence of correlations.

exposition by Rittel (2000). Assuming that at time $t = 0$ the barriers are uniformly and randomly distributed in the interval $[f_0, 1]$, the site with the lowest barrier is located and updated together with $K - 1$ other, distinct sites. [11] On square lattices with nearest neighbour interaction $K = 2d + 1$, whereas in the following $K = 2$.

As mentioned above, because the system is assumed to be infinitely large, the probability that a site is revisited, i.e. a new barrier chosen for a site that has barrier height $f_n < f_0$ already, vanishes, so that *correlations can be neglected*. If there are any **active** sites **n**, i.e. sites with barrier height $f_n < f_0$, the lowest barrier is located and updated together with one freshly chosen random neighbour. Since sites never interact, the order of updates is irrelevant and therefore their actual barrier height does not enter either. For every updated site, all that matters is whether or not the new barrier height is above or below f_0. Strictly speaking, in a finite system, the freshly chosen random neighbour might be active already, i.e. $f_n < f_0$, so that its update, even when still active, $f_n < f_0$, afterwards, might not actually increase the number of active sites. This effect is neglected in the following.

Since $f_n \in [0, 1]$, the probability that a newly assigned barrier height is below f_0 is f_0 and therefore the number $a(t)$ of active sites evolves according to

$$a(t + 1) = a(t) - 1 \qquad \text{with probability } (1 - f_0)^2 \qquad (8.10a)$$

$$a(t + 1) = a(t) \qquad \text{with probability } 2f_0(1 - f_0) \qquad (8.10b)$$

$$a(t + 1) = a(t) + 1 \qquad \text{with probability } f_0^2. \qquad (8.10c)$$

Initially $a(0) = 1$ and the f_0-avalanche terminates after t_x updates, as soon as $a(t_x) = 0$, i.e. when no active site is left. According to Eq. (8.10), the **activity** $a(t)$ performs a random walk along an absorbing wall, [12] as captured in the master equation for the probability $P(a, t)$ of observing activity $a \geq 1$ at time $t \geq 0$,

$$P(a, t + 1) = (1 - f_0)^2 P(a + 1, t) + 2f_0(1 - f_0)P(a, t) + f_0^2 P(a - 1, t). \qquad (8.11)$$

With initial condition $P(a, 0) = \delta_{a,a_0}, a_0 = 1$ and Dirichlet boundary condition $P(0, t) = 0$, this difference equation, Eq. (8.11), can be solved using the method of image charges for the discrete case considered here, as well as in the continuum limit. To do so, one first gauges away the drift, which is due to the difference between $1 - f_0$ and f_0, by defining

$$P(a, t) = \left(\frac{f_0}{1 - f_0} \right)^{a-1} \tilde{P}(a, t), \qquad (8.12)$$

which leaves the boundary and the initial conditions intact, $\tilde{P}(0, t) = 0$ and $\tilde{P}(a, 0) = \delta_{a,1}$. The **Green function** $\tilde{P}_0(a, t)$ with $\tilde{P}_0(a, 0) = \delta_{a,0}$ of

$$\tilde{P}_0(a, t + 1) = f_0(1 - f_0) \left(\tilde{P}_0(a + 1, t) + 2\tilde{P}_0(a, t) + \tilde{P}_0(a - 1, t) \right) \qquad (8.13)$$

for $a \in \mathbb{Z}$ is essentially a binomial

$$\tilde{P}_0(a, t) = (f_0(1 - f_0))^t \binom{2t}{t + a} \qquad (8.14)$$

---

[11] Daerden and Vanderzande (1996) use K for $K - 1$.
[12] Thereby generating Dyck paths (Stanley, 1999).

with the convention that the binomial coefficient vanishes for $t + a < 0$ and $t + a > 2t$. The absorbing boundary is implemented by $\tilde{P}(a, t) = \tilde{P}_0(a - 1, t) - \tilde{P}_0(a + 1, t)$, so that Eq. (8.11) is solved by

$$P(a, t) = f_0^{t+a-1}(1 - f_0)^{t-a+1} \left\{ \binom{2t}{t + a - 1} - \binom{2t}{t + a + 1} \right\}. \tag{8.15}$$

The probability $D_t(t)$ for the activity to cease at $t \geq 1$ therefore is the flux into the absorbing wall, i.e. the probability for the activity to be at $a = 1$ at $t - 1$ times the probability of decreasing by one

$$
\begin{aligned}
D_t(t) &= (1 - f_0)^2 P(1, t - 1) \\
&= (f_0(1 - f_0))^t \frac{1 - f_0}{f_0} \left\{ \binom{2t - 2}{t - 1} - \binom{2t - 2}{t + 1} \right\} \\
&= (f_0(1 - f_0))^t \frac{1 - f_0}{f_0} \left\{ \binom{2t}{t} - \binom{2t}{t - 1} \right\}. \tag{8.16}
\end{aligned}
$$

The expression in the bracket is also known as the Catalan number $C_t = \binom{2t}{t} - \binom{2t}{t-1}$ (Knuth, 1997b; Stanley, 1999) and $C_0 = 1$ by definition. The generating function is $\sum_{n=0}^{\infty} z^n C_n = (1 - \sqrt{1 - 4z})/(2z)$, so that the **moment generating function** of $D_t(t)$ is given by

$$
\begin{aligned}
\mathcal{M}^{(D)}(x) &= \sum_{t=1}^{\infty} D_t(t) x^t = \frac{1 - f_0}{f_0} \sum_{t=1}^{\infty} C_t \left(x f_0 (1 - f_0) \right)^t \\
&= \frac{1 - f_0}{f_0} \frac{1 - \sqrt{1 - 4x f_0 (1 - f_0)} - 2x f_0 (1 - f_0)}{2x f_0 (1 - f_0)} \\
&= \frac{1 - f_0}{f_0} \frac{1 - \sqrt{1 - 4x f_0 (1 - f_0)}}{1 + \sqrt{1 - 4x f_0 (1 - f_0)}}. \tag{8.17}
\end{aligned}
$$

The construction of $D_t(t)$ did not rely on the probability being normalised and in fact $\sum_{t=1}^{\infty} D_t(t) = \mathcal{M}^{(D)}(1)$ is the probability that the activity will cease after a finite time. Evaluating $\mathcal{M}^{(D)}(1)$ gives the termination probability

$$
\begin{aligned}
T_t &= \mathcal{M}^{(D)}(1) \\
&= \frac{1 - f_0}{f_0} \frac{1 - |1 - 2f_0|}{1 + |1 - 2f_0|} = \begin{cases} 1 & \text{for } f_0 \leq 1/2 & \text{(8.18a)} \\ \left(\dfrac{1 - f_0}{f_0} \right)^2 & \text{for } f_0 > 1/2, & \text{(8.18b)} \end{cases}
\end{aligned}
$$

where $\sqrt{1 - 4f_0(1 - f_0)} = |1 - 2f_0|$ has been used. Equation (8.18) is illustrated in Fig. 8.2(a). If it were not for the properties of this final result, the derivation would hardly be worth mentioning. The question that arises immediately after inspecting Fig. 8.2(a) is: how is $\mathcal{M}^{(D)}(1)$ apparently non-analytic at $f_0 = 1/2$, given that the derivation seemingly does not contain a single non-analytic ingredient? The origin of the non-analyticity is the infinite sum in Eq. (8.17), as illustrated by the approach of

$$\mathcal{M}^{(D)}(1; T) = \sum_{t=1}^{T} D_t(t) \tag{8.19}$$

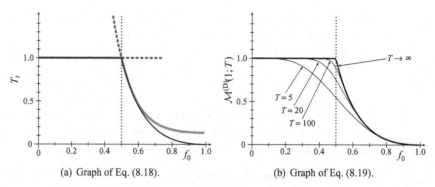

(a) Graph of Eq. (8.18).   (b) Graph of Eq. (8.19).

Figure 8.2   Termination probability T_t. (a) The solid line shows Eq. (8.18), the dashed lines their analytic continuations which are unstable and/or unphysical, Eq. (8.35). For $f_0 \leq 1/2$ ultimate extinction is certain, while for $f_0 > 1/2$ only a finite fraction of realisations terminates eventually. The vertical dotted line marks the critical point $f_0 = 1/2$. The thick grey line shows the continuum approximation Eq. (8.27), which deviates visibly from the exact discrete result for $f_0 \gtrsim 0.6$. (b) $\mathcal{M}^{(D)}(1; T)$ evaluated for $T = 5, 20, 100$ as indicated. The kink at $f_0 = 1/2$ develops in $\mathcal{M}^{(D)}(1; T)$ only in the limit $T \to \infty$, shown as a thick line.

to the full $\mathcal{M}^{(D)}(1)$ with increasing T in Fig. 8.2(b). Its presence signals the phase transition of the MFT at the **critical point** $f_0 = f_0^c = 1/2$. In the **subcritical phase**, $f_0 < f_0^c$, the process is bound to die off, whereas in the **supercritical phase**, $f_0 > f_0^c$, it might continue forever.

Given the definition of time as the number of sites updated, the avalanche size is simply the termination time, and the avalanche size distribution therefore is given by $D_t(t)$, Eq. (8.16), with t being the avalanche size. Because the thermodynamic limit has been taken at the beginning of the derivation, any finite size scaling will effectively be imposed, for example by assuming that all avalanches must have terminated after a certain time. Instead of doing that, in the following the scaling is discussed as a function of the temperature-like parameter f_0. From this point of view, the present MFT looks very much like an AS transition and one might want to consider a feedback loop that tunes the system to its critical point, along the lines of the mechanism introduced by Vespignani *et al.* (1998), see Sec. 9.3. However, the random neighbour BS Model, even for a finite number of sites, does not self-organise by balancing external drive and dissipation due to avalanches.[13] In fact, avalanches are filtered according to the parameter f_0, which is not selected by the system itself. The self-organisation of the system into a barrier height distribution like the one shown in Fig. 5.8(a) is not directly reflected in the avalanche size distribution considered here, which emerges only once it is assumed that all barriers except one are above f_0.

The avalanche size distribution $D_t(t)$, Eq. (8.16), is the difference between two binomial distributions, which can be approximated by two Gaussians, both with mean $2tf_0$ and

---

[13] This is, of course, not to say that the random neighbour BS Model would not evolve into the self-organised state as described earlier. Also, there are random neighbour, finite size models which, after prescribing a suitable feedback mechanism for the temperature-like variable, evolve into the critical state (Zapperi *et al.*, 1995), see also Sec. 9.3, p. 329.

variance $2tf_0(1 - f_0)$, but one evaluated at 'position' t and the other at 'position' $t - 1$:

$$f_0^t(1 - f_0)^t \binom{2t}{t} \approx \frac{1}{\sqrt{4\pi t f_0(1 - f_0)}} e^{-\frac{(t - 2tf_0)^2}{4tf_0(1 - f_0)}} \tag{8.20a}$$

$$f_0^{t-1}(1 - f_0)^{t+1} \binom{2t}{t - 1} \approx \frac{1}{\sqrt{4\pi t f_0(1 - f_0)}} e^{-\frac{(t - 1 - 2tf_0)^2}{4tf_0(1 - f_0)}}. \tag{8.20b}$$

For this approximation to be of any relevance, the position t should not deviate too much from the mean $2tf_0$, i.e. f_0 is very close to $1/2$, so that $f_0 = 1/2 + \epsilon$ appears as a more appropriate parametrisation and to leading order in large t and small ϵ

$$D_t(t) \approx \frac{1}{\sqrt{\pi}} e^{-4\epsilon^2 t} t^{-3/2}. \tag{8.21}$$

The deviation of f_0 from $1/2$ can be identified as a **drift or convection term** in the continuum approximation of Eq. (8.11), which, however, is not unique,

$$P(a, t + 1) - P(a, t)$$
$$= (1 - f_0)^2 \Delta P(a, t) - v(P(a, t) - P(a - 1, t)) \tag{8.22a}$$
$$= f_0^2 \Delta P(a, t) - v(P(a + 1, t) - P(a, t)) \tag{8.22b}$$

where $\Delta P(a, t) = (P(a + 1, t) - 2P(a, t) + P(a - 1, t))$ is the lattice Laplacian operating on $P(a, t)$ and $v = 2f_0 - 1$ is the drift in front of a first order spatial derivative of the form $P(a, t) - P(a - 1, t)$. Neither of the two equations is invariant under the transformation $a \rightarrow -a$ with $f_0 \rightarrow 1 - f_0$, while a linear combination of the two is:

$$P(a, t + 1) - P(a, t) = f_0(1 - f_0)\Delta P(a, t)$$
$$+ (1 - 2f_0)\Big[(1 - f_0)(P(a + 1, t) - P(a, t)) + f_0(P(a, t) - P(a - 1, t))\Big]. \tag{8.23}$$

This ambiguity in the continuum approach is due to $1 - 2f_0 \neq 0$, i.e. for $f_0 = 1/2$ Eq. (8.22) and Eq. (8.23) coincide. The long time asymptotes of the above equations are found for $vt \gg a_0$ (with a_0 the initial distance from the absorbing wall, as introduced after Eq. (8.11)) for non-vanishing drift and $Dt \gg a_0^2$ in the absence of drift.

After Taylor expansion, one arrives at

$$\partial_t P(a, t) = \left(D\partial_a^2 - v\partial_a\right) P(a, t) \tag{8.24}$$

with coefficients $v = 2f_0 - 1$ for the drift velocity and $D = 1/2 - f_0(1 - f_0)$ for the diffusion constant, which is non-zero for all $f_0 \in [0, 1]$ due to the second order contributions from terms of the form $P(a + 1, t) - P(a, t)$. [8.3]

Equation (8.24) can be solved by applying the method of image charges to the problem without drift term (i.e. $v = 0$) and absorbing boundary conditions, which gives, say, $P_0(a, t)$. Imposing $P_0(a, t) = b(a, t)P(a, t)$ with $P(a, t)$ solving Eq. (8.24) gives two differential equations for b, which are solved by $b(a, t) = \exp\left(\frac{v}{2D}\left(\frac{tv}{2} - a\right)\right)$. After some algebra, one has after normalising $\lim_{t \to 0} \int_0^\infty da\, P(a, t) = 1$

$$P(a, t) = \frac{1}{\sqrt{4\pi Dt}} \left(e^{-\frac{(a - a_0 - vt)^2}{4Dt}} - e^{-\frac{(a + a_0 - vt)^2}{4Dt} - \frac{va_0}{D}}\right). \tag{8.25}$$

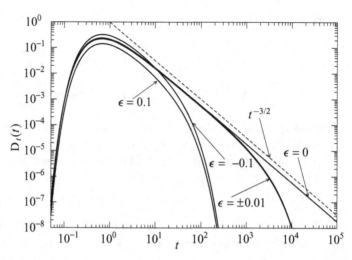

Figure 8.3 Illustration of the distribution of termination times $D_t(t)$ as a function of time t, according to Eq. (8.26), for different values of ϵ, where $f_0 = 1/2 + \epsilon$, so that $D = 1/4 - \epsilon^2$ and $v = 2\epsilon$. The sign of ϵ has little bearing on the general shape of $D_t(t)$, in particular for small values of ϵ. The asymptote $t^{-3/2}$ is shown as a dashed line.

With $\lim_{t\to 0} P(a, t) = \delta(a - a_0) + \delta(a + a_0) \exp(-va_0/D)$, $P(a, t)$ is the Green function on $[0, \infty[$, so that the random walk starts at a_0 with $a_0 = 1$ above.

To find the termination probability (density) one can use the differential equation Eq. (8.24),

$$D_t(t) = -\frac{d}{dt} \int_0^\infty da\, P(a, t) = -\left[(D\partial_a - v) P(a, t) \right]_0^\infty = \frac{1}{\sqrt{4\pi Dt}} \frac{a_0}{t} e^{-\frac{(a_0+vt)^2}{4Dt}} \quad (8.26)$$

as $P(0, t) = 0$ due to the boundary condition. This PDF is discussed in Sec. 2.1.3 (in particular Fig. 2.4).

Equation (8.26) displays the well-known asymptote (in large t) $t^{-3/2}$ for the **return time** of a random walker (Redner, 2001). In the present interpretation, this exponent is the **(classical) mean-field avalanche size exponent** $\tau = 3/2$. The exponential in Eq. (8.26) cuts off the distribution towards small avalanches $t \to 0$, even for $v = 0$, thereby preventing an unphysical singularity at $t = 0$. The distribution $D_t(t)$ is shown in Fig. 8.3 for different values of v, in particular for the critical case $v = 0$. For $v \leq 0$ it is normalised,

$$T_t = \int_0^\infty dt\, D_t(t) = e^{-a_0 \frac{v + a_0 \left| \frac{v}{a_0} \right|}{2D}} = \begin{cases} 1 & \text{for } v \leq 0 \quad (8.27a) \\ e^{-a_0 v/D} & \text{for } v > 0, \quad (8.27b) \end{cases}$$

for $v > 0$ a finite fraction of walkers never returns. Equation (8.27) is shown in Fig. 8.2(a), using $a_0 = 1$, $v = 2f_0 - 1$ and $D = 1/2 - f_0(1 - f_0)$. As it is based on the lowest orders of the Taylor series expansion of the original difference equation, it deviates visibly from the result on the lattice for $f_0 \gtrsim 0.6$. Equation (8.27b) reproduces Eq. (8.18b) to leading order of f_0 about $1/2$.

It is interesting to retrace the above result using **dimensional analysis**, which fixes only $D_t(t) = t^{-1} \mathcal{G}(a_0/\sqrt{Dt}, v\sqrt{t/D})$, without allowing any statement about the scaling of the function $\mathcal{G}$ in large t. The dimensionless variables $a_0/\sqrt{Dt}$ and $v\sqrt{t/D}$ have been chosen so that the sign of a_0 and v can enter, whereas $D, t > 0$. If $a_0 \geq 0$ vanishes, then $D_t(t)$ vanishes on its entire domain $t > 0$, i.e. only for $a_0 \neq 0$ is non-trivial behaviour possible at all. This is illustrated by the scaling of $\mathcal{G}(x, 0) \propto x \exp(-x^2/4)$ if $v = 0$ according to Eq. (8.26), so that $D_t(t) \propto t^{-3/2} \exp(-a_0^2/(4Dt))$, i.e. any occurrence of a non-trivial exponent such as $3/2$ is due to the presence of a (additional) microscopic scale, such as a_0, which restores the dimension of $D_t(t)$ but does not directly enter its asymptote in large t. For $v \neq 0$, the termination probability $D_t(t)$ is dominated by $a_0/(t\sqrt{4\pi Dt}) \exp(-v^2t/(4D))$ at $|vt| \gg a_0$, i.e. $\mathcal{G}(x, y) \propto x \exp(-y^2/4)$, regardless of the sign of v. For positive v the random walkers drift away from the absorbing wall, while for negative v most of them are absorbed after time $t \approx |a_0/v|$, so that the termination rate at later times is reduced in both cases.

Using again $f_0 = 1/2 + \epsilon$ so that $v = 2\epsilon$ and $D = 1/4 + \epsilon^2$, to leading order in small ϵ and large t one finds

$$D_t(t) \approx \frac{1}{\sqrt{\pi t}} \frac{1}{t} e^{-4\epsilon - 4\epsilon^2 t}. \tag{8.28}$$

Apart from the factor $\exp(-4\epsilon)$, this asymptote is identical to the 'cruder' Eq. (8.21), which is, however, invariant under a change $\epsilon \to -\epsilon$. Positive ϵ means $f_0 > 1/2$, i.e. the system tends to produce more activity as time goes by, Eq. (8.10), so after a transient determined by $vt \gg \sqrt{Dt}$ very little termination occurs. Similarly for negative ϵ, as most of the termination occurs within a short time interval of duration $|a_0/v|$. Nevertheless, the factor $\exp(-4\epsilon)$ makes the two regimes distinguishable even on the large time scale.

The above analysis explains why $f_0 = f_0^c = 1/2$ is a critical point. For any $\epsilon \neq 0$ (i.e. $f_0 \neq 1/2$) the drift governs the asymptote. Only at $f_0 = 1/2$ does the diffusive behaviour prevail, i.e. $f_0 = 1/2$ is a separating point between two asymptotic regimes, the subcritical and the supercritical regimes.

8.1.4 The branching process

The present setup of the MFT lends itself naturally to an interpretation in terms of a Galton–Watson branching process (Harris, 1963; Alstrøm, 1988). In the current context, the branching process merely provides a different perspective on the same dynamics and allows the derivation of the PDF of a different observable. Yet, because it plays an important conceptual rôle (for further applications see for example Caldarelli, Tebaldi, and Stella, 1996b; Ivashkevich, 1996; Bröker and Grassberger, 1997; Vergeles *et al.*, 1997; Adami and Chu, 2002; Slanina, 2002), it deserves to be presented in further detail.

During the evolution of the random neighbour BS Model, each site can be assigned a **generation** g_n whenever it is updated and becomes active, i.e. either due to an active random neighbour or by staying active after an update. Following the terminology introduced in Sec. 5.4.5.6, an active site may be called a particle. Starting from $g_n = 0$, a site $\mathbf{n}'$ activated by the update of a given site $\mathbf{n}$ belongs to the next generation $g_{\mathbf{n}'} = g_{\mathbf{n}} + 1$, with $\mathbf{n}' = \mathbf{n}$ as a special case where $g_{\mathbf{n}'}$ on the left hand side is the 'updated' generation. Because the thermodynamic limit has already been taken, sites do not interfere by updating the same (neighbouring) site, for example by retiring a site that has been activated previously by

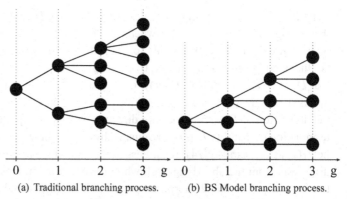

(a) Traditional branching process.	(b) BS Model branching process.

Figure 8.4 The branching process with active sites (particles) indicated by filled circles, as it evolves on a generational time scale g, shown as dotted lines. (a) Non-interacting particles produce offspring independently in a Bernoulli process. Population sizes are, for example, $m(1) = 2$ and $m(3) = 7$. (b) In the BS branching process (Sec. 5.4.5.6), particles reside on sites. In every generation the particle with the globally lowest barrier branches, all other particles are copied from the previous generation. A branching particle might produce new particles or remove existing ones (empty circle).

another site. In the MFT, generations never clash, i.e. every activated site belongs uniquely to a generation. While some of the theoretical insights of the branching process can be carried over to the fully spatial version of the BS Model, the latter feature is particular to the MFT, see Fig. 8.4.

The interpretation of the branching process presented above provides a link to the random walker picture discussed in the previous section. But as sites do not interact, it is equally valid to think of generations of particles that are all updated at once. Updating all sites of the same generation simultaneously leads to a sequence of population sizes $m(g)$ for each generation g with $m(0) = 1$. It is convenient to use a generating function (Harris, 1963; Wilf, 1994) to find the moments of the population size $m(g+1)$ given a certain population size $m(g)$. According to Eq. (8.10), each active site can produce 0, 1 or 2 **offspring** with probabilities $(1 - f_0)^2$, $2f_0(1 - f_0)$ and f_0^2 respectively. [14] Each of these probabilities can be 'decorated' with a power of x which reflects the number of offspring produced,

$$f(x) = (1 - f_0)^2 x^0 + 2f_0(1 - f_0)x^1 + f_0 x^2 = (1 - f_0 + xf_0)^2. \qquad (8.29)$$

If m particles reproduce in generation g, then the coefficient in front of x^n in $(f(x))^m$ is the probability p'_n of producing n offspring, accounting for all possible ways this can happen. The function $(f(x))^m = \sum_n p'_n x^n$ is in fact the moment generating function, which produces the normalisation $f(1)^m = 1$ when evaluated at $x = 1$, and the first moment $\sum_n p'_n n = mf'(1)$ when differentiated with respect to x and evaluated at $x = 1$. If $m = 0$, then the empty generation g is simply carried over to $g + 1$.

The population size m in the gth generation is itself a random variable, size m occurring with probability p_m. The **moment generating function** of the population size in generation

---

[14] In this language, a site remaining active is accounted for as well by producing an offspring.

g is therefore

$$\mathcal{M}^{(g)}(x) = \sum_m p_m x^m. \tag{8.30}$$

As $f(x)^m$ is the moment generating function of the population size in generation $g+1$ *given* m particles in generation g, the full moment generating function of generation $g+1$ is

$$\mathcal{M}^{(g+1)}(x) = \sum_m p_m f(x)^m = \mathcal{M}^{(g)}(f(x)), \tag{8.31}$$

which holds for all $g \geq 0$. The hierarchy is started by $\mathcal{M}^{(g=0)}(x) = x$. The first moment, i.e. the expected size $\langle n \rangle_g$ of the gth generation, is given by $\frac{d}{dx}\big|_{x=1} \mathcal{M}^{(g)}(x)$ (Sec. 2.2.3), so that with $f(1) = 1$

$$\langle n \rangle_{g+1} = \frac{d}{dx}\bigg|_{x=1} \mathcal{M}^{(g+1)}(x) = f'(1) \frac{d}{dx}\bigg|_{x=f(1)} \mathcal{M}^{(g+1)}(x) = f'(1) \langle n \rangle_g, \tag{8.32}$$

and therefore $\langle n \rangle_g = (f'(1))^g$. The **branching ratio** $\sigma = \langle n \rangle_{g+1} / \langle n \rangle_g = f'(1)$ is generally independent of g, so that, for example, the total cumulative population (i.e. the avalanche size) for $\sigma < 1$ is

$$\sum_{g=0}^{\infty} \langle n \rangle_g = \sum_{g=0}^{\infty} \sigma^g = \frac{1}{1-\sigma}, \tag{8.33}$$

essentially reproducing Eq. (5.4). In the BS branching process discussed in Sec. 5.4.5.6, the (effective) branching ratio depends on the state of the system, because two neighbouring individuals have some potential offspring in common (Fig. 8.4(b)). As suggested in Sec. 5.4.5.6, calling it a branching process has led to some disagreement in the literature (also p. 131).

Since $f'(1) = 2f_0$ the present branching process has branching ratio $\sigma = 2f_0$ and on average $(2f_0)^g$ individuals in the gth generation, starting with 1 in generation $g = 0$. For $f_0 = 1/2 = f_0^c$ (critical point) the average population size is sustained across generations, for $f_0 < 1/2$ it decreases exponentially in g (**subcritical branching process**), whereas for $f_0 > 1/2$ the population increases exponentially (**supercritical branching process**). This result of exponential increase and decay is somewhat different to that derived from the random walker approach because of the different time scale used in the branching process. It is therefore natural to ask for the probability $D_g(g)$ of the branching process to terminate in generation g. To answer this question, the (asymptotic) probability that the branching process runs empty at some finite time is introduced, $T_g = \sum_{g=0}^{\infty} D_g(g)$, similar to Eq. (8.27), with probability $D_g(g)$ of terminating at the reproduction in generation g.

The moment generating function contains the probability that the population has size m as the coefficient of the term x^m, i.e. the probability that there is no population after g generations is

$$\mathcal{M}^{(g)}(0) = \mathcal{M}^{(g-1)}(f(0)) = \mathcal{M}^{(g-2)}(f(f(0))) = \underbrace{f(f(f(\dots f(0)\dots)))}_{g \text{ instances of } f}, \tag{8.34}$$

as $\mathcal{M}^{(1)}(x) = f(x)$. The probability T_g of terminating after a finite number of generations is therefore $T_g = \lim_{g\to\infty} \mathcal{M}^{(g)}(0)$, which is $f(x^*)$ where x^* is the fixed point of the map $f(x)$, reached from $x = x_0 = 0$. The fixed point equation $x^* = f(x^*)$ has two solutions,

$$x^* = 1 \tag{8.35a}$$

$$x^* = \left(\frac{1 - f_0}{f_0}\right)^2 \tag{8.35b}$$

to be compared to Eq. (8.18) with $T_g = f(x^*) = x^*$ for the stable fixed point x^* whose basin of attraction contains $x_0 = 0$. A **linear stability analysis** (Strogatz, 1994) indicates that a fixed point x^* is repulsive if $f'(x^*) > 1$ and attractive if $f'(x^*) < 1$. Given that $f(x)$ is a simple quadratic, Eq. (8.29), it is easy to show explicitly that the map

$$x_{g+1} = f(x_g) = \mathcal{M}^{(g+1)}(x_0) \tag{8.36}$$

does not allow x_g to skip over a fixed point, so that the fixed point reached from $x = 0$ is the smaller one of the two in Eq. (8.35b). This fixed point is apparently stable and, because $f(x)$ is quadratic, there can be at most one stable fixed point. [8.4] The fixed point $x^* = 1$ in Eq. (8.35a) corresponds to the case that every incarnation of the branching process eventually dies off, $T_g = \lim_{g\to\infty} \mathcal{M}^{(g)}(0) = x^* = 1$. For that fixed point, $f'(x^*)$ is exactly the branching ratio, $\sigma = f'(1) = 2f_0$, which means that ultimate termination is an attractive fixed point for $\sigma < 1$ i.e. $f_0 < 1/2$, confirming Eq. (8.18) (see Fig. 8.2(a)). Correspondingly, for $x^* = ((1 - f_0)/f_0)^2$, Eq. (8.35b), the first derivative is $f'(x^*) = 2(1 - f_0)$, which means that this fixed point is stable for $f_0 > 1/2$. In summary,

$$T_g = \begin{cases} 1 & \text{for } f_0 \leq 1/2 \tag{8.37a} \\[2mm] \left(\dfrac{1 - f_0}{f_0}\right)^2 & \text{for } f_0 > 1/2, \tag{8.37b} \end{cases}$$

reproducing Eq. (8.18b).

In both cases, $f_0 < 1/2$ and $f_0 > 1/2$ respectively, the evolution of $\Delta x_g = x_g - x^*$, Eq. (8.36), around the attractive fixed point is essentially exponential, $\Delta x_g = \Delta x_0 (f'(x^*))^g$ by keeping only the leading order in Δx_g on the right hand side of $\Delta x_{g+1} = f(x^* + \Delta x_g) - x^*$. The initial $\Delta x_0 = x_0 - x^*$ cannot be positive, according to Eq. (8.34) $x_0 = 0$, so that $\Delta x_0 = -x^*$. At $f_0 = f_0^c = 1/2$, however, $x^* = 1$ is a double root and $f'(x^*) = 1$, which means that the stability cannot be determined without taking into account the second order. [15] The resulting recurrence relation $\Delta x_{g+1} = f'(x^*)\Delta x_g + (1/2)f''(x^*)\Delta x_g^2$ is easily solved in a continuum approximation

$$\frac{\mathrm{d}}{\mathrm{d}g}\Delta x_g = \frac{1}{4}\Delta x_g^2 \tag{8.38}$$

---

[15] In the present case $f(x)$ does not possess any higher orders, so that the resulting *recurrence relation* (but not the continuum equation) is exact.

using $f'(x^*) = 1$ and $f''(x^*) = 1/2$ at $x^* = 1$ and $f_0 = 1/2$. The continuum solution is

$$\Delta x_g = -\left(\frac{1}{4}g - \Delta x_0^{-1}\right)^{-1}, \tag{8.39}$$

i.e. $x^* = 1$ is stable for $f_0 = 1/2$.

The probability that the population vanishes in generation g is given by $\mathcal{M}^{(g)}(0)$ which evolves according to Eq. (8.34) or equivalently according to Eq. (8.36). The difference

$$D_g(g) = \mathcal{M}^{(g+1)}(x_0) - \mathcal{M}^{(g)}(x_0) = x_{g+1} - x_g = \Delta x_{g+1} - \Delta x_g \tag{8.40}$$

is the probability that extinction and therefore termination of the branching process occurs when generation g attempts to reproduce. Close to the attractive fixed points for $f_0 \neq 1/2$ this difference is $-\Delta x_g(1 - f'(x^*)) \propto (f'(x^*))^g$ and therefore exponential in g. However, at $f_0 = 1/2$ the approach to the fixed point is asymptotically (in large g) a power law $\propto g^{-1}$, see Eq. (8.39), so that the difference (8.40) is $D_g(g) \propto g^{-2}$ (Harris, 1963; Bröker and Grassberger, 1997; Lübeck, 2004). This is the well-known **(classical) mean-field avalanche duration exponent** $\alpha = 2$, where the duration of the avalanche is quite naturally measured as the number of generations. On the short time scale, similar to the small avalanche scale (p. 256), a cutoff prevents undesired singularities.

The two exponents z and D characterise the finite size scaling and thus rely on the introduction of a cutoff or boundary 'by hand'. For example, if the branching process is stopped after G generations, the average total number of sites activated is $\sum_{g=0}^{G}(2f_0)^g = (1 - (2f_0)^{G+1})/(1 - 2f_0)$ and thus $\langle s \rangle = G + 1 \propto G$ at the critical point, $f_0 = 1/2$. If G is identified with the system size L, then $D = 2$ follows as **the (classical) mean-field avalanche dimension** from $\langle s \rangle \propto L^{D(2-\tau)}$ with $\tau = 3/2$, Eq. (8.26), and $z = 1$ as **the (classical) mean-field dynamical exponent** via $D(1 - \tau) = z(1 - \alpha)$, Eq. (2.40), and $\alpha = 2$. Similarly, one can call on boundary conditions imposing $D(2 - \tau) = \sigma_1^{(s)} = 1$ or $\sigma_1^{(s)} = 2$, Sec. 8.1.1.1, the latter producing $D = 4$ and $z = 2$.

8.1.4.1 The microscopic time scales of branching process and random walk

The MFT of the BS Model links the branching process and the random walk as two different representations of the same stochastic process. Either the number of active sites is subject to a random walk or a generation of active sites reproduces and thereby gives rise to a certain number of offspring in the following generation. There is no *unique* one-to-one mapping between the two pictures, i.e. there are many ways to translate a particular realisation of the branching process into a trajectory of a random walker and vice versa (e.g. Harris, 1963; Feller, 1966, 1968; Stapleton, 2007, p. 131 contains a very instructive illustration). Yet, since the events in both pictures, reproduction at a node in the branching process and a single **microscopic time step** in the random walk, are independent, their order is irrelevant and, for that matter, one translation between the two pictures is as good as any other.

The nature of the microscopic time, however, is fundamentally different in the two pictures. In a (fine) microscopic time step of the random walk, a single site is updated, whereas in the (coarse) **generational time step** of the branching process, the notion of which does not even exist in the random walk picture, the whole of sites active at that time attempts to reproduce, similar to a parallel update of all active sites in an SOC model. [16] As a result, the probability of terminating at time t in the random walk scales like $t^{-3/2}$, whereas the probability of terminating at generation g scales like g^{-2}.

The outcome of the reproduction attempt of a generation of size $m(g)$ could be determined by a random walk that starts at distance $m(g)$ away from an absorbing wall. The generation size at time $g+1$ is then given by the distance of the walker from the absorbing wall $m(g)$ time steps later, namely after each individual has been given the opportunity to reproduce. In fact, the probability distribution of sizes $m(1), m(2), \ldots$ of consecutive generations $g = 1, 2, \ldots$ starting with $m(0) = 1$ as described in Eq. (8.31) is *identical* to that of the position $a(t_1), a(t_2), \ldots$ of a random walker along an absorbing wall, Eq. (8.11), starting from $a(t_0) = a_0 = 1$ at time $t_0 = 0$ observed at times $t_1 = a(t_0), t_2 = t_1 + a(t_1), \ldots, t_{g+1} = t_g + a(t_g)$, so that

$$t_g = \sum_{g'=0}^{g-1} a(t_{g'}). \tag{8.41}$$

Shrinking the original time scale t of a random walker's trajectory $a(t)$ by a local scale factor $a(t)$ therefore gives rise to the time evolution trajectory of the branching process on a generational time scale. Adopting the latter time scale for the random walk, it becomes subject to a position dependent effective diffusion constant, which is chosen so that the evolution of the activity a is subject to the same distribution in a single time step, as originally in a time steps. The random walk on a generational time scale $\tilde{a}(g) = a(t_g)$ thus has its diffusion constant rescaled by a factor a relative to the original time scale t.

The mapping of the two time scales can be made exact in discrete (see above) as well as in continuous time, as done in the following. As a result, a Langevin equation with RFT-type noise is mapped to one with annealed noise and absorbing boundaries, which might be better behaved than the former. In a first attempt to translate the problem to the continuum (see also Sec. A.6.3), one might consider the continuous time $g(t)$ derived from the original discrete time t as

$$g(t) = \sum_{i=0}^{t-1} 1/a(i), \tag{8.42}$$

motivated by Eq. (8.41) or by the observation that if $a(t)$ updates are to be made in $\Delta g = 1$ time steps, then each time step has length $1/a(t)$ in units of g. However, $a(t)$ itself is a random variable that changes in time, while the time units as introduced above should be $1/a(t_g)$, i.e. strictly $g(t) = \sum_{i=0}^{t-1} 1/a(t_{\lfloor g(i) \rfloor})$, so that $g(t_n) = n$.

The above can be analysed more elegantly in the framework of Fokker–Planck or Langevin equations. A free, fair random walker $a(t)$ is described in the continuum by [17]

$$\dot{a}(t) = \xi(t) \tag{8.43}$$

---

[16] The generational time scale should be distinguished from the macroscopic time scale in an SOC model, which normally comprises an entire avalanche. In the branching process this is best identified with an entire realisation.

[17] For the following, it might be helpful to think of the random walker as going either up, straight or down with probabilities 1/4, 1/2, 1/4 respectively and of the branching process as producing either 2, 1, 0 offspring with the same probabilities.

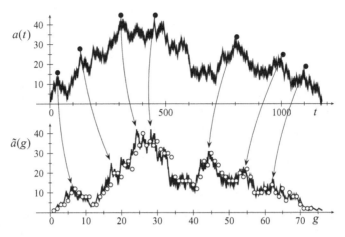

Figure 8.5 Mapping of a (fair) random walk (top) to the evolution of a (critical) branching process (bottom). Both graphs show the same data but on different time scales. In the bottom graph individual generations are shown as empty circles, whereas the solid line shows the data on the approximate continuous time scale Eq. (8.42). For comparison some peaks are marked in both graphs (filled circles in the top graph) and connected by arrows. High activity in the top graph contracts time in the bottom graph, low activity stretches time.

where $\xi(t)$ is Gaussian white noise with vanishing mean and correlator

$$\langle \xi(t)\xi(t')\rangle = 2D\delta(t - t'),\tag{8.44}$$

so that $a(t)$ has variance $2Dt$ starting from $a(0) = a_0$. In the present case, the random walk is subject to the constraint that $\dot{a}(t) = 0$ if $a(t') = 0$ at any time $0 \le t' \le t$, which is most easily accounted for by an absorbing wall at the level of a Fokker–Planck equation, which is the 'equation of motion' of the PDF.

Starting from this random walker $a(t)$, the continuous generational time scale g is introduced implicitly along the same lines as Eq. (8.41),

$$t(g) = \int_0^g \mathrm{d}g' \, \tilde{a}(g'),\tag{8.45}$$

where $\tilde{a}(g) = a(t(g))$ is the generation size in a (critical) branching process, whose equation of motion is derived in the following. A realisation of a random walk trajectory $a(t)$ mapped in this way to a branching process $\tilde{a}(g)$ is shown in Fig. 8.5. In definition (8.45), $t(g)$ is invertible, $g(t) = \int_0^t \mathrm{d}t' \, 1/a(t')$, since $\tilde{a}(g)$ is strictly positive, except for the single point when the random walker touches the wall, which has to be formally excluded from the present analysis. Based on Eq. (8.43), the equation of motion of $\tilde{a}(g)$ is the Langevin equation [18]

$$\frac{\mathrm{d}}{\mathrm{d}g}\tilde{a}(g) = \tilde{a}(g)\xi(t(g)) = \sqrt{\tilde{a}(g)}\tilde{\xi}(g)\tag{8.46}$$

---

[18] Problems arising from Itō versus Stratonovich are boldly ignored.

where $\sqrt{\tilde{a}(g)}\xi(t(g)) =: \tilde{\xi}(g)$ at first looks arbitrary, yet its correlator is that of white noise,

$$\left\langle \tilde{\xi}(g)\tilde{\xi}(g') \right\rangle = 2D\delta(g - g'), \tag{8.47}$$

which can be motivated[19] by evaluating Eq. (8.44) for $t = t(g)$ and $t' = t(g')$ and observing that the derivative $dt/dg = \tilde{a}(g)$ is strictly positive, so that $t(g) = t(g')$ has a unique solution $g = g'$. As illustrated in Fig. 8.5, both Eq. (8.43) and Eq. (8.46) describe the same process. However, the latter implements the condition $\tilde{a}(g) > 0$ not as an additional constraint, but implicitly, since $d\tilde{a}/dg$ vanishes as soon as $\tilde{a} = 0$. This setup can be easier to analyse than the explicit condition on Eq. (8.43) (e.g. Kloster et al., 2001).

As long as $\tilde{a}(g) > 0$ for all g considered, the trajectory $\tilde{a}(g)$ can be mapped uniquely to $a(t)$ and vice versa. If, however, $a(g) = 0$, the time transformation Eq. (8.45) is no longer invertible and the above derivation breaks down either when $\tilde{a}(g)$ vanishes in the denominator or when $t(g) - t(g') = 0$ acquires an additional root other than $g = g'$, which dramatically alters the correlator Eq. (8.47).

The derivation above maps a *fair* random walker to a *critical* branching process. It can be extended further by adding a drift 2ϵ to Eq. (8.43),

$$\dot{a}(t) = 2\epsilon + \xi(t) \tag{8.48}$$

so that Eq. (8.46) becomes

$$\frac{d}{dg}\tilde{a}(g) = 2\epsilon\tilde{a}(g) + \sqrt{\tilde{a}(g)}\tilde{\xi}(g). \tag{8.49}$$

On a discrete time scale, on average $\langle \tilde{a}(g+1) \rangle = \langle \tilde{a}(g) \rangle (1 + 2\epsilon)$, i.e. the branching ratio in this case is $\sigma = 1 + 2\epsilon$. Without the noise term, $\tilde{a}(g)$ is just an exponential, $\exp(2\epsilon g)$, as suggested for example by the probability of extinction, Eq. (8.40) and the discussion thereafter. The mapping can be embellished by calculating the average total population, which is finite in the branching process if $2\epsilon < 0$. Corresponding to Eq. (8.33), one finds

$$\sum_{g=0}^{\infty} \sigma^g = -(2\epsilon)^{-1}. \tag{8.50}$$

In the Langevin equation of the drifting random walker, Eq. (8.48), the drift dominates its behaviour on the large scale. On average, it gets closer to the wall by -2ϵ in every move. If it starts one unit away from the wall, the average number of moves until termination is $-1/(2\epsilon)$, reproducing Eq. (8.50).

The factor $\sqrt{\tilde{a}(g)}$ in front of the Gaussian white noise $\tilde{\xi}(g)$ in Eq. (8.46) or Eq. (8.49) gives rise to the local effective diffusion constant linear in $\tilde{a}(g)$, as suggested above. While Eq. (8.46) contains the noise in the form typical (Janssen, 1981) for **Reggeon field theory** (Gribov, 1967; Brower et al., 1978; Cardy and Sugar, 1980; Liggett, 2000), a treatment of the fully fledged theory with space dependent $a(t, \mathbf{x})$ is hampered by the space dependence of the transform Eq. (8.45) induced by the space dependence of $a(t, \mathbf{x})$.

---

[19] A derivation along these lines was flawed by $t(g)$ being path dependent, while Eq. (8.47) is an ensemble average.

8.1.4.2 The self-organised branching process

The branching process in the form discussed above depends on the control parameter f_0 which has to be tuned to its critical value $f_0^c = 1/2$ to recover scale invariant behaviour. Zapperi *et al.* (1995) devised a self-organised version, where f_0 changes slowly on the macroscopic time scale i, which counts individual realisations of the branching process, which are triggered by an external drive and run for a fixed number of up to $g = n$ generations, involving up to $N = \sum_{i=0}^{n} 2^i = 2^{n+1} - 1$ sites. While the external drive increases f_0 by $1/N$, dissipation occurs when the entire nth generation of size $\tilde{a}(n)$ is dissipated, so that

$$f_0(i+1) = f_0(i) + \frac{1 - \tilde{a}(n)}{N}. \tag{8.51}$$

Various other more concrete interpretations and implementations are possible, none of which alters the following reasoning profoundly.

Ignoring the details of the microscopic dynamics, correlations and finiteness, in the stationary state $1 = \langle \tilde{a}(n) \rangle$ by conservation and therefore $1 \approx (2f_0(i))^n$, so that $f_0(i)$ will fluctuate about $f_0^c = 1/2$, which is the critical point of the process. Although trivial, the model therefore is an exactly solvable model of SOC and its setup makes it particularly interesting for the study of the relation between absorbing states and SOC (Bonachela, 2008). In particular, it has been studied for the effect of non-conservation in the microscopic dynamics, i.e. **bulk-dissipation** (Lauritsen, Zapperi, and Stanley, 1996), which induces a characteristic length scale that eventually suppresses scale invariance. This effect might be altered if the dissipation scales with the system size. Contrary to the suggestions made by Dickman *et al.* (2000), compensating for the dissipation by a re-charging mechanism (Pruessner and Jensen, 2002b; Juanico, Monterola, and Saloma, 2007b) requires tuning of the latter and thus 'cannot be called *bona fide* self-organization' (Bonachela and Muñoz, 2009).

In line with the expectations based on the AS mechanism (Sec. 9.3), the observables of the SOBP are expected to display finite size scaling in n, which is in fact that of the ordinary branching process at the critical point. One might wonder, however, to what extent this is a coincidence, due to the solution $f_0 = 1/2$ of $1 = (2f_0)^n$ being independent of n. This is not expected in other SOC models (Pruessner and Peters, 2006) where, for example, the dissipation rate might depend on the ratio of boundary sites to bulk sites. If only a fraction $\alpha = An^{-\mu} \leq 1$ is dissipated after n generations, f_0 fluctuates about $(1/2)(A^{-1}n^\mu)^{1/n}$ and the scaling of the moments $\langle \tilde{a}(n)^m \rangle$ changes, in particular $\langle \tilde{a}(n) \rangle \propto n^\mu$ is imposed. While this is undesired but in line with the AS mechanism, suitable moment ratios seem to be unaffected (Walker, 2009). This, in turn, is in line with what is expected from an SOC model, but remains unexplained by the AS mechanism.

8.2 Renormalisation

A number of attempts has been made to apply renormalisation group (RG) ideas to SOC models (e.g. Secs. 4.2.2, 5.2.3, 5.4.5.2, 6.2.1.3). They are based on the general notion that degrees of freedom can be integrated out and captured by changing some **bare couplings** to new effective values: the system on a fine scale and one set of parameters corresponds to the system on a coarser scale and **renormalised** parameters. While Kadanoff's (1966) original insight into the origin of scaling behaviour in the Ising Model changed the understanding of critical phenomena forever, the real space technique used to *illustrate* this insight later met with much less success and many hitherto unsolved puzzles (e.g. Burkhardt and van Leeuwen, 1982). Similar in spirit but far more successful is Wilson's integration of **k**-shells, and slightly more removed but more effectively applied is renormalised field theory (Le Bellac, 1991; Barabási and Stanley, 1995; Zinn-Justin, 1997; McComb, 2004; Amit and Martín-Mayor, 2005).

In the following, only those renormalisation schemes are discussed that have been applied directly to SOC models. Beyond that, there is a wide range of renormalisation group approaches to effective (field) theories of SOC models and their mappings (e.g. Dickman *et al.*, 2000; Muñoz *et al.*, 2001) using established (field theoretic) methods (e.g. Doi, 1976; Peliti, 1985; Nattermann *et al.*, 1992; Lee and Cardy, 1995; Pastor-Satorras and Vespignani, 2000d; Chauve and Le Doussal, 2001; Täuber *et al.*, 2005), for example those discussed in Sec. 6.2.1.2, Sec. 6.2.1.3, Sec. 6.3.4.2 or Sec. 9.2. Hwa and Kardar (1989a) were the first to propose a **dynamic renormalisation group scheme** for an SOC model, namely the BTW Model, deriving its *effective* equation of motion from symmetry considerations and physical arguments. The resulting set of equations is generally known as a **driven diffusive system** (Janssen and Schmittmann, 1986; Schmittmann and Zia, 1995). During the course of their analysis, Hwa and Kardar identified a number of key ingredients of SOC models, such as bulk conservation of mass and dissipative relaxation at boundaries, resulting in a net transport of grains. A number of other works, most notably by Garrido, Lebowitz, Maes, and Spohn (1990), Grinstein, Lee, and Sachdev (1990), Grinstein (1991) as well as Socolar *et al.* (1993), were built on these insights and completed the picture of the origin and effect of various symmetries. Further details of these theories can be found in Sec. 9.2. When this type of analysis became more widely accepted, there was some hope that it would provide a *complete* picture on the basis that all interactions allowed by symmetry had been taken care of.

Given, however, that the MANNA universality class is, apparently, generically characterised by a Langevin equation with either RFT-type noise, $\sqrt{g(\mathbf{x}, t)}\xi(\mathbf{x}, t)$, or quenched noise, $\xi(g(\mathbf{x}, t), t)$, neither of which enters the renormalisation schemes discussed here, this hope has since proved unfounded. Moreover, models developed subsequently, such as the BS Model, suggested that bulk conservation was not a necessity and anisotropy was a matter of interpretation (p. 293).

Díaz-Guilera (1992, 1994, also Corral and Díaz-Guilera, 1997) took a similarly traditional approach, applying a Wilson-type renormalisation scheme to the stochastic equation of motion supposed to govern the ZHANG and BTW Models, which was derived by direct inspection of the models' rules. In comparison to the work by Hwa and Kardar (1989a), Díaz-Guilera's (1992) scheme does not make use of anisotropy, has an unusual, time independent noise correlator and has to cope with infinitely many relevant non-linearities. In contrast to the more ad hoc real space renormalisation schemes employed later, his and

Hwa and Kardar's (1989a) approaches are rooted in the established dynamic renormalisation group (Ma, 1976; Forster, Nelson, and Stephen, 1977; Hohenberg and Halperin, 1977; Medina, Hwa, Kardar, and Zhang, 1989; Vasil'ev, 2004; Täuber, 2005), as used in the analysis of growth equations (Barabási and Stanley, 1995).

While the (exact) equation of motion is a solid starting point for an RG approach, Díaz-Guilera's (1992) approach suffers from a number of technical difficulties. First of all, an equation of motion is often not easy to come by (for example Eq. (6.11) on p. 176, Eq. (6.32) on p. 204 and Sec. 8.3.2.1) and constructing them by mere inspection can be very difficult. Secondly, taking the continuum limit can generate or suppress interaction terms, or might require them to be included explicitly, such as an effective diffusion which is not obviously present on the lattice. Thirdly, standard SOC observables, such as avalanches, are not easily identified in the equation of motion, a problem that has been resolved to some extent with the advent of the AS interpretation of SOC. Finally, the presence of thresholds typical in SOC (Cafiero *et al.*, 1995) can translate into a Heaviside step function, which is notoriously difficult to handle (Pérez *et al.*, 1996; Corral and Díaz-Guilera, 1997, also Sec. 6.3.4.2). As discussed in Sec. 4.3.4, the regularisation of the step function is rather arbitrary, for example Eq. (4.11), yet it is far from obvious whether the results are independent of such details.

Compared to Pietronero *et al.*'s (1994) real space renormalisation group (RSRG) discussed below, Díaz-Guilera's (1992) renormalisation of the equation of motion is technically more involved and, because of its level of abstraction, lacks some of the physical appeal. While the RSRG approach has been applied to many different models, Díaz-Guilera's (1992) RG has received less attention. However, it might be seen as an important link between SOC and traditional non-equilibrium critical phenomena, by expressing an SOC model in terms of an equation of motion and applying established techniques to determine its scaling behaviour. In fact, using these techniques to analyse the OSLO Model, amounts to the same analysis as that of the quenched Edwards–Wilkinson equation (see Sec. 6.2.1.2), which was done around the same time by Nattermann *et al.* (1992). As discussed in Sec. 6.3.4.2, instead of a Heaviside step function the quenched Edwards–Wilkinson equation contains a quenched noise, which can be handled using more advanced RG techniques, such as functional or exact renormalisation (Leschhorn *et al.*, 1997; Chauve and Le Doussal, 2001; Le Doussal *et al.*, 2002).

The majority of renormalisation approaches to SOC aim to develop an RSRG (e.g. Vespignani *et al.*, 1997). Since SOC models evolve according to a set of rules which are not readily expressible in terms of equations parameterised by couplings, any RSRG attempt first has to capture the bulk dynamics in such parameters, which is often done on an ad hoc basis. Drawing on the **fixed-scale transformation** developed earlier (Pietronero, Erzan, and Evertsz, 1988; Pietronero and Schneider, 1991; Erzan, Pietronero, and Vespignani, 1995), Pietronero *et al.* (1994, also Vespignani and Zapperi, 1995; Lin and Hu, 2002) were the first to develop a dedicated renormalisation scheme for SOC models, parameterising the two-dimensional MANNA Model as well as the BTW Model by the respective probabilities that a site will charge 1, 2, 3 or 4 sites at toppling. Even when this bold approach ignores most correlations, it produced estimates for exponents that are stunningly close to numerical values even after further simplifications (Moreno, Gómez, and Pacheco, 1999), possibly hinting at the irrelevance of (some) correlations (see Sec. 4.1, in particular p. 99 and p. 101). Ben-Hur and Biham (1996, also Paczuski and Bassler, 2000) suggested early that multiple topplings were missing as a key ingredient from the scheme, to explain why their numerics indicated distinct universality classes for the BTW and MANNA Models in contrast to the RSRG result. Using a triangular lattice (Papoyan and Povolotsky, 1997) or increasing the block size of the renormalisation scheme to 3×3 (Lin and Hu, 2002) produces exponents deviating more from simulation results than those derived originally. Lin and Hu proposed two possible explanations: either multiple topplings are to be included in the procedure, or the exponents of the BTW Model scaling are not well defined to start with.

Marsili (1994a) applied the RSRG to the BS Model (also Mikeska, 1996), and Loreto *et al.* (1995) applied it to the FFM. The latter went on to develop the so-called **dynamically driven renormalisation group** (DDRG) (Vespignani, Zapperi, and Loreto, 1996; Vespignani *et al.*, 1997), discussed and illustrated in great detail by Jensen (1998). Because of technical difficulties, many renormalisation schemes for SOC models ignore correlations in the form of 'higher order proliferations' (Vespignani *et al.*, 1997) of activity, mediated by multiple topplings (Lin, Cheng, and Liaw, 2007) and backward avalanches (Sec. 8.4.4). Ivashkevich (1996) was able to improve the scheme systematically in this respect and re-analyse (Ivashkevich, Povolotsky, Vespignani, and Zapperi, 1999) the ASM originally studied by Pietronero *et al.* (1994). Hasty and Wiesenfeld (1997) applied this scheme to the exactly solvable one-dimensional (non-Abelian) BTW Model, which allowed them to develop a clearer understanding of the scheme, in particular with respect to the rôle and the origin of the 'balance condition'. Applied to a class of directed models (Dhar and Ramaswamy, 1989), Hasty and Wiesenfeld (1998a, also Ben Hur, Hallgass, and Loreto, 1996) found further evidence for the general validity of the renormalisation scheme, while the method does not seem to cope well when faced with anisotropy. Given that the proposed scheme was designed to handle the notoriously difficult problem of correlations, passing tests based on models that lack such correlations represents a minimal requirement. It is a regrettable fact that no non-trivial exactly solvable SOC model is known, which could serve as a testbed for any such method.

While the physically relevant, non-trivial fixed points in Wilsonian RG schemes for traditional critical phenomena have an unstable direction, SOC models flow to non-trivial, *stable* fixed points, as one would expect for a system that develops naturally into a critical state:

> In fact, the fixed point is attractive in the whole phase space, so that the parameters [. . .] evolve spontaneously toward their critical value. Therefore, in this perspective we are able to understand the self-organized nature of the critical stationary state of sandpile models. (Vespignani and Zapperi, 1995)

Similarly, Paczuski and Boettcher (1996) pointed out that the OSLO Model drives the quenched Edwards–Wilkinson equation to the critical point in a very different way compared to the traditional approach. In the latter a *critical* force is applied, whereas in the former, a vanishingly small, constant velocity drives the system to the critical point. In line with the AS mechanism, the velocity is the order parameter of the depinning transition, i.e. the system approaches the critical point by making the order parameter arbitrarily small.

8.3 Abelian symmetry and the operator approach

In the context of SOC, the attribute **Abelian** [20] generally indicates that the order of updates is irrelevant for the final configuration of the system. It is used very loosely, regardless of whether or not operators and the space they operate on are actually defined or form a (semi-)group. In the following, some details of the operator formalism are discussed using the Abelian BTW Model, a special case of the ASM, and the MANNA Model as examples.

The term 'Abelian' was introduced by Dhar (1990a, also Sec. 4.2) together with operators $a_\mathbf{n}$ which charge a *stable* configuration (i.e. all $z_\mathbf{n} \leq z^c$) of the ASM by adding a single slope unit at site $\mathbf{n}$ and fully relaxing the system according to the toppling rules, so that the system reaches a new stable configuration. The operators therefore map one stable configuration to a new stable configuration. The operators do *not* fix the microscopic dynamics, nor can they be used to describe them. Rather, they are the *product* of these toppling rules.

The final state of an Abelian model is independent of the order of external (driving) charges with subsequent relaxations, represented by the operators $a_\mathbf{n}$ mentioned above. In that sense, the Abelian property is a **symmetry**, an invariance of the final state (or its distribution) under permutation of the order of charges, i.e. the operators commute. As discussed below, in a model with deterministic dynamics this state is a concrete configuration, in a stochastic model it is a distribution thereof. *Strictly, the Abelian symmetry of the operators is a property on the* macroscopic *time scale only, which says nothing about how the model is to be updated on the microscopic time scale, i.e. within an avalanche.* [21] Nevertheless, this is where the Abelianness is commonly used (e.g. Biham *et al.*, 2001; Hughes and Paczuski, 2002; Jo and Ha, 2008) – an Abelian model is generally understood to have a microscopic dynamics where the order of updates of toppling sites is irrelevant, and this is how it is used in the present context, where suitable. Clearly the effect on the microscopic dynamics is what is numerically and, more importantly, physically most relevant, as the Abelianness should manifest in each individual avalanche rather than in the fact that the order of external charging can be swapped.

Typically, Abelianness can be extended to the microscopic time scale, because it is derived from the observation that the final configuration of a system is independent of the order of *microscopic* updates (e.g. Dhar, 1990a, 1999a) *within* a single avalanche. For the same reason it is consistent to regard an unstable configuration containing a number of sites that are about to topple as a stable configuration subject to a number of charges represented by the repeated application of the operators $a_\mathbf{n}$, see Sec. 4.2.1. Each of those would actually produce a new *stable* configuration, whereas the update of the original

---

[20] Known as Abelian symmetry, the Abelian property or Abelianness.

[21] In this strict sense, Abelianness is not a concern for any model which is only ever charged on one site only, such as the oOM or the one-dimensional BTW Model.

unstable configuration would become stable only at the end of the avalanche. This approach therefore makes strong use of the Abelian symmetry.

In principle, however, there could be an Abelian model, with a particular, fixed dynamics, whose final state would not depend on the order in which external charges were processed, whose final state, however, would change if the microscopic dynamics was changed. In other words, one cannot claim that a model is not Abelian because its dynamics is 'obviously' not Abelian, unless a link is established between microscopic updates and the macroscopic level of charge and relaxation.

The oMM illustrates this conundrum. By its definition it is clear that two charges arriving simultaneously at a site will produce a different outcome than their sequential arrival. Changing the order of (microscopic) updates or not updating in parallel would therefore generally change the final state. But it is unclear a priori how this translates to the dependence of the final state on the order of charges for a system that is charged externally at two different sites. The way out is to construct an example (see Fig. 6.1, p. 166) which explicitly shows the lack of Abelianness on the macroscopic level. This works in particular in situations where the microscopic dynamics in the definition of the model does not leave any ambiguity in the order of updates, so that proving that something depends on the order of updates is strictly not a statement about the particular model.

The opposite case of a model whose final state is unaffected by a change in the order of updates during an avalanche, while its final state depends on the order in which external charges are processed, has a prominent example in the BTW Model, see Sec. 4.1.2. Here, only the *driving* is non-Abelian, whereas the microscopic updates are identical to that of the ASM and therefore commute.

Small perturbations in Abelian models do not spread but rather persist indefinitely, as illustrated by Stapleton *et al.* (2004), who compared the temporal evolution of two realisations of the OSLO Model, which differed by a single grain added in the bulk.[22] The 'damage' is measured as the total absolute height differences of the two systems (Gleiser, 2001) and is initially unity. Given that the systems are finite, the damage is bounded, yet the authors argued that its scaling in time and system size as well as the frequent returns to its initial value are a consequence of the Abelian nature of the updating, which keeps the model at 'the edge of chaos'.

Using the Abelian symmetry to change the order of updates during an avalanche generally affects all observables that depend on the microscopic time, whereas certain 'static' observables, such as the avalanche size, are invariant under the change of updates, see for example Sec. 4.2.1.[23] The latter feature, however, does *not* extend to the macroscopic time scale, i.e. charging and relaxing an Abelian model in two different orders generally produces two different pairs of avalanche sizes. In deterministic cases, however, their sum is invariant under a change of order of charging (see endnote [4.2] on p. 391), and one might expect the statistics of such sums to be invariant in stochastic cases (see Sec. 6.3.4.1, p. 202).

---

[22] Remarkably, such a perturbation amounts itself to a *non*-Abelian operation.

[23] Transport properties, the residence time and the configuration of the system in terms of heights of stacks of *distinguishable* grains, generally are *not* invariant, see Sec. 6.3.3.

As assumed in the following, generally an Abelian model has operators that commute with respect to *all states*. This condition can be relaxed to some extent, i.e. for a certain (small) set of states the final state might depend on the order in which operators are applied. Some of the results would then need to be adapted, in particular those making general statements about transient and recurrent states, as derived in the next section, as well as in Sec. 8.3.2.2 and in Sec. 4.2.1 (e.g. p. 96). The operator formalism discussed below in general terms has been introduced in preceeding chapters for specific models, such as the BTW Model (Sec. 4.1, specifically Sec. 4.2.1 and Fig. 4.5), the MANNA Model (Sec. 6.1, in particular Sec. 6.2.1.1) and the OSLO Model (Sec. 6.3, especially Sec. 6.3.4.1).

8.3.1 Explicit operator formalism

Unfortunately, the Abelianness of a model is normally demonstrated on the basis of verbal arguments involving its rules, rather than derived in a mathematically rigorous manner. Even when it *really is* 'obvious', 'clear' and 'easy to see', a more formal approach might allow the Abelian nature to be traced back to a particular feature of the model. 'Obviously', a model cannot be Abelian if the number of particles shed in a toppling depends on the number of particles that have been deposited there in the previous time step. However, it is, not at all obvious that such an 'obviously non-Abelian' rule can lead to different final states, if the order of external charges is changed. Yet, it does indeed spoil any attempt to prove Abelianness on the basis of microscopic updates. Similarly, it is 'easy to see' that discharging sites should not render any other unstable site stable. [24] What remains to be discussed is a mathematical property of the operator algebra which gives rise to their being Abelian.

A first step to prove the Abelian symmetry 'from first principles', is to re-express the rules of the model in terms of operators. Representing them explicitly as matrices might help, but, of course, should not *add* any relevant properties, see Sec. 6.2.1.1. One major obstacle to overcome is presented by the configuration space of the system, which a priori includes unstable configurations and might therefore have infinite size; a time evolution operator U would evolve the model according to its rules from one unstable configuration to another (stable or unstable) configuration. The major advantage of this approach is the explicit implementation of time.

A different path has been taken by Pruessner (2004b, also Stapleton and Christensen, 2006) for the TAOM (Sec. 6.2.1.1). Here, unstable sites are effectively represented by operators acting on a stable configuration. Along these lines, the one-dimensional Abelian BTW Model with stable configurations $\mathcal{C} = \{z_1, z_2, \ldots, z_L\}$ and $z_\mathbf{n} \in \{0, 1\}$ for $\mathbf{n} = 1, 2, \ldots, L$ and $z^c = 1$ would be written in terms of operators

$$a_\mathbf{n} = s_\mathbf{n} + a_{\mathbf{n}+1} a_{\mathbf{n}-1} t_\mathbf{n}, \tag{8.52}$$

where $s_\mathbf{n}$ is an operator depositing a slope unit at $z_\mathbf{n}$ and $t_\mathbf{n}$ removes a slope unit from $z_\mathbf{n}$. The open boundary conditions are realised by $a_0 = a_{L+1} = \mathbf{1}$. The operators $a_\mathbf{n}$ evolve a stable configuration $\mathcal{C}$ by charging site $\mathbf{n}$ and relaxing it fully, giving rise to a new stable configuration $\mathcal{C}' = a_\mathbf{n} \mathcal{C}$. The effect of the operators is illustrated for $L = 2$ in Fig. 4.5. For

---

[24] Equation (8.60), discussed in Sec. 8.3.1.3, illustrates the problem more clearly.

some of the following derivations it is more convenient to introduce (Markovian) operators $\hat{a}_n$, which charge site n and fully relax the system just like a_n, but operate on the space of *all* stable configurations, i.e. their distribution, represented by a vector of size 2^L, and again (see Eq. (6.19) on p. 200)

$$\hat{a}_n = S_n + \hat{a}_{n+1}\hat{a}_{n-1}T_n. \tag{8.53}$$

The matrix S_n selects all states where site n has slope $z_n = 0$ and changes it to $z_n = 1$, the matrix T_n does the opposite, in which case the site topples by discharging two particles. The relation between the representation of the dynamics in terms of a_n and $\hat{a}_n$ is identical to that between a Langevin equation of motion and the Fokker–Planck equation. A specific configuration is then represented by having 1 at one particular entry of that vector and 0 everywhere else. Using the direct product $\otimes$ the matrices S_n and T_n can be written down explicitly for the Abelian BTW Model,

$$S_n = \mathbf{1}_2^{\otimes(n-1)} \otimes \begin{pmatrix} 0 & 0 \\ 1 & 0 \end{pmatrix} \otimes \mathbf{1}_2^{\otimes(L-n)} \tag{8.54a}$$

$$T_n = \mathbf{1}_2^{\otimes(n-1)} \otimes \begin{pmatrix} 0 & 1 \\ 0 & 0 \end{pmatrix} \otimes \mathbf{1}_2^{\otimes(L-n)} \tag{8.54b}$$

where $\mathbf{1}_2$ is the 2×2 identity matrix. The resulting operators $\hat{a}_n$ are therefore represented by binary matrices. They are Markovian, but invertible only if restricted to the subspace spanned by the basis vectors of recurrent states (see footnote 10 on p. 95), in which case they are **permutation matrices**, which implies the existence of an inverse and that all recurrent states of a given operator $\hat{a}_n$ appear with the same frequency, which is proven in the next section on the basis of the operators a_n.

One might be tempted to derive the Abelian symmetry $\hat{a}_n\hat{a}_{n'} = \hat{a}_{n'}\hat{a}_n$ using a **cancellation mechanism** of the form [8.5]

$$S_n^2 = 0, \qquad\qquad T_n^2 = 0, \quad S_nT_n + T_nS_n = \mathbf{1}, \tag{8.55a}$$

$$[S_n, S_{n'}] = S_nS_{n'} - S_{n'}S_n = 0, \qquad [T_n, T_{n'}] = 0, \qquad [S_n, T_{n'}] = 0, \tag{8.55b}$$

as suggested by Eq. (8.53) and Eq. (8.54).

This argument, however, is flawed. Defining $\hat{a}_n = S_n + \hat{a}_{n+1}\hat{a}_{n-1}T_n$ recursively already makes use of the Abelian nature of the operator $\hat{a}_n$, by assuming that charging and fully relaxing the neighbours *sequentially*, $\hat{a}_{n+1}\hat{a}_{n-1}$, is a faithful representation of the model, which actually prescribes [25] that sites are updated in parallel, i.e. all sites that are unstable relax simultaneously. In other words, the model defined through Eq. (8.53) and Eq. (8.54) might be Abelian, but it might not be the BTW Model. Writing down operators in the form described above predisposes the model to be Abelian.

8.3.1.1 Stationary state

The external drive strikes with probability [26] c_n at site n. The set $\mathfrak{R} = \{\mathcal{C}, \mathcal{C}', \dots\}$ considered in the following contains all states $\mathcal{C}$ which are recurrent for all operators with $c_n > 0$ (jointly

---

[25] Not always as explicit as one might wish.

[26] If the external drive is applied in a certain pattern, rather than randomly, the proof might either not hold (Sec. 8.3.2.2, p. 281) or have to be adapted accordingly.

recurrent states, see p. 96), i.e. none of the states is transient with respect to any operator. For convenience, none of the operators is trivially $a_n = 1$. This condition can be lifted later and does not change any of the results below, provided the identity operator is applied with equal probability to any of the configurations.

In Abelian systems, a state that is recurrent with respect to one operator might be transient with respect to another, so the union of recurrent states may contain transients of some operators. Yet, the intersection of the sets of recurrent states of each operator is closed under these operators, i.e. an operator $a_{n'}$ acting on a recurrent state of a_n cannot bring about a transient state of a_n, see inaccessible transients, p. 96.

At **stationarity** all C fulfil

$$\sum_{C' \in \mathfrak{R}} P(C)T(C \to C') = \sum_{C'' \in \mathfrak{R}} P(C'')T(C'' \to C), \tag{8.56}$$

where $T(C \to C')$ denotes the probability of a transition from C to C' and $P(C)$ is the probability that C occurs. The terms $C' = C$ and $C'' = C$ can be discounted on both sides as they cancel, so that the left hand side is the outflux from C (otherwise it is $P(C)$) and the right hand side the influx into C in the spirit of a **master equation**. If $\sum_{C'} T(C \to C') = \sum_{C''} T(C'' \to C)$, then $P(C) = \text{constant} = 1/|\mathfrak{R}|$ is a solution of Eq. (8.56), where $|\mathfrak{R}|$ is the cardinality of the set $\mathfrak{R}$ of recurrent states. If the system has N sites, there are N different operators a_n and therefore N new states $C'_n = a_n C$ the given state C can evolve to. In general, they are degenerate, i.e. the C'_n are *not necessarily distinct*. Notably, there might even be some $\mathbf{n} = \mathbf{n}^*$ for which a_{n^*} leaves C unchanged, $C'_{n^*} = a_{n^*}C = C$.

If all C'_n were distinct for each C, then either $T(C \to C') = 0$ (namely when C'_n is recurrent but not reachable from C) or there is, for the initial state C given, exactly one $\mathbf{n}$ with $C' = a_n C$ so that $T(C \to C') = c_n$. Applying the same argument to $C'' = a_n^{-1}C$, means that there is at most one $\mathbf{n}$ so that $a_n C'' = C$ and thus $T(C'' \to C) = c_n$, if it does not vanish. The existence of an inverse is discussed in detail in Sec. 8.3.2.2. The terms in the sums $\sum_{C'} T(C \to C')$ and $\sum_{C''} T(C'' \to C)$ are thus pairwise equal and so are the sums. As a result $P(C) = \text{constant}$ solves Eq. (8.56).

The problem addressed in the following is that the C'_n are not necessarily distinct and thus there is no immediate path to the pairwise equality of terms. A priori, there could be operators a_{n^*} that contribute to the influx into C, i.e. $\sum_{C'' \neq C} T(C'' \to C)$, but not to the outflux, i.e. $\sum_{C' \neq C} T(C \to C')$, because they act as identities in C, so that $a_{n^*}C = C$. Equally, there could be operators that contribute to the outflux, but not to the influx, because there is no C'' such that $a_n C'' = C$.

For each successor C' the transition probability $T(C \to C')$ is the sum of c_n with $C' = a_n C$. The proof of the existence of the solution $P(C) = \text{constant}$ hinges crucially on the fact that if any pair a_n, a_m share the same successor when operating on a recurrent state C, then they also share the same predecessor and vice versa [27]

$$a_n C = a_m C \quad \Leftrightarrow \quad \exists\, C'' \in \mathfrak{R} : a_n C'' = a_m C'' = C. \tag{8.57}$$

The proof of $\Leftarrow$ uses the Abelian symmetry, $a_n C = a_n a_m C'' = a_m a_n C'' = a_m C$, the proof of $\Rightarrow$ uses the existence of an inverse a_n^{-1} as well as Abelianness, which becomes clearer by

---

[27] In practice Eq. (8.57) means that a_n and a_m are identical, although this is not assumed in the following.

writing the right hand side as $a_m a_n^{-1} C = C$ with $a_n^{-1} C = C''$ recurrent, $C = a_n^{-1} a_m C = a_m C''$ and $C = a_m^{-1} a_n C = a_n C''$. Equation (8.57) means that the set of sites $\mathbf{n}$ for which each $a_n C$ equals a given C' is the same set of sites for which a (recurrent) predecessor C'' exists with $C = a_n C''$.

The sum over the c_n of the latter gives $\mathcal{T}(C'' \to C)$ and the sum over the c_n of the former gives $\mathcal{T}(C \to C')$, so that $\mathcal{T}(C \to C') = \mathcal{T}(C'' \to C)$ for all C and C', with C'' fixed by $a_n C'' = C = a_n^{-1} C'$ if $\mathcal{T}(C \to C')$ is non-zero and thus contains the contribution c_n from an operator a_n. If $\mathcal{T}(C \to C')$ vanishes, there is a C'' so that $\mathcal{T}(C'' \to C)$ vanishes as well: as every C'' can be uniquely mapped to a C' if $\mathcal{T}(C \to C') \neq 0$, because then a_n exists such that $a_n C'' = C = a_n^{-1} C'$, and because both C'' and C' are taken from the same set $\mathfrak{R}$, there are precisely as many C'' as there are C' for which $\mathcal{T}(C'' \to C)$ and $\mathcal{T}(C \to C')$ respectively vanish. Each $\mathcal{T}(C'' \to C)$ can thus be paired up with an equal $\mathcal{T}(C \to C')$.

Having identified the transition probabilities in each sum, the sums $\sum_{C'} \mathcal{T}(C \to C')$ and $\sum_{C''} \mathcal{T}(C'' \to C)$ must be identical, so that $P(C) = 1/|\mathfrak{R}|$ is a solution of Eq. (8.56). The key ingredients are the existence of unique predecessors and successors among recurrent states, i.e. the $\hat{a}_n$ being permutation matrices or, equivalently, the system being deterministic and possessing inverse operators (but see Sec. 10.1, p. 352). Every operator a_n that contributes to the influx to C contributes equally to the outflux from C, rather than to $\mathcal{T}(C \to C)$. This suggests a different approach to prove the equality of influx and outflux (see inaccessible transients, p. 96): there is no Abelian operator a_n with $a_n C'' = C$ for a recurrent $C'' \neq C$, which at the same time has $a_n C = C$. Equally, there is no operator with recurrent C and $a_n C \neq C$, but no recurrent C'' such that $a_n C'' = C$.

8.3.1.2 Properties of the Markov matrices

The actual (macroscopic) time evolution operator of stochastically driven, but otherwise deterministic models in a recurrent states, is a linear combination of permutation matrices, i.e. it is itself not a permutation matrix. The full, one-dimensional Abelian BTW Model, where all sites are equally likely to be charged by the external drive, is represented by the matrix

$$\mathcal{O} = \frac{1}{L} \sum_{\mathbf{n}}^{L} \hat{a}_{\mathbf{n}}. \tag{8.58}$$

Generally, the **stationary state** of the Markov matrix $\mathcal{O}$ is a (not necessarily unique) **distribution vector** $|0\rangle$ with eigenvalue 1, i.e.

$$\mathcal{O} |0\rangle = |0\rangle, \tag{8.59}$$

which implies $\mathcal{O}^n |0\rangle = |0\rangle$ for all powers $n \in \mathbb{N}$.[8.6] By the Markov property, a left eigenvector of $\mathcal{O}$ with eigenvalue unity is a column of 1s, $\langle 0|$, whose transpose is also a right eigenvector if $\mathcal{O}$ is symmetric, corresponding to all states being equally likely. If the $\hat{a}_n$ are permutation matrices, as is the case in deterministic models such as the ASM, then their transpose (equal to their inverse, as they are orthogonal) is Markovian as well, so that a row of 1s is an eigenvector with eigenvalue one of every such $\hat{a}_n$. As a result, the uniform distribution is a stationary state, as shown above on the basis of the existence of an inverse.

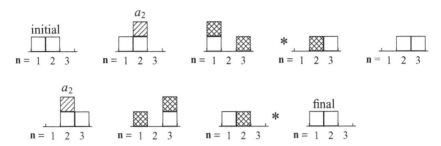

Figure 8.6 Applying $a_2 a_2$ to a recurrent configuration of the Abelian BTW Model in one dimension with $z^c = 1$ and $L = 3$. A box represents a slope unit, a hatched box is a slope unit added by the external drive, a cross-hatched box is a slope unit deposited by a toppling neighbour. A star indicates a slope unit dissipated at one of the open boundaries. The upper row shows the sequence of intermediate (unstable) states after the first charge a_2, the lower row the second sequence. Both sequences show at the end the stable state reached.

The (probability) normalisation $\langle 0|i \rangle$ of every eigenvector $|i\rangle$ of $\mathcal{O}$ with an eigenvalue $\lambda_i \neq 1$ vanishes by orthogonality, or $\langle 0|i \rangle = \langle 0| \mathcal{O} |i \rangle = \lambda_i \langle 0|i \rangle$. The Markov property ensures that $-1 \leq \lambda_i \leq 1$. Not much more can be said regarding completeness and orthogonality of eigenvectors (see, for example, Sec. 6.3.4.1). If they span the entire vector space, then correlations to the initial value of any observable that can be written as a linear operator are dominated by the largest eigenvector $\lambda_i \neq 1$. For example, if the system is prepared in a particular state distribution $|P_0\rangle = \sum_i b_i |i\rangle$, then its evolution is given by $|P_n\rangle = \mathcal{O}^n |P_0\rangle = \sum_i b_i \lambda_i^n |i\rangle$. In this sum all contributions with $|\lambda_i| < 1$ vanish with increasing n, while terms with $\lambda_i = e^{i\phi}$, $\phi \in (0, 2\pi)$ oscillate and those with $\lambda_i = 1$ remain constant. In the simplest case there is a single eigenvector with $\lambda_0 = 1$ and $\mathcal{O}^n |P_0\rangle - |0\rangle$ decays essentially as $\exp(-n/\tau)$, where $\tau = -1/\ln(|\lambda_1|)$ is the correlation time derived from the largest eigenvalue $|\lambda_1| < 1$. This argument extends to 'linear' observables of the form $Q_n = \langle Q|P_n\rangle$. It breaks down if $|P_0\rangle$ cannot be expressed in terms of $|i\rangle$, but might be restored by introducing generalised eigenvectors.

8.3.1.3 An operator identity

The operator identity

$$a_{\mathbf{n}}^{\Delta_{\mathbf{nn}}} = \prod_{\mathbf{n}' \neq \mathbf{n}} a_{\mathbf{n}'}^{\Delta_{\mathbf{nn}'}} \tag{8.60}$$

found by Dhar (1990a) for the ASM is another highly non-trivial property of the operators, which is similarly found for the one-dimensional OSLO Model, as $\hat{a}_{\mathbf{n}}^3 = \hat{a}_{\mathbf{n}+1} \hat{a}_{\mathbf{n}} \hat{a}_{\mathbf{n}-1}$ (Dhar, 2004).[28] Translating it to the one-dimensional Abelian BTW Model with $z^c = 1$,

$$a_{\mathbf{n}}^2 = a_{\mathbf{n}+1} a_{\mathbf{n}-1} \tag{8.61}$$

---

[28] The $a_{\mathbf{n}}$ operate on states, whereas the $\hat{a}_{\mathbf{n}}$ are Markov matrices operating on their distribution and thus do not generally possess an inverse. This is why the $\hat{a}_{\mathbf{n}}$ cannot be cancelled to give $\hat{a}_{\mathbf{n}}^2 = \hat{a}_{\mathbf{n}+1} \hat{a}_{\mathbf{n}-1}$ even when they commute. The power three reflects the fact that it takes up to three charges to make a site topple in the OSLO Model.

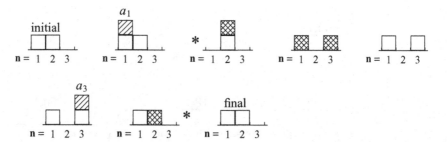

As Fig. 8.6, but applying a_3a_1 to the configuration. The final state is the same as in Fig. 8.6 and Fig. 8.8, and in all three figures one slope unit is dissipated on each boundary. The upper row shows the application of a_1, the lower row the subsequent effect of a_3.

which is confirmed by the formalism developed above, Eq. (8.52) and Eq. (8.54):

$$\hat{a}_n^2 = S_n^2 + \hat{a}_{n+1}\hat{a}_{n-1}(T_nS_n + S_nT_n) + \hat{a}_{n+1}^2\hat{a}_{n-1}^2T_n^2 = \hat{a}_{n+1}\hat{a}_{n-1}, \qquad (8.62)$$

using the cancellation mechanism Eq. (8.55). The order of charges in Eq. (8.61) does not matter, as the model is Abelian. For the same reason, it is irrelevant that the model is fully relaxed after both charges; in principle the first charge could be added and the relaxation performed only after the second external charge.

At first, the interpretation of Eq. (8.61) (and Eq. (8.60) generally) seems clear: Adding two slope units to site n, 'the site is bound to topple once and one [slope unit] is added to each of its nearest neighbours' (Dhar, 1999c). It is tempting to interpret this as: charging a site twice is just as good as leaving it untouched and charging each of its neighbours once. Either the site has $z_n = 0$, then the first charge leads to $z_n = 1$ and the second to its toppling to the neighbours, or it has $z_n = 1$, so that the first charge leads to $z_n = 0$ and a toppling to the neighbours and the second back to $z_n = 1$.

This latter, naïve interpretation of Eq. (8.61) is wrong. As illustrated in Fig. 8.6, it can easily happen that *both* charges a_n^2 lead to the site toppling.[29] That, of course, does not mean that a_n^2 is equivalent to $(a_{n+1}a_{n-1})^2$, because the charging and subsequent toppling of site n after applying a_n adds one and removes two slope units from there.

It seems straightforward, although messy, to generalise the matrix representation used in Eq. (8.62) to prove the generalised statement Eq. (8.60). In any case, it is 'easy to see' (or at least 'easy to agree') that Eq. (8.60) is correct by consulting the microscopic dynamics of the model. Using again the one-dimensional Abelian BTW Model, Figs. 8.7 and 8.8 show the result of operating with a_3a_1 and a_1a_3 respectively on the initial configuration of Fig. 8.6. Both operations lead to the same final configuration and the same dissipation, both go through 3 topplings, namely one less than for a_2a_2. Yet, each of the three operations a_2a_2, a_3a_1 and a_1a_3 reaches an intermediate configuration that does not feature in the other two, even though there is no ambiguity due to sites toppling simultaneously.

---

[29] It is bound to topple at least once.

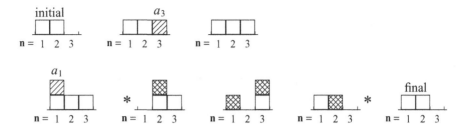

Figure 8.8 As Fig. 8.6 and Fig. 8.7, but applying $a_1 a_3$ to the configuration. Again, the final state is identical to that shown in Fig. 8.6. The upper row shows the application of a_3, the lower row the subsequent effect of a_1.

8.3.2 Abelian operators in stochastic models

Originally, stochasticity was thought to be incompatible with Abelianness (Milshtein *et al.*, 1998; Pastor-Satorras and Vespignani, 2000e). The operators $\hat{a}_n$ have been introduced above as acting on the *distribution* of configurations. In the BTW Model and the ASM, which have deterministic relaxation rules, these operators can therefore be written as **permutation matrices**. As they fulfil the **Markov condition**, they can equally be regarded as **Markov matrices**. The operator $\mathcal{O}$ introduced in Eq. (8.58) is a weighted sum of such permutations and therefore a non-trivial [30] Markov matrix.

The operators characterising stochastic, Abelian models are non-trivial Markov matrices even for charging a single site. The Abelianness of such models again is reflected in the vanishing commutator of these Markov matrices, i.e. the probabilities of the final states are independent of the order of external charges (Dhar, 1999b). As for models with deterministic relaxation rules, the operators act on the *distribution* of configurations, which, again, has a stationary solution of the form Eq. (8.59).

All of the above precautions apply, for example that the Abelian nature of the operators have not been derived formally from a particular feature of their representation, but rather are a matter of very convincing arguments.

Another problem, however, is specific to stochastic systems: the proof of their Abelian nature effectively fixes the random numbers, and therefore does not immediately translate to the associated Markov matrices. This discord is very instructively illustrated by Dhar (1999a, cf. Dhar, 1999b), where the MANNA Model is shown to be Abelian (Dhar, 1999c, 2004, for the OSLO Model), by assuming that each site **n** has its own independent **fixed infinite sequence of random numbers** attached to it (also Sec. 10.1, p. 352). In other words, what is shown to be Abelian are the fully deterministic (Markovian) operators which act on the state of the model including the positions within the sequence of random numbers drawn for each site. However, the Markov matrices which operate directly on the distribution of states *not* including the position in the local sequence of random numbers, are much more interesting and convenient to study. The Markov matrices of the one-dimensional MANNA

---

[30] It is not just a permutation matrix.

Model, for example, obey

$$\hat{a}_{\mathbf{n}}^2 = \frac{1}{4}(\hat{a}_{\mathbf{n}+1} + \hat{a}_{\mathbf{n}-1})^2 \tag{8.63}$$

similar to Eq. (8.61) for the one-dimensional BTW Model, see Sec. 6.2.1.1. Equation (8.63) means that the distribution due to a double charge at site $\mathbf{n}$ equals the distribution due to charges of the neighbours with certain probabilities (a quarter for each possible combination of outcomes). Equation (8.63) does not apply to the fully deterministic Abelian operators mentioned above. If the sequence of random numbers is fixed, charging or not charging site $\mathbf{n}$ completely alters the final outcome, so that charging a particular site can never be replaced by charging neighbours. Deterministic operators incorporate the fixed sequence of random numbers explicitly, which clashes with Eq. (8.63) wrapping up their effect in the prefactor $1/4$. Equation (8.63) therefore applies only to the Markov matrices $\hat{a}_{\mathbf{n}}$ acting on the distribution of states without fixed sequence of random numbers.

8.3.2.1 Equation of motion

The time evolution of Abelian models can be expressed as an equation of motion in terms of the number of (slope) charges $g(\mathbf{n}, t)$ received by site $\mathbf{n}$ up to microscopic time t, thereby fixing the microscopic dynamics. In the presence of stochasticity, the equation of motion is a Langevin equation (Pruessner, 2003a). It is based on the observation that the number of charges at site $\mathbf{n}$ is given by the number of topplings $G(\mathbf{n}', t)$ of its neighbouring sites $\mathbf{n}'$, because in an Abelian model, the same number of particles or slope units are shed in every toppling, see Eq. (8.64). The number of topplings, again by the Abelian property, is determined by the number of charges and the initial value of a site, see Eq. (8.65) below. Without the Abelian symmetry, the number of charges due to toppling neighbours might depend on the order of topplings and no simple equation relates the number of charges to the number of topplings. One might speculate that the existence of a *local* equation of motion like Eq. (8.66) below presupposes Abelianness.

For example, in the one-dimensional Abelian BTW Model with $z^c = 1$ and $z_{\mathbf{n}} = 0$ initially, the number of charges is

$$g(\mathbf{n}, t) = G(\mathbf{n} + 1, t) + G(\mathbf{n} - 1, t), \tag{8.64}$$

ignoring boundary conditions. The number of topplings of a site is [31]

$$G(\mathbf{n}, t + 1) = \left\lfloor \frac{g(\mathbf{n}, t)}{2} \right\rfloor \tag{8.65}$$

ignoring the external drive and taking into account the initial condition. From Eq. (8.64) and Eq. (8.65) one derives

$$g(\mathbf{n}, t + 1) - g(\mathbf{n}, t) = \left\lfloor \frac{g(\mathbf{n} + 1, t)}{2} \right\rfloor - g(\mathbf{n}, t) + \left\lfloor \frac{g(\mathbf{n} - 1, t)}{2} \right\rfloor \tag{8.66}$$

with $g(\mathbf{n}, t)$ subtracted on both sides to make the expression resemble a diffusion equation, $\partial_t g = \partial_x^2 g / 2$.

---

[31] Equation (8.65) prescribes (arbitrarily) the microscopic dynamics in that it fixes the number of topplings performed by a site by time $t + 1$ having received a certain number of charges up to time t.

The configuration of a system effectively encodes the difference between charges and topplings, i.e. the net number of charges at a site is given by

$$z_{\mathbf{n}}(t) = g(\mathbf{n}, t) - 2G(\mathbf{n}, t) = g(\mathbf{n}, t) - 2 \left\lfloor \frac{g(\mathbf{n}, t)}{2} \right\rfloor. \tag{8.67}$$

8.3.2.2 Existence of an inverse

The term 'Abelian' might suggest that the operators generally form a group (rather than a semi-group), which means that an inverse exists and is part of the group. This is in fact the case for the ASM, which is deterministic and has a finite number of stable states, so that there must be a period $p_{\mathbf{n}} \geq 1$, after which the system returns to its original state after a number of charges at the same site, i.e. $a_{\mathbf{n}}^{p_{\mathbf{n}}} \mathcal{C} = \mathcal{C}$ for all *recurrent* states $\mathcal{C}$, so that $a_{\mathbf{n}}^{p_{\mathbf{n}}-1} = (a_{\mathbf{n}})^{-1}$ is the inverse of $a_{\mathbf{n}}$ (Dhar, 1990a; Wiesenfeld *et al.*, 1990), *on the set of recurrent states* (Dhar, 1999b).[32]

If $\mathcal{C}$ is not recurrent, then no power of $a_{\mathbf{n}}$ can invert the action of $a_{\mathbf{n}}$ on $\mathcal{C}$. For example, no power of the operators $a_{\mathbf{n}}$ in the Abelian BTW Model can undo any $a_{\mathbf{n}}$ on the empty lattice of size greater than 1, because the empty lattice is not recurrent. Generally, none of the matrices introduced in Eq. (8.52) is invertible, because they are not restricted to recurrent states. In this respect, Abelianness crucially differs from invertibility. The former applies to an operator operating on any state, the latter only if it is restricted to act on recurrent states.

Starting from any *transient* state $\mathcal{C}$, the sequence $a_{\mathbf{n}}\mathcal{C}, a_{\mathbf{n}}^2\mathcal{C}, a_{\mathbf{n}}^3\mathcal{C}, \ldots$ at some point will hit a recurrent state $a_{\mathbf{n}}^t\mathcal{C}$, so that $a_{\mathbf{n}}^{t+p_{\mathbf{n}}}\mathcal{C} = a_{\mathbf{n}}^t\mathcal{C}$. Clearly, its preceding state is not uniquely determined, as $a_{\mathbf{n}}(a_{\mathbf{n}}^{t-1}\mathcal{C}) = a_{\mathbf{n}}^t\mathcal{C}$ as well as $a_{\mathbf{n}}(a_{\mathbf{n}}^{p_{\mathbf{n}}+t-1}\mathcal{C}) = a_{\mathbf{n}}^t\mathcal{C}$, but $a_{\mathbf{n}}^{p_{\mathbf{n}}+t-1}\mathcal{C} \neq a_{\mathbf{n}}^{t-1}\mathcal{C}$. Because $a_{\mathbf{n}}^{p_{\mathbf{n}}}\mathcal{C} = \mathcal{C}$ for *every* recurrent state $\mathcal{C}$, a similar property follows for the Markovian operators $\hat{a}_{\mathbf{n}}$, which act on the distribution of recurrent states, namely $\hat{a}_{\mathbf{n}}^{p_{\mathbf{n}}} = \mathbf{1}$.

The situation is more complicated for stochastic models, because the operators $\hat{a}_{\mathbf{n}}$ are general Markov matrices and not simply permutations, not even if restricted to recurrent states. While the number of states is finite, their distribution vector $|P\rangle$, which the operator $\hat{a}_{\mathbf{n}}$ acts on, can take infinitely many values, so that there is generally no $p_{\mathbf{n}}$ such that $\hat{a}_{\mathbf{n}}^{p_{\mathbf{n}}}|P\rangle = |P\rangle$. In fact, the inverse of $\hat{a}_{\mathbf{n}}$ does not necessarily exist at all (let alone in the form $\hat{a}_{\mathbf{n}}^{p_{\mathbf{n}}}$) and if it does, it might not be Markovian itself, having negative entries. In the fully stochastic case (sequence of random numbers not fixed), there is generally no unique predecessor for a given final state.

One would think that this would be resolved once a stochastic model is made deterministic by **fixing the sequence of random numbers** (see Sec. 8.3.2, also Paczuski and Bassler, 2000). The state of the model then incorporates the position in a sequence of random numbers at each site. This position is incremented after each toppling so that strictly no state is recurrent. Normally, however, recurrence is the key ingredient to prove the existence of a (unique) inverse: if a subset $\{\mathcal{C}'_1, \mathcal{C}'_2, \ldots\}$ of all stable states are predecessors of $\mathcal{C}$ under $a_{\mathbf{n}}$ such that $a_{\mathbf{n}}\mathcal{C}'_1 = a_{\mathbf{n}}\mathcal{C}'_2 = \ldots = \mathcal{C}$, and all $\mathcal{C}'_i$ are recurrent with the same period $p_{\mathbf{n}}$, then multiplying by $a_{\mathbf{n}}^{p_{\mathbf{n}}-1}$ means $\mathcal{C}'_1 = \mathcal{C}'_2 = \ldots$, i.e. all potential predecessors are identical

---

[32] The period $p_{\mathbf{n}}$ is the smallest $p_{\mathbf{n}} > 0$ such that $a_{\mathbf{n}}^{p_{\mathbf{n}}}\mathcal{C} = \mathcal{C}$. The period generally differs among different operators $a_{\mathbf{n}}$, see Sec. 4.2.1, in particular p. 97.

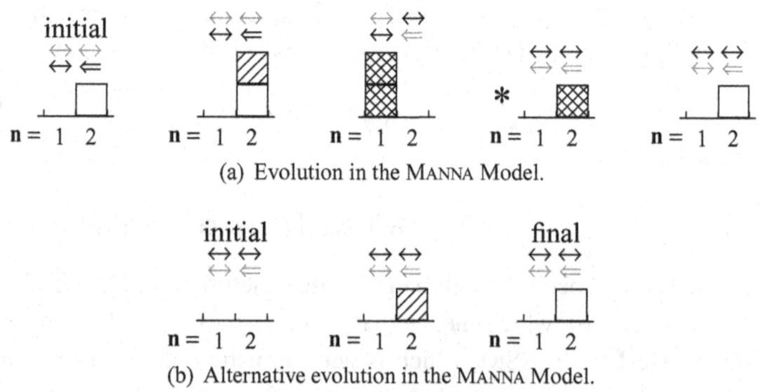

(a) Evolution in the MANNA Model.

(b) Alternative evolution in the MANNA Model.

Figure 8.9 | The evolution in the one-dimensional AMM with fixed sequence of random numbers (cf. Fig. 6.1). The arrows indicate the (randomly chosen) directions of toppling slope units: $\Leftarrow$ means both to the left, $\Rightarrow$ means both to the right and $\leftrightarrow$ means one to either side. The next direction to be used is shown in black, past and future ones are shown in grey. They are traversed from bottom to top. Hatched boxes are added by the external drive, cross-hatched boxes are added by neighbours toppling. (a) An evolution involving two topplings, eventually ending up in the same state as (b), where the two directions used in (a) have been used in the previous avalanche. It turns out that this is impossible, see text.

and the inverse operator of $a_\mathbf{n}$ is $a_\mathbf{n}^{p_\mathbf{n}-1}$. Without recurrence, a 'brute force inverse' of $a_\mathbf{n}$ can still be constructed by retracing the history of the model from the start, which requires knowledge of the initial state, the random numbers used and the sites charged. In general, it is unclear which ingredients are *necessary* to construct an inverse in stochastic models. It certainly is not just the current state, yet how much of the sequence of random numbers is needed is not obvious and maybe even the history of sites charged is required.

The problem is illustrated in Fig. 8.9, which shows two different initial states of a MANNA Model which both end up in the same final state and includes the position in the sequence of random numbers (shown as arrows). Nevertheless, Fig. 8.9 is *not* a counter-example for the existence of an inverse in the MANNA Model, because one of the two possible histories, Fig. 8.9(b), has an initial state which cannot be reached using the sequence of random numbers shown as the past (grey arrows). In other words, $\mathcal{C}'_a$ and $\mathcal{C}'_b$ both lead to the same final state, $a_2\mathcal{C}'_a = a_2\mathcal{C}'_b = \mathcal{C}$. However, there is no operator $a_\mathbf{n}$ and prior state $\mathcal{C}''$ such that $\mathcal{C}'_b = a_\mathbf{n}\mathcal{C}''$, while there is one for $\mathcal{C}'_a$, reminiscent of recurrent states in deterministic models. The initial state can therefore be determined uniquely in the present example, but only by probing which of the two possible initial states is compatible with the history of random numbers used. One might wonder whether this holds for other stochastic sandpiles and for larger systems. Paczuski and Bassler (2000) showed the explicit reconstruction of the initial state in a directed stochastic model.

As outlined in Sec. 4.2.1, Abelianness and the existence of an inverse, $a_\mathbf{n}^{-1} = a_\mathbf{n}^{p_\mathbf{n}-1}$, means that all limit cycles have the same length, i.e. all recurrent configurations appear with the same frequency, in each limit cycle of an operator [33] $a_\mathbf{n}$. Even without constructing an inverse, in a system with finite number of states, a *unique preceding state guarantees that*

---

[33] But of course $p_\mathbf{n}$ generally depends on $\mathbf{n}$.

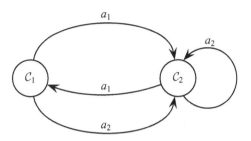

Figure 8.10   Example of two operators a_1 and a_2, each with an inverse within its limit cycle. The encircled symbols are states, the labelled arrows connecting them are the operators. See also Fig. 4.5.

every state features either once or never in every limit cycle of an operator, so that, again, all recurrent configurations appear with the same frequency or not at all. [34] Although maybe less obvious, *the same applies if the succeeding state is unique – in fact, the uniqueness of succeeding and preceding states is equivalent.* If either applies, a limit cycle closes as soon as any state is repeated, making (recurrent) succeeding and preceding states unique with respect to any given operator. As Fig. 4.5, p. 97, illustrates (bold arrows), *equal frequency of all states in the limit cycle does not guarantee that all these states feature in the stationary state (at all or with the same frequency)*, see also Sec. 8.3.1.1 and the comment on p. 352, in particular Fig. 10.1(a).

In this discussion the limit cycle of a single operator has to be distinguished from the stationary state of a system evolving under possibly many operators and maybe starting from many different initial configurations. Depending on the setup, many different distributions might be invariant under the evolution by the operator considered (i.e. stationary), the most important being a uniform distribution, Sec. 8.3.1.1. If more than one operator is concerned, deterministic evolution by individual operaors does not imply unique successors and predecessors and thus an inverse operator might not necessarily exist, not even among recurrent states. As illustrated in Fig. 8.10, for example $a_1 C_1 = a_2 C_1 = C_2 = a_2 C_2$ and $a_1 C_2 = C_1$ is deterministic and each operator has an inverse within its limit cycle, but as soon as both operators are considered simultaneously, the inverse of a_2 no longer exists, as $a_2 C_2 = a_2 C_1 = C_2$. Again, Abelianness comes to the rescue by inducing *jointly* recurrent states: C_1 is transient with respect to a_2 and thus cannot be accessed from C_2 via a_1 if it commutes with a_2. If $a_1 C_2 = C_1$ then $a_2 C_1 = a_2 a_1 C_2 = a_1 a_2 C_2 = a_1 C_2 = C_1$ as $a_1 a_2 = a_2 a_1$, contradicting $a_2 C_1 = C_2$. Generally, if $a_n C' = a_n C'' = C$ (i.e. the predecessor of C is not unique), then either C' or C'' is transient with respect to a_n. However, this transient state cannot be accessed from C using any operator a_n commutes with (see inaccessible transients, p. 96). If there was such a (generally composite) operator X, so that $XC = C'$, then recurrence of C under a_n guarantees recurrence of C' under a_n, as $a_n^{p_n} C = C$ implies $a_n^{p_n} C' = a_n^{p_n} XC = X a_n^{p_n} C = XC = C'$, using Abelianness. Repeating the argument for C'' shows that both C' and C'' are recurrent under a_n, whose effect on both states is therefore invertible, contradicting $a_n C' = a_n C'' = C$. *In Abelian models a jointly recurrent set of*

---

[34] A system with infinitely many states might not have any recurrent states at all.

states exists where the uniqueness of the successor ensures the uniqueness of the predecessor and thus the existence of an inverse.

In stochastic-turned-deterministic models, states are strictly not recurrent, because the non-periodic, fixed sequence of random numbers is processed sequentially, and none of the above arguments therefore applies. If the sequence of random numbers is infinite, then in the space of all possible configurations (local slopes and local positions in the sequences of random numbers), every state is visited either once or never, i.e. all occurring (rather than recurring) states are equally likely for trivial reasons. Yet, this is a statement about states including the position in the fixed sequences of random numbers and therefore says nothing about the marginal probability of certain slope configurations $\{z_n\}$.

More formally (Paczuski and Bassler, 2000), if for each C all transition probabilities $T(C \to C')$ from C to C' vanish except one for which it is unity, then imposing stationarity, $\sum_{C'} P(C)T(C \to C') = \sum_{C''} P(C'')T(C'' \to C)$, Eq. (8.56), immediately gives $P(C) = P(C'')$ for the probability $P(C)$ that C occurs, if C is the state reached after C'', i.e. $T(C'' \to C) = 1$. In that sense, all states (namely those including the random numbers) are equally likely. Otherwise, considering slope configurations $\{z_n\}$ without random numbers, not much can be said about $T(C \to C')$ and thus the $P(C)$ remain unknown (see Sec. 8.3.1).

8.3.2.3 Definition of time

That Abelian models do not require a detailed definition of the microscopic dynamics is a major convenience when implementing a model numerically. Unstable (active) sites can be stored on and processed from a last-in-first-out stack (Sec. A.1). As opposed to non-Abelian models, macroscopic, geometric observables, such as the avalanche size and the number of distinct sites toppling during an avalanche, are independent of the details of the dynamics (i.e. the order in which unstable sites are updated), allowing many different implementations. However, without defining the detailed dynamics of a model, it makes little sense to measure any of its time dependent observables, such as the average activity in AS-type studies, the two-time correlation function or even the duration of an avalanche (e.g. Grassberger, 2002). Moreover, different implementations of time might lead to different weightings of time-averaged observables. The situation is comparable to the Ising Model, which itself does not possess a dynamics. Once it is chosen, its characteristics are nevertheless regarded as features of the Ising Model subject to a particular dynamics. Different dynamics produce different **dynamical exponents**, but **static exponents** remain unaffected (e.g. Hohenberg and Halperin, 1977).

The freedom of choice of the particular updating procedure of an Abelian model quickly turns into a curse (Pruessner, 2003a; Christensen *et al.*, 2004). For example, studies on the connection between the MANNA Model and absorbing state phase transitions frequently omit the details of the microscopic rules. First of all, any observable based on the number of sites about to topple (active sites) is undefined, because their number might change significantly with the order of updates, allowing multiple discharges (see footnote 5 on p. 170) in a single relaxation etc. The most common form of **continuous** time in Abelian models is to allow all (relaxation) processes to occur concurrently with certain rates, i.e. to implement them as a Poisson process (Sec. A.6.3). Because the order of updates determines the intermediate, unstable states visited, such a **random sequential updating** is preferable over parallel updates, which might reduce the accessible phase space, or deterministic sequential sweeps, which might introduce a bias. It is a sign of great robustness of the model and the strong effect of universality, that amid this uncertainty, most studies of Abelian models find the same universal behaviour nevertheless (but see Paczuski *et al.*, 1996, p. 440).

8.3.3 Abelian versus non-Abelian variants

One might speculate on the reason why Abelian models seem, frequently, better behaved than non-Abelian models. One reason certainly is a matter of perception, namely that the former, lacking the constraint of a detailed microscopic dynamics, are more easily accessible through rigorous analysis, such as an operator approach or an equation of motion, which allows their mapping to known problems, such as (loop erased) random walks (Sec. 8.1) or the quenched Edwards–Wilkinson equation (e.g. Sec. 6.3.4.2). The other reason, one might think, is that the Abelian nature of a model implies that it is less constrained by a particular dynamics chosen (being equivalent to so many alternative versions), allowing it to explore phase space more comprehensively. The OFC Model, which is clearly not Abelian, explores a much larger amount of phase space if an open boundary breaks its periodicity (Sec. 5.3.1.2, in particular p. 134). Similarly, driving the Abelian BTW Model on a single site will move it along one of its limit cycles (Creutz, 1991, 2004), whereas a random drive allows it to explore the entire torus of states, as illustrated in Fig. 4.5, p. 97.

The question whether the breakdown of Abelian symmetry triggers a change of universality class is discussed prominently in the literature, because it is still unclear whether, and if so which, symmetries determine the scaling behaviour of SOC models (Hughes and Paczuski, 2002). It is even unclear under what circumstances the Abelianness of a model affects its particular scaling behaviour. It is not easy to address the question in a sensible way. To start with, picking a model that displays particularly clear scaling means that it is particularly robust against changes in its definition, i.e. switching from Abelian to non-Abelian probably has little effect. Also, there is more than one way of making an otherwise Abelian model non-Abelian and breaking the Abelian symmetry always has side-effects and unwanted implications. For example, in the height restricted MANNA Model (Dickman *et al.*, 2002), the partial suppression of **multiple topplings** (by suppressing multiple discharges) renders the model non-Abelian. However, it has been argued (Paczuski and Bassler, 2000) that multiple topplings play a crucial rôle in non-trivial sandpile models (see also Sec. 8.4.4). If therefore this non-Abelian variant of the MANNA Model turns out to be in a different universality class,[35] that might be caused by the lack of multiple topplings, rather than by it being non-Abelian.

It is interesting to note that the MANNA Model has been studied (Lübeck and Heger, 2003a) in great detail using its original non-Abelian definition, where the number of particles shed at a toppling site depends on the number present at toppling, i.e. $h_n \to h_n = 0$, all being

---

[35] This is an unfortunate example, as the height restricted MANNA Model is widely expected to be in the same universality class as the MANNA Model.

redistributed. This version (as well as the non-Abelian, height restricted variant) are fully consistent with its Abelian version (e.g. Chessa *et al.*, 1999b). In most publications on the MANNA Model the presence or absence of the Abelian symmetry does not feature as a particular concern. Nevertheless, the conclusion that it is irrelevant might go too far: in these models, the Abelian symmetry might only be weakly violated, i.e. the processes which destroy the strict Abelian symmetry are rare compared to those effectively obeying it.

On the other hand, Hughes and Paczuski (2002) showed very clearly that in the *directed* MANNA Model (Paczuski and Bassler, 2000; Pastor-Satorras and Vespignani, 2000e), the spatial structure changes significantly from completely uncorrelated in the Abelian version (two particles redistributed to downward neighbours at a time), to algebraically correlated in the non-Abelian version corresponding to the oMM (all particles being redistributed to downward neighbours). They studied the model in three different co-dimensions $d_\perp = 0, 1, 2$ and did *not* find a change in the avalanche statistics for $d_\perp > 0$, i.e. the Abelian and the non-Abelian versions displayed the same scaling behaviour for avalanches, as observed in the undirected MANNA Model (AMM and oMM) and similarly in the 'local, unlimited' and 'local, limited' versions [36] of a ricepile model (Amaral and Lauritsen, 1997; Markošová, 2000a,b). The one model diverting from this behaviour, $d_\perp = 0$, is defined slightly differently compared to $d_\perp > 0$ and can therefore be regarded as a special case. Hughes and Paczuski's (2002) findings seem to suggest that the Abelian symmetry is altogether irrelevant, a view that was challenged by Zhang *et al.* (2005a) who found a different scaling behaviour of the non-Abelian model compared to the Abelian version at $d_\perp = 1$. Similarly, studying the oOM with bulk drive, which is non-Abelian like the original BTW Model, Zhang (1997) and Bengrine *et al.* (1999a) found exponents clearly inconsistent with established values (e.g. Bonachela, 2008). The deviation but not the actual values of the exponents was confirmed independently by Zhang *et al.* (2005b), who also identified a lack of scaling. Zhang (2000) found that the roughness of h_n in the oOM scales differently when switching from non-Abelian bulk drive to (Abelian) drive at $n = 1$.

Jo and Ha (2008) associated the robustness against a breakdown of Abelian symmetry with the stochasticity of the MANNA Model. Breaking it in a *deterministic directed* sandpile leads to a change in scaling behaviour. This is consistent with the study by Milshtein *et al.* (1998), who attributed the difference in scaling behaviour of the (*deterministic*) ZHANG Model compared to the (*deterministic*) BTW Model to the lack of the Abelian symmetry in the former. Similarly, Biham *et al.* (2001) drew the dividing line primarily between deterministic and stochastic models, the latter including Abelian and non-Abelian models (see Sec. 6.1.1 for the relation of the (Abelian) BTW Model and the (Abelian) MANNA Model). Breaking the Abelian symmetry in deterministic models, however, renders their scaling behaviour non-universal.

It is difficult to see why the Abelian symmetry plays a less important rôle in stochastic models than in deterministic ones. Some of the evidence (e.g. Hughes and Paczuski,

---

[36] Yet, 'non-local' versions of this model, similar to those studied by Kadanoff *et al.* (1989, also Lübeck and Usadel, 1993), are affected by changes of the toppling rules. However, introducing next nearest neighbour interaction in the OSLO Model does not change its scaling behaviour (Pruessner, 2003b).

2002) is based on directed models, because the exponents of many directed models are known exactly and therefore more easily identified in numerical studies. The significance of directed models, however, is limited by their lack of multiple topplings (Paczuski and Bassler, 2000, also Sec. 8.4). It remains an open, very important question to what extent the Abelian symmetry is relevant in the field-theoretic sense, i.e. determines the universality class.

8.3.3.1 Weak breakdown of Abelian symmetry

As mentioned above, some established Abelian models can be modified slightly, so that their Abelian symmetry is formally broken, but in a way that rarely shows during the evolution of the model. The AMM and its height restricted variant (Dickman *et al.*, 2002; Dickman, 2006) fall in this category. As long as the breakdown of the symmetry does not actually come to bear, or only in a statistically irrelevant fraction of events, one might argue that it cannot be relevant for the scaling features. The same argument applies in cases where the Abelian symmetry is broken when all configurations are considered, but not so if configurations are restricted to a smaller subset. One could construct models that are Abelian only with respect to recurrent states (the ASM is *not* one of them). Similarly, it is not obvious for which recurrent states in the BTW Model the operators do not commute (see footnote 5 on p. 90). Neither is it obvious whether they have any statistical relevance.

The MANNA Model was originally considered non-Abelian because of its intrinsic stochasticity (Milshtein *et al.*, 1998). If one considers a macroscopic time evolution operator, that charges the system and subsequently relaxes it, then charging the same system in a different order will most likely produce a different outcome. This point of view stresses the fact that the Abelianness of a model is actually a property of its operators, which need to be defined first. Milshtein *et al.* (1998) implicitly considered operators akin to Dhar's (1990a) operators $a_\mathbf{n}$ in the ASM with the randomness being part of the model rather than the operators and concluded that the MANNA Model cannot be Abelian. Using Markov matrices $\hat{a}_\mathbf{n}$, where the stochasticity is built into the operators, restores the Abelianness of the AMM. But even using Dhar's original, deterministic operators $a_\mathbf{n}$, the AMM is still **Abelian on average** (Dhar, 1999a), as the *probability* of a state is independent of the order of charges. If a model passes as Abelian, because it is so on average, none of the counter-examples, demonstrating a model being susceptible to a change of charging order in one particular case, seems to carry much weight.

8.4 Directed models

The first exact solution (Dhar and Ramaswamy, 1989) of an SOC model was a directed version of the BTW Model (reviewed in Dhar, 1999a,c, 2006). In fact all *spatial* exactly solved models are directed and any other exactly solved model either lacks any notion of space, or is trivial in that respect, i.e. is spatially uncorrelated. One of the merits of directed models is their rôle as a testbed for analytical schemes, such as renormalisation procedures

(Ben Hur *et al.*, 1996; Hasty and Wiesenfeld, 1998a). Kadanoff *et al.* (1989) suggested very early that any small degree of anisotropy would change a model's universality class to one which is typically represented by its directed version, as illustrated in the anisotropic version of the OSLO Model (Pruessner and Jensen, 2003). In the following, 'directed' refers to **totally asymmetric** local dynamics, where particles move in one direction only, but most arguments apply as well to merely anisotropic dynamics, with particles moving in a preferred direction. Some ambiguities surrounding the meaning of directedness are discussed in Sec. 8.4.4.

The various attempts to classify directed models along the same lines as undirected models (Kadanoff *et al.*, 1989) have led to a vast number of models, some of which are very similar or equal. Unfortunately, no generally accepted taxonomy has ever been established. For example, Jo and Ha (2008) use DSM, AD, AS, aND, bND, cND to distinguish their models, depending on, for example, whether they are stochastic or deterministic and Abelian or not. Pan, Zhang, Li, *et al.* (2005a) use A-DDM, NA-DDM, A-SDM and NA-SDM for a similar distinction. Pastor-Satorras and Vespignani (2000e) use ADS (DDS in Pastor-Satorras and Vespignani, 2000c), SDS, ESDS, NESDS, and Hughes and Paczuski (2002) use A-SDM for an entire class of directed models, not to be confused with Dhar's (1999b) SASM and DASM.

The two main classes of directed SOC models have been characterised comprehensively by Dhar and Ramaswamy (1989) and by Paczuski and Bassler (2000); both of these are discussed briefly in the following. It is impossible to discuss all models in detail, so the following focuses on general and typical features.

8.4.1 Deterministic directed models

The directed SOC model introduced and solved by Dhar and Ramaswamy (1989) is deterministic, in fact a directed version of the Abelian BTW Model (see Sec. 4.2.3), often referred to as the DR Model (also 'directed sandpile model'). Its dynamics in two dimensions is illustrated in Fig. 8.11. Particles are added on the top row $T = 0$ and toppling sites charge only downstream nearest neighbours, i.e. towards increasing T. In two dimensions (one **transverse dimension**, $d_\perp = d - 1$), the toppling sites form a singly connected domain and the sum of particles in each row $T > 0$ has conserved parity, because the number of particles entering and leaving any row $T > 0$ is even. Starting from any given configuration, all recurrent states are equally frequent (p. 281), given that the model is deterministic and the number of recurrent states finite. Because of conserved quantities and the dependence on the driving location, the latter has to be picked at random and initial conditions have to be averaged over to render *all* (stable) states recurrent. Assuming that all states are recurrent, the uniform distribution of states is invariant under the dynamics, in which case a site is occupied, $z_\mathbf{n} = 1$, or empty, $z_\mathbf{n} = 0$, with equal probability. The envelope shown as dashed lines in Fig. 8.11 is the trajectory of annihilating (vicious) random walkers (Fisher, 1984). The enclosed area is the number of toppling sites and therefore the avalanche size, characterised by an avalanche size exponent $\tau = 4/3$ and avalanche dimension $D = 3/2$, as found in many one-dimensional *stochastic* models (see below) for example in the TAOM (Sec. 6.3.5, Pruessner, 2004b). In $d \geq 3 = d_c$ ($d_\perp \geq 2$) Dhar and Ramaswamy (1989)

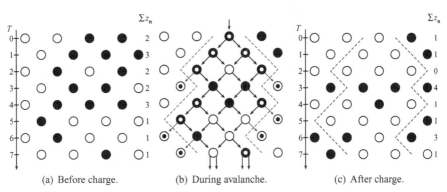

(a) Before charge.   (b) During avalanche.   (c) After charge.

Figure 8.11   An avalanche in the DR Model for $d = 2$, i.e. $d_\perp = 1$. Filled circles indicate sites with $z_\mathbf{n} = 1$, empty circles those with $z_\mathbf{n} = 0$. The initial setup, (a), is charged once in the top row, $T = 0$, triggering an avalanche in (b). Toppling sites charge both downstream (increasing T) nearest neighbours (periodic boundary conditions apply horizontally), with all charges being indicated by arrows. Singly charged sites change state, with a changed state indicated by small, inner circles. The final state of the system is shown in (c) with the (compact) region of toppling sites (its area is the avalanche size) traced out in dashed lines. The parity of the sum of $z_\mathbf{n}$ at constant T is preserved for all $T > 0$, as indicated in (a) and (c).

found $\tau = 3/2$ and $\mathsf{D} = 2$ with additional logarithmic corrections at $d = 3$. In $d = 1$ ($d_\perp = 0$) the particles move through the system ballistically, so that every avalanche has the same size identical to the system size, which means that the nth moment is simply the nth power of the system size, i.e. $\tau = 1$ and $\mathsf{D} = 1$. The exponents generally obey the **scaling law**

$$\mathsf{D} = 1 + \frac{d_\perp}{2} \quad \text{where} \quad 0 \le d_\perp = d - 1 \le 2, \tag{8.68}$$

and using the scaling relations in Sec. 8.4.3, $\tau = 2 - \frac{2}{2+d_\perp}$, $\alpha = \mathsf{D}$ and $\mathsf{z} = 1$. As discussed in Sec. 8.1.3, mean-field theory predicts $\tau = 3/2$ and $\alpha = 2$ and as $\mathsf{z} = 1$ still, $\mathsf{D} = 2$. These values are summarised in Table 8.1.

As mentioned earlier, a number of similar deterministic directed models have been devised to probe, for example, the relevance of Abelian symmetry. Pan *et al.* (2005a) introduced a variant of the DR Model where toppling sites shed *all* particles to downstream neighbours. The exponents found in this study were confirmed by Jo and Ha (2008), who studied in addition two other variants of the DR Model. Breaking the Abelian symmetry this way led to mean-field exponents ($\tau = 3/2$ and $\mathsf{D} = 2$) in all three cases. Pan, Zhang, Yin, and He (2006) investigated the effect of quenched randomness in the toppling matrix of the DR Model and found that the exponents depend on whether or not a 'local flow balance' is satisfied (Karmakar *et al.*, 2005). It remains unclear to what extent numerical results are able firmly to establish universal scaling behaviour different from that of the DR Model and its mean-field theory.

Tadić, Nowak, Usadel, *et al.* (1992) studied the effect of bulk dissipation implemented as random defects (quenched or annealed) in the lattice where particles would 'vanish'. In two dimensions, they found the same exponents $\tau = 4/3$ and $\alpha = 3/2$ as without dissipation, but with cutoffs given by the concentration of impurities which dominate the scaling if exceeded by the system size. Similar observations were made for impurities in the local critical slopes. Theiler (1993) was able to explain some of the observations based on earlier mean-field arguments by Alstrøm (1988). This equivalence of bulk and boundary dissipation is in sharp contrast to the observation in undirected, isotropic models (Sec. 9.3.1, in particular p. 334).

8.4.2 Stochastic directed models

A range of *stochastic* directed models has been studied and for the Abelian case classified comprehensively by Paczuski and Bassler (2000), who derived the exponents τ and D as a function of the dimensionality $d_\perp + 1 = d \le 3 = d_c$ of the system

$$D = \frac{3}{2} + \frac{d_\perp}{4} \qquad (8.69)$$

and $\tau = 2 - 4/(6 + d_\perp)$, again in line with Eq. (8.72). As in the deterministic case, logarithmic corrections are expected at $d = d_c$. Comparing Eq. (8.69) to Eq. (8.68), with a little bit of persuasion the different effects of the $d_\perp$ transverse directions in both setups can be established (also Bunzarova, 2010). The avalanche dimension can be thought of as determining the volume of the system typically covered by an avalanche. The term $d_\perp/2$ in Eq. (8.68) suggests a diffusive exploration in every transverse direction, with the mean squared displacement growing linearly in time, i.e. the system size, the maximum value of T in Fig. 8.11. However, in the deterministic case, particles do not move independently and are rather expelled outwards from the core of the avalanche, where they are subjected to a random environment, so the $d_\perp/2$ is a signature of the environment and not of the dynamics. In the stochastic case particles can move freely and would explore a fixed environment diffusively. The $d_\perp/4$ in Eq. (8.69) can be regarded as the net effect of diffusive exploration of a random environment. At $d_\perp = 0$ the deterministic case displays purely ballistic motion, $D = 1$ (achieved only for $d_\perp = -2$ in the stochastic case), while the stochastic setting displays **multiple topplings** due to accumulated activity, so that $D = 3/2$. The values of the exponents of stochastic models, including the mean-field theory which coincides with that of deterministic models, are tabulated in Table 8.1.

Paczuski and Bassler's result Eq. (8.69) was confirmed independently by Kloster *et al.* (2001) studying the same model, but taking a slightly different analytical approach.[37] It explains and classifies earlier numerical results by Maslov and Zhang (1995, also Priezzhev *et al.*, 2001) as well as those of Pastor-Satorras and Vespignani (2000e), who noted the relevance of the stochasticity in $d = 2$, i.e. at $d < 3$ making a deterministic model stochastic changes its exponents. In $d = 3 = d_c$ and above the same MFT describes the behaviour of deterministic and of stochastic models.

Hughes and Paczuski (2002) considered a set of non-Abelian stochastic directed sandpile models (Sec. 8.3.3, in particular p. 284) and found numerically that avalanche exponents remain unchanged, even when the spatial properties change fundamentally. Zhang *et al.*

---

[37] Like Dhar and Ramaswamy (1989) these authors also considered the triangular lattice as well as a partially directed square lattice.

Table 8.1 Exponents of most directed sandpile models.
$\mathcal{P}^{(s)}(s) = a^{(s)}s^{-\tau}\mathcal{G}^{(s)}(s/(b^{(s)}L^{D}))$ is the avalanche size
distribution and $\mathcal{P}^{(T)}(T) = a^{(T)}T^{-\alpha}\mathcal{G}^{(T)}(T/(b^{(T)}L^{z}))$ is the
avalanche duration distribution, see Sec. 1.3. In both cases
$d_{\perp} = d - 1$ and $d_c = 3$, above which MFT applies. Scaling relations:
$(1 - \tau)D = (1 - \alpha)z$ in all dimensions (see Sec. 2.2.2), $z = 1$
and $\alpha = D$ (see Sec. 8.4.3).

d	$d_\perp$	τ	D	α	z
Deterministic case					
1	0	1	1	1	1
2 (a)	1	4/3	3/2	3/2	1
3 (a)	2	3/2	2	2	1
MFT		3/2	2	2	1
Stochastic case					
1 (b)	0	4/3	3/2	3/2	1
2 (b)	1	10/7	7/4	7/4	1
3 (b)	2	3/2	2	2	1
MFT		3/2	2	2	1

a Dhar and Ramaswamy (1989), b Paczuski and Bassler (2000).

(2005a) questioned these findings, concluding that non-Abelian and Abelian stochastic directed sandpile models are in different universality classes.

Stapleton and Christensen (2005) generalised the one-dimensional (stochastic) TAOM (Sec. 6.3.5, also Pruessner, 2004b) to sites with n states. After being charged, a site has a certain probability of toppling a certain number of times. If n is very big, the developing avalanche resembles a simple branching process characterised by exponents $D = 2$ and $\tau = 3/2$ (p. 256), asymptotically in large n, which gives rise to a characteristic length ξ_n:

> However, as the avalanche propagates through the system, fluctuations in the number of topplings increase and [...] subsequent site[s] [...] will start to feel the fact that n is finite.

Thus, in the limit of $L \gg \xi_n$, the exponents cited in Table 8.1 are recovered, $D = 3/2$ amd $\tau = 4/3$. Jettestuen and Malthe-Sørenssen (2005) considered grain dissipation in the OSLO Model which amounts to an upward net drift of slope units. In that case the average avalanche size approaches a constant (but see Manna, Chakrabarti, and Cafiero, 1999).

Alcaraz and Rittenberg (2008) took an algebraic route to generalising Abelian stochastic directed models. Not all of their results are readily translatable to regular sandpile models. For example, they found that the dynamical exponent is $z = d$, defining it through the 'energy' of the first excited state, which should correspond to the correlation time. In numerical studies, they confirmed $D = 3/2$ in $d = 1$, but found $D = 1.78(1)$ in $d = 2$, which 'at face value [...] contradict[s]' $D = 7/4$.

The undeniable difference between stochastic and deterministic stochastic directed models is seen by some (e.g. Bunzarova, 2010) as further evidence for the BTW and the Oslo universality class being distinct, as the Oslo Model can be seen as the stochastic variant of the BTW Model.

8.4.3 Scaling relations

Directed models are usually set up so that a front propagates through the system ballistically, advancing one step forward at a time, until it reaches the boundary, at which point dissipation extinguishes all activity. [38] Although there might be some transverse motion, the ballistic propagation implies that the upper cutoff of the avalanche duration is linear in the system size, in fact equal to the system size if the driving takes place always at the opposite of the dissipative boundary, as for example in the DR Model, Fig. 8.11. Subsequently, the dynamical exponent is usually (Dhar and Ramaswamy, 1989; Tadić and Dhar, 1997)

$$z = 1, \tag{8.70}$$

even though some authors use the dynamical exponent to relate time to the dynamics in the *transverse* direction, i.e. the horizontal direction in Fig. 8.11 rather than the vertical one, whose size is referred to here as the system size. In that case Paczuski and Bassler (2000) derived $z_\perp = 2$. This use of the dynamical exponent is common in non-equilibrium critical phenomena (Hinrichsen, 2000), but not so much in SOC.

Ignoring boundary conditions and using the same argument as for Eq. (8.70) (Christensen, 1992; Christensen and Olami, 1993),

$$\sigma_1^{(s)} = D(2 - \tau) = 1, \tag{8.71}$$

since every particle added accounts for a number of topplings proportional to the system size until it is dissipated – due to the directedness (ballistic motion), each particle added makes a fixed number of moves until it leaves the system. The same applies even if the particles have only a slight bias, as the ballistic motion asymptotically always dominates. In non-directed models with bulk drive and diffusive motion of particles, $\sigma_1^{(s)} = 2$, so that any small amount of anisotropy can be detected by a change in the scaling of the average avalanche size (Tsuchiya and Katori, 1999b; Pruessner and Jensen, 2003), which also implies a different upper critical dimension (Tsuchiya and Katori, 1999a,b). In many directed models, namely those that are **totally asymmetric**, particles cannot return to sites visited previously, so that the number of distinct sites visited *per particle* equals the number of topplings per particle, i.e. the avalanche size. This, however, does not mean that the average *avalanche* area equals the average avalanche size, because multiple particles take part in an avalanche, which might visit the same sites; the average area per particle (number of distinct sites visited) is different from the average area per avalanche.

Assuming narrow joint distributions of size and duration (Sec. 2.2.2), $z(\alpha - 1) = D(\tau - 1)$, Eq. (8.70) and Eq. (8.71) imply (Dhar and Ramaswamy, 1989; Tadić and Dhar,

---

[38] Bulk conservation is assumed in the following.

1997; Kloster *et al.*, 2001),

$$\alpha = \mathsf{D} \quad \text{and therefore} \quad \tau = 2 - \frac{1}{\mathsf{D}}. \tag{8.72}$$

In directed models, the effect that driving location and boundary conditions have on scaling is different compared to fully isotropic models. If particles are dissipated at a boundary towards which particles are transported due to the net drift, then $\langle s \rangle$, the average number of topplings, is asymptotically proportional to the distance from that boundary to where particles enter the system. If that is chosen to be linear in the system size, as for example in the DR Model, where particles enter only at $T = 0$ and leave after toppling in $T = L$, then $\sigma_1^{(s)} = 1$. If the distance is constant, however, all moments of the avalanche size converge to a constant and $\sigma_n^{(s)} = 0$, suggesting $\mathsf{D} = 0$, provided that scaling holds at all (cf. Jettestuen and Malthe-Sørenssen, 2005). If particles drift towards a reflecting boundary where they 'pile up' (Sec. 6.2.1.1, p. 175), then the avalanche size and duration diverge exponentially in the system size, formally $\sigma_1^{(s)} \to \infty$.

In most totally asymmetric, i.e. purely directed, models the above argument can be made by inspection. The situation is more complicated if the model is merely anisotropic, but often a similar case can be made. For example, in models where particles perform random walks, a small degree of anisotropy effectively amounts to ballistic motion (Pruessner and Jensen, 2003; Pruessner, 2004a). In turn, a model should be regarded as undirected if it is 'nondirected on average' (Ben-Hur and Biham, 1996) or 'isotropic [...] on average' (Biham *et al.*, 2001). Sometimes there remains some ambiguity as to whether a model is directed or not, see Sec. 8.4.4.

Given its great sensitivity to drift, one might argue that SOC is not robust to such perturbations. Of course, this argument extends to many other models in statistical mechanics (Pruessner, 2004a) and thus puts the term 'robustness' rather than 'universality' in perspective. The effect of a drift depends on the time and length scales considered. However, generally it cannot be discounted only because it vanishes asymptotically. If the net displacement scales sublinearly in time, the effective drift vanishes asymptotically but might still dominate over other transport mechanisms, such as diffusion.

8.4.4 The rôle of backward avalanches

One-dimensional equilibrium systems are often solvable because interaction lacks loops (Fisher, 1984) and any *instantaneous* interaction is mediated via a small, fixed set of intermediate sites whose configuration is subject to the same interaction. In non-equilibrium, the interaction between any two sites is complicated by the time dependence which allows many more sites to mediate interaction temporarily. Directed models lack this feature and are thus effectively loop-free. Because of that and the additional absence of correlations, directed non-equilibrium models often develop into a product state at stationarity and thus are very easily solvable, although still not representing an *equilibrium* system.

In some directed models the assumption of no spatial correlations requires further qualification, as for example in the case of the sticky sandpiles, Sec. 8.4.5.1. Generally, an avalanche evolving in time possibly encounters (uncorrelated) remnants of earlier avalanches, as for example in the TAOM and the DR Model, but the avalanche does not interact with itself or its own (correlated) remnants. In addition, as long as activity proceeds in only one direction, a model is fully characterised by the activity in the transverse dimension.

The effect of an avalanche's activity on a region it passed through previously is known as a **backward avalanche** (Kadanoff *et al.*, 1989, also Bouchaud, Cates, Prakash, and Edwards, 1994), which stresses the direction of progression of the avalanche as a whole. Backward avalanches are the signature of interaction and cause sites to topple multiple times due to causally distinguishable charges. **Multiple topplings** (topplings due to charges received after the initial toppling in the same avalanche) mediate backward avalanches which are suppressed as soon as sites are not allowed to topple more than once, which gives rise to the wave decomposition, for example in the BTW Model (Ivashkevich *et al.*, 1994b). Ben-Hur and Biham (1996, also Lin *et al.*, 2007) suggested they were the missing ingredient in the real space renormalisation group of Pietronero *et al.* (1994). Therefore, some authors identify multiple topplings as the key ingredient of non-trivial models (Paczuski and Bassler, 2000): 'Multiple topplings are a fluctuation effect associated with self-intersections of the avalanche cluster in space and time. . . ' This interpretation should be seen in the context that multiple topplings can be caused by **multiple discharges** (see footnote 5 on p. 170) where sites accumulate surplus particles (activity) and redistribute them among their neighbours in more than one toppling event, *without* any self-interaction of the avalanche. In fact, the models investigated by Paczuski and Bassler (2000) and Pastor-Satorras and Vespignani (2000e, but note Peng, 1992) are directed and thus display multiple topplings for this trivial reason. Interestingly, Hughes and Paczuski (2002) considered a directed non-Abelian model in $d_\perp = 0$ whose scaling deviated from that in Sec. 8.4.2 as soon as multiple topplings were effectively suppressed by the dynamics.

The important distinction between multiple discharges and multiple topplings can be illustrated by Dickman *et al.*'s (2002) height restricted MANNA Model. Here a site cannot accumulate an arbitrary number of particle and thus *discharges* at most two particles, whereas no restriction is applied to the number of times it topples within the an avalanche.

Lübeck (2000) called the exponent determining the scaling of the characteristic avalanche size (mass) s_c with the characteristic area A_c covered, the **multitoppling exponent** γ_{sA}

$$s_c \propto A_c^{\gamma_{sA}}, \tag{8.73}$$

because the ratio of the avalanche size and the number of sites covered gives a measure for the number of times a toppling site topples in an avalanche. The area scaling is bounded by the dimensionality of the system, so that the characteristic number of topplings per toppling site scales at least as fast as L^{D-d}, i.e. $s_c/A_c \in \Omega(L^{D-d})$. In a system like the one-dimensional OSLO Model, where D is significantly larger than d, multiple topplings thus dominate an avalanche.

8.4.4.1 Definition of directedness

Models which can be expressed in a way such that at every point in time activity is localised on a (small) number of sites and progresses in one direction only, are called directed. This is the reason why seemingly directed models, such as the BTW Model in, say, one dimension, are not directed, even when grains (but not activity) move in one direction only, which is solely due to the spurious anisotropy induced by the anisotropic definition of height with respect to slope, Eq. (4.2) on p. 87. As soon as the BTW Model is expressed in terms of slope, the anisotropy is removed. In the (non-Abelian) original BTW Model some anisotropy remains in the driving, $z_n \to z_n + 1$ and $z_{n-1} \to z_{n-1} - 1$, even after the translation into the slope picture. If forward and backward are fully equivalent, the notion of a *backward* avalanche becomes empty, and should be understood as a 'returning avalanche' or 'recurrent toppling' (Paczuski and Bassler, 2000).

Hwa and Kardar (1989a) and a number of other authors (Garrido *et al.*, 1990; Grinstein *et al.*, 1990; Grinstein, 1991; Socolar *et al.*, 1993) regarded anisotropy and the resulting particle transport in systems like the BTW Model as a crucial ingredient, not least because of its link to driven diffusive systems (Schmittmann and Zia, 1995). Garrido *et al.* (1990) pointed out that anisotropy is not a sufficient condition for SOC but is certainly a necessary condition. These findings may need to be revised in the light of, apparently, fully isotropic, robust models such as the AMM.

Furthermore, as anisotropy is a relevant perturbation, like drift in a diffusion convection equation (Pruessner, 2004a; Bonachela, 2008), there are models which feature backward avalanches while belonging to a directed universality class (Kadanoff *et al.*, 1989; Pruessner and Jensen, 2003).

Finally, in the BS Model, the concept of forward and backward avalanches is used differently (Paczuski *et al.*, 1996). Here avalanches are identified within a time series as the minimal fitness returns to a minimal value. In finite systems, the time series and its statistical features are *not* invariant under reversal and thus backward and forward directions are distinguishable.

8.4.5 Directed percolation

It is difficult to overstate the importance of **directed percolation** (DP) in non-equilibrium critical phenomena, as it is so widely applicable (Janssen, 1981; Grassberger, 1982; Hinrichsen, 2000). Suspecting an intricate link, early SOC models, such as the BS Model, were analysed frequently with respect to their connection with DP (Sec. 5.4.5.7 Paczuski *et al.*, 1994a,b) and a lot of the technically more involved analysis was made from this starting point. As discussed in Sec. 6.2.1.2, the MANNA Model is still suspected by some to be in the DP universality class, and this is not ruled out even when the MANNA Model has multiple absorbing states (Dickman, 2002b; Ódor, 2004). A comparison of some SOC exponents in the two universality classes is shown in Table 8.2. The link between SOC and DP arises naturally when using the language of **interfaces** and **absorbing state phase transitions** to describe SOC models. Prima facie, the numerics currently available does not allow for the desired (Bonachela and Muñoz, 2008) firm discrimination of the MANNA Model and DP [39] (see Dickman *et al.*, 2002, for moment ratios), so that other numerical techniques have been devised (Bonachela and Muñoz, 2007), probing the respective systems for the effect of a change of boundary conditions and anisotropy.

---

[39] The exponents listed in Table 8.2 seem to be clearer in this respect compared to the exponents considered by Bonachela and Muñoz (2008).

Table 8.2 Comparison of the exponents characterising the MANNA Model, D and z (from Table 6.1, most recent data), and those from simulations (theory for MFT) of the directed percolation universality class, using Eq. (8.80a), $D = z^{(DP)} + d - \beta^{(DP)}/v_\perp^{(DP)}$ (also Lübeck, 2004). The two universality classes are generally believed to be distinct, as opposed to the MANNA and C-DP universality classes (see Table 8.4). Further exponents, such as τ and α, can be derived from scaling relations (Sec. 2.2.2) and are expected to depend on driving and boundary conditions.

d	D		$d + z^{(DP)} - \frac{\beta^{(DP)}}{v_\perp^{(DP)}}$		z		$z^{(DP)}$	
1	2.2(1)◇	(a)	2.328 673(12)	(b)	1.50(4)	(c)	1.580 745(10)	(b)
2	2.764(10)	(d)	2.979(2)	(e)	1.533(24)♮	(f)	1.765(3)	(e)
3	3.36(1)	(g)	3.509(8)	(h)	1.823(23)♮	(f)	1.901(5)	(h)
MFT	4	(i)	4	(i)	2	(i)	2	(i)

◇ Reported for the oMM.
♮ Reported for variants of the MANNA Model (in particular FES, Sec. 9.3.1).
a Nakanishi and Sneppen (1997), b Jensen (1999), c Dickman (2006), d Lübeck (2000), e Voigt and Ziff (1997), f Lübeck and Heger (2003a), g Pastor-Satorras and Vespignani (2001), h Jensen (1992), i Lübeck (2004).

Some SOC models can be thought of as 'accidentally' belonging to the DP universality class – the MANNA Model and the BS Model are possible candidates as long as the numerical results remain open to interpretation. There are, however, SOC models that were designed to be in this class. [40] What sets *directed* SOC models apart from typical DP models is the rôle of the longitudinal dimension: DP models normally appear to be directed only once time is mapped to an extra spatial dimension and there is no return to earlier times. For example, in DP a given configuration gives rise to a new one in the next time step and only when inspecting a time series of configurations does its nature as a type of percolation with directedness emerge. In (directed) SOC models, the situation is different, because even if there are no back-avalanches, the dynamics has to return to sites visited previously (in preceeding avalanches) to allow any self-organisation to take place. The directedness, however, is an integral part of the model, the configuration of which evolves as a whole, affecting possibly all sites in the longitudinal and the transverse directions. The DR Model shown in Fig. 8.11 is an example of such an ever evolving structure, where avalanches occur again and again in the same *directed* lattice. This obscure feature is the reason why mapping a directed SOC model to DP is often so difficult and, more importantly, why there can be self-organisation towards a critical point at all. The following focuses on SOC models with directedness only in time, corresponding to ordinary DP.

Tang and Leschhorn (1992) and Buldyrev *et al.* (1992a) independently devised models of interface (de)pinning in a quenched random environment, which developed DP-related scaling, the latter inspired by experiments. Both models, however, relied on tuning of a probability to its critical value. Only a few months later, Sneppen (1992) proposed two

---

[40] There is also a version of DP that can be regarded as an instance of SOC (Hansen and Roux, 1987; Grassberger and Zhang, 1996), even though it was not intended to be one, Sec. 9.1.

closely related solid-on-solid models, which are the earliest examples of DP-type SOC models, with that link being established only in a comment by Tang and Leschhorn (1993, also Maslov and Paczuski, 1994; Leschhorn and Tang, 1994). Olami, Procaccia, and Zeitak (1994, 1995) later derived a number of scaling relations in the Sneppen Model (Sneppen, 1992). There, the longitudinal dimension has a dual rôle of providing a random (quenched) environment and the degree of freedom of the evolving **interface**. Being inspired by invasion percolation (Wilkinson and Willemsen, 1983), the relation between the Sneppen Model and DP is similar to that of invasion percolation to regular percolation – incidentally Grassberger and Manna (1990) mentioned invasion percolation as an instance of anomalous scaling without a control parameter and thus as a precedent of SOC. Maslov and Zhang (1996) found some numerical support for DP-like behaviour in their variant of the Zhang Model, in $d = 1$ 'probably corresponding to DP with long-range correlated disorder' and in higher dimension displaying scaling more consistent with standard DP. In this model, the longitudinal dimension is that of time.

8.4.5.1 Sticky sandpile

Taking an entirely different route, Tadić and Dhar (1997) developed a two-dimensional, non-Abelian SOC model very similar to the DR Model (Fig. 8.11) but involving features of DP. The connection to DP is based on a direct identification of DP and a few analytical arguments and not on numerics, which however corroborate it. The key ingredient was later coined **stickiness**, namely with tunable probability $1 - p$ *not* to relax a site after whatever number of charges has arrived, i.e. particles can become stuck, not unlike in certain ricepile models (Amaral and Lauritsen, 1996b; Markošová, 2000b), as if grains were **sticky**. The DR Model is recovered exactly at $p = 1$. For any number of simultaneously arriving charges a site attempts only once to relax (thereby shedding two particles) with probability p, so that charging a site twice at once or sequentially gives rise to different behaviour for all $0 < p < 1$, rendering it non-Abelian. At $p < 1$ sites can retain particles without relaxing even when fulfilling the usual toppling condition $z_{\mathbf{n}} > z^c = 1$. If p is very small, the model does not develop into a stationary state as sites continuously accumulate particles and $z_{\mathbf{n}} \geq z^c$ throughout the system, i.e. the density c_0 of sites $\mathbf{n}$ with $z_{\mathbf{n}} = 0$, which cannot topple when being charged once, vanishes. These sites cause complications as they can still topple when charged multiply, but do not if charged only once.

To establish the link to DP, Tadić and Dhar noted that the number of sites toppling below a certain point T in the lattice (Fig. 8.11) is given precisely by the cluster size distribution of directed site percolation at probability p *provided that $c_0 = 0$ everywhere, so that they all topple precisely with probability p*. This is because at $c_0 = 0$ the spreading of an avalanche depends solely on the toppling probability p, corresponding *exactly* to DP. At the DP critical point and beyond, $p \geq p_c^{(\mathrm{DP})}$, the system becomes 'conducting' allowing it to develop into a stationary state. However, as the authors point out, particle conservation prevents the avalanche size distribution coinciding with that of directed percolation. On a two-dimensional square lattice, $d_\perp = 1$, in the stationary state the number of sites toppling in every layer T is $1/2$, shedding 2 particles at each toppling, so that one particle reaches the dissipative boundary per particle added. This is incompatible with DP scaling. The apparent paradox (DP scaling is impossible, although the system is DP-like) is resolved by noting that for $p \geq p_c^{(DP)}$ not all sites fulfil the condition $z_\mathbf{n} \geq z^c$ and the density c_0 is T dependent, so that the system is not solely controlled by a DP-like probability p. For $p < p_c^{(DP)}$ the density c_0 vanishes, but the system never becomes stationary. At the DP critical point it is correlated and inhomogeneous.

Mohanty and Dhar (2002) revised the 'sticky sandpile' by adding bulk dissipation with rate δ, thereby removing the anisotropy induced by the open boundaries which otherwise are needed to sustain the flow of particles through the system. Periodic boundary conditions are applied everywhere and the driving is stochastic, rendering the system translationally invariant. The resulting model displays one-dimensional DP scaling without any apparent tuning provided only that p is large enough, see Fig. 8.12. Arguably, this amounts to one of most desirable achievements, namely devising an SOC model to display the same non-trivial scaling behaviour as an established critical phenomenon that requires tuning (earlier successes by Hansen and Roux, 1987; Grassberger and Zhang, 1996; Sornette and Dornic, 1996, also Sec. 9.1). Based on its generality, they 'argued that the generic behavior of sandpile like models is in the universality class of directed percolation' (Dhar, 2004).

Mohanty and Dhar originally proposed a directed version as well as an undirected (isotropic), the latter being reminiscent of the ricepile model (Amaral and Lauritsen, 1996b) with $S_2 \rightarrow \infty$ ('gravity threshold' removed, Markošová, 2000a,b) and bulk dissipation added. An undirected version of the original model by Tadić and Dhar had been proposed earlier by Vázquez and Sotolongo-Costa (1999, also Vázquez and Sotolongo-Costa, 2000), who found DP scaling below $p_c^{(DP)}$. Because of the model's close link to the MANNA universality class, the DP claim has received some critique (Bonachela et al., 2006; Bonachela and Muñoz, 2007, 2008; Mohanty and Dhar, 2007), not least because it would tie a seemingly typical instance of OSLO-like behaviour to DP, questioning the very existence of the MANNA universality class. This is all the more threatening to the MANNA universality class as the numerical discrimination of DP and MANNA can be difficult. A possible explanation as to why Mohanty and Dhar's (2002) analysis fails, at least in the undirected case, is that correlations cannot be ignored (further details in Bonachela, 2008).

One might question whether the sticky sandpile model, directed or not, is an instance of SOC at all. Two observations motivate this enquiry. Firstly it is surprising that a mechanism as simple (and elegant) as Mohanty and Dhar's 'does the trick' already. Secondly, the bulk dissipation seems to play a rôle similar to regrowing trees in the FFM, this time expunging activity rather than fueling it, but nevertheless supplying the crucial ingredient of empty sites. As mentioned in Sec. 8.1.4.2, such a mechanism seems to require tuning, which is best understood by considering the length scale induced by it: it either dominates the scaling, replacing the system size, in which case its tuning characterises the system's scaling, or is dominated by the system size, rendering it irrelevant.

Mohanty and Dhar argued that SOC and long-range correlations can be expected only in the absence of dissipation, i.e. in the limit $\delta \rightarrow 0$, when the average avalanche size diverges. However, it is reasonable to expect that there is a finite range of $\delta > 0$ where for some value of p the effective site occupation probability, the probability of toppling after charge, $\tilde{p}$ equals its (effective) critical value $p_c^{(DP)}$. This probability has two contributions, namely $p(1-\delta)(1-c_0)$ from the fraction $1-c_0$ of sites with $z_\mathbf{n} \geq z^c$ that are guaranteed to exceed z^c

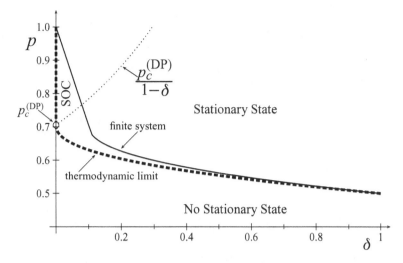

Figure 8.12 Cartoon of the phase diagram for the two-dimensional directed sandpile with sticky grains (cf. Mohanty and Dhar, 2002, Fig. 1). The lines separate a region where a stationary state exists from a region where no stationary state exists. The dotted, updwards pointing line shows the lower bound of p so that $\tilde{p} = p_c^{(DP)}$. The thick dashed line is the phase boundary in the thermodynamic limit, where DP behaviour is expected at vanishing dissipation, $\delta \to 0$, for all $1 \geq p \geq p_c^{(DP)}$ due to SOC. The solid line is the (exaggerated) phase boundary for finite systems, which generally requires a greater dissipation δ to sustain a stationary state at given stickiness $1 - p > 0$.

after being charged, and a contribution from the fraction c_0 of sites with $z_\mathbf{n} < z^c$, for which the probability of toppling depends on the number of charges they receive. These sites might introduce correlations, thereby altering the scaling and the effective critical point. Moreover, the (instantaneous) distribution of such sites is certainly not translationally invariant. Ignoring such complications, for given δ, DP-like behaviour can be expected at $\tilde{p} = p_c^{(DP)}$, where $\tilde{p} \leq p(1 - \delta)$ with the equality holding when $c_0 = 0$, i.e. when the system is not in the stationary state or just on the brink of it. The value of p such that $\tilde{p} = p_c^{(DP)}$ therefore is somewhere above $p_c^{(DP)}/(1 - \delta)$, shown as a dotted line in Fig. 8.12. As it branches immediately into the region where the system develops into a stationary state, $c_0 > 0$, the p required for DP scaling must be clearly bigger than $p_c^{(DP)}/(1 - \delta)$, if accessible at all for finite δ.

Some limited scaling regime can still be expected when $\tilde{p}$ is not precisely at $p_c^{(DP)}$. This can be achieved by tuning $\delta \to 0$ along the phase boundary between stationarity and non-stationarity, which runs along $p = p_c^{(DP)} - A\delta^{1/\gamma^{(DP)}}$ with $\gamma^{(DP)}$ given by ordinary DP, and contains $p_c^{(DP)}$ as the endpoint (Mohanty and Dhar, 2002). Similarly to the work by Tadić and Dhar (1997) this is because avalanches are a product of a directed site percolation process. Clearly, if $\delta = 0$ and $p = p_c^{(DP)}$, in the thermodynamic limit, the model is a realisation of critical DP and just at the brink of stationarity, $c_0 = 0$. The crucial point is that *any* $p > p_c^{(DP)}$ seems to produce DP scaling[41] as $\delta \to 0$, as avalanches are still the

---

[41] At $p \neq p_c^{(DP)}$ this scaling is not critical, i.e. the cutoff is finite.

outcome of a directed site percolation process [42] and their size has to diverge in the limit of vanishing dissipation.

To avoid any tuning, it thus seems one could choose any small δ and any toppling probability $p > p_c^{(DP)}$, corresponding to the dashed line running along the ordinate in Fig. 8.12. As discussed in the following, this procedure is constrained by the system size, which in turn raises the question of the real nature of this 'self-organised critical state'.

In an SOC model, the cutoff in the scaling functions is set by the system size, see Sec. 1.2, in particular p. 9. However, for $\delta > 0$ the cutoff is, at sufficiently large system sizes, determined by $\delta^{-\phi}$ with some exponent $\phi \approx 1.121$ (Mohanty and Dhar, 2002). In turn, δ cannot be made arbitrarily small at a given p, because with decreasing δ the system will eventually fail to develop a stationary state, since in a finite system at $p < 1$ an avalanche has a finite probability of terminating in every time step and will eventually do so. The dissipation during its course, which is proportional to δ, will thus eventually (as $\delta \to 0$) be insufficient to balance the influx of particles due to the drive. The phase boundary of a finite system is therefore slightly different from that in the thermodynamic limit, shown as solid and dashed lines, respectively, in Fig. 8.12.

The limit $\delta \to 0$ corresponds to moving horizontally to the left in Fig. 8.12 and eventually slipping under the solid line where no stationary state (in a finite system) exists. In other words, the limit $\delta \to 0$ cannot be taken without taking the thermodynamic limit $L \to \infty$ first, at which point, however, δ rather than L will determine the cutoff of the scaling functions, i.e. there was no finite size scaling. In principle, one could set $\delta = 0$ and refrain from demanding a stationary state. [43] In this case, the present model is trivially identical to DP, requiring tuning of p to $p_c^{(DP)}$ in order to recover DP's finite size scaling. In other words, the model proposed by Mohanty and Dhar (2002) comes arbitrarily close to the DP critical point, but unless δ is carefully tuned to the minimal value sustaining a stationary state in a finite system of given size, it cannot reach it asymptotically.

This observation is summarised in Fig. 8.12. In a finite system, the entire line $\delta = 0$ for all p lies within the non-stationary region and the avalanche size distribution $\mathcal{P}^{(s)}(s)$ for $\delta = 0$ is *exactly* that given by DP at $\tilde{p} = p$, given $c_0 = 0$, $\delta = 0$ and provided that correlations can be ignored. To recover the critical scaling of DP, p needs to be tuned to the DP critical point $\tilde{p} = p_c^{(DP)}$.

8.5 Surfaces, interfaces, depinning and growth phenomena

SOC models were first analysed from an interface perspective by Hwa and Kardar (1989a). They studied the spatiotemporal evolution of the surface of a sandpile as a continuous height field $h(\mathbf{x}, t)$ over a continuous substrate $\mathbf{x}$, i.e. as a **growth phenomenon**. Their results and those derived in this context (Sec. 9.2, Garrido *et al.*, 1990; Grinstein *et al.*, 1990; Grinstein, 1991; Socolar *et al.*, 1993; Becker and Janssen, 1994) had a profound impact, not least because they were based on a firm mathematical foundation. As epitaxy

---

[42] As $c_0 > 0$ at $p > p_c^{(DP)}$ this is a more complicated process (Domany and Kinzel, 1984).
[43] Wondering then, however, what was really missing in Tadić and Dhar (1997).

is usually performed without tuning to a critical point, even if it is known to exist, the surface growth community had focused on generic scale invariance long before the advent of SOC. In hindsight, it is difficult to understand why their influence has declined over the last decade. One reason could be that this interfacial perspective is somewhat removed from the original models and difficult to relate to a given SOC phenomenon. The BTW Model, the ZHANG Model, and even the MANNA Model, all seem to be indistinguishable, while numerically they clearly are different. A second reason could be the emphasis on bulk conservation in the theory of Hwa, Kardar and others. The advent of the OFC Model, the BS Model and the DS-FFM, all of which violate bulk conservation, challenged this view. Further details of this theory of SOC can be found in Sec. 9.2.

Hwa and Kardar's (1989a) approach belongs to a wider class of studies, which consider SOC from the perspective of surface evolution, which makes it accessible to established techniques (e.g. Frette, 1993; Bouchaud *et al.*, 1994; Mehta, Luck, and Needs, 1996; Chen and den Nijs, 2002), in particular dynamic renormalisation group, applied to Langevin equations of motion with **annealed noise** (Hwa and Kardar, 1989a; Díaz-Guilera, 1992, 1994). All of these methods require taking the continuum limit first, which is not always trivial, as discussed on p. 249 and illustrated in Fig. 8.2(a), Sec. 8.1.3. What is more, the continuum limit induces continuous symmetries, which are present on the lattice at best asymptotically.

Tang and Bak (1988a) related SOC phenomena to what is known as **absorbing state phase transitions** and established the first link between SOC and traditional, *tuned* critical phenomena found at the critical point of a transition between two phases. Their approach was later formalised and greatly extended by Dickman *et al.* (1998). In this setup, it is not the surface of the 'sandpile' which is subject to the stochastic evolution, but the **density of active particles** $\rho_a(\mathbf{x}, t)$, i.e. the **activity**. It can be cast into a stochastic equation of motion with an **RFT-type noise**, i.e. the noise correlator is linear in the activity. Even though the analysis of such equations makes use of similar methods, the analogy of absorbing states and interfaces is incomplete. [44]

Finally, Sneppen (1992), Tang and Leschhorn (1993), Roux and Hansen (1994), Olami *et al.* (1994, 1995) and Amaral *et al.* (1995) established the link between SOC and 'interface growth in a random medium' and thus introduced **quenched noise** (reviewed by Fisher, 1998). In this case, the local degree of freedom is either the number of charges $g(\mathbf{n}, t)$ received at site $\mathbf{n}$ or the number of topplings performed $G(\mathbf{n}, t)$ (Paczuski and Boettcher, 1996; Pruessner, 2003a). Where the mapping has been made for isotropic (undirected) models, it evolves (normally) under the **qEW (quenched Edwards–Wilkinson) equation**, also known as the **linear interface model** (Paczuski *et al.*, 1996; Vespignani *et al.*, 2000; Dickman *et al.*, 2001). The qEW equation was famously introduced as a **depinning transition** by Bruinsma and Aeppli (1984) as well as by Koplik and Levine (1985) and analysed by means of a functional renormalisation group by Nattermann *et al.* (1992, also Leschhorn *et al.*, 1997) also in the context of charge density waves (Narayan and Fisher, 1992a,b, 1993; Narayan and Middleton, 1994). The related qKPZ equation, which has non-conserved bulk dynamics and whose scaling is related to DP (Alava and Muñoz, 2002), has been studied as

---

[44] For example, in absorbing states, the width conditional to survival does not saturate (Dickman *et al.*, 2001), whereas in finite interfaces it normally does (Barabási and Stanley, 1995).

Table 8.3 In the three different interface interpretations of SOC models three different fields (local degrees of freedom) are subject to the evolution by a stochastic partial different equation, with three different types of noise entering.

Interface type	Field	Noise	Example	
Surface	Height $h(\mathbf{x}, t)$	Annealed	BTW Model-like	(a)
Absorbing state	Activity $\rho_a(\mathbf{x}, t)$	RFT	MANNA Model	(b)
Depinning	Charges $g(\mathbf{x}, t)$	Quenched	OSLO Model	(c)

a Hwa and Kardar (1989a), b Vespignani et al. (1998), c Paczuski and Boettcher (1996).

an SOC model as well (Ramasco, López, and Rodríguez, 2001; Szabó, Alava, and Kertesz, 2002).

Linking the qEW equation to absorbing state phase transition on the level of the equation of motion at first seems straightforward, because the degree of freedom in the two pictures is linked by the simple relation $\rho_a(\mathbf{x}, t) \propto \dot{g}(\mathbf{x}, t)$ (Vespignani et al., 2000). Translating the entire equation of motion between the two approaches, however, remains an open challenge (Sec. 6.2.1.2, Sec. 9.3, Alava and Lauritsen, 2001; Alava and Muñoz, 2002), because of the entirely different nature of the noise terms and the complicated effect that the change of variables $\rho_a \propto \dot{g}$ has on them. Bonachela et al. (2007, also Kockelkoren and Chaté, 2003) pointed out that subtle changes in the choice of the degree of freedom can have a profound impact on the scaling. As for the avalanche statistics, both $g(\mathbf{x}, t)$ and $G(\mathbf{x}, t)$ are equally useful, even when for the avalanche size

$$s(t) = \int d^d x \, G(\mathbf{x}, t + \Delta t) - G(\mathbf{x}, t) \tag{8.74}$$

(with suitable microscopic time difference Δt) no corresponding equation exists in terms of $g(\mathbf{x}, t)$.

As listed in Table 8.3, there are thus three different approaches to SOC in the interface representation: surface evolution, where the height field is subject to a stochastic equation of motion; absorbing states, where the activity field evolves in time; and interface depinning, where the number of charges is the local degree of freedom. Being instances of SOC, all three differ fundamentally from traditional non-equilibrium or dynamical *phase transitions* (Hohenberg and Halperin, 1977; Täuber, 2005; Henkel et al., 2008), in that they do not require tuning of a control parameter to produce non-trivial scaling. Paczuski and Boettcher (1996, also Pruessner, 2003a; Corral and Paczuski, 1999, and endnote [6.3], p. 393) noted that the order parameter of the depinning transition is subject to a separation of time scales and thus *forced* to an arbitrarily small value, which ties in with the mechanism proposed by Dickman et al. (1998, see Sec. 9.3).

8.5.1 Growth phenomena

Scaling without tuning is not unheard of, particularly in (well-established) growth equations, such as those governing the dynamics on surfaces discussed above (also Sec. 9.1). In fact, **diffusion limited aggregation** (Witten and Sander, 1981) has been suggested as a precedent of SOC by Grassberger and Manna (1990). The **KPZ universality class** (Kardar, Parisi, and Zhang, 1986) contains an enormous number of models (Aegerter et al., 2003; Dhar, 2004) that display non-trivial scaling without tuning, such as the Eden Model and various solid-on-solid models (Barabási and Stanley, 1995; Vicsek, 1999). Yet, these models require a scale invariant, that is white noise in the driving, whereas many SOC models, for example the Manna Model and the Oslo Model, can be driven deterministically on the same site and any stochasticity is internal (Sec. 8.5.2, Frette, 1993). On the other hand, white noise is actually not hard to come by, as on sufficiently large scale any noise source with finite correlations asymptotically becomes δ-correlated and Gaussian.[45]

One might argue that growth models are merely non-linear transformations, which turn one scale invariant object, the noise, into another scale invariant object, namely the observed field (also Eliazar, 2008, for a similar phenomenon in shot noise). While the non-trivial exponents are the fingerprint of the non-linear growth equation, the scale invariance itself is due to the external noise. What is more, in standard growth phenomena, for example the KPZ equation, drive and growth occur *on the same time scale*, whereas in SOC these are infinitely separated (Grinstein, 1995). The slow drive and the fast relaxation naturally suggests the notion of avalanches in SOC. As Sornette, Johansen, and Dornic (1995) pointed out, diffusion limited aggregation is frozen and even when growing does not naturally display any form of avalanching.

Except for the points raised above, there is indeed little difference between growth models and SOC models. Of course, this insight provides little comfort: the apparent 'self-organised scaling' in such models is not significantly better understood than in SOC. The field of growth models provides tools and examples to study the phenomenon, but it does not provide an explanation. To make matters worse, some surface growth models can undergo roughening transitions, which is a genuine phase transition controlled by a temperature-like variable (Barabási and Stanley, 1995; Pimpinelli and Villain, 1998). Adding anisotropy takes these models into the realm of driven diffusive systems, normally subject to a control parameter (Schmittmann and Zia, 1995; Marro and Dickman, 1999) and conservative noise.

The interfacial language lends itself naturally to experimental situations, in particular the imbibition (fluid invasion) experiments by Buldyrev et al. (1992a, but Maunuksela, Myllys, Kähkönen, et al., 1997 for combustion) and similarly the surface measurements by Frette et al. (1996) and Aegerter et al. (2003, also Welling et al., 2003). Interestingly, the link to invasion percolation (Wilkinson and Willemsen, 1983) made by Sneppen (1992), was made earlier by Cieplak and Robbins (1988, also Martys, Robbins, and Cieplak, 1991; Rubio, Edwards, Dougherty, and Gollub, 1989) in their numerical imbibition study.

In the following, a few more details of the relation between SOC and interfaces are briefly discussed, in particular the origin of the various types of noise, the rôle of boundary conditions and finally the relation between the scaling in SOC and that in growth phenomena. For a general review of interfacial and growth phenomena see Fisher (1986), Krug and Spohn (1991), Godrèche (1992), Barabási and Stanley (1995), Halpin-Healy and Zhang (1995), Krug (1995, 1997), Kardar (1998) and Meakin (1998), for random media in particular see Lässig (1998).

---

[45] What happens to the non-linearities of equations of motion (painstakingly derived on the microscopic scale) when changing to a larger scale, is an entirely different kettle of fish.

8.5.2 Types of noise

As introduced at the beginning of Sec. 8.5, the three different interfacial representations of SOC models, surface, absorbing states and depinning, come with three different types of noise, namely annealed (thermal), RFT-type, and quenched respectively, see Table 8.3. The RFT-type noise, whose correlator is *linear* in the order parameter and enters in the form $+\sqrt{\tilde{a}(g)}\tilde{\xi}(g)$, Eq. (8.46) on p. 263, can be distinguished from multiplicative noise, which is sometimes understood to have a correlator which is *quadratic* in the order parameter, i.e. a noise amplitude linear in the order parameter (Muñoz, 2003), such as $+\tilde{a}(g)\tilde{\xi}(g)$. Not many authors make this distinction and most therefore call RFT-type noise multiplicative as well (e.g. Hinrichsen, 2000), as is done in most of the mathematical literature.

While the relation between models with quenched noise and models with RFT-type noise is riddled with difficulties, RFT-type noise can be linked to annealed noise using a change of time scale at least in a very basic, zero-dimensional model, see Sec. 8.1.4.1. A transition from quenched noise to annealed noise occurs at large interface velocities (Kessler, Levine, and Tu, 1991; Meakin, 1998). In fact Corral and Paczuski (1999) observed a qualitative change in the scaling of the OSLO Model in the continuous flow regime.

SOC models as envisaged at the conception of the OSLO Model should not depend on external scale-free noise sources (Frette, 1993). This desire has been met only partially by the OSLO Model, as its internal randomness remains equally unexplained. The fully deterministic train model by de Sousa Vieira (2000) comes closer to the ideal. Alternatively, one could invoke another deterministic but **chaotic process** (Bhagavatula, Grinstein, He, and Jayaprakash, 1992; Cule and Hwa, 1996; Paczuski and Boettcher, 1996) in the background which effectively produces a noise feeding into a given SOC model. Alava and Lauritsen (2001, also Alava, 2002) took a completely different approach to the noise, effectively wrapping *all* features of a given SOC model in the characteristics of the noise, which radically changes the perspective. Different universality classes of SOC models are then a matter of different ensembles of quenched noise, which are widely independent of any microscopic details, at least where Middleton's (1992) no-passing guarantees a unique interface configuration.

8.5.3 Boundary conditions and driving

On the level of a lattice model evolving under certain rules, the rôle of different boundary conditions remains somewhat obscure compared to the interface picture, not least owing to the unfortunate confusion of conservation and dissipation with respect to height and slope units, p. 94. Once an equation of motion is formulated (e.g. Sec. 6.3.4.2 or Sec. 4.1, p. 87) standard nomenclature can be applied. Boundaries fixed to a particular value are subject to **Dirichlet or absorbing** boundary conditions, those whose first derivative is fixed are subject to **Neumann or reflecting** boundary conditions. When applied to the evolution of the number of charges $g(\mathbf{x}, t)$, an absorbing boundary means that charges applied there have no repercussions inside the system, i.e. particles are lost. A reflecting boundary means that charges and thus particles arrive back at the site they were originating from. The external drive in conjunction with the boundaries induces a current of charges and thus particles in

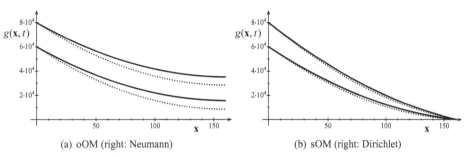

(a) oOM (right: Neumann)   (b) sOM (right: Dirichlet)

Figure 8.13 The interface $g(\mathbf{x}, t)$ at quiescence in the one-dimensional OSLO Model with $L = 160$ sites (solid lines). The distinct large scale features of the average shape render the roughness imperceptible. The dotted lines show the interface of the 'deterministic OSLO Model' with $z^c = 2$ constant (Abelian BTW Model). The external drive charges a 'virtual site' at $\mathbf{n} = 0$ with constant z_0^c, twice for every charge to be received at site $\mathbf{n} = 1$. Other than that, the virtual site does not receive any charges, i.e. Dirichlet boundary conditions apply at $\mathbf{n} = 0$. The number of avalanches triggered up to continuous, microscopic, conditional (see endnote [8.7], p. 395) time $\tilde{t}$ (Poissonian evolution) stated below is $g(0, t)/2$, here $3 \cdot 10^4$ and $4 \cdot 10^4$, fixing the unconditional time t via $g(0, t) = 2\dot{E} \left\lceil \frac{t}{\delta t} \right\rceil \delta t$, see below. (a) In the oOM a Neumann boundary condition applies at the right, implemented by toppling site $\mathbf{n} = L$ like every bulk site, see after Eq. (6.19) and Eq. (8.52), and moving any particle received at the virtual site $\mathbf{n} = L + 1$ back to $\mathbf{n} = L$. At stationarity $\langle g(\mathbf{x}, t + \delta t) - g(\mathbf{x}, t) \rangle = 2\dot{E}\delta t$ with δt taken on the unconditional time scale. Lower profile taken at $\tilde{t} \approx 3.8 \cdot 10^5$, upper profile at $\tilde{t} \approx 5.6 \cdot 10^5$ respectively. (b) In the sOM, a Dirichlet boundary condition applies at the right, $g(L + 1, t) = 0$ (profiles taken at $\tilde{t} \approx 3.5 \cdot 10^5$ and $\tilde{t} \approx 4.8 \cdot 10^5$), so that $\langle g(\mathbf{x}, t + \delta t) - g(\mathbf{x}, t) \rangle = 2\dot{E}\delta t(L - \mathbf{x})$.

the system, which might be responsible for the resulting long-range correlations (Sec. 9.2, also Schmittmann and Zia, 1995).

Figure 8.13 illustrates the effect of different boundary conditions in the interface representation of the OSLO Model, which counts the number of charges received by a site, $g(\mathbf{x}, t)$. In the oOM the right hand boundary is closed with respect to particles, which translates to a Neumann boundary condition. As a result the entire interface lifts off when pulled upwards on the left hand boundary, with roughly the same gap between two snapshots. The number of charges received by all sites is, up to boundary effects, twice the number of topplings performed, so that the avalanche size is, at least asymptotically, $\langle s \rangle = L$, Eq. (8.4b) on p. 247. In the sOM both boundary conditions are fixed which, in the case of boundary drive, results in a wedge-shaped space between consecutive profiles and thus $\langle s \rangle = L/2$, Eq. (8.4a).

The arguments developed in Sec. 8.1.1.1 (also Sec. 6.3.4.1) translate directly to the interface. For example, a combination of one reflecting and one absorbing boundary has the same effect as doubling the size of a system with absorbing boundaries on both ends (also after Eq. (6.19) on p. 200).

A striking feature of Fig. 8.13 is that the roughness of the interface is practically invisible. Given the avalanche size scaling of the OSLO Model and its direct relation to the roughness of the interface, to be discussed in Sec. 8.5.4, this must clearly be an artefact. One might wonder whether some trivial background can be subtracted to reveal the rough surface as its modulation. This is discussed briefly in the following and it will turn out that *the average shape of the interface is caused by and is characteristic of the quenched noise.*

The equation of motion governing the 'OSLO interface' $g(\mathbf{x}, t)$ in the continuum limit, Eq. (6.34) on p. 205, is

$$\partial_t g = D\partial_\mathbf{x}^2 g + \tilde{\xi}(\mathbf{x}, g) \tag{8.75}$$

with $g(0, t) = 2\dot{E}t$, **boundary drive**[46] linear in microscopic time (or, rather, $g(0, t) = 2\dot{E}\lceil\frac{t}{\delta t}\rceil \delta t$, if the driving advances in jumps, with δt much bigger than the avalanche durations), and at $\mathbf{x} = L$ either $\partial_\mathbf{x}|_{\mathbf{x}=L} g(\mathbf{x}, t) = 0$ or $g(L, t) = 0$. Shifting the data, $g(\mathbf{x}, t) = f(\mathbf{x}, t) + \tilde{g}(\mathbf{x}, t)$ with

$$\partial_t f = D\partial_\mathbf{x}^2 f \tag{8.76}$$

and the boundary conditions applied to f, the latter accounts for the deterministic part of the differential equation, in particular the driving. Because of the non-linearity of the noise, any such shift requires a suitable transformation of the noise correlator.

The solution of Eq. (8.76) is easily found to be

$$f(\mathbf{x}, t) = \begin{cases} \dot{E}\left(2t + \dfrac{x(x - 2L)}{D}\right) & \text{Neumann} \tag{8.77a} \\ \dot{E}\left(2t\dfrac{L - x}{L} + \dfrac{x(x - 2L)(L - x)}{3LD}\right) & \text{Dirichlet.} \tag{8.77b} \end{cases}$$

In the light of Fig. 8.13 this might look promising, suggesting that $f(\mathbf{x}, t)$ governs the overall appearance of the interface, i.e. its average value. However, this deterministic part assumes any non-trivial, curved shape only if the driving $\dot{E}$ is fast enough. In the limit of arbitrarily slow $\dot{E}$ as found in SOC, for fixed $t\dot{E}$ (i.e. in the limit of slow drive and long times) the deterministic interface is asymptotically dominated by the non-curved part:

$$f(\mathbf{x}, t) = \begin{cases} 2t\dot{E} & \text{Neumann} \tag{8.78a} \\ 2t\dot{E}\dfrac{L - \mathbf{x}}{L} & \text{Dirichlet.} \tag{8.78b} \end{cases}$$

Thus, in the slow driving limit the deterministic part is trivial and the interface $g(\mathbf{x}, t)$ assumes a non-trivial shape[47] as visible in Fig. 8.13 because it gets stuck in the disorder. The average shape of $g(\mathbf{x}, t)$ is *not* dominated by the deterministic part of the differential equation but by the quenched noise. How the roughness and shape enter into the avalanche size is discussed in Sec. 8.5.4.

8.5.4 Observables, roughness and exponents

What can be learnt about SOC models using their mapping to interfacial phenomena? Which observables in SOC models can be mapped or related to observables in interfacial processes?

As discussed further in Sec. 9.2, Hwa and Kardar (1989a) analysed a **driven diffusive system** formulated as a Langevin equation similar to a KPZ equation, yet conservative,

[46] Bulk driving is more difficult to handle in the present formalism, see endnote [6.3], p. 393.
[47] 'Shape' here means a deviation of the order $t\dot{E}$ from linear. The limit $t\dot{E} \to \infty$ is not to be taken, as it would dominate over any deviation from a straight line.

which describes the surface evolution of a sandpile with a preferred (downhill, longitudinal) direction and subject to a non-conservative noise. The height fluctuations are the essential measure of the avalanche size (see p. 205), which allows the derivation of the avalanche dimension D and exponent τ from the scaling of the fluctuations with the longitudinal size of the system. At this point, their derivation deviates slightly from the BTW Model, which in higher dimension has no single downhill direction. Later versions of this approach (such as Grinstein et al., 1990; Grinstein and Lee, 1991; Socolar et al., 1993, also Aegerter et al., 2003) focused on the more general phenomenon of SOC (conservation, periodicity etc.) rather than the scaling of typical SOC observables.

The absorbing states (AS) approach establishes a fundamental relation between absorbing state phase transitions and SOC, which is discussed in further detail in Sec. 9.3. To some extent, the SOC 'ensemble' can be identified with a mode of driving an absorbing state model, the only difference being the particle density which self-organises in SOC (via dissipative boundaries and external drive) and is fixed in the typical AS setup. The SOC 'mode' can be considered an AS spreading experiment (Marro and Dickman, 1999; Muñoz et al., 2001; Lübeck, 2002b, 2004; Lübeck and Heger, 2003a; Bonachela and Muñoz, 2008), which maps the survival time of the latter to the avalanche duration of the former, so that $\alpha - 1 = \delta^{(AS)}$ (survival exponent) and [48] $z = z^{(AS)}$. Without going into further detail about the meaning of the exponents, scaling relations (Muñoz et al., 1999; Lübeck and Heger, 2003a; Lübeck, 2004) for avalanche sizes follow from a dimensional analysis on the basis of anomalous dimensions. The avalanche size is the time integrated total conditional [49] activity, so that its moments follow (in the same notation as above)

$$L^{D(1+n-\tau)} \propto \langle s^n \rangle \propto \langle (T L^d \rho_a)^n \rangle \propto \langle T^n \rangle L^{nd} \langle \rho_a{}^n \rangle$$
$$\propto L^{z^{(AS)}(n-\delta^{(AS)})} L^{nd} L^{-n\beta^{(AS)}/\upsilon_\perp^{(AS)}} \tag{8.79}$$

for $1 + n - \tau > 0$, which implies (cf. Sec. 8.5.4.1)

$$D = z^{(AS)} + d - \frac{\beta^{(AS)}}{\upsilon_\perp^{(AS)}} \tag{8.80a}$$

$$D(1 - \tau) = -z^{(AS)}\delta^{(AS)}. \tag{8.80b}$$

Table 8.4 shows exponents of the MANNA universality class side by side with those derived from spreading experiments in AS models in the C-DP universality class (cf. Table 8.2), supporting the identity of the two.

In general, however, there are not many observables and exponents common to both AS and SOC, mainly because observations in AS are usually made on the microscopic time scale and observations in SOC are made on the macroscopic time scale (Muñoz et al., 2001). Studies comparing AS and SOC directly are few and far between (e.g. Chessa

---

[48] The exponents with superscript (AS) are measured in AS spreading experiments, characterising the scaling of the probability of surviving, in the notation of, for example, Lübeck (2004). Boundary conditions in SOC normally differ from those in AS. The exponents found in AS with periodic boundary conditions and any drive normally correspond to those found in SOC with open boundary conditions and bulk drive.

[49] Conditional activity is the spatially averaged instantaneous density of active sites averaged over the time during which there was activity at all. The latter is precisely the avalanche duration T.

Table 8.4 Comparison of the exponents characterising the MANNA Model, D and z (from Table 6.1, most recent data for the MANNA Model simulated in SOC mode), and those from simulations (theory for MFT) of the C-DP universality class (CTTP), confirming Eq. (8.80a), $D = z^{(AS)} + d - \beta^{(AS)}/v_\perp^{(AS)}$, via the scaling relation $z^{(AS)} + d - \beta^{(AS)}/v_\perp^{(AS)} = z^{(AS)}(\theta + \delta + 1)$ (Lübeck, 2004). The two classes are therefore generally believed to coincide, as opposed to the MANNA and DP universality classes (see Table 8.2). Other exponents, like α and τ, are expected to depend on driving and boundary conditions.

d	D		$d + z^{(AS)} - \frac{\beta^{(AS)}}{v_\perp^{(AS)}}$		z		$z^{(AS)}$	
1	2.2(1)◇	(a)	2.12(8)	(b)	1.47(7)◇	(a)	1.393(37)	(b)
2	2.764(10)	(c)	2.79(7)	(b)	1.540(10)	(c)	1.533(24)	(b)
3	3.36(1)	(d)	3.47(8)	(b)	1.76(1)	(d)	1.823(23)	(b)
MFT	4	(e)	4	(e)	2	(e)	2	(e)

◇ Reported for the oMM.
a Nakanishi and Sneppen (1997), *b* Lübeck and Heger (2003a), *c* Lübeck (2000), *d* Pastor-Satorras and Vespignani (2001), *e* Lübeck (2004).

et al., 1998; Dickman *et al.*, 1998, 2000; Muñoz *et al.*, 1999; Lübeck, 2004). In order to trigger an avalanche, the archetypical observable of SOC, in AS, the system must be in a quiescent state and fall back into it. For this to happen regularly, the AS system must be in the subcritical phase, where it is fixed by its control parameter. Activated random walkers (Dickman *et al.*, 2000), which are, up to the boundary conditions, an implementation of the MANNA Model, illustrate the problem. Due to a lack of dissipation of the control parameter, the particle density, activity ceases only because of fluctuations (Christensen *et al.*, 2004). Once the system is inactive, there is no inherent way to trigger another avalanche.

So called **fixed-energy sandpiles (FES)** are a popular means of turning sandpile-like models into AS systems (Chessa *et al.*, 1998) by applying periodic boundary conditions and thus removing the dissipation mechanism. If avalanches are triggered by adding particles, this is to be complemented by the introduction of bulk dissipation (Dickman *et al.*, 2000; Malcai *et al.*, 2006; Lin *et al.*, 2006, also Sec. 9.3). Lübeck (2002b) devised a method of creating activity by reshuffling otherwise immobile particles which helps them to overcome the absorbing state, without the need of particle influx or outflux. This method, which unfortunately generally destroys their Abelianness (for the same reason as the original BTW Model is non-Abelian), resembles the application of an external field in equilibrium critical phenomena, which is reflected in the resulting scaling behaviour (Lübeck, 2004). The procedure of triggering spells of activity in AS models is known as a spreading experiment (Hinrichsen, 2000). One might argue that this procedure assimilates AS models to SOC systems and thus does not add to the understanding as to why the two display, apparently, the same scaling behaviour.

Because boundary conditions are generally different in AS (periodic) and SOC (open, dissipative), they do not induce the same scaling relations, so that, for example, the avalanche size exponent τ might no longer be determined solely by D and a scaling relation [50] like

[50] The scaling relation $D(1 - \tau) = z(1 - \alpha)$ (Sec. 2.2.2) is still expected to hold.

$D(2 - \tau) = 1$. The avalanche size and duration exponents, τ and α respectively, derived for example by Lübeck and Heger (2003a) nevertheless reproduce the values found in the MANNA Model with stochastic bulk drive and open boundary conditions. A similar result was found by Chessa *et al.* (1998) as well as Christensen *et al.* (2004) studying avalanches in FES. This is consistent with the view that any driving in periodic systems amounts to bulk driving and that there is no difference between bulk driving with open and periodic boundaries.

The opposite path, namely studying AS-like observables in SOC models, is similarly awkward. The order parameter in AS transitions is the activity, i.e. the instantaneous local density of toppling sites per time unit. In Abelian models, the time unit is a priori undefined (Sec. 8.3.2.3) and so is, strictly, the activity. Most studies so far (e.g. Dickman *et al.*, 1998; Vespignani *et al.*, 2000) do not actually investigate AS-like observables in SOC models, but turn the SOC model into AS models first (FES) and then study the new model.

Finite size scaling (FSS) is the most common scaling behaviour considered in SOC models, but is somewhat ambiguous in AS models, which can be probed more readily in the *approach* to, rather than right *at* the critical point. For FSS, one school measures activity conditional to activity (e.g. Marro and Dickman, 1999), while the other applies an external field (e.g. Lübeck and Heger, 2003a). Both schemes can be reconciled as two interpretations of essentially the same procedure (Pruessner, 2007).

8.5.4.1 SOC and depinning

Relating SOC and extremal dynamics (Boettcher and Paczuski, 1996; Alava and Lauritsen, 2001; Alava, 2002) to growth phenomena in random environments, or more specifically, **depinning** generates a number of elegant scaling relations, comprehensively discussed by Paczuski *et al.* (1996). In this mapping, the interface is formed by the number of charges $g(\mathbf{x}, t)$ received at site $\mathbf{x}$ up to time t, so that a measure s_i' of the size s_i of the ith avalanche [51] is given by the difference of the spatial total of charges, measured at two consecutive times of quiescence t_{i-1} and t_i,

$$s_i' = \int d^d x \left(g(\mathbf{x}, t_i) - g(\mathbf{x}, t_{i-1}) \right), \tag{8.81}$$

see Fig. 8.14. Provided the driving is slow enough, the interface is more likely to be found in a quiescent state, so that the configurations found at times t_i describe its *characteristic* state. [8.7]

Equation (8.81) can be related to the avalanche size discussed in the context of AS, by introducing the activity as $\rho_a(\mathbf{x}, t) = \partial_t g(\mathbf{x}, t)/q'(\mathbf{x})$ (Vespignani *et al.*, 2000), where $q'(\mathbf{x})$ is the local ratio of the number of charges and the number of topplings, in the bulk $q'(\mathbf{x}) = q$, introduced in Sec. 8.1.1.1. The avalanche size is then given by $\int d^d x \int_{t_{i-1}}^{t_i} dt \, \rho_a(\mathbf{x}, t)$. What makes Eq. (8.81) particularly appealing is that the avalanche size can be represented by the area enclosed by two consecutive interface configurations at quiescence, see the hatched

---

[51] In the following, s_i is the total number of topplings, whereas s_i' is the total number of charges. Up to boundary effects (dissipation) and the initial charge (included in s_i', which are therefore *always* non-zero), the two are proportional; in the continuum, the number of charges equals the number of removals discussed above, Sec. 8.1.1.1.

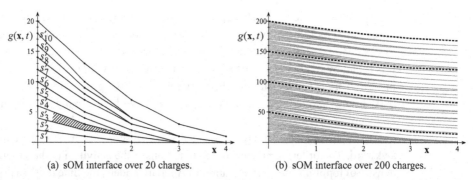

(a) sOM interface over 20 charges.   (b) sOM interface over 200 charges.

Figure 8.14 The sequence of interface configurations in the one-dimensional sOM at quiescence over the course of a number of charges, for illustration purposes in an extremely small system of size $L = 4$. Once two interfaces merge towards large $\mathbf{x}$, they do not depart again, producing a tree-like structure. (a) Charging the system at $\mathbf{n} = 0$ moves the interface up, dragging the rest behind it. The areas s_i' between two consecutive interface configurations, such as the hatched area, measure the total number of charges received throughout the system. (b) After more charges, the interface lifts off. At sufficiently large time steps, a sequence of almost perfectly parallel interfaces appears, shown as bold, dashed lines.

area in Fig. 8.14(a). The vertical fluctuations of the interface are given by its roughness which has characteristic scale $L^{\chi^{(\text{int})}}$. The *characteristic volume encapsulated is therefore* $L^d L^{\chi^{(\text{int})}}$ *and thus* $d + \chi^{(\text{int})} = D$ *(Paczuski et al., 1996, also Eq. (8.80a))*. As long as the scaling of the vertical fluctuations is determined as a spatial average, the compactness of the avalanching area does not enter. It does, however, if the fluctuations are averaged only over the avalanching area (but see Paczuski *et al.*, 1996, after Eq. (55)).

One might think that a more sturdy derivation of $d + \chi^{(\text{int})} = D$ is easy to come by. The following attempt, however, fails until it utilises the remarkable property of the interfacial mapping that *all moments* $\langle s^n \rangle$ *with* $n \geq \tau - 1$ *(or more precisely their leading orders in L) of the avalanche size for **multiple (initial) charges** are linear in the number of charges*, which is subtly linked to the Abelian property. This feature is counter-intuitive, because the number of initial charges M turns out to have a more subtle effect than just rescaling the avalanche size by a factor M, which would result in moments $\langle s^n \rangle \sim M^n$. In fact, as discussed further after Eq. (8.91), consecutive avalanches are anti-correlated. For simplicity, in the following, only the oOM is discussed and only for boundary drive, the traditional form of driving the OSLO Model.

As discussed in Sec. 8.5.3, the average avalanche size with Neumann boundary conditions is linear in the system size. The same applies to the average number of charges received in the system per avalanche attempt, as can be seen in Fig. 8.14(b), which shows four interface configurations each 25 avalanche attempts apart. If the interfaces start with a spacing of $2\dot{E}\delta t$ at $\mathbf{x} = 0$, then the average total area encapsulated is $\langle s' \rangle = 2\dot{E}\delta t L^d$, where δt is the (unconditional, see endnote [8.7], p. 395) waiting time between two consecutive avalanche attempts much bigger than the characterstic avalanche duration, $t_i = i\delta t$. Using $\langle g(\mathbf{x}, t) - g(\mathbf{x}, t + \delta t) \rangle = 2\dot{E}\delta t$ (given a Neumann boundary condition on the right and $\nabla^2 g$ bounded on average, see Fig. 8.13(a) and Sec. 6.3.4.2, in particular Eq. (6.30)), this is trivially confirmed by Eq. (8.81), so that the FSS exponent of the first moment of the

avalanche size in this description is $\sigma_1^{(s)} = D(2 - \tau) = d$, the dimension of the substrate. [52]
To recover the result $\sigma_1^{(s)} = 1$, Eq. (8.5a), it follows $\dot{E}\delta t \propto L^{d-1}$.

The second moment $\langle s'^2\rangle$, which is essentially the second moments of the avalanche size, can be derived similarly, but has some rather unexpected features. Based on Eq. (8.81), one has

$$\langle s'^2\rangle = \int d^d x \, d^d x' \, \langle \big(g(\mathbf{x}, t_i) - g(\mathbf{x}, t_{i-1})\big) \big(g(\mathbf{x}', t_i) - g(\mathbf{x}', t_{i-1})\big)\rangle \tag{8.82}$$

which can be calculated on the basis of the correlation function

$$\begin{aligned} c(\mathbf{x}, \mathbf{x}', t, t') &= \langle g(\mathbf{x}, t)g(\mathbf{x}', t')\rangle - \langle g(\mathbf{x}, t)\rangle \, \langle g(\mathbf{x}', t')\rangle \\ &= c(\mathbf{x}', \mathbf{x}, t', t) \end{aligned} \tag{8.83}$$

(see Eq. (2.48)) using

$$\int d^d x \, d^d x' \, \langle g(\mathbf{x}, t_i) - g(\mathbf{x}, t_{i-1})\rangle \, \langle g(\mathbf{x}', t_i) - g(\mathbf{x}', t_{i-1})\rangle = (2\dot{E}\delta t L^d)^2 = \langle s'\rangle^2, \tag{8.84}$$

so that

$$\begin{aligned} \langle s'^2\rangle - \langle s'\rangle^2 &= \int d^d x \, d^d x' \, c(\mathbf{x}, \mathbf{x}', t_i, t_i) - c(\mathbf{x}, \mathbf{x}', t_i, t_{i-1}) \\ &\quad + \int d^d x \, d^d x' \, c(\mathbf{x}, \mathbf{x}', t_{i-1}, t_{i-1}) - c(\mathbf{x}, \mathbf{x}', t_{i-1}, t_i) \\ &= 2 \int d^d x \, d^d x' \, \big(c(\mathbf{x}, \mathbf{x}', t_i, t_i) - c(\mathbf{x}, \mathbf{x}', t_i, t_{i-1})\big) \end{aligned} \tag{8.85}$$

with the last equality because $\mathbf{x}$ and $\mathbf{x}'$ can be swapped in the integration, so that $c(\mathbf{x}, \mathbf{x}', t, t') = c(\mathbf{x}', \mathbf{x}, t', t)$ can be used, and at stationarity $c(\mathbf{x}, \mathbf{x}', t_i, t_j) = c(\mathbf{x}, \mathbf{x}', t_{i+1}, t_{j+1})$. The expected scaling form of the correlation function (Sec. 8.5.4.2) is

$$c(\mathbf{x}, \mathbf{x}', t, t') = a|\mathbf{x} - \mathbf{x}'|^{2\chi^{(\mathrm{int})}} \mathcal{G}\left(\frac{t - t'}{b|\mathbf{x} - \mathbf{x}'|^{z^{(\mathrm{int})}}}\right) \tag{8.86}$$

with scaling function $\mathcal{G}$, metric factors [53] a and b as well as dynamical exponent $z^{(\mathrm{int})}$ and wandering or roughness exponent $\chi^{(\mathrm{int})}$. Because of the boundary conditions, strictly the positions relative to the boundaries should enter as arguments of the scaling function individually, $\mathbf{x}/L$ and $\mathbf{x}'/L$, but in the following the integral Eq. (8.85) is calculated on the basis of dimensional analysis, which is not affected by that.

The resulting scaling form of the variance Eq. (8.85) thus is

$$\langle s'^2\rangle - \langle s'\rangle^2 = aL^{2d+2\chi^{(\mathrm{int})}} \mathcal{F}\left(\frac{2\delta t}{bL^{z^{(\mathrm{int})}}}\right). \tag{8.87}$$

---

[52] With the first moment linear in δt, the linearity in δt of higher moments follows already from the assumption that the characteristic avalanche size is independent of δt, which is the case when it is given by the substrate volume L^d times the vertical fluctuations, $L^{\chi^{(\mathrm{int})}}$. This argument hinges, however, on the assumption that δt does not enter into the scaling function, which cannot hold in the limit of large δt.

[53] The symbols a, b, e are used as generic metric factors and $\mathcal{G}$ and $\mathcal{F}$ as generic scaling functions in the following, even when they attain different values and different forms in each of the following cases.

Assuming that $\langle s' \rangle^2$ is subleading and that the scaling function $\mathcal{F}$ converges to a non-vanishing value for small arguments gives $\sigma_2^{(s)} = D(3 - \tau) = 2d + 2\chi^{(\text{int})}$ for the finite size scaling exponent of the second moment. Regardless of whether one allows for the scaling of $\dot{E}\delta t$, such that $D(2 - \tau) = 1$, or not, in which case $D(2 - \tau) = d$, together with the result for the first moment, the avalanche dimension is either $D = 2d - 1 + 2\chi^{(\text{int})}$ or $D = d + 2\chi^{(\text{int})}$, compared to the theoretically expected and numerically confirmed $D = d + \chi^{(\text{int})}$ (Paczuski et al., 1996). The derivation cannot be salvaged by appealing to the asymptotic behaviour $\mathcal{G}(0) - \mathcal{G}(x) \propto x^{2\beta^{(\text{int})}}$ of the integrand [54] in Eq. (8.85) for small arguments, which gives $\sigma_2^{(s)} = 2d$ and therefore $D = 2d - 1$ or $D = d$ from $\beta^{(\text{int})}z^{(\text{int})} = \chi^{(\text{int})}$, where $\beta^{(\text{int})}$ is the roughening or growth exponent.

It is rather the presence of a second length scale in Eq. (8.86) which 'corrects' the argument above. The external driving adds $2\dot{E}\delta t$ to $g(0, t)$ each time a new avalanche is triggered. If this gap were so big that the old and the new interface were detached, like the bold configurations in Fig. 8.14(b), then each and every avalanche would have roughly the size $2\dot{E}\delta t L^d$. The driving has to be compared to the width $L^{\chi^{(\text{int})}}$ of the fluctuations (see endnote [8.8], p. 395) of the interface to decide whether or not new configurations are effectively entirely new incarnations of the interface. This is suggested in Eqs. (8.81) and (8.85) where the integrands vanish in the entire area where the new and the old configurations coincide.

Adding the argument $|t - t'|\dot{E}/L^{\chi^{(\text{int})}}$ to the scaling function $\mathcal{G}$ in Eq. (8.86) must be accompanied by a statement about its behaviour in small arguments, given that this ratio in the present case vanishes in large L. The asymptote can be derived from the observation that all moments of the avalanche size distribution scale *linearly* in the number of initial charges. This has been shown analytically for the TAOM (Pruessner, 2004b; Welinder et al., 2007) and can be verified numerically for the OSLO Model and the MANNA Model as an asymptotic behaviour in the limit of large system sizes.

Denoting $t - t' = \Delta t$, charging the model M times instead of once for each avalanche means $\Delta t = M\delta t$. The variance thus becomes

$$\langle s'^2 \rangle - \langle s' \rangle^2 = aL^{2d+2\chi^{(\text{int})}} \mathcal{F}\left(\frac{2\Delta t}{bL^{z^{(\text{int})}}}, \frac{2\dot{E}\Delta t}{eL^{\chi^{(\text{int})}}} \right), \tag{8.88}$$

with metric factor e and scaling function $\mathcal{F}(x, y)$ scaling linearly in y for small x and y, so that asymptotically

$$\langle s'^2 \rangle - \langle s' \rangle^2 = 2ae^{-1}\dot{E}\Delta t L^{2d+\chi^{(\text{int})}} \mathcal{F}_y\left(\frac{2\Delta t}{bL^{z^{(\text{int})}}}, \frac{2\dot{E}\Delta t}{eL^{\chi^{(\text{int})}}} \right), \tag{8.89}$$

with $\mathcal{F}_y(x, y)$ finite in small y. If that holds for small x simultaneously, then the second moment $\langle s'^2 \rangle$ scales like $M\dot{E}\delta t L^{2d+\chi^{(\text{int})}}$. Regardless of the scaling of $\dot{E}\delta t$, since $\langle s' \rangle \propto \dot{E}\delta t L^d$, this confirms (Paczuski et al., 1996)

$$\sigma_2^{(s)} - \sigma_1^{(s)} = D = d + \chi^{(\text{int})}, \tag{8.90}$$

as $L^{\sigma_2^{(s)} - \sigma_1^{(s)}} \propto \langle s'^2 \rangle / \langle s' \rangle \propto L^{d+\chi^{(\text{int})}}$.

---

[54] Such a scaling makes sense only for the difference $\mathcal{G}(0) - \mathcal{G}(x)$, since $\mathcal{G}(x) \propto x^{2\beta^{(\text{int})}}$ contradicts the expected scaling $c(\mathbf{x}, \mathbf{x}', t, t') \propto |\mathbf{x} - \mathbf{x}'|^{2\chi^{(\text{int})}}$ at $t = t'$.

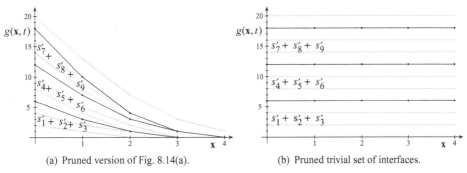

(a) Pruned version of Fig. 8.14(a).   (b) Pruned trivial set of interfaces.

Figure 8.15   (a) The pruned network of interfaces shown in Fig. 8.14(a). Avalanches generated by multiple initial charges (here 3) can be derived from those of single charges, by removing the corresponding number of intermediate interface configurations, here shown as dotted lines (two within each new, bigger avalanche). The resulting, combined areas represent the new, bigger avalanche sizes. (b) Same for a set of trivial, ballistic trajectories. As interfaces do not merge, the moments of the areas encapsulated after removing $k - 1$ intermediate interface configurations scale like k^n (here $k = 3$).

If the model is Abelian, multiple charges can be processed sequentially and the sum of the resulting avalanches (including those of vanishing size) is the size of the single avalanche triggered by the multiple initial charges. The second moment of the avalanche size distribution due to M charges is

$$\langle s^2 \rangle (2M\dot{E}\delta t) = \langle (s_1 + s_2 + \cdots + s_M)^2 \rangle (2\dot{E}\delta t)$$

$$= M \langle s^2 \rangle (2\dot{E}\delta t) + \sum_{i \neq j}^{M} \langle s_i s_j \rangle (2\dot{E}\delta t) \tag{8.91}$$

where the arguments $2M\dot{E}\delta t$ and $2\dot{E}\delta t$, respectively, indicate the amounts of driving received at (virtual) site $\mathbf{n} = 0$. The indexed avalanches s_i are generated sequentially and $\langle s_i s_j \rangle (2\dot{E}\delta t)$ denotes the correlation function in a system with charges of size $2\dot{E}\delta t$. Naïvely, one expects $\sum_{i \neq j}^{M} \langle s_i s_j \rangle$ to scale at least like $M(M-1) \langle s \rangle^2$. For sufficiently large M this must indeed be the case, because $\langle s_i s_j \rangle$ approaches $\langle s \rangle^2$ for large $|i - j|$. Given, however, that $\langle s^2 \rangle (2M\dot{E}\delta t)$ is $M \langle s^2 \rangle (2\dot{E}\delta t)$ to leading order in an intermediate regime of M, the sum scales sublinearly in M. This is due to **anti-correlations**, $\langle s_i s_j \rangle < \langle s \rangle^2$, i.e. large avalanches are followed by small ones and vice versa. Eventually, for sufficiently large M, $\langle s^2 \rangle (2M\dot{E}\delta t)$ must scale at least like M^2 to prevent $\langle s \rangle^2 (2M\dot{E}\delta t)$ from exceeding its value, thereby creating a negative variance.

As illustrated in Fig. 8.15(a), combining avalanches corresponds to pruning the network of quiescent interfaces, thereby joining avalanches. The moments of the areas generated by removing every second interface, are all, to leading order in L, twice the moment in the unpruned network. While this seems to be a general feature it is not a geometrical triviality. [8.8] For example, if all interfaces were horizontal lines (ballistic rather than diffusive), so that all enclosed areas were rectangles of equal size, then removing every second interface would double the areas, so that the nth moment would increase by a factor 2^n, as illustrated in Fig. 8.15(b). In the present case, as long as $\dot{E}\Delta t \ll L^{\chi^{(int)}}$ the vast majority of the interface configurations removed by pruning are those that merge with other interfaces before reaching the right hand side boundary, i.e. fluctuations $L^{\chi^{(\text{int})}}$ in g are big compared to the gap $2\dot{E}\delta T$ at $\mathbf{x} = 0$. Otherwise, a sequence of horizontal patterns develops such as that shown in Fig. 8.14(b) and the linear relationship $\langle s^n \rangle \propto \dot{E}\Delta t$ turns into the naïve scaling $\langle s^n \rangle \propto (\dot{E}\Delta t)^n$.

The scaling of the area (i.e. avalanche) size distribution for multiple charges can be derived alternatively by assuming the usual scaling form $\mathcal{P}^{(\text{s})}(s, L, \dot{E}\Delta t) = as^{-\tau}\hat{\mathcal{G}}(s/(bL^{\text{D}}), \dot{E}\Delta t/e)$ at fixed Δt and in addition that the area size distribution is invariant under simultaneous rescaling of s, L and Δt,

$$\mathcal{P}^{(\text{s})}\left(s, L, \lambda^{\chi^{(\text{int})}}\dot{E}\Delta t\right) \, ds = \mathcal{P}^{(\text{s})}\left(\tilde{s}, \tilde{L}, \dot{E}\Delta t\right) \, d\tilde{s} \tag{8.92}$$

with $\tilde{s} = s/\lambda^{d+\chi^{(\text{int})}}$ as the rescaled avalanche size and $\tilde{L} = L/\lambda$ as the rescaled system size. Combining both scaling symmetries gives

$$\mathcal{P}^{(\text{s})}(s, L, \dot{E}\Delta t) = as^{-\tau}\dot{E}\Delta t e^{-1}L^{d-1}\mathcal{G}\left(\frac{s}{bL^{\text{D}}}, \frac{\dot{E}\Delta t}{eL^{\chi^{(\text{int})}}}\right) \tag{8.93}$$

using Eq. (8.92) at $\lambda = L$ as well as $\text{D} = d + \chi^{(\text{int})}$ and $\text{D}(2 - \tau) = \sigma_1^{(\text{s})} = 1$, as normally found in one dimension. This result looks somewhat surprising because of the additional prefactor $\dot{E}\Delta t L^{d-1}$. As long as $\dot{E}\Delta t$ is adjusted such that $\langle s \rangle = 2\dot{E}\Delta t L^d \propto L$, the prefactor is constant. Assuming more generally $\langle s \rangle = 2\dot{E}\delta t L^d \propto L^{\sigma_1^{(\text{s})}}$ without constraint on $\sigma_1^{(\text{s})}$ and $\Delta t = M\delta t$, then using $\text{D} = \chi^{(\text{int})} + d$ and $\sigma_1^{(\text{s})} = \text{D}(2 - \tau)$ gives

$$\mathcal{P}^{(\text{s})}(s, L, M\dot{E}\delta t) = \tilde{\mathcal{P}}^{(\text{s})}(s, L, M) = \tilde{a}s^{-\tau}M\mathcal{G}\left(\frac{s}{bL^{\text{D}}}, \frac{M}{\tilde{e}L^{\text{D}(\tau-1)}}\right) \tag{8.94}$$

as the driving $\dot{E}\delta t$ is 'enslaved' by the requirement $\langle s \rangle \propto L^{\sigma_1^{(\text{s})}}$. While Eq. (8.93) suggests that standard finite size scaling $\mathcal{P}^{(\text{s})}(s) = as^{-\tau}\mathcal{G}(s/(bL^{\text{D}}))$, Eq. (1.1), is recovered only by the lucky coincidence that $\dot{E}\Delta t \propto L^{-(d-1)}$, Eq. (8.94) clarifies that this scaling form applies for *any* $\sigma_1^{(\text{s})}$ with τ *defined* by $\sigma_1^{(\text{s})} = \text{D}(2-\tau)$. It is reassuring that $L^{\text{D}(\tau-1)}$ has the anomalous dimension of the inverse normalisation, i.e. a (dimensionless) number, to be compared to the (dimensionless) number M. As long as that is small compared to $L^{\text{D}(\tau-1)}$, the second argument does not enter, $\lim_{y\to 0} \mathcal{G}(x, y) = \mathcal{K}(x)$, recovering the simple form of Eq. (1.1) for fixed M. To recover the behaviour in large $M \gg L^{\text{D}(\tau-1)}$, when areas encapsulated by trajectories fluctuate very little, so that $\langle s^n \rangle \propto \left(ML^{\text{D}(2-\tau)}\right)^n \propto \langle s \rangle^n$, Eq. (8.94) can be rewritten as

$$\tilde{\mathcal{P}}^{(\text{s})}(s, L, M) = \tilde{a}'s^{-1}\tilde{\mathcal{G}}\left(\frac{s}{\tilde{e}'ML^{\text{D}(2-\tau)}}, \frac{M}{\tilde{e}L^{\text{D}(\tau-1)}}\right) \tag{8.95}$$

with $\tilde{\mathcal{G}}(x, y)$ asymptotically independent of large y and scaling like $x^{1-\tau}y^{2-\tau}\mathcal{K}(xy)$ in small y to recover the corresponding asymptote of Eq. (8.94).

8.5.4.2 Definition of the roughness

There is significant confusion in the literature about the definition of the roughness and thus of the roughness exponent $\chi^{(\text{int})}$ (also Krim and Indekeu, 1993). Defining it through the width of the interface (the ensemble average of the spatial variance, Barabási and Stanley, 1995),

$$w^2(L,t) = \left\langle \overline{g^2(\mathbf{x},t)} - \overline{g(\mathbf{x},t)}^2 \right\rangle = \left\langle \overline{\left(g(\mathbf{x},t) - \overline{g(\mathbf{x},t)} \right)^2} \right\rangle \qquad (8.96)$$

where the overbar denotes the spatial average

$$\overline{g(\mathbf{x},t)} = L^{-d} \int_{L^d} d^d x \, g(\mathbf{x},t), \qquad (8.97)$$

leads to the standard **Family and Vicsek** (1985) **scaling** form,

$$w^2(L,t) = a L^{2\chi^{(\mathrm{int})}} \mathcal{G}\left(\frac{t}{bL^{z^{(\mathrm{int})}}} \right) \qquad (8.98)$$

to leading order in large L, with metric factors a, b and scaling function $\mathcal{G}(x)$, and universal exponents $\chi^{(\mathrm{int})}$ and $z^{(\mathrm{int})}$ characterising the interface. For small arguments, this scaling function behaves like a power law $\mathcal{G}(x) \propto x^{2\beta^{(\mathrm{int})}}$, replacing the system size dependence by a time-dependence, $\beta^{(\mathrm{int})} z^{(\mathrm{int})} = \chi^{(\mathrm{int})}$.

The roughness, Eq. (8.96), is related to the two-point equal time correlation function through the sum rule [55] (Barabási and Stanley, 1995; Krug, 1997)

$$w^2(L,t) = \frac{1}{2} L^{-2d} \int d^d x_1 \, d^d x_2 \left\langle (g(\mathbf{x}_1,t) - g(\mathbf{x}_2,t))^2 \right\rangle \qquad (8.99)$$

so that any scaling of the latter translates immediately to the scaling of the roughness, which leads to the alternative scaling assumption

$$\left\langle (g(\mathbf{x}_1,t) - g(\mathbf{x}_2,t))^2 \right\rangle = \tilde{a} |\mathbf{x}_1 - \mathbf{x}_2|^{2\chi^{(\mathrm{int})}} \tilde{\mathcal{G}}\left(\frac{t}{b|\mathbf{x}_1 - \mathbf{x}_2|^{z^{(\mathrm{int})}}} \right) \qquad (8.100)$$

normally defined after taking the thermodynamic limit (Krug and Spohn, 1991; Meakin, 1998). As clarified by López (1999, also Das Sarma, Ghaisas, and Kim, 1994; Ramasco, López, and Rodríguez, 2000; Dickman *et al.*, 2001) the roughness exponent in Eq. (8.100) does not necessarily coincide with that in the original *definition* Eq. (8.98). In the argument presented in Sec. 8.5.4.1 it was the roughness exponent of the correlation function in Eqs. (8.83) and (8.86) rather than that of the width that determined the scaling of the avalanche size distribution.

Some authors invoke a different scaling form, namely (e.g. Barabási and Stanley, 1995)

$$\begin{aligned}
\langle g(\mathbf{x}_1,t_1) g(\mathbf{x}_2,t_2) \rangle &- \langle g(\mathbf{x}_1,t_1) \rangle \langle g(\mathbf{x}_2,t_2) \rangle \\
&= \left\langle \left(g(\mathbf{x}_1,t_1) - \langle g(\mathbf{x}_1,t_1) \rangle \right) \left(g(\mathbf{x}_2,t_2) - \langle g(\mathbf{x}_2,t_2) \rangle \right) \right\rangle \\
&= \tilde{a} |\mathbf{x}_1 - \mathbf{x}_2|^{2\chi^{(\mathrm{int})}} \tilde{\mathcal{G}}\left(\frac{(t_1 - t_2)}{b|\mathbf{x}_1 - \mathbf{x}_2|^{z^{(\mathrm{int})}}} \right)
\end{aligned} \qquad (8.101)$$

in the stationary state after taking the thermodynamic limit, which makes it difficult to relate it back to the original scaling form Eq. (8.98) where time t is measured from initialisation with a flat interface.

---

[55] The correlation function $\langle g(\mathbf{x},t)g(\mathbf{x},t) - g(\mathbf{x},t)g(\mathbf{x}',t)\rangle$ used here of course differs from $\langle g(\mathbf{x},t)g(\mathbf{x}',t)\rangle - \langle g(\mathbf{x},t)\rangle \langle g(\mathbf{x}',t)\rangle$ introduced in Eq. (8.83).

The common perception that **super-rough** (Das Sarma *et al.*, 1994) processes, i.e. those with $\chi^{(\text{int})} > 1$, suffer from overhangs, is probably based on the scaling of the correlation function rather than that of the roughness (Fisher, 1986; Krug and Spohn, 1991; Roux and Hansen, 1994; Paczuski *et al.*, 1996, but Ramasco *et al.*, 2000). The interface of the OSLO Model is characterised by a roughness exponent of $\chi^{(\text{int})} \approx 1.25$ in one dimension, however, without any such overhangs. The large exponent is, to some extent, an artefact caused by the definition of the roughness as the ensemble averaged second spatial central moment, Eq. (8.96), which is not invariant under tilting (**Galilei transform**). Even a 'deterministic shape' like $g(\mathbf{x}) = |\mathbf{x}|^{\mu}$ therefore gives rise to a roughness exponent $\chi^{(\text{int})} = \mu$ according to Eq. (8.96) and Eq. (8.98) and similarly for the integral of the correlation function, such as Eq. (8.85). Figure 8.13 shows the interfaces of a deterministic version of the OSLO Model, which are almost indistinguishable from their stochastic counterpart. On the basis of Eq. (8.98), the trivial one-dimensional BTW Model would be rough with exponent $\chi^{(\text{int})} = 1$.

The same argument applies to any non-trivial, noisy $g_0(\mathbf{x}, t)$ that is modulated by a possibly time dependent amplitude; if g_0 gives rise to roughness exponent $\chi_0^{(\text{int})}$, then $g(\mathbf{x}, t) = |\mathbf{x}|^{\mu} g_0(\mathbf{x}, t)$ gives rise to $\chi^{(\text{int})} = \chi_0^{(\text{int})} + \mu$. In the OSLO Model the characteristic shape of the interfaces has $\mu = 1$. From this perspective, the main contribution to the roughness of the interface representation of the OSLO Model is due to its characteristic shape, which in turn is due to the boundary conditions. In studies of depinning transitions, these boundary conditions are periodic and the characteristic shape is fundamentally different, so that one might wonder how the 'roughness' in the interface representations of SOC models such as the OSLO Model is physically related to the roughness observed in depinning – why is there a correspondence at all?

Some authors (e.g. Krug *et al.*, 1992; Lübeck and Usadel, 1993; Zhang, 2000) account for the characteristic shape by measuring the roughness as the spatial average of the ensemble averaged local variance,

$$\tilde{w}^2(L, t) = \overline{\langle g^2(\mathbf{x}, t)\rangle - \langle g(\mathbf{x}, t)\rangle^2} = \overline{\langle (g(\mathbf{x}, t) - \langle g(\mathbf{x}, t)\rangle)^2\rangle} \qquad (8.102)$$

in contrast to Eq. (8.96). While this object is less frequently studied in the interface literature, it is reasonable to assume that it follows Family–Vicsek scaling, Eq. (8.98). It does not suffer from the characteristic shape unduly entering the roughness, but after swapping spatial and ensemble average, the roughness is merely a matter of the ensemble, rather than the individual interface. [8.9] In this definition an ensemble of parallel-displaced, perfectly flat and straight interface can appear rough.

8.5.4.3 Exponents of the qEW equation

In the following, some scaling features of the qEW equation are briefly reviewed. These results all are based on *tuned* non-equilibrium systems at a transition, namely precisely at the depinning transition, where an external force just overcomes the pinning due to a random environment. The qEW equation is of particular importance because it represents the large MANNA universality class, Sec. 6.2.1.2. Dimensional analysis of the qEW equation,

Eq. (8.75), gives

$$w^2(L, t) = \left(\frac{\Gamma}{D}\right)^{\frac{4}{3}} L^{2\frac{4-d}{3}} \mathcal{G}\left(\frac{tD}{L^2}\right) \tag{8.103}$$

assuming δ-correlated noise with amplitude Γ,

$$\left\langle \tilde{\xi}(\mathbf{x}, g)\tilde{\xi}(\mathbf{x}', g')\right\rangle = 2\Gamma^2\delta(\mathbf{x} - \mathbf{x}')\delta(g - g'), \tag{8.104}$$

cf. Eq. (6.35). The exponents are thus

$$\chi^{(\text{int})} = \frac{4 - d}{3} \quad \text{and} \quad z^{(\text{int})} = 2, \tag{8.105}$$

provided that L, t, D and Γ are the only *parameters of the problem.* If this condition is fulfilled, Eq. (8.103) is exact and has no corrections. The upper critical dimension is predicted correctly. Above $d_c = 4$ the diffusive term overcomes the background noise, so that the interface is flat and even the weakest external force will move the interface, i.e. the quenched noise becomes irrelevant as it is unable to pin the interface. Strictly, Eq. (8.103) remains valid, but is normally replaced by scaling in the ultraviolet cutoff (Krug, 1997). Below $d_c = 4$ the interface is flat only below the Larkin-length (Bruinsma and Aeppli, 1984).

Narayan and Fisher (1993) found the roughness exponent Eq. (8.105) to all orders in perturbation theory. Numerically, its value is clearly different, which hints at the existence of an additional scale omitted above – any additional scale renders the derivation above invalid. The most obvious origin for such an additional scale are any non-homogeneous boundary conditions and similarly a finite driving rate $\dot{E}$. More significantly, Nattermann *et al.* (1992) found that the quenched noise cannot be δ-correlated in the longitudinal direction, i.e. in g, which has been confirmed several times since (Leschhorn *et al.*, 1997; Le Doussal *et al.*, 2002). Any non-trivial noise correlator necessarily introduces additional scales. Under renormalisation, it develops a non-analyticity (Wiese, 2002), which might be the explanation why **dimensional reduction** (Efetov and Larkin, 1977) does not apply.

The exponents measured and calculated for the qEW equation have been extensively reviewed by Le Doussal *et al.* (2002, also Leschhorn, 1994). The numerics of the qEW equation, originally due to Leschhorn (1993, 1994) are very consistent with the numerics of the OSLO Model and the MANNA Model for the avalanche dimension $\mathsf{D} = d + \chi^{(\text{int})}$ and the dynamical exponent $\mathsf{z} = z^{(\text{int})}$ in $d = 1, 2, 3$ (but Dickman *et al.*, 2001) as well as in MFT which becomes valid at $d = d_c = 4$, see Table 8.5. The avalanche size exponent τ and the avalanche duration exponent α, sometimes described as secondary to D and z (Christensen, 2004), follow from scaling laws, such as $\mathsf{D}(2 - \tau) = \sigma_1^{(\text{s})} = 1$, depending on the driving, and more generally $\mathsf{D}(1 - \tau) = \mathsf{z}(1 - \alpha)$. Paczuski *et al.* (1996) found that dynamical exponents can differ between SOC and their tuned counterparts. This is a particular problem either when the dynamics of the SOC model is not represented by an equation of motion in the (dynamical) universality class of the tuned critical system or when the dynamics is ambiguous, as is the case in most Abelian models. The situation is somewhat reminiscent of the dynamics of the Ising Model, which can be implemented in many ways,

Table 8.5 Comparison of the exponents characterising the MANNA Model, D and z (from Table 6.1, most recent data), and those from simulations (theory for MFT) of interfaces in the qEW equation universality class, $\chi^{(int)}$ and $z^{(int)}$, confirming $D = d + \chi^{(int)}$ and $z = z^{(int)}$. Other exponents, like α and τ, are expected to depend on driving and boundary conditions.

d	D		$d + \chi^{(int)}$		z		$z^{(int)}$	
1	2.2(1)$\diamond$	(a)	2.25(1)	(b)	1.50(4)♮	(c)	1.42(4)	(b)
2	2.764(10)	(d)	2.75(2)	(b)	1.533(24)♮	(e)	1.56(6)	(b)
3	3.36(1)	(f)	3.35(1)	(b)	1.823(23)♮	(e)	1.75(15)	(b)
MFT	4	(g)	4	(b)	2	(g)	2	(b)

$\diamond$ Reported for the oMM.
♮ Reported for variants of the MANNA Model (in particular FES, Sec. 9.3.1).
a Nakanishi and Sneppen (1997), b Leschhorn *et al.* (1997), c Dickman (2006), d Lübeck (2000), e Lübeck and Heger (2003a), f Pastor-Satorras and Vespignani (2001), g Lübeck (2004).

characterised by different dynamical exponents, but the same static exponents (Hohenberg and Halperin, 1977). The relation between the qEW equation and the OSLO Model does not suffer from this arbitrariness, as the former is precisely a continuum approximation of the equation of motion of the latter. This explains why even the dynamical exponents are reproduced.

9   Mechanisms

How does SOC work? What are the necessary and sufficient conditions for the occurrence of SOC? Can the mechanism underlying SOC be put to work in traditional critical phenomena? These questions are at the heart of the study of SOC phenomena. The hope is that an SOC mechanism would not only give insight into the nature of the critical state in SOC and its long-range, long-time correlations, but also provide a procedure to prompt this state in other systems. In the following, SOC is first placed in the context of ordinary critical phenomena, focusing on the question to what extent SOC has been preceded by phenomena with very similar features. The theories of these phenomena can give further insight into the nature of SOC. In the remainder, the two most successful mechanisms are presented, the second of which, the Absorbing State Mechanism (AS mechanism), is the most recent, most promising development. A few other mechanisms are discussed briefly in the last section.

SOC mechanisms generally fall in one of three categories. Firstly, there are those that show that SOC is an instance of **generic scale invariance**, by showing that SOC models cannot avoid being scale invariant, because of their characteristics, such as bulk conservation and particle transport. The mechanism developed by Hwa and Kardar (1989a), Sec. 9.2, is the most prominent example of this type of explanation. This approach focuses solely on criticality and dismisses any self-organisation. Maybe because it suggests a degree of triviality, it has not always been welcomed very positively. The second category fares better in that respect by attempting to identify the mechanism by which SOC systems place themselves right at the critical point of an underlying **ordinary phase transition**. In the AS mechanism, Sec. 9.3, which identifies the scaling found in SOC with that at the critical point of a non-equilibrium phase transition into an absorbing state and explains how SOC models tune themselves to that critical point, this programme has been realised very successfully. It is an explanation that leaves most of the original notion of 'self-organisation to a critical point' intact. Finally, there is the category of alternative mechanisms, which bring about scale invariance and which are only 'mistaken' for SOC. Some of them are mentioned briefly in Sec. 9.4. Further details can be found in Sornette's (2006) comprehensive review. These suggested mechanisms exemplify the general problem that models can be found which display scale invariance, but can be understood in terms very different from 'self-organisation' and the traditional understanding of critical phenomena.

9.1 SOC and ordinary criticality

Scale invariance characterised by non-trivial exponents is well known from ordinary critical phenomena (Stanley, 1971), where a **temperature-like control parameter** is tuned to trigger a **phase transition** between two phases with and without **broken symmetry**. However, as Grinstein (1995) summarised, the equilibrium is 'a rotten place to hunt for generic scale invariance', i.e. (non-trivial) generic scale invariance is generally expected only in non-equilibrium (but see Sec. 9.4.1). The power of equilibrium lies in **detailed balance** which allows the probability of a particular state to be represented by a **Boltzmann–Gibbs factor** (p. 323), whose value depends solely on the spatial configuration. [1]

What makes SOC fascinating is the appearance of non-trivial spatiotemporal scale invariance, i.e. algebraically decaying correlations in space and time, which indicates that the entire system interacts across time and space making it impossible to split it into finite 'compartments'. While temporal long-range correlations are not that unusual (Hohenberg and Halperin, 1977; Grinstein *et al.*, 1990) and occur without spatial scale invariance, [2] the converse is not true; spatial scale invariance is rare and almost always comes with temporal scale invariance. Nevertheless, as discussed briefly in the following, there are a number of instances of scale invariance in systems not originally associated with SOC, which can serve as a reference point for it.

Firstly, established generically scale invariant phenomena can be used as a template in an attempt to integrate SOC into an existing theory of scale invariance (e.g. Sec. 9.4.1). One very large class of such generically scale invariant phenomena is that of many (but certainly not all, e.g. Schmittmann, Pruessner, and Janssen, 2006) **growth phenomena** (Roland and Grant, 1989, linking growth and SOC at an early time), in particular **diffusion limited aggregation** (Sec. 9.2) and the **KPZ equation** (also pattern formation, Cross and Hohenberg, 1993). As discussed in Sec. 8.5 they arguably differ from typical SOC models by lacking avalanches and time scale separation [3] and requiring external, scale-free noise, which is subject to propagation or generally a transform by the (non-linear) growth equation (p. 301). It is the external nature of the noise that was subject to some early critique (Frette, 1993, also Sec. 6.3), rather than its scale invariance, not least because any noise with finite correlations becomes white on a sufficiently large time scale (but see footnote 45 on p. 301).

Chaos and fractals, sometimes referred to as a predecessor or historic backdrop of SOC, provide a rich variety of scale invariant phenomena. This is a very mature field, with a fully developed phenomenology, detailed mathematical understanding, experimental as well as empirical evidence and various mechanisms. In fact, SOC could be regarded as

---

[1] One consequence of this is the impossibility of a one-dimensional equilibrium system to have a phase transition in one dimension (Fisher, 1984), which can be lifted by allowing interactions to propagate in an additional dimension, time, rendering the system non-equilibrium.

[2] A fair random walker along an absorbing wall is an example. As long as there is no self-organisation to the fair state, this is hardly an instance of SOC.

[3] If one reduces SOC to the mere phenomenon of generic scale invariance resembling an ordinary phase transition, the lack of separation of time scales is of course not a problem, but might water down the definition of the field as a whole.

one of them, spawning a 'physics of fractals' (Kadanoff, 1986; Bak *et al.*, 1987). Bak (1996) illustrated the 'ubiquity' of power laws in nature using a fractal coastline. What makes SOC very different from **dynamical systems**, the systems normally studied for their chaotic behaviour, are the large number of degrees of freedom in SOC, low dimensional observables and the frequent presence of noise (but Hastings, Holm, Ellner, *et al.*, 1993). Regarding initial conditions and small changes, most models remain in the same *statistical ensemble* (but see the OFC Model, Sec. 5.3), but share with dynamical systems their high sensitivity in the immediate response to small external perturbations. Bak (1990) placed SOC in the context of chaos and dynamical systems, introducing the notion of the 'border of chaos' also referred to as the 'edge of chaos'. Langton (1990, also Crutchfield, Nauenberg, and Rudnick, 1981; Kauffman and Johnsen, 1991; Gisiger, 2001; Beggs, 2008) introduced this term in a context very closely related to SOC and it resonated widely in the community (e.g. Socolar *et al.*, 1993; Ray and Jan, 1994; Stapleton *et al.*, 2004). Yet, Bak and Sneppen (1993, also Bak, 1996) refused it as misleading, possibly because it suggests a tuning to one particular point, the 'edge', rather than a dynamical evolution around it, or because it points to an origin of SOC different from phase transitions.

Another large class of systems display generic but **trivial scale invariance**. This can be either due to a lack of interaction or due to a lack of competing length scales. The most prominent example is the Gaussian model, whose fixed points govern the scaling of ordinary φ^4-theory either away from the critical point or at the transition at $d \geq d_c$ (Le Bellac, 1991; Täuber, 2005). In both cases the system is free, i.e. there are (asymptotically) no (non-trivial) correlations. While possibly suffering from any number of the other shortcomings discussed in this section, non-interacting systems are obviously not of great relevance for SOC. In systems that lack competing length scales, correlations exist but scaling is trivial and can be identified by dimensional analysis, as found, for example, in the Edwards–Wilkinson equation (see the discussion on p. 9 about competing versus dominating length scales).

The following list comprises phenomena that display non-trivial scaling in some sense apparently without the need of tuning of a control parameter, but which normally are not regarded as cases of SOC. All of them were known prior to SOC and are therefore sometimes used to illustrate the mechanisms described in this chapter or to demonstrate that SOC is not strictly a new phenomenon. If one accepts these models as instances of SOC, possibly solely on the basis that they do not require explicit tuning of a parameter which otherwise controls a transition, they can only help to elucidate the nature of this peculiar phenomenon, which is not made to go away by claiming it has always been around.

Percolation (Broadbent, 1954; Broadbent and Hammersley, 1957) Various observables display non-trivial scaling, such as different types of hulls (Saleur and Duplantier, 1987), apparently without the need for local interaction. Percolation configurations factorise and can therefore be generated randomly and independently 'on the fly' without the need to employ a Markov chain as happens in other equilibrium critical phenomena, which allows percolation algorithms as proposed by Ziff (1982, also Newman and Ziff, 2000; Ziff, Cummings, and Stells, 1984). Simply *imposing the definition of a cluster afterwards* brings scaling about, on some lattices at 'trivial' values of the (site or bound) occupation probability and in the case of gradient percolation without the need for tuning at all (Sapoval, Rosso, and Gouyet, 1985). One might argue that this imposed interaction is the ultimate external control and thus the opposite of self-organisation.

Turbulence (Kolmogorov, 1991) The scaling in flow at large but not infinite Reynolds numbers is a matter of dimensional analysis (Frisch, 1995) once the relevant quantities are determined. While this is strictly an instance of trivial scaling, other celebrated laws of physics, for example the *universal* law of gravitation have the same 'shortcoming' – great physical insight lies, of course, in the identification of the relevant quantities. Although turbulence displays the desired *separation of time scales*, there is no natural definition of an avalanche. Grinstein (1995, p. 291) pointed out that the separation of time scales occurs in its observables (its 'output') rather than in its driving and rejected turbulence as SOC on these grounds (but Dhar and Majumdar, 1990).

Diffusion limited aggregation (Witten and Sander, 1981) This growth phenomenon requires external, scale-free noise with relaxation and driving occurring on the same time scale, very different from the separation of time scales normally implemented in SOC. Beyond the growth due to random particle influx, it is frozen and does not rearrange structurally, so that avalanches are not readily identified. Also, it grows continuously and does not strictly develop a steady state, although the same can be said about the qEW equation (Sec. 6.3.4.2). It was suggested as possibly an early instance of SOC by Grassberger and Manna (1990), rejected by Grinstein (1995) and discussed along similar lines by Sornette *et al.* (1995, but note Kadanoff, 1990; also Sec. 8.5).

Invasion percolation (Wilkinson and Willemsen, 1983) As pointed out by Grassberger and Manna (1990), this 'scheme' allows the generation of critical percolation clusters without the need for tuning of the bond (or site) occupation probability. By implementation a growth phenomenon, it can be seen as an instance of extremal dynamics, which, in turn, is regarded by some as a general mechanism generating scale-free behaviour (Miller, Miller, and McWhorter, 1993). Invasion percolation can be related to the BS Model, from which it might borrow the definition of avalanches as well as a recipe for removing the need for global control, namely by an exponentially slow dynamics, p. 144. In principle, the noise in invasion percolation is quenched which makes it reminiscent of depinning in disordered media and the qEW equation. However, it suffers from a problem similar to diffusion limited aggregation in that it does not rearrange (Sornette *et al.*, 1995). In fact, it only *passes through* the critical point once, when filling an entire (finite) lattice with a single cluster. This is similar for the **parallel update BS Model** of Sornette and Dornic (1996); instead of generically displaying the scaling of directed percolation (DP), in finite systems the gap function is bound eventually to increase arbitrarily closely to unity. The argument that infinite avalanches develop in the supercritical phase (beyond the percolation threshold) does not apply to finite systems.

KPZ equation (Kardar *et al.*, 1986) While other **growth models** display generic scaling as well, such as the Edwards–Wilkinson equation (Edwards and Wilkinson, 1982), the KPZ equation universality class is remarkably large (Medina *et al.*, 1989; Krug, 1995) and therefore is the most prominent representative of this class of generically scale invariant phenomena. Alava (2002) suggested **interface depinning**, which is a different perspective on growth phenomena, as a general framework to understand scaling in

SOC. As a growth process, the KPZ equation resembles diffusion limited aggregation and thus suffers from similar 'deficiencies' with regard to its self-organised critical nature (Grinstein, 1995, also Paczuski, 1995 and Sec. 8.5): essentially frozen, the need for external noise, a lack of separation of time scales and absence of avalanches. A steady state, at least, could be established through appropriate boundary condition.

DP as a minmax problem (Hansen and Roux, 1987) Similar to invasion percolation, this algorithm generates DP scaling without the need for tuning. As in diffusion limited aggregation, however, avalanches are not naturally defined by the dynamics (Sornette *et al.*, 1995), which points at a lack of separation of time scales. Nevertheless, this model is generally accepted as an instance of SOC. In fact, Grassberger and Zhang (1996, also Dickman *et al.*, 2000) argued that it is an instance of ordinary critical phenomena at the same time, which SOC cannot be distinguished from.

9.2 Dissipative transport in open systems

The first significant mechanism of SOC to be proposed (Hwa and Kardar, 1989a) was largely based on growth phenomena and driven diffusive systems (Krug, 1995; Schmittmann and Zia, 1995, also Garrido *et al.*, 1990). A number of central themes in SOC developed out of it, most notably **bulk conservation** and **boundary dissipation** as well as **transport and anisotropy** as discussed in the following. An early, highly influential review by one of its proponents (Grinstein, 1995) contributed to the widespread success of this explanation of SOC. Hwa and Kardar (1989a) summarised their concept as *dissipative transport in open systems with bulk conservation*, the effect of which Paczuski and Bassler (2000) spelt out as

> [T]he fact that the dissipation process is confined to the boundary, which forces the system to self-organize, is an important and subtle point because the boundary cannot be scaled out in the limit of large system sizes as is usually done in statistical physics. In principle, the boundary is always important, because the incoming sand grains must be transported to it, no matter how large the system size. The broken translational invariance associated with the boundary often leads to long-range boundary effects in the metastable states (see, for example Middleton and Tang, 1995; Drossel, 2000).

Perhaps undeservedly, the life of this theory of SOC was cut short by the arrival of non-conservative models of SOC, notably the 'Game of life' (Bak *et al.*, 1989a, also Bak, 1992), Forest Fire Models of various kinds (Bak *et al.*, 1990; Drossel and Schwabl, 1992a), the OFC Model (Olami *et al.*, 1992, also Sec. 5.3.1) and the BS Model (Bak and Sneppen, 1993), among others (Bak *et al.*, 1989a; Feder and Feder, 1991a; Sneppen, 1992). The mere existence of these models was in direct contradiction to what the present theory derived as necessary conditions of SOC. So much so, that some authors (cf. Socolar *et al.*, 1993, also Feder and Feder, 1991a) suggested the concurrent existence of two types of SOC with two underlying mechanisms, one for the conservative case and one for the (apparently more general and physically more relevant) non-conservative case (Grinstein *et al.*, 1990). That might appear far fetched, in particular as none of the non-conservative models displays scaling as robust as, say, the MANNA Model and the OSLO Model. One can only guess what would have come of this theory of SOC if it had not been 'refuted' so early and so fiercely as mere 'speculation' (Bak *et al.*, 1989a; Bak, 1992).

One key ingredient of SOC, however, is markedly absent in the following Langevin formalism: the separation of time scales (Grinstein, 1995). The relaxational, deterministic part of the dynamics and the stochastic driving by the noise both operate on the same time scale. This is also a 'deficiency' of many growth models, such as the KPZ equation mentioned above. However, even if a separation of time scales is *characteristic* for SOC, it does not mean it is *necessary* and one might therefore add it after performing the following analysis, hoping that it remains unchanged (Grinstein *et al.*, 1990). This view was further supported by a more detailed analysis (Hwa and Kardar, 1992) of the effect of different driving speeds, but has since been challenged numerically (Corral and Paczuski, 1999; Woodard *et al.*, 2007; Lőrincz and Wijngaarden, 2008). Regardless of that, the following analysis 'may at least be taken as a useful guide in the study of SOC' (Socolar *et al.*, 1993).

The natural language of **dissipative transport in open systems** is that of Langevin equations, i.e. stochastic equations of motion of the general form

$$\partial_t h(\mathbf{x}, t) = f([h]; \mathbf{x}, t) + \xi(\mathbf{x}, t) \tag{9.1}$$

where the field $h(\mathbf{x}, t)$ is the height of the surface of an evolving sandpile over a substrate, $f([h]; \mathbf{x}, t)$ is a functional of h (the deterministic part of the Langevin equation) and $\xi(\mathbf{x}, t)$ is a noise. Generally, $f([h]; \mathbf{x}, t)$ is thought to be local in time, i.e. to evaluate $h(\mathbf{x}, t)$ at t and not to contain any temporal derivatives. The noise $\xi(\mathbf{x}, t)$ is defined by its correlator, often chosen to represent a Gaussian white noise,

$$\langle \xi(\mathbf{x}, t)\xi(\mathbf{x}', t') \rangle = 2D\delta(t - t')\delta(\mathbf{x} - \mathbf{x}'), \tag{9.2}$$

with vanishing mean and strength D.

The central ideas of Hwa and Kardar (1989a) and what followed can be illustrated on the basis of a linear theory, which is the programme pursued below. Adding non-linearities can affect the long-range behaviour, as known from all the interesting non-trivial physics of φ^4-theory. Typically, generic scale invariance persists even when the scaling may (Hwa and Kardar, 1989a) or may not (Grinstein *et al.*, 1990) be affected by the non-linearity. Their effect is important, since the typical threshold triggered dynamics of SOC models is *highly* non-linear.

The equation considered by Hwa and Kardar (1989a) is closely related to that analysed by Díaz-Guilera (1992) as an equation of motion not of the surface of a sandpile but of the local energy in the ZHANG Model. He did not consider anistropy and found that *all* terms of the form $\nabla^2 h^n$ for $n > 1$ were either irrelevant in the presence of conserved noise, or all (equally) relevant below $d_c = 4$ in the presence of time independent, non-conserved noise. [4]

---

[4] The usual, time dependent noise correlator Eq. (9.2) gives $d_c = 2$.

9.2.1 The linear theory

A particularly simple linear Langevin equation is given by Eq. (9.1) with

$$f([h]; \mathbf{x}, t) = \nu \nabla^2 h - \epsilon h \tag{9.3}$$

and ξ a Gaussian white noise as defined in Eq. (9.2), describing the surface evolution with surface tension[5] ν and dissipation ϵ under the bombardment by noise. Integrating over space, ignoring surface terms and taking the ensemble average illustrates the rôle of the dissipation ϵ, as $\partial_t \int d^d x \langle h \rangle = -\epsilon \int d^d x \langle h \rangle$ has the solution $\int d^d x \langle h \rangle \propto \exp(-\epsilon t)$. It is a matter of dimensional analysis to determine the equal time correlation function in the thermodynamic limit, considering Eq. (9.1) at stationarity and (thus) disregarding any initial condition (temporal correlations are discussed by Grinstein, Hwa, and Jensen, 1992). In d spatial dimensions, $\mathbf{x}, \mathbf{x}' \in \mathbb{R}^d$, one finds

$$G_d(r, \epsilon) = \langle h(\mathbf{x}, t) h(\mathbf{x}', t) \rangle = r^{-(d-2)} \frac{D}{\nu} \mathcal{G}\left(r\sqrt{\epsilon/\nu}\right) \tag{9.4}$$

where $r = |\mathbf{x} - \mathbf{x}'|$ and $\mathcal{G}(x)$ is a scaling function. After Fourier transforming, the cases $d = 1$ and $d = 3$ are straightforward to integrate so that $\mathcal{G}(x)$ can easily be determined in closed form, recovering the usual Ornstein–Zernike-type correlation functions,

$$G_1(r, \epsilon) = \frac{D\pi}{\sqrt{\epsilon \nu}} e^{-r\sqrt{\epsilon/\nu}} \tag{9.5a}$$

$$G_3(r, \epsilon) = \frac{D}{2\nu r} e^{-r\sqrt{\epsilon/\nu}}. \tag{9.5b}$$

The occurrence of an exponential cutoff or **characteristic scale** is the signature of the (dissipative) **mass term** $-\epsilon h$, a name it carries in the field theoretic analysis of *equilibrium* critical phenomena.[6] Grinstein *et al.* (1990) argued that **equilibrium systems** always generate such a term in the correlation function, because equilibrium systems are subject to **detailed balance** and are thus governed by the exponential of a Boltzmann distribution, which 'makes them incapable of generic scale invariance' (Grinstein, 1995). The problem extends to the current **non-equilibrium system**, because $f([h]; \mathbf{x}, t)$ can be written as a functional derivative of a (bilinear) Hamiltonian, $f = -\delta \mathcal{H}/\delta h$, so that the Langevin equation describes an **out-of-equilibrium** system relaxing back to equilibrium (in contrast to a **far-from-equilibrium** system), which is ultimately governed by detailed balance and the Boltzmann–Gibbs factor (Hohenberg and Halperin, 1977; Zinn-Justin, 1997).

The first step for generic scale invariance is thus to invoke a strong argument as to why ϵ vanishes. Hwa and Kardar (1989a) pointed out that it suffices to demand that h is **conserved** up to the noise. Alternatively, spatial symmetries might demand that f contains only spatial derivatives (Grinstein, 1995). In both cases f can be written as the divergence of a current,

---

[5] For historic reasons, variable naming follows the interface terminology, even when sandpile surface evolution might better be regarded a growth (or roughening) phenomenon. In the former, ν in Eq. (9.1) is a surface tension and D in Eq. (9.2) a diffusion constant, in the latter ν would better be called a diffusion constant (D) and D a noise amplitude or source strength (Γ^2).

[6] The term 'mass' comes from particle physics. In the context of the inverse propagator, e.g. Eq. (9.9), the mass is also known as the 'gap', namely the gap ϵ between the parabola $\epsilon + \nu \mathbf{k}^2$ and the origin.

$f = \nabla \mathbf{j}$, which does not allow for a term of the form $-\epsilon h$. The total of h over the volume V then changes in time like

$$\partial_t \int_V d^d r \, h(\mathbf{x}, t) = \int_V d^d r \, \xi(\mathbf{x}, t) + \int_{\partial V} ds \, \mathbf{nj} \qquad (9.6)$$

where the contribution from the noise scales like $V^{1/2}$ and the last integral over the surface ∂V with normal $\mathbf{n}$ can be made to vanish by suitable boundary conditions. If that is not possible but $\mathbf{j}$ is bounded, it scales at most like $V^{1-1/d}$, i.e. is non-extensive and h is thus (asymptotically) conserved.

Without considering non-linearities, this argument for $\epsilon = 0$ might appear at first somewhat delicate. It is well known that non-linearities can generate an **effective mass**, as happens in φ^4-theory (Täuber, 2005) and even a negative one as can happen in driven diffusive systems (Schmittmann and Zia, 1995). However, by forcing the non-linearity to be a gradient itself it is conservative and thus cannot give rise to any effective mass.

Hwa and Kardar (1989a) included **anisotropy** in the diffusion, ν, thereby implementing a preferred transport direction, $\|$, as well as non-linearities, such as the term with coupling λ in the simplest equation they considered,

$$\partial_t h(\mathbf{x}, t) = (\nu_\| \partial_\|^2 + \nu_\perp \nabla_\perp^2) h(\mathbf{x}, t) - \frac{\lambda}{2} \partial_\| h(\mathbf{x}, t)^2 + \xi(\mathbf{x}, t), \qquad (9.7)$$

where $\|$ and $\perp$ refer to different subspaces of $\mathbf{x}$ with co-dimensions 1 and $d - 1$ respectively. This equation was constructed on the basis of symmetries and conservation laws. Hwa and Kardar showed, using renormalisation group methods (Forster *et al.*, 1977; Medina *et al.*, 1989), that the resulting *conservative Langevin equation with non-conserved noise, Eq. (9.2), displays generic scale invariance* with non-trivial exponents (in the presence of suitable non-linearities, such as in Eq. (9.7)), i.e. non-trivial scaling at the *stable* fixed point.[7] The exponents were derived in closed form and to all orders of perturbation theory. Interestingly, the roughness exponent $\chi^{(\text{int})}$ turns out to be negative, in line with the assumption of a flat surface but suggesting an ultraviolet divergence.

Their findings are all the more a significant, as non-trivial exponents necessarily require competing length scales, which could (but do not) give rise to the need for fine tuning of couplings.[8] They coined the underlying mechanism *dissipative* **transport in open systems** because of the loss of potential energy in sandpile-like models when particles move through the system towards the boundary. This terminology is slightly at odds with the conventions in SOC: dissipation normally refers to particle loss, which is crucially absent in the present systems as it would generate a mass-like term, and the anisotropic transport might be regarded as spurious (Puhl, 1992; Peng and Herrmann, 1993, also p. 293). In their derivation, anisotropy is not generally a necessity for scale invariance but reflects the observed nature of the models.

---

[7] Due to the presence of the non-linearity, the couplings $\nu_\|$ and $\nu_\perp$ renormalise, in principle, differently. No anisotropy needs to be present on the bare level, but can be generated by renormalisation.

[8] Without competing length scales, exponents are given by dimensional analysis, with competing length scales, a dimensionless variable exists, which might necessitate tuning to produce scaling, p. 9.

Grinstein *et al.* (1990) extended the analysis considering **conserved, anisotropic noise**. This was an important improvement, because driving the conservative dynamics with non-conserved noise is a rather 'violent type of driving' (Grinstein, 1995, p. 268).[9] This can be seen very clearly in **k**-space, because for $f = \nabla \mathbf{j}$ in Eq. (9.1) the $\mathbf{k} = \mathbf{0}$ mode performs a random walk, $\partial_t h(\mathbf{0}, t) = \xi(\mathbf{0}, t)$, and cannot ever relax. What is more, if the Langevin equation is to capture the microscopic dynamics, then the noise should represent the randomness of the bulk dynamics, which is conservative in many models, such as the MANNA Model and the OSLO Model. Of course, if the Langevin equation and noise operate on the macroscopic time scale, then it is non-conservative, compensating for dissipation at open boundaries, and might have rather peculiar correlations with the avalanching (Grinstein *et al.*, 1990).

Grinstein (1995, also 1991) popularised the standpoint that far-from-equilibrium systems *typically* are scale invariant, which needs to be put in perspective by the large number of non-equilibrium phase transitions, which require tuning to display (non-trivial) scale invariance, as found in growth processes, absorbing state phase transitions and driven diffusive systems (Schmittmann and Zia, 1995; Henkel *et al.*, 2008). Moreover, while the 'removal of the detailed balance constraint allows for spatial scale invariance more broadly than in equilibrium' (Grinstein, 1995, p. 269), temporal scale invariance without long-range spatial correlations is not uncommon in equilibrium either (Forster *et al.*, 1977; Grinstein *et al.*, 1990).

If the noise is conserved,

$$\langle \xi(\mathbf{x}, t) \xi(\mathbf{x}', t') \rangle = -2D\nabla^2 \delta(\mathbf{x} - \mathbf{x}') \delta(t - t') \tag{9.8}$$

then the algebraic decay, Eq. (9.4), derived above at $\epsilon = 0$ for an isotropic deterministic part, Eq. (9.3), is replaced by a Dirac δ function. This can be seen most clearly in Fourier space, where Eq. (9.1) with Eq. (9.3) and Eq. (9.8) gives

$$\langle h(\mathbf{k}, t) h(\mathbf{k}', t) \rangle = D \frac{(2\pi)^d \delta(\mathbf{k} + \mathbf{k}')}{\nu k^2 + \epsilon} k^2. \tag{9.9}$$

At $\epsilon = 0$ the k^2 cancel and the Fourier transform of the resulting constant is a δ function, indicating the complete lack of correlations, even when it can be rewritten in the form $\delta(\mathbf{x} - \mathbf{x}') = L^{-d}\delta((\mathbf{x} - \mathbf{x}')/L)$ suggesting scaling in the system size.

To recover algebraic correlations, Grinstein *et al.* (1990) considered anisotropy, which can be illustrated rather nicely in two dimensions. The Langevin equation

$$\partial_t h(\mathbf{x}, t) = (\nu_\| \partial_\|^2 + \nu_\perp \partial_\perp^2) h(\mathbf{x}, t) + \xi(\mathbf{x}, t) \tag{9.10}$$

where $\|$ and $\perp$ refer to the two components of $\mathbf{x}$, is the linear part of what Hwa and Kardar (1989a) used originally, Eq. (9.7), and what Grinstein *et al.* (1990) used with $\nu_\| = \nu_\perp = \nu$. In their derivation, the noise is anisotropic and conserved,

$$\langle \xi(\mathbf{x}, t) \xi(\mathbf{x}', t') \rangle = -2(D_\| \partial_\|^2 + D_\perp \partial_\perp^2) \delta(\mathbf{x} - \mathbf{x}') \delta(t - t'), \tag{9.11}$$

---

[9] One might argue that this 'violence' is not uncommon. The linear theory at stationarity described above, Eqs (9.3)–(9.5), is the equilibrium Gaussian model which has its critical point at $\epsilon = 0$ (Le Bellac, 1991).

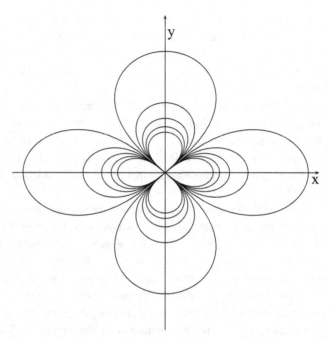

Figure 9.1 Contour plot of Eq. (9.13) with slight anisotropy $D_\parallel/v_\parallel \neq D_\perp/v_\perp$.

with $D_\parallel, D_\perp > 0$. Switching to Fourier space and integrating over ω, the equal time correlator reads

$$\langle h(\mathbf{k}, t) h(\mathbf{k}', t) \rangle = \frac{(2\pi)^2 \delta(\mathbf{k} + \mathbf{k}')(D_\parallel k_\parallel^2 + D_\perp k_\perp^2)}{v_\parallel k_\parallel^2 + v_\perp k_\perp^2}, \quad (9.12)$$

which has a pole only at $k_\parallel = k_\perp = 0$, given that $v_\parallel, v_\perp > 0$. Provided that $D_\parallel/v_\parallel \neq D_\perp/v_\perp$, the squares of the components of the $\mathbf{k}$-vector no longer cancel. After Fourier transforming, the two point correlator is found to be

$$\langle h(\mathbf{x}, t) h(\mathbf{x}', t) \rangle = \frac{1}{2} \left(\frac{D_\parallel}{v_\parallel} + \frac{D_\perp}{v_\perp} \right) \delta(\mathbf{x} - \mathbf{x}')$$
$$+ \frac{\sqrt{v_\parallel v_\perp}}{2\pi |\mathbf{x} - \mathbf{x}'|^2} \left(\frac{D_\parallel}{v_\parallel} - \frac{D_\perp}{v_\perp} \right) \frac{v_\parallel \sin^2 \theta - v_\perp \cos^2 \theta}{\left(v_\parallel \sin^2 \theta + v_\perp \cos^2 \theta \right)^2}, \quad (9.13)$$

where θ is the polar angle, $\tan \theta = r_\perp / r_\parallel$. The last, complicated term in θ simplifies to $-v^{-1} \cos(2\theta)$ for $v_\parallel = v_\perp = v$. The contour plot in Fig. 9.1 shows the 'quadrupole-type' structure (Garrido et al., 1990) of the algebraic correlations with a slight anisotropy $D_\parallel/v_\parallel \neq D_\perp/v_\perp$. As expected, the amplitude of the second, algebraic term in Eq. (9.13) vanishes only at $D_\parallel/v_\parallel = D_\perp/v_\perp$, i.e. to see algebraic correlations $\propto |\mathbf{x} - \mathbf{x}'|^{-2}$ it suffices to have anisotropy in either the noise or the deterministic part of the Langevin equation. The results by Hwa and Kardar (1989a), as summarised after Eq. (9.7), who considered a preferred transport direction, thus remain valid in the presence of conserved noise (Garrido *et al.*, 1990). *A conservative Langevin equation displays generic scale invariance even in the presence of conserved noise, provided noise and transport display different anisotropies.*

The lowest order non-linearity allowed by conservation (but perhaps not by symmetry) is $\gamma \nabla h^2$, which has upper critical dimension 2, as $\gamma D^{1/2} \nu^{-3/2} k^{(d-2)/2}$ is dimensionless (with generic noise strength D and surface tension ν). In one dimension the anisotropy cannot be implemented, simply due to a lack of directions, and in two dimensions the non-linearity is marginally irrelevant (Hwa and Kardar, 1989a). As a result, the anisotropy mechanism gives rise only to trivial exponents in the case of full conservation (including noise).

In summary, a Langevin equation (9.1) with a deterministic part $f([h]; \mathbf{x}, t)$ that vanishes in $\mathbf{k}$-space for vanishing momentum [10] $\mathbf{k}$, can be expected to produce an algebraically decaying correlation function. This is a necessary consequence in the presence of conservation in the bulk dynamics and non-conservation in the noise (Hwa and Kardar, 1989a). If f is linear in h, $f([h]; \mathbf{k}, \omega) = g(\mathbf{k})h$, the equal time correlation function is generally of the form $1/|g(\mathbf{k})|$ and therefore algebraic in $\mathbf{k}$ for a large class of $g(\mathbf{k})$. Mirror symmetry prevents odd derivatives, so that conservative bulk dynamics means $g(\mathbf{k}) \in \mathcal{O}(\mathbf{k}^2)$. If the noise is conservative, its Fourier transform contains momenta squared, which might cancel those in the denominator of the propagator. This does not happen, if noise or bulk dynamics is anisotropic, i.e. algebraic correlations occur also in the presence of bulk conservation, conserved noise and anisotropy (Grinstein *et al.*, 1990). As a rule, a conservative Langevin equation displays generic scale invariance in the presence of non-conserved noise with either isotropic or anisotropic dynamics, generating non-trivial exponents in the presence of non-linearities, but also in the presence of conserved noise, in which case anisotropy is required and non-linearities are irrelevant.

9.2.2 Conservation and dissipation

The importance of **transport**, **anisotropy** and **conservation** had been highlighted very early by Kadanoff *et al.* (1989), who discussed various types of conservation laws. Even when implemented locally, bulk mass conservation can be regarded as a 'nonlocal component of the algorithm' (Frette, 1993), as it enforces transport to boundaries (Paczuski and Bassler, 2000), so that particles explore the extent of the system and are disposed only when reaching its 'end' (Zapperi *et al.*, 1995). This transport is not caused by anisotropy or drift and might not lead to a net current, as is vividly illustrated by the one-dimensional oOM, which displays robust finite size scaling, although it is driven right next to the *only* open boundary, with the other boundary reflective. Bulk dissipation (e.g. Manna *et al.*, 1990, also Sec. 4.1, in particular p. 88 and Sec. 9.3.1) or effective drift (e.g. Jettestuen and Malthe-Sørenssen, 2005, also Sec. 8.4.2) mask or suppress that current and introduce, generally, a different cutoff in the scaling.

Conservation remains a major theme in SOC; some models, such as the BS Model and the FFM, lack any notion of conservation, so that it might be inappropriate to call

---

[10] For example, when f in real space is the divergence of a flux.

these models dissipative. In other models, such as fixed-energy sandpiles (Sec. 9.3.1), conservation is a distinctive feature and they rely crucially on it. The terms 'conservation', 'non-conservation' or equivalently 'dissipation' have led to some confusion. Normally, they apply to the bulk dynamics of particles, and indicate whether their total amount changes at a local update. However, models that conserve the particle number (**mass conservation**) are dissipative in the sense that they lose potential energy when the system is interpreted as a pile of grains, which move downhill, i.e. when particles are subject to dissipative transport. The notion of 'fixed-energy sandpiles', used to suggest a microcanonical ensemble, is particularly confusing in the present context, as the 'energy' refers to the particles, whose number is not allowed to fluctuate but whose potential energy is not fixed.

The OFC Model (Sec. 5.3.1) is a lattice model with a local variable and microscopic dynamics not dissimilar from that in the BTW Model, yet it was designed from the outset to allow for bulk dissipation. Its scaling was challenged by Socolar *et al.* (1993), who tried to reign it in and reconcile it with Hwa and Kardar's (1989a) mechanism (and variants thereof), which suggests that generic scale invariance requires conservation. Their findings were confirmed and complemented by Middleton and Tang (1995), who described a mechanism, **marginal (phase) locking** (Sec. 5.3.1.2), by which the OFC Model seems to develop into a scale invariant state. While numerical results remain open to interpretation, Lise and Jensen's (1996) mean-field-like analysis of the OFC Model suggests that bulk dissipation and critical behaviour are not mutually exclusive. Ultimately, the derivation was shown to be wrong (Sec. 5.3.1, Bröker and Grassberger, 1997; Chabanol and Hakim, 1997; Kinouchi *et al.*, 1998; Pinho *et al.*, 1998), but that cannot take away from the fact that the derivation would work if the force distributions could be made self-consistent, seemingly a mere technicality.

In systems with conservative dynamics and external drive, e.g. the MANNA Model and the OSLO Model, dissipation at some level is a necessary requirement to sustain a steady state. Whether bulk dissipation is equivalent to or can mimic boundary dissipation, continues to be an important question (Sec. 9.3.1, in particular p. 334). For some models, overwhelming evidence exists that suitably scaled bulk dissipation does *not* introduce a characteristic scale beyond that induced by a boundary (e.g. Vespignani and Zapperi, 1998; Pastor-Satorras and Vespignani, 2000c; Malcai *et al.*, 2006), consistent with analytical results for the self-organised branching process (Lauritsen *et al.*, 1996; Bonachela and Muñoz, 2009), even when 'any degree of bulk dissipation makes the system become subcritical' (Bonachela, 2008). This is not to say that bulk dissipation does not introduce a **scale in addition** to the finite size of the system ('additional scales', p. 348). The situation is comparable to an ordinary phase transition, where the system size L is changed in tune with the temperature, $|T - T_c|^{-\nu} \propto L$, so that standard finite size scaling in L is recovered, although moment ratios have **non-universal**, almost arbitrary values (Zheng and Trimper, 2001). The cutoff induced by bulk dissipation does not disappear by compensating for the latter, for example by a random influx of particles. Moreover, such an external field destroys correlations and necessitates tuning (Bröker and Grassberger, 1995; Dickman *et al.*, 2000; Pruessner and Jensen, 2002b; Juanico *et al.*, 2007b; Bonachela and Muñoz, 2009), unless implemented as (conservative) mixing or re-shuffling of particles.

9.2.3 Discussion

The Langevin equations used above describe systems in non-equilibrium, where scale invariance seems to be the rule rather than the exception, whereas in equilibrium, detailed balance almost always necessitates fine-tuning for long-range, spatially invariant correlations to occur. Generic temporal scale invariance, on the other hand, is not uncommon in both (Grinstein, 1995).

Even when many of these findings were initially based on linear equations, they have been extended to non-linear problems. Since then this has led to the popular belief that non-equilibrium systems generally display scale invariance, not in the form of a phase transition, i.e. a critical phenomenon at a singular point in phase space, but as a generic feature. As mentioned above, there are many non-equilibrium systems for which this is not the case. For example, the FES version of the MANNA Model needs to be tuned. On the other hand, it is apparently not governed by a Langevin equation of the type Eq. (9.1), but by one with RFT-type noise. The qEW equation (Leschhorn et al., 1997) has a rich phase diagram incorporating KPZ and Edwards–Wilkinson equations, as well as developing its own, special features which characterise the qEW universality class, if the external force is tuned to the critical value. The model studied by Schmittmann et al. (2006) is a more colourful example of the alternatives. Although the correlator is generically algebraic, characterised by trivial exponents, and thus conforming with the present theory, it develops singularities with non-trivial exponents at a critical point.

Three main questions remain for the present theory of SOC. Firstly, most of the exponents derived are trivial, the only exception being the fractions calculated by Hwa and Kardar (1989a). According to the present theory, algebraic correlations are typically governed by integer valued power laws. This does not correspond to what is found numerically for most SOC models. Secondly, the lack of a time scale separation (and thus avalanching) suggests that a major feature has not been incorporated in the present theory. While one might argue that it is irrelevant for some models and in some limits, models that can be cast in the language of absorbing states (see below) seem to *rely* on it as the mechanism by which activity can cease forever. Thirdly, there is numerical evidence for SOC without bulk conservation, for example in the BS Model. This could mean that an *additional* mechanism of SOC would need to be found for these models. More generally, it challenges the understanding of bulk dissipation as a mass term in a propagator.

9.3 The AS mechanism

As suggested by its name, 'self-organised criticality' is not *intended* to consist of models that are *generically* scale invariant, but 'somehow manage to tune themselves' to an underlying critical point (but see Socolar et al., 1993). To some, a more appealing approach to SOC is thus to trace the observed scale invariance back to an underlying tuned phase transition, focusing on the reasons for the absence of explicit tuning.

The history of analysing SOC from the point of view of **absorbing state** transitions goes through a small number of milestones before culminating in the Absorbing State mechanism (AS mechanism). Tang and Bak (1988a,b) were the first to apply the terminology of critical phenomena to SOC and in particular to pinpoint the concept of an activity. Sornette (1992, also Sornette, 1994) explicitly considered a *non-linear feedback of the order parameter on the control parameter*, which drives an SOC model to an ordinary critical point (also Grassberger and Zhang, 1996). Sornette *et al.* (1995) discussed an entire range of early SOC-like phenomena in this context and Fraysse, Sornette, and Sornette (1993) proposed a number of *experiments* which would make use of the mechanism, some of which have been realised numerically (Pruessner and Peters, 2006; Peters and Girvan, 2009). Vespignani *et al.* (1998, also Dickman *et al.*, 1998) distilled these earlier studies into the single mechanism discussed in the following as the AS mechanism.

Absorbing state (AS) phase transitions (Marro and Dickman, 1999; Hinrichsen, 2000; Lübeck, 2004; Henkel *et al.*, 2008) take place in systems which stop evolving once they hit a so-called absorbing state. [11] The transition is normally driven by a control parameter, say ζ, which needs to be tuned relative to a critical value, ζ_c, above which activity is sustained. The **order parameter** is the **activity**, ρ_a. As illustrated in Fig. 9.2, it vanishes in the **inactive phase**, $\zeta < \zeta_c$ and picks up like a power law at ζ_c, attaining a finite value in the **active phase**, i.e. for any $\zeta > \zeta_c$. Unfortunately, this simple narrative is spoiled by finite size, which normally means that activity ceases eventually for any ζ, even in the supposed active phase. These details are discussed below. The paradigmatic model of AS is **directed percolation**, a form of percolation with directed bonds defining the 'time' direction.

While directed percolation has a single, unique absorbing state, AS models relevant to SOC have **multiple absorbing states**, namely any configuration after an avalanche has terminated. While the number and density of active sites is not conserved, the total number of particles in the system normally is. For example, the AMM can be described as an **activated random walker model** (Dickman *et al.*, 2000, also Rossi *et al.*, 2000; Dickman, Rolla, and Sidoravicius, 2010), where particles move randomly to nearest neighbouring sites only provided that there is a second particle on the same site which they can temporarily pair up with and that makes an independent jump as well. The overall **average particle density** ζ changes only due to particles being lost at open boundaries, but can be made to remain constant by implementing **periodic boundary conditions**. The **activity** ρ_a, the (spatially averaged) density of particle pairs, on the other hand, fluctuates but is always non-negative.

The AS mechanism as formulated by Dickman *et al.* (1998, also Vespignani *et al.*, 1998, 2000; Vespignani and Zapperi, 1998; Dickman *et al.*, 2000) is the most popular and most widely accepted explanation of SOC (e.g. Sornette, 2006, ch. 15.4.5; Meester and Quant, 2005; Rolla and Sidoravicius, 2009), linking SOC explicitly to AS phase transitions by a simple feedback mechanism very much in the spirit of Sornette (1992) and using some of the terminology introduced by Tang and Bak (1988b). The basic idea is that by making the order parameter of an ordinary phase transition ρ_a vanishingly small, $\rho_a \to 0^+$, the system's

---

[11] This term is sometimes used synonymously with recurrent states, but in the present context 'absorbing' entails 'inactive' (van Kampen, 1992).

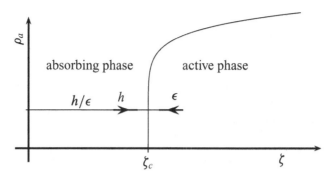

The order parameter ρ_a (activity) as a function of the control parameter ζ (particle density) in an AS transition. In the AS mechanism, the particle density is increased by the external driving, h, and reduced by dissipation, ϵ (bold arrows). It is thus subject to an effective equation of motion, such as Eq. (9.14), which at stationarity fixes the activity to h/ϵ. Tuning $h/\epsilon \rightarrow 0^+$ makes the particle density approach the critical point ζ_c.

(tunable) control parameter ζ is forced to its critical point. From a thermodynamic point of view, this might be considered a change of variables or a change of ensemble (Chessa *et al.*, 1998). However, instead of tuning the order parameter explicitly by imposing a particular value, the rules governing the system set the order parameter implicitly and only *on average* to a particular value. This is implemented by the competition of two mechanisms: the external drive, h, and the internal dissipation, ϵ, both of which act directly on the order parameter.[12] The dissipation is curbed by the system hitting an absorbing state, at which point the instantaneous order parameter and therefore the dissipation strictly ceases. The external drive is constrained by the separation of time scales, which makes it in effect arbitrarily small (but see Sec. 9.3.2). Bonachela and Muñoz (2008, p. 1) summarised:

> It can be argued that slow driving and dissipation acting together at infinitely separated time scales constitute a mechanism able to pin a generic system with absorbing states and a conservation law to its critical point (Grinstein, 1995; Vespignani and Zapperi, 1997, 1998; Dickman *et al.*, 1998; Vespignani *et al.*, 1998, 2000; Dickman *et al.*, 2001; Tang and Bak, 1988a; Dickman *et al.*, 2000; Muñoz *et al.*, 2001).

Figure 9.2 shows the order parameter, the activity ρ_a, of an AS transition as a function of some temperature-like control parameter ζ, such as the particle density. It bears the hallmarks of a traditional continuous phase transition: vanishing order parameter ρ_a in the subcritical (high temperature, disordered phase), a sudden, power law pickup around a critical value and a non-vanishing value in the supercritical (low temperature, ordered) phase. Imposing a particular, non-vanishing (ensemble averaged) value $\langle \rho_a \rangle$ of the order parameter forces the system's control parameter ζ to attain any value in the supercritical regime, down to $\zeta \rightarrow \zeta_c$ as $\langle \rho_a \rangle \rightarrow 0$. This is achieved by constantly increasing the control parameter by an **external field** or **driving** h (a current density, i.e. a number of particles added per time unit and site) and decreasing it *in the presence of activity*, which can be

---

[12] Vespignani *et al.* (1998) make the distinction between SOC models and the corresponding system with fixed control parameter as 'driven dissipative sandpiles' versus 'fixed-energy sandpiles'.

summarised, at least on mean-field level, in a non-linear equation of motion (Dickman *et al.*, 1998):

$$\langle \dot{\zeta} \rangle = \langle h - \epsilon \rho_a(\zeta) \rangle, \tag{9.14}$$

where the time scale is microscopic and ϵ is a **dissipation rate**, which in Eq. (9.14) is constant throughout the bulk,[9.1] whereas a more accurate description allows it to be space dependent, in particular local to the bondary.[13] Assuming that the system reaches stationarity, $\langle \dot{\zeta} \rangle = 0$, gives (Dickman *et al.*, 1998)

$$\langle \rho_a \rangle = \frac{h}{\epsilon}. \tag{9.15}$$

The separation of time scales suggests $h \to 0^+$ (but see Sec. 9.3.2), i.e. $\langle \rho_a \rangle = \langle \rho_a \rangle (\zeta) \to 0^+$ and thus ζ converges to ζ_c where $\langle \rho_a \rangle (\zeta)$ vanishes, $\langle \rho_a \rangle (\zeta_c) = 0$. In fact, the distribution of the control parameter around the critical point collapses to a δ function (see endnote [9.1]). This convergence to the critical point is what distinguishes the AS mechanism from Sornette's (1994) 'sweeping of an instability', where the control parameter has some *fixed* distribution around the critical point.

Strictly, the particle density is a random variable and the result $\langle \zeta \rangle \to \zeta_c$ has to be supported by the assumption $\langle \rho_a(\zeta) \rangle = \rho_a(\langle \zeta \rangle)$. The key ingredient of the AS mechanism, however, is that the function on the right of any such expression is the average activity as a function of the particle density *as found in the AS phase transition* without the external drive and dissipation. If that is the case, then ζ converges to the critical point *of the AS transition*, $\zeta \to \zeta_c^+$. In other words, the SOC model explores the critical point of an underlying AS model, sampling a small region of ζ it is forced into by vanishing h, ϵ. High activity leads to dissipation and thus to a reduction in particle density, which in turn reduces the activity. Low activity is eventually compensated by the external drive h. The system can never venture too deep into the inactive phase, even for very high values of ϵ as it 'gets stuck', i.e. the dynamics and thus the fluctuations stop entirely, when hitting an absorbing state.[14]

Melby, Kaidel, Weber, and Hübler (2000) have applied a mechanism very similar to the present one to the *deterministic* **logistic map**. In that case, the order parameter entering an equation of motion similar to Eq. (9.14) is a low frequency Fourier coefficient of the dynamical variable. Depending on the initial value, the system experiences a chaotic transient until it settles at a periodic state 'at the edge of chaos'.

9.3.1 Fixed-energy sandpiles

The mounting numerical evidence in support of the AS mechanism was mostly found in fixed-energy sandpiles (FES, Dickman *et al.*, 1998; Vespignani *et al.*, 1998, early FES ideas also by Tang and Bak, 1988a). Along the ideas laid out above, these are ordinary SOC models with conservative bulk dynamics, but with their usual dissipative boundaries

---

[13] In that case, the loss term in Eq. (9.14) is an integral over the local activity $\int d^d x\, \bar{\epsilon}(\mathbf{x}) \rho_a(\zeta \mathbf{x})$ with dissipation rate density $\bar{\epsilon}(\mathbf{x})$. The integral can be dropped if the particle density ζ is allowed to vary spatially, $\zeta(\mathbf{x}, t)$, producing a fully local equation of motion (Dickman *et al.*, 1998).

[14] This feature might not be expected to be fully reflected in a continuum formalism.

replaced by periodic ones. This slight modification turns them into AS models, which would (probably) display DP critical behaviour, if it were not for the additional symmetry (bulk conservation) and the existence of *multiple* absorbing states. Without dissipation ($\epsilon = 0$) and external drive ($h = 0$), such a system is set up in a random fashion with a certain number of particles and allowed to evolve in time. Typical observables are the avalanche size and duration, as well as the instanteneous activity (density of topplings sites), analysed in a similar fashion as done in ordinary AS models (Hinrichsen, 2000). Because the particle density is no longer subject to fluctuations as in its original SOC version, the change in the model's definition resembles some features of changing from a canonical to a microcanonical ensemble, hence the suggestive name 'fixed-energy', rather than 'fixed particle number'. Although these fluctuations are often restored by implementing bulk dissipation and external drive, most authors regard them as *additions* to FES models.

Dickman *et al.* (1998) were the first to show that the FES version of the MANNA Model displays an absorbing state phase transition different from DP. Vespignani *et al.* (2000) compared the scaling of the interface made up from the local number of topplings in an FES to the avalanche exponents of the corresponding SOC model and found very good agreement (Sec. 8.5.4). From this point of view, SOC is a 'self-organised absorbing state phase transition'. A large number of numerical studies of FES models (e.g. Chessa *et al.*, 1998; Montakhab and Carlson, 1998; Vespignani *et al.*, 2000; Dickman *et al.*, 2001; Muñoz *et al.*, 2001; Christensen *et al.*, 2004; Ramasco *et al.*, 2004; Dantas and Stilck, 2006) support these findings (despite some anomalies, e.g. Vespignani *et al.*, 1998, 2000; Pastor-Satorras and Vespignani, 2000d; Rossi *et al.*, 2000; Bagnoli, Cecconi, Flammini, and Vespignani, 2003; Dall'Asta, 2006).

A detailed mean-field theory (Vespignani and Zapperi, 1998, also Lübeck, 2003) confirms the picture that *SOC behaviour can be induced in any AS phase transition firstly by adding an external drive h that increases the control parameter, secondly, coupling the order parameter to that control parameter so as to drive the system into the absorbing phase via dissipation ϵ, and thirdly by taking the double limit $\epsilon \to 0, h/\epsilon \to 0$.* The double limit is somewhat reminiscent to that in the DS-FFM, Sec. 5.2.1 and motivated by a similar reasoning. The first, $\epsilon \to 0$, is required to remove the dissipation from the system, which would otherwise constrain avalanches. The second, $h/\epsilon \to 0$, is the mechanism by which the system approaches the critical point in the first place, Fig. 9.2. The dimensional inconsistencies of the limits, manifest by the fact that a change in unit allows both ϵ and h/ϵ to be arbitrarily small, is fixed by comparing them to suitable characteristic scales or observables. *Defining the activity as the density of toppling sites*, the average avalanche size is given by the activity's spatiotemporal integral divided by the number of particles added. Assuming translational invariance of the averaged activity in a system of volume $L^d = \int d^d x$ and replacing the ensemble average by an average over total time T_{tot}, gives $\int d^d x \int_0^{T_{tot}} \rho_a(\mathbf{x}, t) \to L^d T_{tot} \langle \rho_a \rangle$. This is the accumulated total of $\int d^d x \int_0^{T_{tot}} h = L^d T_{tot} h$ avalanches triggered, with average duration $\langle T \rangle = T_{tot}/(L^d T_{tot} h)$ (Sec. 9.3.2). Using Eq. (9.15) then gives

$$\langle s \rangle = \frac{\int d^d x \int_0^{T_{tot}} \rho_a(\mathbf{x}, t)}{\int d^d x \int_0^{T_{tot}} h} = \frac{\langle \rho_a \rangle}{h} = L^d \langle \rho_a \rangle \langle T \rangle = 1/\epsilon. \tag{9.16}$$

The result $\langle s \rangle \epsilon = 1$ is exact provided dissipation takes place homogenously in the bulk of an otherwise conservative system. The characteristic scale ϵ^{-1} might therefore be regarded the characteristic avalanche size s_{ch}, not to be confused with the cutoff of the avalanche size distribution, which has first moment $\langle s \rangle = \epsilon^{-1}$. Similarly, the characteristic time scale of the external driving current, h, is one particle in the entire system per avalanche, $h = L^{-d}T_{\mathrm{c}}^{-1}$.

The tuning of both (effective) quantities, ϵ and h, occurs naturally in SOC systems. Firstly, dissipation takes place only at boundary sites which have density $\propto L^{-1}$. If the activity were homogeneous, this would correspond to a bulk dissipation rate $\epsilon \propto L^{-1}$. However, in the case of bulk drive, the activity at the boundary scales like L^{-1} with respect to the average activity, so that $\epsilon \propto L^{-2}$. Generally $\epsilon \propto L^{-\sigma_1^{(s)}}$ is required to imitate boundary dissipation with bulk dissipation (Barrat, Vespignani, and Zapperi, 1999). Secondly, the separation of time scales means in principle that h is arbitrarily small, since there are arbitrarily long pauses between any two avalanches triggered. This point is discussed further in Sec. 9.3.2.

Dickman *et al.* (2000) even adapted the AS mechanism to dynamics where ζ is reduced by an inbuilt dissipation which cannot be tuned with ϵ. In that case, an external driving is needed to sustain a stationary state and this driving will compensate any dissipation, so that, in effect, Eq. (9.14) holds. A random neighbour model (Pruessner and Jensen, 2002b, also Juanico *et al.*, 2007b; Juanico, Monterola, and Saloma, 2007a) seems to support this suggestion, which has since been questioned by Bonachela and Muñoz (2009) on the grounds that such a compensation mechanism requires tuning.

The **bulk dissipation** implemented in FES models deserves special attention, because dissipation at the boundaries has been considered a fundamental ingredient of SOC, as it forces a particle current through the system (Sec. 9.2.2, Hwa and Kardar, 1989a; Paczuski and Bassler, 2000). This might be the mechanism by which correlations are 'transported', similar to driven diffusive systems (Schmittmann and Zia, 1995) or it could be seen as an effective long-range interaction. While some numerical and analytical studies confirm the equivalence of bulk and boundary dissipation (e.g. Chessa *et al.*, 1998; Vespignani and Zapperi, 1998; Barrat *et al.*, 1999; Pastor-Satorras and Vespignani, 2000c), others seem to question it (e.g. Manna *et al.*, 1990; Vespignani and Zapperi, 1995; Lauritsen *et al.*, 1996; Ghaffari *et al.*, 1997; Jettestuen and Malthe-Sørenssen, 2005; Lin *et al.*, 2006; Casartelli, Dall'Asta, Vezzani, and Vivo, 2006; Bonachela, 2008), often simply on the grounds that bulk dissipation replaces the upper cutoff of the avalanche size distribution as a function of the system size by one determined by the dissipation rate (Bonachela, 2008, also Sec. 9.2.1).

Without the equivalence of the two types of dissipation, SOC systems cannot be considered FES models with a certain level of bulk dissipation, seemingly questioning the entire mapping (see below). Malcai *et al.* (2006, also Vespignani and Zapperi, 1998; Pastor-Satorras and Vespignani, 2000c) reconciled contradicting results by decreasing the dissipation level in an FES model as a function of system size precisely so as to reproduce the average avalanche size of the open SOC system. While the avalanche area distribution is affected by the change of dissipation mechanism, the scaling generally agrees between the open SOC system and the bulk dissipative FES model. The bulk dissipation might well therefore introduce a characteristic scale, but as long as it is chosen to coincide with that induced by dissipative boundaries, bulk dissipation can replace it. Their finding is in line with that by Barrat *et al.* (1999) who found that the correlation length ξ in the BTW Model with bulk dissipation scales approximately like $\epsilon^{-0.5}$. Since $\epsilon \propto L^{-2}$ mimics boundary dissipation, $\xi \propto L$, which is confirmed numerically, and is precisely the scaling expected in a system whose only characteristic length scale is the system size.

However, the problem of bulk dissipation in the AS mechanism might be regarded a red herring. The effective equation of motion Eq. (9.14) does not change fundamentally by making ϵ space dependent, thus implementing boundary dissipation, i.e. the AS mechanism is fully compatible with boundary dissipation (Vespignani and Zapperi, 1998; Vespignani et al., 1998). The only shortcoming of the AS mechanism in this respect is that it does not explain why boundary dissipation is a necessary condition for SOC (without it, the usual finite size scaling disappears). One might argue that considering SOC models as FES models with boundary dissipation is plainly tautological. As long as boundary dissipation and activity in FES models is not understood any better than in SOC, the identity of the two cannot further the understanding of the latter.

The **self-organised branching process** (SOBP) (Sec. 8.1.4.2) is an entirely different model displaying many of the features described by the AS mechanism, in particular Eq. (9.14). First proposed by Zapperi et al. (1995), it is in effect an explicit mean-field model of the AS mechanism and of particular importance because of the paradigmatic rôle the branching process has in SOC (Sec. 8.1.4). Without bulk dissipation, the model displays the usual finite size scaling, the 'size' denoting the maximum number of generations the model evolves until it reaches the dissipative boundary. Lauritsen et al. (1996) as well as Bonachela and Muñoz (2009, but Juanico et al., 2007a,b) analysed in detail the effect of bulk dissipation which was found to introduce an additional length scale which dominates the scaling. In the limit of vanishing bulk dissipation the original behaviour is recovered. According to Bonachela and Muñoz (2009), compensating the bulk dissipation by an external drive to recover finite size scaling requires careful tuning of the latter (see above), so that it 'cannot be called *bona fide* self-organization'. Nevertheless, a concrete implementation of Eq. (9.14) leads to a self-organised critical system, in support of the AS mechanism at least on mean-field level, even when a subsequent implementation of bulk dissipation proves problematic. Yet, manipulating the 'boundary' dissipation of the original SOBP (without bulk dissipation) can affect its scaling significantly (Walker and Pruessner, 2009): for finite system sizes, it no longer evolves to the bulk critical point and the scaling of the relevant observables depends on the scaling of the dissipation, whereas moment ratios remain unaffected. This hints at a level of universality that is beyond what is explained by the equation of motion, Eq. (9.14).

9.3.2 Microscopic time scales

There is no physical difference between running an FES model (or, for that matter, any SOC model) with long waiting times between avalanches and no waiting time between avalanches, as long as no new avalanche is triggered while another one is running. This can be ensured without 'global supervision' of the model by making h arbitrarily small, which Dickman et al. (2000) dubbed 'firing the babysitter'. [15] However, incorporating waiting periods into h affects $\langle \rho_a \rangle$ as it is a temporal average. For example, if h is made to drop by

---

[15] Bröker and Grassberger (1999) call the mechanism 'farmer' (but Dickman et al., 2000).

a factor $1/N$ by increasing the microscopic time that passes between two avalanches being triggered, $\langle \rho_a \rangle$ drops by the same factor (consistent with Eq. (9.15)) without having any physical consequences. In fact, strictly there is no microscopic time as long as the system does not evolve.

Separation of time scales thus allows h and therefore $\langle \rho_a \rangle$ to be arbitrarily small, but this is inconsistent with the density ζ being determined by $\langle \rho_a \rangle$ alone – waiting longer between avalanches decreases $\langle \rho_a \rangle$ but cannot have an effect on ζ. To avoid this complication, a conditional time can be used, which progresses only in the presence of activity and therefore eliminates all waiting times. [16] Consequently, the density ζ can be determined again solely on the basis of a *unique* $\langle \rho_a \rangle$. In turn, h is no longer tunable but is fixed implicitly to $L^d h = \langle T \rangle^{-1}$, so that one particle enters the system per avalanche, see Eq. (9.16) and thereafter. Even when this definition of time is a mere formality one might object to it as confusing AS and SOC. As Pruessner (2007, also 2008) showed, the conditional time scale is a prerequisite for using standard finite size scaling such as Eq. (9.17).

9.3.3 AS phase transitions and SOC models

Some peculiarities of AS phase transitions are important for the interpretation of SOC in terms of AS. Firstly, the probability for a finite system to enter the absorbing state has a non-vanishing lower bound except for some (few, extreme) values of the control parameter, which implies that every finite system enters the absorbing state almost surely. The stationary spatially averaged activity $\overline{\rho_a(t;L)}$ is thus bound to vanish in all finite systems, $\lim_{t\to\infty} \overline{\rho_a(t;L)} = 0$, and there is no phase transition. This is reminiscent of equilibrium phase transitions, which require the thermodynamic limit to be taken first, suggesting $\lim_{t\to\infty} \lim_{L\to\infty} \overline{\rho_a(t;L)} \neq 0$.

The problem is normally circumvented by tuning the control parameter away from the transition, so that the absorbing state is entered either very quickly or not at all within the simulation time available. Alternatively, the absorbing state can be avoided completely, either by introducing an external field (Lübeck and Heger, 2003a) or by considering observables conditional to activity (Marro and Dickman, 1999; de Oliveira and Dickman, 2005). The problem becomes more pressing when studying finite size scaling (as in SOC), with the temperature-like control parameter set to its critical value. This point in phase space can also be studied on the basis of **spreading experiments**, which are closely related to the way of driving SOC models (Muñoz *et al.*, 1999; Hinrichsen, 2000).

The second problem concerns the lack of symmetry of the order parameter of AS phase transitions. In equilibrium phase transitions, the Hamiltonian normally obeys a symmetry which is reflected in the vanishing of the order parameter in the disordered phase (subcritical, high temperature phase, below the transition). In the ordered phase (supercritical, low temperature phase, above the transition) this symmetry is broken, i.e. the ensemble does not reflect the symmetry of the Hamiltonian any longer and the order parameter is finite. This leads to a very tangible picture of the nature of fluctuations in equilibrium phase

---

[16] The time scales with and without waiting times were disputed in the context of AS (Marro and Dickman, 1999; Lübeck and Heger, 2003a; de Oliveira and Dickman, 2005) but unified later (Pruessner, 2007).

transitions. Below the transition the order parameter 'averages out' across the system, above the transition the order parameter is biased giving rise to a finite average.

Below the transition the order parameter (density) $\langle m \rangle$ of a finite system of linear size L therefore behaves like the average over uncorrelated patches of size ξ, of which there are $(L/\xi)^d$ in d dimensions. The variance of the order parameter (density) within each correlated patch is given by the susceptibility divided by the area, $\xi^{\gamma/\nu}\xi^{-d}$, so that typically $\langle m \rangle \propto L^{-d/2}\xi^{\gamma/(2\nu)}$ consistent with standard finite size scaling, $L \approx \xi$, as $\nu d = 2\beta + \gamma$ (e.g. Pfeuty and Toulouse, 1977). [17] This argument cannot be applied to AS phase transitions, because the non-negative activity cannot 'average out' across the lattice. Instead, one finds $\langle \rho_a \rangle \propto L^{-d}\xi^{d-\beta^{(AS)}/\nu_\perp^{(AS)}}$ for the conditional activity below the transition which is an artefact of enforcing that at least one site is active (Marro and Dickman, 1999, p. 172, also Vespignani *et al.*, 2000; Dickman *et al.*, 2001). [9.2]

The third problem is the implementation of the external field, which in AS phase transitions often adds *particles* to the system. This is indeed the case in the AS mechanism, where the external field h affects ζ. However, an external field in ordinary phase transitions as well as in AS increases the order parameter. In this strict sense h is *not* an external field. That has been introduced by Lübeck (2002b, also Lübeck, 2002a; Lübeck and Heger, 2003a,b) as an activation mechanism, which moves otherwise immobile particles to other sites in an attempt to create activity.

9.3.4 A critique

Pruessner and Peters (2006) pointed out that that the graph in Fig. 9.2 is idealised and appears more rounded in finite systems, Figure 9.3, which at first seems a mere technicality. However, imposing $\langle \rho_a \rangle = h/\epsilon$ in a finite system results in a system size dependent ζ which is determined by $\rho_a(\zeta; L) = h/\epsilon$ in the spirit of the AS mechanism; the function on the left hand side is given by the scaling form as observed in AS, namely [18]

$$\rho_a(\zeta \, L) = a^{(AS)}L^{-\beta^{(AS)}/\nu_\perp^{(AS)}} \mathcal{G}\left(\frac{\zeta - \zeta_c}{b^{(AS)}L^{-1/\nu_\perp^{(AS)}}}\right) \tag{9.17}$$

with $a^{(AS)}$ and $b^{(AS)}$ metric factors and $\beta^{(AS)}$ and $\nu_\perp^{(AS)}$ the usual critical exponents. The scaling function $\mathcal{G}(x)$ behaves like

$$\mathcal{G}(x) = \begin{cases} x^{\beta^{(AS)} - d\nu_\perp^{(AS)}} & \text{for } x \to -\infty \tag{9.18a} \\ \text{constant} > 0 & \text{for } x \to 0 \tag{9.18b} \\ x^{\beta^{(AS)}} & \text{for } x \to +\infty \tag{9.18c} \end{cases}$$

so that the usual asymptotes are recovered in the various limits (Marro and Dickman, 1999; Vespignani *et al.*, 2000). The asymptote of the order parameter in the disordered, subcritical phase, $x \to -\infty$, is $L^{-d}|\zeta - \zeta_c|^{\beta^{(AS)} - d\nu_\perp^{(AS)}}$ and thus different from that found

---

[17] The exponents β, γ and ν are the usual order parameter exponent, the susceptibility exponent and the correlation length exponent, respectively.

[18] As mentioned above, this type of finite size scaling is found for the conditional activity (Pruessner, 2007, 2008); on this time scale the external driving is $h = 1/(L^d \langle T \rangle)$.

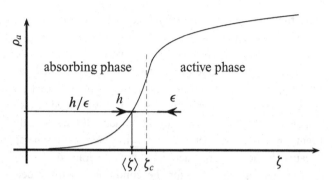

Figure 9.3 Similar to Fig. 9.2, the order parameter ρ_a (activity) as a function of the control parameter ζ (particle density) in an AS transition as measured in a *finite* system. The approach of the effective (average) particle density $\langle\zeta\rangle$ to the critical value ζ_c (dashed vertical line) depends on the way h/ϵ is tuned to 0^+ with increasing system size.

in ordinary equilibrium critical phenomena (Privman *et al.*, 1991), where it scales like $L^{-d/2}|\zeta - \zeta_c|^{(\beta - d\nu/2)}$ with $\beta - d\nu/2 = -\gamma/2$, see Sec. 9.3.3. This difference is caused by imposing activity, i.e. measuring observables conditional to activity.[19]

In SOC, the activity ρ_a is forced to take a particular value, h/ϵ, Eq. (9.15), which in turn produces a particular value of the particle density ζ by solving for it in Eq. (9.17). In an FES model, the dissipation ϵ and external drive h can be implemented as free parameters. In an SOC model, however, h and ϵ change implicitly with the system size L and can be expected to approach 0^+ as a power law. *Assuming* that $\rho_a(\zeta)$ in Eq. (9.14) is determined by the corresponding AS model, ζ will quite generally approach the critical point of the AS transition, ζ_c (but Fey *et al.*, 2010b,c; Park, 2010). However, whatever way h/ϵ scales in L, it dictates the scaling of $\zeta - \zeta_c$, which feeds through to all other observables. It controls in particular the effective correlation length (which generally diverges, but not linearly with the system size) and thus enters into all cutoffs and most moments.[20] *Only if h/ϵ scales like $L^{-\beta^{(AS)}/\nu_\perp^{(AS)}}$ do the other observables display standard finite size scaling (Pruessner and Peters, 2006).* If h/ϵ does not scale like $L^{-\beta^{(AS)}/\nu_\perp^{(AS)}}$, the system nevertheless ends up at the critical point of the AS transition for a wide range of asymptotes of h/ϵ. Although the correlation length ξ will diverge as a result, the ratio ξ/L will vanish, giving rise to (almost) arbitrary finite size scaling (similar to Zheng and Trimper, 2001). This was confirmed by applying the AS mechanism naïvely to the two-dimensional Ising Model and finding that its finite size depended on the way h/ϵ vanished in the thermodynamic limit. The comparison is somewhat bold, because the Ising Model does not 'get stuck' in the inactive (disordered) phase when it hits an absorbing state. Nevertheless, one might conclude that the AS mechanism *explains* SOC only once an explanation is found as to why h/ϵ scales as it should. This is, in fact, equivalent to asking why ρ_a scales like $L^{-\beta^{(AS)}/\nu_\perp^{(AS)}}$, i.e. why it displays standard finite size scaling – precisely what the AS mechanism sets out to explain in the first place.

---

[19] The scaling function in Eq. (9.17) acquires an additional dependence on an external (activity creating) field if ρ_a is not measured conditional to activity (Lübeck and Heger, 2003a).

[20] But not directly into τ (Alava *et al.*, 2008), which is, however, generally found to be less universal than D.

There are three possible outcomes to the issue raised by Pruessner and Peters (2006). Firstly, the AS mechanism as described above might not apply at all, secondly h/ϵ might in fact be arbitrary and thus SOC non-universal and thirdly, there might be an explanation as to why h/ϵ scales like $L^{-\beta^{(AS)}/\nu_\perp^{(AS)}}$. The first option conflicts with the seemingly obvious link between AS and SOC, but hinges crucially on the (over-) simplification Eqs. (9.14) and (9.17). Describing the equation of motion of ζ in greater detail (Dickman *et al.*, 1998; Vespignani *et al.*, 1998) might be a way to enhance the AS mechanism and thus to invalidate the critique above (see also p. 335 for space dependent dissipation). The second option is not what is generally observed or desired (Sec. 1.4), SOC is generally regarded as a phenomenon displaying universality. In particular the MANNA Model and the OSLO Model display scaling behaviour so robust that the case for non-universality is difficult to make. Where non-universality occurs, for example in the exponent τ which depends on driving and boundary conditions, it is well understood on the basis of conservation laws (Nakanishi and Sneppen, 1997; Pruessner and Peters, 2008, but Alava *et al.*, 2008). The third option, namely that $h/\epsilon \propto L^{-\beta^{(AS)}/\nu_\perp^{(AS)}}$ has not been very widely addressed. One might wonder whether finding a mechanism for this would amount to finding the 'actual' mechanism of SOC.

The SOC model's approach to the AS critical point can be measured directly by the scaling of $|\langle\zeta\rangle(L) - \langle\zeta\rangle(\infty)| \propto L^{-1/\nu_\perp^{(SOC)}}$, with $\langle\zeta\rangle(\infty)$ the asymptotic value of $\langle\zeta\rangle(L)$. If the SOC model experiences the AS critical point through the AS mechanism then $\langle\zeta\rangle(\infty)$ is expected to coincide with the critical value ζ_c found in AS and $\nu_\perp^{(SOC)}$ should have the same value as $\nu_\perp^{(AS)}$. Although both universal and non-universal quantities are generally expected to be affected by boundary conditions (Privman *et al.*, 1991), which are different in SOC (at least partially open) and AS (periodic, normally even in spreading experiments), this is generally confirmed by numerics. The asymptotic particle density in the two-dimensional oMM was extrapolated to $\langle\zeta\rangle(\infty) = 0.6832(10)$ (Manna, 1991b), while Lübeck (2004, citing Lübeck and Heger, 2003b) found for the AS variant $\zeta_c = 0.68333(3)$. Manna (1990, 1991b) found $\nu_\perp^{(SOC)}$ not too far from 1 in the two-dimensional BTW Model and MANNA Model respectively, which was later confirmed by Pradhan and Chakrabarti (2001), compared to $\nu_\perp^{(AS)} = 0.799(14)$ measured in the steady state of the FES MANNA Model (Lübeck and Heger, 2003b). Christensen *et al.* (2004) found $\nu_\perp^{(SOC)} \approx 1.4$ for the one-dimensional OSLO Model, compared to $\nu_\perp^{(AS)} = 1.347(91)$ (Lübeck and Heger, 2003b) for the AS version of the oMM. Dickman *et al.* (2001) on the other hand found $\nu_\perp^{(AS)} = 1.81(1)$, maybe due to significant logarithmic corrections (Dickman and Campelo, 2003).

The case for the AS mechanism finds further support in the scaling of both ϵ and h which are fixed by the dynamics (Sec. 9.3.1, Eq. (9.16) and Sec. 9.3.2): $\epsilon \propto \langle s\rangle^{-1} \propto L^{-D(2-\tau)}$ and $h \propto L^{-d}\langle T\rangle^{-1} \propto L^{-d}L^{-z(2-\alpha)}$. The scaling of ϵ is due to the scaling of the average avalanche size and is thus set by bulk conservation and boundary dissipation, and the scaling of h relies on the separation of time scales, [21] which prevents avalanches from overlapping. On the conditional time scale they are thus separated on average by $\langle T\rangle$ (see footnote 18 on

---

[21] If the scaling form Eq. (9.17) applies with Eq. (9.18), i.e. variables are measured on a conditional time scale, then $h = 1/(L^d\langle T\rangle)$ and vice versa. Most importantly, Eq. (9.17) precludes arbitrary scaling of h by padding time with quiescent periods.

Table 9.1 Comparison of $\beta^{(AS)}/\nu_\perp^{(AS)}$ (Lübeck, 2004, based on Lübeck and Heger, 2003b) and $d + z - D$ using the data in Table 6.1.

d	$\beta^{(AS)}/\nu_\perp^{(AS)}$	$d + z - D$
1	0.28(3)	0.30(5)
2	0.80(3)	0.79(3)
3	1.42(6)	1.44(5)
MFT	2	2

p. 337). With $D(1 - \tau) = z(1 - \alpha)$, Eq. (2.40) on p. 43, this behaviour gives $h/\epsilon \propto L^{D-z-d}$ and standard scaling arguments [22] yield indeed (Lübeck and Heger, 2003a) [23]

$$D - z - d = -\beta^{(AS)}/\nu_\perp^{(AS)}. \tag{9.19}$$

In other words, regardless of the boundary conditions (but see footnote 48 on p. 305 and p. 307), SOC models considered here are driven so that $h/\epsilon \propto L^{-\beta^{(AS)}/\nu_\perp^{(AS)}}$, which is precisely the scaling needed to recover standard finite size scaling according to Pruessner and Peters (2006), provided only that $D - z - d = -\beta^{(AS)}/\nu_\perp^{(AS)}$. It is reassuring for the AS mechanism to see three fundamental aspects of SOC, bulk conservation, boundary dissipation and separation of time scales, entering this way.

The AS mechanism has been studied numerically mostly for the MANNA Model (e.g. Dickman *et al.*, 2001; Vespignani *et al.*, 2000; Lübeck and Heger, 2003a; Ramasco *et al.*, 2004; Dantas and Stilck, 2006). Fewer studies are available for the BTW Model (e.g. Chessa *et al.*, 1998; Vespignani *et al.*, 2000; Dantas and Stilck, 2006) and some for the OSLO Model (e.g. Christensen *et al.*, 2004). Using the exponents published by Lübeck (2004), Table 9.1 shows a comparison of $\beta^{(AS)}/\nu_\perp^{(AS)}$ measured directly in the *steady state* [24] of the FES MANNA Model (Dickman *et al.*, 2001) and $d + z - D$ based on the SOC exponents of the MANNA Model as listed in Table 6.1 on p. 168. The remarkable correspondence corroborates AS as the critical phenomenon underlying SOC and lends further support to the AS mechanism.

However, the argument just presented is circular, because the scaling relation $D - z - d = -\beta^{(AS)}/\nu_\perp^{(AS)}$ is derived by assuming that both AS and SOC probe the same critical point, which is the only way to link the exponents on the right to the exponents on the left. A priori they need to be considered **different ensembles** (Dickman *et al.*, 1998). None of the exponents characterising SOC and AS are *necessarily* the same, not even z which at least probes closely related observables in the two ensembles. The central problem is best

---

[22] This is easily seen by a dimensional analysis based on the anomalous dimensions involved: $\langle s^n \rangle \propto \langle \rho_a^n \rangle \langle T^n \rangle L^{dn}$ means $D(1 + n - \tau) = -\beta^{(AS)}/\nu_\perp^{(AS)} n + z(1 + n - \alpha) + dn$. The constants on both sides cancel because of Eq. (2.40) on p. 43 and the terms in n produce Eq. (9.19).

[23] Similar results are due to Muñoz *et al.* (1999), who use D_f for D and a different definition of z, and Lübeck (2004, A.3), who uses D_f for what is $D - z$ in the present notation. The scaling law A.3, $z(\theta + \delta) = D - z\delta$, cited in the latter work holds only for $\beta' = \beta$ in his notation.

[24] Rather than in spreading experiments, which could be considered as an 'SOC mode' of AS.

illustrated by noting that $h/\epsilon = L^{-d} \langle s \rangle / \langle T \rangle$ is merely a trivial consequence of stationarity, $\langle \dot{\zeta} \rangle = 0$ in Eq. (9.14), so that $\epsilon \langle s \rangle = hL^d \langle T \rangle$. Bulk conservation, boundary dissipation and separation of time scales enter only when determining the scaling of ϵ and h in L individually. That h/ϵ scales at all, let alone that it scales with $D - z - d$, which can be linked to AS, remains the *proposition* to be proven. Only if one *assumes* that h/ϵ scales as needed, is the scaling of all other observables recovered (Pruessner and Peters, 2006).

9.3.4.1 Discussion and conclusion

The problem becomes manifest as soon as one tries to *apply* the AS mechanism to any other phase transition, say the **contact process** (CP, Harris, 1974; Liggett, 2000), which is essentially governed by a single parameter λ, the colonisation rate over the extinction rate. An external mechanism would activate a single site, triggering an avalanche. At the same time, λ increases by a small amount,[25] whereas activity reduces λ for example whenever a boundary site attempts to colonise a (virtual) site beyond the border. Such a setup finds a stationary state, with $\langle \lambda \rangle$ converging to λ_c with increasing system size, but $h/\epsilon = L^{-d} \langle s \rangle / \langle T \rangle$ seems to display a scaling different from $L^{-\beta^{(CP)}/\nu_\perp^{(CP)}}$. The resulting avalanche exponents D and z in the SOC setup may or may not differ from their values when measured or derived in the AS model at the critical point (Muñoz *et al.*, 1999). Nothing stops the scaling relation Eq. (9.19) from being violated, if the exponents on the left and on the right are measured in different ensembles. Accepting that, to what extent can the AS mechanism explain SOC ?

There is a confusing dichotomy in the AS mechanism regarding the underlying AS model. The two parameters, ϵ and h, can firstly be thought of as being two control parameters in a (or rather added to a) standard AS model which can be set to any value, or secondly as being set implicitly by the dynamics of an SOC model. It has been shown (Pruessner and Peters, 2006; Alava *et al.*, 2008) that an arbitrary choice of h and ϵ leads to arbitrary scaling. Thus, contrary to the original claim (Dickman *et al.*, 1998), SOC is *not* an AS model with ζ subject to the equation of motion Eq. (9.14) and any arbitrary implementation of the double limit $\epsilon \to 0$, $h/\epsilon \to 0$.

The second interpretation, where the two parameters h and ϵ are set implicitly by the SOC model, which is recast in the language of AS, remains unchallenged. An SOC model can be seen as an AS model where ζ is tuned through an equation of motion like Eq. (9.14) and h and ϵ are chosen implicitly. Why they scale in precisely the 'right way' (Pruessner and Peters, 2006) so that SOC models probe the critical point of an AS phase transition remains unknown. To complete the AS mechanism, an explanation must be found for the scaling of the terms in $\epsilon \langle s \rangle = hL^d \langle T \rangle$, Eq. (9.19), and why $\rho_a(\zeta)$ in Eq. (9.14) is the same function in both AS and SOC.

The SOC ensemble where driving and dissipation are set implicitly and the AS ensemble where they are (if added in the form of ϵ and h) a matter of choice, must be carefully

---

[25] The effect of the external driving on λ is not as immediate as in, say, the FES MANNA Model, which can be remedied by considering λ as the density of perfectly mixed particles, which "catalytically" mediate colonisation.

distinguished and explicitly linked. The AS mechanism cannot be confirmed by measuring AS quantities, say $\beta^{(AS)}/\nu_{\perp}^{(AS)}$ in Eq. (9.19), in the SOC setup, i.e. by spreading experiments (Muñoz *et al.*, 1999), possibly even with open boundaries as in SOC models (see footnote 24, on p. 340). It would be pointless to state that SOC probes an AS phase transition, if this particular transition can only be probed in the SOC setup.

Recently a very different critique was raised by Fey *et al.* (2010b,c, also Park, 2010), who *proved* that the average slope of the ASM on certain lattices in the stationary state exceeds, minutely, the critical value of the corresponding FES (but Jo and Jeong, 2010). This minuscle effect can also be shown numerically to exist on the two-dimensional square lattice. If this is confirmed for other models, it puts into question the most basic assumptions of the AS mechanism.

9.4 Other mechanisms

Three more mechanisms of SOC are discussed in the following. There are many more 'mechanisms for power laws', comprehensively reviewed by Sornette (2006, chapter 14; also Sornette, 1994, 2002; Newman and Sneppen, 1996), but the following have gained some popularity explicitly as explanations for SOC.

The line between an explanation and a description of SOC is blurred; an explanation usually contains a description but whether it should contain a *prescription* is a contentious epistemological question. Even what amounts to a prescription is unclear. Peters and Girvan (2009) suggested that including in the system's definition *any* recipe by which a numericist or an experimentalist tunes a system to its critical point renders the system an instance of SOC. Of course, such an explanation does not apply to the models in Part II.

9.4.1 Goldstone bosons

Obukhov (1990) proposed very early that SOC models were governed by **interacting Goldstone bosons**, which are massless excitations in the ordered phase of a phase transition, i.e. excitations of arbitrarily low energy and arbitrarily long wavelength. The fact that the excitations are massless is important to explain the apparent long-range behaviour in SOC models, in particular their scale invariance. Due to the lack of mass, small local perturbations are propagated through the entire system. Mathematically, this is reflected in an algebraic divergence of the propagator in the limit of small momenta, as in the algebraic form of the correlation functions discussed in Sec. 9.2, for example $\langle h(\mathbf{k}, t)h(\mathbf{k}', t)\rangle = \delta(\mathbf{k}+\mathbf{k}')/(\nu \mathbf{k}^2)$, Eq. (9.9).

Obukhov's (1990) approach differs from that of Hwa and Kardar (1989a) by tracing the origin of the masslessness of the propagator to Goldstone bosons, rather than symmetries or conservation laws. Goldstone bosons appear in equilibrium phase transitions of systems with continuous symmetries below the critical temperature. [26] There, transversal

---

[26] That extends to the Kosterlitz–Thouless transition, which is not a second order phase transition in the strict sense.

spin-waves can modulate the local order parameter without changing its spatial average, whose direction represents the broken symmetry. Below the critical temperature, the transversal susceptibility diverges, as it takes an arbitrarily small external field perpendicular to the magnetisation to turn it towards the field. The divergent susceptibility signals long-range correlations (Parisi, 1998). Interestingly, Obukhov (1990) appealed to conservation and stationarity to justify the need for a massless propagator.

In ferromagnetic phase transitions, transversal correlation functions in the ordered phase are dominated by the 'trivial' divergence $1/k^2$ (Täuber and Schwabl, 1992), which is consistent with the picture that asymptotically, the non-linearity becomes irrelevant and the system's behaviour is dominated by the trivial fixed point. Similar observations have been made in other systems (Dhar and Thomas, 1992; Zürcher, 1994, non-trivial exponents in Newman, Bray, and Moore, 1990). Nevertheless Obukhov (1990) suggested that the non-trivial exponents observed in SOC could be a product of the non-linear interaction of Goldstone bosons.

This explanation of SOC is somewhat phenomenological and might be reduced to the statement that long-range correlations and scale invariance can be observed in Goldstone bosons in ordinary equilibrium critical phenomena below the critical temperature. From this perspective, SOC is rooted in classical phase transition, without the need of precise tuning, as long as the system remains in the ordered phase. Around the same time as Obukhov (1990), the rôle of Goldstone bosons was discussed by other authors (Dhar and Majumdar, 1990; Newman *et al.*, 1990; Nowak and Usadel, 1990, 1991; Hwa and Kardar, 1992). The Goldstone boson approach to SOC illustrates that scale invariance is more common in nature than one might expect at first. It remains open whether Goldstone bosons are the dominating excitation, not suppressed by other effects, such as the formation of interacting domains in ferromagnets. As for its relevance to SOC, the Goldstone boson approach fails to incorporate some of its key ingredients, such as boundary conditions, conservation, separation of time scales, thresholds and avalanches.

9.4.2 Singular diffusion

Carlson, Chayes, Grannan, and Swindle (1990a, extensions by Montakhab and Carlson, 1998) proposed another mechanism for SOC based on the characteristics of the propagator, similar to the approach by Hwa and Kardar (1989a). They studied the *effective* diffusion constant of particles in a diffusive, sandpile-like SOC model and found that it contains a singularity in the local particle density.

Kadanoff, Chhabra, Kolan, *et al.* (1992) repeated some of the analysis adding a second model and confirmed the results. One important ingredient is the notion of **troughs**, [27] originally introduced by Carlson, Chayes, Grannan, and Swindle (1990b) which delimit the domains avalanches extend over. The effective diffusion constant of the particles diverges as the density of troughs vanishes so that particles become trapped less frequently (one might argue that this is a matter of choosing an appropriate time scale for the microscopic dynamics). At the same time diffusion fills up the troughs.

---

[27] Similar to the 'sinks' used in the analysis of the OSLO Model by Chua and Christensen (2002).

According to Grinstein (1995), particle conservation and mass balance causes the divergence of the effective diffusion constant, which controls the flow of particles through the system. This divergence implies the existence of a diverging length, which explains why the system is scale invariant. The proposed mechanism incorporates separation of time scales, conservation and open boundaries, although it makes use of results derived for systems with closed boundaries (Carlson *et al.*, 1990a). While phenomenologically consistent (Sornette, 2006), its explanatory power hinges on the precise origin of the singularity which remains somewhat unclear. Illustrating the rôle and the functioning of the singularity, Machta, Candela, and Hallock (1993) suggested an experimental setup which would maintain liquid helium at its superfluid critical point throughout the bulk, which has divergent thermal conductivity.

9.4.3 Extremal dynamics

Models with a dynamics which depends on a local quantity being globally extremal are clearly distinct from sandpile-like models, which have purely local dynamics. Global extremality requires global 'supervision' or infinitely many separated time scales, as in the BS Model (Sec. 5.4, p. 144). Even within the class of extremal models, the BS Model is different from, say, the OFC Model, as the former is always updated at the extremal site, whereas the latter is only driven by an amount determined by the extremal site. Models of the former type are thus generally non-Abelian (provided the attribute applies), whereas the latter could in principle have an Abelian dynamics (even when that is not the case for the OFC Model). On a more basic level, extremality without a high degree of degeneracy requires continuous local variables, which is a second distinguishing feature of extremal models. It therefore would not be surprising if extremal SOC models were governed by a mechanism very different from that underlying sandpile-like models.

A mechanism of SOC based on extremal dynamics was proposed by Miller *et al.* (1993), whose notion of extremal models, however, deviates from the narrow outline above. They noticed that *exceeding a threshold* is a matter of extreme value statistics – the probability that any element in a sequence of random numbers exceeds a certain limit equals the probability that the maximum within this sequence exceeds the limit. With thresholds being a key ingredient of SOC from the outset (Cafiero *et al.*, 1995), Miller *et al.*'s (1993) mechanism applies to SOC generally.

If the dynamics of a system is determined by extreme values of a dynamical variable, classical results of the statistics of extremes (Gumbel, 1958; Galambos, 1978; Leadbetter, Lindgren, and Rootzén, 1983) show its distribution under very broad conditions to be universal and essentially exponential. The power spectrum consequently is close to a Lorentzian, giving rise to $1/f^2$ behaviour in the *ultraviolet* and converging to a non-zero value in the infrared. Miller *et al.* (1993) argued that at intermediate values, the scaling would approximate $1/f$ and that the presence of *multiple* thresholds would broaden this region dramatically.

Although this approach to SOC is rather phenomenological and does seem to incorporate very many of the details and characteristics of the model, it is underpinned by a widely accepted *mathematical* theory which features, in particular, universality. Miller *et al.* (1993)

were not the only ones who put the statistics of extremes at the centre of their theory. Paczuski *et al.* (1996) popularised the notion of **extremal dynamics** as the distinctive feature of many SOC models, including those that are in the most robust MANNA universality class (Sec. 6.2.1.2). Gabrielli, Cafiero, Marsili, and Pietronero (1997) proposed a transformation of the quenched noise governing the extremal dynamics of certain models into annealed (thermal) noise, which allowed them to apply methods successfully used in fractal growth phenomena.

'Highly optimised tolerance'[28] (Carlson and Doyle, 1999, 2000, also Langton, 1990) can be thought of as an extremisation procedure or a design principle, by which systems optimise a yield, which is a measure of their tolerance to external perturbations. SOC models can be altered to fall in this category, by introducing some tunable variable and a yield function. As a result, they display very structured spatial patterns that are clearly different from those found in SOC models, while still generating scale invariance. Although the authors clearly distinguish 'highly optimised tolerance' from SOC, it is regarded by others as an explanation for it.

---

[28] Incidentally, HOT was later contrasted with COLD, 'constrained optimisation with limited deviations' (Newman, Girvan, and Farmer, 2002). TEPID is yet to be discovered.

Summary and discussion

In his review of SOC, Jensen (1998) asked four central questions paraphrased here.

1. Can SOC be defined as a distinct phenomenon?
2. Are there systems that display SOC?
3. What has SOC taught us?
4. Does SOC have any predictive power?

As discussed in the following, the answers are positive throughout, but slightly different from what was expected ten years ago, when the general consensus was that the failure of SOC experiments and computer models to display the expected features was merely a matter of improving the setup or increasing the system size. Firstly, this is not true: larger and purer systems have, in many cases, not improved the behaviour. Secondly, truly *universal* behaviour is not expected to be prone to tiny impurities or to display such dramatic finite size corrections. If the conclusion is that this is what generally happens in systems studied in SOC over the last twenty years, critical phenomena may not be the most suitable framework to describe them.

Can SOC be defined as a distinct phenomenon?

In the preceding chapters, SOC was regarded as the observation that some systems with spatial degrees of freedom evolve, by a form of self-organisation, to a critical point, where they display intermittent behaviour (avalanching) and (finite size) scaling as known from ordinary phase transitions (Bak *et al.*, 1987, also Ch. 1). This definition makes it clearly distinct from other phenomena, although generic scale invariance has been observed elsewhere. Responding to this by including these phenomena in SOC *by definition* renders the first question above meaningless. Jensen (1998) therefore turned his attention to the necessary and sufficient conditions for SOC, some of which seem to be slow drive and highly non-linear interaction (thresholds). Within SOC, the identification of these conditions remains the biggest puzzle.

One discussion that has been resolved over the last ten years is the dispute over the meaning of the term 'tuning'. If SOC did not allow for any tuning at all, then a separation of time scales would strictly not be allowed. However, it is now widely accepted that tuning that ultimately removes a scale, namely that of the driving rate, is qualitatively different from tuning which provides a scale, such as the finite critical temperature in the Ising Model. It is the latter that is absent in SOC (Sec. 1.2).

Are there systems that display SOC?

SOC was envisaged as an explanation for supposedly ubiquitous phenomena, namely $1/f$ noise and scaling in natural systems, which in many detailed experiments and careful analysis of observations turn out to be elusive. It seems nature is much too rich and complicated to be categorised in universality classes and characterised by scaling. Yet, there are a number of experiments, most notably those in granular media (Sec. 3.1) and superconductivity (Sec. 3.2), and observations such as the Gutenberg–Richter law (Sec. 3.4) and the scaling in the statistics of precipitation (Sec. 3.7.1), which suggest that robust scaling exists in nature and governs at least some systems. Perhaps, some of the observed 'dirty power laws' are spoilt by experimental shortcomings, which would imply that scaling and SOC had some, although limited, application.

A breathtaking number of SOC models has been developed and at least partially analysed. Setting up models has helped to identify certain important features that may or may not be present in SOC. Unfortunately, it has also led to a perception of models as ends in themselves. What is worse, many of the models do not actually display any of the desired features, such as avalanching and scale invariance. Looking at the tables at the beginning of Part II, pp. 83, 84, suggests that the general answer to the question 'Is it SOC?' is 'Probably not.' Some models, such as the MANNA Model (Sec. 6.1) and the OSLO Model (Sec. 6.3), however, have it all: robust, universal, non-trivial scaling behaviour with local interactions, simple rules and nothing but a separation of time scales to drive them.

What has SOC taught us?

The most important lesson SOC has taught is that it exists as a robust phenomenon. The two models just mentioned are perfect candidates to put to the test any future theory of emergence, which complexity is wating for.

No complete theory of SOC exists. Some explanations of SOC, in particular the AS mechanism (Sec. 9.3), are widely accepted, but curiously have so far failed to fulfil the promise of a recipe that allows the conversion of a traditional phase transition into an instance of SOC. It has *not yet* taught us how to trigger phase transitions without tuning.

Does SOC have any predictive power?

If universality were to apply and SOC were as widespread as initially suspected, the predictive power of SOC would be immense, as witnessed by the enormous range of experiments (Ch. 3). Where it applies, SOC has immediate implications for the *generic* behaviour of a system, because scaling is not confined to a small region around a critical point, but is the *only* behaviour accessible to the system. The question of how broadly SOC applies therefore remains of great importance and one can only hope that the critical perspective adopted above will turn out to be too pessimistic.

What is needed is a theory that rationalises the plethora of observations and models and provides rules for how to identify and even produce SOC. To cite Kadanoff's (1986) appeal for a physics of fractals: 'Without that underpinning much of the work [. . .] seems somewhat superficial and even slightly pointless. It is easy, too easy, to perform computer simulations upon all kinds of models and to compare the results with each other and with real-world outcomes. But without organizing principles, the field tends to decay into a zoology of interesting specimens and facile classifications.'

Some important questions in SOC and complexity are discussed further in Sec. 10.2. The following section contains, in a nutshell, a list of misunderstandings, wrong assumptions and wrong conclusions, that have plagued SOC in the past.

10.1 Convenient fallacies and inconvenient truths

There are a number of popular *technical* misbeliefs, assumptions and non sequiturs shrouding the perception and analysis of SOC. [1] They are listed in the following *without* referencing publications where they have been disseminated. They vary in popularity, level of technicality and impact.

Additional scales (context: Chs. 1, 2, Sec. 5.3.1)

Misbelief: 'Any scale imposed onto a scale invariant system will necessarily destroy its scale invariant behaviour.'

This misbelief might be traced back to another one, according to which there can only be one scale in a scale invariant system, such that adding a scale means that scaling is destroyed. This is incorrect – if the 'additional scale' is to be understood as a *dimensionful* quantity, then the presence of several of them is necessary to form a dimensionless variable and thus to generate scaling beyond what is dictated by dimensional analysis (Sec. 1.2, in particular p. 9). But even if 'scale' means some additional *dimensionless* variable, the statement that its mere presence destroys scaling in this generality is wrong for at least two reasons. Firstly, to have any asymptotic effect, the new scale needs to replace the old dominating scale. Secondly, even when that happens, it does not necessarily imply that scaling breaks down, only that scaling in the *old* scaling variable is replaced or amended by scaling in the new variable. For example, adding a drift term in the MANNA Model makes it change universality class (Secs. 8.4, 8.4.3), yet scaling itself is maintained. Scale invariance does not mean the absence of any scale or cutoff, but the presence of a scaling symmetry (Stanley, 1971; Barenblatt, 1996, also p. 13).

Moreover, according to the universality hypothesis, adding a certain microscopic interaction often has no or very little effect on the scaling behaviour, which is of particular interest precisely because it is expected to be robust against such perturbations. However, certain interactions, such as bulk dissipation, seem to be particularly destructive, because they naturally give rise to some length scale comparable to the original cutoff (such as the system size, e.g. Eq. (9.16) on p. 333) and seem to impose an effective cutoff on the observables, e.g. Sec. 9.2.1. Both are very difficult to prove (Sec. 5.3.1). In principle, the dominating scale can enslave the additional scale (i.e.

---

[1] I would like to thank Darryl Holm for suggesting this chapter.

making the latter a function of the former, say $\epsilon(L)$ as in the AS mechanism, p. 334), possibly generating critical behaviour of a different universality class.

Exponential scaling (context: Ch. 2, p. 36, Sec. 2.1.3)

Misbelief: 'An exponential distribution does not display scaling.'

The presence of an exponential cutoff in a relevant PDF is frequently interpreted as a sign for the absence of scaling. This is incorrect. If $\mathcal{P}^{(s)}(s) \propto \exp(-s/s_{\mathrm{c}})$, then after normalising

$$\mathcal{P}^{(s)}(s) \propto s^{-1} \frac{s}{s_{\mathrm{c}}} e^{-s/s_{\mathrm{c}}} \tag{10.1}$$

which fulfils simple scaling with $\tau = 1$. One-dimensional percolation close to the percolation threshold is a prominent example of exponential scaling.

The origin of the misinterpretation that exponentials indicate the absence of scaling is twofold. Firstly, an exponential is the signature of a Poisson process. An event of size s might be generated and sustained by a resource the system loses with a certain rate and eventually runs out of, at which point the event stops. In a bulk dissipative model, this could be the amount of 'energy' available at a site. Secondly, and more importantly, the exponential cutoff might be mediated by an upper cutoff s_{c} *that does not scale with system size*. In other words, what is important in Eq. (10.1) is not the particular form of the scaling function, but the fact that the cutoff depends on some independent control parameter, which ultimately dominates the long-range behaviour of the system, and does not diverge with the system size, thus failing to produce finite size scaling.

First moment (context: Ch. 2, Sec. 2.2)

Misbelief: 'The first moment coincides with the characteristic scale as there is only one in a system displaying scaling.'

If there were only one scale (i.e. only one dimensionful quantity), then the first moment, say $\langle s \rangle$, would indeed coincide with the characteristic scale s_{c} and consequently (Secs. 2.2, 7.4.2, p. 240), $\tau = 1$. However, if $\tau \neq 1$, then the first moment necessarily differs from the characteristic scale, $\langle s \rangle \propto s_{\mathrm{c}}^{2-\tau}$. This apparent dimensional inconsistency is mediated by a second scale, whose physical origin and effect is irrelevant in the long-range limit, but which nevertheless is a dimensional necessity (see for example Sec. 8.1.3, p. 257). The statement 'there can only be one scale' ignores such a second scale and should be replaced by 'there can only one *dominating* scale'.

Lower cutoff (context: Ch. 2)

Misbelief: 'The lower cutoff of a distribution is irrelevant for its asymptotes and can therefore be ignored.'

The lower cutoff of the PDF of an observable like the cluster size distribution is constant and scaling is therefore normally considered as a function of the upper cutoff, which in turn is, in the case of finite size scaling, a function of the system size. The rôle of the lower cutoff is therefore normally reduced to that of a threshold above which scaling is clearly visible. The PDFs as shown in Fig. 7.12(a), p. 238, display a clear power law only beyond that lower cutoff.

That, however, does not mean the lower cutoff can be ignored. Firstly, it is often determined by the same microscopic details of the system that enter the dimensionful metric factor that makes exponents $\tau \neq 1$ possible at all (additional scales as discussed above, e.g. a_0 in Eq. (8.26) as discussed on p. 37 and p. 257). Such scales therefore enter themselves as a power law prefactor, as can be seen in the metric factor Eq. (2.20) of the simple scaling form Eq. (2.19). In both cases, the initial gap a_0 between a random walker (with effective diffusion constant D) and an absorbing wall provide a suitable scale in the form $a_0^2/(4D)$. In that sense, the lower cutoff itself gives rise to power laws.

What is more, if the lower cutoff is dropped completely from the simple scaling form, this is the same as making it vanish, which forces $\tau = 1$. In particular in the early days of SOC, $\tau = 1$ was frequently 'derived' on the tacit, implicit assumption that the lower cutoff vanishes. To avoid inconsistencies the lower cutoff has to be kept as an essential part of the scaling ansatz.

Narrow joint distributions (context: Ch. 2, Sec. 2.2.2)

Misbelief: 'The scaling law $D(1 - \tau) = z(1 - \alpha)$ is a mathematical necessity.'

Based on the observation that D is the anomalous dimension of the avalanche size s in terms of the system size L and correspondingly z the anomalous dimension of the avalanche duration T, the two observables are sometimes regarded as proportional to each other $s^{1/D} \propto T^{1/z}$. Consequently the scaling of the PDF of one implies the scaling of the PDF of the other, as $\mathcal{P}^{(s)}(s)\,ds = \mathcal{P}^{(T)}(T)\,dT$ and as $s \propto T^{D/z}$

$$\mathcal{P}^{(T)}(T) = \mathcal{P}^{(s)}(T^{D/z})\frac{D}{z}T^{D/z-1}. \tag{10.2}$$

Based on the simple scaling form expected for the avalanche size and duration distribution, Eqs. (1.1) and (1.2) respectively, one can identify z as the exponent with which the cutoff of the duration exponent scales and read off the exponent $-\alpha = (D/z)(1-\tau)-1$, from

$$\mathcal{P}^{(T)}(T) = \frac{D}{z}a^{(s)}T^{-\tau D/z+D/z-1}\,\mathcal{G}^{(s)}\left(\left(\frac{T}{s_c^{z/D}}\right)^{D/z}\right) \tag{10.3}$$

reproducing the scaling law mentioned above, $D(1 - \tau) = z(1 - \alpha)$, see Eq. (2.40). Moreover, the scaling function of the avalanche duration distribution is essentially identical to that of the avalanche size.

This derivation is deeply flawed. It suggests that the scaling law $D(1-\tau) = z(1-\alpha)$ is a mathematical necessity on the basis that observables are, on the grounds of dimensionality, functions of each other. They are not – although not independent, they are merely correlated random variables, except for a few hard constraints. For example, the avalanche duration is normally bounded from above by the avalanche size, and the size is bounded from below by the duration, but normally these constraints become asymptotically irrelevant. [2]

The dimensional argument presented above applies by definition to the cutoffs, $s_c^{1/D} \propto T_c^{1/z}$. Based on the *assumption* of sufficiently narrow distributions, it holds

---

[2] For example, $s \geq T$, but asymptotically the avalanche size is so much greater than the duration that $s \geq T$ does not actually pose a constraint. In practice, the borderline case $s = T$ does not occur at any reasonable system size.

for conditional expectation values, for example $\langle s|T \rangle^{1/D} \propto T^{1/z}$, Eq. (2.35a), which is very close in spirit to observables being functions of each other. But basing that on dimensional consistency is misleading, as it suggests, for example, that the first moment $\langle s \rangle$ should scale like L^D, see 'first moment', p. 349.

Even when the scaling law $D(1 - \tau) = z(1 - \alpha)$ therefore has to be tested for each and every model, it is very widely assumed to hold in general. So much so, it is difficult to untangle the numerical results in the literature in a way that would allow the validation of this scaling law on the basis of *independently* determined exponents.

Temporal correlations by spatial correlations (context: Ch. 1, Sec. 9.2)

Misbelief: 'Temporal correlations are stored in and maintained by spatial correlations.'

It seems natural to assume that the memory required to display temporal correlations is provided by the state of a system and as this state is spatially extended, by spatial correlations. This point is commonly made, arguing that algebraic temporal correlations signal algebraic spatial correlations, which can be difficult to measure, but are ultimately at the heart of a cooperative phenomenon.

However, temporal correlations can easily emerge from systems with product states, i.e. with no spatial correlations whatsoever. Because SOC systems normally have two time scales, the microscopic and the macroscopic, correlations can be measured for both. The TAOM (Sec. 6.3.5) is a very instructive example. It has a product state and after every update and on the microscopic time scale, correlations between states at a given site are trivial, because the local state remains unchanged until a new state is chosen randomly and independently. On the macroscopic time scale, macroscopic observables such as the avalanche size, however, are clearly correlated. This result might appear counter-intuitive: how can there be memory, if the state is 'completely random'? The answer is that even a spatially uncorrelated state needs time to evolve in time and will thus display possibly non-trivial temporal correlations. A product measure state thus does not preclude (temporal) correlations, at least not on the macroscopic time scale.

Grinstein (1995) almost dismisses temporal correlations as 'difficult to avoid' ('in the presence of a local conservation law') on the grounds that they can be present without signalling non-trivial cooperation. Considering time as another, yet, anisotropic direction in space, it is difficult to see why time and space are so fundamentally different. One reason might be the different rôles they play in the presence of symmetries such as conservation, when the *time* derivative of a *spatially* integrated quantity vanishes. Also, equations of motion often involve higher order derivatives in space, notably a gradient squared term in the presence of diffusive transport, but only first derivatives in time. In particular, there are *no* second derivatives in time, which produce momentum effects.

Criticality pre-condition for scaling (context: Ch. 2, Sec. 9.2)

Misbelief: 'Only critical processes display scaling and thus a data collapse.'

This misbelief has its origin probably in the perception of equilibrium critical phenomena. For example, in percolation, away from the critical point the cluster size distribution has (asymptotically) a cutoff independent of the system size, so that finite size scaling is precluded. One might conclude that there is no scaling at all. However, at a sufficiently coarse scale, trivial scaling is recovered and data can be made to collapse.

In fact, such scaling without long-range order is very common. For example, by the central limit theorem, the sum of N independent, identically distributed random variables with finite variance and vanishing mean produces a distribution that collapses increasingly well (in N) onto a Gaussian when rescaled appropriately. Less trivial examples of *generic scale invariance* have been presented, for example, by Hwa and Kardar (1989a, also Sec. 9.2). In these systems, the two point correlation function can be shown to display scaling. This is not to say, however, that these systems are necessarily good representatives of SOC models, as they lack certain characteristic features, in particular a separation of time scales.

SOC as non-equilibrium critical phenomenon (context: Secs. 9.1, 9.2)

Misbelief: 'SOC is just an instance of non-equilibrium critical phenomena, which generally do not require tuning of a control parameter, in contrast to equilibrium critical phenomena.'

A more refined version of this misbelief distinguishes (correctly) SOC as a critical phenomenon far from equilibrium from an out of equilibrium critical phenomenon, which is non-equilibrium as well, but relaxes back to equilibrium (Hohenberg and Halperin, 1977). However, equilibrium is not possible in SOC, as the constant particle flux required to sustain stationarity breaks detailed balance. What is more, Hwa and Kardar (1989a) discovered entire classes of such non-equilibrium systems that display scale invariance without tuning of a control parameter. As Grinstein (1995, p. 269) put it:

> [The] removal of the detailed balance constraint allows for spatial scale invariance more broadly than in equilibrium.

One might conclude that SOC does not require parameter tuning *because* it is a non-equilibrium critical phenomenon.

This conclusion is incorrect. Even when the universality classes identified by Hwa and Kardar (1989a) are very big, they by no means exhaust the space of non-equilibrium systems that SOC models are part of, i.e. many of those that display scaling, do so only at a certain, tunable parameter value. In fact, the underlying phase transition of the most solid, best understood SOC models, such as the MANNA Model, is known and controlled by a temperature-like variable. Tuning is very commonly required in non-equilibrium. Just because something is non-equilibrium and capable of displaying a phase transition it certainly does not have to be generically scale invariant.

Operator algebra (context: Secs. 8.3.1, 8.3.2)

Misbelief: 'Existence of an inverse implies uniform probability of all states.'

It seems obvious at first: if a model has a unique preceding state then all states appear with the same frequency. That is because if an inverse exists, only one preceding state can give rise to a certain next state. But it seems that the only way that some states can appear more often than others is that they have more preceding states, illustrated in Fig. 10.1(a).

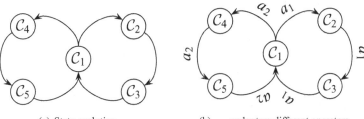

(a) State evolution.   (b) ... under two different operators.

Figure 10.1 The frequency of recurrent states (encircled) can vary, even in deterministic systems where each individual operator has an inverse. (a): If state C_1 has more than one preceding state, here C_3 and C_5, then it might occur more frequently than the others. If there is only one operator evolving the system, then it does not possess an inverse. (b): Each of the deterministic, invertible operators a_1 and a_2 generates the states on its limit cycle with the same frequency. Yet, a mixed operation like $a_2^3 a_1^3 a_2^3 a_1^3 \dots$ applied to state C_1, produces a sequence of states with state C_1 appearing twice as frequently as any other state.

This is incorrect or at least incomplete or misleading. Firstly, there is no need to *construct* an inverse. If the evolution among recurrent states is deterministic, i.e. evolution *can take only one path*, then an inverse necessarily exists and vice versa. Secondly, a deterministic evolution by the same operator[3] of the system with finite state space is sufficient for uniform probability, as every state on the single, unique limit cycle is visited precisely once in every revolution, i.e. all states appear with the same frequency or not at all, such as the limit cycle of either a_1 or a_2 in Fig. 10.1(b). Yet, limit cycles vary between operators and if more than one operator is applied, then uniform probability does not necessarily follow (regardless of whether or not each individual operator has an inverse). This is shown in Fig. 10.1(b): applying to a system in state C_1 operators a_1 and a_2 each three times before switching to the other operator, means that all states appear with frequency $1/6$ except C_1, which occurs with frequency $1/3$. Uniform probability does follow, however, when the operators are applied at random, as discussed further in Sec. 8.3.1, p. 272. The criterion is that the operator applied to the system next does not depend on its state, i.e. the state evolution is Markovian. A *deterministic*, Markovian matrix is a permutation matrix, which leaves the uniform distribution invariant as well.

The arguments above rely strongly on the finiteness of state space, which entails necessarily finite limit cycles. It breaks down as soon as the number of states considered is infinite, simply because none of the states might ever be repeated. Also, none of the arguments above extends to the density of the system's trajectory in (continuous) phase space. The state space is infinite when, for example, an infinite sequence of random numbers is 'attached' to each site. This is a popular route used to turn an essentially stochastic model into a formally deterministic one.

[3] If the system were allowed to evolve under different operators, then a sequence like $a_2^3 a_1^3 a_2^3 a_1^3 \dots$ in Fig. 10.1(b) would be allowed. In addition to the state C_i of the system, the position in the sequence would be required to determine the operator to be applied and thus the state in the next time step, i.e. the evolution of the state (without information about the position in the sequence) would not be Markovian.

Abelianness of operator algebras (context: Secs. 4.2, 8.3)

Misbelief: 'The Abelian symmetry of the operator algebras representing some SOC models derives from their (matrix) representation.'

Dhar's (1990a) reformulation of the sandpile model was a brilliant success that provided for the first time an algebraic access route to SOC models. Some authors seem to assume that the derivation of the Abelian symmetry of the algebra is based on the properties of its representation in terms of Markov matrices. This is not the case. To start with, Dhar's operators originally act on the state of the system, rather than on their probability distribution (vector) as a Markov matrix would. Secondly, the proof of Abelianness does not rely on a particular representation, but on general considerations of the rules of the model.

Abelianness means arbitrary update order (context: Sec. 8.3)

Misbelief: 'In Abelian models, the order in which sites are updated is irrelevant.'

It seems to be a widespread belief that the order of updates in an Abelian model is irrelevant for its statistical and scaling properties and therefore a mere technicality. This is incorrect. Firstly, time is defined through the microscopic dynamics, i.e. time dependent observables such as the avalanche duration depend crucially on the definition of the microscopic update rules. Secondly, the Abelian property of the operator algebra applies to the macroscopic time scale. It states that the configuration of a system does not depend on the order in which external charges are applied, rather than the order in which a system is updated. This is a somewhat pedantic remark, because the proof of Abelianness normally relies on and establishes the invariance of the evolution on the microscopic time scale as well. Thirdly, even observables on the macroscopic time scale that do not depend on the specifics of the microscopic time, still normally depend on the order in which charges are carried out. For example, the avalanche sizes generated by charging two distinct sites depend on the order of these charges; only their sum is invariant (p. 98, p. 270).

SOC optimises (context: Ch. 3, Sec. 9.4)

Misbelief: 'SOC optimises, as signalled by power laws.'

Where considered outside the realm of statistical mechanics, SOC is often regarded as the signature of an optimisation mechanism, which makes itself visible through the appearance of power laws. The origin of this notion could be the occurrence of distributions with a cutoff determined by a single scale, beyond which event sizes cannot possibly stretch. In that sense, events extend as far as possible, which amounts to a form of optimisation. Another perspective is that system-wide avalanching in SOC models hints at extreme and thus 'optimised' susceptibility, as found at the critical point of an ordinary phase transition. [4] Such a large susceptibility could be caused (via a sum rule) by algebraic correlations, signalled by power laws.

Beyond such handwaving, there is no justification for regarding SOC as a form of (self-) optimisation. SOC might optimise something, but there is no general rule *what*. In fact, where power laws are used as evidence for optimisation of one feature, one

---

[4] But also in the ordered phase, if massless Goldstone bosons exist.

might find equally good evidence for the optimisation of the opposite. SOC does not optimise everything and a power law is generally not the signature of an optimisation.

Power laws imply unpredictability (context: Sec. 3.4)

Misbelief: 'An event size distribution that follows a power law means that the occurrence of (large) events cannot be predicted.'

Power law distributions are broad, so that large events might occur more frequently as one might naturally tend to expect. If, for example, they follow an exponential distribution, certain large events practically never occur, and adding a dimensionful (characteristic) constant to a given event size reduces its probability by a constant factor. In the case of a power law distribution, the event size has to be *multiplied* by a dimensionless number to achieve the same. However, frequent occurrences of large events does not imply that they are unpredictable, and neither does the lack of an upper cutoff (or, for that matter, a very large upper cutoff). If caused by cooperative phenomena, then the presence of long time temporal correlations suggests the opposite, namely that power law distributed events are strongly correlated and thus very predictable (e.g. Pepke and Carlson, 1994). Models like the TAOM (Sec. 6.3.5) illustrate that: strong anti-correlations prevent the occurrence of sequences of large avalanches (Sec. 8.5.4.1).

10.2 Outlook

The phenomenology is understood, the models are studied, the theory developed. The questions what SOC is, where it can be found and how to analyse it have been answered at least partially in the preceding chapters. A number of important questions in their specific context have been raised throughout. What are the relevant questions for the field as a whole?

On the technical side, *analytical results* would probably have the greatest impact, such as bringing to bear in SOC established theories and methods, producing an *exact solution* of any of the models or deriving the algebraic features of the models from first principles. Although there has been a lot of progress, many mathematical and physical features of the best known models, such as Abelianness or the rôle of dissipation, are still not fully understood. One of the most promising routes is field theory, which requires as input, for example, a Langevin equation of motion, a Fokker–Planck equation or a reaction-diffusion description of the underlying processes. It remains a great puzzle why SOC models are so resistant to analysis by these techniques.

At the heart of all research into SOC is the question, *what are the necessary and sufficient conditions for SOC?*. Even when the possible answers are far less clear cut than the question, this has been answered at least partially and approximately in the present book. The central ingredients in SOC are probably a separation of time scales and thus avalanching as well as an internal transport mechanism, possibly induced by bulk conservation. An underlying critical point is triggered, possibly by a feedback loop of the order parameter to the (temperature-like) control parameter, or by avalanches being forced to spread through the system until they reach the boundaries. An Abelian symmetry ensures that the scale invariant features are independent of the choice of the specific dynamics and seems to have important implications for the interfacial interpretation of a model. How all these physical and mathematical features are joined together to produce SOC remains to be understood – again, a full, analytical description is still missing.

Is it possible to 'apply' SOC, like a recipe, to other critical phenomena? It seems that the known mechanisms should be applicable to broad classes of well-understood models, most notably absorbing state phase transitions. However, a successful example of such an 'implementation' does not yet exist. What features of working SOC models need to be implanted into an ordinary critical phenomenon to 'make it SOC'?

Denying flatly the existence of SOC, as its fiercest critics do, is difficult to uphold, given the MANNA and the OSLO Models, which display all the desired features so clearly. Unfortunately, both models belong to the same universality class and no model outside that class displays SOC as indisputably as they do. If that remains unchanged, one might well argue that SOC is a feature of this particular universality class, just as generic scale invariance is a feature of many growth models. Experimental results cannot be used to diffuse this argument, as the most solid experimental evidence (Secs. 3.1 and 3.2) produces scaling close to the very same universality class.

Finally, there is an enormous amount of unexplained experimental and numerical data that display *almost scaling*. Claiming that these systems are not instances of SOC or that SOC is confined to a single universality class does not help to explain such 'dirty' power laws which are produced so often in complex systems. This is probably the biggest challenge of the years to come: *what is the cause of* approximate *scaling in nature?* Even if SOC should turn out not to be part of the answer, it remains a crucial part of the question.

Appendix A The OLAMI–FEDER–CHRISTENSEN Model in C

Most computational physicists try to strike a balance between a number of conflicting objectives. Ideally, *a model is quickly implemented, easy to maintain, readily extensible, fast and demands very little memory.* A few general rules can help to get closer to that ideal. Well written code that uses proper indentation, comments and symmetries (see for example PUSH and POP below), helps to avoid bugs and improves maintainability. How much tweaking and tuning can be done without spoiling readability and maintainability of the code is a matter of taste and experience. Sometimes an obfuscated implementation of an obscure algorithm makes all the difference.[1] Yet, many optimisations have apparent limits where any reduction of interdependence and any improvement of data capture is compensated by an equal increase in computational complexity and thus runtime. Often a radical rethink is necessary to overcome such an ostensible limit of maximum information per CPU time, as examplified by the Swendsen–Wang algorithm (Swendsen and Wang, 1987) for the Ising Model, which represents a paradigmatic change from the classic Metropolis algorithm (Metropolis, Rosenbluth, Rosenbluth, *et al.*, 1953).

Nevertheless, one should not underestimate the amount of real time as well as CPU time that can be saved by opting for a slightly less powerful code in favour of one that is more stable and correct from the start. On the same account, it usually pays to follow up even a little hunch that something is not working correctly. Failing programs, instances of which break down, question all the results obtained by that code, unless they can be shown to be unaffected by the error – a proof that very rarely exists.

Of all program failures, a **segmentation fault** (also known as segfault) is the most common and also most corrosive problem. On UN*X systems, a program is terminated (by default) via the signal SIGSEGV, when the operating system detects that a program operates beyond its declared memory limits, which does not imply that it was working correctly at any point prior to the detection and subsequent termination. A segmentation fault should be regarded as a sign that either the code is not a faithful implementation of the algorithm or that the latter is flawed. Both can be tested using **invariants**, i.e. verifying that a certain structure or order is maintained during runtime, as well as **assertions** (see for example l. 129), i.e. probing assumptions about the state of the algorithm and the data it operates on. Using macros and compiler flags, such tests can be turned off in the code used for production without further interference with the source files. A revision control system, such as RCS and Subversion, is often the best way to organise code across time and authors.

---

[1] The 'fast inverse square root' (Wikipedia, 2010a) is one of the most famous examples of hardly readable code which uses the very details of the C language specification and is responsible for huge technical and commercial success.

It is generally advisable to store data together with the executable and the source code used for their production [2] and the flags set at compilation. Comparison with exactly known as well as published and widely accepted results is an obvious means of quality control.

Even though most of the discussion below is particular to the OFC Model, many techniques are easily carried over to other models. A more general discussion on simulation techniques, in particular **Monte Carlo**, can be found in various superb textbooks on the subject (e.g. Binder and Heermann, 1997; Newman and Barkema, 1999; Berg, 2004; Landau and Binder, 2005). Unfortunately, many powerful techniques, say the single histogram method (Ferrenberg and Swendsen, 1988), do not apply in SOC, where the aim of the 'simulation' is not to realise a string of configurations according to a known probability density function, as normally aimed for using a Markov chain, but to evolve a set of rules that gives rise to that very distribution.

The famous review by Press *et al.* (2007) contains many scientific algorithms and plenty of technical details. Cormen *et al.* (1996) have written probably the most widely used text on the theory of algorithms, introducing the basic concepts, data structures, cost analysis etc. from a computer science perspective. Knuth's (1997a) famous work takes a slightly more mathematical perspective on often needed algorithms. Many implementations of common algorithms in C, in particular on graphs, have been collected by Sedgewick (1990). Dowd and Severance (1998) addressed the relation between code and hardware and practical problems that can arise. Kernighan and Ritchie's (1988) original work remains the most important guide to C programming, whereas van der Linden (1994) highlighted many of its less known quirks. Stevens and Rago (2005) wrote a new edition of the famous reference for programming in the UN*X environment (Nemeth, Snyder, Seebass, and Hein, 1995). For post-processing of data, UN*X-tools such as grep, sed and awk are frequently used and are described in great detail by Dougherty (1992). Brandt (1998) discussed most if not all of the standard techniques used to derive estimates for observables from measurements. Anderson (1971) wrote a classic text on time series analysis; Kantz and Schreiber (2005) took a more modern approach to the same subject.

The following discusses, almost line-by-line, a faithful, fast implementation of the OFC Model, [3] which has received a lot of attention (e.g. Grassberger, 1994; Ceva, 1998; Drossel, 2002), because of the extensive numerics required to study it and the immediate benefit of a good algorithm. In many SOC models, the CPU spends the vast majority of its time in very few lines of the code. This is different in the OFC Model, which requires a larger framework to run efficiently.

The basic setup of the OFC Model can be found in Box 5.3, p. 127, and a brief outline of the most important numerical techniques is given in Sec. 5.3.4. Below, some additional improvements and simplifications are discussed which would speed it up quite significantly, but which rely on hitherto untested assumptions. In contrast to other published code, the implementation presented here can in fact be used for practical purposes, i.e. it is not just pseudocode or the bare bones of the OFC Model which would be impractical for

---

[2] Adding a function which allows the executable to write out its own source code and possibly adding it to the data itself can help organising data and code.

[3] All lines of code that are part of the implementation are numbered sequentially from l. 1 to l. 368. Line breaks due to page size limitations are indicated by ↳.

real applications. The code is presented and discussed linearly, which is unfortunately not always in line with its logic, as some details occasionally need to be anticipated. Two readings might be required to digest it fully. Except for some examples in Sec. A.6, the code is written in C, which should easily translate to C++.

The first few lines,

```
1    #include <stdio.h>                                    /* For standard I/O. */
2    #include <stdlib.h>                                    /* For malloc and exit. */
3    #include <time.h>                                      /* For time and ctime. */
4    #include <limits.h>                              /* For ULLONG_MAX, if it exists. */
5    #ifndef ULLONG_MAX
6    #define ULLONG_MAX (~0ULL)
7    #endif
8    #include <fpu_control.h>                           /* For setting the precision. */
9    /* More convenient interface to set the precision: */
10   #define FPSETPREC(a) {fpu_control_t flags; _FPU_GETCW(flags); flags &=~(
     ↳    _FPU_EXTENDED | _FPU_DOUBLE | _FPU_SINGLE); flags |=(a); _FPU_SETCW(
     ↳    flags);}
11   #include <signal.h>                                  /* For signal handling. */
12   #include <sys/types.h>                   /* For determining the pid of the program. */
13   #include <unistd.h>                             /* For pid and hostname. */
```

contain the usual header files, for example that of standard I/O, some details of which will be discussed below. The following few lines contain all the parameters of the simulation as macros

```
14   #define LENGTH (64)                                   /* System size. */
15   #define OBC                                      /* Boundary condition. */
16   #define NUM_BOXES (LENGTH*32)                     /* Number of boxes. */
17   #define ALPHA (0.07)                           /* Level of conservation. */
18   #define NUM_CHUNKS (5)                            /* Number of chunks. */
19   #define ITERATIONS_PER_CHUNK (1000000LL)                    /* ... */
20   #define SEED (1L)                     /* Seed of the random number generator. */
```

which by convention appear in all capitals. Most of them could be exchanged for variables set at runtime. The advantage of using a macro is a potential increase in performance and the ability to store the source file including all parameters with the results. The latter option is watered down by the possibility of defining macros at compile time. Moreover, macros are (usually) not typed, i.e. no strict type is assigned to them. The use of const variables is an alternative [4] preferred by many (Kernighan and Pike, 2002; van der Linden, 1994), unless when used in arithmetics at compile time (error: initializer element is not constant).

```
21                    /* 64 bit long long */
22   #if (1)                                      /* For quick commenting out. */
23   typedef unsigned long long int force_type;              /* Type of force. */
24   /* Name of the force type for result file: */
25   #define FORCE_TYPE_NAME "unsigned_long_long"
26   #define F_SLIP ((force_type)1<<40)                    /* Force threshold. */
```

---

[4] In ANSI C const int can be used for array bounds, so that there is no reason to use enum as an alternative (Schildt, 2000; Kernighan and Pike, 2002).

```
27   /* Minimal force at which slipping takes place: */
28   #define INIT_OFFSET (ULLONG_MAX − 64*F_SLIP)
29   /* Criterion to trigger readjustment of forces: */
30   #define READJUST_CRITERION(a) ((a)<F_SLIP)
31   #define FORCE_OUT_FMT "%llu"                        /* Output format for forces. */
32   #define PRECISION _FPU_EXTENDED                     /* Precision setting. */
33   /* Calculate charge in accordance with force_type: */
34   #define CHARGE(a) (force_type)((long double)ALPHA*(long double)(a))
35   #endif
36
37                   /* 64 bit double */
38   #if (0)                                            /* For quick commenting out. */
39   typedef double force_type;
40   #define FORCE_TYPE_NAME "double"
41   #define F_SLIP ((force_type)1.0)
42   #define INIT_OFFSET ((force_type)(1<<24))
43   #define READJUST_CRITERION(a) ((a)<F_SLIP)
44   #define FORCE_OUT_FMT "%g"
45   #define PRECISION _FPU_DOUBLE
46   /* Note: Precision clashes with MOMENT_TYPE below (long double). */
47   #define CHARGE(a) ((force_type)ALPHA*(a))
48   #endif
49
50   force_type f_threshold;
```

These two sections provide alternative choices for the **variable type** the forces in the OFC Model are stored in, see l. 23 and l. 39, the corresponding values of the static friction, F_SLIP, and various auxiliary macros, such as a name for the type for later reference and its formatting string. For quick commenting in and out #if (0) and #if (1), l. 38 and l. 22 respectively, are used. In proper production code this would be done better using other macros as flags, such as #define FORCE_LL after l. 20 and #ifdef FORCE_LL instead of l. 22. The different macro definitions are discussed further in Sec. A.4.4. Once force_type is defined, the threshold f_threshold is declared in l. 50.

Each site on the lattice has a number of properties associated with it, which are conveniently (but not necessarily most efficiently [5]) organised in the form of a struct

```
51   struct lattice {
52     force_type f;                                    /* Local force. */
53     force_type charge;                               /* Force to be charged. */
54   #define NUM_NEIGHBOURS (4)
55     struct lattice *nb[NUM_NEIGHBOURS];              /* Pointers to neighbours. */
56     struct lattice **bx_left;
57     /* ... the address of (neighbour's) pointer to me. */
58     struct lattice *bx_right;                        /* Box: pointer to the next site. */
59   } *site;
```

The field charge is necessary so that charges by neighbouring sites can be stored and accounted for *after* the charged site has toppled (see p. 369, red-black approach, for further

---

[5] Access to an array by index takes one addition, namely pointer plus index, access to members of a struct beyond the first takes one more addition.

details). Pointers to neighbouring sites are stored in nb, see the initialisation after l. 219. The other two fields are used for organising the sites according to their force (Sec. A.3).

```
60    #define NUM_SITES (LENGTH*LENGTH)
61    #define COO2INDEX(x,y) ((x)*LENGTH+(y))
```

Grassberger (1994) suggested that indexing sites linearly is more efficient than using d-dimensional arrays. Over the last 50 years or so, **addressing of sites and their neighbours** has experienced a number of changes (see the discussion on boundary conditions by Binder and Heermann, 1997; Newman and Barkema, 1999, Ch. 13) and there is no unique, best way of doing it. If large neighbourhoods are addressed frequently, their on-the-fly computation, using, for example, a macro like COO2INDEX defined in l. 61, is probably detrimental to the overall performance. On the other hand, determining the neighbouring sites by integer arithmetics sometimes benefits from more sophisticated techniques available on modern processors, such as out-of-order processing. When calculating them on the fly, a conditional assignment left=((x-1)<0)? (LENGTH-1): (x-1) can be faster than a modulo operation left=(x-1+LENGTH)%LENGTH. [6]

Using a map of neighbours as introduced in l. 55 corresponds to storing results of integer arithmetics in arrays, with the intention of swapping memory for CPU time. Nowadays, memory access is often a bottleneck compared to what can be done within the CPU. A comparison would take into account problem size, frequency of use, compiler and its settings, as well as the platform.

The array nb[4] of l. 55 is therefore a matter of convenience, as it greatly improves the readability and facilitates maintenance of the code. In general 4 neighbours are assumed to exist, see l. 307, while sites with fewer neighbours (in the case of open boundary conditions, at the edge and in corners) acquire NULL pointers to neighbours which do not exist. A loop like the one in l. 306 could break as soon as a NULL pointer is encountered, which would require all non-NULL neighbours to be stored in a contiguous sequence when they are calculated in l. 219–245. Instead of using NULL pointers for non-existing neighbours and testing for it, one could also point to a **virtual site** which does not interact with the rest of the lattice, but can receive arbitrary amounts of force. This type of **padding** is commonly used in lattice models to avoid complicated if-statements to account for boundaries. Occasional resetting of the local force prevents accidental toppling as it crosses the threshold due to (integer) overflow.

In the present code, neighbours, the content of the stack and sites generally are accessed through **pointers**. While pointers are often regarded as the greatest blessing of C, they are probably the biggest curse for any highly optimising compiler, because pointers impede the ability of the compiler to 'recognise' the intended effect of a piece of code (Dowd and Severance, 1998, p. 84). For example, depending on platform, compiler, the compiler settings and the problem size, copying an array by [7]

```
for (i=0; i<LENGTH; i++) a[i]=b[i];
```

---

[6] The addition +LENGTH avoids the ambiguity of negative operands in modulo operations if x-1 is negative.

[7] The library function memcpy is much faster at performing that operation than any of the two codes discussed here.

can be much faster than the apparent equivalent

> **for** (p=a, q=b; p!=a+LENGTH; p++,q++) *p=*q;

Again, the choice between the two alternatives often comes down to readability and maintainability of the code.

A.1 Stacks

Another, rather simple optimisation pointed out by Grassberger (1994) is the use of **stacks** to maintain lists of sites that are subject to future update, instead of rescanning the entire lattice to find such sites. Using stacks, however, means that the lattice structure, i.e. the relative position of active sites, is not immediately known at the time of the update, which makes it difficult to distinguish sites that have been visited and updated already from sites that will be updated in the future. In the simplest application of a stack, only one such list exists and all sites to topple are treated equivalently. This is often the method of choice where many **Poisson processes** take place simultaneously across the lattice. A single stack is the most convenient facility when the order of updates is irrelevant, such as in Abelian models if observables do not depend on the definition of microscopic time. Contrary to what the name suggests, stacks can (usually) be accessed in an arbitrary order, see below.

The first complication that arises for the OFC Model is the sequential nature of the updates, because all sites whose forces exceed the threshold at the beginning of a time step must have toppled and distributed their current force before any freshly charged site does with the newly acquired force. A stack, by its very design, is a **last in, first out (LIFO)** mechanism: whatever goes last on the stack by PUSH, is taken out first by POP. Yet, sites to be toppled in the future need to be processed only after all other sites have been updated.

This problem can be resolved either by using a **first in, first out (FIFO)** queue (circular buffer) or two independent stacks. The latter is the method used here:

```
62                    /* Stacks */
63    int stackheightCurr, stackheightNext;
64    struct lattice **stackCurr, **stackNext;
65
66    #define PUSH(a) stackNext[stackheightNext++]=(a)
67    #define POP(a) (a)=stackNext[--stackheightNext]
```

Sites that are updated are taken from `stackCurr`, sites that will be updated in the future reside in `stackNext`. As soon as `stackCurr` has been worked through, the two entities are swapped, see l. 329–333. The macro POP is never used and is stated here for completeness only. As a final remark, random access can be realised by code of the form `a=stackNext[random_pos]; POP(stackNext[random_pos]).`

A.2 Moments

The scaling of an observable is most conveniently determined by means of moments (see Sec. 2.2, p. 37 and Sec. 7.3). Over the course of the run, observables are taken, their nth power calculated, accumulated and later normalised and printed in the result file. If the observable spans m orders of magnitude, its nth power spans nm orders of magnitude. Each order of magnitude requires about 3.3 bits to be resolved. The results will be distorted if the size of the mantissa of the type chosen for the moment does not exceed $3.3nm$. Given that the moments are at first **accumulated**, i.e. many measurements are taken and summed, the size of the mantissa should by far exceed $3.3nm$. The smaller the mantissa, the more easily small contributions drown in large ones, even when the former actually dominate; if they occur, small additions might not actually change a variable's value if its resolution is low and its current accumulated value is too large. For very high moments one might therefore observe a saturation effect, when observables are effectively cut off below a certain size. If the observable is integer valued, then a sufficiently large integer type provides an exact representation. Otherwise, if performance allows, one should use a floating point type with the biggest mantissa. This choice, however, needs to be in tune with the FPU precision (see Sec. A.4.5) and therefore should be made together with it in the sections starting at l. 22 and l. 38.

```
68   #define MOMENT_TYPE long double
69   #define MOMENT_OUT_FMT "%10.20Lg"
70
71   MOMENT_TYPE mom_pow;
72   int mom_cnt;
73
74   #define MOMENTS_DECL(n,m) MOMENT_TYPE mom_ ##n[(m) + 1]; const int
  ↳     mom_max_ ##n = m;
75   #define MOMENTS_INIT(n) {for (mom_cnt=0; mom_cnt<=mom_max_ ##n; mom_cnt++)
  ↳     mom_ ##n [mom_cnt]=0;}
76   #define MOMENTS(n,x) {mom_ ##n [0]++; for (mom_pow=1, mom_cnt=1; mom_cnt<=
  ↳     mom_max_ ##n; mom_cnt++) {mom_pow*=((MOMENT_TYPE)(x)); mom_ ##n[
  ↳     mom_cnt]+=mom_pow;}}
77   #define MOMENTS_OUT(n) {force_type chunk_time; \
78     chunk_time =total_time_readjusted − chunk_start_readjusted; \
79     chunk_time +=(total_time − chunk_start); \
80     printf("#M_%s_%i_%i_%g_%lli_" FORCE_OUT_FMT "_" FORCE_OUT_FMT "_"
  ↳       FORCE_OUT_FMT "_%i_" MOMENT_OUT_FMT "_", #n, chunk, LENGTH,
  ↳       ALPHA, ITERATIONS_PER_CHUNK, chunk_time, total_time +
  ↳       total_time_readjusted, f_threshold, mom_max_ ##n, mom_ ##n [0]); \
81     for (mom_cnt=0; mom_cnt<=mom_max_ ##n; mom_cnt++) \
82     printf("_" MOMENT_OUT_FMT, mom_ ##n [mom_cnt]/((mom_ ##n [0]) ? mom_ ##n [0]
  ↳       : −1)); fputc('\n', stdout);}
83
84   MOMENTS_DECL(siz, 5);                          /* Avalanche size. */
85   MOMENTS_DECL(dur, 5);                          /* Avalanche duration. */
86   MOMENTS_DECL(deg, 2);                          /* Degeneracy of the maximum. */
```

For illustration purposes the **variable type** for moments is chosen through a macro rather than a `typedef`, which makes it easier for the compiler to pick up output-formatting mistakes, but possibly more difficult to detect clashes in assignments of that variable type than it would be using `typedef`. All macros regarding moments have prefix `MOMENTS_` and are meant to be used throughout the rest of the code, whereas all variables associated with moments have prefix `mom_` and should not be touched in any other way than through the macros. This is one of many instances where an object oriented language would be more suitable.

To keep the output format consistent with the type, a macro `MOMENT_OUT_FMT` is defined. The two global variables `mom_pow` and `mom_cnt` are needed for the internal workings of the macros `MOMENTS_INIT`, `MOMENTS` and `MOMENTS_OUT`.

The macro `MOMENTS_DECL(n,m)` is the interface used to define an array that will accumulate the **powers** of an observable, identified by the name n. This is used as a suffix for variables associated with that moment, such as the largest power `const int mom_max_ ##n = m`. The macro `MOMENTS_INIT(n)` initialises all moments with suffix n, see l. 257–259.

The macro `MOMENTS(n,x)` records the moments of the various observables. First, the total weight is increased by 1, `mom_ ##n [0]++`, then the moments are generated in a loop, which the compiler might unroll. It is often beneficial to have a large number of moments (see Sec. 7.3) without taking, however, moments of too high an order, which suggests using **fractional powers** (but note Sec. 7.3.2, in particular p. 226). Because the library functions `pow` and `powl` are often too expensive, one can instead generate one or a few fractional powers, say $pow(x,1./8.)$, $pow(x,1./4.)$ and $pow(x,1./2.)$, and generate any power $m/8$ by multiplication. Because each outcome of a `pow` call has only limited accuracy and precision, their combination might result in undesirable overall numerical error. The same objection applies to `MOMENTS(n,x)`, where repeated multiplication of a floating point number rather than an integer produces the higher powers. Higher powers are produced more reliably by `pow`. If that is not feasible, very high powers $n = 2, 4, 8, 16\ldots$ with good precision can be produced by repeated squaring.

The last macro, `MOMENTS_OUT` is a routine for the **output** of the results. The macro begins with a careful calculation of the total amount of force applied per site. Like all other output, the data are preceded by a **tag**, such as `#M_siz` which facilitates access to the various moments and other information via standard tools, such as `awk`, `grep` and `sed`. [8] The tag is followed by all parameters that are thought to be relevant to the outcome, concluding with the total accumulated weight, which is of great importance for the calculation of averages across sets of results, i.e. independent runs or repeated output such as the one discussed here (chunks, Sec. 7.2). Precautions have to be taken in case the total weight `mom_ ##n [0]` vanishes, here indicated by a minus sign.

In the present code, `stdout` is used as the sole data output stream, which helps keep the output in sync (even with error messages, which are written to both `stderr` and `stdout`, e.g. l. 181) and avoids a wide range of problems, such as descriptor leaks by repeated opening of the same file or failure to open files due to non-existing files and directories,

---

[8] A typical command would be `grep '#M_siz' output.dat | sed 's/#M_siz//'`, which picks avalanche size moments in the file `output.dat`.

incorrect permissions, stale NFS handles etc. Some potential pitfalls remain, as stdout is redirected to a file (see p. 371), but at least it is, by default, guaranteed to exist. The processing of the chunks written out, in particular the calculation of statistical errors, is discussed in Sec. 7.2.

A.3 Boxes

The two declarations

```
87                         /* Signalling */
88    void signal_handler(int signo);
89    int global_stop_flg=0;
```

to be discussed in Sec. A.4.1 are followed by three prototypes and the definitions of functions handling boxes, a facility suggested by Grassberger (1994):

```
90    #define BOX_POS(f) ((int)((((unsigned long long)((double)(NUM_BOXES−1)*(double)(f)
      ↳    /(double)F_SLIP))%(unsigned long long)NUM_BOXES))
91
92    struct lattice *box[NUM_BOXES];
93
94    int box_insert(struct lattice *a);
95    int box_delete(struct lattice *a);
96    int box_getmaxima(struct lattice **list, int *num);
```

Each site, when it is not about to topple and is therefore maintained on the stack, is held together with other sites in a box. The reason for the distinction between stack and boxes is purely logical – as explained in the following, the algorithm works as long as all sites are in boxes after an avalanche has finished.

Grassberger's (1994) **box-technique** is a means of keeping track of the maximum site with a certain resolution, F_SLIP/(NUM_BOXES-1), whose value is discussed below. From the point of view of computing, the method uses the force as a **hash value** to access subsets of unsorted forces within a particular range. At first sight, this technique, to be detailed below, appears a bit cumbersome. It is obviously much better than a naïve sweep over the lattice, but a more traditional approach using a **binary search tree** seems to be much faster. In fact, another extremal model, the BS Model, can be implemented very successfully using a binary search tree (also endnote [5.7], p. 392). The key difference is that, most of the time, the fitness distribution in the BS Model is very sparse for the small fitnesses which are accessed at *every* update (as opposed to the OFC Model, where the extremal site is located only at the beginning of the avalanche). Finding the extremal site, when the density of such sites is sparse, is relatively expensive using the box-technique. In the OFC Model, however, the distribution is (fairly) continuous (but note Torvund and Frøyland, 1995), even close to F_{th}, with only a small degree of degeneracy. Insertion of a site is very fast using boxes (constant time) and comparatively slow using binary trees (about logarithmic in the number

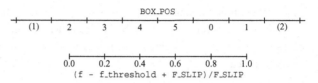

How the force at a site, f, maps to a box. The upper line represents the boxes, the lower line the forces. The box numbers in brackets are the continuation of those without brackets, which are the outcomes of BOX_POS (see l. 90) for the force ranges indicated. The box numbers are periodic due to modulo operation in BOX_POS. Even when the range of possible forces covers at most a range of size F_SLIP, dividing F_SLIP into 5 ranges will map the whole to 6 boxes.

of sites). As a result, using the box-technique in the OFC Model is about two orders of magnitude faster [9] than using a binary search tree.

The boxes are divided evenly into ranges of forces. At any point, the maximum force throughout the system is f_threshold or less and the minimum is f_threshold-F_SLIP or more. The threshold f_threshold can be a large multiple of F_SLIP, which is discussed in detail in Sec. A.4.4.

If there are NUM_BOXES boxes, see l. 16, then each box contains a range of forces of size F_SLIP/(NUM_BOXES-1), so that there is effectively one more box than actually needed, given that all forces are within a range of F_SLIP after an avalanche has finished. The reason for the extra box is that the ranges are generally misaligned with the boxes, so that the first and the last ranges of size F_SLIP/(NUM_BOXES-1) generally spill over and thus require one more box, as illustrated in Fig. A.1. The macro BOX_POS(f), l. 90, maps every force within the range from f_threshold-F_SLIP to f_threshold to a box, which wrap around periodically due to the modulo operation in the macro BOX_POS(f). According to Hergarten (2002) it is advantageous to change the range of the forces maintained in boxes not after every avalanche but rather only once all forces are outside the range, i.e. no forces are left in the boxes, at which point the overall range is readjusted and all forces are sorted back into boxes (not implemented in the following).

The accuracy and precision needed in the macro BOX_POS(f) depends mostly on the number of boxes and the magnitude of F_SLIP. Because of the extra box, it cannot be plagued by rounding errors, even when the force is implemented as a floating point number. [10] The force f is rescaled to a multiple of F_SLIP by (double)(f)/(double)F_SLIP. Because of the initial offset INIT_OFFSET, and the effective offset of the vanishing force f_threshold-F_SLIP, neither maximum nor minimum force will generally align with any boundary of a box, Fig. A.1. More importantly, (double)(f)/(double)F_SLIP will generally be a very large number. Scaled to the number of boxes available minus one, the number is even larger. For that reason, the result is cast to the largest available integer, here

---

[9] 10^6 avalanches per second are easily achievable on today's (2008) hardware, depending on boundary conditions and level of conservation.

[10] The theoretical worst case scenario is a minimum aligned with a box boundary, and the difference between minimum and maximum force being exactly F_SLIP *in the precision chosen* in the macro. This is impossible with sufficient precision, because a site with force equal to the minimum force plus F_SLIP should always have slipped during the avalanche.

unsigned long long, before the modulo operation finally determines the box number. In l. 179 is a test to determine whether the limit of the largest integer is exceeded. Similarly, the test in l. 186 probes whether the floating point precision suffices for the given settings, checking directly whether two effectively neighbouring boxes (the one for the largest and the one for the smallest forces) blur into the same.

The three functions prototyped in l. 94–96 maintain the forces in the boxes (box_insert and box_delete) and find the maxima (box_getmaxima). They could quite easily be implemented as macros, but for readability and because the compiler is likely to inline them anyway, in the present code they are implemented as regular functions. By default, all functions return 0 to report their success and non-zero otherwise, which is never evaluated and is likely to be removed by the compiler.

Each box[0...NUM_BOXES-1] consists of a pointer to a site, see l. 92. Being global variables, pointers are initialised to 0 which in ANSI C corresponds to NULL.[11] To make the behaviour well defined, also in case the memory for box is allocated (from the heap) via malloc, they are explicitly initialised in l. 203. A NULL pointer for a box corresponds to having no site in the respective box, Fig. A.2.

The function box_insert, defined as

```
97    int box_insert(struct lattice *a)
98    {
99      int b;
100
101      b=BOX_POS(a->f);
102      if ((a->bx_right=box[b])!=NULL) box[b]->bx_left=&(a->bx_right);
103      box[b]=a;
104      a->bx_left=&box[b];
105    return(0);                                              /* Success. */
106    }
```

places a site in a box, by first determining the position in the box array, see l. 101. Whatever address is currently stored in box[b] is written in the .bx_right pointer of the site to be inserted. This is illustrated in Fig. A.3, where site[1] is **inserted** in the construct shown in Fig. A.2. If the pointer box[b] is non-NULL, i.e. the box is not empty, then the object pointed to by a->bx_right is to be informed about that change in the pointing pointer. For when a site is later **removed** from the box, the site needs to know the address of the pointer pointing at it. This address is stored in .bx_left, of the site pointed to (Ceva, 1998). Every site that is maintained in the boxes has a valid .bx_left, containing the address of the pointer pointing at it. The value at the address a->bx_left is the address of a, i.e. whenever the data structures of the boxes are intact, *(a->bx_left)==a is true. In Fig. A.3, this is ensured by bending the pointer .bx_left of site[8] to point to .bx_right of site[1] pointing at site[8].

The arrangement of sites in a list, Fig. A.2 and Fig. A.3, strictly is not a **doubly linked list** (Cormen *et al.*, 1996) as suggested by Ceva (1998), because the pointer to the left points to the pointer pointing at itself, rather than to the whole object of type struct lattice.

---

[11] For portability, readability and cleanliness of the code some prefer expressions of the form if (p!=NULL) over if (p).

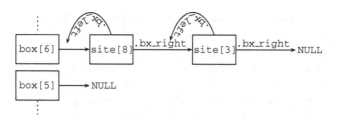

Figure A.2 The arrangement of sites in boxes. box[5] is empty by containing a NULL pointer. box[6] contains two sites, by pointing to site[8], which in turn points at site[3] via its .bx_right pointer. The .bx_left pointers of every site contain the address of the pointer pointing at them.

That would be disadvantageous, not only because box[...] would need to be a **struct lattice** rather than just a pointer, but also because more information than actually needed about the pointing site would be disclosed to the site pointed at, which is detrimental to the maintainability of the code.

In l. 102 the existing chain is connected to site a. The expression box[b]=a in the next line makes box[b] point to the site and the address of that pointer is subsequently stored in a's .bx_left, see the rewired pointer coming out of box[6] and .bx_left coming out of site[1] in Fig. A.3. This finishes the insertion, the computational complexity of which is apparently bounded.

The function box_delete,

```
107    int box_delete(struct lattice *a)
108    {
109        if ((*(a−>bx_left)=a−>bx_right)!=NULL)
110            a−>bx_right−>bx_left=a−>bx_left;
111        return(0);                                          /* Success. */
112    }
```

removes a site from a list, see Fig. A.4. This is done by skipping it, first by assigning to the pointer that points to it, *(a->bx_left), the address of the site it points to, a->bx_right. If a is located at the end of the chain, a->bx_right is NULL. Otherwise the address that stores the pointer to a->bx_right has to be updated, which happens in l. 110, restoring the invariant mentioned above. The pointers a->bx_right and a->bx_left still contain valid data, but are not used henceforth and therefore do not need to be reset.

Finally, box_getmaxima,

```
113    int box_getmaxima(struct lattice **list, int *num)
114    {
115        int b;
116        struct lattice *p;
117
118        for (b=BOX_POS(f_threshold); ; b=(b−1+NUM_BOXES)%NUM_BOXES) {
119            if (box[b]==NULL) continue;
120            list[0]=box[b];
121            *num=1;
122            for (p=box[b]−>bx_right; (p!=NULL); p=p−>bx_right) {
123                if (p−>f>list[0]−>f) list[0]=p, *num=1;        /* New max found. */
124                else if (p−>f==list[0]−>f) list[(*num)++]=p;   /* Another, same...*/
```

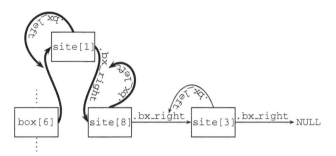

Starting with the arrangement shown in Fig. A.2, site[1] is inserted by rewiring the pointers shown in bold.

```
125                                                                  /* ... max found. */
126         }
127      return(0);
128      }
129                              /* Example for an assertion: Code cannot reach this point. */
130   return(−1);
131   }
```

finds the entire set of largest forces. This degeneracy poses a major problem for the most naïve implementation of the OFC Model based on a **red-black approach** (also p. 107 for the ZHANG Model and p. 129 for the OFC Model, Dowd and Severance, 1998). If the maximum is unique and provided boundary conditions are either open, or, for periodic boundary conditions, the system size is even, then at any given time step, all sites currently toppling, i.e. those on stackCurr, are located on the same sublattice, which together generate a checkerboard-like pattern. Consequently, toppling sites could never charge each other while toppling, so that any charge arriving at a site could be added onto site[...].f instantly. Instead, all charges have to be accumulated first in site[...].charge and added after all sites have discharged, see l. 321.[A.1]

The search for the maxima starts at the box that carried the maximum force in the last search, i.e. the box of f_threshold. Stepwise decreasing the box-index, the maxima are found in the first non-empty box. The first candidate for the maximum is the first entry in the box, see l. 120. Starting from its right neighbour in the box, box[b]->bx_right, the for loop visits all elements in the chain and either identifies a new maximum, p->f> list[0]->f, or finds further, degenerate maxima, p->f==list[0]->f. A test of floating point numbers for equality is often complained about by compilers, but in the present case it makes sense, because the maxima actually *can* be degenerate. It is a genuine feature of the model, due to sites being reset to vanishing force simultaneously producing the checkerboard pattern mentioned above.

A.4 Main

The main function starts with some variable declarations, before printing all parameters thought to be relevant to the simulation, including hostname, time and PID (see below).

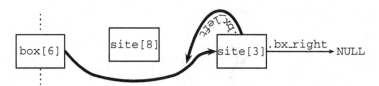

Figure A.4   Removal of a site from a box. All changes relative to Fig. A.2 are shown in bold. `site[8]` is removed from the chain by re-routing the pointer previously pointing at it, identified by its `.bx_left`, to what it used to point to, `.bx_right`, and informing that site about the change as well.

```
132    int main(int argc, char *argv[])
133    {
134      long long int it;                              /* Iteration counter. */
135      int chunk;                                     /* Chunk counter. */
136      int i, j;                                      /* Auxiliary variables. */
137      struct lattice *nb;                            /* Pointer to neighbour. */
138      force_type d, charge;                          /* Derivation of charges. */
139      int s, t;                                /* Avalanche size and duration. */
140      force_type f_threshold_start;          /* Force threshold at start of time. */
141      force_type total_time;                         /* Total time counter. */
142      force_type total_time_readjusted;            /* Total at readjustment. */
143      force_type chunk_start, chunk_start_readjusted;  /* Chunk start time. */
144      sigset_t sigmask;                              /* Signalling. */
145
146
147      { char hostname[256];                   /* HOST_NAME_MAX is 255. */
148        gethostname(hostname, sizeof(hostname)-1);
149        hostname[sizeof(hostname)-1]=0;
150        printf("#Info_Running_on:_%s\n", hostname);
151      }
152      { time_t tm;
153        time(&tm);
154        printf("#Info_Started_at_time_%s", ctime(&tm));
155      }
156      printf("#Info_PID:_%i\n", (int)getpid());
157    #define PRINT_PARAM(a,f) printf("#Info_" #a ":_" f "\n", a)
158      PRINT_PARAM(LENGTH, "%i");
159      PRINT_PARAM(NUM_BOXES, "%i");
160      PRINT_PARAM(ALPHA, "%g");
161      PRINT_PARAM(ITERATIONS_PER_CHUNK, "%lli");
162      PRINT_PARAM(SEED, "%li");
163      PRINT_PARAM(F_SLIP, FORCE_OUT_FMT);
164      PRINT_PARAM(INIT_OFFSET, FORCE_OUT_FMT);
165      PRINT_PARAM(FORCE_OUT_FMT, "%s");
166      PRINT_PARAM(FORCE_TYPE_NAME, "%s");
167      PRINT_PARAM(PRECISION, "%i");
168      printf("#Info_force_type_has_%i_bytes\n", (int)sizeof(force_type));
169
170      setlinebuf(stdout);                       /* Avoid buffering of output. */
171
```

```
172    FPSETPREC(PRECISION);                                    /* Set precision. */
173
174    srand48(SEED);                         /* Initialise random number generator. ONLY for */
175                              /* illustration purposes, the library functions are used. To be REPLACED. */
```

The call of `setlinebuf`, l. 170, changes the buffering behaviour of `stdout`. By default `stdio` buffers several kilobytes of output if that is written to a file. All output of the present code is written to `stdout`, which in an actual run would be redirected to a file using UN*X redirection, such as `./ofc > output.dat`. It is often sensible to redirect `stderr` as well. In the Bourne shell (`sh` and supersets, such as `bash` and the Korn shell `ksh`), this is done via `./ofc > output.dat 2> output.err` or, if tagging is used in the `stderr` output stream as well, to the same file, `./ofc > output.dat 2>&1`.

In line l. 172, the precision of the FPU is set to the value specified in the configuration blocks, l. 22 and l. 38. Finally, the (pseudo) random number generator (RNG) is seeded, which is used (only) during the initialisation. In the present example, for simplicity, the RNG of the C-library is used, which in its earlier incarnation (`rand`) was notoriously unreliable, but has since been improved and standardised in the `rand48` family of RNGs. In actual production code a more solid RNG should be used (Knuth, 1997c; Newman and Barkema, 1999; Press *et al.*, 2007; Matsumoto, 2008; Galassi, Davies, Theiler, *et al.*, 2009, also Gentle, 1998).

In stochastic models with simple local rules, involving mainly integer operations, the RNG often dominates the overall performance. There is a clear trade-off between speed and quality of an RNG and it is worthwhile, not least to test the robustness of code and results, to try out a number of different RNGs.

A.4.1 Signals

The three lines

```
176    signal(SIGHUP, signal_handler);                       /* Register signal handler. */
177    sigemptyset(&sigmask);                                 /* Prepare signal mask. */
178    sigaddset(&sigmask, SIGHUP);
```

set up the **signal** SIGHUP to be used for communication with the running program. Signals are one form of inter-process communication in the UN*X environment (Stevens and Rago, 2005) and the most convenient form of instructing a simulation to terminate gracefully at the next convenient point in time. In the present implementation a 'chunk' (data written out in regular intervals, Sec. A.4.3) is finished and the data written before the program terminates.

In l. 176 `signal(SIGHUP, signal_handler)` tells the operating system that the function `signal_handler` is to be called, whenever the signal SIGHUP is sent to the program, which can be done using for example the command `kill -HUP <pid>`, where `<pid>` is the pid of the program, found via `ps` or `top` or printed at the start of the program, l. 156. SIGHUP is also sent automatically for a variety of (good) reasons; SIGUSR1 and SIGUSR2 are alternatives. The signal handler is executed asynchronously, i.e. the program stops whatever it is doing, jumps to the handler, and after its completion continues where it had previously left off. It might not, however, continue some system calls, [12] which as a matter of caution should therefore be shielded from incoming signals. The two functions `sigemptyset` and `sigaddset` prepare a signal mask by first clearing it and then setting SIGHUP. Any section in the code that produces output is wrapped in calls of `sigprocmask`, see l. 340 and l. 352. The first call blocks the signal, so that it is not processed during output, the second unblocks it again. A signal sent while it is blocked is (nowadays) delivered as soon as it is unblocked.

When the signal arrives the signal handler is called, l. 365, which in the present case simply sets a global flag `global_stop_flg` to 1, see l. 367. If the flag `global_stop_flg` is set, the program terminates directly after the output, l. 354. Using signalling releases the simulator from fixing the number of iterations before the program is started; it is a way of communicating with the process in a controlled way. In a more sophisticated setup, output is produced on a regular basis in real time (say hourly). Terminating the process while it is still collecting statistics and using only the output produced so far, possibly wastes a significant amount of CPU time.

A.4.2 Initialisation

The first line of the initialisation section tests whether INIT_OFFSET/F_SLIP + 1 does not exceed the maximum unsigned long long, which is vital for the macro BOX_POS, defined in l. 90, to work, as discussed in Sec. A.3. A similar test is done in line l. 186.

```
179    if ((long double)(NUM_BOXES−1)*(long double)(INIT_OFFSET+F_SLIP)
180         /(long double)F_SLIP > (long double)ULLONG_MAX) {
181      fprintf(stderr, "Box␣position␣overspills.␣Reduce␣INIT_OFFSET.\n");
182      fprintf(stdout, "#ERROR␣"
183                      "Box␣position␣overspills.␣Reduce␣INIT_OFFSET.\n");
184      exit(0);
185    }
186    if (BOX_POS(INIT_OFFSET+F_SLIP)==BOX_POS(INIT_OFFSET)) {
187      fprintf(stderr, "Box␣position␣ambiguous.␣Reduce␣INIT_OFFSET.\n");
188      fprintf(stdout, "#ERROR␣"
189                      "Box␣position␣ambiguous.␣Reduce␣INIT_OFFSET.\n");
190      exit(0);
191    }
192
193
194    if ((site=malloc(sizeof(*site)*NUM_SITES))==NULL)
195      exit(EXIT_FAILURE);                                    /* Malloc has failed. */
196    if ((stackNext=malloc(sizeof(*stackNext)*NUM_SITES))==NULL)
197      exit(EXIT_FAILURE);                                    /* Malloc has failed. */
198    if ((stackCurr=malloc(sizeof(*stackCurr)*NUM_SITES))==NULL)
199      exit(EXIT_FAILURE);                                    /* Malloc has failed. */
200
201    /* Initialise boxes (although that should have happened automatically,
```

---

[12] Notably, `write` operations to pipes then lead to interrupted system calls which return with error EINTR (Stevens and Rago, 2005).

```
202     * as it is a global variable): */
203     for (i=0; i<NUM_BOXES; i++) box[i]=NULL;
204
205     total_time=0;
206     total_time_readjusted=0;
207     f_threshold=INIT_OFFSET;
208
209     /* Initialise stacks: */
210     stackheightCurr=0;
211     stackheightNext=0;
212
213     /* Initialise lattice: */
214     for (i=0; i<LENGTH; i++) for (j=0; j<LENGTH; j++) {
215       site[COO2INDEX(i,j)].f=INIT_OFFSET+(force_type)((double)F_SLIP*drand48());
216       if (site[COO2INDEX(i,j)].f>f_threshold) f_threshold=site[COO2INDEX(i,j)].f;
217       site[COO2INDEX(i,j)].charge=0;
218
219     #ifdef OBC
220       /* Left neighbour: */
221       if (i>0) site[COO2INDEX(i,j)].nb[0]=&site[COO2INDEX(i−1,j)];
222       else site[COO2INDEX(i,j)].nb[0]=NULL;
223       /* Right neighbour: */
224       if (i<LENGTH−1) site[COO2INDEX(i,j)].nb[1]=&site[COO2INDEX(i+1,j)];
225       else site[COO2INDEX(i,j)].nb[1]=NULL;
226       /* Lower neighbour: */
227       if (j>0) site[COO2INDEX(i,j)].nb[2]=&site[COO2INDEX(i,j−1)];
228       else site[COO2INDEX(i,j)].nb[2]=NULL;
229       /* Upper neighbour: */
230       if (j<LENGTH−1) site[COO2INDEX(i,j)].nb[3]=&site[COO2INDEX(i,j+1)];
231       else site[COO2INDEX(i,j)].nb[3]=NULL;
232     #define BC_NAME "OBC"
233     #elif defined(PBC)
234       /* Left neighbour: */
235       if (i>0) site[COO2INDEX(i,j)].nb[0]=&site[COO2INDEX(i−1,j)];
236       else site[COO2INDEX(i,j)].nb[0]=&site[COO2INDEX(LENGTH−1,j)];
237       /* Right neighbour: */
238       if (i<LENGTH−1) site[COO2INDEX(i,j)].nb[1]=&site[COO2INDEX(i+1,j)];
239       else site[COO2INDEX(i,j)].nb[1]=&site[COO2INDEX(0,j)];;
240       /* Lower neighbour: */
241       if (j>0) site[COO2INDEX(i,j)].nb[2]=&site[COO2INDEX(i,j−1)];
242       else site[COO2INDEX(i,j)].nb[2]=&site[COO2INDEX(i,LENGTH−1)];
243       /* Upper neighbour: */
244       if (j<LENGTH−1) site[COO2INDEX(i,j)].nb[3]=&site[COO2INDEX(i,j+1)];
245       else site[COO2INDEX(i,j)].nb[3]=&site[COO2INDEX(i,0)];
246     #define BC_NAME "PBC"
247     #else
248     #error "No_boundary_condition_defined"
249     #endif
250
251       /* bx_right and bx_left are initialised automatically with box_insert: */
252       box_insert(&site[COO2INDEX(i,j)]);
```

```
253    }
254    printf("#Info␣Boundary␣condition:␣%s\n", BC␣NAME);
255    f␣threshold␣start=f␣threshold;
```

The **memory** for the lattice and the two stacks is both taken from the heap via malloc. The estimates for the maximal memory requirements for the two stacks are very conservative, only in a very bizarre setup can all sites topple at the same time. If memory is scarce, the stacks can be chosen to be smaller, but exceeding them needs to be monitored (again trading CPU time for memory requirements).

The malloc calls above (l. 194, l. 196, l. 198) are bracketed in an if statement to terminate the program immediately if the return values are not valid. It is of great importance to catch any error thrown by the operating system when invoking (directly or indirectly) a system call, most notoriously malloc and fopen. Endless hours of debugging and enumerable hours of CPU time can be wasted when the program behaves in some erratic, undefined way, because assumptions about the value of some variables, the memory available or the files open do not match reality. A detailed analysis of the problem usually starts with evaluating errno or strerror(errno).

All boxes are initialised with NULL, l. 203, indicating that they are empty. The global time is determined from total_time and total_time_readjusted, which are set to 0, l. 205 and l. 206, see Sec. A.4.4. The force threshold f_threshold is first set to its minimal value, INIT_OFFSET, l. 207, and set to the maximum throughout the system at initialisation of the individual sites, l. 206. The rôle of f_threshold and INIT_OFFSET is discussed in detail in Sec. A.4.4. The two stacks stackCurr and stackNext are initialised by setting both their heights to 0.

In the loop starting in l. 214, randomly chosen forces are assigned in l. 215 before f_threshold is updated. This is the only place in the code where random numbers are used. After the field for the cumulative charge is initialised, the neighbourhood of each site is set up, see the discussion about the neighbourhood, p. 361. The macros OBC and PBC decide how boundary sites are connected. Each macro also triggers the definition of the macro BC_NAME, the value of which is written to stdout after the initialisation loop in l. 254. Finally, all sites are placed into boxes by box_insert(&site[COO2INDEX(i,j)]).

A.4.3 Generating chunks

The main loop iterating the OFC Model is divided in two. The outer loop runs over **chunks**,

```
256    for (chunk=1; ((chunk<=NUM␣CHUNKS) || (NUM␣CHUNKS<0)); chunk++) {
257        MOMENTS␣INIT(siz);
258        MOMENTS␣INIT(dur);
259        MOMENTS␣INIT(deg);
260        chunk␣start=total␣time;
261        chunk␣start␣readjusted=total␣time␣readjusted;
```

the inner loop runs over iterations within each chunk. In the present case, iterations refer to avalanches, which might translate into a varying influx of force per site. At the beginning of each chunk, all cumulative observables are initialised using MOMENTS_INIT, l. 257–259, and the starting time of a chunk is recorded, l. 260 (see also Sec. A.4.4), to be used in the output. At the end of each chunk, i.e. after ITERATIONS_PER_CHUNK avalanches, all observables are written to stdout, see l. 340–352, before being reinitialised for the next chunk. The total number of chunks generated by the program is NUM_CHUNKS, defined in l. 18. A negative value causes an infinite loop, see l. 256.

The motivation for producing chunks is the generation of a continuous stream of results instead of a single, final outcome at the end. Correlated or not, multiple outputs allow a very simple estimation of the **statistical error**, Sec. 7.2, and easy control of the **transient** by probing for **equilibration** (convergence) in the output and omitting a suitable number of chunks. Moreover, the program can be terminated at any convenient point using the signalling facility. Because all variables are reinitialised after the output, overruns and problems with precision are less likely to occur or are less serious, see the discussion in Sec. A.2. Also, **checkpointing** [13] can be implemented just behind the data output and amounts to saving the configuration of the lattice and a few internal variables such as total_time. [14] All observables are reinitialised anyway and therefore need not be restored when the checkpoint file is used.

Finally, on the basis of a regular output in the form of chunks, bugs in the code can be identified much earlier and easier, if their effect becomes immediately apparent in the observable (negative avalanche sizes due to integer overflow etc.), which would otherwise drown in too many other, possibly valid, measurements.

```
262    for (it=1; it<=ITERATIONS_PER_CHUNK; it++) {
263        /* Find sites with maximum force: */
264        box_getmaxima(stackCurr, &stackheightCurr);
265        MOMENTS(deg, stackheightCurr);
266        for (i=0; i<stackheightCurr; i++)
267            /* Delete from boxes (sites in stack are not in boxes): */
268            box_delete(stackCurr[i]);
269        /* Adjust force threshold of OFC model; every site with that force
270         * or greater should topple. */
271        f_threshold=stackCurr[0]->f;
272        /* Total macroscopic time without readjustment: */
273        total_time=(f_threshold_start−f_threshold);
```

The actual iteration over avalanches starts with a search for the possibly **degenerate maxima**, l. 264. The moments of the degeneracy are recorded in the following line, l. 265. Because box_getmaxima places all maximal sites on the stack, they are removed from the boxes in the following loop. According to the scheme discussed in the following section, the force threshold f_threshold is set to the maximum across the lattice and the total force influx per site, total_time, is derived as the difference between the current maximum and the previous maximum.

---

[13] 'Checkpointing' means to store the state of the simulation in a file, from where it can be restarted if the program terminates too early, for example in the case of hardware failure or if more data after the transient are needed.

[14] Such internal variables can be omitted if the simulation is to be continued without *exactly* restoring the internal state of the program (which would manifest itself by slight deviations of the observables, caused by rounding etc.).

A.4.4 Receding force threshold

This section discusses one of the most important performance enhancing techniques used in the OFC Model, which was first suggested by Grassberger (1994). While it is straightforward to implement, it has to be analysed with respect to its impact on precision, the relevance of which has been highlighted by Drossel (2002).

The options in l. 22 and l. 38 choosing the type of dynamical variable (the force F_n at a site n, see Box 5.3) are rather unusual. The reason why it was implemented in this form was to enable various performance and precision tests. The perfomance can increase dramatically when floating point operations are avoided, see Sec. A.4.5. In the present case, the rôle of precision is much more important.

Grassberger (1994) proposed to implement the global, uniform driving in the OFC Model by **lowering the force threshold** F_{th} (variable f_threshold in the code) rather than **sweeping** through the lattice and adding a certain amount of force to every site. Sweeping through a lattice is enormously expensive in terms of CPU time, as every site has to be fetched from memory and processed. It should generally be avoided, unless the computational effort of fetching every site can be justified by the amount of computation done on every site. Yet, lowering F_{th} has a significant impact on the precision of the computation, the cause and the consequences of which are discussed in the following.

In the code above, the macro INIT_OFFSET in l. 28 and l. 42 respectively defines the initial offset value of vanishing force, i.e. the force F_n at each site n is initialised in l. 215 with INIT_OFFSET plus an amount taken randomly from the interval $[0, F_{th}[$. In its original definition, the OFC Model is driven by identifying the (possibly degenerate) maximum force F_m and increaseing all forces by $F_{th} - F_m$. Grassberger (1994) suggested decreasing F_{th} instead, and (effectively) reducing the force offset (vanishing force) by the same amount. Using this technique, F_{th} might become very negative, so one might as well start with very large F_n throughout and see them gradually reduced towards 0. If the forces are stored in an unsigned type a positive offset is necessary, if they are stored in a floating point type, then it is a matter of consistency to have a positive offset.

When site n topples, it transfers a force αF_n to its nearest neighbours. In the code, the actual force F_n is determined as the difference site[...].f-(f_threshold-F_SLIP) between the actual value site[...].f and the current, effective offset f_threshold-F_SLIP, see l. 300. Here, F_SLIP is the width of the range of stable forces, i.e. the difference between f_threshold, where sites start slipping, and the offset of vanishing force. The value of F_SLIP corresponds to F_{th} in the original definition with fixed F_{th}. If the type chosen for force_type, see l. 39, is a floating point type, then F_SLIP is sensibly chosen to be unity.

Depending on f_threshold, the difference f_threshold-F_SLIP in l. 300 is calculated from very big numbers, either at early stages if INIT_OFFSET was chosen big and positive, or at later stages during the run, when f_threshold becomes very negative, if INIT_OFFSET is small. If the forces F_n are implemented as floating point numbers that implies a loss of **precision**, [15] when determining the difference in l. 300 as well as when

---

[15] The larger a floating point, the larger the smallest number that changes its value when added (Press *et al.*, 2007, p. 10).

adding the charge by neighbours, l. 308. The effective precision is the size of the mantissa minus the number of bits that all sites have in common, which increases with increasing magnitude of f_threshold. In particular when the level of conservation is low, it is not unusual to see values of f_threshold around 10^8, which corresponds to 27 bits and would reduce the precision available in a double to a mere 26 bits, just over single precision.

Regardless of the value of INIT_OFFSET, using **floating point numbers** to represent the forces will make the (effective) precision *change* as the value of f_threshold changes. Given the subtle tuning and synchronisation in the OFC Model, Sec. 5.3.1.2, p. 133, this effect is highly undesirable.

One way of controlling this change of precision it is to introduce a **readjustment mechanism**, which is triggered by the macro READJUST_CRITERION, written in tune with the type chosen for force_type, l. 30 or l. 43. It probes whether f_threshold has dropped below some minimal value and thus triggers fewer readjustments if the initial value INIT_OFFSET of f_threshold is large or its minimal allowed value, here F_SLIP, is small.

```
274      if (READJUST_CRITERION(f_threshold)) {
275          /* Readjust all forces, as they have fallen below the threshold: */
276          printf("#Info_Readjusting.\n");
277          /* Charge to all sites that pushes minimum to INIT_OFFSET: */
278          charge=INIT_OFFSET+F_SLIP-f_threshold;
279          /* Reset force threshold, using the same arithmetic operation
280           * as for all forces: */
281          f_threshold+=charge;
282          f_threshold_start=f_threshold;                      /* Reset time keeping. */
283          total_time_readjusted+=total_time;
284          total_time=0;
285          for (i=0; i<NUM_BOXES; i++) box[i]=NULL;            /* Flush boxes. */
286          for (i=0; i<NUM_SITES; i++) {
287              site[i].f+=charge;
288              site[i].bx_right=NULL;
289              site[i].bx_left=NULL;
290              box_insert(&site[i]);                           /* Insert all sites into boxes. */
291          }
292          for (i=0; i<stackheightCurr; i++)
293              box_delete(stackCurr[i]);                       /* Delete toppling sites from boxes. */
294      }
```

Starting in l. 278, all forces are ramped up to values such that the current offset (value of vanishing force), f_threshold-F_SLIP is INIT_OFFSET again, l. 28 or l. 42, the value at initialisation. For consistency, all forces have to be changed in the same fashion, including f_threshold, which could in principle be set to INIT_OFFSET+F_SLIP, a value that might differ slightly from f_threshold+(INIT_OFFSET+F_SLIP-f_threshold). After the readjustment, the largest force is equal to f_threshold. Because total_time is affected by precision problems just as much as any of the forces, it needs to be set back to 0, l. 284, and the accumulated force influx is charged to total_time_readjusted, l. 283, to be added as an offset whenever the correct total time is needed. All times, including chunk_start and chunk_start_readjusted, e.g. l. 260 and l. 261 respectively, are generally affected by limited precision and therefore are unreliable.

The place of the sites within the boxes has to be readjusted according to their new force, so they are first flushed in l. 285 and refilled in l. 290. Only the sites that have been identified as current maxima have to be taken out again in l. 293.

When using floating point numbers, to maintain the same precision throughout the run, the readjustment procedure would need to be called after every update, which corresponds to waiving the huge benefit of a receding F_{th} and replacing it by a uniform increase in force across the lattice. To make maximum use of that technique, readjustment is to be triggered as rarely as possible, i.e. INIT_OFFSET needs to be as large as possible, which, however, clashes with the need for high precision (see above).

In order to keep the precision constant throughout the run, alternatively one can use **integers** to represent the forces, switched on in the section starting in l. 22, which has the additional benefit that the arithmetic is in some cases faster then for floating point numbers. The minimal 'resolution' of an integer is 1 and fractions $0 < \alpha < 1$ of this unit effectively vanish. Compared to floating point numbers, the forces therefore have to be rescaled, i.e. the value of F_SLIP has to be increased drastically. The accuracy with which the force received by neighbouring sites is determined is given by the number of bits needed to represent the toppling force, which is of the order of F_SLIP. While large F_SLIP is therefore desirable, readjustment, as discussed above, is triggered earlier with increasing F_SLIP. The macroscopic time until readjustment is given by INIT_OFFSET/F_SLIP, so that INIT_OFFSET is to be chosen as large as possible, but not so large that forces can overflow and wrap around at the early stages during the run. The minimal largest force is F_SLIP and the maximum is a small multiple of that, so if the largest possible integer value INIT_OFFSET can take is ULLONG_MAX, then INIT_OFFSET should be ULLONG_MAX -n*INIT_OFFSET with $n > 1$. The choice of $n = 64$ in l. 28 is certainly far beyond any observed maximum value. Using an unsigned type for the forces, see l. 23, means that the minimum force in the system should not fall below 0, which is guaranteed by keeping the maximum at or above F_SLIP.[16]

It is worthwhile summarising the rôles of the relevant macros defined in l. 26–l. 30 and l. 41–43 respectively:

F_SLIP, l. 26, l. 41 Force level at which slipping occurs, the force difference between vanishing force and threshold force. It is the order of magnitude of the force redistributed at toppling. If forces are integers, the effective precision is 1 in F_SLIP, in the present case 40 bits, l. 26.

INIT_OFFSET, l. 28, l. 42 Minimal initial value of all forces and therefore of force_threshold. If forces are floating point numbers, the magnitude of f_threshold determines the effective precision. It is the difference between the FPU's setting (here 53 bits, l. 45) and the number of bits necessary to represent f_threshold/F_SLIP, which in the present case only very briefly exceeds 24 bits (initially f_threshold is equal to INIT_OFFSET, l. 42, and is increased by less than F_SLIP in l. 216), so that the effective precision is most of the time at least 29 bits.

---

[16] Allowing for negative forces, one could let the force threshold run from (about) INIT_OFFSET to (about) -INIT_OFFSET. This doubling in range corresponds to the extra bit available when using unsigned types.

READJUST_CRITERION, l. 30, l. 43 Macro to trigger readjustment of forces. The total external drive per site until all forces are to be adjusted is INIT_OFFSET/F_SLIP, about 2^{24} in the present case. To avoid frequent (costly) readjustments, INIT_OFFSET should be as big as possible. In the case of integers, it needs to be small enough so that avalanches at forces around this value do not produce local forces exceeding the largest allowed integer. In the case of floating point numbers, the value of INIT_OFFSET is bounded from above by the desired precision. In turn, F_SLIP needs to be as small as possible, which is, however, bounded from below by precision requirements (see above).

A.4.5 Floating point operations and their precision

A few general comments about floating point numerics are in order (for further details see Goldberg, 1991). On most platforms, basic arithmetic on floating point numbers is (almost) as fast as on integers. Multiple pipelines and new instruction sets, such as SSE and MMX, might render the occasional use of floating point numbers beneficial to the overall performance. However, library functions, such as pow or sqrt, are usually very slow and should be avoided wherever possible.

Some operations are superfluous and can be avoided altogether. For example, most random number generators return naturally random integers. Their floating point relatives are usually simple wrappers which rescale a random integer in its natural range to a random floating point number in the desired interval. Where random numbers are used heavily, say in the form of int rand(), in order to perform a certain task with a fixed probability, say double probab, it pays significantly to use

int probab_i=probab*(1.+RAND_MAX);

...

if (rand()<probab_i) {...}

instead of if (rand()/(1.+RAND_MAX)<probab)..., where RAND_MAX is the maximum random number returned by rand().[17] Special care must be taken if probab is 1.0, in which case probab*(1.+RAND_MAX) might exceed the range of probab_i. Replacing the factor by RAND_MAX and the if condition by rand()<=probab_i is similarly problematic if probab vanishes. If probab is 1/2, rand() should be replaced by a macro which returns random bits using a bitmask that slides over a random, unsigned integer, for example

#define RNG_MT_BITS (32)
unsigned long mt_bool_mask=1UL<<(RNG_MT_BITS−1);
unsigned long mt_bool_rand;
#define RNG_MT_BOOLEAN ((mt_bool_mask==(1UL<<(RNG_MT_BITS−1))) ? ((
↳   mt_bool_rand=genrand_int32()) & (mt_bool_mask=1UL)) : (mt_bool_rand & (
↳   mt_bool_mask+=mt_bool_mask)))

Here RNG_MT_BOOLEAN returns either 0 or a power of 2, by masking all but a single bit from mt_bool_rand using mt_bool_mask. The number of random bits stored in

---

[17] Again, rand() of libc is used for illustration purposes only.

`mt_bool_rand` after every call of `unsigned long genrand_int32()` (Matsumoto and Nishimura, 1998; Matsumoto, 2008) is `RNG_MT_BITS`. The macro first tests whether the mask has reached the highest bit which is also the value it is initialised with. If so, a new random number is drawn, stored in `mt_bool_rand` and the first bit read out using the reset mask. Otherwise, the mask is updated to extract the next random bit. The bit shift of the mask is based on an addition, `mt_bool_mask+=mt_bool_mask` rather than a bit shift `mt_bool_mask<<=1`, because the latter requires an extra CPU cycle to write the constant 1 into one of the registers.

Given that floating point numbers on a computer have only a finite mantissa, the internal representation of a real number suffers from a lack of precision. That applies all the more to (natural) decimal floating point numbers, such as 0.1 which cannot be faithfully represented in a binary floating point type, like, say, $0.25 = 2^{-2}$. Improved precision is often costly in terms of CPU time. In this light, the only advantage of a floating point number over integers is readability of the code. Precision and accuracy are much more controlled with integers, as the smallest number that can be added to or subtracted from an integer (so that it actually changes its value) is 1. As a result, the new value differs from the old one by exactly 1. The price of this feature is the limited range of an integer – within a limited range a floating point number has the same properties, but the minimal unit addable, such as `DBL_EPSILON` if the initial value is unity, is neither obvious, nor natural, nor easily controlled across the enormous range spanned by floating point numbers.

One of the more subtle problems with floating point numbers is the precision setting of the FPU. The IEEE 754-1985 standard (Dowd and Severance, 1998; Press *et al.*, 2007) and more recently the IEEE 754-2008 standard define floating point numbers with 24 (`float`, 4 byte), 53 (`double`, 8 byte) and 64 bit mantissas (`long double`, 10 to 16 byte, sometimes larger mantissas, normally no hidden bit) and exponents of similarly increasing size. The largest type, `long double`, was added to the C programming language long after its original publication and the first implementation of its arithmetics, even its IEEE 754 conformity depends heavily on the compiler, the libraries and the platform.[18] Contrary to common belief, the type of a variable has little to do with the precision with which they are processed, most notably on most i386 platforms.[19] Most operating systems, such as many Linux distributions, default to extended precision, whereas others, for example FreeBSD, default to double precision, which is the native floating point type of the mathematical library. While this has little or no impact on the performance, the wrong precision setting of the FPU can have unforeseeable and erratic consequences, as described below. It is therefore highly recommended to set the floating point precision explicitly to the value corresponding to the floating point type used in the code (if they cannot be avoided), using interfaces such as `fpu_control`[20] (see l. 8 and the wrapper macro FPSETPREC(a) in l. 10, used in l. 172).

---

[18] The type `long double` was introduced with ANSI C and further standardised in the C99 revision of the C language definition (Schildt, 2000) by adding `long double` versions to the usual mathematical functions (Wikipedia, 2010b). Some non-compliant compilers treat `long double` as `double`.

[19] The PowerPC architecture generally seems to be less prone to this type of problem.

[20] Some C compilers provide a facility to set the precision at compile time. As of C99 (Schildt, 2000) the floating point environment is to some extent accessible at runtime via `fenv(3)`. FreeBSD provides a particularly wide and convenient interface via `ieeefp.h` with library functions such as `fpsetprec(3)`.

Testing for equality of floating point numbers is generally regarded as bad style, because of the finite precision of floating point numbers. [21] Unexpectedly, the problem extends to relations and is exacerbated by finite precision. For example, the code

```
/* Illustration of precision/accuracy issues. Not part of OFC code. */
double a, b, c1, c2;

/* gcc 3.4.6 20060404 (Red Hat 3.4.6−11) on Linux 2.6.9−89.0.18.ELsmp */
a=1.;
b=3.;
c1=a/b;
if (c1<(c2=a/b)) printf("less\n");
else printf("not_less\n");
if (c1<c2) printf("less\n");
else printf("not_less\n");
```

prints first 'less' and then 'not less', depending on the platform, the compiler and its settings. [22] While it is difficult to provoke this erratic behaviour in a small, isolated piece of code, it is not rare in more complex codes. For example, when searching for the smallest value among a set of floating point numbers, a particular value that is identified as smaller than another at one point in the code might be identified as not smaller at a later stage. As a result, invariants of ordered trees holding data with floating point numbers as keys, such as binary search trees, can unexpectedly break down, which leads to wrong results and unexpected terminations.

The cause of this behaviour is a discrepancy between the precision of a variable type and the precision (or accuracy) of the arithmetics. As long as a result of a calculation, a/b in the example above, resides in a register of the FPU, its (effective) accuracy is determined by the FPU's precision and not by the type of the variable it is assigned to. As soon as the result is written to memory, however, the precision is determined by the type. In the example above (c1<(c2=a/b)), therefore is (incorrectly) true, even when (c1<c2) is (correctly) false two lines further down. In the first case, c1 was fetched from memory, while c2 was in a register, in the second case both variables were read from memory and were therefore equal. To avoid such behaviour, one has to set the precision of the FPU, as well as any other floating point unit (such as SSE or MMX) if that is used as well, to the value corresponding to the floating point type used in the code. This is what happens in l. 172 of the OFC Model code. A more radical measure is to use a compiler flag like -ffloat-store in gcc, which might, however, significantly impact performance.

While the problem might be relevant only to some operations and variables, the setting of the precision affects all operations. Choosing the precision of the FPU to be lower than that of the largest floating point type makes little sense, unless the latter has been chosen for the size of its exponent. Generally, only one type of floating point number should be used. In the current example, the choice in l. 39 and therefore in l. 45 undermines the choice for the moments, long double, in l. 68. For consistency, MOMENT_TYPE should be chosen in the blocks starting at l. 22 and l. 38.

---

[21] In most applications, floating point numbers are appropriately treated as identical if they differ by less than a certain, small amount.

[22] In fact, optimisation might need to be curtailed to prevent the compiler from optimising the if statements away.

A.4.6 Charging neighbours

After an occasional force readjustment, the following piece of code implements the **toppling** part of an avalanche:

```
295      s=t=0;                                              /* Avalanche size and durations. */
296      do {
297        t++;                                              /* Time ticks away. */
298        for (i=0; i<stackheightCurr; i++) {
299          /* Force at toppling site above current effective 0: */
300          d=stackCurr[i]->f - (f_threshold - F_SLIP);
301          stackCurr[i]->f=f_threshold - F_SLIP;           /* Reset to current 0. */
302          /* Insert toppled site back into boxes, as site was taken out
303           * of box when placed on stack: */
304          box_insert(stackCurr[i]);
305          charge=CHARGE(d);                               /* To be added later. */
306          for (j=0; j<NUM_NEIGHBOURS; j++)                /* Visit all neighbours. */
307            if (stackCurr[i]->nb[j]!=NULL)                /* Neighbour exists. */
308              stackCurr[i]->nb[j]->charge+=charge;
309        }                                                 /* End of loop over stackCurr. */
```

After initialising observables, l. 295, a do-loop runs until the stack is empty, incrementing the time first, l. 297, before looping through the list of toppling sites stored in stackCurr, l. 298. The total amount of force subject to redistribution is calculated in l. 300, so that the force at the current site can be reset and the toppling site moved back to one of the boxes. The charge of neighbouring sites is derived in l. 305 in accordance with the choice of the force type, using the macro CHARGE(). In the following loop, l. 306–308, the neighbours are charged, not by adding the charge directly to their current force, but by adding it to the charge field, so that charges do not arrive at sites that are about to topple, which would contravene the definition of the OFC Model. [23]

After the for-loop, l. 298–309, none of the forces exceeds the threshold any longer and all sites are placed back into boxes by box_insert after the force is reset in l. 301 (so they all arrive in the box of lowest force). The forces received from toppling neighbours still need to be added:

```
310            /* All sites in boxes at this stage. Now charge. */
311        for (i=0; i<stackheightCurr; i++) {
312          /* Visit all neighbours: */
313          for (j=0; j<NUM_NEIGHBOURS; j++)
314            if (stackCurr[i]->nb[j]!=NULL) {              /* Neighbour exists. */
315              nb=stackCurr[i]->nb[j];                     /* To ease notation. */
316              /* Ignore sites visited already: */
317              if (nb->charge==0) continue;
318              box_delete(nb);                             /* Delete from box before update. */
319              /* On the stack goes >= not just >, because f=f_threshold has
```

---

[23] The order of updates is fixed by imposing that sites exceeding the force threshold in one time step topple in the next. The force redistributed at toppling does not include the force received in the time step the site topples. The latter point necessitates the present charging mechanism and is an issue when maxima are degenerate or the lattice does not decompose into even and odd sublattices, see Sec. A.3 p. 369.

```
320                      * initiated the avalanche: */
321                      if ((nb->f+=nb->charge)>=f_threshold) PUSH(nb);
322                      else box_insert(nb);
323                      nb->charge=0;
324                  }
325              }
326          s+=stackheightCurr;                              /* Avalanche size. */
```

If more than one neighbour of a site has toppled, the site is visited multiple times. This could be avoided by an additional stack but only at the expense of additional overhead, which might outweigh any potential perfomance gain. After charging, l. 321, charge is reset in l. 323 which serves an indicator to detect multiple visits, l. 317, and prevents superfluous operations in l. 318 and l. 322.

Before the update of the force, a site is removed from the boxes, see l. 318, and either placed on the stack stackNext using PUSH, l. 321, or put back into the appropriate box after the update of the force, l. 322. Every site whose force exceeds the threshold enters the stack exactly once, so that the height of the current stack, stackheightCurr, is the number of currently topplings sites, which makes up the avalanche size, l. 326.

As mentioned before, two stacks are maintained: stackCurr contains the sites currently toppling, stackNext contains those toppling in the next time step. Before that can happen, the stacks are swapped, which concludes the do-loop starting in l. 296.

```
327                      /* Swap stacks. */
328          { struct lattice **stackTemp;
329              stackheightCurr=stackheightNext;
330              stackheightNext=0;
331              stackTemp=stackCurr;
332              stackCurr=stackNext;
333              stackNext=stackTemp;
334          }
335      } while (stackheightCurr);
```

The loop terminates if there are no more sites to topple, stackheightCurr==0, i.e. the avalanche ceases. At this point, statistics is collected about the avalanche duration t and size s, using the MOMENTS macro discussed above:

```
336                      /* Avalanche has finished, collect statistics. */
337      MOMENTS(siz, s);
338      MOMENTS(dur, t);
339  }                        /* End of iterations over s aingle chunk; output follows. */
```

A.5 Output and termination

Finally, after each loop over ITERATIONS_PER_CHUNK avalanches (starting in l. 262), data are written out using the macro MOMENTS_OUT, preceding each such chunk by a short preamble, which contains its index and the total force influx (including the force influx in units of F_SLIP). All regular **output** (in the form of chunks) is bracketed in calls for blocking and unblocking signals, as discussed in Sec. A.4.1.

```
340     sigprocmask(SIG_BLOCK, &sigmask, NULL);
341     { time_t tm;
342       time(&tm);
343       printf("#Info_Chunk_%i_at_total_time_" FORCE_OUT_FMT
344             "_(%Lg_F_SLIP)_finished_at_%s",
345         chunk, total_time+total_time_readjusted,
346         ((long double)(total_time+total_time_readjusted))/((long double)F_SLIP),
347         ctime(&tm));
348     }
349     MOMENTS_OUT(siz);
350     MOMENTS_OUT(dur);
351     MOMENTS_OUT(deg);
352     sigprocmask(SIG_UNBLOCK, &sigmask, NULL);
353                                                        /* End of output. */
354     if (global_stop_flg) {
355       printf("#Info_Termination_because_of_global_stop_flg.\n");
356       exit(0);
357     }
```

The last piece of code contains the end of the loop over chunks and the definition of the signal handler discussed earlier.

```
358   }                                                  /* End of chunk loop. */
359   printf("#Info_Termination_at_the_end_of_the_loop.\n");
360   return(0);
361   }
362
363
364   /* For documentation reasons this is at the end. */
365   void signal_handler(int signo)
366   {
367     global_stop_flg=1;
368   }
```

This concludes the code for the OFC Model and its discussion.

A.6 Other techniques

There are a number of other techniques worth mentioning. The first is the method of choice for measuring the **force distribution**. Unfortunately, it cannot be applied to the OFC Model. This is because the algorithm as discussed below is geared twoards *local* rather than *global* changes, as they occur in the OFC Model due to the uniform driving (regardless of how the latter is implemented). Furthermore, the technique does not cope well with *continuous changes* either. Both these aspects generally pose a problem for any averaging technique. Yet, Grassberger's (1994) boxes effectively maintain a histogram of forces and averaging over it provides a quick route to a reliable estimate of the force distribution.

The second technique discussed below is used for producing **histograms** of quantities such as avalanche size and duration, only a few moments of which have been determined in the code above. 'Histogram' is used here synonymously for a (weighted) probability density function (PDF), and a histogram therefore is an estimator for the probability that a certain observable has a value within a certain range. Estimating a **probability** amounts to averaging over an indicator function, being unity whenever the observable is in the correct range. In that sense, producing a histogram in a numerical simulation is technically closely related to producing moments. However, plenty of CPU time can be saved by determining the range of an observable efficiently.

In Sec. A.6.3 some implementation details of **concurrent Poisson processes** are discussed, which are frequently used in random sequential updates of Abelian models. The last procedure to be presented here is a particularly elegant way of presenting and analysing strongly time dependent observables. This approach ties in with some of the discussion in Ch. 7.

A.6.1 Force distributions by integration by parts

Some observables, such as the force histogram across the OFC Model, require a significant amount of CPU time to be collected, but every **sweep** contains very little new information compared to a sweep made in the previous time step. Given that **correlations** are so strong in the observable, one might take new measurements with a lower frequency, effectively skipping intermediate measurements altogether. The frequency needs to be tuned carefully. Every sweep is equally costly but the gain in new information depends on the time that has passed since the last measurement. Generally, *every* measurement improves the estimate, but potentially only very little.

It often pays to include all measurements and to avoid the sweeps, (for example Pruessner and Jensen, 2004, also for details on correlations). This can be done very efficiently by keeping track only of *changes of the observable*. The technique is based on the observation

$$\overline{f} = \frac{1}{T} \int_0^T dt\, f(t) = f(T) - \frac{1}{T} \int_0^T dt\, t f'(t) \tag{A.1}$$

for taking the average over the continuous time observable $f(t)$. In a numerical simulation, even in continuous time, the observable $f(t)$ usually changes by finite jumps. Figure A.5 shows an example for such a temporal signal, the average of which can be written as

$$\overline{f} = \frac{1}{T} \sum_{i=1}^{N} (t_i - t_{i-1}) f_i = f_N - \frac{1}{T} \sum_{i=1}^{N-1} t_i \Delta f_i \tag{A.2}$$

where $t_0 = 0$ is the time at start and $\Delta f_i = f_{i+1} - f_i$ as illustrated in Fig. A.5.

The method applies where changes occur infrequently and repeated polling is expensive. For example, in the BTW Model, to estimate the ensemble averaged slope $\overline{z_n}$ for each site **n**, one could sweep over the entire lattice after every avalanche, corresponding to the left hand side of Eq. (A.2), and normalise at the end:

/* *Latest avalanche has terminated.* */
for all sites s **do** {

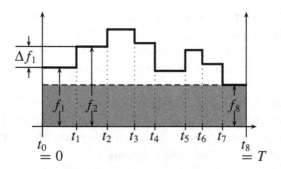

Figure A.5 The average of an observable $f(t)$ (*thick line*) that changes in discrete jumps at times t_i can be calculated in various ways, see Eq. (A.2). One is based on the sum over all columns $(t_i - t_{i-1})f_i$ with $i = 1, 2, \ldots, N$ and $N = 8$. Alternatively, the grey area $f_N T$ is added to the area enclosed by the signal relative to the last measurement (dashed line), which is $-\sum_{i=1}^{N-1} t_i \Delta f_i / T$ where $\Delta f_i = f_{i+1} - f_i$.

```
        average[s] += slope[s];
    }
    T++;
    ...
    /* After T avalanches. */
    for all sites s do {
        average[s] /= T;
    }
```

Alternatively, the change of slope is recorded whenever the site is charged (by c units below). After the final avalanche, the local slope is added according to the right hand side of Eq. (A.2) (the grey background in Fig. A.5):

```
    /* Site s is charged c units during the nth avalanche. */
    average[s] -= (c * n);
    slope[s] += c;
    ...
    /* Site s discharges c units during the nth avalanche. */
    average[s] += (c * n);
    slope[s] -= c;
    ...
    /* After T avalanches. */
    for all sites s do {
        average[s] = slope[s] + average[s]/T;
    }
```

Whether or not this algorithm is more efficient than the brute force (sweeping) method above or any type of occasional sampling, is a matter of the observable,[24] correlations and computational overhead, to name but a few.

---

[24] For example, in the present case, if sites often topple more than once during the same avalanche, the computational cost caused has no statistical benefit.

A.6.2 Histograms

Traditionally, in SOC, the most important observable is the avalanche size distribution, i.e. the **histogram** of avalanche sizes that converges to the exact **probability density function** (PDF) with increasing sample size. This will serve as an example for any histogram in the following. During a numerical simulation, histograms should be coarse grained and rounded as little as possible and systematic binning (Sec. 7.4.1) should be confined to the stage of data analysis (post-processing).

By the very nature of SOC models, the avalanche sizes cover many orders of magnitude, with an overwhelming majority of events being comparatively small. Two conflicting demands arise from this. On the one hand, **variable types** must be large enough to resolve the number of events even if there are very many of them; on the other hand the resolution of the histogram should be as fine as possible, requiring, at given memory size, small variable types for the storage of events.

Assuming discrete avalanche sizes s, the simplest algorithm for histogram estimation is of the form `histogram[s]++` (referred to as the *naïve* algorithm below), which, however, is often not an option because the range of s may be significantly larger than could be covered by the memory available. The avalanche size s therefore must be mapped to a coarser grid using what is effectively a **hash function** $h(s)$ that assigns to every size s an index $h(s)$ of a slot in the histogram. [25] The histogram is updated through this hash function, `histogram[h(s)]++`. The total count for each slot is written out at the end of the simulation, perhaps normalised by its size or, if that is done during post-processing, at least together with its size. In the following different such hash functions are discussed. Two types can be distinguished: those that need constant CPU time and are essentially based on a single mathematical operation, and those that require a more complicated procedure, in particular a loop over ranges that terminates when the correct range is found. In practice, a combination of both is almost always needed (multiple ranges with individual hashes), not least to accommodate small events which should be fully resolved to allow inspection of the non-universal part of the PDF estimated.

The simplest hash function is the integer part of the division by a constant b, which means $h(s) = \lfloor s/b \rfloor$. Similarly straightforward is the hash function $h(s) = \lfloor \log(s)/b \rfloor$ or $h(s) = \lfloor \log(s+1)/b \rfloor$ which maps a linearly increasing range of avalanche sizes into the same slot of the histogram. An increased range reduces memory wastage, as it decreases the number of empty slots towards larger event sizes as found in the naïve histogram.

A number of caveats apply. Hash functions provide a form of binning and are thus prone to any of the artefacts mentioned in Sec. 7.4.1. For example, it is advisable to refrain from using power law slot sizes. In terms of CPU time, using library calls such as `pow` and `log` on most platforms requires considerably more CPU time than simple mathematical manipulations such as division or a small number of `if` statements. Finally,

---

[25] In the following, a slot of a histogram is the entry in an array associated with a certain range of event sizes, whose extent is referred to as the size of the slot.

modern (virtual) memory management can mean that the naïve (but fast) `histogram [s]++` is not so wasteful after all, as memory is effectively provided only when it is used.

Even when the hash function provides wider slots for larger event sizes, in most applications, a slot for small avalanches will receive many more entries than one for large avalanches. In order to account for the former and prevent overflows, large variable types are necessary for small event sizes which are not justifiable for the larger ones – if the ranges were big enough towards larger event sizes to make up for the decrease in frequency, the histogram would become far too coarse. In order to reduce memory requirements, one can resort to a more complex hash procedure, which gives access to different variable types depending on the event size. To this end a number of thresholds is introduced, which the event sizes are compared to in a way that minimises the expected number of comparisons, for example

if (s < threshold_low) { histogram_low[hash_low(s)]++; }
else if (s < threshold_med) { histogram_med[hash_med(s)]++; }
else if (s < threshold_hig) { histogram_hig[hash_hig(s)]++; }
else { histogram_max[hash_max(s)]++; }

Within each such range, again a (simple) hash function is used. Anticipating a large number of entries for small events, `histogram_low` could be of type `unsigned long long`, whereas `int` or even `short` might be sufficient for the larger events. Each comparison costs CPU time and the number of such comparisons should be kept at a minimum. Looping over a large number of such thresholds is generally not advisable. If many ranges are used, a 'divide and conquer' technique can be adopted which arranges the range tests in a tree-like structure, so that the expected number of comparisons (on the basis of a simplified estimate of the PDF) is minimised (Pruessner, 2009). This might result in a degenerate tree, however, such as the algorithm above, which makes the least comparisons for the most likely event sizes and vice versa.

There is usually no need to stretch the coverage of the histogram so far as to handle the largest event theoretically possible (which practically never occurs). Very large events can be captured in a **minority report**, a dynamically growing list of exceptionally large events which is converted into a histogram on the fly at the time of output, or stored separately for post-processing.

It is good practice to compare moments derived from the histogram with moments generated directly within the code, in order to gauge the numerical error of both.

A.6.3 Implementing concurrent Poisson processes

In models that can be cast in the form of a continuous time Markov process,[26] the relaxation of individual sites can be thought of as **concurrent Poisson processes**, e.g. all sites topple with the same rate. Each (local) relaxation process i has a certain rate λ_i associated with it, so that the probability of event i not to occur over the time interval $[0, t]$ is $\exp(-t\lambda_i)$ and

---

[26] Usually in Abelian models, if the order of updates is irrelevant for the final state, see Sec. 8.3.

the probability for it to take place on the interval $[t, t + dt]$ is $dt\, \lambda_i$. The product of the two probabilities describes a 'decay' with no event until t, at which point it occurs, so that t is the **waiting time**.

As the processes themselves do not interact, they behave as naïvely expected, for example, the rate with which *any* event takes place is $\Lambda = \sum_i \lambda_i$. The probability that none of the events takes place is simply the product $\prod_i \exp(-t\lambda_i) = \exp(-t\Lambda)$. The average waiting time for the next event to occur therefore is Λ^{-1}. If there are N processes all occurring with the same rate λ, time is usually rescaled such that $\lambda = 1$ and the average weighting time between processes is $1/N$.

In a naïve, ideal implementation of N concurrent Poisson processes, time would be incremented in infinitesimally small units Δt and each process would occur with the small probability $\Delta t\, \lambda_i$ within the time interval. Such an implementation is obviously very time consuming, because time increments are to be chosen as small as possible and every process must be given a chance to occur, which rarely happens. Many authors therefore resort to a much larger Δt, such that on average one event occurs in that interval, $\Delta t = \Lambda^{-1}$ or $\Delta t = 1/N$ for $\lambda_i = 1$. In this approximation, after waiting time Δt event i occurs with probability λ_i/Λ. On average, in M such time steps, corresponding to time $M\Lambda^{-1}$, event i occurs $M\lambda_i/\Lambda$ times, i.e. with frequency λ_i as it should.

The choice $\Delta t = \Lambda^{-1}$ is a rather drastic oversimplification, because it suppresses all fluctuations in the waiting time. In principle, any number of the N processes can take place within any finite time interval. Only in the limit of infinitely many concurrent processes does the procedure become exact (Liggett, 2005). It is, however, not an option where an accurate implementation of time is needed, for example when measuring time correlation functions in a system with few active sites. Also, expressing the model in terms of probabilities often leads to events that are, within a time step, mutually exclusive rather than concurrent which can lead to bias and undesired correlations.

A simple but exact representation of multiple concurrent Poisson processes is implemented by **randomising the waiting time** Δt after which process i is picked with probability λ_i/Λ. The distribution the waiting time is drawn from is $\mathcal{P}(\Delta t) = \Lambda \exp(-\Lambda \Delta t)$ which is easily generated from a uniform distribution $\mathcal{P}^{(u)}(u) = 1$ on the interval $]0, 1]$ using the transformation $\Delta t = -\ln(1 - u)/\Lambda$ (or, of course, $\Delta t = -\ln(u)/\Lambda$).

Picking an event with probability λ_i/Λ is difficult if the rates λ_i vary widely, although there are very elegant solutions for the general case (the 'method of aliases' Knuth, 1997c, p. 120). When they are all equal, event $i \in \{1, 2, \ldots, N\}$ can be chosen by rounding up uN where u is taken from a uniform distribution as above. Sometimes all events can be arranged in pairs (or generally tuples) i, j so that the sum of their rates is identical $\lambda_i + \lambda_j = \lambda$. Each pair is picked with equal probability using $\lceil uN/2 \rceil$. The random choice within a pair is based on a comparison of λ_i/λ with a random number drawn from a uniform distribution. If the range of λ_i is narrow, the largest frequency among them can be chosen for λ and each event i is paired up with a 'washer', a null event that occurs with frequency $\lambda_i' = \lambda - \lambda_i$. This is in effect **rejection sampling** (Press *et al.*, 2007), which is easy to implement but can be very wasteful. It is not an option if null events $N\lambda - \Lambda$ would dominate the statistics. In that case, more sophisticated methods might be needed, which are subject to ongoing research (e.g. Loison, Qin, Schotte, and Jin, 2004, also Knuth, 1997c, p. 120).

A.6.4 Characterising the transient

Dickman (2004b) devised a particularly convenient technique to analyse time dependent observables, especially observables displaying a strong transient in *microscopic* time. Time dependent observables, such as the particle density in the Oslo Model with suitable dynamics, are usually recorded in arrays, where each element is associated with a certain time frame. If moment [n] [t] denotes the *n*th accumulated (non-normalised) moment of an observable at time *t*, then the naïve output of the first moment has the form moment [1] [t] /moment [0] [t]. By fitting against an exponential, comparison with exact results or any other method, the **transient time** τ_t can be established, beyond which the system is assumed to be at stationarity. To derive an estimate of the first moment from the present data, a temporal average has to be taken from τ_t to some maximum time tmax, using (pseudo) code of the form

```
/* Initialise: */
for (n=0; n<=mom_max; n++) total_moment[n]=0;

/* Accumulate: */
for (t=τ_t; t<=tmax; t++) {
  for (n=0; n<=mom_max; n++) {
    total_moment[n]+=moment[n][t];
  }
}
/* Output (normalised) total_moment[n] for given τ_t. */
...
```

In this implementation the loop has to be repeated for each estimate of the transient time τ_t, which is normally determined iteratively during post-processing by testing for convergence observables such as particle density or activity. Convergence can be verified very elegantly by considering observables for all possible values of τ_t. These can be computed without repeated summation by re-using the result[27] for $\tau_t + 1$ in the summation up to τ_t.

```
/* Initialise: */
for (n=0; n<=mom_max; n++) total_moment[n]=0;

/* Accumulate incrementally for each τ_t: */
for (t=tmax; t>=0; t--) {
  for (n=0; n<=mom_max; n++) {
    total_moment[n]+=moment[n][t];
  }
  /* Output (normalised) total_moment[n] for each t, i.e. every possible τ_t. */
  ...
}
```

If added to an existing output routine, which prints the raw estimates moment [n] [t] (so that they can be accumulated later), in principle this can be done during the simulation (rather than in post-processing).

---

[27] Note that below the first loop runs from the maximum time tmax down to 0.

Notes

Chapter 2

[2.1] (page 41) Using the scaling law Eq. (2.40) on p. 43, moment ratios can combine different observables, such as

$$\frac{\langle s^2 \rangle \langle T \rangle^2}{\langle s \rangle^2 \langle T^2 \rangle} = \frac{a^{(\mathrm{T})} (b^{(\mathrm{T})})^{1-\alpha}}{a^{(s)} (b^{(s)})^{1-\tau}} \frac{g_2^{(s)}(0)(g_1^{(\mathrm{T})}(0))^2}{g_2^{(\mathrm{T})}(0)(g_1^{(s)}(0))^2}.$$

In this case, the metric factors do not cancel, but combine to a dimensionless quantity, which is expected to be universal as well.

Chapter 4

[4.1] (page 96) As illustrated (Fig. 4.4), that does not mean that a configuration which is transient with respect to $a_{\mathbf{n}}$ has to be transient with respect to $a_{\mathbf{n}'}$. Similarly, $\mathcal{C}'$ being transient under $a_{\mathbf{n}}$ does not contradict $\mathcal{C} = a_{\mathbf{n}'}\mathcal{C}'$ being recurrent under $a_{\mathbf{n}}$, so that $\mathcal{C} = a_{\mathbf{n}}^{p_{\mathbf{n}}}\mathcal{C} = a_{\mathbf{n}}^{p_{\mathbf{n}}} a_{\mathbf{n}'}\mathcal{C}' = a_{\mathbf{n}'} a_{\mathbf{n}}^{p_{\mathbf{n}}}\mathcal{C}'$, and therefore $a_{\mathbf{n}'}\mathcal{C}' = a_{\mathbf{n}'} a_{\mathbf{n}}^{p_{\mathbf{n}}}\mathcal{C}'$. This is because Abelianness (at least in the ASM) applies to recurrent and non-recurrent states. As long as transient states such as $\mathcal{C}'$ are not excluded, there is generally no inverse of $a_{\mathbf{n}'}$, so that $a_{\mathbf{n}'}\mathcal{C}' = a_{\mathbf{n}'} a_{\mathbf{n}}^{p_{\mathbf{n}}}\mathcal{C}'$ does *not* entail $\mathcal{C}' = a_{\mathbf{n}}^{p_{\mathbf{n}}}\mathcal{C}'$, which would contradict the assumption of $\mathcal{C}'$ being transient under $a_{\mathbf{n}}$. In other words, if $\mathcal{C}'$ is transient under $a_{\mathbf{n}}$ then $a_{\mathbf{n}'}\mathcal{C}'$ might nevertheless be recurrent under $a_{\mathbf{n}}$, provided that $\mathcal{C}'$ is not recurrent under $a_{\mathbf{n}'}$ (because if $\mathcal{C}'$ was recurrent under $a_{\mathbf{n}'}$, then there is an inverse such that $a_{\mathbf{n}'}\mathcal{C}' = a_{\mathbf{n}'} a_{\mathbf{n}}^{p_{\mathbf{n}}}\mathcal{C}'$ entailed $\mathcal{C}' = a_{\mathbf{n}}^{p_{\mathbf{n}}}\mathcal{C}'$, i.e. $\mathcal{C}'$ recurrent under $a_{\mathbf{n}}$).

[4.2] (page 98) An example is easily constructed: one site with $z_{\mathbf{n}} = z_{\mathbf{n}}^c$ next to a site with $z_{\mathbf{n}'} = z_{\mathbf{n}'}^c - 1$. Charging and relaxing the former first (and assuming no further topplings among any other sites) leaves the neighbour at $z_{\mathbf{n}'}^c$, so that it will topple at the next charge and the sequence of avalanche sizes will be 1, 1. Inverting the sequence gives 0, 2.

Chapter 5

[5.1] (page 112) In a realistic model of forest fires one would not assume that the forest is essentially perpetually burning with reforestation occurring on a much longer yet comparable time scale, so that fires sweep constantly through a regrowing forest. In that sense, the name 'forest fire model' is misleading, although the model was never actually intended to represent real forests. Rather, the name is meant as a caricature of the basic dynamic process. Nevertheless, the naming received some (maybe unjustified) critique (Grassberger and Kantz, 1991).

[5.2] (page 118) One has to be somewhat careful to compare like with like, because the *order parameter* in percolation *is* in fact characterised by $\tau = 1$ (Christensen *et al.*, 2008), while the cluster size distribution, indeed, is not. This is important, because some of Drossel and Schwabl's arguments actually *do* apply to percolation, but they apply to the order parameter distribution (characterised by a trivial exponent 1), rather than the cluster size distribution (characterised by the non-trivial percolation exponent $187/91$ in two dimensions).

[5.3] (page 131) For example, the distribution of F^+ must be self-consistent given a certain distribution of stable sites that are charged, as the former results from the convolution of the distribution of stable sites with the distribution of F^+ itself. Given that the range of stable and toppling sites available to contribute to a small resulting F^+ is larger than the corresponding range contributing to large F^+, one would expect that the overall distribution of forces cannot be uniform simultaneously with that of F^+.

[5.4] (page 145) It is tempting to define f_c as the fitness so that on average one fitness arrives below f_c in every update, along the lines of a branching ratio (Sec. 5.4.5.6) of 1. This is wrong or misleading for several reasons: first of all, because an update affects the nearest neighbours, which might be located below f_c as well, the true branching ratio is the number of sites arriving below f_c divided by the number of sites departing from there. Secondly, the stationary state is exactly defined by having such a branching ratio exactly equal to unity for *any* f, corresponding to vanishing net flux across any f. To make use of the branching ratio, the existence of a stationary state has to be established first. This can be done by taking the thermodynamic limit first and preparing the system in a certain fitness distribution, say a step function rising at f^*. If all fluxes vanish (asymptotically), this initial fitness distribution is the stationary state. At the first update the flux across f^* is $-(2d + 1)f^*$, yet the average can vanish.

Without fluctuations and correlations, in the thermodynamic limit, f_c is given by the probability that K new fitnesses per site (and neighbours) updated contain on average 1 site below f_c, i.e. $Kf_c = 1$ for a uniform distribution of updated fitnesses. In a one-dimensional system as described in Box 5.4, $K = 3$, corresponding to $f_c = 1/3$ in the random neighbour version, yet numerically the threshold turns out to be about 0.667 02, as if $K_{\text{eff}} \approx 1.5$.

[5.5] (page 145) This is an approximate result, assuming a certain 'free path' (the small range just above $G(t)$ devoid of any particles after the avalanche) in a system with N fitnesses on an interval of size $1 - G(t)$. The probability that $f_{\min}(t)$ becomes f^* after the update equals the probability that there is precisely one fitness with $f_{\mathbf{n}} = f^*$ and that all other fitnesses are greater than this value. The probability density for this event is $1/(1 - G(t))$ times the probability that all other fitnesses are above f^*, which is $((1 - f^*)/(1 - G(t)))^{N-1}$, times N, the number of possible realisations. The resulting average fitness at termination therefore is

$$\int_{G(t)}^1 df^* \, f^* \frac{1}{1 - G(t)} \left(\frac{1 - f^*}{1 - G(t)} \right)^{N-1} N = \frac{NG(t) + 1}{N + 1},$$

so that the expected increase is $\Delta G = (NG(t) + 1)/(N + 1) - G(t) = (1 - G(t))/(N + 1)$. The correctness of this result is illustrated by the special cases $N = 0$ and $N = 1$. In the latter case, the average increase is half the remaining gap.

[5.6] (page 147) The situation is actually slightly more complicated. The probability that no fitness arrives in the interval conditional to a fixed (given) number of updated sites n_{cov} is $((1 - f_0 - df_0)/(1 - f_0))^{n_{\text{cov}}}$. Using the complement, to leading order in df_0, the probability that one fitness arrives in the interval is $n_{\text{cov}} \, df_0/(1 - f_0)$. Averaging over this result gives $\langle n_{\text{cov}} \rangle \, df_0/(1 - f_0)$ as stated, on the basis that n_{cov} is independent of f_0 and therefore does not need to be conditioned to it, i.e. n_{cov} is independent of the fact that the last fitness updated arrived in the interval $[f_0, f_0 + df_0]$.

Since the BS Model generally has correlated fitnesses, some subsets of which are mutually independent, it is very often difficult to decide where assumptions of independence enter and whether they are met.

[5.7] (page 160) The BS Model can make use of some of the techniques developed for the OFC Model, such as the boxes (Sec. 5.3.4, Sec. A.3). There are, however, two significant differences between the OFC Model and the BS Model. Firstly, there is no uniform driving and no need for or benefit in letting the range of fitnesses slide slowly in the BS Model. The bracket of fitnesses, which contains all likely candidates to be recruited for the minimum, is constant in the stationary state. Secondly, while the extremal site in the OFC Model is to be found only when driving it, it is required in *every* update of the BS Model. This puts a strong emphasis on being able to find the minimum fitness efficiently, normally in a sparse part of the distribution.

In the OFC Model, on the other hand, while an avalanche is running, there is no performance gain from being able to locate the minimum force quickly, but computational costs if they are to be maintained in a tree.

Chapter 6

[6.1] (page 175) Algebraically, this problem is *not* solely caused by $\{C, C^{\dagger}\} = \mathbf{1}_2$, which can similarly be found in the ASM, Eq. (8.55) on p. 272, yet does not suffer from this problem. On the same note, it cannot be avoided by C and $C^{\dagger}$ both being nilpotent, $C^2 = (C^{\dagger})^2 = 0$.

[6.2] (page 190) The average number of topplings per particle added in the one-dimensional sOM can generally be derived most easily in the height picture. Whenever site $\mathbf{n} = L$ topples $L + 1$ particles are dissipated, namely the particle falling off plus an entire row of particles because of the boundary condition, see Fig. 4.3, p. 93. The particle falling off has made L topplings while in the pile, the particle removed from $\mathbf{n} = L$ has made $L - 1$ topplings, the one at $\mathbf{n} = L - 1$ has made $L - 2$ and so on, up to $\mathbf{n} = 1$ which has only been added and has made no topplings at all. This amounts in total to $\sum_{i=0}^{L} i = L(L + 1)/2$ topplings and therefore $\langle s \rangle = L/2$ per particle dissipated or equivalently per particle added. This reasoning is identical for the ASM. No condition has been imposed that the pile relaxes at all after particle addition so that the average $\langle s \rangle$ contains a contribution from avalanches of size 0.

[6.3] (page 205) In the (boundary driven) one-dimensional sOM, the right boundary condition can be represented by an additional site at $\mathbf{n} = L + 1$, which never topples and dissipates all charges, $g(L + 1, t) = 0$. The boundary conditions can be homogenised by shifting the data, $g(\mathbf{n}, t) = f(\mathbf{n}, t) + \tilde{g}(\mathbf{n}, t)$, where $f(\mathbf{n}, t)$ is a simple, deterministic expression, so that $\tilde{g}$ essentially obeys Eq. (6.34) on p. 205 with $\tilde{g}(0, t) = \tilde{g}(L + 1, t) = 0$.

In the boundary driven version of the OSLO Model, the external drive $E(t)$ fixes the value of $g(0, t)$, which increases slowly in time (see also Fig. 8.13, p. 303). In the traditional version of the qEW equation, on the other hand, the driving enters as a constant additional velocity $\dot{E}$, usually called a 'force', so that $\partial_t g = \cdots + \dot{E}$. At the critical force $\dot{E} = \dot{E}_c$ the interface starts moving with a finite average velocity, considered as the order parameter. In SOC, this order parameter is set by the external drive (Paczuski and Boettcher, 1996).

The different nature of the driving in the SOC model compared to the traditional qEW equation becomes more apparent by allowing **bulk drive**, which is implemented naturally in the sOM (Bonachela, 2008, confirming earlier results for the linear interface model by Paczuski *et al.*, 1996).

In the sOM with bulk drive, both boundaries are dissipative and fixed to 0, $g(0, t) = g(L + 1, t) = 0$. Without noise and external force, $g(\mathbf{n}, t)$ would relax diffusively to $g(\mathbf{n}, t) = 0$ according to Eq. (6.34) in the limit of large t. However, with external drive (and noise) the number of charges is bound to increase throughout the interface, so that $g(\mathbf{n}, t)$ is in fact monotonically increasing in t and has constant average velocity in the stationary state. Being pinned at two ends, the increasing tension by the Laplacian cannot be quelled by the noise, which is bounded on the lattice, Eq. (6.27) on p. 204. It is instead balanced (in fact, created) by an increasing *source term* $E(\mathbf{n}, t)$, which enters as an additional term on the right of Eq. (6.29) on p. 204,

$$g(\mathbf{n}, t) = G(\mathbf{n} + 1, t) + G(\mathbf{n} - 1, t) + E(\mathbf{n}, t),$$

and turns into a *monotonically increasing source* in the equation of motion

$$g(\mathbf{n}, t + 1) - g(\mathbf{n}, t) = \frac{1}{2}\Big(g(\mathbf{n} - 1, t) - 2g(\mathbf{n}, t) + g(\mathbf{n} + 1, t)$$
$$+ \xi(\mathbf{n} + 1, g(\mathbf{n} + 1, t)) + \xi(\mathbf{n} - 1, g(\mathbf{n} - 1, t))\Big) + E(\mathbf{n}, t).$$

With Neumann boundary conditions, i.e. no particles being able to leave the system, the external drive $E(\mathbf{n}, t)$ could be ramped up until it reached roughly the critical force $\dot{E}_c$ mentioned above, at which point the interface would lift off and move indefinitely, reminiscent of an absorbing state phase transition into the active phase.

Chapter 7

[7.1] (page 231) It is generally advisable to keep a simulation as simple as possible and move as much of the more difficult data handling to post-processing. Mistakes can be easily corrected at the stage of data analysis, while an error in the simulation might ruin the result of thousands of hours of CPU time (see Appendix A).

[7.2] (page 236) Only a continuum description allows a dimensional analysis in the form of continuous symmetries of (partial) differential equations. The present example is parameterised by a_0, L and D which can be combined to a dimensionless variable $a_0^2/(DL)$. If it were not for this quantity, the PDF would be fully determined by dimensional analysis, which means that $\tau = 1$, see Sec. 2.1, p. 27. Only the presence of $a_0^2/(DL)$ allows for $\tau = 4/3$, which implies that a_0 cannot be ignored or taken to some convenient limit anywhere in the scaling regime. While small a_0 means early termination of the random walker and fewer survivors after long times, the characteristic size of the area under their trajectory after long times is simply $L\sqrt{DL}$ independent of a_0.

[7.3] (page 237) This view can be *very* misleading, suggesting that a sample of size N typically does not contain events of a size greater than s^* with $\mathcal{P}^{(s)}(s^*) = 1/N$. This is of course incorrect, which can be seen immediately by considering small sample sizes, say $N = 1$. The expected value of any unbiased estimator is still its expectation value, for example $\bar{s} = s_1$ and $\langle \bar{s} \rangle = \langle s \rangle$.

Chapter 8

[8.1] (page 250) The term 'mean-field theory' is frequently abused in SOC, as generally in statistical mechanics. It implies that spatial correlations are neglected in one way or another, for example by accounting for the presence of neighbours merely by some effective, local field, as is done in a molecular field approximation, where the local field is replaced by the average magnetisation (e.g. McComb, 2004). Or, similarly, nearest neighbours are replaced by randomly chosen interaction partners, as is done in the MFT of the BS Model, Sec. 8.1.3. In the thermodynamic limit, at finite time, the probability that a site is chosen twice vanishes, so that correlations can be neglected. This is usually not the case for finite, random neighbour models. Another form of MFT is to assume that correlations vanish (more generally, probabilities factorise) beyond a certain length scale, thereby effectively truncating a hierarchy of differential equations of the n-body joint PDFs (ben-Avraham and Köhler, 1992; Ferreira and Mendiratta, 1993, also Opper and Saad, 2001). Any combination of these three approaches is possible. Many of them can be improved systematically.

[8.2] (page 251) If an avalanche of size s occurs in a system of N sites, so that s sites are visited at random, the expected fraction of sites visited more than once is $1 - ((N+s-1)/N)(1 - 1/N)^{s-1}$, which to leading order in N at fixed s is $s(s-1)/(2N^2)$. If s is a power law in N, say $s = aN^\alpha$, the asymptote in large N is

$$1 - ((N+s-1)/N)(1-1/N)^{s-1} \sim \begin{cases} \dfrac{a(a-1)}{2N^2} & \text{for } \alpha = 0 \\[2mm] \dfrac{a^2}{2N^{2(1-\alpha)}} & \text{for } 0 < \alpha < 1 \\[2mm] 1 - \dfrac{1+a}{\exp(a)} & \text{for } \alpha = 1 \\[2mm] 1 & \text{for } \alpha > 1. \end{cases}$$

[8.3] (page 255) This result is somewhat counter-intuitive, because from Eq. (8.11) on p. 252 it is clear that $f_0 = 0$ or $f_0 = 1$ means that the evolution of $P(a, t)$ is purely ballistic, which is not the case for Eq. (8.24) on p. 255. Similarly, the variance of an individual step of the random walker is $\sigma^2 = 2f_0(1 - f_0)$, suggesting $D = f_0(1 - f_0)$ according to footnote 2 on p. 244. The root of the problem lies in the terms ignored when making the continuum approximation of the difference equation, Eq. (8.23).

[8.4] (page 260) More generally, since the generating function $f(x)$ is continuous and contains only positive coefficients, $f(x)$ and all its derivatives are non-negative for any $x \geq 0$ and there can only be 0, 1, 2 or infinitely many roots $f(x) = x \in [0, 1]$. Since $f(1) = 1$ by normalisation and $f''(x) \geq 0$ on $x \in [0, 1]$, if there is only one other fixed point $f(x^*) = x^* < 1$, then $f'(x^*) < 1 < f'(1)$, so that the fixed point $x^* < 1$ is the unique stable one. If $x^* = 1$ is the only root, then this must be the stable one.

[8.5] (page 272) Pursuing such an attempt of a more formal proof very quickly becomes very messy and not much more convincing (e.g. Pruessner, 2004c, for the OSLO Model). The aim of the proof is to show that every representation of the dynamics which fullfils certain conditions (see below) generates *all possible* terms, i.e. every possible scenario of particle creation and annihilation is realised in every representation. The proof identifies in one representation the terms generated in another representation. The key ingredients are firstly, a representation of the model in terms of operators that correspond to local 'atomic' updates and commute for different sites. In the example of Sec. 8.3.1, Eq. (8.54) on p. 272, the matrices $S_\mathbf{n}$ and $T_\mathbf{n}$ fulfil this condition, $[S_\mathbf{n}, S_{\mathbf{n}'}] = S_\mathbf{n} S_{\mathbf{n}'} - S_{\mathbf{n}'} S_\mathbf{n} = 0$, $[T_\mathbf{n}, T_{\mathbf{n}'}] = 0$ and $[S_\mathbf{n}, T_{\mathbf{n}'}] = 0$ in particular for $\mathbf{n} \neq \mathbf{n}'$. Secondly, '[m]ultiple *local* charges must be equivalent to powers of [. . .] operators' (Pruessner, 2004c). Unfortunately, this last condition is little less than demanding that operators be Abelian to start with, so that, say, two slope units arriving at the same time can be dealt with sequentially. There seems little merit in demonstrating Abelianness on the basis of this condition.

[8.6] (page 274) Contrary to common belief, this does *not* imply that $\lim_{n \to \infty} \mathcal{O}^n$ exists, as can be easily seen in the case $\mathcal{O} = \begin{pmatrix} 0 & 1 \\ 1 & 0 \end{pmatrix}$, which has eigenvectors $(1/2, 1/2)$ and $(-1/2, 1/2)$ with eigenvalues 1 and -1 respectively. If the eigenvalue with magnitude 1 is unique and the eigenvectors span the entire vector space, then $\lim_{n \to \infty} \mathcal{O}^n$ exists and each of its columns is the stationary distribution vector.

[8.7] (page 307) As for the notion of time used here, t_i is introduced as a time when the system is quiescent, which implies that time evolves without the interface moving. The external drive $2\dot{E}t$, Sec. 8.5.3, very slowly ramps up on this time scale, in fact arbitrarily slowly. One might then introduce a second microscopic time scale, the **conditional time**, with all quiescent periods removed – conditional time ticks provided the system is active. Such a change of time scale apparently does not affect the physics of the interface, but it is crucial for a clear theoretical understanding, in particular with respect to the scaling of the activity (Pruessner, 2007). The activity measured in conditional time is always non-zero and obeys the scaling form Eq. (9.17).

At closer inspection, even this carefully constructed time scale reveals some flaws. The driving discussed above takes place in the continuum and assuming that no surprising non-analyticities develop, it will always affect some small part of the interface (possibly even when advancing in jumps, $2\dot{E}\lceil t/\delta t \rceil \delta t$), which means that it never rests and the conditional time scale coincides with the non-conditional one. In order to restore the distinction, a cutoff would need to be introduced.

[8.8] (page 311) Of essentially geometrical origin are the anti-correlations that can be detected in the statistical error when estimating $\langle s \rangle$. Considering a sequence of avalanches $s'_1, s'_2, \ldots, s'_M$, their average size is $\bar{s'} = M^{-1} \sum_i^M s'_i$. The sum $\sum_i^M s'_i$ is the total area encapsulated by the first and the last interface configurations, such as the one between the dashed trajectories shown in Fig. 8.14(b), p. 308, which depart from $\mathbf{x} = 0$ with vertical distance $(2\dot{E}\delta t)M$. The total area encapsulated, $S = M\bar{s'} = \sum_i^M s'_i$, is essentially $(2\dot{E}\delta t)ML$ plus fluctuations due to the upper and lower boundaries, which do not depend on M. The variance

$$\sigma^2(S) = \sum_{ij}^M \langle s'_i s'_j \rangle - \langle s'_i \rangle \langle s'_j \rangle$$

of the total area is therefore constant in the limit of large M and the standard deviation of the estimate of the average avalanche size $\overline{s'}$ vanishes as fast as $1/M$, rather than $1/\sqrt{M}$, as would be the case for fully uncorrelated measurements $s'_1, s'_2, \ldots, s'_M$, see Sec. 7.1. Of course, asymptotically in large k avalanche sizes s'_i and s'_{i+k} are independent, but that does not imply that the sum $\sum_k \langle s'_i s'_{i+k} \rangle - \langle s'_i \rangle \langle s'_{i+k} \rangle$ in the present, *correlated* case is at least as big as the *fully uncorrelated* case $\langle s'_i s'_{i+k} \rangle - \langle s'_i \rangle \langle s'_{i+k} \rangle \propto \delta_{0k}$. In fact, for any model which possesses an interface representation so that the geometrical arguments above apply, the correlation function is negative for some intermediate k, so that the sum is *less* than what is obtained for uncorrelated random variables s'_i (Welinder *et al.*, 2007; Morand, Pruessner, and Christensen, 2010).

The characteristic number M_c for which the variance of S approaches a constant is the number of charges beyond which the consecutive interface configurations typically detach, Fig. 8.14(b), p. 308. This happens when the initial gap $(2\dot{E}\delta t)M$ between first and last interfaces is of the order of its fluctuations, $M_c \propto L^{\chi^{(\text{int})}}$. This is hence the *macroscopic* correlation time, relating consecutive avalanches, which is linked to the (supposed) *microscopic* correlation time $L^{z^{(\text{int})}}$, via $z^{(\text{int})} = \chi^{(\text{int})}/\beta^{(\text{int})}$ (Krug, 1995).

In Abelian models, where the sum of consecutive avalanche sizes equals the avalanche size caused by a corresponding number of multiple charges, M_c is the number of charges required to 'flush' the system. Charging it at once more than M_c times, $M = M_c + Q$, amounts to (effectively) adding a determinstic number of topplings (or charges, for that matter) to the avalanche size, namely for the Q charges beyond those that caused correlated avalanches. In the one-dimensional oOM, such additional charges simply topple L times each until they leave the system on the right end.

[8.9] (page 314) For example, the flat interface $g(\mathbf{x}, t) = \xi$, constant in space and time, but with random ξ having finite variance across the ensemble, would have finite roughness. If the variance increases with system size, this might even result in a positive roughness exponent. Similarly, if $g(\mathbf{x}, t) = \xi|\mathbf{x}|$ and $\sigma^2(\xi) = \langle \xi^2 \rangle = $ constant, then $\chi^{(\text{int})} = 1$.

Chapter 9

[9.1] (page 332) The averaging $\langle \cdot \rangle$ in Eq. (9.14) is somewhat delicate in that it contains a spatial as well as an ensemble average, which cannot be replaced by a time average, in order to keep the expression time dependent. Equation (9.14) does not apply without averaging, even when adding a noise term. Even if $\rho_a(\zeta(\mathbf{x}, t))$ denotes the *local* response of the activity to a *local* particle density, $\dot{\zeta}(\mathbf{x}, t) = h - \epsilon \rho_a(\zeta(\mathbf{x}, t)) + \xi(\mathbf{x}, t)$ still lacks (multiplicative ?) noise in the dissipation as well as spatial interaction (see Dickman *et al.*, 1998, for details). Identifying the microscopic equation of motion, coupling the *local, instantaneous* particle density to the *local, instantaneous* activity is not trivial, and linking the latter to the *spatially averaged* activity of the AS model and its scaling is at best a tall order.

Yet, ignoring the space dependence and linearising this expression about $\zeta = \zeta^*$ with $\rho_a(\zeta^*) = h/\epsilon$ generates an Ornstein–Uhlenbeck process,

$$\dot{\Delta\zeta}(t) = -\epsilon \Delta\zeta(t)\rho_a'(\zeta^*) + \xi(t)$$

where $\Delta\zeta = \zeta - \zeta^*$. This equation describes the dynamics of the SOC as it 'hovers' around ζ_c and together with the scaling of the noise correlator determines the distribution of ζ values sampled. If the AS mechanism applies in SOC, one would expect its width to vanish like $L^{-1/\nu_\perp^{(\text{AS})}}$. As the slope $\rho_a'(\zeta^*)$ of the activity at the critical point increases, the particle density is increasingly localised or 'pinched'.

[9.2] (page 337) At closer inspection a similar problem can be identified in equilibrium phase transitions. Firstly, the symmetry of the Hamiltonian is reflected in the ensemble, not in space. The average over the former rather than the latter displays the phase transition. The averaging across space therefore has as little justification for equilibrium phase transitions as for those into absorbing states. In equilibrium, the ensemble averaged order parameter could display a phase transition without a change in *magnitude* of the order parameter measured in individual realisations. However, secondly, even the *absolute* order parameter in these equilibrium models signals a phase transition, very similar to the conditional order parameter in AS which is forced to be non-zero.

Appendix A

[A.1] (page 369) The code needs to be carefully revised if it is decided, for the sake of perfomance, to pick one of the degenerate maxima at random, rather than handling all of them in parallel. This is a questionable approach, as degenerate maximum forces are likely to reside in the same patch and their interaction therefore is an integral part of the model. The technical key problem with using a randomly selected maximum is the criterion used in l. 321 to decide whether a site is on the stack already. Given that a site with force `f==f_threshold` triggers the avalanche, it is assumed that any site with force `f>=f_threshold` before being charged, must be on the stack already. That is not the case if such a site is a degenerate maximum which has been omitted. The criterion to enter the stack could be changed to `f>f_threshold`, leading, however, to two different criteria for toppling sites: one criterion for it to be on the stack and topple (`f>f_threshold`) and another criterion to trigger an avalanche, (`f==f_threshold`, l. 269 – l. 271).

References

Key references are marked as follows: (●) selected reviews and overview articles about SOC and closely related topics, (▶) definitions of some selected (typical or important) SOC models, (⇒) proposed mechanisms and theories of SOC.
The page numbers following the arrow (→) at the end of each entry refer to the location of the citation, with *quotations* being indicated by italics.
For the sake of consistency name prefixes are regarded part of the names.

Adami, C., and J. Chu, 2002, Critical and near-critical branching processes, *Phys. Rev. E* **66**(1), 011907 (8 pp), arXiv:cond-mat/9903085. → p 257.

Aegerter, C. M., 1998, Evidence for self-organized criticality in the Bean critical state in superconductors, *Phys. Rev. E* **58**(2), 1438–1441. → p 60.

Aegerter, C. M., 2003, A sandpile model for the distribution of rainfall?, *Physica A* **319**, 1–10. → pp 187, 189, 192, 209.

Aegerter, C. M., R. Günther, and R. J. Wijngaarden, 2003, Avalanche dynamics, surface roughening, and self-organized criticality: experiments on a three-dimensional pile of rice, *Phys. Rev. E* **67**(5), 051306 (6 pp), arXiv:cond-mat/0305592. → pp 58, 192, 193, 301, 305.

Aegerter, C. M., K. A. Lőrincz, M. S. Welling, and R. J. Wijngaarden, 2004a, Extremal dynamics and the approach to the critical state: experiments on a three dimensional pile of rice, *Phys. Rev. Lett.* **92**(5), 058702 (4 pp). → pp 58, 192.

Aegerter, C. M., M. S. Welling, and R. J. Wijngaarden, 2004b, Self-organized criticality in the Bean state in YBa$_2$Cu$_3$O$_{7-x}$ thin films, *Europhys. Lett.* **65**(6), 753–759. → pp 60, 61.

Aegerter, C. M., M. S. Welling, and R. J. Wijngaarden, 2005, Dynamic roughening of the magnetic flux landscape in YBa$_2$Cu$_3$O$_{7-x}$, *Physica A* **347**, 363–374. → p 61.

Ahlgren, P., M. Avlund, I. Klewe, J. N. Pedersen, and Á. Corral, 2002, Anomalous transport in conical granular piles, *Phys. Rev. E* **66**(3), 031305 (5 pp). → pp 58, *192*, 193, 197.

Alava, M., 2002, Scaling in self-organized criticality from interface depinning?, *J. Phys.: Condens. Matter* **14**(9), 2353–2360, arXiv:cond-mat/0204226. → pp 150, 180, 203, 302, 307, 320.

● Alava, M., 2004, Self-organized criticality as a phase transition, in *Advances in Condensed Matter and Statistical Physics*, edited by E. Korutcheva and R. Cuerno (Nova Science Publishers, New York, NY, USA), pp. 69–102, arXiv:cond-mat/0307688. → p 6.

Alava, M., M. Dubé, and M. Rost, 2004, Imbibition in disordered media, *Adv. Phys.* **53**(2), 83–175. → p 155.

Alava, M. J., and K. B. Lauritsen, 2001, Quenched noise and over-active sites in sandpile dynamics, *Europhys. Lett.* **53**(5), 563–569, arXiv:cond-mat/0002406v2. → pp 165, 171, 178, 180, 182, 300, 302, 307.

Alava, M. J., L. Laurson, A. Vespignani, and S. Zapperi, 2008, Comment on 'self-organized criticality and absorbing states: lessons from the Ising model', *Phys. Rev. E* **77**(4), 048101 (2 pp), comment on (Pruessner and Peters, 2006), reply (Pruessner and Peters, 2008). → pp 170, 190, 227, 338, 339, 341, 445.

Alava, M., and M. A. Muñoz, 2002, Interface depinning versus absorbing-state phase transitions, *Phys. Rev. E* **65**(2), 026145 (8 pp), arXiv:cond-mat/0105591. → pp 61, 168, 178–180, 209, 299, 300.

Albert, R., I. Albert, D. Hornbaker, P. Schiffer, and A.-L. Barabási, 1997, Maximum angle of stability in wet and dry spherical granular media, *Phys. Rev. E* **56**(6), R6271–R6274. → p 56.

Alcaraz, F. C., and V. Rittenberg, 2008, Directed Abelian algebras and their application to stochastic models, *Phys. Rev. E* **78**(4), 041126 (13 pp). → pp 208, *289*.

Alessandro, B., C. Beatrice, G. Bertotti, and A. Montorsi, 1990a, Domain-wall dynamics and Barkhausen effect in metallic ferromagnetic materials. I. Theory, *J. Appl. Phys.* **68**(6), 2901–2907. → p 62–64.

Alessandro, B., C. Beatrice, G. Bertotti, and A. Montorsi, 1990b, Domain-wall dynamics and Barkhausen effect in metallic ferromagnetic materials. II. Experiments, *J. Appl. Phys.* **68**(6), 2908–2915. → p 62.

Ali, A. A., and D. Dhar, 1995a, Breakdown of simple scaling in Abelian sandpile models in one dimension, *Phys. Rev. E* **51**(4), R2705–R2708. → p 94.

Ali, A. A., and D. Dhar, 1995b, Structure of avalanches and breakdown of simple scaling in the Abelian sandpile model in one dimension, *Phys. Rev. E* **52**(5), 4804–4816. → p 94.

Alstrøm, P., 1988, Mean-field exponents for self-organized critical phenomena, *Phys. Rev. A* **38**(9), 4905–4906. → pp 130, 251, 257, 288.

• Alstrøm, P., T. Bohr, K. Christensen, H. Flyvbjerg, M. H. Jensen, B. Lautrup, and K. Sneppen (editors), 2004, *Complexity and Criticality*, *Physica A* **340**(4), Proceedings of the Symposium Complexity and Criticality: in Memory of Per Bak (1947–2002), Copenhagen, Denmark, August 21–23, 2003. → p 6.

Altshuler, E., and T. H. Johansen, 2004, Colloquium: experiments in vortex avalanches, *Rev. Mod. Phys.* **76**(2), 471–487. → pp 53, 61.

Altshuler, E., T. H. Johansen, Y. Paltiel, P. Jin, K. E. Bassler, O. Ramos, Q. Y. Chen, G. F. Reiter, E. Zeldov, and C. W. Chu, 2004, Vortex avalanches with robust statistics observed in superconducting niobium, *Phys. Rev. B* **70**(14), 140505(R) (4 pp), early preprint arXiv:cond-mat/0208266. → pp 60, 61.

Altshuler, E., O. Ramos, E. Martínez, A. J. Batista-Leyva, A. Rivera, and K. E. Bassler, 2003, Sandpile formation by revolving rivers, *Phys. Rev. Lett.* **91**(1), 014501 (4 pp). → p 192.

Altshuler, E., O. Ramos, C. Martínez, L. E. Flores, and C. Noda, 2001, Avalanches in one-dimensional piles with different types of bases, *Phys. Rev. Lett.* **86**(24), 5490–5493. → pp 56, 134.

Amaral, L. A. N., A.-L. Barabási, S. V. Buldyrev, S. T. Harrington, S. Havlin, R. Sadr-Lahijany, and H. E. Stanley, 1995, Avalanches and the directed percolation depinning model: experiments, simulations, and theory, *Phys. Rev. E* **51**(5), 4655–4673. → pp 203, 299.

Amaral, L. A. N., and K. B. Lauritsen, 1996a, Energy avalanches in a rice-pile model, *Physica A* **231**(4), 608–614. → p 192.

▶ Amaral, L. A. N., and K. B. Lauritsen, 1996b, Self-organized criticality in a rice-pile model, *Phys. Rev. E* **54**(5), R4512–R4515. → pp 33, 178, *184*, 185, 186, 188, 189, 191, 192, 208, 295, 296, 485.

Amaral, L. A. N., and K. B. Lauritsen, 1997, Universality classes for rice-pile models, *Phys. Rev. E* **56**(1), 231–234. → pp 189, 192, 284.

Amit, D. J., and V. Martín-Mayor, 2005, *Field Theory, the Renormalization Group, and Critical Phenomena* (World Scientific, Singapore), 3rd edition. → p 266.

Anderson, P. E., H. J. Jensen, L. P. Oliviera, and P. Sibani, 2004, Evolution in complex systems, *Complexity* **10**(1), 49–56. → p 149.

Anderson, P. W., 1972, More is different, *Science* **177**(4047), 393–396. → pp *4*, 6, *13*, 19.

Anderson, T. W., 1971, *The Statistical Analysis of Time Series* (John Wiley & Sons, New York, NY, USA). → p 358.

Andrade, R. F. S., H. J. Schellnhuber, and M. Claussen, 1998, Analysis of rainfall records: possible relation to self-organized criticality, *Physica A* **254**(3–4), 557 – 568. → p 71.

Arakawa, A., and W. H. Schubert, 1974, Interaction of a cumulus cloud ensemble with the Large-Scale environment, Part I, *J. Atmos. Sci.* **31**(3), 674–701. → p 72.

Avnir, D., O. Biham, D. Lidar, and O. Malcai, 1998, Is the geometry of nature fractal?, *Science* **279**(5347), 39–40. → p 53.

Axtell, R. L., 2001, Zipf distribution of U.S. firm sizes, *Science* **293**(5536), 1818–1820. → p 76.

Azimi-Tafreshi, N., H. Dashti-Naserabadi, and S. Moghimi-Araghi, 2008, The spatial asymmetric two-dimensional continuous Abelian sandpile model, *J. Phys. A: Math. Theor.* **41**(43), 435002 (14 pp). → p 108.

Azimi-Tafreshi, N., H. Dashti-Naserabadi, S. Moghimi-Araghi, and P. Ruelle, 2010, The Abelian sandpile model on the honeycomb lattice, *J. Stat. Mech.* **2010**(02), P02004, arXiv:0912.3331v2. → pp 23, 99.

Babcock, K. L., R. Seshadri, and R. M. Westervelt, 1990, Coarsening of cellular domain patterns in magnetic garnet films, *Phys. Rev. A* **41**(4), 1952–1962. → p 62.

Babcock, K. L., and R. M. Westervelt, 1989, Elements of cellular domain patterns in magnetic garnet films, *Phys. Rev. A* **40**(4), 2022–2037. → p 62.

Babcock, K. L., and R. M. Westervelt, 1990, Avalanches and self-organization in cellular magnetic-domain patterns, *Phys. Rev. Lett.* **64**(18), 2168–2171. → p 62.

Bagnoli, F., F. Cecconi, A. Flammini, and A. Vespignani, 2003, Short-period attractors and non-ergodic behavior in the deterministic fixed-energy sandpile model, *Europhys. Lett.* **63**(4), 512–518. → p 333.

Bak, P., 1990, Is the world at the border of chaos?, *AIP Conf. Proc.* **213**(1), 110–118, Proceedings of Frontiers in Condensed Matter Theory, New York, NY, USA, July 2, 1990. → pp 16, *319*.

Bak, P., 1992, Self-organized criticality in non-conservative models, *Physica A* **191**(1–4), 41 – 46. → pp 81, 321, 322.

● Bak, P., 1996, *How Nature Works* (Copernicus, New York, NY, USA). → pp xii, *4*, 5, *6*, *17*, 319.

Bak, P., 1998, Life laws, *Nature* **391**(6668), 652–653. → p *68*.

● Bak, P., and K. Chen, 1991, Self-organized criticality, *Sci. Am.* **264**(1), 26–33. → p 6.

▶ Bak, P., K. Chen, and M. Creutz, 1989a, Self-organized criticality in the 'Game of Life', *Nature* **342**(6251), 780–782. → pp 81, 321, *322*, 485.

Bak, P., K. Chen, J. Scheinkman, and M. Woodford, 1993, Aggregate fluctuations from independent sectoral shocks: self-organized criticality in a model of production and inventory dynamics, *Ric. Econ.* **47**(1), 3 – 30. → p 75.

▶ Bak, P., K. Chen, and C. Tang, 1990, A forest-fire model and some thoughts on turbulence, *Phys. Lett. A* **147**(5–6), 297–300. → pp 72, 82, *112*, 113–116, 321, 485.

Bak, P., K. Christensen, L. Danon, and T. Scanlon, 2002a, Unified scaling law for earthquakes, *Phys. Rev. Lett.* **88**(17), 178501 (4 pp). → pp 67, 68.

Bak, P., K. Christensen, L. Danon, and T. Scanlon, 2002b, Unified scaling law for earthquakes, *Proc. Natl. Acad. Sci. USA* **99**(Suppl. 1), 2509–2513, colloquium. → p 68.

Bak, P., and H. Flyvbjerg, 1992, Self-organization of cellular magnetic-domain patterns, *Phys. Rev. A* **45**(4), 2192–2200. → p 62.

● Bak, P., and M. Paczuski, 1993, Why nature is complex, *Phys. World* **6**(12), 39–43. → p 6.

● Bak, P., and M. Paczuski, 1995, Complexity, contingency, and criticality, *Proc. Natl. Acad. Sci. USA* **92**(15), 6689–6696. → pp 6, 141.

▶ Bak, P., and K. Sneppen, 1993, Punctuated equilibrium and criticality in a simple model of evolution, *Phys. Rev. Lett.* **71**(24), 4083–4086. → pp 17, 68, 82, 141–144, 151, 152, 161, 319, 321, 485.

Bak, P., and C. Tang, 1989a, Earthquakes as a self-organized critical phenomenon, *J. Geophys. Res.* **94**(B11), 15635–15637. → pp 66, 67, 135, 139, 141.

● Bak, P., and C. Tang, 1989b, Self-organized criticality, *Phys. Today* **42**(1), S-27–S-28, American Institute of Physics Special Report 'Physics News in 1988'. → p 6.

▶ Bak, P., C. Tang, and K. Wiesenfeld, 1987, Self-organized criticality: an explanation of $1/f$ noise, *Phys. Rev. Lett.* **59**(4), 381–384. → pp xi, *3*, 4, 5, 9, 16, 18, 22, 54, 62, 82, 85–87, 89, 92, 102, 104, 204, 319, 346, 485.

Bak, P., C. Tang, and K. Wiesenfeld, 1988, Self-organized criticality, *Phys. Rev. A* **38**(1), 364–374. → pp 16, 62, 87–89, 102.

Bak, P., C. Tang, and K. Wiesenfeld, 1989b, Are earthquakes, fractals and $1/f$ noise self-organized critical phenomena?, in *Cooperative Dynamics in Complex Physical Systems, Proceedings of the Second Yukawa International Symposium, Kyoto, Japan, August 24–27, 1988*, edited by H. Takayama (Springer-Verlag, Berlin, Germany), volume 43 of *Springer Series in Synergetics*, pp. 274–279. → p *66*.

Ball, P., 2001, Long waits will always try patients, *Nature News* (online journal), 5 April 2001, doi:10.1038/news010404-16, comments (Smethurst and Williams, 2001), accessed 15 Jun 2010, URL http://www.nature.com/news/2001/010404/full/news010404-16.html. → p 75.

Banavar, J. R., A. Maritan, and A. Rinaldo, 1999, Size and form in efficient transportation networks, *Nature* **399**(6732), 130–132. → p 74.

Bandt, C., 2005, The discrete evolution model of Bak and Sneppen is conjugate to the classical contact process, *J. Stat. Phys.* **120**(3/4), 685–693. → p 158.

Barabási, A.-L., and H. E. Stanley, 1995, *Fractal Concepts in Surface Growth* (Cambridge University Press, Cambridge, UK). → pp 47, 178, 195, 203, 207, 266, 267, 299, 301, 313.

Barbay, J., and C. Kenyon, 2001, On the discrete Bak-Sneppen model of self-organized criticality, in *Proceedings of the twelfth annual ACM-SIAM symposium on Discrete Algorithms 2001, Washington, DC, USA, January 7–9, 2001*, pp. 928–944. → p 157.

Barber, M. N., 1983, Finite-size scaling, in *Phase Transitions and Critical Phenomena*, edited by C. Domb and J. L. Lebowitz (Academic Press, New York, NY, USA), volume 8, pp. 145–266. → pp 12, 27.

Barenblatt, G. I., 1996, *Scaling, Self-Similarity, and Intermediate Asymptotics* (Cambridge University Press, Cambridge, UK). → pp 25, 27, 348.

Barker, G. C., and A. Mehta, 1996, Rotated sandpiles: the role of grain reorganization and inertia, *Phys. Rev. E* **53**(6), 5704–5713. → p 56.

Barker, G. C., and A. Mehta, 2000, Avalanches at rough surfaces, *Phys. Rev. E* **61**(6), 6765–6772. → p 56.

Barkhausen, H., 1919, Zwei mit Hilfe der neuen Verstärker entdeckte Erscheinungen, *Phys. Z.* **20**, 401–403. → p 62.

Barrat, A., A. Vespignani, and S. Zapperi, 1999, Fluctuations and correlations in sandpile models, *Phys. Rev. Lett.* **83**(10), 1962–1965. → p 334.

Bartolozzi, M., D. B. Leinweber, and A. W. Thomas, 2005, Self-organized criticality and stock market dynamics: an empirical study, *Physica A* **350**(2–4), 451 – 465. → p 75.

Bassler, K. E., and M. Paczuski, 1998, Simple model of superconducting vortex avalanches, *Phys. Rev. Lett.* **81**(17), 3761–3764. → p 61.

Batterman, R. W., 2002, *The Devil in the Detail* (Oxford University Press, New York, NY, USA). → p 6.

Batty, M., and P. Longley, 1994, *Fractal Cities* (Academic Press, London, UK). → p 76.

Batty, M., and Y. Xie, 1999, Self-organized criticality and urban development, *Discrete Dyn. Nat. Soc.* **3**, 109–124. → p 76.

Bean, C. P., 1962, Magnetization of hard superconductors, *Phys. Rev. Lett.* **8**(6), 250–253. → p 58.

Bean, C. P., 1964, Magnetization of high-field superconductors, *Rev. Mod. Phys.* **36**(1), 31–39. → pp 58, 62.

Becker, V., and H. K. Janssen, 1994, Current-current correlation function in a driven diffusive system with nonconserving noise, *Phys. Rev. E* **50**(2), 1114–1122. → p 298.

Bédard, C., H. Kröger, and A. Destexhe, 2006, Does the $1/f$ frequency scaling of brain signals reflect self-organized critical states?, *Phys. Rev. Lett.* **97**(11), 118102 (4 pp). → p 70.

Beggs, J. M., 2008, The criticality hypothesis: how local cortical networks might optimize information processing, *Philos. Trans. R. Soc. London, Ser. A* **366**(1864), 329–343. → pp 70, 319.

Beggs, J. M., and D. Plenz, 2003, Neuronal avalanches in neocortical circuits, *J. Neurosci.* **23**(35), 11167–11177. → pp 70, 71.

Beggs, J. M., and D. Plenz, 2004, Neuronal avalanches are diverse and precise activity patterns that are stable for many hours in cortical slice cultures, *J. Neurosci.* **24**(22), 5216–5229. → p 70.

Behnia, K., C. Capan, D. Mailly, and B. Etienne, 2000, Internal avalanches in a pile of superconducting vortices, *Phys. Rev. B* **61**(6), R3815–R3818. → p *60*.

Behnia, K., C. Capan, D. Mailly, and B. Etienne, 2001, Avalanches in a pile of superconducting vortices, *J. Magn. Magn. Mater.* **226–230**(Part 1), 370 – 371. → p 60.

ben-Avraham, D., and S. Havlin, 2000, *Diffusion and Reactions in Fractals and Disordered Systems* (Cambridge University Press, Cambridge, UK). → pp 243, 248, 249.

ben-Avraham, D., and J. Köhler, 1992, Mean-field (n,m)-cluster approximation for lattice models, *Phys. Rev. A* **45**(12), 8358–8370. → pp 170, 181, 250, 394.

Ben-Hur, A., and O. Biham, 1996, Universality in sandpile models, *Phys. Rev. E* **53**(2), R1317–R1320. → pp 19, 23, 43, 92, 101, 164, 165, 167, 168, 171, 174, 177, 183, 268, *291*, 292.

Ben Hur, A., R. Hallgass, and V. Loreto, 1996, Renormalization procedure for directed self-organized critical models, *Phys. Rev. E* **54**(2), 1426–1432. → pp 268, 286.

Bengrine, M., A. Benyoussef, A. E. Kenz, M. Loulidi, and F. Mhirech, 1999a, A sharp transition between a trivial 1D BTW model and self-organized critical rice-pile model, *Eur. Phys. J. B* **12**(1), 129–133, numerics may not be independent from Bengrine *et al.* (1999b). → pp 189, 193, 198, 200, 284, 403.

Bengrine, M., A. Benyoussef, F. Mhirech, and S. D. Zhang, 1999b, Disorder-induced phase transition in a one-dimensional model of rice-pile, *Physica A* **272**(1–2), 1–11, numerics may not be independent from Bengrine *et al.* (1999a). → pp 189, 198, 403.

Bentley, R. A., and H. D. G. Maschner, 1999, Subtle nonlinearity in popular album charts, *Adv. Complex Syst.* **2**(3), 197–208. → p 76.

Bentley, R. A., and H. D. G. Maschner, 2001, Stylistic change as a self-organized critical phenomenon: an archaeological study in complexity, *J. Archaeol. Method Theory* **8**(1), 35–66. → p 76.

Benton, M. J., 1993, *The Fossil Record 2* (Chapman and Hall, London, UK). → p 69.

Benton, M. J., 1995, Diversification and extinction in the history of life, *Science* **268**(5207), 52–58. → pp 68, 69.

Berg, B. A., 1992, Double Jackknife bias-corrected estimators, *Comput. Phys. Commun.* **69**(1), 7–14. → p 216.

Berg, B. A., 2004, *Markov Chain Monte Carlo Simulations and Their Statistical Analysis* (World Scientific, Singapore). → pp 220, 358.

Bertotti, G., G. Durin, and A. Magni, 1994, Scaling aspects of domain wall dynamics and Barkhausen effect in ferromagnetic materials, *J. Appl. Phys.* **75**(10), 5490–5492, Proceedings of the 38th Annual Conference On Magnetism and Magnetic Materials, Minneapolis, MN, USA, November 15–18, 1993. → p 64.

Bhagavatula, R., G. Grinstein, Y. He, and C. Jayaprakash, 1992, Algebraic correlations in conserving chaotic systems, *Phys. Rev. Lett.* **69**(24), 3483–3486. → p 302.

Bhattacharjee, S. M., and F. Seno, 2001, A measure of data collapse for scaling, *J. Phys. A: Math. Gen.* **34**(33), 6375–6380. → p 238.

Biham, O., E. Milshtein, and O. Malcai, 2001, Evidence for universality within the classes of deterministic and stochastic sandpile models, *Phys. Rev. E* **63**(6), 061309 (8 pp). → pp 23, 100, 106, 165, 168, 178, 183, 269, 284, *291*.

Biham, O., E. Milshtein, and S. Solomon, 1998, Symmetries and universality classes in conservative sandpile models, arXiv:cond-mat/9805206 (unpublished). → pp 23, 106.

Binder, K., 1981a, Critical Properties from Monte Carlo coarse graining and renormalization, *Phys. Rev. Lett.* **47**(9), 693–696. → pp 39, 49, 69, 134, 227.

Binder, K., 1981b, Finite Size Scaling Analysis of Ising Model Block Distribution Functions, *Z. Phys. B* **43**, 119–140. → pp 39, 227.

Binder, K., and D. W. Heermann, 1997, *Monte Carlo Simulation in Statistical Physics* (Springer-Verlag, Berlin, Germany), 3rd edition. → pp 213, 220, 358, 361.

Birkeland, K. W., and C. C. Landry, 2002, Power-laws and snow avalanches, *Geophys. Res. Lett.* **29**(11), 1554 (3 pp). → p 194.

Blanchard, P., B. Cessac, and T. Krüger, 1997, A dynamical system approach to SOC models of Zhang's type, *J. Stat. Phys.* **88**(1/2), 307–318. → p 110.

Blanchard, P., B. Cessac, and T. Krüger, 2000, What can one learn about self-organized criticality from dynamical systems theory?, *J. Stat. Phys.* **98**(1/2), 375–404. → p 141.

Boettcher, S., and M. Paczuski, 1996, Exact results for spatiotemporal correlations in a self-organized critical model of punctuated equilibrium, *Phys. Rev. Lett.* **76**(3), 348–351. → pp 152, 157, 307.

Boettcher, S., and M. Paczuski, 1997a, Aging in a model of self-organized criticality, *Phys. Rev. Lett.* **79**(5), 889–892. → p 149.

Boettcher, S., and M. Paczuski, 1997b, Broad universality in self-organized critical phenomena, *Physica D* **107**(2–4), 171–173, Proceedings of the 16th Annual International Conference of the Center for Nonlinear Studies, Los Alamos, NM, USA, May 13–17, 1996, largely identical to Paczuski and Boettcher (1996). → p 440.

Boettcher, S., and M. Paczuski, 2000, $d_c = 4$ is the upper critical dimension for the Bak-Sneppen model, *Phys. Rev. Lett.* **84**(10), 2267–2270, arXiv:cond-mat/9911273v2. → pp 144, 152, 154, 155, 161.

Boffetta, G., V. Carbone, P. Giuliani, P. Veltri, and A. Vulpiani, 1999, Power laws in solar flares: self-organized criticality or turbulence?, *Phys. Rev. Lett.* **83**(22), 4662–4665. → p 72.

Boguñá, M., and Á. Corral, 1997, Long-tailed trapping times and Lévy flights in a self-organized critical granular system, *Phys. Rev. Lett.* **78**(26), 4950–4953. → pp 193, 195, 196.

Bonabeau, E., and P. Lederer, 1994, On $1/f$ power spectra, *J. Phys. A: Math. Gen.* **27**(9), L243–L250. → p 15.

● Bonachela, J. A., 2008, *Universality in Self-Organized Criticality*, Ph.D. Thesis, Departmento de Electromagnetismo y Física de la Materia & Institute Carlos I for Theoretical and Computational Physics, University of Granada, Granada, Spain, accessed 12 September 2009, URL http://hera.ugr.es/tesisugr/17706312.pdf. → pp 44, 92, 168, 177, 184, 187–189, 209, 247, 265, 284, 293, 296, *328*, 334, 393.

Bonachela, J. A., H. Chaté, I. Dornic, and M. A. Muñoz, 2007, Absorbing states and elastic interfaces in random media: two equivalent descriptions of self-organized criticality, *Phys. Rev. Lett.* **98**(15), 155702 (4 pp). → pp 176, 179, 180, 183, 300.

Bonachela, J. A., and M. A. Muñoz, 2007, How to discriminate easily between directed-percolation and Manna scaling, *Physica A* **384**(1), 89–93, Proceedings of the International Conference on Statistical Physics, Raichak and Kolkata, India, January 5–9, 2007. → pp 177, 179, 293, 296.

Bonachela, J. A., and M. A. Muñoz, 2008, Confirming and extending the hypothesis of universality in sandpiles, *Phys. Rev. E* **78**(4), 041102 (8 pp), arXiv:0806.4079. → pp 177–179, 183, 188, 293, 296, 305, *331*.

Bonachela, J. A., and M. A. Muñoz, 2009, Self-organization without conservation: true or just apparent scale-invariance?, *J. Stat. Mech.* **2009**(09), P09009 (37 pp). → pp 111, *265*, 328, 334, *335*.

Bonachela, J. A., J. J. Ramasco, H. Chaté, I. Dornic, and M. A. Muñoz, 2006, Sticky grains do not change the universality class of isotropic sandpiles, *Phys. Rev. E* **74**(5), 050102(R) (4 pp). → pp 178, 179, 296.

Bons, P. D., and B. P. van Milligen, 2001, New experiment to model self-organized critical transport and accumulation of melt and hydrocarbons from their source rocks, *Geology* **29**(10), 919–922. → pp 65, 193.

Bottani, S., and B. Delamotte, 1997, Self-organized criticality and synchronization in pulse coupled relaxation osillator system; the Olami, Feder and Christensen and the Feder and Feder model, *Physica D* **103**(1–4), 430–441. → pp 133, 138, 140.

Bouchaud, J.-P., M. E. Cates, J. R. Prakash, and S. F. Edwards, 1994, A model for the dynamics of sandpile surfaces, *J. Phys. I (France)* **4**, 1383–1410. → pp 292, 299.

Bouchaud, J.-P., M. E. Cates, J. R. Prakash, and S. F. Edwards, 1995, Hysteresis and metastability in a continuum sandpile model, *Phys. Rev. Lett.* **74**(11), 1982–1985. → p 56.

Boulter, C. J., and G. Miller, 2003, Nonuniversality and scaling breakdown in a nonconservative earthquake model, *Phys. Rev. E* **68**(5), 056108 (6 pp). → p 137.

Bramwell, S. T., K. Christensen, J.-Y. Fortin, P. C. W. Holdsworth, H. J. Jensen, S. Lise, J. M. López, M. Nicodemi, J.-F. Pinton, and M. Sellitto, 2000, Universal fluctuations in correlated systems, *Phys. Rev. Lett.* **84**(17), 3744–3747, arXiv:cond-mat/991225. → pp 134, 149, 458.

Bramwell, S. T., K. Christensen, J.-Y. Fortin, P. C. W. Holdsworth, H. J. Jensen, S. Lise, J. M. López, M. Nicodemi, J.-F. Pinton, and M. Sellitto, 2001, Bramwell et al. reply:, *Phys. Rev. Lett.* **87**(18), 188902 (1 p), reply to comment (Zheng and Trimper, 2001). → p 458.

Bramwell, S. T., P. C. W. Holdsworth, and J.-F. Pinton, 1998, Universality of rare fluctuations in turbulence and critical phenomena, *Nature* **396**, 552–554. → p 134.

Brandt, S., 1998, *Data Analysis* (Springer-Verlag, Berlin, Germany). → pp 214, 358.

Bretz, M., J. B. Cunningham, P. L. Kurczynski, and F. Nori, 1992, Imaging of avalanches in granular materials, *Phys. Rev. Lett.* **69**(16), 2431–2434. → pp *55*, 57.

Brézin, E., and J. Zinn-Justin, 1985, Finite size effects in phase transitions, *Nucl. Phys. B* **257** [FS14], 867–893. → p 251.

British Library Science Technology and Business (STB), 2001, *Welfare Reform on the Web (June 2001): National Health Service – Reform – General*, Welfare Reform Digest No. 22, http://www.bl.uk/welfarereform/issue22/nhs-rfrm.html, summary of Smethurst and Williams (2001), accessed 4 March 2010. → p *75*.

Broadbent, S. R., 1954, Discussion on Symposium on Monte Carlo Methods, *J. R. Stat. Soc. B, Stat. Method.* **16**(1), 68. → p 319.

Broadbent, S. R., and J. M. Hammersley, 1957, Percolation processes I. crystals and mazes, *Proc. Cambridge. Philos. Soc.* **53**(03), 629–641. → p 319.

Bröker, H.-M., and P. Grassberger, 1995, Mean-field behaviour in a local low-dimensional model, *Europhys. Lett.* **30**(6), 319–324. → pp 250, 328.

Bröker, H.-M., and P. Grassberger, 1997, Random neighbor theory of the Olami-Feder-Christensen earthquake model, *Phys. Rev. E* **56**(4), 3944–3952. → pp 129, 131, *133*, 251, 257, 261, 328.

Bröker, H.-M., and P. Grassberger, 1999, SOC in a population model with global control, *Physica A* **267**(3–4), 453 – 470, arXiv:cond-mat/9902195. → p 335.

Brower, R. C., M. A. Furman, and M. Moshe, 1978, Critical exponents for the Reggeon quantum spin model, *Phys. Lett. B* **76**(2), 213 – 219. → pp 142, 264.

Brown, J. H., and G. B. West (editors), 2000, *Scaling in Biology*, Santa Fe Institute Studies on the Sciences of Complexity (Oxford University Press, New York, NY, USA). → p 53.

Brown, S. R., C. H. Scholz, and J. B. Rundle, 1991, A simplified spring-block model of earthquakes, *Geophys. Res. Lett.* **18**(2), 215–218. → p 126.

Bruinsma, R., and G. Aeppli, 1984, Interface motion and nonequilibrium properties of the random-field Ising model, *Phys. Rev. Lett.* **52**(17), 1547–1550. → pp 205, 299, 315.

Buldyrev, S. V., A.-L. Barabási, F. Caserta, S. Havlin, H. E. Stanley, and T. Vicsek, 1992a, Anomalous interface roughening in porous media: experiment and model, *Phys. Rev. A* **45**(12), R8313–R8316. → pp 149, 203, 294, 301.

Buldyrev, S. V., N. V. Dokholyan, A. L. Goldberger, S. Havlin, C. K. Peng, H. E. Stanley, and G. M. Viswanathan, 1998, Analysis of DNA sequences using methods of statistical physics, *Physica A* **249**(1–4), 430 – 438. → p 74.

Buldyrev, S. V., J. Ferrante, and F. R. Zypman, 2006, Dry friction avalanches: experiment and theory, *Phys. Rev. E* **74**(6), 066110 (12 pp). → p 65.

Buldyrev, S. V., A. L. Goldberger, S. Havlin, C.-K. Peng, M. Simons, F. Sciortino, and H. E. Stanley, 1993, Long-range fractal correlations in DNA, *Phys. Rev. Lett.* **71**(11), 1776, comment on Voss (1992). → pp 74, 455.

Buldyrev, S. V., S. Havlin, and H. E. Stanley, 1992b, Anisotropic percolation and the d-dimensional surface roughening problem, *Physica A* **200**(1–4), 200–211. → p 203.

Bunzarova, N. Z., 2010, Statistical properties of directed avalanches, *Phys. Rev. E* **82**(3), 031116 (14 pp). → pp 288, 290.

Burkhardt, T. W., and J. M. J. van Leeuwen (editors), 1982, *Real-Space Renormalization*, Topics in Current Physics (Springer-Verlag, Berlin, Germany). → pp 101, 121, 266.

Burlando, B., 1990, The fractal dimension of taxonomic systems, *J. Theor. Biol.* **146**(1), 99–114. → p 69.

Burlando, B., 1993, The fractal geometry of evolution, *J. Theor. Biol.* **163**(2), 161–172. → p 69.

Burridge, R., and L. Knopoff, 1967, Model and theoretical seismicity, *Bull. Seismol. Soc. Am.* **57**(3), 341–371. → pp 125, 127, 128, 206, 485.

Butkovskiy, A. G., 1982, *Green's Functions and Transfer Functions Handbook* (Ellis Horwood, Chichester, UK). → p 249.

Cafiero, R., P. De Los Rios, F.-M. Dittes, A. Valleriani, and J. L. Vega, 1998, Power-law behavior in a nonextremal Bak-Sneppen model, *Phys. Rev. E* **58**(3), 3993–3996. → p 150.

Cafiero, R., A. Gabrielli, M. Marsili, and L. Pietronero, 1996, Theory of extremal dynamics with quenched disorder: invasion percolation and related models, *Phys. Rev. E* **54**(2), 1406–1425. → p 150.

Cafiero, R., V. Loreto, L. Pietronero, A. Vespignani, and S. Zapperi, 1995, Local rigidity and self-organized criticality for avalanches, *Europhys. Lett.* **29**(2), 111–116. → pp *109*, 204, 267, 344.

Cafiero, R., A. Valleriani, and J. L. Vega, 1999, Damage-spreading in the parallel Bak-Sneppen model, *Eur. Phys. J. B* **7**(4), 505–508. → pp 156, 157.

Caldarelli, G., F. D. Di Tolla, and A. Petri, 1996a, Self-organization and annealed disorder in a fracturing process, *Phys. Rev. Lett.* **77**(12), 2503–2506. → p 65.

Caldarelli, G., C. Tebaldi, and A. L. Stella, 1996b, Branching processes and evolution at the ends of a food chain, *Phys. Rev. Lett.* **76**(26), 4983–4986. → p 257.

Cannelli, G., R. Cantelli, and F. Cordero, 1993, Self-organized criticality of the fracture processes associated with hydrogen precipitation in niobium by acoustic emission, *Phys. Rev. Lett.* **70**(25), 3923–3926. → p 64.

Canning, D., L. A. N. Amaral, Y. Lee, M. Meyer, and H. E. Stanley, 1998, Scaling the volatility of GDP growth rates, *Econ. Lett.* **60**(3), 335 – 341. → p 75.

Carbone, V., R. Cavazzana, V. Antoni, L. Sorriso-Valvo, E. Spada, G. Regnoli, P. Giuliani, N. Vianello, F. Lepreti, R. Bruno, E. Martines, and P. Veltri, 2002, To what extent can dynamical models describe statistical features of turbulent flows?, *Europhys. Lett.* **58**(3), 349–355. → p 72.

Cardy, J. L. (editor), 1988, *Finite-Size Scaling* (North-Holland, Amsterdam, The Netherlands). → pp 12, 27.

Cardy, J. L., and R. L. Sugar, 1980, Directed percolation and Reggeon field theory, *J. Phys. A: Math. Gen.* **13**(12), L423–L427. → p 264.

Carlson, J. M., 1991, Time intervals between characteristic earthquakes and correlations with smaller events: an analysis based on a mechanical model of a fault, *J. Geophys. Res.* **96**(B3). → p 66.

⇒ Carlson, J. M., J. T. Chayes, E. R. Grannan, and G. H. Swindle, 1990a, Self-orgainzed criticality and singular diffusion, *Phys. Rev. Lett.* **65**(20), 2547–2550. → pp 343, 344.

Carlson, J. M., J. T. Chayes, E. R. Grannan, and G. H. Swindle, 1990b, Self-organized criticality in sandpiles: nature of the critical phenomenon, *Phys. Rev. A* **42**(4), 2467–2470. → p 343.

Carlson, J. M., and J. Doyle, 1999, Highly optimized tolerance: a mechanism for power laws in designed systems, *Phys. Rev. E* **60**(2), 1412–1427. → p 345.

Carlson, J. M., and J. Doyle, 2000, Highly optimized tolerance: robustness and design in complex systems, *Phys. Rev. Lett.* **84**(11), 2529–2532. → p 345.

Carlson, J. M., and J. S. Langer, 1989, Properties of earthquakes generated by fault dynamics, *Phys. Rev. Lett.* **62**(22), 2632–2635. → p 125.

Carlson, J. M., J. S. Langer, and B. E. Shaw, 1994, Dynamics of earthquake faults, *Rev. Mod. Phys.* **66**(2), 657–670. → p 66.

Carreras, B. A., V. E. Lynch, D. E. Newman, and G. M. Zaslavsky, 1999, Anomalous diffusion in a running sandpile model, *Phys. Rev. E* **60**(4), 4770–4778. → pp 193, 195.

Carreras, B. A., D. E. Newman, I. Dobson, and A. B. Poole, 2004, Evidence for self-organized criticality in a time series of electric power system blackouts, *IEEE Trans. Circuits Syst. I* **51**(9), 1733–1740. → p 76.

Carreras, B. A., D. Newman, V. E. Lynch, and P. H. Diamond, 1996, A model realization of self-organized criticality for plasma confinement, *Phys. Plasmas* **3**(8), 2903–2911. → pp 72, 193.

Carrillo, L., L. Mañosa, J. Ortín, A. Planes, and E. Vives, 1998, Experimental evidence for universality of acoustic emission avalanche distributions during structural transitions, *Phys. Rev. Lett.* **81**(9), 1889–1892. → p 63.

Casartelli, M., L. Dall'Asta, A. Vezzani, and P. Vivo, 2006, Dynamical invariants in the deterministic fixed-energy sandpile, *Eur. Phys. J. B* **52**(1), 91–105, arXiv:cond-mat/0502208v2. → p 334.

Černák, J., 2002, Self-organized criticality: robustness of scaling exponents, *Phys. Rev. E* **65**(4), 046141 (6 pp). → pp 88, 135.

Cessac, B., P. Blanchard, and T. Krüger, 2001, Lyapunov exponents and transport in the Zhang model of self-organized criticality, *Phys. Rev. E* **64**(1), 016133 (16 pp). → p 110.

Ceva, H., 1995, Influence of defects in a coupled map lattice modeling earthquakes, *Phys. Rev. E* **52**(1), 154–158. → p 134.

Ceva, H., 1998, On the asymptotic behavior of an earthquake model, *Phys. Lett. A* **245**(5), 413–418. → pp 128, 129, 134–136, 140, 358, 367.

Chabanol, M.-L., and V. Hakim, 1997, Analysis of a dissipative model of self-organized criticality with random neighbors, *Phys. Rev. E* **56**(3), R2343–R2346. → pp 131, 328.

Chaikin, P. M., and T. C. Lubensky, 1995, *Principles of Condensed Matter Physics* (Cambridge University Press, Cambridge, UK). → p 47.

Chang, T., 1992, Low-dimensional behavior and symmetry breaking of stochastic systems near criticality – can these effects be observed in space and in the laboratory?, *IEEE Trans. Plasma Sci.* **20**(6), 691–694. → p 72.

Chang, T., 1999, Self-organized criticality, multi-fractal spectra, sporadic localized reconnections and intermittent turbulence in the magnetotail, *Phys. Plasmas* **6**(11), 4137–4145. → p 72.

Chapman, S. C., 2000, Inverse cascade avalanche model with limit cycle exhibiting period doubling, intermittency, and self-similarity, *Phys. Rev. E* **62**(2), 1905–1911. → p 193.

Chapman, S. C., R. O. Dendy, and B. Hnat, 2001, Sandpile model with Tokamaklike enhanced confinement phenomenology, *Phys. Rev. Lett.* **86**(13), 2814–2817. → pp 72, 193, 209.

Chapman, S. C., R. O. Dendy, and G. Rowlands, 1999, A sandpile model with dual scaling regimes for laboratory, space and astrophysical plasmas, *Phys. Plasmas* **6**(11), 4169–4177. → p 72.

Chatto, A., R. Lee, R. Duncan, and D. Goodstein, 2007, Experiments on the self-organized critical state of ^{4}He, *J. Low Temp. Phys.* **148**(5), 519–526. → p 62.

Chau, H. F., 1994, Generalized Abelian sandpile model, *Physica A* **205**(1–3), 292–298. → p 101.

Chau, H. F., and K. S. Cheng, 1992, Relations of $1/f$ and $1/f^2$ power spectra to self-organized criticality, *Phys. Rev. A* **46**(6), R2981–R2983. → p 185.

Chau, H. F., and K. S. Cheng, 1993, Generalized Abelian sandpile model, *J. Math. Phys.* **34**(11), 5109–5117. → p 101.

Chauve, P., and P. Le Doussal, 2001, Exact multilocal renormalization group and applications to disordered problems, *Phys. Rev. E* **64**(5), 051102 (27 pp). → pp 266, 267.

Che, X., and H. Suhl, 1990, Magnetic domain patterns as self-organizing critical systems, *Phys. Rev. Lett.* **64**(14), 1670–1673. → p 62.

Chen, C.-C., and M. den Nijs, 2002, Interface view of directed sandpile dynamics, *Phys. Rev. E* **65**(3), 031309 (4 pp). → p 299.

Chen, C.-F., and C.-Y. Lin, 2009, Scalings of a modified Manna model with bulk dissipation, *Int. J. Mod. Phys. C* **20**(2). → p 169.

Chen, D.-M., S. Wu, A. Guo, and Z. R. Yang, 1995, Self-organized criticality in a cellular automaton model of pulse-coupled integrate-and-fire neurons, *J. Phys. A: Math. Gen.* **28**(18), 5177–5182. → p 70.

Chen, K., P. Bak, and M. H. Jensen, 1990, A deterministic critical forest fire model, *Phys. Lett. A* **149**(4), 207–210. → pp 114, 116.

Chen, K., P. Bak, and S. P. Obukhov, 1991, Self-organized criticality in a crack-propagation model of earthquakes, *Phys. Rev. A* **43**(2), 625–630. → pp 66, 129.

Chen, Y., M. Ding, and J. A. S. Kelso, 1997, Long memory processes ($1/f^\alpha$ type) in human coordination, *Phys. Rev. Lett.* **79**(22), 4501–4504. → p 76.

Chessa, A., E. Marinari, and A. Vespignani, 1998, Energy constrained sandpile models, *Phys. Rev. Lett.* **80**(19), 4217–4220. → pp 171, 305–307, 331, 333, 334, 340.

Chessa, A., H. E. Stanley, A. Vespignani, and S. Zapperi, 1999a, Universality in sandpiles, *Phys. Rev. E* **59**(1), R12–R15, numerics may not be independent from Chessa *et al.* (1999b). → pp 30, 31, 39, 43, 44, 92, 103, 164, 166, 168, 183.

Chessa, A., A. Vespignani, and S. Zapperi, 1999b, Critical exponents in stochastic sandpile models, *Comput. Phys. Commun.* **121–122**, 299–302. → pp 168, 183, 284, 409.

Chialvo, D. R., 2004, Critical brain networks, *Physica A* **340**(4), 756–765, Proceedings of the Symposium Complexity and Criticality: in Memory of Per Bak (1947–2002), Copenhagen, Denmark, August 21–23, 2003. → p 71.

Chianca, C. V., J. S. Sá Martins, and P. M. C. de Oliveira, 2009, Mapping the train model for earthquakes onto the stochastic sandpile model, *Eur. Phys. J. B* **68**(4), 549–555. → pp 178, 206.

Christensen, K., 1992, *Self-Organization in Models of Sandpiles, Earthquakes, and Flash-ing Fireflies*, Ph.D. Thesis, Institute of Physics and Astronomy, University of Aarhus, Denmark. → pp 88, 99, 104, 128, 129, 134, 290.

Christensen, K., 1993, Christensen replies, *Phys. Rev. Lett.* **71**(8), 1289, reply to comment (Klein and Rundle, 1993). → pp 136, 428.

Christensen, K., 2004, On self-organised criticality in one dimension, *Physica A* **340**(4), 527–534, Proceedings of the Symposium Complexity and Criticality: in Memory of Per Bak (1947–2002), Copenhagen, Denmark, August 21–23, 2003. → pp 184, 188–190, 209, 236, 315.

Christensen, K., 2008, personal communication. → p 185.

▶ Christensen, K., Á. Corral, V. Frette, J. Feder, and T. Jøssang, 1996, Tracer dispersion in a self-organized critical system, *Phys. Rev. Lett.* **77**(1), 107–110. → pp 23, 58, 82, *180*, 183, 186, 189, 192, *193*, 194–196, *197*, 198, 485.

Christensen, K., R. Donangelo, B. Koiller, and K. Sneppen, 1998, Evolution of random networks, *Phys. Rev. Lett.* **81**(11), 2380–2383. → p 159.

Christensen, K., N. Farid, G. Pruessner, and M. Stapleton, 2008, On the scaling of probability density functions with apparent power-law exponents less than unity, *Eur. Phys. J. B* **62**(3), 331–336. → pp 27, 29, 32, 57, 58, 192, 391.

Christensen, K., H. Flyvbjerg, and Z. Olami, 1993, Self-organized critical forest-fire model: mean-field theory and simulation results in 1 to 6 dimensions, *Phys. Rev. Lett.* **71**(17), 2737–2740. → pp 120, 121, 124, 125, 251.

Christensen, K., H. C. Fogedby, and H. J. Jensen, 1991, Dynamical and spatial aspects of sandpile cellular automata, *J. Stat. Phys.* **63**(3/4), 653–684. → pp 16, 42, 92, 93, 102, 104.

Christensen, K., D. Hamon, H. J. Jensen, and S. Lise, 2001, Comment on 'self-organized criticality in the Olami-Feder-Christensen model', *Phys. Rev. Lett.* **87**(3), 039801, comment on (de Carvalho and Prado, 2000), reply (de Carvalho and Prado, 2001). → pp 131, 413.

● Christensen, K., and N. R. Moloney, 2005, *Complexity and Criticality* (Imperial College Press, London, UK). → pp 6, 128, 129, 136, 138, 139, 220.

Christensen, K., N. R. Moloney, O. Peters, and G. Pruessner, 2004, Avalanche behavior in an absorbing state Oslo model, *Phys. Rev. E* **70**(6), 067101 (4 pp), arXiv:cond-mat/0405454. → pp 248, 282, 306, 307, 333, 339, 340.

Christensen, K., and Z. Olami, 1992a, Scaling, phase transitions, and nonuniversality in a self-organized critical cellular-automaton model, *Phys. Rev. A* **46**(4), 1829–1838. → pp *68*, 128, 129, 133, *135*, 136, 138, 139, 141.

Christensen, K., and Z. Olami, 1992b, Variation of the Gutenberg-Richter *b* values and nontrivial temporal correlations in a spring-block model of earthquakes, *J. Geophys. Res.* **97**(B6), 8729–8735. → pp 67, 68, 76, 128.

Christensen, K., and Z. Olami, 1993, Sandpile models with and without an underlying spatial structure, *Phys. Rev. E* **48**(5), 3361–3372. → pp 43, 103, 131, 247, 290.

Christensen, K., Z. Olami, and P. Bak, 1992, Deterministic $1/f$ noise in nonconservative models of self-organized criticality, *Phys. Rev. Lett.* **68**(16), 2417–2420. → pp 15, 135.

Chua, A., 2003, Creation and annihilation operators for the Manna model, personal communication (unpublished). → p 173.

Chua, A., and K. Christensen, 2002, Exact enumeration of the critical states in the Oslo Model, arXiv:cond-mat/0203260v2 (unpublished). → pp 186, *187*, 343.

Chua, A., and G. Pruessner, 2003, The Markov matrix of the Oslo rice pile (unpublished). → p 202.

Cieplak, M., and M. O. Robbins, 1988, Dynamical transition in quasistatic fluid invasion in porous media, *Phys. Rev. Lett.* **60**(20), 2042–2045. → p 301.

Ciliberto, S., and C. Laroche, 1994, Experimental evidence of self organized criticality in the stick-slip dynamics of two rough elastic surfaces, *J. Phys. I (France)* **4**, 223–235. → p 65.

Clar, S., B. Drossel, and F. Schwabl, 1994, Scaling laws and simulation results for the self-organized critical forest-fire model, *Phys. Rev. E* **50**(2), 1009–1018. → pp 120, 121, 123–125.

Clar, S., B. Drossel, and F. Schwabl, 1996, Forest fires and other examples of self-organized criticality, *J. Phys.: Condens. Matter* **8**(37), 6803–6824. → p 118.

Clauset, A., C. R. Shalizi, and M. E. J. Newman, 2009, Power-law distributions in empirical data, *SIAM Rev.* **51**(4), 661–703, arXiv:0706.1062v2. → pp 220, 229.

Coniglio, A., and W. Klein, 1980, Clusters and Ising critical droplets: a renormalisation group approach, *J. Phys. A: Math. Gen.* **13**(8), 2775–2780. → p 115.

Cormen, T. H., C. E. Leiserson, and R. L. Rivest, 1996, *Introduction to Algorithms* (MIT Press, Cambridge, MA, USA). → pp 123, 160, 358, 367.

Corral, Á., 2003, Local distributions and rate fluctuations in a unified scaling law for earthquakes, *Phys. Rev. E* **68**(3), 035102(R) (4 pp). → p 68.

Corral, Á., 2004a, Calculation of the transition matrix and of the occupation probabilities for the states of the Oslo sandpile model, *Phys. Rev. E* **69**(2), 026107 (12 pp), arXiv:cond-mat/0310181v1. → pp 17, 93, 187, 200–203.

Corral, Á., 2004b, Long-term clustering, scaling, and universality in the temporal occurrence in earthquakes, *Phys. Rev. Lett.* **92**(10), 108501 (4 pp). → p 68.

Corral, Á., 2004c, Universal local versus unified global scaling laws in the statistics of seismicity, *Physica A* **340**(4), 590–597. → p 68.

Corral, Á., and K. Christensen, 2006, Comment on 'earthquakes descaled: on waiting time distributions and scaling laws', *Phys. Rev. Lett.* **96**(10), 109801 (1 p), comment on (Lindman *et al.*, 2005), reply (Lindman *et al.*, 2006). → pp 68, 431.

Corral, Á., and A. Díaz-Guilera, 1997, Symmetries and fixed point stability of stochastic differential equations modeling self-organized criticality, *Phys. Rev. E* **55**(3), 2434–2445, arXiv:cond-mat/9612100. → pp 101, 109, 203, 266, 267.

Corral, Á., and M. Paczuski, 1999, Avalanche merging and continuous flow in a sandpile model, *Phys. Rev. Lett.* **83**(3), 572–575. → pp 193, *194*, 300, 302, 322.

Corral, Á., C. J. Pérez, A. Díaz-Guilera, and A. Arenas, 1995, Self-organized criticality and synchronization in a lattice model of integrate-and-fire oscillators, *Phys. Rev. Lett.* **74**(1), 118–121. → p 70.

Corral, Á., L. Telesca, and R. Lasaponara, 2008, Scaling and correlations in the dynamics of forest-fire occurrence, *Phys. Rev. E* **77**(1), 016101 (7 pp). → p 73.

Corté, L., P. M. Chaikin, J. P. Gollub, and D. J. Pine, 2008, Random organization in periodically driven systems, *Nat. Phys.* **4**(5), 420–424. → p 58.

Corté, L., S. J. Gerbode, W. Man, and D. J. Pine, 2009, Self-organized criticality in sheared suspensions, *Phys. Rev. Lett.* **103**(24), 248301 (4 pp). → pp 58, *61*.

Costello, R. M., K. L. Cruz, C. Egnatuk, D. T. Jacobs, M. C. Krivos, T. S. Louis, R. J. Urban, and H. Wagner, 2003, Self-organized criticality in a bead pile, *Phys. Rev. E* **67**(4), 041304 (9 pp). → pp 56, 134.

Cote, P. J., and L. V. Meisel, 1991, Self-organized criticality and the Barkhausen effect, *Phys. Rev. Lett.* **67**(10), 1334–1337. → p 62.

Cote, P. J., and L. V. Meisel, 1993, Self-organized criticality and the Barkhausen effect in amorphous and polycrystalline metals, *Int. J. Mod. Phys. B* **7**(1–3), 934–937, Proceedings of the International Conference on the Physics of Transition Metals, Darmstadt, Germany, July 20–24, 1992. → p 63.

Creed, M., 2004, Work 370: Balls (image of the work), multiple parts; overall dimensions variable. Installation view at Hauser & Wirth, London. Courtesy of the artist and Hauser & Wirth. Photo: Hugo Glendinning. → p xvi.

Creutz, M., 1991, Abelian sandpiles, *Comput. Phys.* **5**(2), 198–203. → p 283.

Creutz, M., 2004, Playing with sandpiles, *Physica A* **340**(4), 521 – 526, Proceedings of the symposium Complexity and Criticality: in Memory of Per Bak (1947–2002), Copenhagen, Denmark, August 21–23, 2003. → pp 97, 283.

Crisanti, A., M. H. Jensen, A. Vulpiani, and G. Paladin, 1992, Strongly intermittent chaos and scaling in an earthquake model, *Phys. Rev. A* **46**(12), R7363–R7366. → p 141.

Cross, M. C., and P. C. Hohenberg, 1993, Pattern formation outside of equilibrium, *Rev. Mod. Phys.* **65**(3), 851. → p 318.

Crovella, M. E., and A. Bestavros, 1997, Self-similarity in World Wide Web traffic: evidence and possible causes, *IEEE Trans. Network.* **5**(6), 835–846. → p 76.

Crutchfield, J., M. Nauenberg, and J. Rudnick, 1981, Scaling for external noise at the onset of chaos, *Phys. Rev. Lett.* **46**(14), 933–935. → p 319.

Cule, D., and T. Hwa, 1996, Tribology of sliding elastic media, *Phys. Rev. Lett.* **77**(2), 278–281. → pp 178, 206, 302.

da Silva, L., A. R. R. Papa, and A. M. C. de Souza, 1998, Criticality in a simple model for brain functioning, *Phys. Lett. A* **242**(6), 343–348. → pp 70, 149.

Daerden, F., and C. Vanderzande, 1996, $1/f$ noise in the Bak-Sneppen model, *Phys. Rev. E* **53**(5), 4723–4728, comment (Davidsen and Lüthje, 2001). → pp 152, 154, 252.

Daerden, F., and C. Vanderzande, 1998, Sandpiles on a Sierpinski gasket, *Physica A* **256**(3–4), 533–546. → pp 24, 101.

Dahlstedt, K., and H. J. Jensen, 2001, Universal fluctuations and extreme-value statistics, *J. Phys. A: Math. Gen.* **34**(50), 11193–11200. → p 149.

Dahmen, K., S. Kartha, J. A. Krumhansl, B. W. Roberts, J. P. Sethna, and J. D. Shore, 1994, Disorder-driven first-order phase transformations: a model for hysteresis, *J. Appl. Phys.* **75**(10), 5946–5948, Proceedings of the 38th Annual Conference On Magnetism and Magnetic Materials, Minneapolis, MN, USA, November 15–18, 1993. → p 63.

Dahmen, K., and J. P. Sethna, 1993, Hysteresis loop critical exponents in 6-ϵ dimensions, *Phys. Rev. Lett.* **71**(19), 3222–3225. → p 63.

Dahmen, K., and J. P. Sethna, 1996, Hysteresis, avalanches, and disorder-induced critical scaling: a renormalization-group approach, *Phys. Rev. B* **53**(22), 14872–14905. → p 63.

Dall'Asta, L., 2006, Exact solution of the one-dimensional deterministic fixed-energy sandpile, *Phys. Rev. Lett.* **96**(5), 058003 (4 pp). → p 333.

Dantas, W. G., and J. F. Stilck, 2006, Generalized Manna sandpile model with height restrictions, *Braz. J. Phys.* **36**(3A). → pp 333, 340.

Das Sarma, S., S. V. Ghaisas, and J. M. Kim, 1994, Kinetic super-roughening and anomalous dynamic scaling in nonequilibrium growth models, *Phys. Rev. E* **49**(1), 122–125. → pp 313, 314.

Datta, A. S., K. Christensen, and H. J. Jensen, 2000, On the physical relevance of extremal dynamics, *Europhys. Lett.* **50**(2), 162–168. → pp 150, 161.

Davidsen, J., and C. Goltz, 2004, Are seismic waiting time distributions universal?, *Geophys. Res. Lett.* **31**(21), L21612 (4 pp). → p 68.

Davidsen, J., and N. Lüthje, 2001, Comment on '1/f noise in the Bak-Sneppen model', *Phys. Rev. E* **63**(6), 063101 (2 pp), arXiv:cond-mat/0201201. → pp 15, 154, 412.

Davidsen, J., and M. Paczuski, 2005, Analysis of the spatial distribution between successive earthquakes, *Phys. Rev. Lett.* **94**(4), 048501 (4 pp), comment (Werner and Sornette, 2007). → pp 68, 456.

Davidsen, J., and M. Paczuski, 2007, Davidsen and Paczuski reply:, *Phys. Rev. Lett.* **99**(17), 179802, reply to comment (Werner and Sornette, 2007). → pp 68, 456.

de Arcangelis, L., C. Perrone-Capano, and H. J. Herrmann, 2006, Self-organized criticality model for brain plasticity, *Phys. Rev. Lett.* **96**(2), 028107 (4 pp). → pp 70, 71.

de Boer, J., B. Derrida, H. Flyvbjerg, A. D. Jackson, and T. Wettig, 1994, Simple model of self-organized biological evolution, *Phys. Rev. Lett.* **73**(6), 906–909. → pp 143, 152, 158, 251.

de Boer, J., A. D. Jackson, and T. Wettig, 1995, Criticality in simple models of evolution, *Phys. Rev. E* **51**(2), 1059–1074. → pp 152, 153, 251.

de Carvalho, J. X., and C. P. C. Prado, 2000, Self-organized criticality in the Olami-Feder-Christensen model, *Phys. Rev. Lett.* **84**(17), 4006–4009, comment (Christensen *et al.*, 2001). → pp 131, 410.

de Carvalho, J. X., and C. P. C. Prado, 2001, de Carvalho and Prado reply:, *Phys. Rev. Lett.* **87**(3), 039802 (1 p), reply to comment (Christensen *et al.*, 2001). → p 410.

de Gennes, P. G., 1966, *Superconductivity of Metals and Alloys* (Physica B, Reading, MA, USA), translated by P. A. Pincus. → p 58.

De Los Rios, P., M. Marsili, and M. Vendruscolo, 1998, High-dimensional Bak-Sneppen Model, *Phys. Rev. Lett.* **80**(26), 5746–5749. → pp 143, 150, 152, 161.

De Los Rios, P., and Y.-C. Zhang, 1999, Universal 1/f noise from dissipative self-organized criticality models, *Phys. Rev. Lett.* **82**(3), 472–475. → p 15.

De Menech, M., and A. L. Stella, 2000, From waves to avalanches: two different mechanisms of sandpile dynamics, *Phys. Rev. E* **62**(4), R4528–R4531. → pp 92, 100, 170.

De Menech, M., A. L. Stella, and C. Tebaldi, 1998, Rare events and breakdown of simple scaling in the Abelian sandpile model, *Phys. Rev. E* **58**(3), R2677–R2680. → pp 37, 51, 92, 102, 103, 220, 229.

de Oliveira, M. M., and R. Dickman, 2005, How to simulate the quasistationary state, *Phys. Rev. E* **71**(1), 016129 (5 pp). → p 336.

▶ de Sousa Vieira, M., 1992, Self-organized criticality in a deterministic mechanical model, *Phys. Rev. A* **46**(10), 6288–6293. → pp 126, 178, 485.

de Sousa Vieira, M., 2000, Simple deterministic self-organized critical system, *Phys. Rev. E* **61**(6), R6056–R6059. → pp 126, 178, 302.

de Sousa Vieira, M., 2002, Breakdown of self-organized criticality in sandpiles, *Phys. Rev. E* **66**(5), 051306 (5 pp). → p 192.

Dendy, R. O., and P. Helander, 1998, Appearance and nonappearance of self-organized criticality in sandpiles, *Phys. Rev. E* **57**(3), 3641–3644. → pp 55, 193.

Dendy, R. O., P. Helander, and M. Tagger, 1998, On the role of self-organised criticality in accretion systems, *Astron. Astrophys.* **337**(3), 962–965, updated in Dendy *et al.* (1999). → p 414.

Dendy, R. O., P. Helander, and M. Tagger, 1999, Self-organised criticality in astrophysical accretion systems, *Phys. Scripta* **T82**, 133–136, Topical Issue, Nonlinear Plasma Science: Special Issue in Honour of Professor Lennart Stenflo on the Occasion of his 60th Birthday, essentially identical to Dendy, Helander, and Tagger (1998). → pp *4*, 72, 193, 414.

Derrida, B., and M. R. Evans, 1997, The asymmetric exclusion model: exact results through a matrix approach, in *Nonequilibrium Statistical Mechanics in One Dimension*, edited by V. Privman (Cambridge University Press, Cambridge, UK), pp. 277–304. → p 202.

Derrida, B., M. R. Evans, V. Hakim, and V. Pasquier, 1993, Exact solution of a 1D asymmetric exclusion model using a matrix formulation, *J. Phys. A: Math. Gen.* **26**(7), 1493–1517. → p 176.

▶ Dhar, D., 1990a, Self-organized critical state of sandpile automaton models, *Phys. Rev. Lett.* **64**(14), 1613–1616, see erratum (Dhar, 1990b). → pp 82, 85–87, 90, 91, *94*, 95, 96, 103, 129, 187, 198, 243, 246, 248, 249, 269, 275, 279, 285, 354, 485.

Dhar, D., 1990b, Self-organized critical state of sandpile automaton models, *Phys. Rev. Lett.* **64**(23), 2837, erratum regarding numbering of references. → p 414.

▶ ● Dhar, D., 1999a, The Abelian sandpile and related models, *Physica A* **263**(1–4), 4–25, arXiv:cond-mat/9808047. → pp 6, 82, 91, 95, 99–102, 163, 164, 167, 169, 183, 199, 250, 269, 277, 285, 485.

Dhar, D., 1999b, Some results and a conjecture for Manna's stochastic sandpile model, *Physica A* **270**(1–2), 69–81, arXiv:cond-mat/9902137. → pp 167, 169, 181, 277, 279, 286.

▶ ● Dhar, D., 1999c, Studying self-organized criticality with exactly solved models, arXiv:cond-mat/9909009 (unpublished). → pp 6, 94–97, 101, 135, 201, *276*, 277, 285, 414.

Dhar, D., 2004, Steady state and relaxation spectrum of the Oslo rice-pile model, *Physica A* **340**(4), 535–543, arXiv:cond-mat/0309490. → pp 95, 163, 186–188, 198–200, *201*, 275, 277, *296*, 301.

● Dhar, D., 2006, Theoretical studies of self-organized criticality, *Physica A* **369**(1), 29–70, Proceedings of the 11th International Summerschool on Fundamental Problems in Statistical Physics, Leuven, Belgium, September 4 – 17, 2005; updated from Dhar (1999c). → pp 6, 101, 193, 197, 285.

Dhar, D., and S. N. Majumdar, 1990, Abelian sandpile model on the Bethe lattice, *J. Phys. A: Math. Gen.* **23**(19), 4333–4350. → pp 100, 101, 243, 251, 320, 343.

Dhar, D., and S. S. Manna, 1994, Inverse avalanches in the Abelian sandpile model, *Phys. Rev. E* **49**(4), 2684–2697. → p 99.

Dhar, D., and P. Pradhan, 2004, Probability distribution of residence times of grains in sand-pile models, *J. Stat. Mech.* **2004**(05), P05002 (12 pp), includes erratum. → p 196.

► Dhar, D., and R. Ramaswamy, 1989, Exactly solved model of self-organized critical phenomena, *Phys. Rev. Lett.* **63**(16), 1659–1662. → pp 22, 24, 101, 206, 249, 268, 285, 286, 288–290, 485.

Dhar, D., P. Ruelle, S. Sen, and D.-N. Verma, 1995, Algebraic aspects of Abelian sandpile models, *J. Phys. A: Math. Gen.* **28**(4), 805–831. → pp 87, 95, 96.

Dhar, D., and P. B. Thomas, 1992, Hysteresis and self-organized criticality in the $O(N)$ model in the limit $N \to \infty$, *J. Phys. A: Math. Gen.* **25**(19), 4967–4984. → p 343.

Diamond, P. H., and T. S. Hahm, 1995, On the dynamics of turbulent transport near marginal stability, *Phys. Plasmas* **2**(10), 3640–3649. → p 72.

Díaz-Guilera, A., 1992, Noise and dynamics of self-organized critical phenomena, *Phys. Rev. A* **45**(2), 8551–8558. → pp 104, 106, 108, 109, 243, 266, 267, 299, 322.

Díaz-Guilera, A., 1994, Dynamic renormalization group approach to self-organized critical phenomena, *Europhys. Lett.* **26**(3), 177–182, arXiv:cond-mat/9403036v1. → pp 101, 106, 109, 110, 203, 266, 299, 420.

Dickman, R., 1994, Numerical study of a field theory for directed percolation, *Phys. Rev. E* **50**(6), 4404–4409. → p 182.

Dickman, R., 2002a, *n*-site approximations and coherent-anomaly-method analysis for a stochastic sandpile, *Phys. Rev. E* **66**(3), 036122 (6 pp). → p 181.

Dickman, R., 2002b, Nonequilibrium phase transitions in epidemics and sandpiles, *Physica A* **306**, 90–97, arXiv:cond-mat/0110043. → pp 171, 177, 293.

Dickman, R., 2004a, Fractal rain distributions and chaotic advection, *Braz. J. Phys.* **34**(2A), 337–346, arXiv:physics/0311083. → p 71.

Dickman, R., 2004b, personal communication. → p 390.

Dickman, R., 2006, Critical exponents for the restricted sandpile, *Phys. Rev. E* **73**(3), 036131 (5 pp). → pp 44, 168, 170, 177–180, 285, 294, 316.

Dickman, R., M. Alava, M. A. Muñoz, J. Peltola, A. Vespignani, and S. Zapperi, 2001, Critical behavior of a one-dimensional fixed-energy stochastic sandpile, *Phys. Rev. E* **64**(5), 056104 (7 pp). → pp 168, 177, 179, 299, 313, 315, 331, 333, 337, 339, 340.

Dickman, R., and J. M. M. Campelo, 2003, Avalanche exponents and corrections to scaling for a stochastic sandpile, *Phys. Rev. E* **67**(6), 066111 (5 pp). → pp 32, 165, 168, 169, 171, 180, 339.

Dickman, R., and J. Kamphorst Leal da Silva, 1998, Moment ratios for absorbing-state phase transitions, *Phys. Rev. E* **58**(4), 4266–4270. → p 39.

► ● Dickman, R., M. A. Muñoz, A. Vespignani, and S. Zapperi, 2000, Paths to self-organized criticality, *Braz. J. Phys.* **30**(1), 27–41, arXiv:cond-mat/9910454v2. → pp 6, *10*, 53, 56, 63, 64, 111, 170–172, 265, 266, 306, 321, 328, 330, 331, 334, *335*, 485.

Dickman, R., L. Rolla, and V. Sidoravicius, 2010, Activated random walkers: facts, conjectures and challenges, *J. Stat. Phys.* **138**(1), 126–142. → p 330.

Dickman, R., T. Tomé, and M. J. de Oliveira, 2002, Sandpiles with height restrictions, *Phys. Rev. E* **66**(1), 016111 (8 pp). → pp 40, 168, 170, 171, 177, 178, *180*, 181, 182, 283, 285, 292, 293.

▶ ⇒ Dickman, R., A. Vespignani, and S. Zapperi, 1998, Self-organized criticality as an absorbing state phase transition, *Phys. Rev. E* **57**(5), 5095–5105. → pp 10, 165, 170–172, 299, 300, 306, 307, 330–333, 339–341, 396.

Dickman, R., and R. Vidigal, 2002, Path-integral representation for a stochastic sandpile, *J. Phys. A* **35**(34), 7269–7285. → p 182.

Dimri, V. P., and M. R. Prakash, 2001a, Reply to comments by Dr. James W. Kirchner on 'scaling of power spectrum of extinction events in the fossil record', *Earth Planet. Sci. Lett.* **192**(4), 623 – 625, reply to Kirchner (2001). → pp 69, 428.

Dimri, V. P., and M. R. Prakash, 2001b, Scaling of power spectrum of extinction events in the fossil record, *Earth Planet. Sci. Lett.* **186**(3–4), 363 – 370, comment (Kirchner, 2001). → pp 69, 428.

Ding, E. J., Y. N. Lu, and H. F. Ouyang, 1992, Theoretical sandpile with stochastic slide, *Phys. Rev. A* **46**(10), R6136–R6139. → pp 56, 184.

Diodati, P., F. Marchesoni, and S. Piazza, 1991, Acoustic emission from volcanic rocks: an example of self-organized criticality, *Phys. Rev. Lett.* **67**(17), 2239–2243. → p 64.

Doi, M., 1976, Second quantization representation for classical many-particle system, *J. Phys. A: Math. Gen.* **9**(9), 1465–1477. → p 266.

Domany, E., and W. Kinzel, 1984, Equivalence of cellular automata to Ising models and directed percolation, *Phys. Rev. Lett.* **53**(4), 311–314. → p 298.

Dorn, P. L., D. S. Hughes, and K. Christensen, 2001, On the avalanche size distribution in the BTW model, preprint from `http://www.cmth.ph.ic.ac.uk/kim/papers/preprints/preprint_btw.pdf`, accessed 19 October 2010 (unpublished). → p 103.

Dornic, I., H. Chaté, and M. A. Muñoz, 2005, Integration of Langevin equations with multiplicative noise and the viability of field theories for absorbing phase transitions, *Phys. Rev. Lett.* **94**(10), 100601 (4 pp). → pp 177, 182.

Dougherty, D., 1992, *sed & awk* (O'Reilly, Sebastopol, CA, USA). → p 358.

Dowd, K., and C. Severance, 1998, *High Performance Computing* (O'Reilly, Sebastopol, CA, USA), 2nd edition. → pp 358, 361, 369, 380.

Drossel, B., 1997, Renormalization group approach to the critical behavior of the forest-fire model, *Phys. Rev. Lett.* **78**(7), 1392. → p 119.

Drossel, B., 1999a, An alternative view of the Abelian sandpile model, `arXiv:cond-mat/9904075v1` (unpublished). → p 104.

Drossel, B., 1999b, On the scaling behavior of the Abelian sandpile model, `arXiv:cond-mat/9904075v2` (unpublished). → p *104*.

Drossel, B., 2000, Scaling behavior of the Abelian sandpile model, *Phys. Rev. E* **61**(3), R2168–R2171, `arXiv:cond-mat/9904075v2`. → pp 55, 103, 104, 137, 169, 321.

Drossel, B., 2002, Complex scaling behavior of nonconserved self-organized critical systems, *Phys. Rev. Lett.* **89**(23), 238701 (4 pp), `arXiv:cond-mat/0205658`. → pp 134, 137, *140*, 358, 376.

Drossel, B., S. Clar, and F. Schwabl, 1993, Exact results for the one-dimensional self-organized critical forest-fire model, *Phys. Rev. Lett.* **71**(23), 3739–3742. → pp 116, 120, *121*, 124.

▶ Drossel, B., and F. Schwabl, 1992a, Self-organized critical forest-fire model, *Phys. Rev. Lett.* **69**(11), 1629–1632, largely identical to Proceedings article (Drossel and Schwabl, 1992b). → pp 82, 112, 115, 116, *117*, 118–121, 124, 321, 391, 417, 485.

Drossel, B., and F. Schwabl, 1992b, Self-organized criticality in a forest-fire model, *Physica A* **191**(1–4), 47–50, Proceedings of the International Conference on Fractals and Disordered Systems, Hamburg, Germany, July 29–31, 1992, largely identical to Drossel and Schwabl (1992a). → pp 115, 417.

Duarte, J. A. M. S., 1990, cited by Manna (1990) as private communication for studying the BTW model on a triangular lattice. → p 23.

Durin, G., G. Bertotti, and A. Magni, 1995, Fractals, scaling and the question of self-organized criticality in magnetization processes, *Fractals* **3**(2), 351–370. → p 64.

Durin, G., and S. Zapperi, 2000, Scaling exponents for Barkhausen avalanches in polycrystalline and amorphous ferromagnets, *Phys. Rev. Lett.* **84**(20), 4705–4708. → p 64.

Dutta, P., and P. M. Horn, 1981, Low-frequency fluctuations in solids: $1/f$ noise, *Rev. Mod. Phys.* **53**(3), 497–516. → p 4.

Edwards, S. F., and D. R. Wilkinson, 1982, The surface statistics of a granular aggregate, *Proc. R. Soc. London, Ser. A* **381**(1780), 17–31. → p 320.

Efetov, K. B., and A. I. Larkin, 1977, Charge-density wave in a random potential, *Sov. Phys. JETP* **45**, 1236, original *Zh. Eksp. Teor. Fiz.* **72**, 2350–2361 (1971), translated by J. G. Adashko. → p 315.

Efron, B., 1982, *The Jackknife, the Bootstrap and Other Resampling Plans* (SIAM, Philadelphia, PA, USA). → p 216.

Eldredge, N., and S. J. Gould, 1972, Punctuated equilibria: an alternative to phyletic gradualism, in *Models in Paleobiology*, edited by T. J. M. Schopf (Freeman, Cooper & Company, San Francisco, CA, USA), pp. 82–115. → pp 68, 141.

Eliazar, I., 2008, Intrinsic fractality of classic shot noise, *Phys. Rev. E* **77**(6), 061103 (5 pp). → p 301.

Erzan, A., L. Pietronero, and A. Vespignani, 1995, The fixed-scale transformation approach to fractal growth, *Rev. Mod. Phys.* **67**(3), 545–604. → p 267.

Evernden, J. F., 1970, Study of regional seismicity and associated problems, *Bull. Seismol. Soc. Am.* **60**(2), 393–446. → p 66.

Evesque, P., 1991, Analysis of the statistics of sandpile avalanches using soil-mechanics results and concepts, *Phys. Rev. A* **43**(6), 2720–2740. → p 56.

Evesque, P., and J. Rajchenbach, 1989, Instability in a sand heap, *Phys. Rev. Lett.* **62**(1), 44–46. → pp 53, 54.

Faillettaz, J., F. Louchet, and J.-R. Grasso, 2004, Two-threshold model for scaling laws of noninteracting snow avalanches, *Phys. Rev. Lett.* **93**(20), 208001 (4 pp). → p 57.

Family, F., and T. Vicsek, 1985, Scaling of the active zone in the Eden process on percolation networks and the ballistic deposition model, *J. Phys. A: Math. Gen.* **18**(2), L75–L81. → p 313.

▶ Feder, H. J. S., and J. Feder, 1991a, Self-organized criticality in a stick-slip process, *Phys. Rev. Lett.* **66**(20), 2669–2672, see erratum (Feder and Feder, 1991b). → pp 65, 81, 126, 129, 321, 485.

Feder, H. J. S., and J. Feder, 1991b, Self-organized criticality in a stick-slip process, *Phys. Rev. Lett.* **67**(2), 283, erratum. → pp 126, 418.

Feder, J., 1988, *Fractals* (Plenum Press, London, UK). → p 17.

Feder, J., 1995, The evidence for self-organized criticality in sandpiles dynamics, *Fractals* **3**(3), 431–443. → pp 55, 57.

Feller, W., 1966, *An Introduction to Probability Theory and its Applications*, volume II (John Wiley & Sons, New York, NY, USA). → pp 37, 261.

Feller, W., 1968, *An Introduction to Probability Theory and its Applications*, volume I (John Wiley & Sons, New York, NY, USA), 3rd edition. → p 261.

Fenger, N. P., 1976, Dimensional analysis of swirl atomizers, *Engineering* **216**(12), 896–899. → p 22.

Ferreira, A. L. C., and S. K. Mendiratta, 1993, Mean-field approximation with coherent anomaly method for a non-equilibrium model, *J. Phys. A: Math. Gen.* **26**(4), L145–L150. → pp 181, 250, 394.

Ferrenberg, A. M., D. P. Landau, and K. Binder, 1991, Statistical and systematic errors in Monte Carlo sampling, *J. Stat. Phys.* **63**(5/6), 867–882. → p 225.

Ferrenberg, A. M., and R. H. Swendsen, 1988, New Monte Carlo technique for studying phase transitions, *Phys. Rev. Lett.* **61**(23), 2635–2638. → p 358.

Fey, A., L. Levine, and Y. Peres, 2010a, Growth rates and explosions in sandpiles, *J. Stat. Phys.* **138**(1), 143–159, arXiv:0901.3805v2. → p 101.

Fey, A., L. Levine, and D. B. Wilson, 2010b, Approach to criticality in sandpiles, *Phys. Rev. E* **82**(3), 031121 (14 pp), arXiv:1001.3401v1. → pp 101, 338, 342.

Fey, A., L. Levine, and D. B. Wilson, 2010c, Driving sandpiles to criticality and beyond, *Phys. Rev. Lett.* **104**(14), 145703 (4 pp), arXiv:0912.3206v3. → pp 101, 338, 342, 426.

Fey, A., R. Meester, and F. Redig, 2009, Stabilizability and percolation in the infinite volume sandpile model, *Ann. Probab.* **37**(2), 654–675, arXiv:0710.0939v2. → p 101.

Fey-den Boer, A., and F. Redig, 2005, Organized versus self-organized criticality in the Abelian sandpile model, *Markov Process. Relat. Fields* **11**(3), 425–442, arXiv:math-ph/0510060. → p 101.

Fey-den Boer, A., and F. Redig, 2008, Limiting shapes for deterministic centrally seeded growth models, *J. Stat. Phys.* **130**(3), 579–597. → p *110*.

Feynman, R. P., R. B. Leighton, and M. Sands, 1964, *The Feynman Lectures on Physics*, volume 2 (Addison-Wesley, Reading, MA, USA). → p 62.

Field, S., J. Witt, F. Nori, and X. Ling, 1995, Superconducting vortex avalanches, *Phys. Rev. Lett.* **74**(7), 1206–1209. → pp 59, 60.

Fisher, D. S., 1998, Collective transport in random media: from superconductors to earthquakes, *Phys. Rep.* **301**(1–3), 113 – 150. → pp 58, 62, 299.

Fisher, M. E., 1984, Walks, walls, wetting, and melting, *J. Stat. Phys.* **34**(5/6), 667–729. → pp 243, 249, 286, 291, 318.

Fisher, M. E., 1986, Interface wandering in adsorbed and bulk phases, pure and impure, *J. Chem. Soc., Faraday Trans. 2* **82**, 1596–1603 (Faraday Symposium 20). → pp 301, 314.

Fisher, M. E., and M. F. Sykes, 1959, Excluded-volume problem and the Ising model of ferromagnetism, *Phys. Rev.* **114**(1), 45–58. → p 23.

Flyvbjerg, H., and H. G. Petersen, 1989, Error estimates on averages of correlated data, *J. Chem. Phys.* **91**(1), 461–466. → p 217.

Flyvbjerg, H., K. Sneppen, and P. Bak, 1993, Mean field theory of a simple model of evolution, *Phys. Rev. Lett.* **71**(24), 4087–4090. → pp 152, 251.

Forster, D., D. R. Nelson, and M. J. Stephen, 1977, Large-distance and long-time properties of a randomly stirred fluid, *Phys. Rev. A* **16**(2), 732–749. → pp 267, 324, 325.

Fraysse, N., A. Sornette, and D. Sornette, 1993, Critical phase transitions made self-organized: proposed experiments, *J. Phys. I (France)* **3**, 1377–1386. → p 330.

Freckleton, R. P., and W. J. Sutherland, 2001, Hospital waiting-lists (communication arising): do power laws imply self-regulation?, *Nature* **413**(6854), 382–382. → p 76.

Frette, V., 1993, Sandpile models with dynamically varying critical slopes, *Phys. Rev. Lett.* **70**(18), 2762–2765. → pp 58, 87, 184, 185, 190, *205*, 206, 299, 301, 302, 318, *327*.

Frette, V., 2009, personal communication. → p 185.

Frette, V., K. Christensen, A. Malthe-Sørenssen, J. Feder, T. Jøssang, and P. Meakin, 1996, Avalanche dynamics in a pile of rice, *Nature* **379**, 49–52. → pp 55, 57, 184–186, 190, 191, *192*, 193, 195, 203, 301.

Frette, V., J. Feder, T. Jøssang, and P. Meakin, 1993, 'Stick-slip' motion during slow fluid migration in porous media, personal comunication from Vidar Frette, March 2009 (unpublished). → p 185.

Frigg, R., 2003, Self-organised criticality – what it is and what it isn't, *Stud. Hist. Philos. Sci.* **34**, 613–632. → p 6.

Frisch, U., 1995, *Turbulence* (Cambridge University Press, Cambridge, UK). → pp 112, 320.

Gabaix, X., P. Gopikrishnan, V. Plerou, and H. E. Stanley, 2003, A theory of power-law distributions in financial market fluctuations, *Nature* **423**(6937), 267–270. → p 75.

Gabrielli, A., R. Cafiero, M. Marsili, and L. Pietronero, 1997, Theory of self-organized criticality for problems with extremal dynamics, *Europhys. Lett.* **38**(7), 491–496, arXiv: cond-mat/9702176. → p 345.

Gabrielli, A., M. Marsili, R. Cafiero, and L. Pietronero, 1996, Comment on the run time statistics in models of growth in disordered media, *J. Stat. Phys.* **84**(3/4), 889–893, erratum of Marsili (1994b). → pp 150, 152, 434.

Galambos, J., 1978, *The Asymptotic Theory of Extreme Order Statistics* (John Wiley & Sons, New York, NY, UK). → p 344.

Galassi, M., J. Davies, J. Theiler, B. Gough, G. Jungman, P. Alken, M. Booth, and F. Rossi, 2009, *GNU Scientific Library Reference Manual* (Network Theory Ltd.), 3rd (v1.12) edition, http://www.network-theory.co.uk/gsl/manual/, accessed 18 August 2009. → p 371.

Garcia, G. J. M., and R. Dickman, 2004, On the thresholds, probability densities, and critical exponents of Bak-Sneppen-like model, *Physica A* **342**(1–2), 164–170. → p 151.

García-Pelayo, R., 1994a, Dimension of branching processes and self-organized criticality, *Phys. Rev. E* **49**(6), 4903–4906, erratum (García-Pelayo, 1994b). → p 131.

García-Pelayo, R., 1994b, Erratum: dimension of branching processes and self-organized criticality, *Phys. Rev. E* **50**(6), 5146. → p 420.

Garcimartín, A., A. Guarino, L. Bellon, and S. Ciliberto, 1997, Statistical properties of fracture precursors, *Phys. Rev. Lett.* **79**(17), 3202–3205. → p 64.

Gardiner, C. W., 1997, *Handbook of Stochastic Methods* (Springer-Verlag, Berlin, Germany), 2nd edition. → p 15.

Garrido, P. L., J. L. Lebowitz, C. Maes, and H. Spohn, 1990, Long-range correlations for conservative dynamics, *Phys. Rev. A* **42**(4), 1954–1968. → pp 266, 293, 298, 321, 326.

Gattass, R., and R. Desimone, 1996, Responses of cells in the superior colliculus during performance of a spatial attention task in the macaque, *Rev. Bras. Biol.* **56**(Su 1 Pt 2), 257–279. → p 70.

Gaunt, D. S., M. E. Fisher, M. F. Sykes, and J. W. Essam, 1964, Critical isotherm of a ferromagnet and of a fluid, *Phys. Rev. Lett.* **13**(24), 713–715. → p 23.

Gaveau, B., and L. S. Schulman, 1991, Mean-field self-organized criticality, *J. Phys. A: Math. Gen.* **24**(9), L475–L480. → pp 250, 251.

Gawlinski, E. T., and S. Redner, 1983, Monte-Carlo renormalisation group for continuum percolation with excluded-volume interactions, *J. Phys. A: Math. Gen.* **16**(5), 1063–1071. → p 113.

Gentle, J. E., 1998, *Random Number Generation and Monte Carlo Methods* (Springer-Verlag, Berlin, Germany). → p 371.

Geoffroy, O., and J. L. Porteseil, 1991a, On the fractal nature of Barkhausen noise in magnetically soft materials, *J. Magn. Magn. Mater.* **97**(1–3), 205 – 209. → p 62.

Geoffroy, O., and J. L. Porteseil, 1991b, Sandpile simulation of Barkhausen noise in soft magnetic materials, *J. Magn. Magn. Mater.* **97**(1–3), 198 – 204. → p 62.

Geoffroy, O., and J. L. Porteseil, 1994, Scaling properties of irreversible magnetization and their consequences on the macroscopic behaviour of soft materials, *J. Magn. Magn. Mater.* **133**(1–3), 1 – 5. → p 62.

Ghaffari, P., and H. J. Jensen, 1996, Comment on dynamical renormalization group approach to self-organized critical phenomena, *Europhys. Lett.* **35**(5), 397–398, comment on Díaz-Guilera (1994). → pp 106, 110.

Ghaffari, P., S. Lise, and H. J. Jensen, 1997, Nonconservative sandpile models, *Phys. Rev. E* **56**(6), 6702–6709. → pp 88, 109, 334.

Ghashghaie, S., W. Breymann, J. Peinke, P. Talkner, and Y. Dodge, 1996, Turbulent cascades in foreign exchange markets, *Nature* **381**(6585), 767–770. → p 75.

Giacometti, A., and A. Díaz-Guilera, 1998, Dynamical properties of the Zhang model of self-organized criticality, *Phys. Rev. E* **58**(1), 247–253. → pp 104–106, 109.

Gilden, D. L., T. Thornton, and M. W. Mallon, 1995, $1/f$ noise in human cognition, *Science* **267**(5205), 1837–1839. → p 76.

Gillet, F., 2003, Asymptotic behaviour of watermelons, arXiv:math/0307204v1 (unpublished). → p 249.

Gisiger, T., 2001, Scale invariance in biology: coincidence or footprint of a universal mechanism?, *Biol. Rev.* **76**, 161–209. → pp 3, *4*, 53, 68, 70, 71, *158*, 159, 319.

Gleiser, P. M., 2001, Damage spreading in a rice pile model, *Physica A* **295**(1–2), 311–315. → pp 192, 270.

Gleiser, P. M., S. A. Cannas, F. A. Tamarit, and B. Zheng, 2001, Long-range effects in granular avalanching, *Phys. Rev. E* **63**(4), 042301 (4 pp). → p 192.

Godrèche, C. (editor), 1992, *Solids Far From Equilibrium* (Cambridge University Press, Cambridge, UK). → p 301.

Goldberg, D., 1991, What every computer scientist should know about floating-point arithmetic, *ACM Comput. Surv.* **23**(1), 5–48. → p 379.

Gopal, A. D., and D. J. Durian, 1995, Nonlinear bubble dynamics in a slowly driven foam, *Phys. Rev. Lett.* **75**(13), 2610–2613. → p 62.

Gopikrishnan, P., V. Plerou, L. A. N. Amaral, M. Meyer, and H. E. Stanley, 1999, Scaling of the distribution of fluctuations of financial market indices, *Phys. Rev. E* **60**(5), 5305–5316. → p 75.

Gould, S. J., 2002, *The Structure Of Evolutionary Theory* (Belknap Press of Harvard University Press, Cambridge, MA, USA). → p 141.

Gould, S. J., and N. Eldredge, 1993, Punctuated equilibrium comes of age, *Nature* **366**, 223–227. → pp 68, 141.

Grassberger, P., 1982, On phase transitions in Schlögels second model, *Z. Phys. B* **47**, 365–374. → pp 155, 177, 293.

Grassberger, P., 1986, Toward a quantitative theory of self-generated complexity, *Int. J. Theor. Phys.* **25**(9), 907–938. → pp 16, *17*.

Grassberger, P., 1992, Spreading and backbone dimensions of 2D percolation, *J. Phys. A: Math. Gen.* **25**(21), 5475–5484. → p 117.

Grassberger, P., 1993, On a self-organized forest-fire model, *J. Phys. A: Math. Gen.* **26**(9), 2081–2089. → pp 116, *117*, 118–121, 123–125.

Grassberger, P., 1994, Efficient large-scale simulations of a uniformly driven system, *Phys. Rev. E* **49**(3), 2436–2444. → pp 127, 128, 133–136, 140, 141, 358, 361, 362, 365, 376, 384.

Grassberger, P., 1995, The Bak-Sneppen model for punctuated evolution, *Phys. Lett. A* **200**(3–4), 277–282. → pp 143, 151, 154–158, 160, 161.

Grassberger, P., 2002, Critical behaviour of the Drossel-Schwabl forest fire model, *New J. Phys.* **4**(1), 17 (15 pp), arXiv:cond-mat/0202022. → pp 116, 119–122, 124, 125, 282.

Grassberger, P., and H. Kantz, 1991, On a forest-fire model with supposed self-organized criticality, *J. Stat. Phys.* **63**(3–4), 685–700. → pp 113, *114*, 116, 391.

Grassberger, P., and S. S. Manna, 1990, Some more sandpiles, *J. Phys. France* **51**, 1077–1098. → pp 91, 92, 99, 101, 104, 156, 163, 233, 243, 295, 301, 320.

Grassberger, P., and Y.-C. Zhang, 1996, 'Self-organized' formulation of standard percolation phenomena, *Physica A* **224**(1–2), 169–179. → pp 294, 296, 321, 330.

Graves, J. P., R. O. Dendy, K. I. Hopcraft, and E. Jakeman, 2002, The role of clustering effects in interpreting nondiffusive transport measurements in tokamaks, *Phys. Plasmas* **9**(5), 1595–1605. → pp 72, 193.

Gribov, V. N., 1967, Reggeon diagram technique, *Zh. Eksp. Teor. Fiz.* **53**, 654–672 (Sov. Phys. JETP **26**, 414–422 (1968)). → pp 142, 264.

Grinstein, G., 1991, Generic scale invariance in classical nonequilibrium systems (invited), *J. Appl. Phys.* **69**(8), 5441–5446, Proceedings of the 35th Annual Conference on Magnetism and Magnetic Materials, San Diego, CA, USA October 29 – Nov 1, 1990. → pp 266, 293, 298, 325.

● Grinstein, G., 1995, Generic scale invariance and self-organized criticality, in *Scale Invariance, Interfaces, and Non-Equilibrium Dynamics*, edited by A. McKane, M. Droz, J. Vannimenus, and D. Wolf (Plenum Press, New York, NY, USA), pp. 261–293, NATO Advanced Study Institute on Scale Invariance, Interfaces, and Non-Equilibrium Dynamics, Cambridge, UK, June 20–30, 1994. → pp 6, 15, 56, 172, 301, *318*, *320*, 321, 322, *323*, *325*, 329, 331, 344, *351*, *352*.

Grinstein, G., T. Hwa, and H. J. Jensen, 1992, $1/f^{\alpha}$ noise in dissipative transport, *Phys. Rev. A* **45**(2), R559–R562. → p 323.

Grinstein, G., and D.-H. Lee, 1991, Generic scale invariance and roughening in noisy model sandpiles and other driven interfaces, *Phys. Rev. Lett.* **66**(2), 177–180. → p 305.

⇒ Grinstein, G., D.-H. Lee, and S. Sachdev, 1990, Conservation laws, anisotropy, and 'Self-organized criticality' in noisy nonequilibrium systems, *Phys. Rev. Lett.* **64**(16), 1927–1930. → pp 266, 293, 298, 305, 318, 321–325, 327.

Grumbacher, S. K., K. M. McEwen, D. A. Halverson, D. T. Jacobs, and J. Lindner, 1993, Self-organized criticality: an experiment with sandpiles, *Am. J. Phys.* **61**(4), 329–335. → p 56.

Guarino, A., A. Garcimartín, and S. Ciliberto, 1998, An experimental test of the critical behaviour of fracture precursors, *Eur. Phys. J. B* **6**(1), 13–24. → p 64.

Guclu, H., G. Korniss, and Z. Toroczkai, 2007, Extreme fluctuations in noisy task-completion landscapes on scale-free networks, *Chaos* **17**, 026104 (13 pp), arXiv: cond-mat/0701301v1. → p 149.

Gumbel, E. J., 1958, *Statistics of Extremes* (Columbia University Press, New York, NY, USA). → pp 132, 149, 344.

Gutenberg, B., and C. F. Richter, 1954, *Seismicity of the Earth and Associated Phenomena* (Princeton University Press, Princeton, NJ, USA), 2nd edition. → p 66.

Halley, J. M., 1996, Ecology, evolution and $1/f$-noise, *Trends Ecol. Evol.* **11**(1), 33 – 37. → pp 69, 74.

Halpin-Healy, T., and Y.-C. Zhang, 1995, Kinetic roughening phenomena, stochastic growth, directed polymers and all that. Aspects of multidisciplinary statistical mechanics, *Phys. Rep.* **254**(4–6), 215–414. → p 301.

Hansen, A., and S. Roux, 1987, Application of 'logical transport' to determine the directed and isotropic percolation thresholds, *J. Phys. A: Math. Gen.* **20**(13), L873–L878. → pp 294, 296, 321.

Hardner, H. T., M. B. Weissman, M. B. Salamon, and S. S. P. Parkin, 1993, Fluctuation-dissipation relation for giant magnetoresistive $1/f$ noise, *Phys. Rev. B* **48**(21), 16156–16159. → p 63.

Harris, T. E., 1963, *The Theory of Branching Processes* (Springer-Verlag, Berlin, Germany). → pp 154, 251, 257, 258, 261.

Harris, T. E., 1974, Contact interactions on a lattice, *Ann. Prob.* **2**(6), 969–988. → p 341.

Hastings, A., C. L. Holm, S. Ellner, P. Turchin, and H. C. J. Godfray, 1993, Chaos in ecology: is mother nature a strange attractor?, *Annu. Rev. Ecol. Syst.* **24**, 1–33. → p 319.

Hasty, J., and K. Wiesenfeld, 1997, Renormalization of one-dimensional avalanche models, *J. Stat. Phys.* **86**(5/6), 1179–1201. → pp 182, *268*.

Hasty, J., and K. Wiesenfeld, 1998a, Renormalization group for directed sandpile models, *Phys. Rev. Lett.* **81**(8), 1722–1725. → pp 268, 286.

Hasty, J., and K. Wiesenfeld, 1998b, Renormalization of self-organized critical models, *Ann. NY Acad. Sci.* **848**, 9–17. → p 182.

Haussmann, R., 1999, Liquid ⁴He near the superfluid transition in the presence of a heat current and gravity, *Phys. Rev. B* **60**(17), 12349–12372. → p 61.

Head, D. A., and G. J. Rodgers, 1997, Crossover to self-organized criticality in an inertial sandpile model, *Phys. Rev. E* **55**(3), 2573–2579. → pp 192, 203.

Head, D. A., and G. J. Rodgers, 1998, The anisotropic Bak-Sneppen model, *J. Phys. A: Math. Gen.* **31**(17), 3977–3988. → p 157.

Head, D. A., and G. J. Rodgers, 1999, Stretched exponentials and power laws in granular avalanching, *J. Phys. A: Math. Gen.* **32**(8), 1387–1393. → pp 192, 193.

Heiden, C., and G. I. Rochlin, 1968, Flux jump size distribution in low-κ type-II superconductors, *Phys. Rev. Lett.* **21**(10), 691–694. → p 59.

Held, G. A., D. H. Solina, H. Solina, D. T. Keane, W. J. Haag, P. M. Horn, and G. Grinstein, 1990, Experimental study of critical-mass fluctuations in an evolving sandpile, *Phys. Rev. Lett.* **65**(9), 1120–1123. → pp 54–56, 102, 192.

Helmstetter, A., Y. Y. Kagan, and D. D. Jackson, 2006, Comparison of short-term and time-independent earthquake forecast models for southern California, *Bull. Seismol. Soc. Am.* **96**(1), 90–106. → p 67.

Henkel, M., H. Hinrichsen, and S. Lübeck, 2008, *Non-Equilibrium Phase Transitions* (Springer-Verlag, Berlin, Germany). → pp 20, 300, 325, 330.

▶ Henley, C. L., 1989, M18.2: Self-organized percolation: a simpler model, *Bull. Am. Phys. Soc.* **34**(3), 838, abstract of talk M18.2, 23 March 1989, of the 1989 March Meeting of The American Physical Society, St. Louis, MO, USA, March 20–24, 1989. → pp 112, 115, 116, 121.

Henley, C. L., 1993, Statics of a 'self-organized' percolation model, *Phys. Rev. Lett.* **71**(17), 2741–2744. → pp 117, 120, 123.

● Hergarten, S., 2002, *Self-Organized Criticality in Earth Systems* (Springer-Verlag, Berlin, Germany). → pp 6, 66, 139, 366.

● Hergarten, S., 2003, Landslides, sandpiles, and self-organized criticality, *Nat. Hazards Earth Syst. Sci.* **3**, 505–514. → p *139*.

Hergarten, S., and H. J. Neugebauer, 2002, Foreshocks and aftershocks in the Olami-Feder-Christensen model, *Phys. Rev. Lett.* **88**(23), 238501 (4 pp). → p *67*.

Herrmann, H. J., and S. Luding, 1998, Modeling granular media on the computer, *Continuum Mech. Thermodyn.* **10**, 189–231. → p 56.

Herz, A. V. M., and J. J. Hopfield, 1995, Earthquake cycles and neural reverberations: collective oscillations in systems with pulse-coupled threshold elements, *Phys. Rev. Lett.* **75**(6), 1222–1225. → p 70.

Hethcote, H. W., 2000, The mathematics of infectious diseases, *SIAM Rev.* **42**(4), 599–653. → p 115.

Hinrichsen, H., 2000, Non-equilibrium critical phenomena and phase transitions into absorbing states, *Adv. Phys.* **49**, 815–958, arXiv:cond-mat/0001070v2. → pp 155, 157, 176, 202, 290, 293, 302, 306, 330, 333, 336.

Hohenberg, P. C., and B. I. Halperin, 1977, Theory of dynamic critical phenomena, *Rev. Mod. Phys.* **49**(3), 435–479. → pp 19, 177, 267, 282, 300, 316, 318, 323, 352.

Honecker, A., and I. Peschel, 1997, Length scales and power laws in the two-dimensional forest-fire model, *Physica A* **239**(4), 509–530. → pp 116, 119, 120, 123.

Hooge, F. N., and P. A. Bobbert, 1997, On the correlation function of $1/f$ noise, *Physica B* **239**(3–4), 223 – 230. → p 16.

Hooge, F. N., T. G. M. Kleinpenning, and L. K. J. Vandamme, 1981, Experimental studies on $1/f$ noise, *Rep. Progr. Phys.* **44**(5), 479–532. → p 4.

Hopcraft, K. I., E. Jakeman, and R. M. J. Tanner, 1999, Lévy random walks with fluctuating step number and multiscale behavior, *Phys. Rev. E* **60**(5), 5327–5343. → p 195.

Hopcraft, K. I., E. Jakeman, and R. M. J. Tanner, 2001a, Characterization of structural reorganization in rice piles, *Phys. Rev. E* **64**(1), 016116 (10 pp). → p 195.

Hopcraft, K. I., R. M. J. Tanner, E. Jakeman, and J. P. Graves, 2001b, Fractional non-Brownian motion and trapping-time distributions of grains in rice piles, *Phys. Rev. E* **64**(2), 026121 (6 pp). → p 195.

Hopfield, J. J., 1994, Neurons, dynamics and computation, *Phys. Today* **47**(2), 40–46, special issue, Physics and Biology. → pp 69, 70.

Horgan, J., 1995, From complexity to perplexity, *Sci. Am.* **272**(6), 74–79. → p 6.

Hoshen, J., 1997, Percolation and cluster structure parameters: the radius of gyration, *J. Phys. A: Math. Gen.* **30**(24), 8459–8469. → p 124.

Hoshen, J., M. W. Berry, and K. S. Minser, 1976, Percolation and cluster structure parameters: The enhanced Hoshen-Kopelman algorithm, *Phys. Rev. E* **56**(2), 1455–1460. → p 124.

Hoshen, J., and R. Kopelman, 1976, Percolation and cluster distribution. I. Cluster multiple labeling technique and critical concentration algorithm, *Phys. Rev. B* **14**(8), 3438–3445. → p 124.

Houle, P. A., and J. P. Sethna, 1996, Acoustic emission from crumpling paper, *Phys. Rev. E* **54**(1), 278–283. → p 65.

Hu, C.-K., and C.-Y. Lin, 2003, Universality in critical exponents for toppling waves of the BTW sandpile model on two-dimensional lattices, *Physica A* **318**(1–2), 92–100. → pp 23, 99.

Huberman, B. A., and L. A. Adamic, 1999, Internet: growth dynamics of the World Wide Web, *Nature* **401**(6749), 131. → p 76.

Huberman, B. A., P. L. T. Pirolli, J. E. Pitkow, and R. M. Lukose, 1998, Strong regularities in World Wide Web surfing, *Science* **280**(5360), 95–97. → p 76.

Hughes, D., and M. Paczuski, 2002, Large scale structures, symmetry, and universality in sandpiles, *Phys. Rev. Lett.* **88**(5), 054302 (4 pp). → pp 22, 23, 107, 206, 269, 283, 284, 286, 288, 292.

Hughes, D., M. Paczuski, R. O. Dendy, P. Helander, and K. G. McClements, 2003, Solar flares as cascades of reconnecting magnetic loops, *Phys. Rev. Lett.* **90**(13), 131101 (4 pp), arXiv:cond-mat/0210201. → p 72.

Huynh, H. N., L. Y. Chew, and G. Pruessner, 2010, The Abelian Manna model on two fractal lattices, arXiv:1006.5807 (unpublished). → p 24.

⇒ Hwa, T., and M. Kardar, 1989a, Dissipative transport in open systems: an investigation of self-organized criticality, *Phys. Rev. Lett.* **62**(16), 1813–1816, identical to proceedings article (Hwa and Kardar, 1989b). → pp 88, 111, 129, 130, 205, 243, 266, 267, 293, 298–300, 304, 317, 321–329, 334, 342, 343, 352, 425.

Hwa, T., and M. Kardar, 1989b, Fractals and self-organized criticality in dissipative dynamics, *Physica D* **38**(1–3), 198–202, identical to Hwa and Kardar (1989a); Proceedings of a Conference Held in Honour of Benoit B. Mandelbrot's 65th Birthday, Les Mas d'Artigny (Vence), France, October 1 – 4, 1989. → p 425.

Hwa, T., and M. Kardar, 1992, Avalanches, hydrodynamics, and discharge events in models of sandpiles, *Phys. Rev. A* **45**(10), 7002–7023. → pp 193, 194, 322, 343.

Ito, K., 1995, Punctuated-equilibrium model of biological evolution is also a self-organized-criticality model of earthquakes, *Phys. Rev. E* **52**(3), 3232–3233. → p 68.

Ito, K., and M. Matsuzaki, 1990, Earthquakes as self-organized critical phenomena, *J. Geophys. Res.* **95**(B5), 6853–6860. → p 66.

Ivashkevich, E. V., 1994, Boundary height correlations in the two-dimensional Abelian sandpile, *J. Phys. A: Math. Gen.* **27**(11), 3643–3653. → p 99.

Ivashkevich, E. V., 1996, Critical behavior of the sandpile model as a self-organized branching process, *Phys. Rev. Lett.* **76**(18), 3368–3371. → pp 257, 268.

Ivashkevich, E. V., D. V. Ktitarev, and V. B. Priezzhev, 1994a, Critical exponents for boundary avalanches in a two-dimensional Abelian sandpile, *J. Phys. A: Math. Gen.* **27**(16), L585–L590. → pp 100, 178, 248.

Ivashkevich, E. V., D. V. Ktitarev, and V. B. Priezzhev, 1994b, Waves of topplings in an Abelian sandpile, *Physica A* **209**(3–4), 347–360. → pp 99, 100, 170, 292.

Ivashkevich, E. V., A. M. Povolotsky, A. Vespignani, and S. Zapperi, 1999, Dynamical real space renormalization group applied to sandpile models, *Phys. Rev. E* **60**(2), 1239–1251. → p 268.

● Ivashkevich, E. V., and V. B. Priezzhev, 1998, Introduction to the sandpile model, *Physica A* **254**(1–2), 97–116. → p 99.

Jaeger, H. M., C.-h. Liu, and S. R. Nagel, 1989, Relaxation at the angle of repose, *Phys. Rev. Lett.* **62**(1), 40–43. → pp 52, 53, *54*, 55, 57, 102.

Jaeger, H. M., and S. R. Nagel, 1992, Physics of the granular state, *Science* **255**(5051), 1523–1531. → p 56.

Jaeger, H. M., S. R. Nagel, and R. P. Behringer, 1996, Granular solids, liquids, and gases, *Rev. Mod. Phys.* **68**(4), 1259–1273. → pp 53, 197.

Jánosi, I. M., 1990, Effect of anisotropy on the self-organized critical state, *Phys. Rev. A* **42**(2), 769–774. → pp 105, 106, 108.

Jánosi, I. M., and V. K. Horváth, 1989, Dynamics of water droplets on a window pane, *Phys. Rev. A* **40**(9), 5232–5237. → pp 57, 102.

Jánosi, I. M., and J. Kertész, 1993, Self-organized criticality with and without conservation, *Physica A* **200**(1–4), 179–188. → pp 128, 129, 134, 135, 138, 140, 236.

Janowsky, S. A., and C. A. Laberge, 1993, Exact solutions for a mean-field Abelian sandpile, *J. Phys. A: Math. Gen.* **26**(19), L973–L980. → pp 100, 251.

Janssen, H. K., 1981, On the nonequilibrium phase transition in reaction-diffusion systems with an absorbing stationary state, *Z. Phys. B* **42**, 151–154. → pp 155, 177, 264, 293.

Janssen, H.-K., 2005, Survival and percolation probabilities in the field theory of growth models, *J. Phys.: Condens. Matter* **17**(20), S1973–S1993. → p 179.

Janssen, H. K., and B. Schmittmann, 1986, Field theory of long time behaviour in driven diffusive systems, *Z. Phys. B* **63**(4), 517–520. → p 266.

Jeng, M., 2005a, Conformal field theory correlations in the Abelian sandpile model, *Phys. Rev. E* **71**(1), 016140 (12 pp). → p 99.

Jeng, M., 2005b, Four height variables, boundary correlations, and dissipative defects in the Abelian sandpile model, *Phys. Rev. E* **71**(3), 036153 (17 pp). → p 99.

Jeng, M., G. Piroux, and P. Ruelle, 2006, Height variables in the Abelian sandpile model: scaling fields and correlations, *J. Stat. Mech.* **2006**(10), P10015-1–63, arXiv: cond-mat/0609284. → pp 99, 100.

▶ Jensen, H. J., 1990, Lattice gas as a model of $1/f$ noise, *Phys. Rev. Lett.* **64**(26), 3103–3106. → pp 15, 81, 178, 185, 485.

● Jensen, H. J., 1998, *Self-Organized Criticality* (Cambridge University Press, New York, NY, USA). → pp xii, 6, *7*, *11*, 16, 56, 204, 268, 346.

Jensen, H. J., K. Christensen, and H. C. Fogedby, 1989, $1/f$ noise, distribution of lifetimes, and a pile of sand, *Phys. Rev. B* **40**(10), 7425–7427. → pp 12, 15, 16, 42, 54, 55, 62, 102.

Jensen, I., 1992, Critical behavior of the three-dimensional contact process, *Phys. Rev. A* **45**(2), R563–R566. → p 294.

Jensen, I., 1999, Low-density series expansions for directed percolation: I. A new efficient algorithm with applications to the square lattice, *J. Phys. A: Math. Gen.* **32**(28), 5233–5249. → p 294.

Jettestuen, E., and A. Malthe-Sørenssen, 2005, Scaling properties of a one-dimensional sandpile model with grain dissipation, *Phys. Rev. E* **72**(6), 062302 (4 pp). → pp *184*, 207, 289, 291, 327, 334.

Ji, H., and M. O. Robbins, 1992, Percolative, self-affine, and faceted domain growth in random three-dimensional magnets, *Phys. Rev. B* **46**(22), 14519–14527. → p 63.

Jinghua, F., M. Ta-chung, R. Rittel, and K. Tabelow, 2001, Criticality in quark-gluon systems far beyond thermal and chemical equilibrium, *Phys. Rev. Lett.* **86**(10), 1961–1964. → p 72.

Jo, H.-H., and M. Ha, 2008, Relevance of Abelian symmetry and stochasticity in directed sandpiles, *Phys. Rev. Lett.* **101**(21), 218001 (4 pp). → pp 269, 284, 286, 287.

Jo, H.-H., and H.-C. Jeong, 2010, Comment on 'Driving Sandpiles to Criticality and Beyond', *Phys. Rev. Lett.* **105**(1), 019601 (1 p), comment on (Fey *et al.*, 2010c), no reply. → pp 101, 342.

Johansen, A., P. Dimon, C. Ellegaard, J. S. Larsen, and H. H. Rugh, 1993, Dynamic phases in a spring-block system, *Phys. Rev. E* **48**(6), 4779–4790. → p 65.

▶ Jovanović, B., S. V. Buldyrev, S. Havlin, and H. E. Stanley, 1994, Punctuated equilibrium and 'history-dependent' percolation, *Phys. Rev. E* **50**(4), R2403–R2406. → pp 143, 151, 155–157, 485.

Juanico, D. E., C. Monterola, and C. Saloma, 2007a, Dissipative self-organized branching in a dynamic population, *Phys. Rev. E* **75**(4), 045105(R) (4 pp), see Juanico *et al.* (2007b). → pp 334, 335, 427.

Juanico, D. E., C. Monterola, and C. Saloma, 2007b, Self-organized critical branching in systems that violate conservation laws, *New J. Phys.* **9**(4), 92 (18 pp), see Juanico *et al.* (2007a). → pp 265, 328, 334, 335, 427.

Kadanoff, L. P., 1966, Scaling laws for Ising models near T_c^*, *Physics* **2**(6), 263–272. → pp 4, 25, 266.

Kadanoff, L. P., 1986, Fractals: where's the physics?, *Phys. Today* **39**(2), 6–7. → pp *4*, 16, *319*, *347*.

Kadanoff, L. P., 1990, Scaling and universality in statistical physics, *Physica A* **163**(1), 1 – 14. → pp 25, 51, 320.

Kadanoff, L. P., 2000, *Statistical Physics* (World Scientific, Singapore). → p 6.

Kadanoff, L. P., A. B. Chhabra, A. J. Kolan, M. J. Feigenbaum, and I. Procaccia, 1992, Critical indices for singular diffusion, *Phys. Rev. A* **45**(8), 6095–6098. → p 343.

Kadanoff, L. P., S. R. Nagel, L. Wu, and S.-m. Zhou, 1989, Scaling and universality in avalanches, *Phys. Rev. A* **39**(12), 6524–6537. → pp 12, 18, 19, 22, 23, 25, 51, 54, 89, 93, 139, 203, 243, 284, 286, 292, 293, 327.

Kagan, Y. Y., 1991a, Likelihood analysis of earthquake catalogues, *Geophys. J. Int.* **106**(1), 135–148. → p 66.

Kagan, Y. Y., 1991b, Seismic moment distribution, *Geophys. J. Int.* **106**(1), 123–134. → p 66.

Kagan, Y. Y., 2003, Accuracy of modern global earthquake catalogs, *Phys. Earth Planet. Inter.* **135**(2–3), 173–209. → p 66.

Kakalios, J., 2005, Resource letter GP-1: granular physics or nonlinear dynamics in a sandbox, *Am. J. Phys.* **73**(1), 8–22. → pp 53, 85, 193.

Kaneko, K., 1983, Transition from torus to chaos accompanied by frequency lockings with symmetry breaking – in connection with coupled-logistic map –, *Progr. Theor. Phys.* **69**(5), 1427–1442. → p 126.

Kaneko, K., 1984, Period-doubling of kink-antikink patterns, quasiperiodicity in antiferro-like structures and spatial intermittency in coupled logistic lattice – towards a prelude of a 'field theory of chaos' –, *Progr. Theor. Phys.* **72**(3), 480–486. → p 126.

Kaneko, K., 1989, Spatiotemporal chaos in one- and two-dimensional coupled map lattices, *Physica D* **37**(1–3), 60–82. → pp 126, 127.

Kantz, H., and T. Schreiber, 2005, *Nonlinear Time Series Analysis* (Cambridge University Press, Cambridge, UK), 2nd edition. → p 358.

Kardar, M., 1996, Avalanche theory in rice, *Nature* **379**, 22. → pp *19*, *71*, 192.

Kardar, M., 1998, Nonequilibrium dynamics of interfaces and lines, *Phys. Rep.* **301**, 85–112. → p 301.

Kardar, M., G. Parisi, and Y.-C. Zhang, 1986, Dynamic scaling of growing interfaces, *Phys. Rev. Lett.* **56**(9), 889–892. → pp 301, 320.

Karmakar, R., S. S. Manna, and A. L. Stella, 2005, Precise toppling balance, quenched disorder, and universality for sandpiles, *Phys. Rev. Lett.* **94**(8), 088002 (4 pp). → pp 23, 95, 100, 104, 165, 178, 181, 204, 287.

Katori, M., and H. Kobayashi, 1996, Mean-field theory of avalanches in self-organized critical states, *Physica A* **229**(3–4), 461–477. → p 251.

Kauffman, S. A., and S. Johnsen, 1991, Coevolution to the edge of chaos: coupled fitness landscapes, poised states, and coevolutionary avalanches, *J. Theor. Biol.* **149**(4), 467–505. → pp 158, 319.

Kaulke, M., 1999, *Anwendung der Dichtematrix-Renormierung auf nichthermitesche Probleme*, Ph.D. Thesis, Fachbereich Physik, Freie Universität Berlin, Germany. → pp 176, 202.

Keitt, T. H., and P. A. Marquet, 1996, The introduced Hawaiian avifauna reconsidered: evidence for self-organized criticality?, *J. Theor. Biol.* **182**(2), 161 – 167. → p 74.

Kellog, D., 1975, The role of phyletic change in the evolution of *Pseudocubus vema* (Radiolaria), *Paleobiology* **1**(4), 359–370. → p 141.

Kernighan, B. W., and R. Pike, 2002, *The Practice of Programming* (Addison-Wesley, Boston, MA, USA). → p 359.

Kernighan, B. W., and D. M. Ritchie, 1988, *The C Programming Language* (Prentice Hall, Englewood Cliffs, NJ, USA), 2nd edition. → p 358.

Kertész, J., and L. B. Kiss, 1990, The noise spectrum in the model of self-organised criticality, *J. Phys. A: Math. Gen.* **23**(9), L433–L440. → pp 16, 102.

Kessler, D. A., H. Levine, and Y. Tu, 1991, Interface fluctuations in random media, *Phys. Rev. A* **43**(8), 4551–4554. → p 302.

Khfifi, M., and M. Loulidi, 2008, Scaling properties of a rice-pile model: inertia and friction effects, *Phys. Rev. E* **78**(5), 051117 (8 pp). → pp 135, 192.

Kinouchi, O., S. T. R. Pinho, and C. P. C. Prado, 1998, Random-neighbor Olami-Feder-Christensen slip-stick model, *Phys. Rev. E* **58**(3), 3997–4000. → pp 131, 328.

Kirchner, J. W., 2001, Fractal power spectra plotted upside-down: comment on 'scaling of power spectrum of extinction events in the fossil record' by V. P. Dimri and M. R. Prakash, *Earth Planet. Sci. Lett.* **192**(4), 617 – 621, comment on (Dimri and Prakash, 2001b), reply (Dimri and Prakash, 2001a). → pp 69, 416.

Kirchner, J. W., and A. Weil, 1998, No fractals in fossil extinction statistics, *Nature* **395**(6700), 337–338. → pp 52, 69, 158.

Klein, W., and J. Rundle, 1993, Comment on 'self-organized criticality in a continuous, nonconservative cellular automaton modeling earthquakes', *Phys. Rev. Lett.* **71**(8), 1288, comment on (Olami *et al.*, 1992), reply (Christensen, 1993). → pp 128, 136, 410, 439.

▶ Kloster, M., S. Maslov, and C. Tang, 2001, Exact solution of a stochastic directed sandpile model, *Phys. Rev. E* **63**(2), 026111 (4 pp). → pp 22, 24, 107, 172, 206, 208, 249, 264, 288, 291.

Knuth, D. E., 1997a, *The Art of Computer Programming*, Volumes 1–3 (Addison-Wesley, Reading, MA, USA). → p 358.

Knuth, D. E., 1997b, *Fundamental Algorithms*, volume 1 of *The Art of Computer Programming* (Addison-Wesley, Reading, MA, USA), 3rd edition. → p 253.

Knuth, D. E., 1997c, *Seminumerical Algorithms*, volume 2 of *The Art of Computer Programming* (Addison-Wesley, Reading, MA, USA), 2nd edition. → pp 371, 389.

Koba, Z., H. B. Nielsen, and P. Olesen, 1972, Scaling of multiplicity distributions in high energy hadron collisions, *Nucl. Phys. B* **40**, 317–334. → p 221.

Koch, C., 1997, Computation and the single neuron, *Nature* **385**, 207–210. → p 69.

Kockelkoren, J., and H. Chaté, 2003, Absorbing phase transitions with coupling to a static field and a conservation law, `arXiv:cond-mat/0306039v1` (unpublished). → pp 179, 300.

Kolmogorov, A. N., 1991, The local structure of turbulence in incompressible viscous fluid for very large Reynolds number, *Proc. R. Soc. London, Ser. A* **434**(1890), 9–13, original *Dokl. Akad. Nauk SSSR* **30**, 299–303 (1941), translated by V. Levin. → p 320.

Koplik, J., and H. Levine, 1985, Interface moving through a random background, *Phys. Rev. B* **32**(1), 280–292. → pp 205, 299.

Koscielny-Bunde, E., A. Bunde, S. Havlin, H. E. Roman, Y. Goldreich, and H.-J. Schellnhuber, 1998, Indication of a universal persistence law governing atmospheric variability, *Phys. Rev. Lett.* **81**(3), 729–732. → p 71.

Krim, J., and J. O. Indekeu, 1993, Roughness exponents: a paradox resolved, *Phys. Rev. E* **48**(2), 1576–1578. → p 312.

Krug, J., 1995, Statistical physics of growth processes, in *Scale Invariance, Interfaces, and Non-Equilibrium Dynamics*, edited by A. McKane, M. Droz, J. Vannimenus, and D. Wolf (Plenum Press, New York, NY, USA), pp. 1–61, NATO Advanced Study Institute on Scale Invariance, Interfaces, and Non-Equilibrium Dynamics, Cambridge, UK, June 20–30, 1994. → pp 301, 320, 321, 396.

Krug, J., 1997, Origins of scale invariance in growth processes, *Adv. Phys.* **46**(2), 139–282. → pp 251, 301, 313, 315.

Krug, J., 2007, Records in a changing world, *J. Stat. Mech.* **2007**(07), P07001 (13 pp). → p 149.

Krug, J., J. E. S. Socolar, and G. Grinstein, 1992, Surface fluctuations and criticality in a class of one-dimensional sandpile models, *Phys. Rev. A* **46**(8), R4479–R4482. → pp 56, 192, 197, 314.

Krug, J., and H. Spohn, 1991, Kinetic roughening of growing surfaces, in *Solids far from Equilibrium*, edited by C. Godrèche (Cambridge University Press, Cambridge, UK), pp. 479–582. → pp 301, 313, 314.

Ktitarev, D. V., S. Lübeck, P. Grassberger, and V. B. Priezzhev, 2000, Scaling of waves in the Bak-Tang-Wiesenfeld sandpile model, *Phys. Rev. E* **61**(1), 81–92. → p 100.

Kulkarni, R. V., E. Almaas, and D. Stroud, 1999, Evolutionary dynamics in the Bak-Sneppen model on small-world networks, `arXiv:cond-mat/9905066` (unpublished). → p 159.

Kutnjak-Urbanc, B., S. Zapperi, S. Milošević, and H. E. Stanley, 1996, Sandpile model on the Sierpinski gasket fractal, *Phys. Rev. E* **54**(1), 272–277. → pp 24, 101.

Laherrère, J., and D. Sornette, 1998, Stretched exponential distributions in nature and economy: 'fat tails' with characteristic scales, *Eur. Phys. J. B* **2**(4), 525–539. → p 57.

Landau, D. P., and K. Binder, 2005, *A Guide to Monte Carlo Simulations in Statistical Physics* (Cambridge University Press, Cambridge, UK), 2nd edition. → pp 39, 220, 227, 358.

Langton, C. G., 1990, Computation at the edge of chaos: phase transitions and emergent computation, *Physica D* **42**(1–3), 12 – 37. → pp *17, 319*, 345.

Lässig, M., 1998, On growth, disorder, and field theory, *J. Phys.: Condens. Matter* **10**(44), 9905–9950. → p 301.

Lauritsen, K. B., S. Zapperi, and H. E. Stanley, 1996, Self-organized branching processes: avalanche models with dissipation, *Phys. Rev. E* **54**(3), 2483–2488. → pp 265, 328, 334, 335.

Le Bellac, M., 1991, *Quantum and Statistical Field Theory (Phenomenes critiques aux champs de jauge)* (Oxford University Press, New York, NY, USA), translated by G. Barton. → pp 266, 319, 325.

Le Doussal, P., K. J. Wiese, and P. Chauve, 2002, Two-loop functional renormalization group theory of the depinning transition, *Phys. Rev. B* **66**(17), 174201 (34 pp), `arXiv: cond-mat/0205108`. → pp 179, 205, 267, 315.

Leadbetter, M. R., G. Lindgren, and H. Rootzén, 1983, *Extremes and Related Properties of Random Sequences and Processes* (Springer-Verlag, New York, NY, USA). → p 344.

Lee, B. P., and J. Cardy, 1995, Renormalization group study of the $A + B \rightarrow \emptyset$ diffusion-limited reaction, *J. Stat. Phys.* **80**(5/6), 971–1007. → p 266.

Leschhorn, H., 1993, Interface depinning in a disordered medium – numerical results, *Physica A* **195**(3–4), 324–335, `arXiv:cond-mat/9302039`. → pp 207, 315.

Leschhorn, H., 1994, *Grenzflächen in ungeordneten Medien*, Ph.D. Thesis, Ruhr-Universität Bochum, Bochum, Germany. → pp 182, 205, 315.

Leschhorn, H., T. Nattermann, S. Stepanow, and L.-H. Tang, 1997, Driven interface depinning in a disordered medium, *Ann. Phys.* **6**, 1–34, `arXiv:cond-mat/9603114`. → pp 182, 205, 267, 299, 315, 316, 329, 438.

Leschhorn, H., and L.-H. Tang, 1994, Avalanches and correlations in driven interface depinning, *Phys. Rev. E* **49**(2), 1238–1245. → pp 182, 295.

• Leung, K., J. Müller, and J. V. Andersen, 1997, Generalization of a two-dimensional Burridge-Knopoff model of earthquakes, *J. Phys. I (France)* **7**, 423–429. → pp 125, 133.

Levina, A., J. M. Herrmann, and T. Geisel, 2006, Dynamical synapses give rise to a power-law distribution of neuronal avalanches, in *Advances in Neural Information Processing Systems 18*, edited by Y. Weiss, B. Schölkopf, and J. Platt (MIT Press, Cambridge, MA, USA), pp. 771–778. → p 71.

Levina, A., J. M. Herrmann, and T. Geisel, 2007, Dynamical synapses causing self-organized criticality in neural networks, *Nat. Phys.* **3**(12), 857–860. → p 71.

Liggett, T. M., 2000, *Interacting Particle Systems*, volume 276 of *Grundlehren der mathematischen Wissenschaften* (Springer-Verlag, Berlin, Germany). → pp 142, 264, 341.

Liggett, T. M., 2005, *Stochastic Interacting Systems: Contact, Voter and Exclusion Processes* (Springer-Verlag, Berlin, Germany). → p 389.

Lin, C.-Y., C.-F. Chen, C.-N. Chen, C.-S. Yang, and I.-M. Jiang, 2006, Effects of bulk dissipation on the critical exponents of a sandpile, *Phys. Rev. E* **74**(3), 031304 (11 pp). → pp 88, 94, 169, 306, 334.

Lin, C.-Y., A.-C. Cheng, and T.-M. Liaw, 2007, Numerical renormalization-group approach to a sandpile, *Phys. Rev. E* **76**(4), 041114 (7 pp). → pp 268, 292.

Lin, C.-Y., and C.-K. Hu, 2002, Renormalization-group approach to an Abelian sandpile model on planar lattices, *Phys. Rev. E* **66**(2), 021307 (12 pp). → pp 92, 101, 267, 268, 455.

Lindman, M., K. Jónsdóttir, R. Roberts, B. Lund, and R. Bödvarsso, 2005, Earthquakes descaled: on waiting time distributions and scaling laws, *Phys. Rev. Lett.* **94**(10), 108501 (4 pp), comment (Corral and Christensen, 2006). → pp 68, 411.

Lindman, M., K. Jónsdóttir, R. Roberts, B. Lund, and R. Bödvarsso, 2006, Lindman et al. Reply:, *Phys. Rev. Lett.* **96**(10), 109802 (1 p), reply to comment (Corral and Christensen, 2006). → pp 68, 411.

Ling, X. S., D. Shi, and J. L. Budnick, 1991, Self-organized critical state in High-*T*c superconductors, *Physica C* **185–189**(Part 4), 2181 – 2182. → p *59*.

Linkenkaer-Hansen, K., V. V. Nikouline, and R. J. Ilmoniemi, 2000, Critical dynamics in the human brain, *NeuroImage* **11**(5, Supplement 1), S760. → p 70.

Linkenkaer-Hansen, K., V. V. Nikouline, J. M. Palva, and R. J. Ilmoniemi, 2001, Long-range temporal correlations and scaling behavior in human brain oscillations, *J. Neurosci.* **21**(4), 1370–1377. → pp 52, 70.

Lise, S., 2002, Self-organization to criticality in a system without conservation law, *J. Phys. A: Math. Gen.* **35**(22), 4641–4649, arXiv:cond-mat/0204490. → pp 134, *135*, 137, 172, 238.

Lise, S., and H. J. Jensen, 1996, Transitions in nonconserving models of self-organized criticality, *Phys. Rev. Lett.* **76**(13), 2326–2329. → pp *130*, 131, 132, 328.

Lise, S., and M. Paczuski, 2001a, Scaling in a nonconservative earthquake model of self-organized criticality, *Phys. Rev. E* **64**(4), 046111 (5 pp). → pp 128, 134, 137, 141, 169.

Lise, S., and M. Paczuski, 2001b, Self-organized criticality and universality in a non-conservative earthquake model, *Phys. Rev. E* **63**(3), 036111 (5 pp). → pp 128, 136, 141.

Loison, D., C. L. Qin, K. D. Schotte, and X. F. Jin, 2004, Canonical local algorithms for spin systems: heat bath and Hasting's methods, *Eur. Phys. J. B* **41**, 395–412. → p 389.

López, C., and M. A. Muñoz, 1997, Numerical analysis of a Langevin equation for systems with infinite absorbing states, *Phys. Rev. E* **56**(4), 4864–4867. → p 182.

López, J. M., 1999, Scaling approach to calculate critical exponents in anomalous surface roughening, *Phys. Rev. Lett.* **83**(22), 4594–4597. → p 313.

Loreto, V., L. Pietronero, A. Vespignani, and S. Zapperi, 1995, Renormalization group approach to the critical behavior of the forest-fire model, *Phys. Rev. Lett.* **75**(3), 465–468. → pp 119–121, 268.

Loreto, V., L. Pietronero, A. Vespignani, and S. Zapperi, 1997, Loreto et al Reply:, *Phys. Rev. Lett.* **78**(7), 1393, reply to comment (Torcini, Livi, Politi, and Ruffo, 1997). → pp 119, 453.

Loreto, V., A. Vespignani, and S. Zapperi, 1996, Renormalization scheme for forest-fire models, *J. Phys. A: Math. Gen.* **29**(21), 2981–3004. → p 121.

Lőrincz, K. A., and R. J. Wijngaarden, 2007, Edge effect on the power law distribution of granular avalanches, *Phys. Rev. E* **76**(4), 040301(R) (4 pp). → pp 58, 192.

Lőrincz, K. A., and R. J. Wijngaarden, 2008, Influence of the driving rate in a two-dimensional rice pile model, *Phys. Rev. E* **77**(6), 066110 (4 pp). → pp 189, 194, 322.

Lowen, S. B., S. S. Cash, M.-m. Poo, and M. C. Teich, 1997, Quantal neurotransmitter secretion rate exhibits fractal behavior, *J. Neurosci.* **17**(15), 5666–5677. → p 70.

Lowen, S. B., and M. C. Teich, 1993, Fractal renewal processes generate $1/f$ noise, *Phys. Rev. E* **47**(2), 992–1001. → p 15.

Lu, E. T., and R. J. Hamilton, 1991, Avalanches and the distribution of solar flares, *Astrophys. J.* **380**(2), L89–L92. → p 72.

Lübeck, S., 1997, Large-scale simulations of the Zhang model, *Phys. Rev. E* **56**(2), 1590–1594. → pp 105, 106, 108, 109.

Lübeck, S., 1998, Logarithmic corrections of the avalanche distributions of sandpile models at the upper critical dimension, *Phys. Rev. E* **58**(3), 2957–2964. → pp 104, 250.

Lübeck, S., 2000, Moment analysis of the probability distribution of different sandpile models, *Phys. Rev. E* **61**(1), 204–209. → pp 42–44, 92, 101–104, 165, 168, 183, 292, 294, 306, 316.

Lübeck, S., 2002a, Scaling behavior of the conserved transfer threshold process, *Phys. Rev. E* **66**(4), 046114 (6 pp). → pp 170, 337.

Lübeck, S., 2002b, Scaling behavior of the order parameter and its conjugated field in an absorbing phase transition around the upper critical dimension, *Phys. Rev. E* **65**(4), 046150 (7 pp). → pp 179, 305, 306, 337.

Lübeck, S., 2003, Mean-field theory for self-organized critical sandpile models, personal communication (unpublished). → pp 251, 333.

Lübeck, S., 2004, Universal scaling behavior of non-equilibrium phase transitions, *Int. J. Mod. Phys. B* **18**(31/32), 3977–4118. → pp x, 20, 44, 155, 168, 179, 261, 294, 305, 306, 316, 330, 339, 340.

Lübeck, S., and P. C. Heger, 2003a, Universal finite-size scaling behavior and universal dynamical scaling behavior of absorbing phase transitions with a conserved field, *Phys. Rev. E* **68**(5), 056102 (11 pp). → pp 44, 168–170, 178, 179, 183, 283, 294, 305–307, 316, 336–338, 340.

Lübeck, S., and P. C. Heger, 2003b, Universal scaling behavior at the upper critical dimension of nonequilibrium continuous phase transitions, *Phys. Rev. Lett.* **90**(23), 230601 (4 pp). → pp x, 164, 167, 169, 179, 183, 337, 339, 340.

Lübeck, S., and H.-K. Janssen, 2005, Finite-size scaling of directed percolation above the upper critical dimension, *Phys. Rev. E* **72**(1), 016119 (4 pp). → p 39.

Lübeck, S., N. Rajewsky, and D. E. Wolf, 2000, A deterministic sandpile automaton revisited, *Eur. Phys. J. B* **13**(4), 715–721, 10.1007/s100510050090. → pp 97, 134.

Lübeck, S., B. Tadić, and K. D. Usadel, 1996, Nonequilibrium phase transition and self-organized criticality in a sandpile model with stochastic dynamics, *Phys. Rev. E* **53**(3), 2182–2189. → p 135.

Lübeck, S., and K. D. Usadel, 1993, SOC in a class of sandpile models with stochastic dynamics, *Fractals* **1**(4), 1030–1036. → pp 135, 197, 284, 314.

Lübeck, S., and K. D. Usadel, 1997a, Bak-Tang-Wiesenfeld sandpile model around the upper critical dimension, *Phys. Rev. E* **56**(5), 5138–5143. → pp 92, 101, 104.

Lübeck, S., and K. D. Usadel, 1997b, Numerical determination of the avalanche exponents of the Bak-Tang-Wiesenfeld model, *Phys. Rev. E* **55**(4), 4095–4099. → pp 92, 102, 103.

Luding, S., 1996, Langkorn- oder Kurzkornreis?, *Phys. Bl.* **52**(3), 203. → p 192.

Luijten, E., and H. W. J. Blöte, 1995, Monte Carlo method for spin models with long-range interactions, *Int. J. Mod. Phys. C* **6**(3), 359–370. → p 251.

Lux, T., and M. Marchesi, 1999, Scaling and criticality in a stochastic multi-agent model of a financial market, *Nature* **397**(6719), 498–500. → p 75.

Ma, S.-K., 1976, *Modern Theory of critical Phenomena* (Addison-Wesley, Reading, MA, USA). → pp 251, 267.

Machta, J., D. Candela, and R. B. Hallock, 1993, Self-organized criticality in ^{4}He with a heat current, *Phys. Rev. E* **47**(6), 4581. → p 344.

Maddox, J., 1994, Punctuated equilibrium by computer, *Nature* **371**, 197. → pp *68, 158*.

Maddox, J., 1995, The case for great many journals, *Nature* **375**, 11. → p *158*.

Maes, C., A. Van Moffaert, H. Frederix, and H. Strauven, 1998, Criticality in creep experiments on cellular glass, *Phys. Rev. B* **57**(9), 4987–4990. → p 64.

Mahan, G. D., 1990, *Many-Particle Physics* (Plenum Press, New York, NY, USA), 2nd edition. → p 175.

Mahieu, S., and P. Ruelle, 2001, Scaling fields in the two-dimensional Abelian sandpile model, *Phys. Rev. E* **64**(6), 066130 (19 pp), `arXiv:hep-th/0107150`. → p 99.

Main, I. G., and P. W. Burton, 1986, Long-term earthquake recurrence constrained by tectonic seismic moment release rates, *Bull. Seismol. Soc. Am.* **76**(1), 297–304. → p 66.

Maini, P. K., K. J. Paintera, and H. N. P. Chaub, 1997, Spatial pattern formation in chemical and biological systems, *J. Chem. Soc., Faraday Trans.* **93**, 3601–3610. → p 114.

Majumdar, S. N., and A. Comtet, 2004, Exact maximal height distribution of fluctuating interfaces, *Phys. Rev. Lett.* **92**(22), 225501 (4 pp). → pp 149, 208, 232, 250.

Majumdar, S. N., and A. Comtet, 2005, Airy distribution function: from the area under a Brownian excursion to the maximal height of fluctuating interfaces, *J. Stat. Phys.* **119**(3/4), 777–826. → pp 208, 232, 250.

Majumdar, S. N., and D. Dhar, 1991, Height correlations in the Abelian sandpile model, *J. Phys. A: Math. Gen.* **24**(7), L357–L362. → pp 98, 99, 246.

Majumdar, S. N., and D. Dhar, 1992, Equivalence between the Abelian sandpile model and the $q \to 0$ limit of the Potts model, *Physica A* **185**(1–4), 129–145. → pp 95, 99, 100.

Malamud, B. D., J. D. A. Millington, and G. L. W. Perry, 2004, Characterizing wildfire regimes in the United States, *Proc. Natl. Acad. Sci. USA* **102**(13), 4694–4699. → pp 73, 122.

Malamud, B. D., G. Morein, and D. L. Turcotte, 1998, Forest fires: an example of self-organized critical behavior, *Science* **281**(5384), 1840–1842. → pp *73*, 120, 122, 123.

Malcai, O., D. A. Lidar, O. Biham, and D. Avnir, 1997, Scaling range and cutoffs in empirical fractals, *Phys. Rev. E* **56**(3), 2817–2828. → p *53*.

Malcai, O., Y. Shilo, and O. Biham, 2006, Dissipative sandpile models with universal exponents, *Phys. Rev. E* **73**(5), 056125 (5 pp). → pp 169, 246, 306, 328, 334.

Malthe-Sørenssen, A., 1996, Kinetic grain model for sandpiles, *Phys. Rev. E* **54**(3), 2261–2265. → p 178.

Malthe-Sørenssen, A., 1999, Tilted sandpiles, interface depinning and earthquake models, *Phys. Rev. E* **59**(4), 4169–4174. → pp 177, 178, 187, 193, 227.

Malthe-Sørenssen, A., J. Feder, K. Christensen, V. Frette, and T. Jøssang, 1999, Surface fluctuations and correlations in a pile of rice, *Phys. Rev. Lett.* **83**(4), 764–767. → pp 193, 197.

Mandelbrot, B. B., 1983, *The Fractal Geometry of Nature* (Freeman, New York, NY, USA). → pp *4*, 74.

Manna, S. S., 1990, Large-scale simulation of avalanche cluster distribution in sand pile model, *J. Stat. Phys.* **59**(1/2), 509–521. → pp 23, 88, 92, 99, 102, 103, 134, 339, 417.

Manna, S. S., 1991a, Critical exponents of the sand pile models in two dimensions, *Physica A* **179**(2), 249–268. → pp 23, 92, 93, 100, 103.

▶ Manna, S. S., 1991b, Two-state model of self-organized criticality, *J. Phys. A: Math. Gen.* **24**(7), L363–L369. → pp 22, 82, 162, *163*, *164*, 166–169, 185, 201, 339, 485.

Manna, S. S., A. D. Chakrabarti, and R. Cafiero, 1999, Critical states in a dissipative sandpile model, *Phys. Rev. E* **60**(5), R5005–R5008. → p 289.

Manna, S. S., L. B. Kiss, and J. Kertész, 1990, Cascades and self-organized criticality, *J. Stat. Phys.* **61**(3/4), 923–932. → pp 88, 129, 225, 327, 334.

Mantegna, R. N., and H. E. Stanley, 1995, Scaling behaviour in the dynamics of an economic index, *Nature* **376**(6535), 46–49. → p 75.

Mantegna, R. N., and H. E. Stanley, 1997, Stock market dynamics and turbulence: parallel analysis of fluctuation phenomena, *Physica A* **239**(1–3), 255 – 266. → p 75.

Mari, D. D., and S. Kotz, 2001, *Correlation and Dependence* (Imperial College Press, London, UK). → p 212.

Marinari, E., G. Parisi, D. Ruelle, and P. Windey, 1983, Random walk in a random environment and $1/f$ noise, *Phys. Rev. Lett.* **50**(17), 1223–1225. → p 15.

Markošová, M., 2000a, Ricepiles: experiments and models, *Prog. Theor. Phys. Suppl.* **139**, 489–495. → pp 189, 192, 284, 296.

Markošová, M., 2000b, Universality classes for the ricepile model with absorbing properties, *Phys. Rev. E* **61**(1), 253–260. → pp 23, 189, 192, 284, 295, *296*.

Markošová, M., and P. Markoš, 1992, Analytical calculation of the attractor periods of deterministic sandpiles, *Phys. Rev. A* **46**(6), 3531–3534. → p 96.

Marro, J., and R. Dickman, 1999, *Nonequilibrium Phase Transitions in Lattice Models* (Cambridge University Press, New York, NY, USA). → pp 155, 301, 305, 307, 330, 336, 337.

Marsili, M., 1994a, Renormalization group approach to the self-organization of a simple model of biological evolution, *Europhys. Lett.* **28**(6), 385–390. → pp 143, 152, *156*, 161, 268.

Marsili, M., 1994b, Run time statistics in models of growth in disordered media, *J. Stat. Phys.* **77**(3/4), 733–754, erratum (Gabrielli *et al.*, 1996). → pp 149, 150, 152, 419.

Marsili, M., G. Caldarelli, and M. Vendruscolo, 1996, Quenched disorder, memory, and self-organization, *Phys. Rev. E* **53**(1), R13–R16. → pp 149, 157.

Marsili, M., P. De Los Rios, and S. Maslov, 1998, Expansion around the mean-field solution of the Bak-Sneppen model, *Phys. Rev. Lett.* **80**(7), 1457–1460. → pp 143, 152, 157.

Martys, N., M. O. Robbins, and M. Cieplak, 1991, Scaling relations for interface motion through disordered media: application to two-dimensional fluid invasion, *Phys. Rev. B* **44**(22), 12294–12306. → p 301.

Maslov, S., 1995, Time directed avalanches in invasion models, *Phys. Rev. Lett.* **74**(4), 562–565. → p 145.

Maslov, S., 1996, Infinite series of exact equations in the Bak-Sneppen model of biological evolution, *Phys. Rev. Lett.* **77**(6), 1182–1185. → pp 143, 157.

Maslov, S., and M. Paczuski, 1994, Scaling theory of depinning in the Sneppen model, *Phys. Rev. E* **50**(2), R643–R646. → p 295.

Maslov, S., and Y.-C. Zhang, 1995, Exactly solved model of self-organized criticality, *Phys. Rev. Lett.* **75**(8), 1550–1553. → pp 206, 288.

▶ Maslov, S., and Y.-C. Zhang, 1996, Self-organized critical directed percolation, *Physica A* **223**(1–2), 1–6. → pp 81, 178, 179, *295*, 485.

Masuda, N., K.-I. Goh, and B. Kahng, 2005, Extremal dynamics on complex networks: analytic solutions, *Phys. Rev. E* **72**(6), 066106 (6 pp), `arXiv:cond-mat/0508623v1`. → p 159.

Matsumoto, M., 2008, Mersenne Twister Home Page, available from `http://www.math.sci.hiroshima-u.ac.jp/~m-mat/MT/emt.html`, accessed 9 October 2008. → pp 371, 380.

Matsumoto, M., and T. Nishimura, 1998, Mersenne Twister: A 623-Dimensionally Equidistributed Uniform Pseudorandom Number Generator, *ACM Trans. Model. Comput. Sim.* **8**(1), 3–30. → p 380.

Maunuksela, J., M. Myllys, O.-P. Kähkönen, J. Timonen, N. Provatas, M. J. Alava, and T. Ala-Nissila, 1997, Kinetic roughening in slow combustion of paper, *Phys. Rev. Lett.* **79**(8), 1515–1518. → p 301.

Mazzoni, A., F. D. Broccard, E. Garcia-Perez, P. Bonifazi, M. E. Ruaro, and V. Torre, 2007, On the dynamics of the spontaneous activity in neuronal networks, *PLoS ONE* **2**(5), e439. → p 70.

McComb, W. D., 2004, *Renormalization Methods* (Oxford University Press, New York, NY, USA). → pp 266, 394.

● McKane, A., M. Droz, J. Vannimenus, and D. Wolf (editors), 1995, *Scale Invariance, Interfaces, and Non-Equilibrium Dynamics* (Plenum Press, New York, NY, USA), NATO Advanced Study Institute on Scale Invariance, Interfaces, and Non-Equilibrium Dynamics, Cambridge, UK, June 20–30, 1994. → p 6.

Meakin, P., 1998, *Fractals, Scaling and Growth far from Equilibrium* (Cambridge University Press, Cambridge, UK). → pp 205, 301, 302, 313.

Medina, E., T. Hwa, M. Kardar, and Y.-C. Zhang, 1989, Burgers equation with correlated noise: renormalization-group analysis and applications to directed polymers and interface growth, *Phys. Rev. A* **39**(6), 3053–3075. → pp 267, 320, 324.

Meester, R., and C. Quant, 2005, Connections between 'self-organised' and 'classical' criticality, *Markov Process. Relat. Fields* **11**(2), 355–370. → pp 101, 330.

Meester, R., and D. Znamenski, 2002, Non-triviality of the discrete Bak-Sneppen evolution model, *J. Stat. Phys.* **109**(5/6), 987–1004, `arXiv:cond-mat/0301480`. → p 158.

Meester, R., and D. Znamenski, 2003, Limit behavior of the Bak–Sneppen evolution model, *Ann. Probab.* **31**(4), 1986–2002, arXiv:cond-mat/0301479. → pp 151, 157.

Mehta, A., 1992, Real sandpiles: dilatancy, hysteresis and cooperative dynamics, *Physica A* **186**(1–2), 121 – 153. → pp 56, 85.

Mehta, A., 2007, *Granular Physics* (Cambridge University Press, Cambridge, UK). → p 56.

Mehta, A., and G. Barker, 1991, The self-organising sand pile, *New Sci.* **130**(1773), 40–43. → p 56.

Mehta, A., and G. C. Barker, 1994, Disorder, memory and avalanches in sandpiles, *Europhys. Lett.* **27**(7), 501–506. → pp 17, 56, 134, 207.

Mehta, A., J. M. Luck, and R. J. Needs, 1996, Dynamics of sandpiles: physical mechanisms, coupled stochastic equations, and alternative universality classes, *Phys. Rev. E* **53**(1), 92. → p 299.

Mehta, A. P., A. C. Mills, K. A. Dahmen, and J. P. Sethna, 2002, Universal pulse shape scaling function and exponents: critical test for avalanche models applied to Barkhausen noise, *Phys. Rev. E* **65**(4), 046139 (6 pp). → p 64.

Meisel, L. V., and P. J. Cote, 1992, Power laws, flicker noise, and the Barkhausen effect, *Phys. Rev. B* **46**(17), 10822–10828. → p 63.

Melby, P., J. Kaidel, N. Weber, and A. Hübler, 2000, Adaptation to the edge of chaos in the self-adjusting logistic map, *Phys. Rev. Lett.* **84**(26), 5991–5993. → p *332*.

Mendes, J. F. F., 2000, Critical behavior of models with infinitely many absorbing states, *Braz. J. Phys.* **30**(1). → p 177.

Mendes, J. F. F., R. Dickman, M. Henkel, and M. C. Marques, 1994, Generalized scaling for models with multiple absorbing states, *J. Phys. A: Math. Gen.* **27**(9). → pp 170, 178.

Meng, T.-c., R. Rittel, and Y. Zhang, 1999, Inelastic diffraction and color-singlet gluon clusters in high-energy hadron-hadron and lepton-hadron collisions, *Phys. Rev. Lett.* **82**(10), 2044–2047. → p 72.

Metropolis, N., A. W. Rosenbluth, M. N. Rosenbluth, A. H. Teller, and E. Teller, 1953, Equation of state calculations by fast computing machines, *J. Chem. Phys.* **21**(6), 1087–1092. → p 357.

Middleton, A. A., 1992, Asymptotic uniqueness of the sliding state for charge-density waves, *Phys. Rev. Lett.* **68**(5), 670–673. → pp 205, 302.

Middleton, A. A., and C. Tang, 1995, Self-organized criticality in nonconserved systems, *Phys. Rev. Lett.* **74**(5), 742–745. → pp 133–136, 139, 140, 321, 328.

Mikeska, B., 1996, *A Renormalization-Group approach to Self-Organized Criticality*, Ph.D. Thesis, Universität Hamburg, Hamburg, Germany. → p 268.

⇒ Miller, S. L., W. M. Miller, and P. J. McWhorter, 1993, Extremal dynamics: a unifying physical explanation of fractals, $1/f$ noise, and activated processes, *J. Appl. Phys.* **73**(6), 2617–2628. → pp 320, 344.

Milshtein, E., O. Biham, and S. Solomon, 1998, Universality classes in isotropic, Abelian, and non-Abelian sandpile models, *Phys. Rev. E* **58**(1), 303–310. → pp 23, 100, 106, 164, 166, 277, 284, 285.

Mineshige, S., N. B. Ouchi, and H. Nishimori, 1994a, On the generation of $1/f$ Fluctuations in X-rays from black-hole objects, *Publ. Astron. Soc. Jpn.* **46**, 97–105. → p 72.

Mineshige, S., M. Takeuchi, and H. Nishimori, 1994b, Is a black hole accretion disk in a self-organized critical state?, *Astrophys. J.* **435**, L125–L128. → p 72.

Miramontes, O., and P. Rohani, 1998, Intrinsically generated coloured noise in laboratory insect populations, *Proc. R. Soc. London, Ser. B* **265**(1398), 785–792. → p 74.

Moeur, W. A., P. K. Day, F.-C. Liu, S. T. P. Boyd, M. J. Adriaans, and R. V. Duncan, 1997, Observation of self-organized criticality near the superfluid transition in ^{4}He, *Phys. Rev. Lett.* **78**(12), 2421–2424. → p 62.

▶ Mohanty, P. K., and D. Dhar, 2002, Generic sandpile models have directed percolation exponent, *Phys. Rev. Lett.* **89**(10), 104303 (4 pp). → pp 178, 179, 296–298.

Mohanty, P. K., and D. Dhar, 2007, Critical behavior of sandpile models with sticky grains, *Physica A* **384**(1), 34–38, Proceedings of the International Conference on Statistical Physics, Raichak and Kolkata, India, January 5–9, 2007. → pp 178, 179, 296.

Moloney, N. R., and G. Pruessner, 2003, Asynchronously parallelized percolation on distributed machines, *Phys. Rev. E* **67**(3), 037701 (4 pp), arXiv:cond-mat/0211240. → p 124.

Montakhab, A., and J. M. Carlson, 1998, Avalanches, transport, and local equilibrium in self-organized criticality, *Phys. Rev. E* **58**(5), 5608–5619. → pp 333, 343.

Montroll, E. W., and M. F. Shlesinger, 1982, On $1/f$ noise and other distributions with long tails, *Proc. Natl. Acad. Sci. USA* **79**(10), 3380–3383. → p 15.

Morand, J., G. Pruessner, and K. Christensen, 2010, Correlations of avalanche sizes in the Oslo model (unpublished). → p 396.

Moreno, Y., J. B. Gómez, and A. F. Pacheco, 1999, Modified renormalization strategy for sandpile models, *Phys. Rev. E* **60**(6), 7565–7568. → p 268.

Moreno, Y., and A. Vazquez, 2002, The Bak-Sneppen model on scale-free networks, *Europhys. Lett.* **57**(5), 765–771. → p 159.

Moßner, W. K., B. Drossel, and F. Schwabl, 1992, Computer simulations of the forest-fire model, *Physica A* **190**(3–4), 205–217. → pp 113, 115.

Mousseau, N., 1996, Synchronization by disorder in coupled systems, *Phys. Rev. Lett.* **77**(5), 968–971. → p 134.

Muñoz, M. A., 2003, Multiplicative noise in non-equilibrium phase transitions: a tutorial, arXiv:cond-mat/0303650v2 (unpublished). → p 302.

● Muñoz, M. A., R. Dickman, R. Pastor-Satorras, A. Vespignani, and S. Zapperi, 2001, Sandpiles and absorbing-state phase transitions: recent results and open problems, *AIP Conf. Proc.* **574**(1), 102–110, arXiv:cond-mat/0011447. → pp 6, 177–179, 266, 305, 331, 333.

Muñoz, M. A., R. Dickman, A. Vespignani, and S. Zapperi, 1999, Avalanche and spreading exponents in systems with absorbing states, *Phys. Rev. E* **59**(5), 6175–6179. → pp 177, 305, 306, 336, 340–342.

Murray, J. D., 2003, *Mathematical Biology II: Spatial Models and Biomedical Applications* (Springer-Verlag, Berlin, Germany), 3rd edition. → pp 112, 114, 115.

Nagel, K., and H. J. Herrmann, 1993, Deterministic models of traffic jams, *Physica A* **199**(2), 254–269. → p 76.

Nagel, S. R., 1992, Instabilities in a sandpile, *Rev. Mod. Phys.* **64**(1), 321–325. → p 56.

Nakanishi, H., 1990, Cellular-automaton model of earthquakes with deterministic dynamics, *Phys. Rev. A* **41**(12), 7086–7089. → p 126.

Nakanishi, H., and K. Sneppen, 1997, Universal versus drive-dependent exponents for sandpile models, *Phys. Rev. E* **55**(4), 4012–4016. → pp 21, 23, 44, 100, 165, 168, 170, 175, 177, 178, 180, 183, *184*, 186, 187, 198, 227, 245, 248, 294, 306, 316, 339.

Narayan, O., 1996, Self-similar Barkhausen noise in magnetic domain wall motion, *Phys. Rev. Lett.* **77**(18), 3855–3857. → p 64.

Narayan, O., and D. S. Fisher, 1992a, Critical behavior of sliding charge-density waves in 4-epsilon dimensions, *Phys. Rev. B* **46**(18), 11520–11549. → p 299.

Narayan, O., and D. S. Fisher, 1992b, Dynamics of sliding charge-density waves in 4-epsilon dimensions, *Phys. Rev. Lett.* **68**(24), 3615–3618. → p 299.

Narayan, O., and D. S. Fisher, 1993, Threshold critical dynamics of driven interfaces in random media, *Phys. Rev. B* **48**(10), 7030–7042. → pp 299, 315.

Narayan, O., and A. A. Middleton, 1994, Avalanches and the renormalization group for pinned charge-density waves, *Phys. Rev. B* **49**(1), 244–256. → p 299.

Nattermann, T., S. Stepanow, L.-H. Tang, and H. Leschhorn, 1992, Dynamics of interface depinning in a disordered medium, *J. Phys. II (France)* **2**, 1483–1488, for details see Leschhorn *et al.* (1997). → pp 179, 182, 205, 266, 267, 299, 315.

Neelin, J. D., O. Peters, and K. Hales, 2009, The transition to strong convection, *J. Atmos. Sci.* **66**(8), 2367–2384. → p 72.

Nemeth, E., G. Snyder, S. Seebass, and T. R. Hein, 1995, UNIX® System Administration Handbook (Prentice Hall, Upper Saddle River, NJ, USA), 2nd edition. → p 358.

Nerone, N., and S. Gabbanelli, 2001, Surface fluctuations and the inertia effect in sandpiles, *Gran. Matter* **3**(1), 117–120. → p 56.

Newman, D. E., B. A. Carreras, P. H. Diamond, and T. S. Hahm, 1996, The dynamics of marginality and self-organized criticality as a paradigm for turbulent transport, *Phys. Plasmas* **3**(5), 1858–1866. → pp 72, 193.

Newman, M. E. J., 1996, Self-organized criticality, evolution and the fossil extinction record, *Proc. R. Soc. London, Ser. B* **263**(1376), 1605–1610, see Newman (1997a,b). → pp 69, 135, 158, 159, 438.

Newman, M. E. J., 1997a, Evidence for self-organized criticality in evolution, *Physica D* **107**(2–4), 293–296, Proceedings of the 16th Annual International Conference of the Center for Nonlinear Studies, Los Alamos, NM, USA, May 13–17, 1996, see Newman (1996, 1997b). → pp 69, 159, 438.

Newman, M. E. J., 1997b, A model of mass extinction, *J. Theor. Biol.* **189**, 235–252, see Newman (1996, 1997a). → pp 69, 159, 438.

Newman, M. E. J., and G. T. Barkema, 1999, *Monte Carlo Methods in Statistical Physics* (Oxford University Press, New York, NY, USA). → pp 220, 358, 361, 371.

Newman, M. E. J., and G. J. Eble, 1999, Power spectra of extinction in the fossil record, *Proc. R. Soc. London, Ser. B* **266**(1425), 1267–1270. → p *69*.

Newman, M. E. J., M. Girvan, and J. D. Farmer, 2002, Optimal design, robustness, and risk aversion, *Phys. Rev. Lett.* **89**(2), 028301 (4 pp), arXiv:cond-mat/0202330. → p 345.

Newman, M. E. J., and B. W. Roberts, 1995, Mass extinction: evolution and the effects of external influences on unfit species, *Proc. R. Soc. London, Ser. B* **260**(1357), 31–37, arXiv:adap-org/9410004v1. → p 159.

Newman, M. E. J., and K. Sneppen, 1996, Avalanches, scaling, and coherent noise, *Phys. Rev. E* **54**(6), 6226–6231. → pp 159, 192, 193, 342.

Newman, M. E. J., and R. M. Ziff, 2000, Efficient Monte Carlo algorithm and high-precision results for percolation, *Phys. Rev. Lett.* **85**(19), 4104–4107. → p 319.

Newman, T. J., A. J. Bray, and M. A. Moore, 1990, Growth of order in vector spin systems and self-organized criticality, *Phys. Rev. B* **42**(7), 4514–4523. → p 343.

Noever, D. A., 1993, Himalayan sandpiles, *Phys. Rev. E* **47**(1), 724–725. → p 57.

Nowak, E. R., O. W. Taylor, L. Liu, H. M. Jaeger, and T. I. Selinder, 1997, Magnetic flux instabilities in superconducting niobium rings: tuning the avalanche behavior, *Phys. Rev. B* **55**(17), 11702–11705. → p 60.

Nowak, U., and K. D. Usadel, 1990, Slow relaxation of diluted antiferromagnets, *Physica B* **165–166**(Part 1), 211–212, Proceedings of the 19th International Conference on Low Temperature Physics, Brighton, UK, August 16–22, 1990. → p 343.

Nowak, U., and K. D. Usadel, 1991, Nonexponential relaxation of diluted antiferromagnets, *Phys. Rev. B* **43**(1), 851–853. → p 343.

O'Brien, K. P., and M. B. Weissman, 1994, Statistical characterization of Barkhausen noise, *Phys. Rev. E* **50**(5), 3446–3452. → pp 63, 185.

⇒ Obukhov, S. P., 1990, Self-organized criticality: Goldstone modes and their interactions, *Phys. Rev. Lett.* **65**(12), 1395–1398. → pp 342, 343.

Ódor, G., 2004, Universality classes in nonequilibrium lattice systems, *Rev. Mod. Phys.* **76**(3), 663–724, arXiv:cond-mat/0205644. → pp 20, 177, 181, 293.

Ódor, G., and N. Menyhárd, 2008, Crossovers from parity conserving to directed percolation universality, *Phys. Rev. E* **78**(4), 041112 (7 pp). → p 155.

Olami, Z., and K. Christensen, 1992, Temporal correlations, universality, and multifractality in a spring-block model of earthquakes, *Phys. Rev. A* **46**(4), R1720–R1723. → p 67.

▶ Olami, Z., H. J. S. Feder, and K. Christensen, 1992, Self-organized criticality in a continuous, nonconservative cellular automaton modeling earthquakes, *Phys. Rev. Lett.* **68**(8), 1244–1247, comment (Klein and Rundle, 1993). → pp 82, 126–128, *130*, 133–135, 138, 139, 321, 428, 485.

Olami, Z., I. Procaccia, and R. Zeitak, 1994, Theory of self-organized interface depinning, *Phys. Rev. E* **49**(2), 1232–1237. → pp 295, 299.

Olami, Z., I. Procaccia, and R. Zeitak, 1995, Interface roughening in systems with quenched disorder, *Phys. Rev. E* **52**(4), 3402–3414. → pp 295, 299.

Olson, C. J., C. Reichhardt, and F. Nori, 1997, Superconducting vortex avalanches, voltage bursts, and vortex plastic flow: effect of the microscopic pinning landscape on the macroscopic properties, *Phys. Rev. B* **56**(10), 6175–6194. → p 61.

Omori, F., 1894, On the aftershocks of earthquakes, *J. Coll. Sci., Imp. Univ. Tokyo* **7**, 111–200. → p 67.

Onuki, A., 1987, The He I-He II interface in He^4 and He^3-He^4 near the Superfluid transition, *Jpn. J. Appl. Phys.* **26S3**(Supplement 26-3-1), 365–366, proceedings of the 18th

International Conference on Low Temperature Physics, Kyoto, Japan, August 20–26, 1987. → p 61.

Onuki, A., 1996, Superfluid transition in ^{4}He in gravity and heat flow, *J. Low Temp. Phys.* **104**(3), 133–142. → p 61.

Ooi, S., T. Shibaushi, and T. Tamegai, 2000, Vortex avalanches in the vortex lattice phase of $Bi_2Sr_2CaCu_2O_{8+y}$, *Physica B* **284–288**(Part 1), 775 – 776. → p 60.

Opper, M., and D. Saad (editors), 2001, *Advanced Mean Field Methods*, Neural Information Processing Series (MIT Press, Cambridge, MA, USA). → p 394.

Otsuka, M., 1971, A simulation of earthquake occurrence, part 1, A mechanical model (in Japanese), *Jishin (also Zisin, J. Seismol. Soc. Jpn.)* **24**, 13–25. → p 125.

Otsuka, M., 1972a, A chain-reaction-type source model as a tool to interpret the magnitude-frequency relation of earthquakes, *J. Phys. Earth* **20**, 34–45. → p 125.

Otsuka, M., 1972b, A simulation of earthquake occurrence, *Phys. Earth Planet. Inter.* **6**, 311–315. → p 125.

Pacheco, J. F., C. H. Scholz, and L. R. Sykes, 1992, Changes in frequency-size relationship from small to large earthquakes, *Nature* **355**(6355), 71–73. → pp 66, 67, 139.

Paczuski, M., 1995, Dynamic scaling: distinguishing self-organized from generically critical systems, *Phys. Rev. E* **52**(3), R2137–R2140. → p 321.

Paczuski, M., and P. Bak, 1993, Theory of the one-dimensional forest-fire model, *Phys. Rev. E* **48**(5), R3214–R3216. → pp 116, 121, 124.

Paczuski, M., P. Bak, and S. Maslov, 1995, Laws for stationary states in systems with extremal dynamics, *Phys. Rev. Lett.* **74**(21), 4253–4256. → p 149.

▶ Paczuski, M., and K. E. Bassler, 2000, Theoretical results for sandpile models of SOC with multiple topplings, arXiv:cond-mat/0005340v2 (unpublished). → pp 22, 101, 107, 111, 165, 170, 172, 206–208, 249, 251, 268, 279, 280, 282–286, 288–290, *292*, *293*, *321*, 327, 334.

Paczuski, M., and S. Boettcher, 1996, Universality in sandpiles, interface depinning, and earthquake models, *Phys. Rev. Lett.* **77**(1), 111–114, largely identical to proceedings article (Boettcher and Paczuski, 1997b). → pp 125, 126, 162, 177, 178, 180, 182, *183*, 184, 187, 189, 190, 197, 198, 203, 206, 268, 299, 300, 302, 393, 404.

Paczuski, M., and S. Boettcher, 1997, Avalanches and waves in the Abelian sandpile model, *Phys. Rev. E* **56**(4), R3745–R3748. → p 100.

Paczuski, M., S. Boettcher, and M. Baiesi, 2005, Interoccurrence times in the Bak-Tang-Wiesenfeld sandpile model: a comparison with the observed statistics of solar flares, *Phys. Rev. Lett.* **95**(18), 181102 (4 pp). → p 72.

Paczuski, M., S. Maslov, and P. Bak, 1994a, Erratum: field theory for a model of self-organized criticality, *Europhys. Lett.* **28**(4), 295–296. → pp 142, 147, 155, 157, 293.

Paczuski, M., S. Maslov, and P. Bak, 1994b, Field theory for a model of self-organized criticality, *Europhys. Lett.* **27**(2), 97–102. → pp 142, 143, 145, 146, 150, 154, 155, 157, 293.

● Paczuski, M., S. Maslov, and P. Bak, 1996, Avalanche dynamics in evolution, growth, and depinning models, *Phys. Rev. E* **53**(1), 414–443, arXiv:adap-org/9510002. → pp 6, 142–147, 149, 151, 153, 156, 157, 160, 161, 178, 187, 203, 251, 283, 293, 299, 307, 308, 310, 314, 315, 345, 393.

Pan, G.-J., D.-M. Zhang, Z.-H. Li, H.-Z. Sun, and Y.-P. Ying, 2005a, Critical behavior in non-Abelian deterministic directed sandpile, *Phys. Lett. A* **338**(3–5), 163–168. → pp 286, 287.

Pan, G.-J., D.-M. Zhang, H.-Z. Sun, and Y.-P. Yin, 2005b, Universality class in Abelian sandpile models with stochastic toppling rules, *Commun. Theor. Phys.* **44**(3), 483–486. → pp 177, 178, 187.

Pan, G.-J., D.-M. Zhang, Y.-P. Yin, and M.-H. He, 2006, Avalanche dynamics in quenched random directed sandpile models, *Chin. Phys. Lett.* **23**(10), 2811–2814. → p 287.

Papa, A. R. R., and L. da Silva, 1997, Earthquakes in the brain, *Theor. Biosci.* **116**, 321–327. → pp 70, 149.

Papadopoulos, M. C., M. Hadjitheodossiou, C. Chrysostomou, C. Hardwidge, and B. A. Bell, 2001, Is the National Health Service at the edge of chaos?, *J. R. Soc. Med.* **94**(12), 613–616. → p *75*.

Papoyan, V. V., and A. M. Povolotsky, 1997, Renormalization group study of sandpile on the triangular lattice, *Physica A* **246**(1–2), 241 – 252. → p 268.

Parisi, G., 1998, *Statistical Field Theory* (Physica B, Reading, MA, USA). → p 343.

Park, K., S. Kang, and I. Kim, 2005, Absorbing phase transition with a conserved field, *Phys. Rev. E* **71**(6), 066129 (5 pp). → p 177.

Park, S.-C., 2010, Absence of the link between self-organized criticality and deterministic fixed energy sandpiles, `arXiv:1001.3359` (unpublished). → pp 338, 342.

Pastor-Satorras, R., 1997, Multifractal properties of power-law time sequences: application to rice piles, *Phys. Rev. E* **56**(5), 5284–5294. → p 195.

Pastor-Satorras, R., and A. Vespignani, 2000a, Anomalous scaling in the Zhang model, *Eur. Phys. J. B* **18**(2), 197–200. → pp 104, 106, 107, *110*, 178.

Pastor-Satorras, R., and A. Vespignani, 2000b, Corrections to scaling in the forest-fire model, *Phys. Rev. E* **61**(5), 4854–4859. → pp 120, 125.

Pastor-Satorras, R., and A. Vespignani, 2000c, Critical behavior and conservation in directed sandpiles, *Phys. Rev. E* **62**(5), 6195–6205, see Pastor-Satorras and Vespignani (2000e). → pp 286, 328, 334, 441.

Pastor-Satorras, R., and A. Vespignani, 2000d, Field theory of absorbing phase transitions with a nondiffusive conserved field, *Phys. Rev. E* **62**(5), R5875–R5878. → pp 162, 177, 178, 182, 266, 333.

▶ Pastor-Satorras, R., and A. Vespignani, 2000e, Universality classes in directed sandpile models, *J. Phys. A: Math. Gen.* **33**(3), L33–L39, see Pastor-Satorras and Vespignani (2000c). → pp 22, 206, 220, 277, 284, 286, 288, 292, 441.

Pastor-Satorras, R., and A. Vespignani, 2001, Reaction-diffusion system with self-organized critical behavior, *Eur. Phys. J. B* **19**(4), 583–587, `arXiv:cond-mat/0101358`. → pp 44, 168, 178, 294, 306, 316.

Patzlaff, H., and S. Trimper, 1994, Analytical approach to the forest-fire model, *Phys. Lett. A* **189**(3), 187–192. → pp 120, 122.

Peixoto, T. P., and C. P. C. Prado, 2004, Statistics of epicenters in the Olami-Feder-Christensen model in two and three dimensions, *Physica A* **342**(1–2), 171–177. → pp 138, 141.

Peliti, L., 1985, Path integral approach to birth-death processes on a lattice, *J. Phys. (Paris)* **46**, 1469–1483. → p 266.

Peng, C. K., S. V. Buldyrev, A. L. Goldberger, S. Havlin, F. Sciortino, M. Simons, and H. E. Stanley, 1992, Long-range correlations in nucleotide sequences, *Nature* **356**(6365), 168–170. → p 74.

Peng, C.-K., S. Havlin, H. E. Stanley, and A. L. Goldberger, 1995, Quantification of scaling exponents and crossover phenomena in nonstationary heartbeat time series, *Chaos* **5**(1), 82–87. → p 74.

Peng, C.-K., J. Mietus, J. M. Hausdorff, S. Havlin, H. E. Stanley, and A. L. Goldberger, 1993, Long-range anticorrelations and non-Gaussian behavior of the heartbeat, *Phys. Rev. Lett.* **70**(9), 1343–1346. → p 74.

Peng, G., 1992, Self-organized critical state in a directed sandpile automaton on Bethe lattices: equivalence to site percolation, *J. Phys. A: Math. Gen.* **25**(20), 5279–5282. → p 292.

Peng, G., and H. J. Herrmann, 1993, Boundary-induced anisotropy of the avalanches in the sandpile automaton, *Physica A* **199**(3–4), 476 – 484. → p 324.

Pepke, S. L., and J. M. Carlson, 1994, Predictability of self-organizing systems, *Phys. Rev. E* **50**(1), 236–242. → pp 67, 355.

• Pérez, C. J., Á. Corral, A. Díaz-Guilera, K. Christensen, and A. Arenas, 1996, On self-organized criticality and synchronization in lattice models of coupled dynamical systems, *Int. J. Mod. Phys. B* **10**(10), 1111–1151. → pp 88, 109, 126, 128, 129, 203, 267.

Perković, O., K. Dahmen, and J. P. Sethna, 1995, Avalanches, Barkhausen noise, and plain old criticality, *Phys. Rev. Lett.* **75**(24), 4528–4531. → pp *63*, 64.

Peschel, I., X. Wang, M. Kaulke, and K. Hallberg (editors), 1999, Density-Matrix Renormalization, volume 528 of *Lecture Notes in Physics* (Springer-Verlag, Berlin, Germany). → pp 176, 202.

Peskin, C. S., 1975, *Mathematical Aspects of Heart Physiology* (Courant Insititute of Mathematical Sciences, New York, NY, USA). → p 135.

Peters, O., and K. Christensen, 2002, Rain: relaxations in the sky, *Phys. Rev. E* **66**(3), 036120 (9 pp). → p 71.

Peters, O., and M. Girvan, 2009, Universality under conditions of self-tuning, `arXiv: 0902.1956` (unpublished). → pp 330, 342.

Peters, O., C. Hertlein, and K. Christensen, 2002, A complexity view of rainfall, *Phys. Rev. Lett.* **88**(1), 018701 (4 pp), `arXiv:cond-mat/0201468`. → p 71.

Peters, O., and J. D. Neelin, 2006, Critical phenomena in atmospheric precipitation, *Nat. Phys.* **2**(6), 393–396. → p 71.

Petri, A., G. Paparo, A. Vespignani, A. Alippi, and M. Costantini, 1994, Experimental evidence for critical dynamics in microfracturing processes, *Phys. Rev. Lett.* **73**(25), 3423–3426. → p 64.

Pfeuty, P., and G. Toulouse, 1977, *Introduction to the Renormalization Group and to Critical Phenomena* (John Wiley & Sons, Chichester, UK). → pp 38, 337.

Pietronero, L., A. Erzan, and C. Evertsz, 1988, Theory of fractal growth, *Phys. Rev. Lett.* **61**(7), 861–864. → p 267.

Pietronero, L., and W. R. Schneider, 1991, Fixed scale transformation approach to the nature of relaxation clusters in self-organized criticality, *Phys. Rev. Lett.* **66**(18), 2336–2339. → p 267.

Pietronero, L., P. Tartaglia, and Y.-C. Zhang, 1991, Theoretical studies of self-organized criticality, *Physica A* **173**(1–2), 22–44. → pp 108, 126, 133.

Pietronero, L., A. Vespignani, and S. Zapperi, 1994, Renormalization scheme for self-organized criticality in sandpile models, *Phys. Rev. Lett.* **72**(11), 1690–1693. → pp 92, 101, 110, 121, 165, 168, 182, 267, 268, 292.

Pimpinelli, A., and J. Villain, 1998, *Physics of Crystal Growth* (Cambridge University Press, Cambridge, UK). → p 301.

Pinho, S. T. R., and R. F. S. Andrade, 2004, Power law sensitivity to initial conditions for abelian directed self-organized critical models, *Physica A* **344**(3–4), 601–607, Proceedings of the International Workshop on Trends and Perspectives in Extensive and Non-extensive Statistical Mechanics, in Honor of the 60th birthday of Constantino Tsallis, Angra dos Reis, Brazil, Nov 19–21, 2003. → p 17.

Pinho, S. T. R., C. P. C. Prado, and O. Kinouchi, 1998, Absence of self-organized criticality in a random-neighbor version of the OFC stick-slip model, *Physica A* **257**(1–4), 488–494. → pp 131, 328.

Pla, O., and F. Nori, 1991, Self-organized critical behavior in pinned flux lattices, *Phys. Rev. Lett.* **67**(7), 919–922. → p 59.

Pla, O., N. K. Wilkin, and H. J. Jensen, 1996, Avalanches in the Bean critical state: a characteristic of the random pinning potential, *Europhys. Lett.* **33**(4), 297–302. → pp *60*, 61.

Planet, R., S. Santucci, and J. Ortín, 2009, Avalanches and non-Gaussian fluctuations of the global velocity of imbibition fronts, *Phys. Rev. Lett.* **102**(9), 094502 (4 pp), comment (Pruessner, 2010). → pp 240, 445.

Planet, R., S. Santucci, and J. Ortín, 2010, Planet, Santucci, and Ortín reply:, *Phys. Rev. Lett.* **105**(2), 029402 (1 p), reply to comment (Pruessner, 2010). → pp 240, 445.

Plourde, B., F. Nori, and M. Bretz, 1993, Water droplet avalanches, *Phys. Rev. Lett.* **71**(17), 2749–2752. → p 57.

Politzer, P. A., 2000, Observation of avalanchelike phenomena in a magnetically confined plasma, *Phys. Rev. Lett.* **84**(6), 1192–1195. → p 72.

Portugali, J., 2000, *Self-organization and the City* (Springer-Verlag, Berlin, Germany). → p 76.

Pradhan, P., and D. Dhar, 2006, Probability distribution of residence times of grains in models of rice piles, *Phys. Rev. E* **73**(2), 021303 (12 pp), `arXiv:cond-mat/0608144`. → pp 187, 197, 209.

Pradhan, P., and D. Dhar, 2007, Sampling rare fluctuations of height in the Oslo ricepile model, *J. Phys. A: Math. Theor.* **40**(11), 2639–2650, `arXiv:cond-mat/0511237`. → p 197.

Pradhan, S., and B. K. Chakrabarti, 2001, Precursors of catastrophe in the Bak-Tang-Wiesenfeld, Manna, and random-fiber-bundle models of failure, *Phys. Rev. E* **65**(1), 016113 (7 pp). → pp 165, 339.

Prado, C. P. C., and Z. Olami, 1992, Inertia and break of self-organized criticality in sandpile cellular-automata models, *Phys. Rev. A* **45**(2), 665–669. → p 56.

Press, W. H., 1978, Flicker noises in astronomy and elsewhere, *Comments Astrophys.* **7**(4), 103–119. → p 4.

Press, W. H., S. A. Teukolsky, W. T. Vetterling, and B. P. Flannery, 2007, *Numerical Recipes* (Cambridge University Press, Cambridge, UK), 3rd edition. → pp 220–222, 358, 371, 376, 380, 389.

Priezzhev, V. B., 1994, Structure of two-dimensional sandpile. I. Height probabilities, *J. Stat. Phys.* **74**(5/6), 955–979. → pp 99, 100.

Priezzhev, V. B., 2000, The upper critical dimension of the Abelian sandpile model, *J. Stat. Phys.* **98**(3/4), 667–684, arXiv:cond-mat/9904054. → p 104.

▶ Priezzhev, V. B., D. Dhar, A. Dhar, and S. Krishnamurthy, 1996a, Eulerian walkers as a model of self-organized criticality, *Phys. Rev. Lett.* **77**(25), 5079–5082. → pp 81, 485.

Priezzhev, V. B., E. V. Ivashkevich, A. M. Povolotsky, and C.-K. Hu, 2001, Exact phase diagram for an asymmetric avalanche process, *Phys. Rev. Lett.* **87**(8), 084301 (4 pp). → pp 206, 288.

Priezzhev, V. B., D. V. Ktitarev, and E. V. Ivashkevich, 1996b, Formation of avalanches and critical exponents in an Abelian sandpile model, *Phys. Rev. Lett.* **76**(12), 2093–2096. → pp 92, 99, 100, 103.

Privman, V., and M. E. Fisher, 1984, Universal critical amplitudes in finite-size scaling, *Phys. Rev. B* **30**(1), 322–327. → pp 12, 14, 27, 39.

Privman, V., P. C. Hohenberg, and A. Aharony, 1991, Universal critical-point amplitude relations, in *Phase Transitions and Critical Phenomena*, edited by C. Domb and J. L. Lebowitz (Academic Press, New York, NY, USA), volume 14, chapter 1, pp. 1–134. → pp 18, 20, 25, 39, 189, 227, 338, 339.

Pruessner, G., 2003a, Oslo rice pile model is a quenched Edwards-Wilkinson equation, *Phys. Rev. E* **67**(3), 030301(R) (4 pp), arXiv:cond-mat/0209531. → pp 162, 178, 180, 182, 184, 186, 199, 203–205, 278, 282, 299, 300.

Pruessner, G., 2003b, Universality in the Oslo model (unpublished). → pp 107, 229, 284.

Pruessner, G., 2004a, Drift causes anomalous exponents in growth processes, *Phys. Rev. Lett.* **92**(24), 246101 (4 pp), arXiv:cond-mat/0404007v2. → pp 207, 251, 291, 293.

Pruessner, G., 2004b, Exact solution of the totally asymmetric Oslo model, *J. Phys. A: Math. Gen.* **37**(30), 7455–7471, arXiv:cond-mat/0402564. → pp 174, 201, 206, 207, 222, 223, 232, 249, 271, 286, 289, 310.

● Pruessner, G., 2004c, *Studies in Self-Organised Criticality*, Ph.D. Thesis, Imperial College London, UK, accessed 19 November 2009, URL http://www.ma.imperial.ac.uk/~pruess/publications/thesis_final/. → pp 173, 174, 180, 189, 207, *395*.

Pruessner, G., 2007, Equivalence of conditional and external field ensembles in absorbing-state phase transitions, *Phys. Rev. E* **76**(6), 061103 (4 pp), arXiv:0712.0979v1. → pp 39, 181, 307, 336, 337, *395*.

Pruessner, G., 2008, Block scaling in the directed percolation universality class, *New J. Phys.* **10**(11), 113003 (13 pp), arXiv:0706.1144. → pp 21, 49, 134, 336, 337.

Pruessner, G., 2009, Probability densities in complex systems, measuring, in *Encyclopedia of Complexity and Systems Science*, edited by R. A. Meyers (Springer-Verlag, New York, NY, USA), volume 7, pp. 6990–7009. → pp 230, 236, 388.

Pruessner, G., 2010, Comment on 'avalanches and non-gaussian fluctuations of the global velocity of imbibition fronts', *Phys. Rev. Lett.* **105**(2), 029401 (1 p), comment on (Planet *et al.*, 2009), reply (Planet *et al.*, 2010). → pp 240, 443.

Pruessner, G., and H. J. Jensen, 2002a, Broken scaling in the forest-fire model, *Phys. Rev. E* **65**(5), 056707 (8 pp), arXiv:cond-mat/0201306. → pp 116, 119–121, 124, 125, 224.

Pruessner, G., and H. J. Jensen, 2002b, A solvable non-conservative model of self-organised criticality, *Europhys. Lett.* **58**(2), 250–256, arXiv:cond-mat/0104567. → pp 130, 265, 328, 334.

▶ Pruessner, G., and H. J. Jensen, 2003, Anisotropy and universality: the Oslo model, the rice pile experiment and the quenched Edwards-Wilkinson equation, *Phys. Rev. Lett.* **91**(24), 244303 (4 pp), arXiv:cond-mat/0307443. → pp 174, 187–189, 192, 206, 222, 286, 290, 291, 293.

Pruessner, G., and H. J. Jensen, 2004, Efficient algorithm for the forest fire model, *Phys. Rev. E* **70**(6), 066707 (25 pp), arXiv:cond-mat/0309173. → pp 44, 45, 119, 120, 124, 125, 224, 385.

Pruessner, G., D. Loison, and K.-D. Schotte, 2001, Monte Carlo simulation of an Ising model on a Sierpiński carpet, *Phys. Rev. B* **64**(13), 134414 (10 pp). → pp 216, 229.

Pruessner, G., and N. R. Moloney, 2003, Numerical results for crossing, spanning and wrapping in two-dimensional percolation, *J. Phys. A: Math. Gen.* **36**(44), 11213–11228, arXiv:cond-mat/0309126. → p 226.

Pruessner, G., and N. R. Moloney, 2006, Asynchronously parallelised percolation on distributed machines, in *Computer Simulation Studies in Condensed-Matter Physics XVIII; Proceedings of the Eighteens Workshop, Athens, GA, USA, March 7–11, 2005* (Springer-Verlag, Berlin, Germany), volume 105 of *Springer Proceedings in Physics*, pp. 121 – 125. → p 124.

Pruessner, G., and O. Peters, 2006, Self-organized criticality and absorbing states: lessons from the Ising model, *Phys. Rev. E* **73**(2), 025106(R) (4 pp), arXiv:cond-mat/0411709. → pp 52, 265, 330, 337–341, 399.

Pruessner, G., and O. Peters, 2008, Reply to 'Comment on "self-organized criticality and absorbing states: lessons from the Ising model"', *Phys. Rev. E* **77**(4), 048102 (2 pp), reply to comment (Alava *et al.*, 2008). → pp 170, 339, 399.

Puhl, H., 1992, On the modelling of real sand piles, *Physica A* **182**(3), 295 – 319. → pp 56, 85, 324.

Puhl, H., 1993, Sandpiles on random lattices, *Physica A* **197**(1–2), 14 – 22. → p 56.

Quartier, L., B. Andreotti, S. Douady, and A. Daerr, 2000, Dynamics of a grain on a sandpile model, *Phys. Rev. E* **62**(6), 8299–8307. → p 192.

Ramasco, J. J., J. M. López, and M. A. Rodríguez, 2000, Generic dynamic scaling in kinetic roughening, *Phys. Rev. Lett.* **84**(10), 2199–2202. → pp 313, 314.

Ramasco, J. J., J. M. López, and M. A. Rodríguez, 2001, Interface depinning in the absence of an external driving force, *Phys. Rev. E* **64**(6), 066109 (5 pp). → p 300.

Ramasco, J. J., M. A. Muñoz, and C. A. da Silva Santos, 2004, Numerical study of the Langevin theory for fixed-energy sandpiles, *Phys. Rev. E* **69**(4), 045105(R) (4 pp). → pp 177, 182, 183, 333, 340.

Ramos, O., E. Altshuler, and K. J. Måløy, 2009, Avalanche prediction in a self-organized pile of beads, *Phys. Rev. Lett.* **102**(7), 078701 (4 pp). → pp 67, 134.

Raup, D. M., 1976, Species diversity in the Phanerozoic; a tabulation, *Paleobiology* **2**(4), 279–288. → p 68.

Raup, D. M., 1986, Biological extinction in earth history, *Science* **231**(4745), 1528–1533. → pp *68*, 69.

Raup, D. M., 1991, A kill curve for Phanerozoic marine species, *Paleobiology* **17**(1), 37–48. → p 69.

Raup, D. M., and G. E. Boyajian, 1988, Patterns of generic extinction in the fossil record, *Paleobiology* **14**(2), 109–125. → p 68.

Raup, D. M., and J. J. Sepkoski, Jr., 1984, Periodicity of extinctions in the geologic past, *Proc. Natl. Acad. Sci. USA* **81**(3), 801–805. → p 68.

Ray, T. S., and N. Jan, 1994, Anomalous approach to the self-organized critical state in a model for 'life at the edge of chaos', *Phys. Rev. Lett.* **72**(25), 4045–4048. → pp 145, 155, 319.

Redner, S., 2001, *A Guide to First-Passage Processes* (Cambridge University Press, Cambridge, UK). → pp 243, 256.

Reed, W. J., and K. S. McKelvey, 2002, Power-law behaviour and parametric models for the size-distribution of forest-fires, *Ecol. Modell.* **150**, 239–254. → pp 73, 122.

Rhodes, C. J., and R. M. Anderson, 1996, Power laws governing epidemics in isolated populations, *Nature* **381**, 600–602. → p 73.

Rhodes, C. J., H. J. Jensen, and R. M. Anderson, 1997, On the critical behaviour of simple epidemics, *Proc. R. Soc. London, Ser. B* **264**(1388), 1639–1646. → pp 73, 74.

Rhodes, T. L., R. A. Moyer, R. Groebner, E. J. Doyle, R. Lehmer, W. A. Peebles, and C. L. Rettig, 1999, Experimental evidence for self-organized criticality in tokamak plasma turbulence, *Phys. Lett. A* **253**(3–4), 181 – 186. → p 72.

Ricotta, C., M. Arianoutsou, R. Díaz-Delgado, B. Duguy, F. Lloret, E. Maroudi, S. Mazzoleni, J. M. Moreno, S. Rambal, R. Vallejo, and A. Vázquez, 2001, Self-organized criticality of wildfires ecologically revisited, *Ecol. Modell.* **141**, 307–311. → p 73.

Ricotta, C., G. Avena, and M. Marchetti, 1999, The flaming sandpile: self-organized criticality and wildfires, *Ecol. Modell.* **119**, 73–77. → p 73.

• Riste, T., and D. Sherrington (editors), 1991, *Spontaneous Formation of Space-Time Structures and Criticality* (Kluwer, Dordrecht, The Netherlands), NATO Advanced Study Institute on Spontaneous Formation of Space-Time Structures and Criticality, Geilo, Norway, April 2–12, 1991. → p 6.

Rittel, R., 2000, *Selbstorganisierte Kritikalität auf der Ebene von Quarks und Gluonen und deren Manifestation in inelastisch diffraktiven hochenergetischen Streuprozessen*, Ph.D. Thesis, Fachbereich Physik, Freie Universität Berlin, Berlin, Germany. → p 252.

Roberts, D. C., and D. L. Turcotte, 1998, Fractality and self-organized criticality of wars, *Fractals* **6**(4), 351–357. → p 76.

Roering, J. J., J. W. Kirchner, L. S. Sklar, and W. E. Dietrich, 2001, Hillslope evolution by nonlinear creep and landsliding: an experimental study, *Geology* **29**(2), 143–146, comment (van Milligen and Pons, 2002). → p 454.

Roering, J. J., J. W. Kirchner, L. S. Sklar, and W. E. Dietrich, 2002, Hillslope evolution by nonlinear creep and landsliding: an experimental study: comment and reply: reply, *Geology* **30**(5), 482, reply to comment (van Milligen and Pons, 2002). → p 454.

Roland, C., and M. Grant, 1989, Lack of self-averaging, multiscaling, and $1/f$ noise in the kinetics of domain growth, *Phys. Rev. Lett.* **63**(5), 551–554. → p 318.

Rolla, L. T., and V. Sidoravicius, 2009, Absorbing-state phase transition for stochastic sandpiles and activated random walks, `arXiv:0908.1152v1` (unpublished). → p 330.

Rosendahl, J., M. Vekić, and J. Kelley, 1993, Persistent self-organization of sandpiles, *Phys. Rev. E* **47**(2), 1401–1404. → pp 56, *57*.

Rosendahl, J., M. Vekić, and J. E. Rutledge, 1994, Predictability of large avalanches on a sandpile, *Phys. Rev. Lett.* **73**(4), 537–540. → p 57.

Rossi, M., R. Pastor-Satorras, and A. Vespignani, 2000, Universality class of absorbing phase transitions with a conserved field, *Phys. Rev. Lett.* **85**(9), 1803–1806. → pp 170, 171, *177*, 178–180, 330, 333.

Roux, S., and A. Hansen, 1994, Interface roughening and pinning, *J. Phys. I (France)* **4**, 515–538. → pp 178, 299, 314.

Royer, S., and D. Pare, 2003, Conservation of total synaptic weight through balanced synaptic depression and potentiation, *Nature* **422**(6931), 518–522. → p 71.

Rubio, M. A., C. A. Edwards, A. Dougherty, and J. P. Gollub, 1989, Self-affine fractal interfaces from immiscible displacement in porous media, *Phys. Rev. Lett.* **63**(16), 1685–1688. → p 301.

Rudnick, J., and G. Gaspari, 2004, *Elements of the Random Walk* (Cambridge University Press, Cambridge, UK). → p 243.

Ruelle, P., 2002, A $c = -2$ boundary changing operator for the Abelian sandpile model, *Phys. Lett. B* **539**(1), 172–177, `arXiv:hep-th/0203105`. → p 99.

Ruelle, P., and S. Sen, 1992, Toppling distributions in one-dimensional Abelian sandpiles, *J. Phys. A: Math. Gen.* **25**(22), L1257–L1264. → pp 87, 92–94, 174, 202.

Saberi, A. A., S. Moghimi-Araghi, H. Dashti-Naserabadi, and S. Rouhani, 2009, Direct evidence for conformal invariance of avalanche frontiers in sandpile models, *Phys. Rev. E* **79**(3), 031121 (5 pp). → p 99.

Sadhu, T., and D. Dhar, 2008, Emergence of quasiunits in the one-dimensional Zhang model, *Phys. Rev. E* **77**(3), 031122 (6 pp). → p *109*.

Sadhu, T., and D. Dhar, 2009, Steady state of stochastic sandpile models, *J. Stat. Phys.* **134**(3), 427–441. → pp 173, 202.

Salas, J., and A. D. Sokal, 2000, Universal amplitude ratios in the critical two-dimensional Ising model on a torus, *J. Stat. Phys.* **98**(3–4), 551–588, `arXiv:cond-mat/9904038v2`. → pp 20, 39.

Saleur, H., and B. Duplantier, 1987, Exact determination of the percolation hull exponent in two dimensions, *Phys. Rev. Lett.* **58**(22), 2325–2328. → p 319.

Sanchez, R., D. E. Newman, and B. A. Carreras, 2001, Mixed SOC diffusive dynamics as a paradigm for transport in fusion devices, *Nucl. Fusion* **41**(3), 247–256. → pp 72, 193.

Sánchez, R., D. E. Newman, and B. A. Carreras, 2002, Waiting-time statistics of self-organized-criticality systems, *Phys. Rev. Lett.* **88**(6), 068302 (4 pp). → p 76.

Sapoval, B., M. Rosso, and J. F. Gouyet, 1985, The fractal nature of a diffusion front and the relation to percolation, *J. Phys. (Paris) Lett.* **46**(4), 149–156. → p 320.

Sarkar, A., and P. Barat, 2006, Analysis of rainfall records in India: self-organized criticality and scaling, *Fractals* **14**(4), 289–293, arXiv:physics/0512197. → p 71.

Scheidegger, A. E., 1967, A stochastic model for drainage patterns into an intramontane trench, *Bull. Int. Assoc. Sci. Hydrol.* **12**, 15–20. → p 102.

Scheinkman, J. A., and M. Woodford, 1994, Self-organized criticality and economic fluctuations, *Am. Econ. Rev.* **84**(2), 417–421. → p 75.

Schenk, K., B. Drossel, S. Clar, and F. Schwabl, 2000, Finite-size effects in the self-organized critical forest-fire model, *Eur. Phys. J. B* **15**(1), 177–185, arXiv:cond-mat/9904356. → pp 116, 123.

Schenk, K., B. Drossel, and F. Schwabl, 2002, Self-organized critical forest-fire model on large scales, *Phys. Rev. E* **65**(2), 026135 (8 pp), arXiv:cond-mat/0105121. → pp 116, 120–122.

Schick, K. L., and A. A. Verveen, 1974, $1/f$ noise with a low frequency white noise limit, *Nature* **251**(5476), 599–601. → pp 53, 85.

Schildt, H., 2000, *C: The Complete Reference* (McGraw-Hill, Berkeley, CA, USA), 4th edition. → pp 359, 380.

Schmittmann, B., G. Pruessner, and H.-K. Janssen, 2006, Strongly anisotropic roughness in surfaces driven by an oblique particle flux, *Phys. Rev. E* **73**(5), 051603 (10 pp), arXiv:cond-mat/0604363. → pp 318, 329.

Schmittmann, B., and R. K. P. Zia, 1995, Statistical mechanics of driven diffusive systems, in *Phase Transitions and Critical Phenomena*, edited by C. Domb and J. L. Lebowitz (Academic Press, New York, NY, USA), volume 17, pp. 1–220. → pp 266, 293, 301, 303, 321, 324, 325, 334.

Scholz, C. H., 1991, Earthquakes and faulting: self-organized critical phenomena with a characteristic dimension, in *Spontaneous Formation of Space-Time Structures and Criticality*, edited by T. Riste and D. Sherrington (Kluwer, Dordrecht, The Netherlands), pp. 41–56, NATO Advanced Study Institute on Spontaneous Formation of Space-Time Structures and Criticality, Geilo, Norway, April 2–12, 1991. → p 66.

Schotte, K. D., 1999, personal communication. → p *251*.

Schwoll, H., 2004, (Danfoss), personal communication. → p 22.

Sedgewick, R., 1990, *Algorithms in C* (Addison-Wesley, Reading, MA, USA). → p 358.

Segev, R., M. Benveniste, E. Hulata, N. Cohen, A. Palevski, E. Kapon, Y. Shapira, and E. Ben-Jacob, 2002, Long term behavior of lithographically prepared in vitro neuronal networks, *Phys. Rev. Lett.* **88**(11), 118102 (4 pp). → p 70.

Sepkoski, J. J., Jr., 1982, A compendium of fossil marine families, *Milwaukee Public Museum Contrib. Biol. Geol.* **51**, 1–125. → pp 68, 69.

Sepkoski, J. J., Jr., 1993, Ten years in the library: new data confirm paleontological patterns, *Paleobiology* **19**(1), 43–51. → pp 68, 158.

Sepkoski, J. J., Jr., 2002, A compendium of fossil marine animal genera, *Bull. Am. Paleo.* **363**, 1–560, published posthumously, edited by D. Jablonski and M. Foote. → p 68.

Sethna, J. P., K. Dahmen, S. Kartha, J. A. Krumhansl, B. W. Roberts, and J. D. Shore, 1993, Hysteresis and hierarchies: Dynamics of disorder-driven first-order phase transformations, *Phys. Rev. Lett.* **70**(21), 3347–3350. → p 63.

Sethna, J. P., K. A. Dahmen, and C. R. Myers, 2001, Crackling noise, *Nature* **410**(6825), 242–250. → p 62.

Shi, K., and C.-Q. Liu, 2009, Self-organized criticality of air pollution, *Atmos. Environ.* **43**(21), 3301–3304. → p 71.

Shilo, Y., and O. Biham, 2003, Sandpile models and random walkers on finite lattices, *Phys. Rev. E* **67**(6), 066102 (8 pp). → pp 198, 245, 246, 249.

Shlesinger, M. F., and B. J. West, 1991, Complex fractal dimension of the bronchial tree, *Phys. Rev. Lett.* **67**(15), 2106–2108. → p 74.

Slanina, F., 2002, Self-organized branching process for a one-dimensional rice-pile model, *Eur. Phys. J. B* **25**(2), 209–216. → pp 199, 257.

Smethurst, D. P., and H. C. Williams, 2001, Power laws: are hospital waiting lists self-regulating?, *Nature* **410**(6829), 652–653. → pp *75*, 401, 406.

▶ Sneppen, K., 1992, Self-organized pinning and interface growth in a random medium, *Phys. Rev. Lett.* **69**(24), 3539–3542, comment (Tang and Leschhorn, 1993). → pp 142, 149, 182, 294, 295, 299, 301, 321, 452, 485.

Sneppen, K., 1995a, Extremal dynamics and punctuated co-evolution, *Physica A* **221**(1–3), 168–179. → pp 144, 149.

Sneppen, K., 1995b, Minimal SOC: intermittency in growth and evolution, in *Scale Invariance, Interfaces, and Non-Equilibrium Dynamics*, edited by A. McKane, M. Droz, J. Vannimenus, and D. Wolf (Plenum Press, New York, NY, USA), pp. 295–302, NATO Advanced Study Institute on Scale Invariance, Interfaces, and Non-Equilibrium Dynamics, Cambridge, UK, June 20–30, 1994. → pp 149, 151.

Sneppen, K., P. Bak, H. Flyvbjerg, and M. H. Jensen, 1995, Evolution as a self-organized critical phenomenon, *Proc. Natl. Acad. Sci. USA* **92**(11), 5209–5213. → pp *69*, *158*.

Sneppen, K., and M. H. Jensen, 1993, Sneppen and Jensen reply, *Phys. Rev. Lett.* **70**(24), 3833, reply to comment (Tang and Leschhorn, 1993). → pp 142, 151, 452.

▶ Sneppen, K., and M. E. J. Newman, 1997, Coherent noise, scale invariance and intermittency in large systems, *Physica D* **110**(3–4), 209–222. → pp 81, 485.

Socolar, J. E. S., G. Grinstein, and C. Jayaprakash, 1993, On self-organized criticality in nonconserving systems, *Phys. Rev. E* **47**(4), 2366–2376. → pp 133, 134, 266, 293, 298, 305, 319, 321, *322*, 328, 329.

Solé, R. V., and J. Bascompte, 1996, Are critical phenomena relevant to large-scale evolution?, *Proc. R. Soc. London, Ser. B* **263**(1367), 161–168. → pp 69, *158*, 159.

Solé, R. V., and S. C. Manrubia, 1995, Self-similarity in rain forests: evidence for a critical state, *Phys. Rev. E* **51**(6), 6250–6253. → p 74.

Solé, R. V., S. C. Manrubia, M. Benton, and P. Bak, 1997, Self-similarity of extinction statistics in the fossil record, *Nature* **388**(6644), 764–767. → p 69.

Soler, J. M., 1982, Alternative exact method for random walks on finite and periodic lattices with traps, *Phys. Rev. B* **26**(2), 1067–1070. → p 249.

Solomon, T. H., E. R. Weeks, and H. L. Swinney, 1994, Chaotic advection in a two-dimensional flow: Lévy flights and anomalous diffusion, *Physica D* **76**(1–3), 70–84. → p 72.

Somfai, E., A. Czirok, and T. Vicsek, 1994, Power-law distribution of landslides in an experiment on the erosion of a granular pile, *J. Phys. A: Math. Gen.* **27**(20), L757–L763. → p 56.

Song, W., F. Weicheng, W. Binghong, and Z. Jianjun, 2001, Self-organized criticality of forest fire in China, *Ecol. Modell.* **145**, 61–68. → p 73.

Sornette, A., and D. Sornette, 1989, Self-organized criticality and earthquakes, *Europhys. Lett.* **9**(3), 197–202. → p 66.

• Sornette, D., 1991, Self-organized criticality in plate tectonics, in *Spontaneous Formation of Space-Time Structures and Criticality*, edited by T. Riste and D. Sherrington (Kluwer, Dordrecht, The Netherlands), pp. 57–106, NATO Advanced Study Institute on Spontaneous Formation of Space-Time Structures and Criticality, Geilo, Norway, April 2–12, 1991. → pp 66, 139.

⇒ Sornette, D., 1992, Critical phase transitions made self-organized: a dynamical system feedback mechanism for self-organized criticality, *J. Phys. I (France)* **3**, 2065–2073. → p 330.

Sornette, D., 1994, Sweeping of an instability: an alternative to self-organized criticality to get powerlaws without parameter tuning, *J. Phys. I (France)* **4**, 209–221. → pp 7, *11*, 71, 330, *332*, 342.

Sornette, D., 2002, Mechanism for powerlaws without self-orgnization, *Int. J. Mod. Phys. C* **13**(2), 133–136, arXiv:cond-mat/0110426. → pp 76, 342.

• Sornette, D., 2006, *Critical Phenomena in Natural Sciences* (Springer-Verlag, Berlin, Germany), 2nd edition. → pp 6, 317, 330, *342*, 344.

Sornette, D., and J. V. Andersen, 1998, Scaling with respect to disorder in time-to-failure, *Eur. Phys. J. B* **1**(3), 353–357. → p 65.

Sornette, D., and I. Dornic, 1996, Parallel Bak-Sneppen model and directed percolation, *Phys. Rev. E* **54**(4), 3334–3338. → pp *150*, 156, 296, 320.

Sornette, D., A. Johansen, and I. Dornic, 1995, Mapping self-organized criticality onto criticality, *J. Phys. I (France)* **5**, 325–335. → pp 301, 320, 321, 330.

Sornette, D., and M. J. Werner, 2009, Seismicity, statistical physics approaches to, in *Encyclopedia of Complexity and Systems Science*, edited by R. A. Meyers (Springer-Verlag, New York, NY, USA), volume 9, pp. 7872–7891, arXiv:0803.3756v2. → pp 67, 68, 139.

Spasojević, D., S. Bukvić, S. Milošević, and H. E. Stanley, 1996, Barkhausen noise: elementary signals, power laws, and scaling relations, *Phys. Rev. E* **54**(3), 2531–2546. → p 63.

Speer, E. R., 1993, Asymmetric Abelian sandpile models, *J. Stat. Phys.* **71**(1/2), 61–76. → p 95.

Spitzer, F., 2001, *Principles of Random Walk* (Springer-Verlag, Berlin, Germany). → pp 246, 249.

Stanley, H. E., 1971, *Introduction to Phase Transitions and Critical Phenomena* (Oxford University Press, New York, NY, USA). → pp 3, 23, 25, 318, 348.

Stanley, H. E., L. A. N. Amaral, S. V. Buldyrev, A. L. Goldberger, S. Havlin, B. T. Hyman, H. Leschhorn, P. Maass, H. A. Makse, C. K. Peng, M. A. Salinger, M. H. R. Stanley, and G. M. Viswanathan, 1996, Scaling and universality in living systems, *Fractals* **4**(3), 427–451. → p 53.

Stanley, H. E., L. A. N. Amaral, P. Gopikrishnan, V. Plerou, and M. A. Salinger, 2002, Scale invariance and universality in economic phenomena, *J. Phys.: Condens. Matter* **14**(9), 2121–2131. → p 75.

Stanley, H. E., P. J. Reynolds, S. Redner, and F. Family, 1982, Position-space renormalization group for models of linear polymers, in *Real-Space Renormalization*, edited by T. W. Burkhardt and J. M. J. van Leeuwen (Springer-Verlag, Berlin, Germany), Topics in Current Physics, pp. 169–206. → p 113.

Stanley, R. P., 1999, *Enumerative Combinatorics*, Volume II, number 62 in Cambridge Studies in Advanced Mathematics (Cambridge University Press, Cambridge, UK). → pp 252, 253.

● Stapleton, M. A., 2007, *Self-Organised Criticality and Non-Equilibrium Statistical Mechanics*, Ph.D. Thesis, Imperial Collage London, UK, accessed 12 May 2007, URL http://www.matthewstapleton.com/thesis.pdf. → pp 100, 172, 178, 184, 187, 188, 190, 198, 229, 247–249, 261, 451.

► Stapleton, M., and K. Christensen, 2005, Universality class of one-dimensional directed sandpile models, *Phys. Rev. E* **72**(6), 066103 (4 pp). → pp 206, 208, 249, *289*.

Stapleton, M. A., and K. Christensen, 2006, One-dimensional directed sandpile models and the area under a Brownian curve, *J. Phys. A: Math. Gen.* **39**(29), 9107–9126, table B1 seems to contain typos (Stapleton, 2007). → pp 206, 208, 223, 228, 232, 249, 271.

Stapleton, M., M. Dingler, and K. Christensen, 2004, Sensitivity to initial conditions in self-organized critical systems, *J. Stat. Phys.* **117**(5/6), 891–900. → pp 17, *270*, 319.

Stassinopoulos, D., and P. Bak, 1995, Democratic reinforcement: a principle for brain function, *Phys. Rev. E* **51**(5), 5033–5039. → p 71.

Stauffer, D., and A. Aharony, 1994, *Introduction to Percolation Theory* (Taylor & Francis, London, UK). → pp 12, 103, 112, 121, 124.

Stella, A. L., and M. De Menech, 2001, Mechanisms of avalanche dynamics and forms of scaling in sandpiles, *Physica A* **295**(1–2), 101–107. → pp 103, 165.

Stevens, W. R., and S. A. Rago, 2005, *Advanced Programming in the UNIX® Environment* (Addison-Wesley, Reading, MA, USA), 2nd edition. → pp 358, 371, 372.

Stierstadt, K., and W. Boeckh, 1965, Die Temperaturabhängigkeit des magnetischen Barkhauseneffekts, *Z. Phys.* **186**(2), 154–167. → p 62.

Stilck, J. F., R. Dickman, and R. Vidigal, 2004, Series expansion for a stochastic sandpile, *J. Phys. A: Math. Gen.* **37**(4), 1145–1157, arXiv:cond-mat/0306214. → p 173.

Strogatz, S. H., 1994, *Nonlinear Dynamics and Chaos* (Physica B, Cambridge, MA, USA). → p 260.

Suki, B., A.-L. Barabasi, Z. Hantos, F. Petak, and H. E. Stanley, 1994, Avalanches and power-law behaviour in lung inflation, *Nature* **368**(6472), 615–618. → p 74.

Surdeanu, R., R. J. Wijngaarden, E. Visser, J. M. Huijbregtse, J. H. Rector, B. Dam, and R. Griessen, 1999, Kinetic roughening of penetrating flux fronts in high-T_c thin film superconductors, *Phys. Rev. Lett.* **83**(10), 2054–2057. → p 61.

Suzuki, M., and M. Katori, 1986, New method to study critical phenomena – mean-field finite-size scaling theory, *J. Phys. Soc. Jpn.* **55**(1), 1–4. → p *182*.

Suzuki, M., M. Katori, and X. Hu, 1987, Coherent anomaly method in critical phenomena. I., *J. Phys. Soc. Jpn.* **56**(9), 3092–3112, first of five parts. → p 182.

Swendsen, R. H., and J.-S. Wang, 1987, Nonuniversal critical dynamics in Monte Carlo simulations, *Phys. Rev. Lett.* **58**(2), 86–88. → p 357.

Syozi, I., 1972, Transformation of Ising models, in *Phase Transitions and Critical Phenomena*, edited by C. Domb and M. S. Green (Academic Press, London, UK), volume 2, pp. 269–329. → p 23.

Szabó, G., M. Alava, and J. Kertesz, 2002, Self-organized criticality in the Kardar-Parisi-Zhang equation, *Europhys. Lett.* **57**(5), 665–671. → p 300.

Tabelow, K., 2001, Gap function in the finite Bak-Sneppen model, *Phys. Rev. E* **63**(4), 047101 (3 pp). → p 144–146.

Tadić, B., 1999, Time distribution and loss of scaling in granular flow, *Eur. Phys. J. B* **7**(4), 619–625. → p 135.

► Tadić, B., and D. Dhar, 1997, Emergent spatial structures in critical sandpiles, *Phys. Rev. Lett.* **79**(8), 1519–1522. → pp 171, 290, 295–298.

Tadić, B., U. Nowak, K. D. Usadel, R. Ramaswamy, and S. Padlewski, 1992, Scaling behavior in disordered sandpile automata, *Phys. Rev. A* **45**(12), 8536–8545. → p 287.

Takayasu, H., 1989, Steady-state distribution of generalized aggregation system with injection, *Phys. Rev. Lett.* **63**(23), 2563–2565. → p 102.

Takayasu, H., and M. Matsuzaki, 1988, Dynamical phase transition in threshold elements, *Phys. Lett. A* **131**, 244–247. → p 125.

Tang, C., 1993, SOC and the Bean critical state, *Physica A* **194**(1–4), 315 – 320. → p 59.

Tang, C., and P. Bak, 1988a, Critical exponents and scaling relations for self-organized critical phenomena, *Phys. Rev. Lett.* **60**(23), 2347–2350. → pp 59, 92, 171, 194, 299, 330–332.

Tang, C., and P. Bak, 1988b, Mean field theory of self-organized critical phenomena, *J. Stat. Phys.* **51**(5/6), 797–802. → pp 5, 92, 251, 330.

Tang, L.-H., M. Kardar, and D. Dhar, 1995, Driven depinning in anisotropic media, *Phys. Rev. Lett.* **74**(6), 920–923. → p 207.

Tang, L.-H., and H. Leschhorn, 1992, Pinning by directed percolation, *Phys. Rev. A* **45**(12), R8309–R8312. → p 294.

Tang, L.-H., and H. Leschhorn, 1993, Self-organized interface depinning, *Phys. Rev. Lett.* **70**(24), 3832, comment on (Sneppen, 1992), reply (Sneppen and Jensen, 1993). → pp 142, 295, 299, 449.

Täuber, U. C., 2005, Critical dynamics, preprint available at http://www.phys.vt.edu/~tauber/utaeuber.html, accessed 11 February 2010 (unpublished). → pp 177, 267, 300, 319, 324.

Täuber, U. C., M. Howard, and B. P. Vollmayr-Lee, 2005, Applications of field-theoretic renormalization group methods to reaction-diffusion problems, *J. Phys. A: Math. Gen.* **38**(17), R79–R131. → pp 154, 266.

Täuber, U. C., and F. Schwabl, 1992, Critical dynamics of the O(n)-symmetric relaxational models below the transition temperature, *Phys. Rev. B* **46**(6), 3337–3361. → p 343.

Tebaldi, C., M. De Menech, and A. L. Stella, 1999, Multifractal scaling in the Bak-Tang-Wiesenfeld sandpile and edge events, *Phys. Rev. Lett.* **83**(19), 3952–3955, arXiv: cond-mat/9903270. → pp 50, 103, 220.

Teich, M. C., C. Heneghan, S. B. Lowen, T. Ozaki, and E. Kaplan, 1997, Fractal character of the neural spike train in the visual system of the cat, *J. Opt. Soc. Am. A* **14**(3), 529 – 546. → p 70.

Theiler, J., 1993, Scaling behavior of a directed sandpile automata with random defects, *Phys. Rev. E* **47**(1), 733–734. → p 288.

Toib, A., V. Lyakhov, and S. Marom, 1998, Interaction between duration of activity and time course of recovery from slow inactivation in mammalian brain Na^+ channels, *J. Neurosci.* **18**(5), 1893–1903. → p 70.

Toner, J., 1991, Dirt roughens real sandpiles, *Phys. Rev. Lett.* **66**(6), 679–682. → pp 56, 207.

Tononi, G., and G. M. Edelman, 1998, Consciousness and complexity, *Science* **282**(5395), 1846–1851. → p 71.

Torcini, A., R. Livi, A. Politi, and S. Ruffo, 1997, Comment on 'universal scaling law for the largest lyapunov exponent in coupled map lattices', *Phys. Rev. Lett.* **78**(7), 1391, comment on (Yang, Ding, and Ding, 1996), reply (Loreto *et al.*, 1997). → pp 431, 457.

Torvund, F., and J. Frøyland, 1995, Strong ordering by non-uniformity of thresholds in a coupled map lattice, *Phys. Scripta* **52**, 624–627. → pp 129, 134, 139, 365.

Tracy, C. A., and H. Widom, 2007, Nonintersecting Brownian excursions, *Ann. Appl. Probab.* **17**(3), 953–979. → p 249.

Tsuchiya, T., and M. Katori, 1999a, Effect of anisotropy on the self-organized critical states of Abelian sandpile models, *Physica A* **266**(1–4), 358–361, Proceedings of the International Conference on Percolation and Disordered Systems – Theory and Applications, Giessen, Germany, July 14 – 17, 1998, see Tsuchiya and Katori (1999b). → p 290.

Tsuchiya, T., and M. Katori, 1999b, Exact results for the directed Abelian sandpile models, *J. Phys. A: Math. Gen.* **32**(9), 1629–1641. → pp 290, 453.

Tsuchiya, T., and M. Katori, 2000, Proof of breaking of self-organized criticality in a nonconservative Abelian sandpile model, *Phys. Rev. E* **61**(2), 1183–1188. → p 88.

Tullis, T. E., and J. D. Weeks, 1986, Constitutive behavior and stability of frictional sliding of granite, *Pure Appl. Geophys.* **124**(3), 383–414. → p 65.

Turcotte, D. L., 1993, *Fractals and Chaos in Geology and Geophysics* (Cambridge University Press, Cambridge, UK). → p 66.

Turcotte, D. L., 1999, Self-organized criticality, *Rep. Prog. Phys.* **62**, 1377–1426. → pp 53, 112.

Tuszyński, J. A., M. Otwinowski, and J. M. Dixon, 1991, Spiral-pattern formation and multistability in Landau-Ginzburg systems, *Phys. Rev. B* **44**(17), 9201–9213. → p 114.

Urbach, J. S., R. C. Madison, and J. T. Markert, 1995, Interface depinning, self-organized criticality, and the Barkhausen effect, *Phys. Rev. Lett.* **75**(2), 276–279. → pp 62, 63.

Uritsky, V. M., M. Paczuski, J. M. Davila, and S. I. Jones, 2007, Coexistence of self-organized criticality and intermittent turbulence in the solar corona, *Phys. Rev. Lett.* **99**(2), 025001 (4 pp). → p 72.

Utsu, T., 1961, A statistical study on the occurrence of aftershocks, *Geophys. Mag.* **30**, 521–605. → p 67.

Utsu, T., Y. Ogata, and R. S. Matsu'ura, 1995, The centenary of the Omori formula for a decay law of aftershock activity, *J. Phys. Earth* **43**, 1–33. → p 67.

Vallette, D. P., and J. P. Gollub, 1993, Spatiotemporal dynamics due to stick-slip friction in an elastic-membrane system, *Phys. Rev. E* **47**(2), 820–827. → p 65.

van der Linden, P., 1994, *Expert C Programming* (Sunsoft Press, A Prentice Hall Title, Mountain View, CA, USA). → pp 358, 359.

van der Ziel, A., 1950, On the noise spectra of semi-conductor noise and of flicker effect, *Physica* **16**(4), 359–372. → p 16.

van Kampen, N. G., 1992, *Stochastic Processes in Physics and Chemistry* (Elsevier Science, Amsterdam, The Netherlands), 3rd impression 2001, enlarged and revised. → pp 201, 212, 330.

van Milligen, B. P., and P. D. Pons, 2002, Hillslope evolution by nonlinear creep and landsliding: an experimental study: comment and Reply: comment, *Geology* **30**(5), 481–482, comment on (Roering, Kirchner, Sklar, and Dietrich, 2001), reply (Roering, Kirchner, Sklar, and Dietrich, 2002). → pp 193, 447.

van Wijland, F., 2002, Universality class of nonequilibrium phase transitions with infinitely many absorbing states, *Phys. Rev. Lett.* **89**(19), 190602 (4 pp). → p 179.

van Wijland, F., K. Oerding, and H. J. Hilhorst, 1998, Wilson renormalization of a reaction–diffusion process, *Physica A* **251**(1–2), 179–201. → p 177.

Vasil'ev, A. N., 2004, *The Field Theoretic Renormalization Group in Critical Behavior Theory and Stochastic Dynamics* (Chapman and Hall, Boca Raton, FL, USA). → p 267.

Vattay, G., and A. Harnos, 1994, Scaling behavior in daily air humidity fluctuations, *Phys. Rev. Lett.* **73**(5), 768–771. → p 71.

Vázquez, A., 2000, Nonconservative Abelian sandpile model with the Bak-Tang-Wiesenfeld toppling rule, *Phys. Rev. E* **62**(6), 7797–7801. → p 88.

Vázquez, A., and O. Sotolongo-Costa, 1999, Self-organized criticality and directed percolation, *J. Phys. A: Math. Gen.* **32**(14), 2633–2644. → p 296.

Vázquez, A., and O. Sotolongo-Costa, 2000, Universality classes in the random-storage sandpile model, *Phys. Rev. E* **61**(1), 944–947. → p 296.

Vere-Jones, D., 1976, A branching model for crack propagation, *Pure Appl. Geophys.* **114**(4), 711–725. → p 66.

Vere-Jones, D., 1977, Statistical theories of crack propagation, *Math. Geol.* **9**(5), 455–481. → p 66.

Vergeles, M., A. Maritan, and J. R. Banavar, 1997, Mean-field theory of sandpiles, *Phys. Rev. E* **55**(2), 1998–2000. → pp 92, 104, 106, 251, 257.

Vespignani, A., R. Dickman, M. A. Muñoz, and S. Zapperi, 1998, Driving, conservation and absorbing states in sandpiles, *Phys. Rev. Lett.* **81**(25), 5676–5679, arXiv:cond-mat/9806249v2. → pp 162, 165, 171, 254, 300, 330–333, 335, 339.

Vespignani, A., R. Dickman, M. A. Muñoz, and S. Zapperi, 2000, Absorbing-state phase tranistions in fixed-energy sandpiles, *Phys. Rev. E* **62**(4), 4564–4582, arXiv:cond-mat/0003285. → pp 100, 162, 171, 178, 179, 182, 299, 300, 307, 330, 331, 333, 337, 340.

Vespignani, A., and S. Zapperi, 1995, Renormalization approach to the self-organized critical behavior of sandpile models, *Phys. Rev. E* **51**(3), 1711–1724, see Lin and Hu (2002). → pp 92, 101, 121, 165, 168, 182, 267, 268, 334.

Vespignani, A., and S. Zapperi, 1997, Order parameter and scaling fields in self-organized criticality, *Phys. Rev. Lett.* **78**(25), 4793–4796. → pp 167, 171, *251*, 331.

⇒ Vespignani, A., and S. Zapperi, 1998, How self-organized criticality works: a unified mean-field picture, *Phys. Rev. E* **57**(6), 6345–6362. → pp 111, 118, 171, 251, 328, 330, 331, 333–335.

Vespignani, A., S. Zapperi, and V. Loreto, 1996, Renormalization of nonequilibrium systems with critical stationary states, *Phys. Rev. Lett.* **77**(22), 4560–4563. → p 268.

Vespignani, A., S. Zapperi, and V. Loreto, 1997, Dynamically driven renormalization group, *J. Stat. Phys.* **88**(1/2), 47–79. → pp 243, 267, *268*.

Vicsek, T., 1999, *Fractal Growth Phenomena* (World Scientific, Singapore), 2nd edition. → p 301.

Vicsek, T., 2002, Complexity: the bigger picture, *Nature* **418**(6894), 131–131. → p 6.

Vidigal, R., and R. Dickman, 2005, Asymptotic behavior of the order parameter in a stochastic sandpile, *J. Stat. Phys.* **118**(1/2), 1–25. → p 182.

Vlasko-Vlasov, V. K., U. Welp, V. Metlushko, and G. W. Crabtree, 2004, Experimental test of the self-organized criticality of vortices in superconductors, *Phys. Rev. B* **69**(14), 140504(R) (4 pp). → pp 60, 61.

Voigt, C. A., and R. M. Ziff, 1997, Epidemic analysis of the second-order transition in the Ziff-Gulari-Barshad surface-reaction model, *Phys. Rev. E* **56**(6), R6241–R6244. → p 294.

Voss, R. F., 1992, Evolution of long-range fractal correlations and $1/f$ noise in DNA base sequences, *Phys. Rev. Lett.* **68**(25), 3805–3808, comment (Buldyrev *et al.*, 1993). → pp 74, 406.

Voss, R. F., 1993, Voss replies, *Phys. Rev. Lett.* **71**(11), 1777, reply to comment (Buldyrev *et al.*, 1993). → p 74.

Walker, D., 2009, *The Self-Organised Branching Process*, M.Sc. Thesis, Department of Mathematics, Imperial College London. → p 265.

Walker, D., and G. Pruessner, 2009, The self-organised branching process revisted once more (unpublished). → p 335.

Walsh, C. A., and J. J. Kozak, 1981, Exact algorithm for d-dimensional walks on finite and infinite lattices with traps, *Phys. Rev. Lett.* **47**(21), 1500–1502. → p 249.

Walsh, C. A., and J. J. Kozak, 1982, Exact algorithm for d-dimensional walks on finite and infinite lattices with traps. II. General formulation and application to diffusion-controlled reactions, *Phys. Rev. B* **26**(8), 4166–4189. → p 249.

Wang, Z., and D. Shi, 1993, Thermally activated flux avalanches in single crystals of high-T_c superconductors, *Phys. Rev. B* **48**(13), 9782–9787. → pp 59, 60.

Wannier, G. H., 1987, *Statistical Physics* (Dover, Mineola, NY, USA), reprint, originally published by John Wiley & Sons, New York, NY, USA, 1966. → p 175.

Wegner, F. J., 1972, Corrections to scaling laws, *Phys. Rev. B* **5**(11), 4529–4536. → pp 30, 39.

Weichman, P. B., and J. Miller, 2000, Theory of the self-organized critical state in nonequilibrium ^{4}He, *J. Low Temp. Phys.* **119**(1), 155–179. → p 61.

Weinrib, A., 1984, Long-range correlated percolation, *Phys. Rev. B* **29**(1), 387–395. → p 121.

Weiss, J., and J.-R. Grasso, 1997, Acoustic emission in single crystals of ice, *J. Phys. Chem. B* **101**(32), 6113–6117. → p 64.

Weissman, M. B., 1988, $1/f$ noise and other slow, nonexponential kinetics in condensed matter, *Rev. Mod. Phys.* **60**(2), 537–571. → pp 4, 16.

Welinder, P., G. Pruessner, and K. Christensen, 2007, Multiscaling in the sequence of areas enclosed by coalescing random walkers, *New J. Phys.* **9**(5), 149 (18 pp). → pp 18, 206, 208, 223, 225, 232, 249, 251, 310, 396.

Welling, M. S., C. M. Aegerter, and R. J. Wijngaarden, 2003, Structural similarity between the magnetic-flux profile in superconductors and the surface of a $2d$ rice pile, *Europhys. Lett.* **61**(4), 473–479. → pp 187, 189, 193, 197, 209, 301.

Welling, M. S., C. M. Aegerter, and R. J. Wijngaarden, 2004, Noise correction for roughening analysis of magnetic flux profiles in $YBa_2Cu_3O_{7-x}$, *Eur. Phys. J. B* **38**(1), 93–98. → p 61.

Welling, M. S., C. M. Aegerter, and R. J. Wijngaarden, 2005, Self-organized criticality induced by quenched disorder: experiments on flux avalanches in NbH_x films, *Phys. Rev. B* **71**(10), 104515 (5 pp). → p 61.

Werner, M. J., and D. Sornette, 2007, Comment on 'analysis of the spatial distribution between successive earthquakes', *Phys. Rev. Lett.* **99**(17), 179801, comment on (Davidsen and Paczuski, 2005), reply (Davidsen and Paczuski, 2007). → pp 68, 413.

West, G. B., J. H. Brown, and B. J. Enquist, 1997, A general model for the origin of allometric scaling laws in biology, *Science* **276**(5309), 122–126. → p 74.

West, G. B., J. H. Brown, and B. J. Enquist, 1999, The fourth dimension of life: fractal geometry and allometric scaling of organisms, *Science* **284**(5420), 1677–1679. → p 74.

White, E. P., B. J. Enquist, and J. L. Green, 2008, On estimating the exponent of power-law frequency distributions, *Ecology* **89**(4), 905–912. → p 230.

Widom, B., 1965a, Equation of state in the neighborhood of the critical point, *J. Chem. Phys.* **43**(11), 3898–3905. → p 3.

Widom, B., 1965b, Surface tension and molecular correlations near the critical point, *J. Chem. Phys.* **43**(11), 3892–3897. → p 3.

Wiese, K. J., 2002, Disordered systems and the functional renormalization group, a pedagogical introduction, *Acta Phys. Slovaca* **52**(4), 341–351, arXiv:cond-mat/0205116. → p 315.

Wiesenfeld, K., J. Theiler, and B. McNamara, 1990, Self-organized criticality in a deterministic automaton, *Phys. Rev. Lett.* **65**(8), 949–952. → pp 96, 97, 134, 279.

Wijngaarden, R. J., M. S. Welling, C. M. Aegerter, and M. Menghini, 2006, Avalanches and self-organized Criticality in superconductors, *Eur. Phys. J. B* **50**(1), 117–122. → p 61.

Wikipedia, 2010a, Fast inverse square root – Wikipedia, The Free Encyclopedia, accessed 14 Aug 2010, URL http://en.wikipedia.org/w/index.php?title=Fast_inverse_square_root&oldid=378232698. → p 357.

Wikipedia, 2010b, Long double – Wikipedia, The Free Encyclopedia, accessed 1 Oct 2010, URL `http://en.wikipedia.org/w/index.php?title=Long_double&oldid=367785987`. → p 380.

Wilf, H. S., 1994, *Generatingfunctionology* (Academic Press, London, UK). → p 258.

Wilkinson, D., and J. F. Willemsen, 1983, Invasion percolation: a new form of percolation theory, *J. Phys. A: Math. Gen.* **16**(14), 3365–3376. → pp 295, 301, 320.

Willinger, W., R. Govindan, S. Jamin, V. Paxson, and S. Shenker, 2002, Scaling phenomena in the Internet: critically examining criticality, *Proc. Natl. Acad. Sci. USA* **99**(Suppl. 1), 2573–2580. → p 76.

Winfree, A. T., 1980, *The Geometry of Biological Time* (Springer-Verlag, Berlin, Germany). → p 135.

Wissel, F., and B. Drossel, 2005, The Olami-Feder-Christensen earthquake model in one dimension, *New J. Phys.* **7**(1), 5 (19 pp). → p 141.

Wissel, F., and B. Drossel, 2006, Transient and stationary behavior in the Olami-Feder-Christensen model, *Phys. Rev. E* **74**(6), 066109 (9 pp). → pp 134, *137*.

Witten, T. A., and L. M. Sander, 1981, Diffusion-limited aggregation, a kinetic critical phenomenon, *Phys. Rev. Lett.* **47**(19), 1400–1403. → pp 300, 320.

Woodard, R., D. E. Newman, R. Sánchez, and B. A. Carreras, 2007, Persistent dynamic correlations in self-organized critical systems away from their critical point, *Physica A* **373**, 215 – 230. → pp 172, 194, 322.

Worrell, G. A., S. D. Cranstoun, J. Echauz, and B. Litt, 2002, Evidence for self-organized criticality in human epileptic hippocampus, *NeuroReport* **13**(16), 2017–2021. → p 70.

Yang, C. B., 2004, The origin of power-law distributions in self-organized criticality, *Physica A* **37**(42), L523–L529. → p 251.

Yang, C. B., and X. Cai, 2001, Fluctuations and finite-size effect in the Bak-Sneppen model, *Eur. Phys. J. B* **21**(1), 109–114. → pp 141, 155.

Yang, W., E.-J. Ding, and M. Ding, 1996, Universal scaling law for the largest Lyapunov exponent in coupled map lattices, *Phys. Rev. Lett.* **76**(11), 1808–1811, comment (Torcini *et al.*, 1997). → p 453.

Yeh, W. J., and Y. H. Kao, 1984, Measurements of flux-flow and $1/f$ noise in superconductors, *Phys. Rev. Lett.* **53**(16), 1590–1593. → p 60.

▶ Zaitsev, S. I., 1992, Robin Hood as self-organized criticality, *Physica A* **189**(3–4), 411–416. → pp 81, 142, 149, 178, 185, 485.

Zapperi, S., C. Castellano, F. Colaiori, and G. Durin, 2005, Signature of effective mass in crackling-noise asymmetry, *Nat. Phys.* **1**(1), 46–49. → p 64.

Zapperi, S., P. Cizeau, G. Durin, and H. E. Stanley, 1998, Dynamics of a ferromagnetic domain wall: avalanches, depinning transition, and the Barkhausen effect, *Phys. Rev. B* **58**(10), 6353–6366. → p 64.

Zapperi, S., K. B. Lauritsen, and H. E. Stanley, 1995, Self-organized branching processes: mean-field theory for avalanches, *Phys. Rev. Lett.* **75**(22), 4071–4074. → pp 131, 183, 251, 254, 265, 327, 335.

Zapperi, S., P. Ray, H. E. Stanley, and A. Vespignani, 1999, Avalanches in breakdown and fracture processes, *Phys. Rev. E* **59**(5), 5049–5057. → p 64.

Zapperi, S., A. Vespignani, and H. E. Stanley, 1997, Plasticity and avalanche behaviour in microfracturing phenomena, *Nature* **388**(6643), 658–660. → p 65.

Zhang, D.-M., G.-J. Pan, H.-Z. Sun, Y.-P. Ying, and R. Li, 2005a, Moment analysis of different stochastic directed sandpile model, *Phys. Lett. A* **337**(4–6), 285–291. → pp 107, 284, 288.

Zhang, D.-M., H.-Z. Sun, Z.-H. Li, G.-J. Pan, B.-M. Yu, R. Li, and Y.-P. Yin, 2005b, Anomalous scaling behaviors in rice-pile model with two different driving mechanisms, *Commun. Theor. Phys.* **44**(1), 99–102, numerics may not be independent from Zhang *et al.* (2005c). → pp 189, 198, 209, 284, 458.

Zhang, D.-M., H.-Z. Sun, G.-J. Pan, B.-M. Yu, Y.-P. Yin, F. Sun, R. Li, and X.-Y. Su, 2005c, Moment analysis of a rice-pile model, *Commun. Theor. Phys.* **43**(3), 483–486, numerics may not be independent from Zhang *et al.* (2005b). → pp 189, 458.

Zhang, S., 1997, On the universality of a one-dimensional model of a rice pile, *Phys. Lett. A* **233**(4–6), 317–322. → pp 178, 188, 189, 193, 194, 196, 198, 209, 284.

Zhang, S., Z. Huang, and E. Ding, 1996, Predictions of large events on a spring-block model, *J. Phys. A: Math. Gen.* **29**(15), 4445. → p 141.

Zhang, S.-d., 2000, $1/f^{\alpha}$ fluctuations in a ricepile model, *Phys. Rev. E* **61**(5), 5983–5986. → pp 15, 186, 206, 284, 314.

▶ Zhang, Y.-C., 1989, Scaling theory of self-organized criticality, *Phys. Rev. Lett.* **63**(5), 470–473. → pp 22, 82, 104–110, 485.

Zheng, B., and S. Trimper, 2001, Comment on 'universal fluctuations in correlated systems', *Phys. Rev. Lett.* **87**(18), 188901 (1 p), comment on (Bramwell *et al.*, 2000), reply (Bramwell, Christensen, Fortin, *et al.*, 2001). → pp 328, 338, 405.

Zhu, J., G. Zeng, X. Zhao, G. Huang, and Y. Jiang, 2005, Self-organized critical behavior of acid deposition, *Water Air Soil Poll.* **162**(1), 295–313. → p 71.

Zieve, R. J., T. F. Rosenbaum, H. M. Jaeger, G. T. Seidler, G. W. Crabtree, and U. Welp, 1996, Vortex avalanches at one thousandth the superconducting transition temperature, *Phys. Rev. B* **53**(17), 11849–11854. → p 60.

Ziff, R. M., 1982, The perimeter of percolation clusters as a random walk, *J. Stat. Phys.* **28**(4), 838, announcement of a talk at the 47th Statistical Mechanics Meeting, New Brunswick, NJ, USA, May 13–14, 1982. → p 319.

Ziff, R. M., P. T. Cummings, and G. Stells, 1984, Generation of percolation cluster perimeters by a random walk, *J. Phys. A: Math. Gen.* **17**(15), 3009–3017. → p 319.

Zinn-Justin, J., 1997, *Quantum Field Theory and Critical Phenomena* (Oxford University Press, New York, NY, USA), 3rd edition. → pp 266, 323.

Zipf, G. K., 1949, *Human Behavior and the Principle of Least Effort* (Addison-Wesley, Reading, MA, USA). → pp 53, 76.

Zürcher, U., 1994, Scaling behavior of fluctuations in systems with continuous symmetry, *Phys. Rev. Lett.* **72**(21), 3367–3369. → p 343.

Author index

Quotes are indicated by italic page numbers, **bibliographical entries** by bold page numbers. Name prefixes are regarded part of the entries.

Subject index

Page ranges of complete sections on a particular subject are underlined. **Keywords** that appear bold in the main text have bold page numbers. Finally, **key models** discussed in Part II are printed bold and models' EPONYMS and NAMESAKES are shown in small capitals.

Printed in the United States
by Baker & Taylor Publisher Services